EA-ITP-G

TRAINING MANUAL GENERAL SECTION BOOKS 1 THROUGH 7

by
Dale Crane

AVIATION MAINTENANCE PUBLISHERS, INC.

INTEGRATED TRAINING PROGRAM FOR THE AVIATION MAINTENANCE TECHNICIAN

For sale by: Aviation Maintenance Publishers
P.O. Box 36 Riverton, Wyoming 82501
Tel: (307) 856-1582

International Standard Book Number 0-89100-241-3

Printed in the United States of America

Table of Contents

AVIATION
TECHNICIAN
INTEGRATED
TRAINING
PROGRAM

EA-ITP-G1

Mathematics and Physics

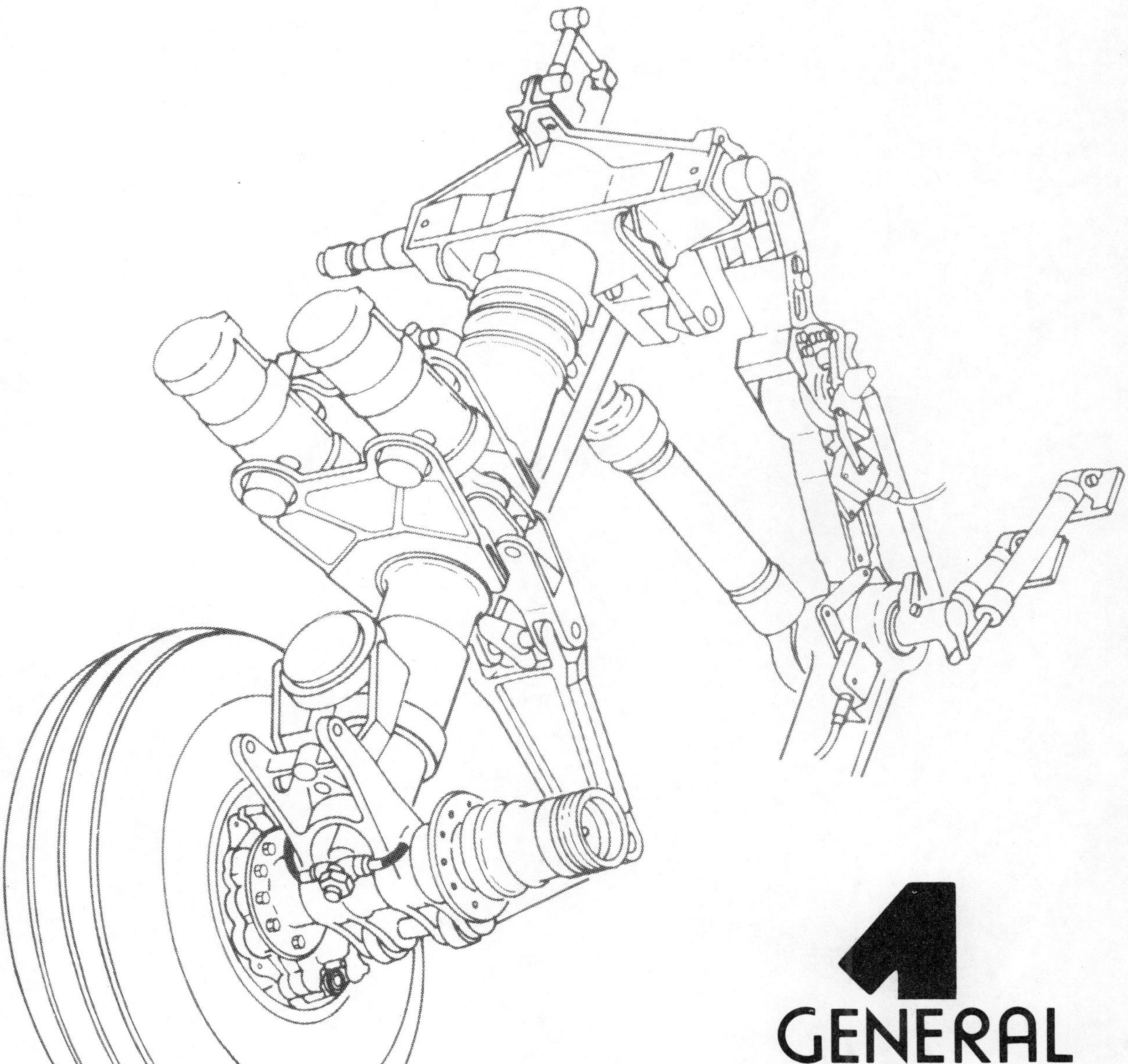

GENERAL SECTION 1

EA-ITP-G1

TRAINING MANUAL
GENERAL SECTION
BOOK 1

MATHEMATICS AND PHYSICS

by
Dale Crane

AVIATION MAINTENANCE PUBLISHERS, INC.

**INTEGRATED TRAINING PROGRAM
FOR THE
AVIATION MAINTENANCE TECHNICAN**

International Standard Book Number 0-89100-116-6

Aviation Maintenance Publishers, Inc.
1000 College View Drive, Riverton, Wyoming 82501

Printed in the United States of America

Table of Contents

PREFACE

Part 65 of the Federal Aviation Regulations requires that an applicant for a Mechanic Certificate with an Airframe or Powerplant rating have a general knowledge of the construction and maintenance of the aircraft covered by the rating he seeks. And each applicant must prove this knowledge by passing a written, an oral, and a practical examination, each given in five sections: General, Airframe Structures, Airframe Systems and Components, Powerplant Theory and Maintenance, and Powerplant Systems and Components.

Certificated Aviation Maintenance Technician Schools all operate under individual Federal Aviation Administration approvals, and all have in their approved curriculum enough information to satisfy the FAA that anyone completing their course will meet the knowledge requirements for certification as a mechanic.

In 1972, the Federal Aviation Administration published a series of three Advisory Circulars providing background information for mechanic certification applicants. These are entitled: *Airframe and Powerplant Mechanics General Handbook,* AC 65-9; *Airframe and Powerplant Mechanics Powerplant Handbook,* AC 65-12; and *Airframe and Powerplant Mechanics Airframe Handbook,* AC 65-15. These three handbooks were written to cover the area over which a mechanic applicant was to be tested, and they were an attempt by the FAA to coordinate the information available to the applicants with the tests that were given for certification. In 1976, the three handbooks were revised into their present A Revision format: AC 65-9A, AC 65-12A, and AC 65-15A.

Individuals attempting to certificate under FAR Part 65, as well as schools operating under FAR Part 147, have found that these handbooks actually do not follow the material used in the FAA examinations, and because of this, plus their lack of study helps, they fail to effectively provide the broad base of knowledge needed for certification and for a foundation on which to build in the aviation maintenance industry.

Realizing the need for updated information, presented in a format that is in keeping with the latest developments in educational technology, the Aviation Maintenance Publishers, Inc., is offering an Integrated Training Program (ITP) that presents the basic information needed for mechanic certification, in a form that will provide a good foundation in aviation maintenance knowledge. Developed in coordination with Crane Aeronautical Activities, Inc., the ITP training manuals, workbooks, study guides, and audiovisuals are designed to follow the FAA written examinations and are written to the level required for the FAR Part 147 schools.

All of this material can be used by an individual preparing for certification on his own, or by schools preparing groups of individuals for certification under FAR Part 65, but it has been prepared primarily for the Aviation Maintenance Technician Schools operating under FAR Part 147. Not only has every attempt been made to meet the requirements of Part 147, in producing this series, but the authors have followed many of the recommendations of the Aviation Mechanic Occupation Study conducted by the University of California, Los Angeles, under a contract from the U.S. Department of Health, Education, and Welfare.

One of the most basic premises of aviation maintenance is that all operations must be carried out according to FAA-approved data, and all information of a specific nature furnished by the manufacturer of an aircraft, powerplant, or component must be followed in detail. These training manuals are, of necessity, general in nature and cover a wide scope of aircraft, powerplants, and components. Because of this, it must be understood that they should *never* be allowed to take precedence over specific information furnished by the manufacturer.

The authors and publishers of this Integrated Training Program wish to express our appreciation to the various manufacturers, Aviation Maintenance Technician Schools, and FAA personnel who have aided our efforts to provide this coordinated training material to meet the challenge of better training for aviation maintenance technicians in the eighties.

PART I
MATHEMATICS

SECTION I

An Introduction to Mathematics

Mathematics is the basic language of science and technology. It is an exact language that has a vocabulary and a meaning for every term. And since it follows very definite rules and behaves in exactly the same way every time, it is one of the true absolutes in our life.

Scientists and engineers use mathematics as their basic tool, and long before any metal is cut on a new aircraft design, there have been literally millions of mathematical computations made.

The electronic computer has opened new doors in engineering, as it can make such a huge volume of computations in such a short time that more and more dependence can be placed on mathematical analysis.

The small hand-held electronic calculator has almost replaced the slide rule for rapid calculations, and its low cost has made it a standard item in almost every well-equipped toolbox. Use of these little calculators relieves us of almost all of the monotonous details of addition, subtraction, multiplication, and division, and makes our calculations less subject to errors.

However, we must not look on these calculators as something to relieve us of the responsibility of learning to handle mathematics, but rather as an aid in getting the most out of our other mathematical tools.

SECTION II

The Number Systems

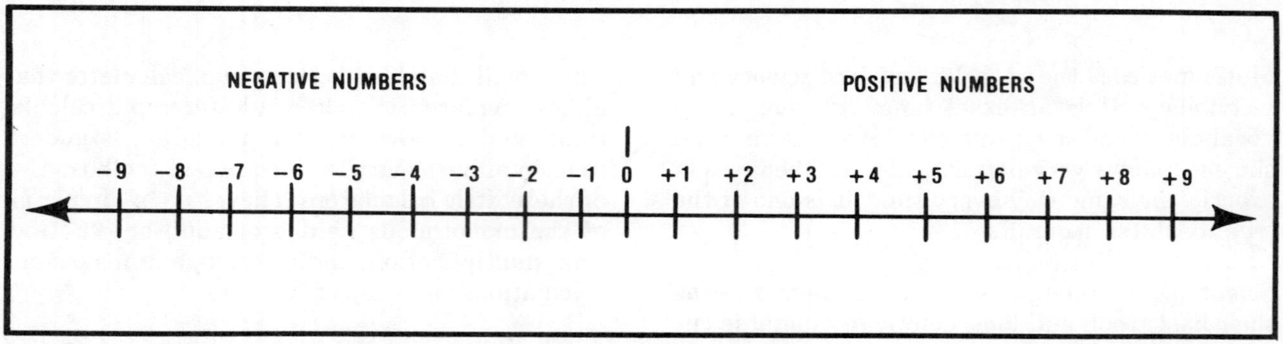

Fig. 2-1 The number line includes all numbers from negative infinity to positive infinity.

A. Decimal System

The decimal, or base-ten, numbering system is the one most generally used in our computations. It is based on ten whole numbers, the digits from zero to nine. These numbers may be either positive or negative, and may be added to or subtracted from each other. Numbers greater than nine may be formed by using columns for units, tens, hundreds, thousands, ten-thousands, hundred-thousands, millions, and even higher.

0 1 2 3 4 5 6 7 8 9

Fig. 2-2 The digits in the decimal number system

B. Binary System

The extreme speed with which electronic calculators work mathematical problems is due to their using the binary number system. In this system, rather than the ten different digits we use in the decimal system, there are only two digits used, zero and one. In the electronic computer or calculator, these values are represented by the condition of the circuit. The circuit conducts for digit one, and does not conduct for digit zero.

Binary numbers can be added or subtracted, and they can be converted into decimal numbers, numbers using ten as the base. Also, decimal numbers can be converted into binary numbers, and to do this, we need a simple table such as the

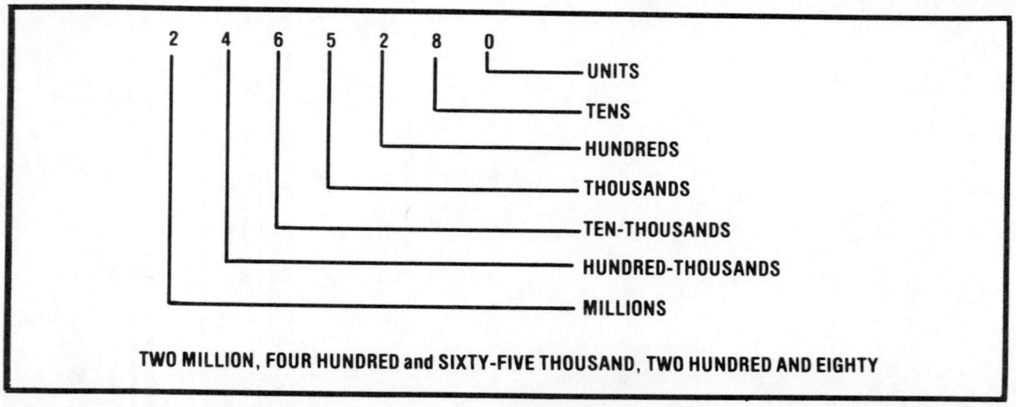

Fig. 2-3 In the decimal number system, the position of the digit indicates its value.

0	1

Fig. 2-4 The digits in the binary number system

one in Fig. 2-5. This table is prepared by starting with the number one and doubling it to get the number to its left. Then we double *this* number, and continue doubling each number until we come up with all of the decimal numbers we need across the top of the chart.

BINARY NUMBER							DECIMAL NUMBER
64	32	16	8	4	2	1	
0	0	1	1	0	1	1	27
0	1	1	0	0	0	0	48
1	0	1	1	1	0	0	92
1	1	1	0	1	0	1	117

Fig. 2-5 Conversion of binary numbers into decimal numbers.

In our chart, beginning on the right and working to the left, we have columns for 1, 2, 4, 8, 16, 32, and 64. Now we work from left to right in order to convert the number 27 into binary numbers. We find that there are no 64's or 32's in 27, but there is one 16 and one 8, giving us 24. For the remaining 3, there is one 2 and one 1.

$$16 + 8 + 2 + 1 = 27$$

The binary equivalent of decimal 27 is therefore 0011011.

The decimal number 117 can be converted in the same manner into the binary 1110101, as $64 + 32 + 16 + 4 + 1 = 117$.

Binary numbers can be converted into decimal numbers by using the same chart. To convert the binary number 1011100 into a decimal, we see by the chart that it contains a 64, a 16, an 8, and a 4. When all of these numbers are added together, we find that the decimal equivalent of 1011100 is 92.

Electronic computers and calculators do their seeming magic by using binary numbers, even though we feed decimal numbers into them. Electronic circuits inside them do the conversions.

SECTION III

Whole Numbers

Most of us use electronic calculators for almost all of our mathematical jobs, but we must understand the fundamental mathematical manipulations in order to get the best use of these handy little tools.

A. Addition

When we add numbers, we must place them in vertical columns so that all of the units are in line, as are all of the tens, hundreds, and thousands. Start at the right, and add the units and then the tens. If the sum of either of these columns equals more than 9, we will have to put part of it in the "sums" column and the rest in the "carry" column. In the example of Fig. 3-1, the units column added up to 9, which we put in the units area of the sum. There are twelve 10's, so we put two of them in the sum, and since the remaining ten 10's are the same as 100, we place 1 in the carry space of the hundreds. There are now 5 hundreds and 9 thousands. The sum of these two numbers is 9 thousands, plus 5 hundreds, plus 2 tens, and 9 units. All of this is equal to 9,529.

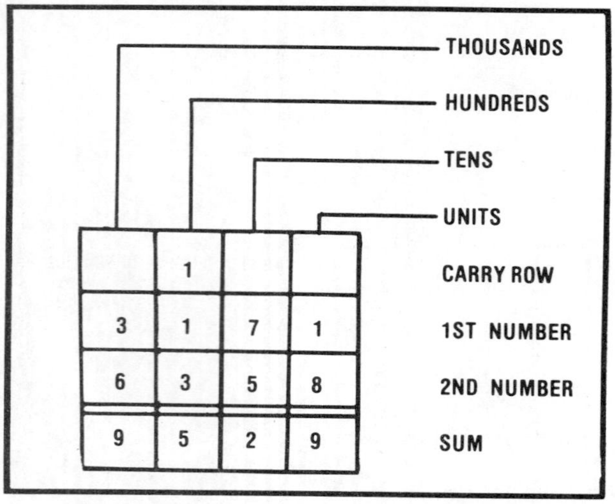

Fig. 3-1 Addition of whole numbers

To work addition problems on a hand calculator, all we need to do is turn it on and press the clear key to set everything to zero. Now press the keys for all of the digits in the first number, starting at the left and moving to the right. We punch the keys in the same order in which we say the numbers. To add the numbers in Fig. 3-1, we press keys 3, 1, 7, and 1, and then press the plus (+) key. Now press 6, 3, 5, and 8. When we press the equal (=) key, the number 9,529 should show up on the display.

The best way to check our addition is to add it again, this time starting at the opposite end of the column. If we started with the bottom number and worked upward the first time, we start with the top number to check our addition and add as we move down the column. It makes no difference in what sequence we add the numbers.

B. Subtraction

In order to find the difference between two whole numbers, we arrange them in the same way we did the numbers for addition; that is, we put all of the units, tens, hundreds, and thousands in vertical rows. But, be sure to place the larger whole number above the smaller one.

Starting at the right-hand side of Fig. 3-2, we subtract 2 from 6, and put 4 in the units area of the difference row. Then we go to the tens column, and find that we cannot subtract 40 from 30, so we must borrow 10 tens from the hundreds column. This gives us 13 tens from which to subtract the 40, and so we have 9 in the tens column of the answer. When we borrow 10 tens from the 4 in the hundreds column, we were left with only 3 hundreds, and since we cannot subtract 7 hundreds from 3 hundreds, we must borrow 10 hundreds from the thousands column. This gives us 13 hundreds from which we can subtract the 7 hundreds, and 6 hundreds is the difference. Since there are no thousands in the subtrahend (the bottom number), we bring the 1 thousand that was left in the minuend (the top number) down into the difference and find that when we subtract 742 from 2,436 we get a difference of 1,694.

On the electronic calculator, press the clear key, then 2, 4, 3, and 6, and then press the subtract, or (−) key. Now, punch in the subtrahend: 7, 4, and 2. When you press the equal (=) key, the difference of 1,694 will show up on the display.

In order to check your subtraction for accuracy, add the difference to the subtrahend and you should get the minuend.

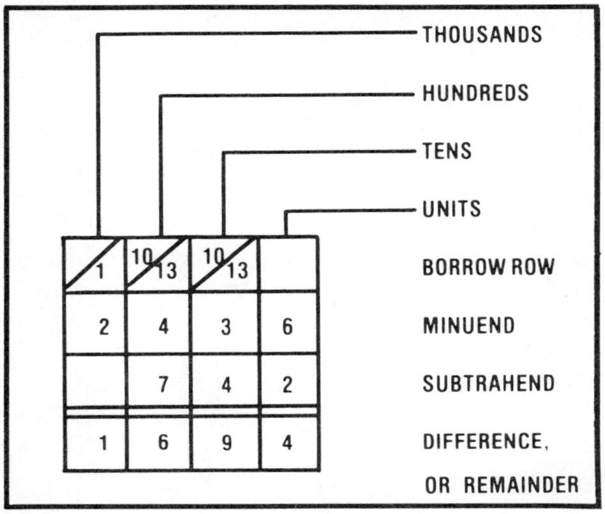

Fig. 3-2 Subtraction of whole numbers

C. *Multiplication*

Multiplication is a process of adding one number to itself a given number of times. For example, when we multiply 4 times 6, the results are the same as if we had added 4 to itself 6 times.

$$4 \times 6 = 24$$

$$4 + 4 + 4 + 4 + 4 + 4 = 24$$

To multiply numbers having more than one digit, we simply multiply all of the digits in the multiplican (Fig. 3-3) by each digit in the multiplier. We must be careful when we do this, however, to keep all of the units, tens, hundreds, and thousands in their correct columns.

In Fig. 3-3, we can see the steps used when we multiply 416 by 32. We first multiply the 416 by the units of the multiplier, the 2. In this step, we have 2 times 6 which is 12, so we put the 2 in the units column of the first partial product and carry the remaining 10 into the tens column of the first operation carry. We know that 2 times 10 is 20, and since we have 1 in the tens carry column, we *add* 20 and 10 and put 3 in the tens column of the first partial product. And finally, 2 times 400 is 800, so 8 goes in the hundreds column of the first partial product.

Our next step is to multiply the multiplicand by the tens digit of the multiplier. We first multiply 6 by 30 and get 180, so we put 8 in the tens column of the second partial product (we may ignore the unit of zero in this step — it will always be zero), and then we carry the 1, which is actually 10 times 10, or 100, in the tens column of the second operation carry. Now, 30 times 10 is 300, and adding the 100 we carried gives us 400. We bring the 4 down into the hundreds column of the second partial product. The last part of this step is to multiply 30 times 400, which gives us 12,000, or 1 in the ten-thousands column, and 2 in the thousands column. Note that the second partial product is *actually* 12,480, as it was derived from 416 multiplied by 30. It is customary to ignore the zero in the units column, but we must be sure that the last digit that we do use — in this case the 8 — is placed in the *tens* column of the second partial product.

When we add the two partial products, we have the final product of 13,312.

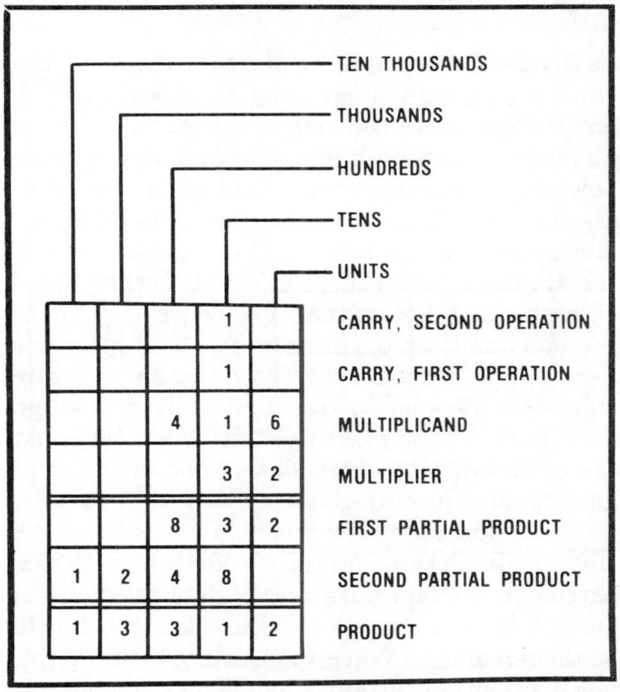

Fig. 3-3 Multiplication of whole numbers

To multiply on an electronic calculator, simply turn it on and clear the display by pressing the clear key. Now press in the digits for the multiplicand (the top number), then press the times (×) key and then each digit for the multiplier (the bottom number). When the equal (=) key is pressed, the product will show up on the display.

In order to check multiplication for accuracy, divide the product by either the multiplier or the multiplicand. You should get the other number as the quotient.

D. Division

Division is the means of finding out how many times one number will go into another, or how many times one number can be subtracted from the other.

$$24 \div 6 = 4$$

$$24 - 6 - 6 - 6 - 6 = 0$$

"Twenty-four divided by 6" means that we can subtract 6 from 24 four times. Try this on your calculator and you will find that you end up with zero; it comes out even.

To divide by longhand, we place the dividend (the number we are dividing) under a division sign (÷), and then we place the divisor (the number we are dividing by) to the left of it. Our first step is to see how many times the divisor will go into the first digits of the dividend. In Fig. 3-4, we see that our divisor of 8 will go into 35, the left two digits of the dividend, four times; so we place the digit 4 above the last digit of the dividend we used, the 5. We now multiply our divisor by 4 and place the results, the number 32, below the 35 and subtract. It is obvious that the divisor of 8 will not go into the 3 we got when we subtracted. We must therefore bring down enough of the dividend to get a number into which we can divide. To do this, we bring down the zero, which gives us 30, and 8 will go into 30 three times. We continue with this partial division until we use up all of the digits in our dividend, and anything that is left over is called the remainder. When we divide 3,504 by eight, our quotient is 438, and there is no remainder; it comes out even.

To check our work, multiply the quotient by the divisor and add any remainder. If your work is correct, your answer will be the dividend.

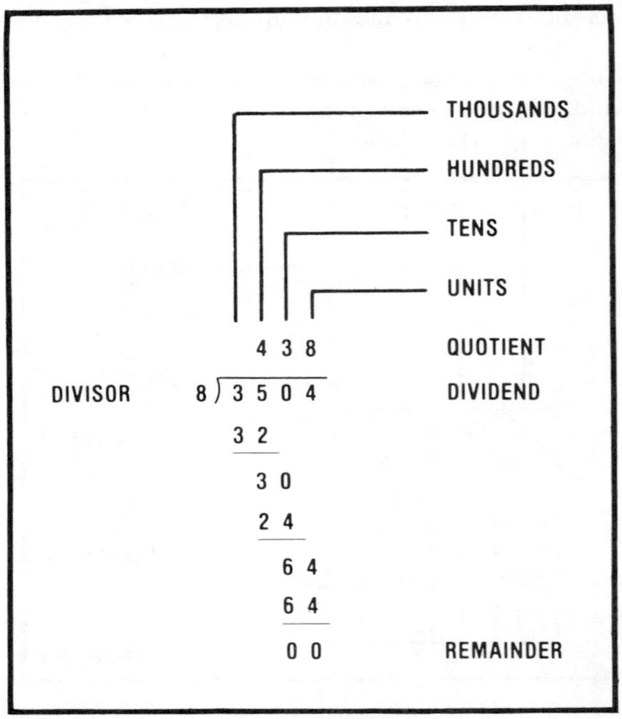

Fig. 3-4 Division of whole numbers

Division problems may be written by placing the dividend under the division sign and the divisor to the left; by putting the dividend on top of a line and the divisor below it; or by separating the dividend and the divisor by a "divided by" (÷) sign. They all mean the same — just find the number of times the divisor will go into the dividend.

8) 362 $\qquad \dfrac{362}{8} \qquad 362 \div 8$

Fig. 3-5 All of these arrangements tell us to divide by eight.

Division by electronic calculator is done by first clearing the display, then pressing the keys for the dividend. Now, press the divide (÷) key and then the keys for the divisor. When the equal (=) key is pressed, the quotient will show up on the display.

SECTION IV

Decimal Fractions

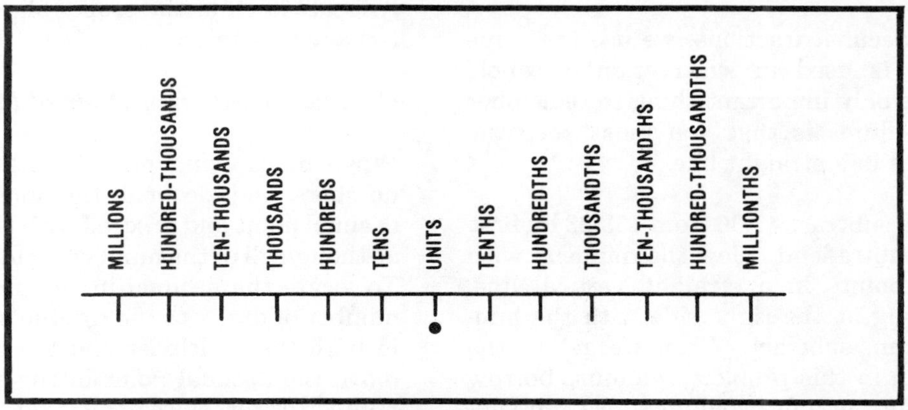

Fig. 4-1 Decimal placement

In the decimal system, we assume that a point resembling a period may be placed after the units column to indicate the end of the whole numbers. This "decimal point" is generally called just the "decimal." For example, the number 126, or one hundred and twenty-six, can also be written 126.0. And we can multiply 126.0 by ten by simply moving the decimal point over one place to the right, and the number becomes 1,260, the correct product.

The decimal point can also be moved to the left, creating a decimal fraction, and every place we move it, the number is divided by ten. When we move the decimal point in our 126.0 one place to the left, the number becomes the decimal fraction 12.60. With the decimal moved one place more to the left, the number is 1.260.

It makes no difference how many zeros we have to the right of the decimal, as long as there are no significant (non-zero) numbers following the zero. Numerically, 126, 126., and 126.000 are all the same.

A. *Addition of Decimal Fractions*

Addition of decimal fractions (numbers with significant figures to the right of the decimal point) is done in exactly the same way as addition of whole numbers. The decimal point of all of the numbers must be kept in a straight vertical line, and the decimal point in the sum will be located in line with those numbers being added. (Columns to the right of the decimal point indicate tenths, hundredths, thousandths, etc.)

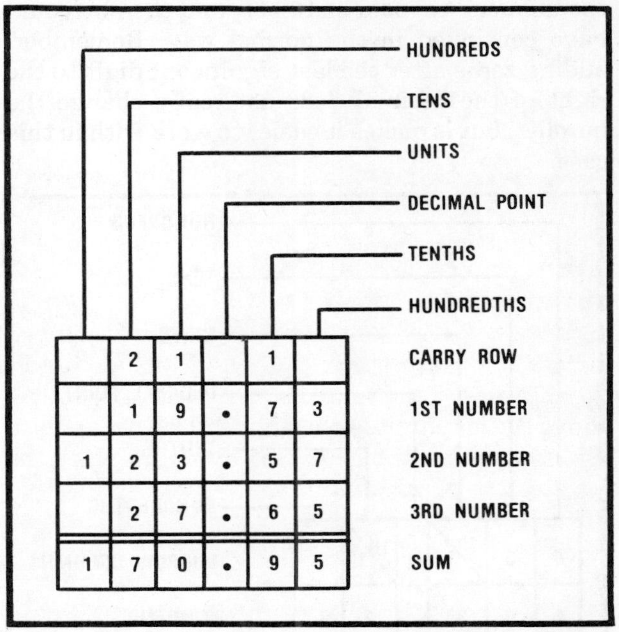

Fig. 4-2 Addition of mixed numbers with decimal fractions

When using the electronic calculator, perform all of the steps in the same way you did for the addition of whole numbers, except you must be sure that the key for the decimal point is pressed in its proper sequence each time you enter a number. The decimal point will be automatically located at the correct place in the answer when the equal key is pressed.

B. Subtraction of Decimal Fractions

To subtract decimal fractions, we use the same process that is used in subtraction of whole numbers. The only important thing to remember in both procedures is that you must keep the decimal points in a straight line.

In Fig. 4-3, we subtract 47.36 from 463.28 by first putting the subtrahend below the minuend with the decimal points in a straight vertical line. Now, beginning at the right side with the hundredths column, subtract. When we get to the tenths column in this problem, we must borrow 10 tenths from the units column so we can subtract. Continue on beyond the decimal point, subtracting in the normal way. The decimal point in the difference will be in line with the decimal point in the subtrahend and the minuend.

If our problem had been to subtract 14.275 from 16.5, we could have just added a couple of zeros to the minuend to make it 16.500, and then we could have continued in the normal way. Remember, adding zeros after the last significant digit to the right of the decimal does not really change the number, but it makes it easier to work with in this case.

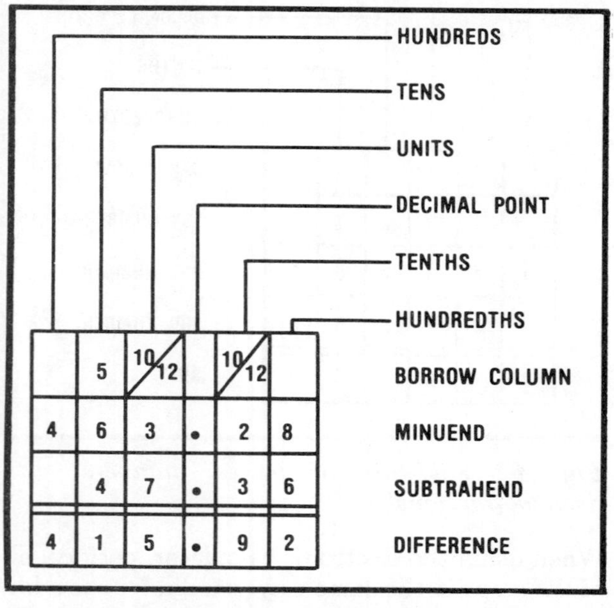

Fig. 4-3 Subtraction of mixed numbers with decimal fractions

Subtraction of decimal fractions by use of the electronic calculator is done in the same way as subtraction of whole numbers, except that you must remember to press the decimal key in the correct sequence each time you enter a number. The decimal point in the answer will automatically locate itself at the proper place when the equal (=) key is pressed.

C. Multiplication of Decimal Fractions

When multiplying numbers having both whole numbers and decimal fractions, disregard the decimal point and proceed with the multiplication as though all of the numbers were whole numbers. To locate the decimal in the product, count the number of digits to the right of the decimal point in both the multiplier and the multiplicand and move the decimal point in the answer the same number of places to the *left* of the last digit. In the example of Fig. 4-4, we multiply 19.76 by 0.275. There are two digits to the right of the decimal point in the multiplicand and three in the multiplier. Our product has an absolute value of 543400, but when we move the decimal point five places to the left, we find the product of the two numbers to be 5.434. (The two zeros beyond the last significant digit had no effect on the answer and, to make it less confusing, we have dropped them.)

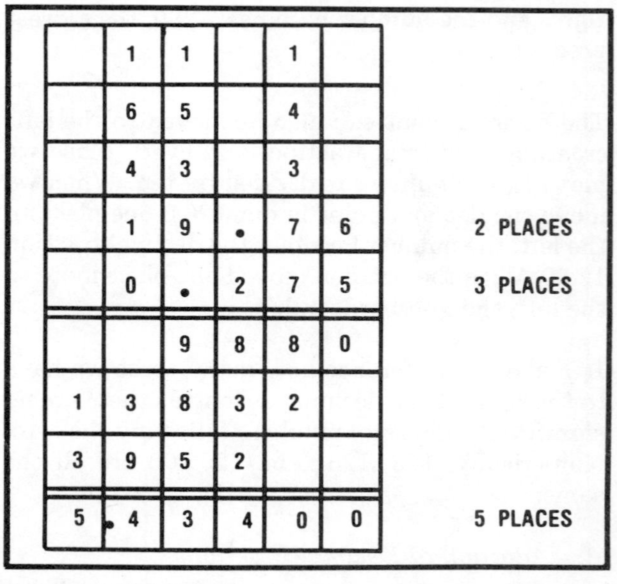

Fig. 4-4 Multiplication of numbers involving decimal fractions

When we must multiply large numbers having several digits to the left of the decimal point, by small numbers with a number of digits to the right of the decimal point, we can make the problem simpler by rearranging the decimal points. In Fig. 4-5, we have the problem of 1,620 multiplied by 0.002. We can greatly simplify the problem by moving the decimal point in the multiplier three places to the right to make it 2. All we have done is to multiply it by one thousand. Now, in order not to change the problem, we must divide the multiplicand by one thousand, by moving its decimal point to the *left* three places, and the 1,620 becomes 1.62. When we multiply 1.62 by 2, we get 3.24. This is exactly the same thing we get if we multiply out our original problem; try it.

$$1{,}620 \times 0.002 = 3.24$$

← 3 PLACES — — 3 PLACES →

$$1.62 \times 2 = 3.24$$

Fig. 4-5 Each place a decimal is moved to the left divides the number by ten, and each place the decimal is moved to the right multiplies the number by ten.

D. Division of Decimal Fractions

Division by a decimal fraction is accomplished in exactly the same way as division of whole numbers, but the problem is much simplified if both the divisor and the dividend are multiplied by some number that will make a whole number of the divisor. In Fig. 4-6, we are dividing 676 by 42.25. We can simplify this by multiplying both the divisor and the dividend by 100. We do this by simply moving the decimal point in both numbers two places to the right. Our problem then becomes 67,600 divided by 4,225. Since we do not have any mixed numbers to contend with, the problem is simply one of dividing whole numbers, and the quotient is 16.

Of course, we do not have to simplify the problem at all if we use an electronic calculator. Just punch out the digits in the numbers, putting the decimal point in the divisor as it comes up, and the decimal point in the answer will be in the proper place.

$$42.25\overline{)\,6\;7\;6} \qquad 4225_{\wedge}\overline{)\,6\;7\;6\;0\;0}_{\wedge}$$

Fig. 4-6 When dividing a number by a decimal fraction, multiply both the divisor and dividend by the same number that changes the decimal fraction in the divisor into a whole number.

SECTION V

Scientific Notation, or Powers of Ten

1 0 0 0 0 0 0 .	=	10^6
1 0 0 0 0 0 .	=	10^5
1 0 0 0 0 .	=	10^4
1 0 0 0 .	=	10^3
1 0 0 .	=	10^2
1 0 .	=	10^1
1 .	=	10^0
. 1	=	10^{-1}
. 0 1	=	10^{-2}
. 0 0 1	=	10^{-3}
. 0 0 0 1	=	10^{-4}
. 0 0 0 0 1	=	10^{-5}
. 0 0 0 0 0 1	=	10^{-6}

Fig. 5-1 Scientific notation, or powers of ten

Handling *very* large or *very* small numbers is made easy, and the chance of error is decreased, when keeping track of the decimal place by using "scientific notation" or the powers of ten.

We have found that all of the numbers we work with can be changed into numbers between 1 and 10 by moving the decimal point in the appropriate direction. In using scientific notation, in order not to change our problem, we multiply the number we have changed by a power of ten equal to the number of places we have moved the decimal point. This is really doing nothing but dividing by the same number we had multiplied by to make the original change. For example, we can change 4,260,000 into 4.26 by dividing by one million (1,000,000). Now, in order not to change the number, we must *multiply* it by one million, and we can do this by multiplying 4.26 by 10^6. This is an easy way of telling us that we must add enough zeros to our number to get six places between the decimal point and the last zero.

$$4.26 \times 10^6 = 4{,}260{,}000$$

$$0.4 \times 0.000426 \times 13.520 \times 560 =$$

$$4 \times 10^{-1} \times 4.26 \times 10^{-4} \times 1.352 \times 10^{4} \times 5.6 \times 10^{2} =$$

$$4 \times 4.26 \times 1.352 \times 5.6 \times 10^{-1} \times 10^{-4} \times 10^{4} \times 10^{2} =$$

$$129.013 \times 10^{1} = 1{,}290.13$$

Fig. 5-2 Multiplication by the use of scientific notation

When the number we are working with is smaller than 1, we must move the decimal point to the right to get a number between 1 and 10. Count the number of places we move the decimal point, and then multiply our number by a *negative* power of ten. If our number were 0.000 000 003, we would move the decimal point nine places to the right, and then, in order not to change the number, we would have to multiply it by 10^{-9}.

$$0.000\ 000\ 003 = 3 \times 10^{-9}$$

A. Multiplication by Scientific Notation

Multiplication of very large and very small numbers becomes simple by the use of scientific notation. All we do is multiply the numbers and then *algebraically* add the powers of ten and move the decimal point the number of places the powers of ten tells us.

For example, let's take a problem that has several large and small numbers: $0.4 \times 0.000426 \times 13{,}520 \times 560 =$ ____________. The first thing we do is convert all of these numbers into powers of ten, and our problem becomes: $4 \times 10^{-1} \times 4.26 \times 10^{-4} \times 1.352 \times 10^{4} \times 5.6 \times 10^{2}$. The numbers multiplied out are $4 \times 4.26 \times 1.352 \times 5.6 = 129.013$, and our powers of ten are -5 and $+6$, or a total of $+1$. So we must multiply our answer by 10^{1}, or 10. The answer to our problem is therefore 1,290.13.

B. Division by Scientific Notation

When we divide by powers of ten, we find the powers in the same way we did for multiplication, and when we have divided the numbers, we *subtract* the powers rather than add them. To divide 5,280 by 0.25, we first convert the numbers into powers of ten and get 5.28×10^{3}, and 2.5×10^{-1}. When we divide, we get 2.112. Subtracting the powers of ten, -1 from $+3$, gives us $+4$. (Subtracting a negative number from a positive number is simply determining the difference between them and is done by adding.) Our answer is then 2.112×10^{4}, or, in decimal values, is 21,120.

You can prove this by punching it out on your electronic calculator: $5{,}280 \div 0.25 = 21{,}120$.

$$5{,}280 \div 0.25 =$$

$$5.28 \times 10^{3} \div 2.5 \times 10^{-1} =$$

$$5.28 \div 2.5 \times 10^{3-(-1)} =$$

$$2.112 \times 10^{4} = 21{,}120$$

Fig. 5-3 Division by the use of scientific notation

SECTION VI

Common Fractions

A common fraction consists of two numbers, one above and one below a line. The number below the line (the denominator) indicates the number of parts the whole unit is divided into, and the number above the line (the numerator) represents the number of parts being considered.

If we divide a square into four equal parts by dividing each side in its center, we will have four-fourths, or four quarters. Each of the smaller squares is called one-fourth, or 1/4. Now, if we divide the entire square into parts the size of one-fourth of each *fourth*, there will be 16 even smaller squares, each of them having an area of 1/16 that of the large square.

We can divide each of these smaller squares into four equal parts, and each will have one sixty-fourth of the area of the entire square. And each of our original *quarters* will contain 16 of these sixty-fourths.

If we divide each of the sixty-fourths into four equal parts, we will have tiny squares of such a small size that it would take 256 of them to equal the original square, and each quarter will contain 64 of these smallest squares. This breakdown of the square sets up a relationship that helps us understand common fractions: one-fourth = 4/16 = 16/64 = 64/256.

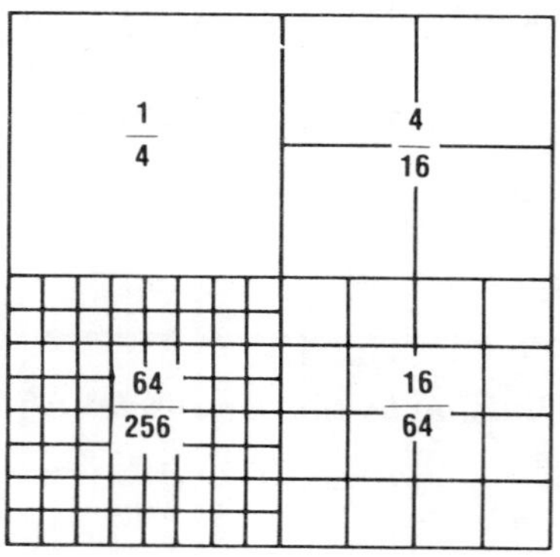

Fig. 6-1 The denominator of a fraction tells the number of parts the unit is divided into, and the numerator tells how many of these parts are being considered.

A. Finding the Common Denominator

We cannot add or subtract common fractions unless all of them are in the same units. If we are to add 1/4, 1/8, and 1/16, we must convert all of them into fractions of the same unit, or the same denominator. In this case, we can change them all into sixteenths and get 4/16, 2/16, and 1/16, and when the numerators of these three fractions are added, the sum will be 7/16.

If all of the original denominators are not divisible by the same number, as this combination was, we may have trouble finding a usable common denominator. One way to find a common denominator is to multiply all of the denominators together. This will give us a common denominator, but it will often be so large that we will have trouble using it.

Let's take a far-out example to illustrate the point. If we want to find a common denominator for 7/15, 5/24, and 13/32, we can multiply them together and get 11,520. This will work, but the number is so large that there will be unnecessary chances for error.

$$\frac{7}{15} \qquad \frac{5}{24} \qquad \frac{13}{32}$$

$$15 \times 24 \times 32 = 11{,}520$$

Fig. 6-2 A common denominator may be found by multiplying together all of the denominators. This number is often large and difficult to work with.

We can find the *least*, or smallest, common denominator by setting up a chart such as the one we have in Fig. 6-3. Across the top, we list the fractions for which we are to find the least common denominator. Below this, we place the three

	7/15	5/24	13/32	FRACTIONS
4	15	24	32	DENOMINATORS
3	15	6	8	
2	5	2	8	
5	5	1	4	
4	1	1	4	
	1	1	1	

$$4 \times 3 \times 2 \times 5 \times 4 = 480$$

$$\frac{7}{15} = \frac{224}{480}$$

$$\frac{5}{24} = \frac{100}{480}$$

$$\frac{13}{32} = \frac{195}{480}$$

Fig. 6-3 Finding the least common denominator

denominators, 15, 24, and 32. Now, we find a number that will go into at least two of these numbers, and place it to the left of the denominators. Four will go into both 24 and 32, so we place the 4 to the left of our denominators and divide, placing the quotients in the squares below the 24 and 32. Four will not go evenly into 15, so we bring the 15 down into the next line. Four will go into 24 six times, and into 32, eight times. Our next line now consists of 15, 6, and 8.

We want a number that will go evenly into at least two of *these* numbers, and so we choose 3. Three will go evenly into 15 five times and into 6, two times, but it will not go evenly into 8. Our third line is therefore 5, 2, and 8.

Now, we use 2 as a divisor and find that it will not go evenly into 5, but it will go into 2, one time, and into 8, four times. There is nothing that will go into 5, 1, and 4 an even number of times, but we can divide by 5, and get for our next line, 1, 1, and 4. Now, the final division is by 4, which goes into itself one time, and we end up with nothing across the bottom of our chart but 1's. Finally, to find the least common denominator, we multiply together all of the numbers we have used to divide by, and we get: $4 \times 3 \times 2 \times 5 \times 4 = 480$.

All of our original denominators will go into 480 an even number of times. Fifteen will go into 480 thirty-two times; 24 will go into 480 twenty times; and 32 will go into it fifteen times. When we convert all of these fractions into common fractions using this common denominator, we find that 7/15 = 224/480; 5/24 = 100/480; and 13/32 = 195/480.

B. Adding Common Fractions

To add common fractions, all we need to do is to be sure that they all have the same denominator. If they do not, change them so they will, as we have just discussed, and then add the numerators.

In Fig. 6-4, we want to add 1/16, 1/8, and 1/4. In order to do this, we must convert all of these fractions into common fractions having the same denominator, and in this case we can use 16. The first step then, is to convert all of the fractions into sixteenths.

1/16 = 1/16; 1/8 = 2/16; and 1/4 = 4/16

When we add all of these together, we find their sum is 7/16.

$$\frac{1}{16} = \frac{1}{16}$$

$$\frac{1}{8} = \frac{2}{16}$$

$$\frac{1}{4} = \frac{4}{16}$$

$$\frac{7}{16}$$

Fig. 6-4 Addition of common fractions

C. Subtracting Common Fractions

We make the same preparation in subtracting common fractions we did when adding. Be sure the subtrahend and minuend have the same denominator, and then subtract the numerators. In Fig. 6-5, we are to subtract 1/8 from 1/3. Our first step is to find a common denominator and convert both the subtrahend and minuend into fractions using this common denominator. In our example, we will use 24, so 1/3 becomes 8/24, and 1/8 becomes 3/24. When we subtract, we find that 1/3 − 1/8 = 8/24 − 3/24 = 5/24.

$$\begin{array}{rcc} & \frac{1}{3} & = & \frac{8}{24} \\ (-) & \frac{1}{8} & = & \frac{3}{24} \\ \hline & & & \frac{5}{24} \end{array}$$

Fig. 6-5 Subtraction of common fractions

SECTION VII

Mixed Numbers

In order to add or subtract "mixed numbers," which contain both whole numbers and common fractions, we must change them into an improper fraction (one having a larger numerator than denominator) by changing the whole number into a fraction using the same denominator as the fraction, and then adding them. For example, to change 3-5/16 into an improper fraction, we change 3 into a common fraction with 16 as its denominator. Its numerator is derived by multiplying the 3 by 16 to get 48 as the numerator. In this way, we find that 3 = 48/16. To this, we add the 5/16 to get 53/16, the improper fraction of our original 3-5/16.

A. Addition of Mixed Numbers

In Fig. 7-1, we want to add the three mixed numbers, 1-5/16 + 3-1/4 + 5-7/8. We must first change all of them into mixed numbers with a denominator that is common to all. We can use 16 as the common denominator. Starting at the top, 1 can be changed into 16/16, and then we add the 5/16 giving us the improper fraction 21/16. We find that 3-1/4 becomes 52/16, and 5-7/8 converts into 94/16. When all of these numbers are added together, we have 167/16. This can be changed back into a mixed number, 10-7/16.

B. Subtraction of Mixed Numbers

In subtracting mixed numbers, we first, of course, convert them into improper fractions with the subtrahend and the minuend both having the same denominator. Now all we have to do is to subtract the numerator of the subtrahend from the numerator of the minuend. Finally, convert the resulting improper fraction into a mixed number.

In Fig. 7-2, 8-1/3 is subtracted from 16-3/4. We first convert them both into improper fractions with 12 as the denominator. Our 16-3/4 becomes 201/12, and 8-1/3 becomes 100/12. When we subtract, we get 101/12 as the difference. In reducing this answer to its lowest terms, we find that 12 will go into 101 eight times, with 5 as the remainder, so the mixed number for our answer is 8-5/12.

$$
\begin{aligned}
1\tfrac{5}{16} &= \frac{16}{16} + \frac{5}{16} = \frac{21}{16} \\
3\tfrac{1}{4} &= \frac{48}{16} + \frac{4}{16} = \frac{52}{16} \\
5\tfrac{7}{8} &= \frac{80}{16} + \frac{14}{16} = \frac{94}{16} \\
& \qquad\qquad \frac{167}{16} = 10\tfrac{7}{16}
\end{aligned}
$$

Fig. 7-1 Addition of mixed numbers

$$
\begin{aligned}
16\tfrac{3}{4} &= \frac{192}{12} + \frac{9}{12} = \frac{201}{12} \\
(-)\ 8\tfrac{1}{3} &= \frac{96}{12} + \frac{4}{12} = \frac{100}{12} \\
& \qquad\qquad \frac{101}{12} = 8\tfrac{5}{12}
\end{aligned}
$$

Fig. 7-2 Subtraction of mixed numbers

SECTION VIII

Reduction of a Common Fraction to Its Lowest Terms

A fraction such as 9/27 can be left in this form, and it is a workable number, but it is often better if we can reduce it to its lowest terms; that is, change it into a fraction whose numerator and denominator are not divisible by the same number.

In Fig. 8-1, we see that the numerator, 9, will go into the denominator, 27, an even number of times. So, in order to reduce 9/27 to a fraction of the lowest terms, we divide both the numerator and denominator by 9. Nine twenty-sevenths then becomes 1/3.

$$\begin{matrix} 9 \\ 9 \end{matrix} \Big| \frac{9}{27} = \frac{1}{3}$$

Fig. 8-1 Reduction of a common fraction to its lowest terms

If we have a fraction whose numerator will not divide evenly into the denominator, but one in which both the numerator and denominator are divisible by the same number, we can reduce the fraction by dividing by a number with which they are both divisible. In Fig. 8-2, both 100 and 125 are divisible by 5, and when each is divided, the fraction becomes 20/25. Both of these numbers are also divisible by five, and when we again divide, we get the fraction 4/5, the lowest terms into which we can reduce 100/125. (Actually, we could have eliminated one step by dividing 100 and 125 by 25. This also would have given us 4/5.)

$$\begin{matrix} 5 \\ 5 \end{matrix} \Big| \frac{100}{125} = \begin{matrix} 5 \\ 5 \end{matrix} \Big| \frac{20}{25} = \frac{4}{5}$$

Fig. 8-2 Reduction of a common fraction to its lowest terms

SECTION IX

Multiplication of Fractions

In order to multiply common fractions, we simply multiply all of the denominators to find the new denominator, and then multiply all of the numerators to find the new numerator. For example, if we want to multiply 3/4 × 2/3, we multiply the denominators and find that the new denominator will be 12. And the numerators, when multiplied together, will be 6. So, 3/4 × 2/3 = 6/12. This reduces to 1/2.

Longer problems are worked in exactly the same way. In Fig. 9-2, we multiply 8/32 × 5/8 × 4/16.

$$\frac{3}{4} \times \frac{2}{3} = \frac{6}{12} = \frac{1}{2}$$

Fig. 9-1 Multiplication of common fractions

The new denominator is 4096 and the numerator is 160. By inspection, we see that both 160 and 4096 are divisible by 16, and this gives us the fraction 10/256. Both numerator and denominator are even numbers so they are both divisible by 2. This division gives us a fraction of 5/128 which is the lowest terms to which this fraction can be reduced.

$$\frac{8}{32} \times \frac{5}{8} \times \frac{4}{16} = \frac{160}{4096}$$

$$\begin{array}{c|c} 16 & 160 \\ 16 & 4096 \end{array} = \begin{array}{c|c} 2 & 10 \\ 2 & 256 \end{array} = \frac{5}{128}$$

Fig. 9-2 Multiplication of common fractions and reduction of the product to its lowest terms

SECTION X

Simplification of Fractions by Cancellation

$$\frac{\overset{1}{\cancel{8}}}{32} \times \frac{5}{\underset{1}{\cancel{8}}} \times \frac{4}{16}$$

$$\frac{1}{32} \times \frac{5}{1} \times \frac{\overset{1}{\cancel{4}}}{\underset{4}{\cancel{16}}}$$

$$\frac{1}{32} \times \frac{5}{1} \times \frac{1}{4} = \frac{5}{128}$$

Fig. 10-1 Simplification of common fractions by cancellation

We do not change a problem if we perform the same operation on both the numerator and the denominator of a fraction. This fact allows us to greatly simplify multiplication problems involving common fractions. In the example of Fig. 10-1, we have the same problem we worked in Fig. 9-2, but this time we are going to simplify it by a process known as cancellation. We look for numbers in the numerator that will go an even number of times into a number in the denominator. It makes no difference which fraction the denominator is in. The 8 in the numerator of 8/32 will cancel the 8 in the denominator of 5/8. This leaves the problem 1/32 × 5/1 × 4/16. We can cancel the 4 in the numerator of 4/16 into its own denominator, and the problem now becomes 1/32 × 5/1 × 1/4. When we multiply all of the numerators, we get 5, and the denominators multiply out to 128. This is exactly the same answer we got by multiplying all of the original numbers as we did in Fig. 9-2.

SECTION XI

Division by Common Fractions

$$6 \div \frac{1}{2} = 6 \times \frac{2}{1} = 12$$

Fig. 11-1 Division by a common fraction

Division by common fractions is performed in the same way as multiplication, *except* that we invert the divisor. For example, if we divide a 6-inch line into 1/2-inch increments, we know that we will have more than 6 increments. In Fig. 11-1, we see that 6 ÷ 1/2 = 6 × 2/1 = 12 half-inch increments.

We must remember that although in multiplication it makes no difference how we arrange the numbers — that is, 2 × 3 × 6 is the same as 6 × 2 × 3 — in division, we must never invert anything but the *divisor*. Dividing 6 by 1/2 is *not* the same as dividing 1/2 by 6.

SECTION XII

Converting Common Fractions into Decimals

The electronic calculator is one of our handiest tools for working almost any kind of mathematical problem, but it is somewhat difficult to use in working problems involving common fractions. It is easy, however, to work these problems when we convert the common fractions into decimal fractions. We can do this simply by dividing the numerator by the denominator. In this way, 13/16 becomes 0.8125.

$$\frac{13}{16} = 0.8125$$

Fig. 12-1 Conversion of a common fraction into a decimal fraction

SECTION XIII

Converting Decimal Fractions into Common Fractions

A decimal fraction can be written as a common fraction whose denominator is one, with the same number of zeros annexed to it as places the decimal point would be moved in order to convert the decimal fraction into a whole number. For example, 0.5 is 5/10; 0.75 is 75/100; and 0.003 may be written as the common fraction 3/1000.

To convert a decimal fraction into a common fraction of the lowest terms, write it in its common fraction form and divide both the numerator and the denominator by any number that will go into both of them an even number of times. Six hundred and twenty-five thousandths, written as 0.625, a decimal fraction, can also be written as 625/1000. Then both numerator and denominator can be divided by 5 to get 125/200. Both can again be divided by 5 to get 25/40. And finally, these numbers can be divided by 5 once more to get the fraction in its lowest terms, 5/8.

$$0.625 = 5/8$$

$$0.625 = \frac{625}{1000} \quad \frac{5}{5} \,\Big|\, \frac{625}{1000} = \frac{5}{5} \,\Big|\, \frac{125}{200} = \frac{5}{5} \,\Big|\, \frac{25}{40} = \frac{5}{8}$$

Fig. 13-1 Conversion of a decimal fraction into a common fraction of the lowest terms

SECTION XIV

Rounding Off Decimal Numbers

It is easy when working with decimal fractions to carry the problem out to an unreasonable degree, and so a little practical thought should be given to the use to which the numbers will be put. For example, if you want to lay out a metal strap to fit around a cylindrical air conditioning duct, you can find the circumference of the duct by multiplying its diameter by pi, or 3.1416. If the duct is 3 inches in diameter, the strap will have to be 9.4248 inches long. This number is highly accurate, but impractical to use for this purpose, as there is no need to measure to one ten-thousandth of an inch when the metal can be cut to an accuracy of less than one-hundredth of an inch. Since you need to measure to an accuracy of only one-hundredth of an inch, you should use 3.14 for the value of pi, and the metal should be cut to a length of 9.42 inches.

To round off numbers in a decimal fraction, determine the accuracy you want—tenths, hundredths, thousandths, or ten-thousandths—and then carry your answer out one decimal place beyond this amount. If the last digit is less than 5, disregard it. But, if it is 5 or greater, add one count to the last digit in the rounded off answer. In our example, we wished to read the circumference measurement to hundredths of an inch, and it came out 9.424. The last digit, 4, is less than 5, so the answer is 9.42. If the number, when multiplied out, had been, for example, 9.426, 1 would have been added to the 2 and the answer would have been rounded off to 9.43.

SECTION XV

Signed Numbers

The whole number scale runs from negative infinity to positive infinity and includes all of the numbers on both sides of zero. Numbers on the positive side are identified with the plus sign, of course, and those on the negative side use the minus sign. Addition, subtraction, multiplication, and division can all be carried out with numbers on either side of zero by following the simple rules for signed numbers.

A. *Addition of Signed Numbers*

When adding numbers having the same sign, find the sum of the absolute values, and place the common sign before the answer. If there is no sign before the number, it is assumed that the number is positive.

If both positive and negative numbers are being added (refer to Fig. 2-1), add all of the positive numbers and then all of the negative numbers. Find the difference between the two and use the sign of the larger as the sign of the answer.

−26	+26	+26
−34	−34	+34
−16	−16	−16
−76	−24	+44

Fig. 15-1 Addition of signed numbers

B. *Subtraction of Signed Numbers*

In order to subtract signed numbers — that is, to find the difference between them — change the sign of the subtrahend and proceed as you do in addition. It makes no difference if the subtrahend is larger than the minuend, since all of the operation is done as though it were addition. If we subtract a positive 48 from a negative 216, it is logical that the answer will be more negative than the minuend — it will be negative 264. But, if we subtract a positive 48 from a positive 216, the difference will still be positive, but it will be less than the 216; it will be positive 168.

−216	+216	−216	+216
(−)+ 48	(+)− 48	(+)− 48	(−)+ 48
−264	+264	−168	+168

Fig. 15-2 Subtraction of signed numbers

C. *Multiplication of Signed Numbers*

Multiplication of signed numbers is carried out in exactly the same way as multiplication of *unsigned* numbers, and if the signs of the numbers are alike, the sign of the answer will be positive. If the signs of the numbers are not alike, however, the sign of the answer will be negative.

$-6 \times -2 = +12$ $\quad$ $+6 \times +2 = +12$

$-6 \times +2 = -12$ $\quad$ $+6 \times -2 = -12$

Fig. 15-3 Multiplication of signed numbers

D. *Division of Signed Numbers*

As in multiplication of signed numbers, the process of division is the same as it is for unsigned numbers, and the sign of the answer will be positive if the signs of the two numbers are alike, and it will be negative if their signs are different.

$-12 \div -3 = +4$ $\quad$ $+12 \div +3 = +4$

$-12 \div +3 = -4$ $\quad$ $+12 \div -3 = -4$

Fig. 15-4 Division of signed numbers

SECTION XVI

Percentage

The percent, or parts in a hundred, is a special type of fraction which uses one hundred as the denominator. The decimal fraction 0.75 is the same as 75/100 and may also be expressed as 75 percent or 75%. Common fractions may be expressed as percentages by dividing the numerator by the denominator and then multiplying by one hundred, by moving the decimal point two places to the right. Five-eighths, changed into a decimal fraction, is 0.625. When it is converted into a percentage by moving the decimal point two places to the right, 5/8 = 0.625 = 62.5%.

A. Finding a Number Which Is a Given Precentage of Another

It is often necessary to determine what number is a given percentage of another, and in order to do this, multiply the known number by the decimal equivalent of the percentage. For example, if 100% on a turbine engine tachometer represents 11,000 RPM, what is the engine RPM at 38%?

$$11{,}000 \times 0.38 = 4{,}180 \text{ RPM}$$

B. Finding the Percentage That One Number is of Another

Determining what percentage one number is of another is done by setting the two numbers up as a common fraction, with the number representing one hundred percent as the denominator.

In order to determine what percentage 4,180 RPM is of 11,000 RPM, we set up the fraction:

$$\frac{4{,}180}{11{,}000} = 4{,}180 \div 11{,}000 = 0.380 = 38\%$$

C. Finding the Total Number When Part of the Number and Its Percentage Are Known

When one number is known to be a certain percentage of an unknown number, this unknown number, representing 100%, can be found. To do this, we change the percentage into a decimal and divide it into the known number. If we know that 4,180 RPM is 38% of the total RPM (100%), we can find the value of 100% by:

$$100\% \text{ RPM} = 4{,}180 \div 0.38 = 11{,}000 \text{ RPM}$$

SECTION XVII

Ratio

$$\text{ASPECT RATIO} = \frac{\text{SPAN}}{\text{CHORD}} = \frac{35}{5} = \frac{7}{1} = 7$$

Fig. 17-1 A ratio is a relationship between two values and is the quotient of one number divided by another.

A ratio is a relationship between two values and is the quotient of one number divided by another. For example, the aspect ratio of an airplane wing is the ratio of its span, or length, to its chord, or width. If the wing has a span of 35 feet and a chord of five feet, it has an aspect ratio of 35:5, which may be expressed as 35/5. Or, when it is reduced to its lowest terms, as 7/1 or 7:1. A ratio such as this, in which the denominator is one, may be expressed simply as an aspect ratio of seven.

SECTION XVIII

Proportion

A proportion is a relationship that shows two ratios to be equal, and is a convenient method of solving problems involving ratios.

If an engine has a reduction gear ratio between the crankshaft and the propeller of 3:2, and the engine is turning at 2700 RPM, what is the speed of the propeller? We can solve this problem of proportions by letting the symbol X represent the unknown value, which in this case is the speed of the propeller. We first set up the proportional statement $3/2 = 2700/X$, as we have it in Fig. 18-1. To solve this equation, we simply cross multiply; that is we find that $3X = 2 \times 2700$, or 5,400. To solve for X, we divide 5,400 by 3 and find the speed of the propeller to be 1800 RPM.

$$\text{REDUCTION RATIO} = \frac{3}{2} = \frac{\text{ENGINE RPM}}{\text{PROPELLER RPM}}$$

ENGINE SPEED = 2700 RPM

PROPELLER SPEED = X

$$\frac{3}{2} = \frac{2700}{X}$$

$$3X = 5400$$

$$X = 1800$$

X = 1800 RPM

Fig. 18-1 A proportion is a relationship between two ratios that shows them to be equal.

Another way we can solve this proportion problem is to convert the known ratio into a percentage. We have the ratio of 3:2 between the engine RPM and the propeller RPM. This means that the propeller is turning only 2/3 the speed of the engine. Two-thirds is roughly 67%, so the propeller is turning at 67% of the engine speed. Sixty-seven percent of 2700 RPM is 1809 RPM. The difference of nine RPM in the two ways of working the problem is caused by rounding off the percentage from 66.6666% to 67%, and since it represents only 1/2 of one percent, it is sufficiently close for practical work. Actually, this computation is more accurate than the average aircraft tachometer.

Proportional problems may also be solved by arranging them as we have in Fig. 18-2, and instead of cross multiplying, multiplying the means (the two middle numbers) by the extremes (the two outside numbers). In this problem, we make the ratios of the engine gear over the propeller gear, equal to the ratio of X which represents the engine RPM we want to find, to 1800, which represents the RPM of the propeller.

Engine gear : Propeller gear =

Engine RPM : Propeller RPM

$$3 : 2 = X : 1800$$

When we multiply the means, we get 2X, and when we multiply the extremes, we get 5,400. Now dividing 5,400 by 2 gives the engine RPM of 2,700.

REDUCTION RATIO = RPM RATIO

ENGINE GEAR : PROPELLER GEAR = ENGINE RPM : PROPELLER RPM

$$3:2 = X : 1800$$

$$2X = 5400$$

X = 2700 RPM

Fig. 18-2 The product of the means (the inner numbers) of a proportion must equal the product of the extremes (the outer numbers).

SECTION XIX

Power and Roots

A. *Raising a Number to a Given Power*

When a number is multiplied by itself, it is said to be raised to a given power. For example, if a number is multiplied by itself one time, it is raised to its second power or, more conveniently, it is squared. (The original number is the first power.) Four squared, for example, is equal to 4×4 which is 16. This is written with the number 4 as the base, and a small 2 is placed above it to represent the power. Four squared is written 4^2.

If a number is multiplied by itself twice, it is raised to its third power, or is cubed. Four cubed is equal to $4 \times 4 \times 4 = 64$, and is written $4^8 = 64$. Any number can be raised to any power, but to raise numbers to fractional powers usually requires the use of logarithms, which is not within the scope of this text.

$$4^2 = 4 \times 4 = 16$$

$$4^3 = 4 \times 4 \times 4 = 64$$

$$4^4 = 4 \times 4 \times 4 \times 4 = 256$$

$$4^5 = 4 \times 4 \times 4 \times 4 \times 4 = 1024$$

$$4^6 = 4 \times 4 \times 4 \times 4 \times 4 \times 4 = 4096$$

Fig. 19-1 Raising a number to a given power

B. *Extracting the Root of a Number*

The root of a number is that value, which, when raised to the indicated power, will produce the number. For example, the square root of 144 is that number which, when squared, will equal 144. This root number is 12, and we can therefore say that 12 is the square root of 144. The symbol used to indicate that we are to extract a root is the radical sign ($\sqrt{\ }$) placed over the number. We can show that we want to extract the square root of 81 by placing the number 81 under the radical sign; $\sqrt{81}$.

We can indicate other roots by placing the number of the root we want above the radical sign. If we want to indicate the cube root of a number such as 175,616, we place the number 3 above the radical sign:

$$\sqrt[3]{175616}$$

$\sqrt{81}$ = SQUARE ROOT OF 81 = 9

$\sqrt[3]{729}$ = SQUARE ROOT OF 729 = 9

Fig. 19-2 The radical sign over a number indicates that the indicated root of the number should be taken. If there is no number over the radical sign, the square root should be taken.

There are many mathematical handbooks and tables, such as that in Fig. 9-3, that give us squares, cubes, square roots, and cube roots for the most commonly used numbers. And the electronic calculator has made the process of finding powers and the square root of almost any number such an easy matter that powers and roots have become handy mathematical tools for us.

It is also an easy matter for us to find the square root of a number by the longhand method if we do not have an electronic calculator.

If we need to find the length of the sides of a square hangar which we know to have an area of 7,225 square feet, we do it by finding the square root of 7,225. We first place 7,225 under a radical sign as we have done in Fig. 9-4. Now, we pair off the digits, starting with the decimal point and moving to the left. This gives us the two pairs, 72 and 25, and means that we will have two digits in our answer.

Our next step is to find the largest number that when squared will go into 72. This number is eight, which when squared is 64. We place the 8

NO.	SQUARE	CUBE	SQUARE ROOT	CUBE ROOT
1	1	1	1.0000	1.0000
2	4	8	1.4142	1.2599
3	9	27	1.7321	1.4422
4	16	64	2.0000	1.5874
5	25	125	2.2361	1.7100
6	36	216	2.4495	1.8171
7	49	343	2.6458	1.9129
8	64	512	2.8284	2.0000
9	81	729	3.0000	2.0801
10	100	1,000	3.1623	2.1544
11	121	1,331	3.3166	2.2240
12	144	1,728	3.4641	2.2894
13	169	2,197	3.6056	2.3513
14	196	2,744	3.7417	2.4101
15	225	3,375	3.8730	2.4662
16	256	4,096	4.0000	2.5198
17	289	4,913	4.1231	2.5713
18	324	5,832	4.2426	2.6207
19	361	6,859	4.3589	2.6684
20	400	8,000	4.4721	2.7144
21	441	9,261	4.5826	2.7589
22	484	10,648	4.6904	2.8020
23	529	12,167	4.7958	2.8439
24	576	13,824	4.8990	2.8845
25	625	15,625	5.0000	2.9240
26	676	17,576	5.0990	2.9625
27	729	19,683	5.1962	3.0000
28	784	21,952	5.2915	3.0366
29	841	24,389	5.3852	3.0723
30	900	27,000	5.4772	3.1072
31	1,961	29,791	5.5678	3.1414
32	1,024	32,768	5.6569	3.1748
33	1,089	35,937	5.7446	3.2075
34	1,156	39,304	5.8310	3.2396
35	1,225	42,875	5.9161	3.2717
36	1,296	46,656	6.0000	3.3019
37	1,369	50,653	6.0828	3.3322
38	1,444	54,872	6.1644	3.3620
39	1,521	59,319	6.2450	3.3912
40	1,600	64,000	6.3246	3.4200
41	1,681	68,921	6.4031	3.4482
42	1,764	74,088	6.4807	3.4760
43	1,849	79,507	6.5574	3.5034
44	1,936	85,184	6.6332	3.5303
45	2,025	91,125	6.7082	3.5569
46	2,116	97,336	6.7823	3.5830
47	2,209	103,823	6.8557	3.6088
48	2,304	110,592	6.9282	3.6342
49	2,401	117,649	7.0000	3.6593
50	2,500	125,000	7.0711	3.6840

NO.	SQUARE	CUBE	SQUARE ROOT	CUBE ROOT
51	2,601	132,651	7.1414	3.7084
52	2,704	140,608	7.2111	3.7325
53	2,809	148,877	7.2801	3.7563
54	2,916	157,464	7.3485	3.7798
55	3,025	166,375	7.4162	3.8030
56	3,136	175,616	7.4833	3.8259
57	3,249	185,193	7.5498	3.8485
58	3,364	195,112	7.6158	3.8709
59	3,481	205,379	7.6811	3.8930
60	3,600	216,000	7.7460	3.9149
61	3,721	226,981	7.8102	3.9365
62	3,844	238,328	7.8740	3.9579
63	3,969	250,047	7.9373	3.9791
64	4,096	262,144	8.0000	4.0000
65	4,225	274,625	8.0623	4.0207
66	4,356	287,496	8.1240	4.0412
67	4,489	300,763	8.1854	4.0615
68	4,624	314,432	8.2462	4.0817
69	4,761	328,509	8.3066	4.1016
70	4,900	343,000	8.3666	4.1213
71	5,041	357,911	8.4261	4.1408
72	5,184	373,248	8.4853	4.1602
73	5,329	389,017	8.5440	4.1793
74	5,476	405,224	8.6023	4.1983
75	5,625	421,875	8.6603	4.2172
76	5,776	438,976	8.7178	4.2358
77	5,929	456,533	8.7750	4.2543
78	6,084	474,552	8.8318	4.2727
79	6,214	493,039	8.8882	4.2908
80	6,400	512,000	8.9443	4.3089
81	6,561	531,441	9.0000	4.3267
82	6,724	551,368	9.0554	4.3445
83	6,889	571,787	9.1104	4.3621
84	7,056	592,704	9.1652	4.3795
85	7,225	614,125	9.2195	4.3968
86	7,396	636,056	9.2376	4.4140
87	7,569	638,503	9.2374	4.4310
88	7,744	681,472	9.3808	4.4480
89	7,921	704,969	9.4340	4.4647
90	8,100	729,000	9.4868	4.4814
91	8,281	753,571	9.5394	4.4979
92	8,464	778,688	9.5917	4.5144
93	8,649	804,357	9.6437	4.5307
94	8,836	830,584	9.6954	4.5468
95	9,025	857,375	9.7468	4.5629
96	9,216	884,736	9.7980	4.5789
97	9,409	912,673	9.8489	4.5947
98	9,604	941,192	9.8995	4.6104
99	9,801	970,299	9.9499	4.6261
100	10,000	1,000,000	10.0000	4.6416

Fig. 19-3 Squares and square roots; cubes and cube roots of numbers.

both above the 72 and to the left of the radical sign. Now, multiply, and put the product 64 below the 72. Subtract 64 from 72, and then bring down the next *pair* of numbers, placing them beside the 8 to get our second-step dividend, 825.

Double the *last digit* in the divisor, which, in the first step is 8. Now, we must add one more digit to this for our second divisor. The divisor will be one hundred and sixty something, and to find what the last digit will be, we must find the number of times 160 will go into 825. This proves to be five times, so the second divisor is 165. Place the 5 both above the second pair of numbers under the radical sign and in the divisor, and then multiply. This process shows us the square root of 7,225 to be 85.

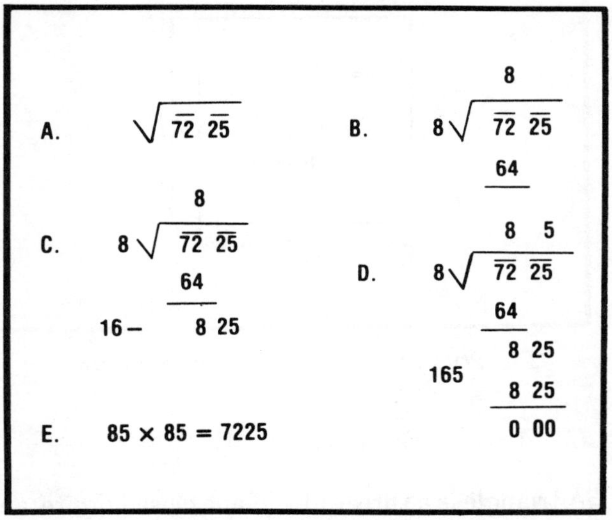

Fig. 19-4 Extraction of the square root

If the number from which we are extracting the square root is a mixed number with a decimal fraction, pair off the numbers on both sides of the decimal point. For example, to find the square root of 16,012.3716, pair the digits off from the decimal point as:

1 60 12 . 37 16

The square root is extracted as shown in Fig. 19-5.

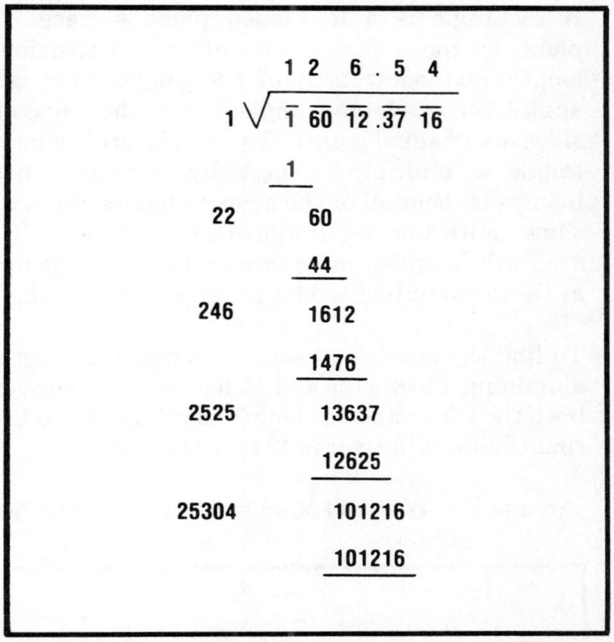

Fig. 19-5 Extraction of the square root of a number containing a decimal fraction

To find the square root with an electronic calculator, all we have to do is press the keys for the number for which we want to find the square root, then press the key with the radical sign (√‾), and the square root will pop up on the display.

SECTION XX
Measurement of Area

A. Area of a Rectangle

A rectangle is a four-sided plane surface. By plane, we mean that it has only two dimensions, length and width. All of the angles are right angles, or 90-degree angles, and the opposite sides are of equal length. To find the area of a rectangle, we multiply its length by its width — first being sure that all of the measurements are in the same units (inches, millimeters, feet, etc.). The area will be given in *square* units, the same unit as the linear units used for its length and width.

To find the number of square inches in a sheet of aluminum 4 feet wide and 12 feet long, we convert both the width and the length into inches and find that the sheet measures 48 by 144 inches.

$$\text{Area} = L \times W = 144 \times 48 = 6{,}912 \text{ square inches}$$

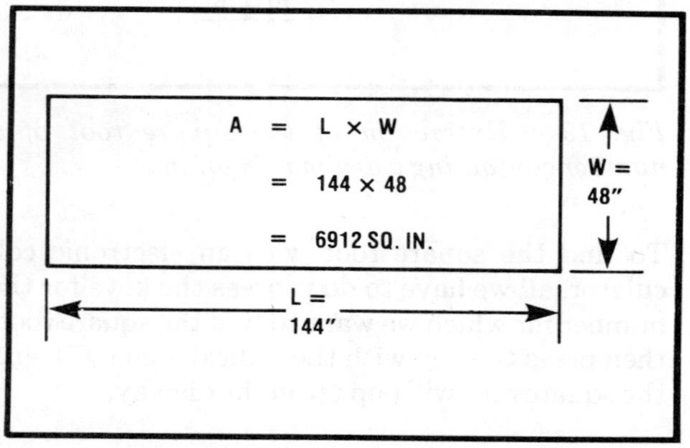

Fig. 20-1 Finding the area of a rectangular figure

B. Area of a Square

A square is a special form of rectangle in which all of the sides are of equal length. The same formula may be used for finding the area of a square as was used for finding the area of a rectangle, but since all of the sides are the same length, the formula may be simplified to:

$$\text{Area} = \text{Side}^2$$

Square the length of one of the sides to get the area in square units.

To find the area of a square whose sides are each 3.25 inches, we may use the formula:

$$\text{Area} = S^2 = 3.25^2 = 10.56 \text{ square inches}$$

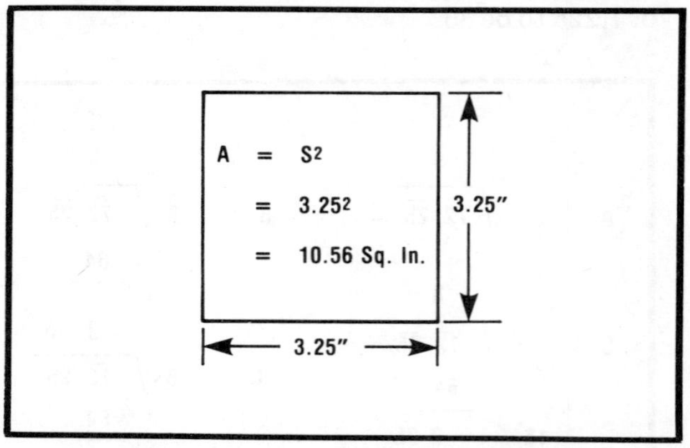

Fig. 20-2 Finding the area of a square

C. Area of a Triangle

A triangle is a three-sided plane closed figure, and as such it has certain characteristics and follows certain rules. For example, the sum of all of the angles in a triangle always equals 180 degrees. If a triangle is set inside a rectangle, with its base being equal to one of the sides and its altitude equal to the other side, the area of the triangle will be exactly one-half that of the rectangle. We can see this in Fig. 20-3 with five different types of triangles. The shaded portion of each of these rectangles is equal in area to the triangle that is not shaded. The formula for the area of a triangle is therefore one-half of the base times the altitude, or:

$$\text{Area} = \frac{B \times A}{2}$$

Triangles are named according to their angles. If two of the angles are equal, it is called an

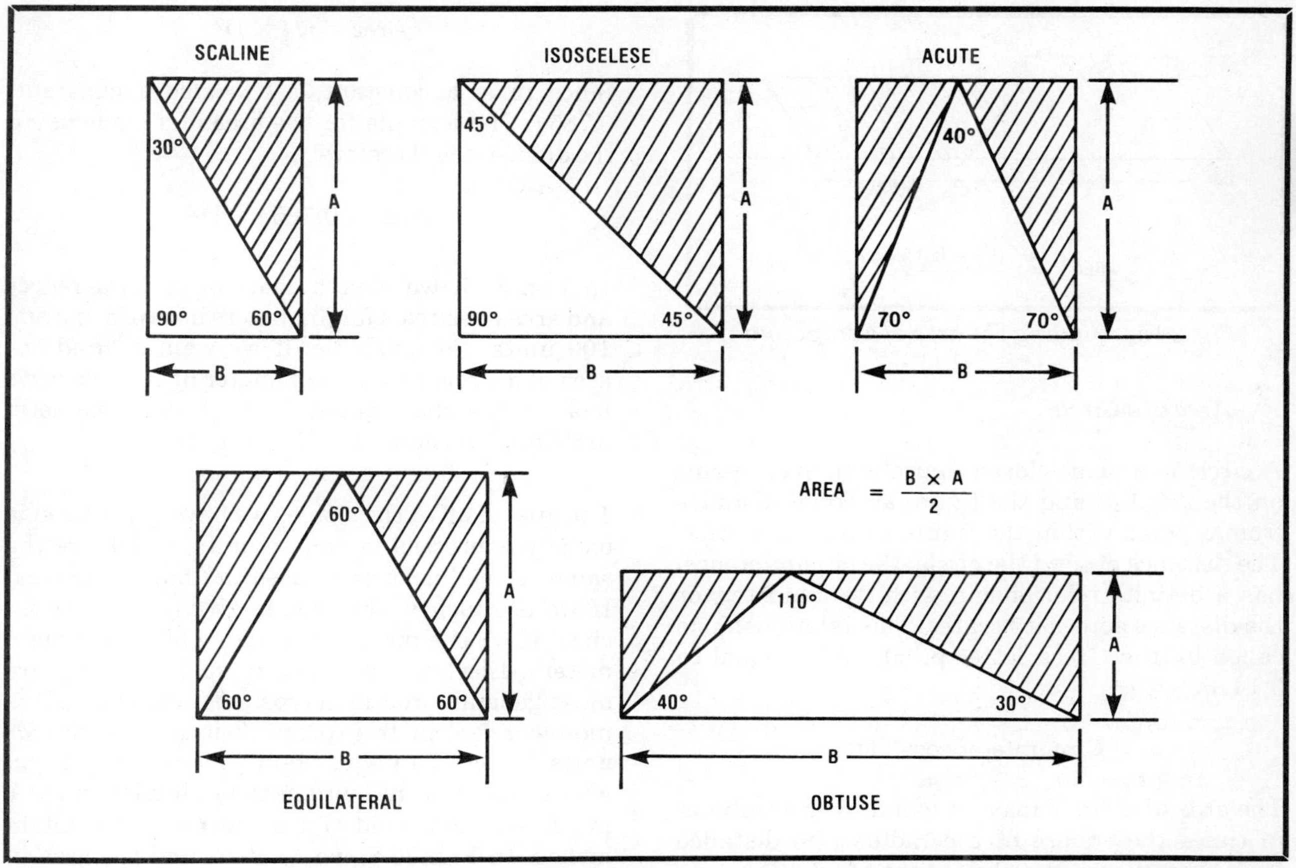

Fig. 20-3 Finding the area of a triangle

isosceles triangle. If all three of the angles are equal, it is an equilateral, or equiangular triangle. And if none of the sides or angles are alike, it is called a scalene triangle. If one angle in the triangle is a 90-degree, or right, angle, the triangle is a right triangle. If one of the angles is greater than 90 degrees, the triangle is an obtuse triangle, but if all of the angles are less than 90 degrees, the triangle is called an acute triangle.

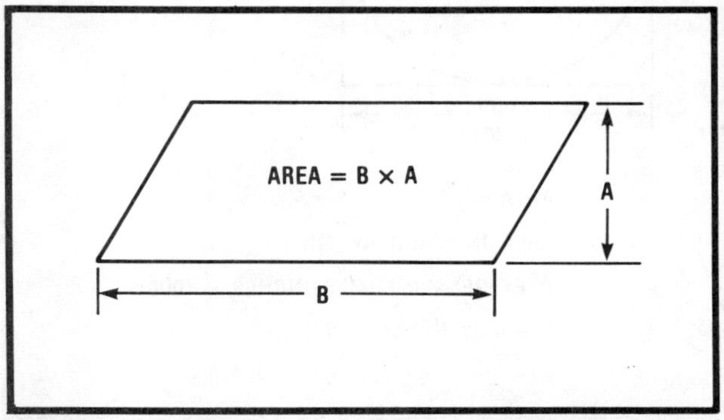

Fig. 20-4 Finding the area of a parallelogram

D. Area of a Parallelogram

A parallelogram is a four-sided plane closed figure, whose opposite sides are parallel, but whose angles are not right angles. The area of a parallelogram is simply the product of the base times the altitude.

E. Area of a Trapezoid

A trapezoid is a four-sided plane closed figure having only two sides parallel. The area of a trapezoid may be found, if we consider two identical trapezoids laid side by side so that their parallel sides form connected straight lines, Fig. 20-5. This forms a parallelogram whose base is equal to the combined lengths of B_1 and B_2 of the trapezoid and whose altitude is the altitude of the trapezoid. The area is therefore one-half of the parallelogram we have just formed:

$$\text{Area} = \frac{(B_1 + B_2) \times A}{2}$$

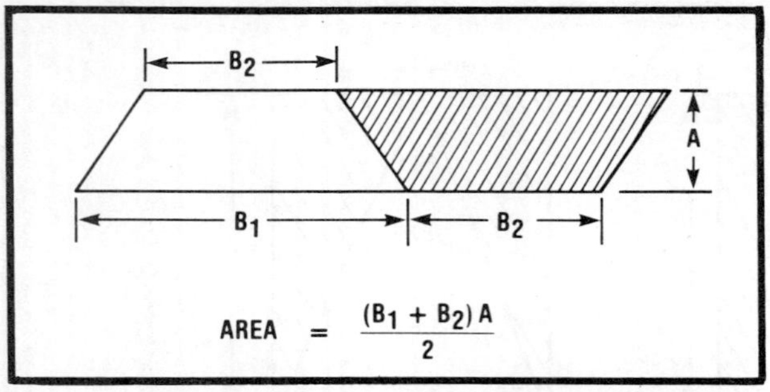

Fig. 20-5 Finding the area of a trapezoid

F. Area of a Circle

A circle is a plane closed figure with every point on the line forming the figure an equal distance from a point within the figure called the center. The distance around the circle, the circumference, has a definite relationship with its diameter, or the distance across its center. This relationship is called by the Greek letter pi (π), and is equal to 3.1416.

$$\text{Circumference} = \pi D$$

The area of a circle may be found by multiplying pi times the square of the radius (the distance from the center of the circle to the edge).

$$\text{Area} = \pi R^2$$

But, it is often more convenient to use the diameter rather than the radius, and since the diameter is two times the length of the radius, the square of the diameter will be *four* times the square of the radius, and we can keep the equation the same as it was originally by dividing the constant pi by 4. This formula is easier to handle:

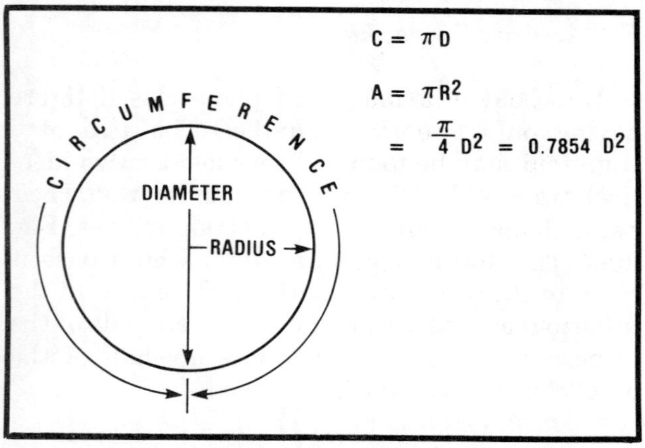

Fig. 20-6 Finding the circumference and the area of a circle

$$\text{Area} = \pi/4 \times D^2$$

Since pi is a constant, $\pi/4$ is also a constant, 0.7854. The formula for the area of a circle using the diameter is therefore:

$$\text{Area} = 0.7854 \times D^2$$

In Fig. 20-7, we have a chart of circumferences and areas of circles for diameters between one and 100 units. For example, if we want to find the area of a circle having a diameter of 15 inches, we look on the chart opposite the 15-unit diameter and find that the area is 176.71 square inches.

The area of all of the shapes we have considered is usually expressed in square units, which are the same as the linear units used in finding the area. If we use inches, the area is given in square inches. If we use meters, the area will be in square meters. Electrical conductors such as wires are most generally round in cross section, and so it is more convenient to express their area in *circular* units rather than square units. For wires, then, we use an area measurement of circular mils. A mil is one thousand times the measurement in inches. In Fig. 20-8, we want to find the area in circular mils of a round conductor whose diameter is 1/4 inch or 0.25 inch. The formula for finding the area in circular mils is simply the square of the diameter of the circle in mils.

$$\text{Area} = D^2 = 250^2 = 62{,}500 \text{ circular mils}$$

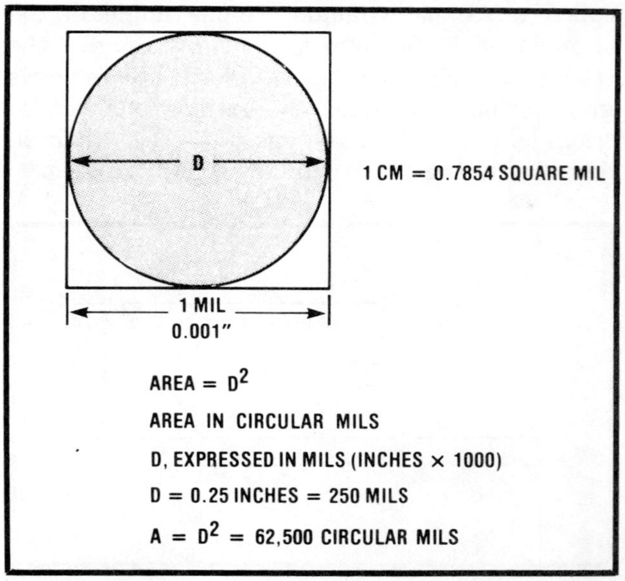

Fig. 20-8 Finding the area of a circle expressed in circular mils

DIAMETER	CIRCUMFERENCE	AREA
1	3.1416	0.7854
2	6.2832	3.1416
3	9.4248	7.0680
4	12.5664	12.5664
5	15.7080	19.635
6	18.850	28.274
7	21.991	38.485
8	25.133	50.266
9	28.274	63.617
10	31.416	78.540
11	34.558	95.033
12	37.699	113.10
13	40.841	132.73
14	43.982	153.94
15	47.124	176.71
16	50.265	201.06
17	53.407	226.98
18	56.549	254.47
19	59.690	283.53
20	62.832	314.16
21	65.973	346.36
22	69.115	380.13
23	72.257	415.48
24	75.398	452.39
25	78.540	490.87
26	81.681	530.93
27	84.823	572.56
28	87.965	615.75
29	91.106	660.52
30	94.248	706.86
31	97.389	754.77
32	100.53	804.25
33	103.67	855.30
34	106.81	907.92
35	109.96	962.11
36	113.10	1,017.88
37	116.24	1,075.21
38	119.38	1,134.11
39	122.52	1,194.59
40	125.66	1,256.64
41	128.81	1,320.25
42	131.95	1,385.44
43	135.09	1,452.20
44	138.23	1,520.53
45	141.37	1,590.43
46	144.51	1,661.90
47	147.65	1,734.94
48	150.80	1,809.56
49	153.94	1,885.74
50	157.08	1,963.50

DIAMETER	CIRCUMFERENCE	AREA
51	160.22	2,042.82
52	163.36	2,123.72
53	166.50	2,206.18
54	169.65	2,290.22
55	172.79	2,375.83
56	175.93	2,463.01
57	179.07	2,551.76
58	182.21	2,642.08
59	185.35	2,733.97
60	188.50	2,827.43
61	191.64	2,922.47
62	194.78	3,019.07
63	197.92	3,117.25
64	201.06	3,126.99
65	204.20	3,381.31
66	207.34	3,421.19
67	210.49	3,525.65
68	213.63	3,631.68
69	216.77	3,739.28
70	219.91	3,848.45
71	233.05	3,959.19
72	226.19	4,071.50
73	229.34	4,185.39
74	232.48	4,300.84
75	235.62	4,417.86
76	238.76	4,536.46
77	241.90	4,656.63
78	245.05	4,778.36
79	248.19	4,901.67
80	251.33	5,026.55
81	254.47	5,153.00
82	257.61	5,281.02
83	260.75	5,410.61
84	263.89	5,541.77
85	267.04	5,674.50
86	270.18	5,808.80
87	273.32	5,944.68
88	276.46	6,082.12
89	279.60	6,221.14
90	282.74	6,361.73
91	285.88	6,503.88
92	289.03	6,647.61
93	292.17	6,792.91
94	295.31	6,939.78
95	298.45	7,088.22
96	301.59	7,283.23
97	304.73	7,389.81
98	307.88	7,542.96
99	311.02	7,697.69
100	314.16	7,853.98

Fig. 20-7 Area and circumference of circles with a given diameter.

SECTION XXI

Measurement of Volume

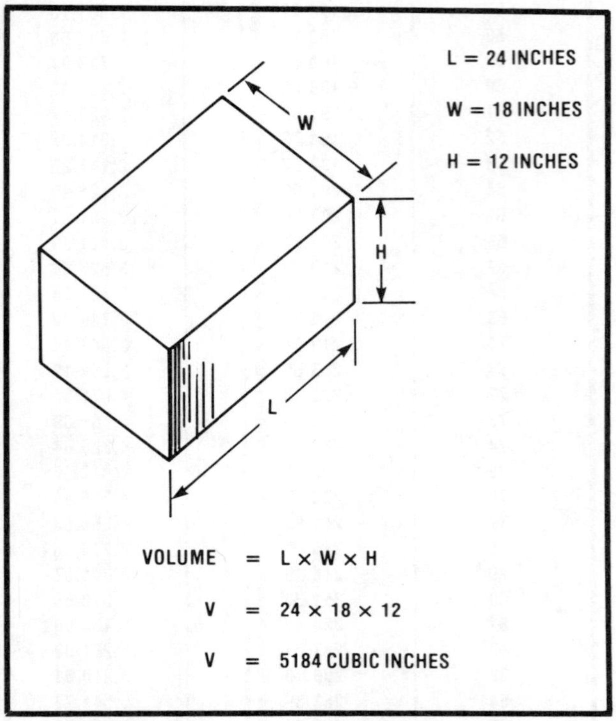

Fig. 21-1 Finding the volume of a rectangular solid

A. *Volume of a Rectangular Solid*

The volume of a rectangular solid is found by multiplying together the dimensions of length, width, and height of the object being considered. It is important that all of the units be in the same terms, and the volume will be given in cubic measurements of the same terms. A rectangular block having dimensions of 24 inches by 18 inches by 12 inches contains 5,184 cubic inches of material.

B. *Volume of a Cube*

A cube is a special form of rectangular solid in which all of the sides have the same length. The formula for the volume of a cube is simply:

$$\text{Volume} = \text{Side}^3$$

A six-inch cube of aluminum contains $6 \times 6 \times 6$, or 216, cubic inches of aluminum.

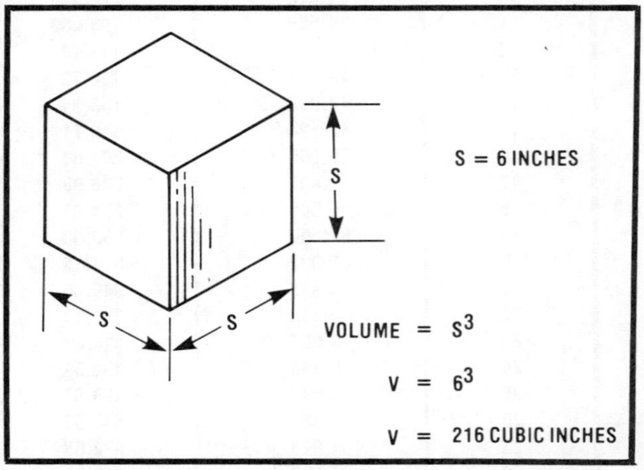

Fig. 21-2 Finding the volume of a cube

C. *Volume of a Cylinder*

A cylinder is a solid form having circular ends and parallel sides. Its volume may be found by multiplying together the area of one of the ends and the height of the side.

In Fig. 21-3, we have an eight-inch-long section of eight-inch-diameter aluminum rod, and we want

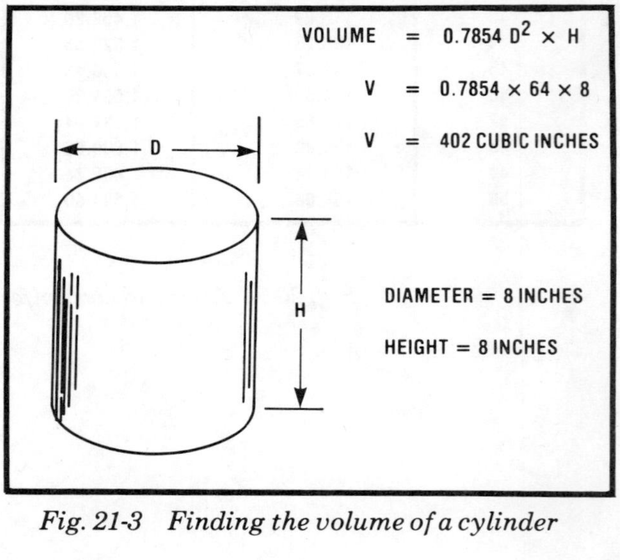

Fig. 21-3 Finding the volume of a cylinder

to know the amount of material it contains. We can find this by the formula:

$$\text{Volume} = 0.7854 \times D^2 \times H$$

By using this formula, we find that the rod contains 402 cubic inches of material.

D. Volume of a Sphere

A sphere is a solid form bounded by a surface on which all points are an equal distance from a point within, called the center. It is the shape which has the greatest volume for its surface area and is often used in aircraft as the shape for liquid oxygen converters and for hydraulic accumulators.

The volume of a sphere is found by multiplying the cube of its diameter by the constant 0.5236, which is one-sixth of pi. In Fig. 21-4, we want to find the volume of a six-inch-diameter sphere. Using the formula:

$$\text{Volume} = 0.5236 \times 6^3 = 0.5236 \times 216 = 113.1$$

we find that this six-inch-diameter sphere contains 113.1 cubic inches of material.

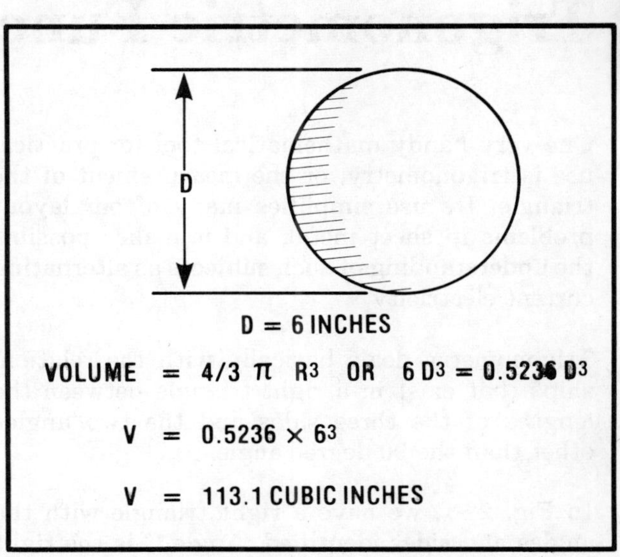

Fig. 21-4 Finding the volume of a sphere

SECTION XXII

Trigonometric Functions

One very handy mathematical tool for practical use is trigonometry, or the measurement of the triangle. Its use simplifies many of our layout problems in sheet metal, and it makes possible the understanding of such subjects as alternating current electricity.

Trigonometry deals basically with the relationships that exist in a right triangle between the lengths of the three sides and the two angles other than the 90-degree angle.

In Fig. 22-1, we have a right triangle with the angles and sides identified. Angle C is the right angle. For this explanation, we will use angle A as the angle for which we are setting up the relationships. Side *c* is the hypotenuse, which, by definition, is the side opposite the right angle. Side *a* is the side opposite angle A, and side *b* is the side adjacent, or next to, angle A.

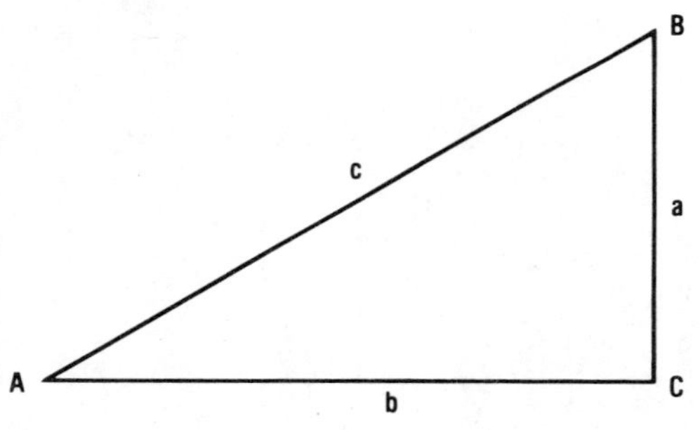

C = RIGHT ANGLE

c = HYPOTENUSE

A = ANGLE BEING CONSIDERED

a = SIDE OPPOSITE

b = SIDE ADJACENT

$$\text{SINE A (SIN A)} = \frac{\text{SIDE OPPOSITE}}{\text{HYPOTENUSE}} = \frac{a}{c}$$

$$\text{COSINE A (COS A)} = \frac{\text{SIDE ADJACENT}}{\text{HYPOTENUSE}} = \frac{b}{c}$$

$$\text{TANGENT A (TAN A)} = \frac{\text{SIDE OPPOSITE}}{\text{SIDE ADJACENT}} = \frac{a}{b}$$

Fig. 22-1 The trigonometric relations of a right triangle

The sine of angle A is the ratio of the length of the side opposite the angle to the length of the hypotenuse. For any degree of angle A, this ratio will be constant, regardless of the size of the triangle. In the chart of Fig. 22-2, we see that the sine of 30°, which is written Sin 30°, is 0.500. This means that the side opposite the thirty-degree angle will be 50%, or one-half, the length of the hypotenuse. For a 45° angle, the ratio is 0.7071. The side opposite the 45° angle is 0.7071 times the length of the hypotenuse.

Trigonometric charts usually list the angles only to 45 degrees in a single column, as we see by the numbers in the left-hand column, but notice that in the extreme right-hand column, the numbers start at 45 and go *upward* to 90 degrees. And the names of the columns at the bottom are opposite the names at the top of the same column. Reading up in the right column, we find 60 degrees, and in the "Sin" column (at the bottom of the chart) we find that Sin 60° = 0.8990.

The Cosine (Cos) of an angle is the ratio of the length of the side adjacent to the angle to the length of the hypotenuse, and is found in the same way we found the sine of the angle. The cosine of 25 degrees is 0.9063, and for 75 degrees, it is 0.2588.

The third ratio we are interested in for this brief introduction to mathematics is the tangent. This is the ratio of the length of the side opposite the angle, to the side adjacent the angle. The tangent of 20 degrees is 0.3640, and for 70 degrees is 2.747.

Remember that in any triangle, the sum of the angles is always 180 degrees, and all of the triangles we are considering in this use of trigonometry are right triangles. When we know the size of angle A, we can find angle B.

$$\text{Angle B} = 90° - \text{Angle A}$$

TRIGONOMETRIC FUNCTIONS					
DEG	SIN	COS	TAN		
0	.0	1.000	.0	+	90
1	.0175	.9999	.0175	57.29	89
2	.0349	.9994	.0349	28.64	88
3	.0523	.9986	.0524	19.08	87
4	.0698	.9976	.0699	14.30	86
5	.0872	.9962	.0875	11.43	85
6	.1045	.9945	.1051	9.514	84
7	.1219	.9926	.1228	8.144	83
8	.1392	.9903	.1405	7.115	82
9	.1564	.9877	.1584	6.314	81
10	.1737	.9848	.1763	5.671	80
11	.1908	.9816	.1944	5.145	79
12	.2079	.9782	.2126	4.705	78
13	.2250	.9744	.2309	4.331	77
14	.2419	.9703	.2493	4.011	76
15	.2588	.9659	.2680	3.732	75
16	.2756	.9613	.2868	3.487	74
17	.2924	.9563	.3057	3.271	73
18	.3090	.9511	.3249	3.078	72
19	.3256	.9455	.3443	2.904	71
20	.3420	.9397	.3640	2.747	70
21	.3584	.9336	.3839	2.605	69
22	.3746	.9272	.4040	2.475	68
23	.3907	.9205	.4245	2.356	67
24	.4067	.9136	.4452	2.246	66
25	.4226	.9063	.4663	2.145	65
26	.4384	.8988	.4877	2.050	64
27	.4540	.8910	.5095	1.963	63
28	.4695	.8830	.5317	1.881	62
29	.4848	.8746	.5543	1.804	61
30	.5000	.8660	.5774	1.732	60
31	.5150	.8572	.6009	1.664	59
32	.5299	.8481	.6249	1.600	58
33	.5446	.8387	.6494	1.540	57
34	.5592	.8290	.6745	1.483	56
35	.5736	.8192	.7002	1.428	55
36	.5878	.8090	.7265	1.376	54
37	.6018	.7986	.7536	1.327	53
38	.6157	.7880	.7813	1.280	52
39	.6293	.7772	.8098	1.235	51
40	.6428	.7660	.8391	1.192	50
41	.6561	.7547	.8693	1.150	49
42	.6691	.7431	.9004	1.111	48
43	.6820	.7314	.9325	1.072	47
44	.6947	.7193	.9657	1.036	46
45	.7071	.7071	1.0000	1.000	45
	COS	SIN		TAN	DEG

Fig. 22-2 Table of trigonometric functions

SECTION XXIII

Charts and Graphs

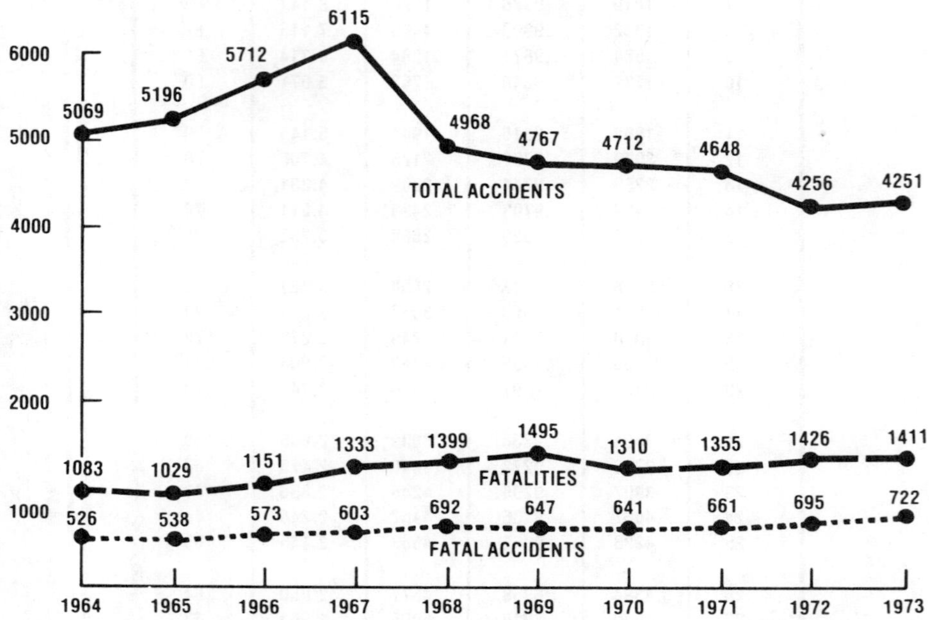

Fig. 23-1 A broken-line graph

Columns of figures and other forms of dry statistics are difficult to interpret. But, if all of these numbers are put into some graphic form, it is easy for us to visualize their significance and to use the information.

One of the most generally used methods of graphical presentation is the broken-line graph such as the one in Fig. 23-1. With this graph it is easy for us to see that while the total number of aircraft accidents has drastically dropped since 1967, when there were more than any previous year, the number of fatalities is gradually, but steadily, increasing.

In order to show comparison, we can use a bar graph such as the one shown in Fig. 23-2. Using this graph, we are able to visualize the portions of a flight in which most of the accidents occur. From this, we can see that the most of the fatal accidents occur during the landing phase of a flight, and the fewest occur during taxiing.

"Pie" charts, or circular graphs, are used to show the relationships that exist between the parts of a whole. For example, in the graphs of Fig. 23-3, we see the total number of instrument hours flown in 1975 and predicted to be flown in 1986, and these numbers are broken down by the type of user. From these two charts we can see that the total number of hours will greatly increase in 1986, and the percentage that will be flown by general aviation aircraft will increase, while the percentage flown by all other types will decrease.

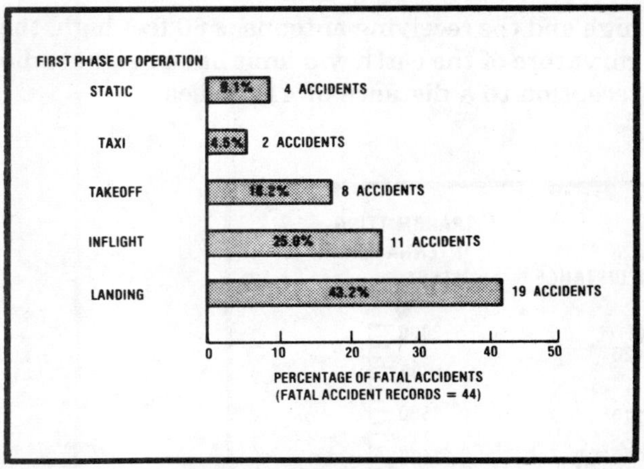

Fig. 23-2 A bar graph.

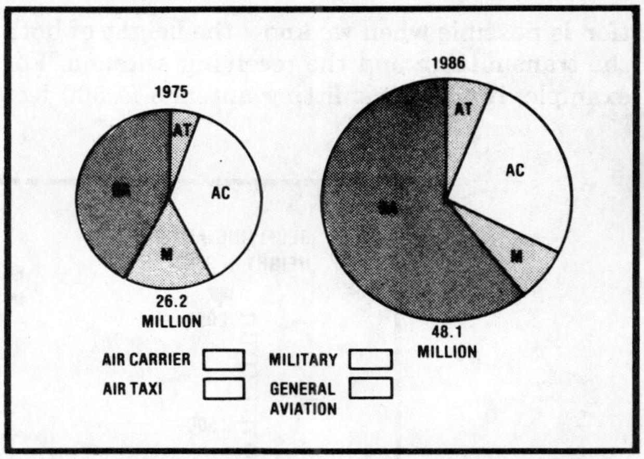

Fig. 23-3 A circular graph, or "pie" chart.

Continuous-curve graphs such as the one in Fig. 23-4 are used to show trends and general variations. Here we see the relationships between the freezing and boiling points of typical aviation fuels. From this graph, we can see that, generally speaking, the lower the freezing point, the lower will be the boiling point. This change is not necessarily uniform, as the separation between the two points widens out with fuels having the higher boiling points.

It is sometimes necessary to show the way *two* variable affect a value. In the nomograph of Fig. 23-5, we are able to find the distance radio recep-

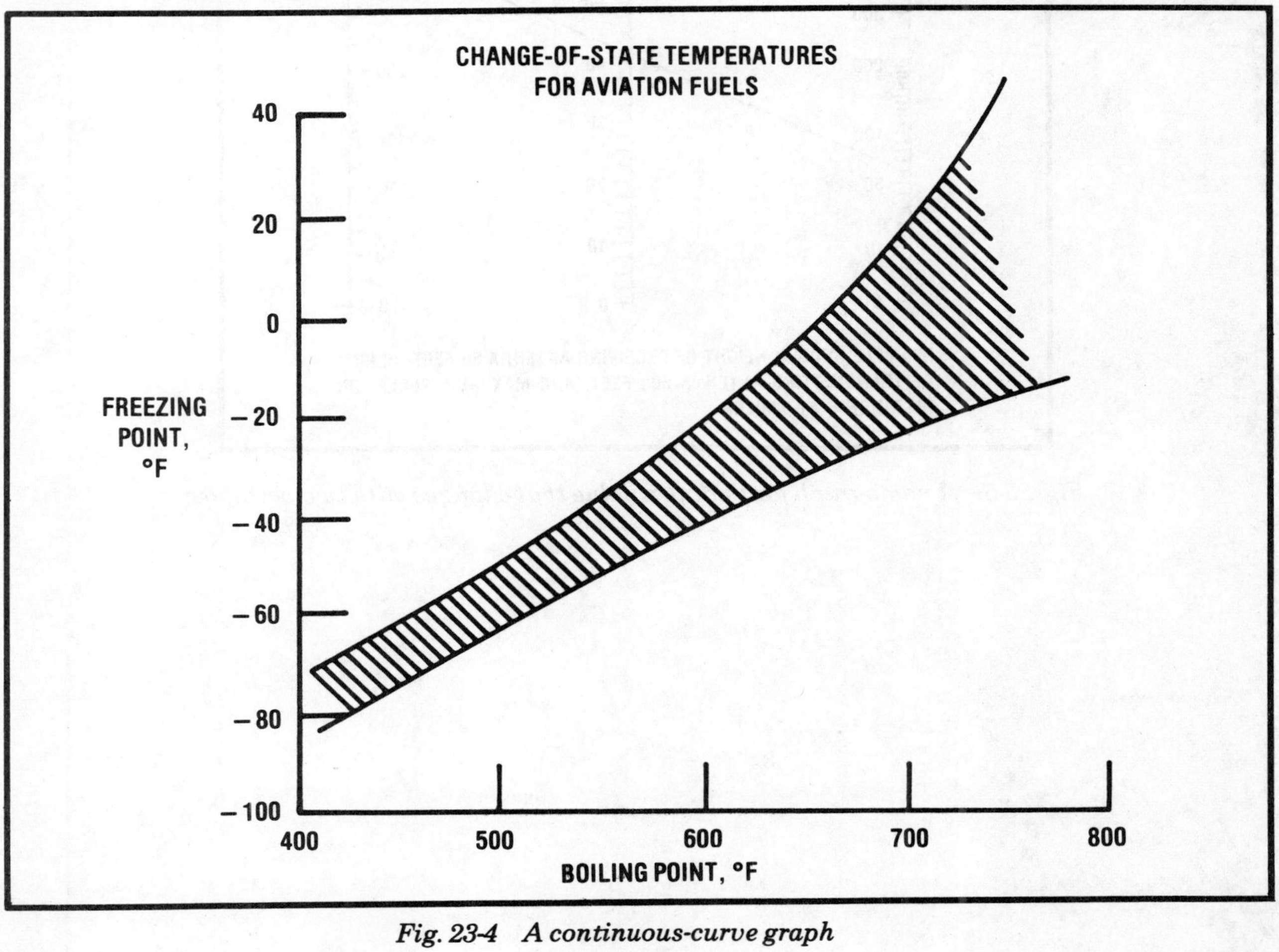

Fig. 23-4 A continuous-curve graph

tion is possible when we know the height of both the transmitting and the receiving antenna. For example, if the transmitting antenna is 500 feet high and the receiving antenna is 60 feet high, the curvature of the earth will limit line-of-sight radio reception to a distance of 41.5 miles.

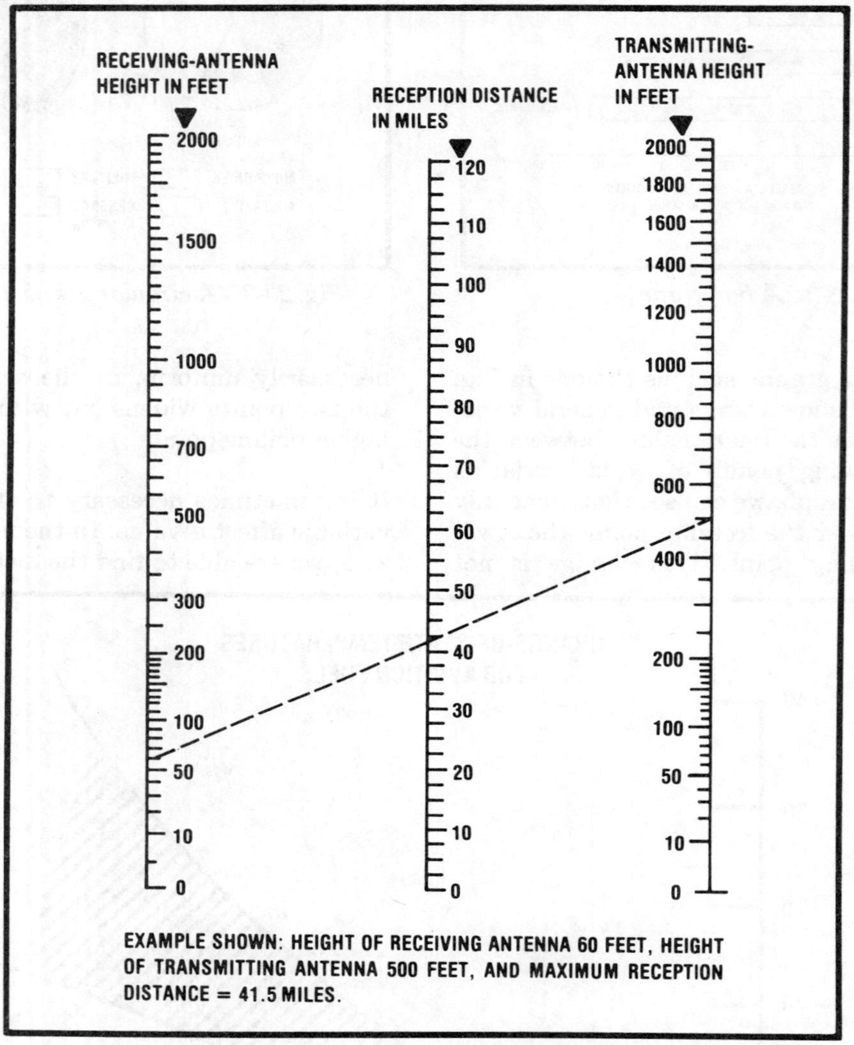

Fig. 23-5 A nomograph for finding a value that changes with two variables.

SECTION XXIV

English-Metric Conversion

Number	Prefix	Symbol	Scientific Notation
1,000,000,000,000	tera	t	1×10^{12}
1,000,000,000	giga	g	1×10^{9}
1,000,000	mega	m	1×10^{6}
1,000	kilo	k	1×10^{3}
100	hecto	h	1×10^{2}
10	deka	dk	1×10^{1}
0.1	deci	d	1×10^{-1}
0.01	centi	c	1×10^{-2}
0.001	milli	m	1×10^{-3}
0.000,001	micro	μ	1×10^{-6}
0.000,000,001	nano	n	1×10^{-9}
0.000,000,000,001	pico	p	1×10^{-12}

Fig. 24-1 Prefixes, symbols, and multiples for basic quantities

The International System of Units, which is now called SI, had its origin when the meter was selected as the unit of length and the kilogram as the unit of mass. These units were created by the Paris Academy of Science and adopted by the National Assembly of France in 1795.

The United States Congress *legalized* the use of the metric system throughout the United States on July 28, 1866, but it was not until December 23, 1975, that the Metric Bill was signed into law in the United States to convert all of our measurements into the metric system. It is becoming more and more important for us to be familiar with this system.

Multiples and Prefixes

One of the great advantages of the metric system is the fact that it is built on decimal units. Each basic unit may be multiplied or divided by ten as many times as it is necessary to get a convenient size. Each of these multiples has a definite prefix, symbol, and name.

The basic unit of length measurement is the meter, but if we want to express a smaller measurement, we can use the centimeter, 1/100 of a meter; or the millimeter, 1/1000 of a meter. If we want a larger unit, we can use the hektometer, 100 meters, or the kilometer, 1000 meters. Fig. 24-1 gives a list of the metric prefixes along with their symbols and scientific notation, or powers of ten, multipliers.

In Fig. 24-2, we have the linear conversion between the English and metric systems, with inches being given in both common fractions and decimals. The metric dimensions are all given in millimeters. (See following page.)

In Fig. 24-3 we have some of the commonly used English-metric conversions. (See following pages.)

Inches			Decimal Equivalent	Millimeter Equivalent
1/64			.0156	0.397
	1/32		.0313	0.794
3/64			.0469	1.191
		1/16	.0625	1.588
5/64			.0781	1.985
	3/32		.0938	2.381
7/64			.1094	2.778
		1/8	.1250	3.175
9/64			.1406	3.572
	5/32		.1563	3.969
11/64			.1719	4.366
		3/16	.1875	4.762
13/64			.2031	5.159
	7/32		.2188	5.556
15/64			.2344	5.953
		1/4	.2500	6.350
17/64			.2656	6.747
	9/32		.2813	7.144
19/64			.2969	7.541
		5/16	.3125	7.937
21/64			.3281	8.334
	11/32		.3438	8.731
23/64			.3594	9.128
		3/8	.3750	9.525
25/64			.3906	9.922
	13/32		.4063	10.319
27/64			.4219	10.716
		7/16	.4375	11.112
29/64			.4531	11.509
	15/32		.4688	11.905
31/64			.4844	12.303
		1/2	.5000	12.700
33/64			.5156	13.097
	17/32		.5313	13.494
35/64			.5469	13.891
		9/16	.5625	14.287
37/64			.5781	14.684
	19/32		.5938	15.081
39/64			.6094	15.478
		5/8	.6250	15.875
41/64			.6406	16.272
	21/32		.6563	16.669
43/64			.6719	17.067
		11/16	.6875	17.463
45/64			.7031	17.860
	23/32		.7188	18.238
47/64			.7344	18.635
		3/4	.7500	19.049
49/64			.7656	19.446

Inches			Decimal Equivalent	Millimeter Equivalent
	25/32		.7813	19.842
51/64			.7969	20.239
		13/16	.8125	20.636
53/64			.8281	21.033
	27/32		.8438	21.430
55/64			.8594	21.827
		7/8	.8750	22.224
57/64			.8906	22.621
	29/32		.9063	23.018
59/64			.9219	23.415
		15/16	.9375	23.812
61/64			.9531	24.209
	31/32		.9688	24.606
63/64			.9844	25.004
		1.0	1.0000	25.400

Fig. 24-2 Decimal and metric equivalent of inches.

To Convert From	To	Multiply By
acres	sq ft	4.356×10^{4}
atmospheres	cm Hg at 0°C	76.00
	in. Hg at 0°C	29.92
	lb/sq in	14.696
	bars	1.013
bars	lb/sq in	14.5
Btu	ft-lb	778.26
	kilowatt-hr	2.931×10^{-4}
	joules	1055
Btu/sec	watts	1055
centimeters	in	0.3937
cm Hg	in. H_2O at 4°C	5.354
	lb/sq in	1.934×10^{-1}
cm/second	Ft/sec	3.281×10^{-2}
circular mils	sq in	7.854×10^{-7}
cu centimeters	cu in	6.102×10^{-2}
	U.S. gal	2.642×10^{-4}
cu ft	cu cm	2.832×10^{4}
	U.S. gal	7.481
	liters	28.32
cu ft H_2O	lb	62.428
cu in.	cu cm	16.39
	liters	1.639×10^{-2}
	U.S. gal	4.329×10^{-3}
cu meters	U.S. gal	264.2
deg [arc]	radians	1.745×10^{-2}
feet	meters	3.048×10^{-1}
ft/min	mph	1.136×10^{-2}
	km/hr	1.829×10^{-2}
ft/sec	mph	.6818
	cm/sec	30.48
	knots	.5925

To Convert From	To	Multiply By
ft-lb/min	hp	3.030×10^{-5}
ft-lb/sec	hp	1.818×10^{-3}
	kilowatts	1.356×10^{-3}
fluid oz	dram	8
gal, Imperial	cu in.	277.4
	U.S. gal	1.201
	liters	4.546
gal, U.S. dry	U.S. gal, liquid	1.164
gal, U.S. liquid	cu in.	231.0
grams	oz avdp	3.527×10^{-2}
	lb avdp	2.205×10^{-3}
grams/cm	lb/ft	6.721×10^{-2}
hectopieze	in. Hg	29.53
horsepower	ft-lb/min	33,000
	ft-lb/sec	550
	m-kg/sec	76.04
	kilowatts	7.457×10^{-1}
	Btu/sec	7.068×10^{-1}
horsepower, metric	hp	9.863×10^{-1}
in.	cm	2.540
in. water at 4°C	in. Hg at 0°C	7.355×10^{-2}
kilograms	lb	2.205
	oz	35.27
kilometers	ft	3.281×10^{3}
	miles	6.214×10^{-1}
	nautical miles	5.400×10^{-1}
km/hr	ft/sec	9.113×10^{-1}
	knots	5.396×10^{-1}
	mph	6.214×10^{-1}
kilowatts	Btu/sec	9.480×10^{-1}
	hp	1.341

Fig. 24-3 Conversion factors

To Convert From	To	Multiply By
knots	nautical mph ft/sec mph	1.0 1.688 1.151
liters	cu cm cu in. U.S. gal	10^3 61.03 2.642×10^{-1}
meters	in. ft	39.37 3.281
meter-kilogram	ft-lb	7.233
meter/sec	ft/sec	3.281
microns	in.	3.937×10^{-5}
miles [stat.]	ft km	5280 1.609
mph	ft/sec km/hr knots	1.467 1.609 8.690×10^{-1}
milibars	in. Hg at 0°C	2.953×10^{-2}
nautical miles [naumiles]	ft miles [stat.] m	6076.1 1.151 1852
ounces, avdp	grams	28.35

To Convert From	To	Multiply By
ounces, fluid	cu in.	1.805
lb/avdp	grams ounces	453.6 16.0
lb/cu in.	grams/cu cm	27.68
lb/sq in.	in. Hg at 0°C	2.036
radians	deg [arc]	57.30
radians/sec	deg/sec rev/min	57.30 9.549
revolutions	radians	6.283 [2π]
rev/min	radians/sec	1.047×10^{-1}
slug	lb	32.174
sq cm	sq in.	1.550×10^{-1}
sq ft	sq cm	929.0
sq in.	sq cm	6.452
sq meters	sq ft	10.76
sq miles	sq km	2.590
watts	Btu/sec	9.481×10^{-4}

PART II

PHYSICS

SECTION I

Introduction to Physics

Aeronautics is the branch of science which deals with the operation of aircraft, and as a science it operates with absolutes whose properties and values behave in the same way every time because they cannot function otherwise.

A knowledge of basic physics gives us the understanding of the laws which determine the operation of our engines, the transmission of power by our electrical and hydraulic systems, and especially the flight of our aircraft. We can gain through this study an appreciation for the forces that act on aircraft structures, so we will be able to design repairs that will give back to damaged aircraft the strength they are required to have.

SECTION II

Matter

A. Chemical Nature of Matter

In the world in which we live, everything we come into contact with may be classified as some form of matter. Matter exists in an untold number of forms. The air we breathe, the water we drink, the food we eat, and the clothes we wear, as well as the airplanes we fly in and work on are all forms of matter.

By a very loose definition, matter is anything that occupies space, and since we are primarily concerned with matter that exists on our earth, we can say that it also has weight, or is attracted by the earth's gravity.

Man has not been given the prerogative to create or to destroy matter, but fortunately for us, we are able to change its form and to extract from it an extremely useful property, energy.

More years ago than we like to think about, energy from our basic source, the sun, reached the earth, and was converted into matter in the form of animals and plant life. In a great upheaval, this living matter and these creatures were crushed beneath countless billions of tons of earth and have become what we know today as fossil fuel: petroleum products and coal. When these forms of matter are brought to the surface, and mixed with the proper amount of another form of matter, oxygen, they will release the energy they have held trapped for so long, and we can use it to produce light and heat, the same forms of energy that were brought to earth so long ago.

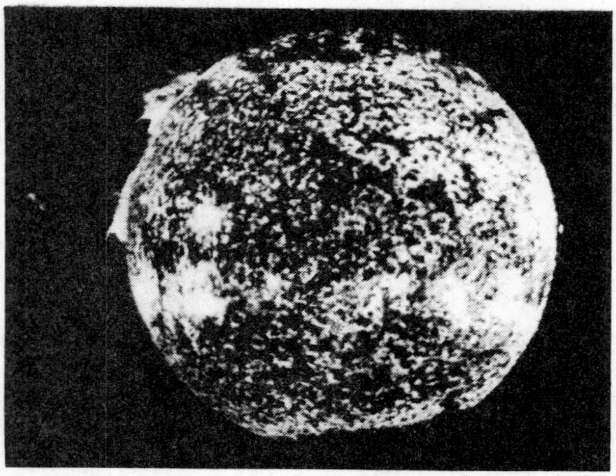

Courtesy of NASA.

Fig. 2-1 The sun is our chief source of energy.

When we break all matter down to its most fundamental form, we see that it is composed of the basic building blocks of the universe: electrons, protons, and neutrons. An atom is the smallest particle of a chemical element which can exist either alone or in combination with other atoms. It consists of a nucleus, containing positive charges of electricity called protons, and neutrons, which have the same mass as the protons, but which are electrically neutral. Spinning around the nucleus in orderly rings, or shells, are negatively charged electrons, which are much lighter than the protons, their mass being only about 1/1847 that of a proton.

Hydrogen is the simplest atom, as it has only one proton in the nucleus, and no neutrons. Spinning around this proton is one electron.

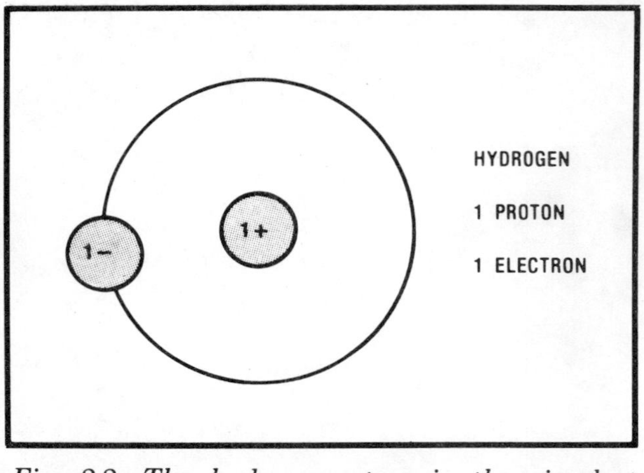

Fig. 2-2 The hydrogen atom is the simplest atom, having only one proton and one electron.

A more complex atom is oxygen which contains eight protons and eight neutrons in the nucleus, and spinning around this nucleus are eight electrons. The protons and electrons in the hydrogen

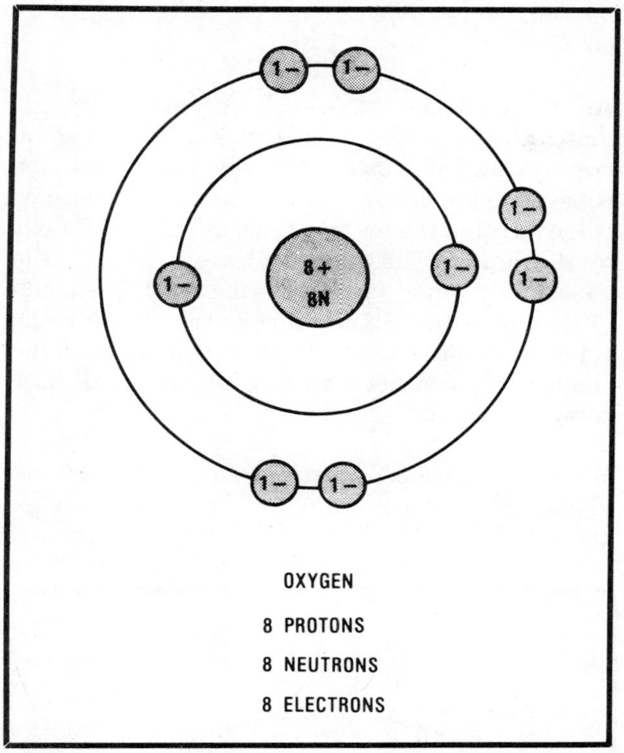

Fig. 2-3 The oxygen atom has eight protons and eight electrons, along with eight neutrons.

and oxygen atoms are exactly the same. It is when they link together in their specific combination that they form matter that has specific characteristics.

The oxygen atom has two electrons in its inner ring, or shell, and in its outer ring, it has six. But, oxygen could accept two more electrons in its outer ring if there were protons to balance them. So, if two atoms of hydrogen are brought near an atom of oxygen, the electrons from the hydrogen will link with electrons in the outer ring of the oxygen atom and form a molecule, the smallest particle of an entirely different substance, the chemical compound water.

B. *Physical Nature of Matter*

1. *Physical states*

All of the atoms and molecules in all forms of matter are constantly in motion, in a motion which is caused by the heat energy in the material. The amount of this motion determines the physical state of the matter.

a. *Solid*

If the amount of heat energy in a certain material is low, the molecular movement is relatively weak, and the cohesive force within the molecule holds the matter in a rigid form which has a definite volume and shape. Water in its solid form is familiar to us as ice.

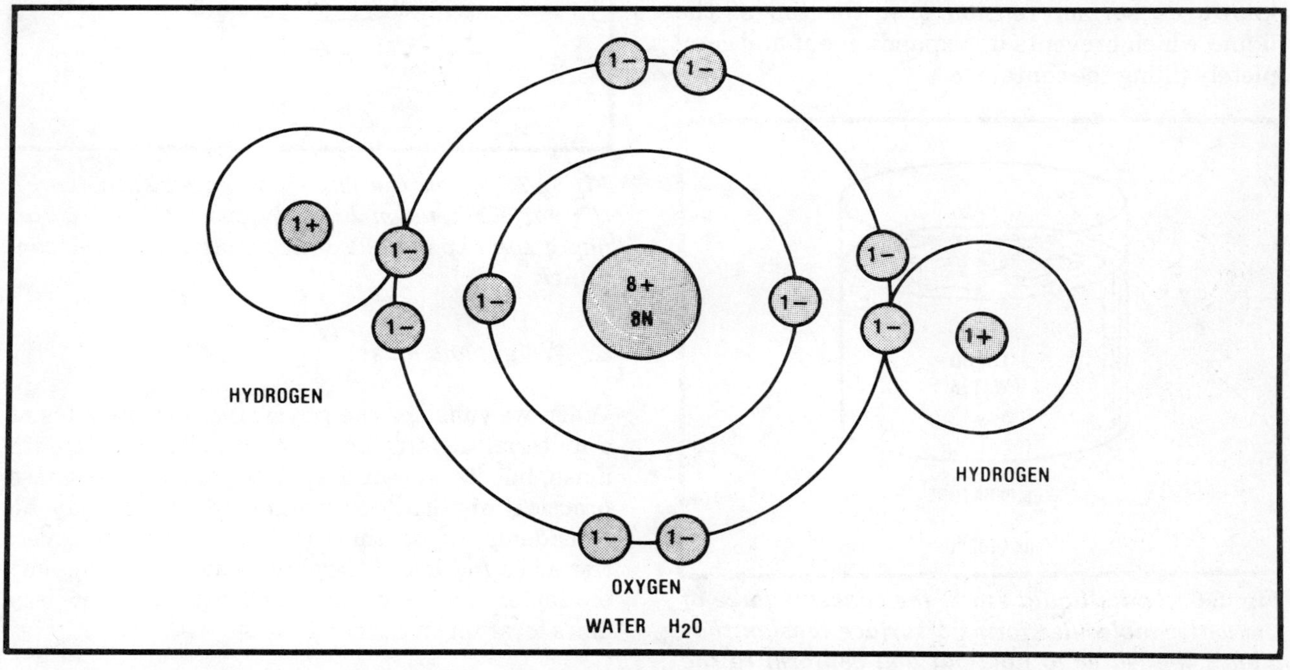

Fig. 2-4 A molecule of water (H_2O) is formed when two atoms of hydrogen join one atom of oxygen.

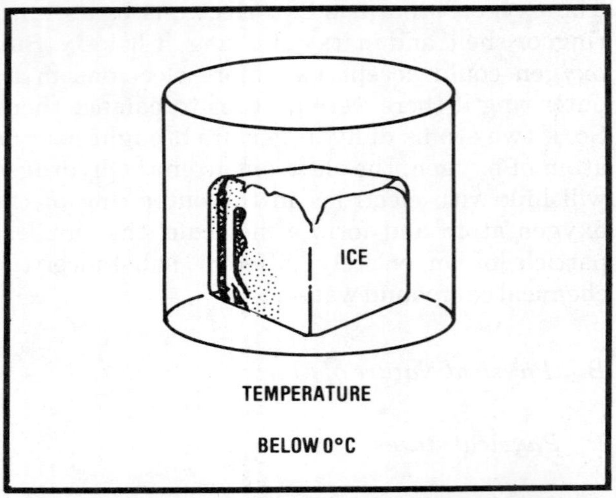

Fig. 2-5 Water exists in its solid state, ice, when the cohesive force of the molecules holds them in a rigid shape.

b. Liquid

When heat energy is added to solid matter, the molecular movement is speeded up, until it becomes great enough to overcome the cohesive force that held the solid in its rigid shape. The volume of the material does not greatly change, but it will conform to the shape of the container that holds it. When this happens, it has become a liquid, which is the most familiar form for water. The cohesion of the molecules forms a force known as surface tension over the top of the liquid which prevents its expanding out and completely filling the container.

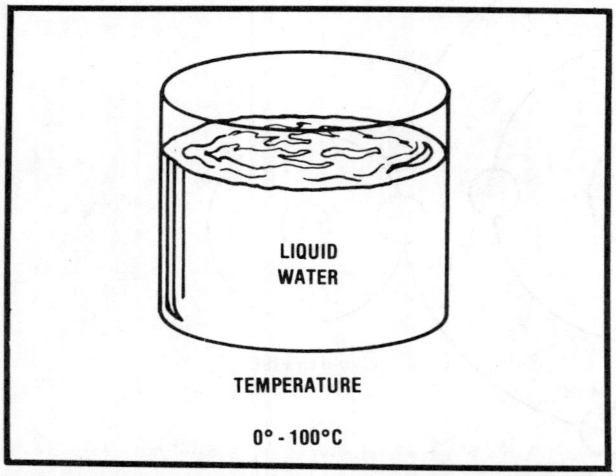

Fig. 2-6 In its liquid state, the cohesive force of the water molecules forms a surface tension that allows the water to flow out and conform to the shape of the container, but prevents its escaping from the surface.

c. Gas

As we continue to add heat energy to our material, the molecular movement is further increased, and it will reach a point at which the cohesive force is so weakened that the molecules will overcome the surface tension and will fly out of the liquid. These molecules are now in the gaseous, or vapor, form of the material. Chemically they are exactly the same as they were in the solid or liquid state, but they will assume the shape of the container and will expand until they completely fill it.

Water in its gaseous form is water vapor, an invisible gas, slightly more than half as heavy as air.

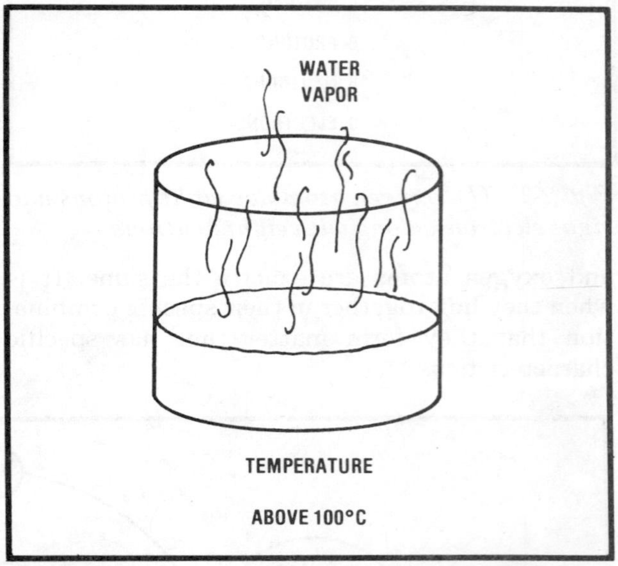

Fig. 2-7 When the water is heated sufficiently, the surface tension breaks down and the water vapor can expand out and completely fill the container.

2. Weight and mass

When we consider the physical characteristics of a material, it is difficult to actually measure its mass, but its weight is easy to find, and for most practical applications weight and mass may be considered to be the same. Actually, however, there is a marked difference. Mass is the amount of matter in a body, and it is constant, regardless of its location on the earth.

Weight, on the other hand, is the force with which the gravity of the earth attracts the mass, and it

Courtesy of NASA.

Fig. 2-8 A body can exist in a weightless condition even though its mass is not changed.

varies with the distance between the body and the center of the earth. Mass and weight are both measured in pounds in the English system and in grams or kilograms in the metric system. Mass may be found by dividing the weight of the object by the acceleration due to gravity.

$$\text{Mass} = \frac{\text{Weight}}{\text{Acceleration due to gravity}}$$

$$= \frac{\text{Weight}}{32.2}$$

The acceleration figure of 32.2 is obtained from the fact that under standard conditions the force of gravity will cause a freely falling body to accelerate, or speed up, 32.2 feet per second each second it falls.

3. Density

The density of a material is its mass per unit volume, and it varies with temperature in both liquids and gases.

$$\text{Density} = \frac{\text{Mass}}{\text{Volume}}$$

The density of pure water, expressed in the English system is 62.3 pounds per cubic foot, and in the metric system it is one gram per cubic centimeter.

4. Specific gravity

The ratio of the density of a solid or a liquid to the density of pure water at 4°C is known as its specific gravity, and it is expressed as a dimensionless number; that is a number having no units. Any material having a specific gravity of less than one will float in water, while materials having specific gravities higher than one will sink. An example of a practical use of specific gravity with which aviation maintenance technicians are quite familiar is the specific gravity of the electrolyte of a lead-acid battery. When the battery is fully charged, the float in the hydrometer will sink only enough to indicate that the electrolyte has a specific gravity of about 1.275.

Specific gravity is concerned with the density of the material, and so the temperature must be considered. For this reason, the specific gravity should be measured at a specified temperature. Battery electrolyte is measured at 80° Fahrenheit. If the electrolyte is at a different temperature, a correction must be applied.

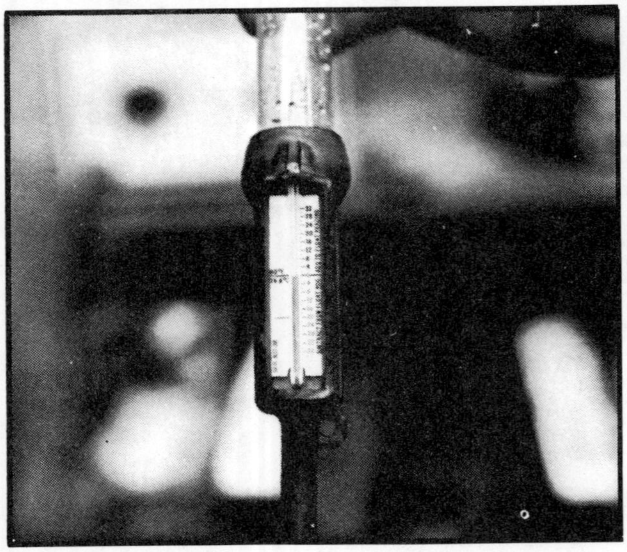

Fig. 2-9 The specific gravity of a liquid is measured with a hydrometer.

The specific gravity of gases is measured using air as the reference. Hydrogen has a specific gravity of 0.0695 and was therefore used for inflating many of the early balloons and dirigibles. Nonflammable helium is favored for this use today for safety reasons, even though its specific gravity is only 0.138 and much more is needed to lift an equivalent weight.

LIQUID		SOLIDS	
Gasoline	0.72	Ice	0.917
Jet fuel JP-4	0.785	Aluminum	2.7
Jet fuel JP-5	0.871	Titanium	4.4
Alcohol (ethyl)	0.789	Zinc	7.1
Kerosine	0.82	Iron	7.9
Lubricating oil	0.89	Brass	8.4
Synthetic oil	0.928	Copper	8.9
Water	1.000	Lead	11.4
Sulfuric acid	1.84	Gold	19.3
Mercury	13.6	Platinum	21.5

GASES	
Hydrogen	0.0695
Helium	0.138
Acetylene	0.898
Nitrogen	0.967
Air	1.000
Oxygen	1.105
Carbon dioxide	1.528

Pure water weighs 62.4 pounds per cubic foot or 1 gram per cubic centimeter.

Standard air weighs 0.07651 pounds per cubic foot or 1.225 kilograms per cubic meter.

Fig. 2-10 Specific gravity of various solids, liquids, and gases

SECTION III

Energy

In 1905, the great scientist Albert Einstein stated that there was a relationship between mass and energy. And without having to dig into the inner depths of matter, we can see this relationship. Our chief source of energy is the sun, and energy from the sun is transmitted to the earth and absorbed by trees as they grow into solid matter. These trees can be cut up into firewood and burned in our fireplace where they are converted back into energy in the form of heat and light, and a little solid residue known as ash. The energy that is in the burned fuel is not used up, but it is rather changed into some other forms. Some of these are usable to us, and others, because of our lack of technological knowledge, are at the present unusable.

Energy, as we know it in its practical form, can be thought of as the capacity of an object or a body to perform work, and it is classified into two rather broad types; potential and kinetic.

A. Potential Energy

Potential energy is that energy stored in a material because of its position or configuration, and while doing no work, it is capable of doing work.

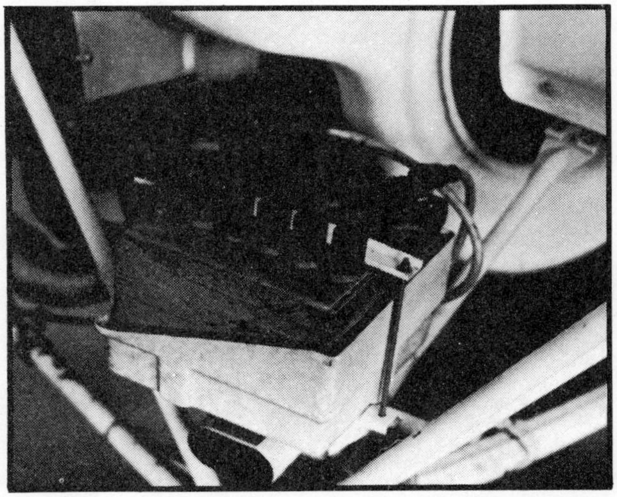

Fig. 3-1 An aircraft battery stores potential energy in the form of chemical energy.

Chemical energy may be stored in an aircraft battery and is there to produce mechanical work when the starter switch is pressed. Electrical energy may be stored in a capacitor to produce light when a strobe light is fired. Mechanical energy may be stored in a compressed spring or in the tremendous weight of water stored behind a dam.

B. Kinetic Energy

When potential energy is released in such a way that it causes motion, it has been changed into kinetic energy.

When you raise a hammer above your head, you convert potential energy from the food you have eaten into kinetic energy which is stored in the hammer because of its position, or height. Then, as you bring the hammer down, the potential energy stored in it, plus the kinetic energy put into it by your muscles speeding up its movement, gives it a great deal of kinetic energy. When the hammer strikes a nail, work is done forcing the nail into the wood, and some of the energy is dissipated in the form of heat caused by friction between the nail and the wood.

Fig. 3-2 The hammer forces the nail into the wood by the kinetic energy due to its motion.

C. *Units of Energy*

Our entire lifestyle and economy are based on energy, and one of the greatest searches of our modern age is for more sources of this important commodity.

In order to better understand energy, we must have units with which to express it, and in Fig. 3-3 we have several of the most generally used terms.

Unit	Value in Foot-Pounds
British Thermal Unit (Btu)	778.
Small Calorie (cal.)	3.087
Large Calorie (Cal.)	3,087.
Kilowatt Hour (Kw-hr)	2,655,000.
Joule	0.7376
Erg	7.376×10^{-8}

Fig. 3-3 Relationship between heat energy and work.

SECTION IV

Work

The end result of our use of energy and of almost all of our machines is for them to produce work. By definition, work is the product of a force applied to an object, times the distance the force causes the object to move. If an engine weighing 400 pounds is lifted 10 feet, 4000 foot-pounds of work has been done. In the metric system, the unit of work is the meter-kilogram and is the amount of work done when one kilogram of weight is lifted through a distance of one meter.

SECTION V

Power

When we speak of work, we do not take into consideration the amount of time required to do the work. A very small, low-powered motor can be geared to lift a large weight if the time is not critical, but if it is important to lift it quickly, more power is required.

Power may be defined as the time-rate of doing work, and in the English system is expressed in foot-pounds per second. In the metric system, the unit of power is the meter-kilograms per second. Power may be found by the formula:

$$\text{Power} = \frac{\text{Force} \times \text{Distance}}{\text{Time}}$$

1 Joule	=	0.7376 foot-pounds
1 Watt	=	1 Joule per minute
1 Kilowatt	=	1.34 Horsepower
1 Horsepower	=	33,000 foot-pounds per minute
1 Horsepower	=	746 Watts

Fig. 5-1 Relationship between work and power

For practical measurement of power, the horsepower is used. This unit was first used by James Watt in the comparison of the performance of his steam engine with the typical English dray horse. One horsepower is the amount of power required to do 33,000 foot-pounds of work in one minute, or 550 foot-pounds of work in one second. Electrical power is expressed in terms of watts, with one horsepower being equal to 746 watts.

We can find the amount of power required to do a given job by using the horsepower equation. For example, to find the electrical power needed to raise a 12,000-pound airplane six feet off of the hangar floor in one-half minute, we would use the formula in Fig. 5-2.

AIRPLANE WEIGHT = 12,000 POUNDS

HEIGHT = 6 FEET

TIME = 1/2 MINUTE

$$\text{HORSEPOWER} = \frac{12{,}000 \times 6}{33{,}000 \times 1/2}$$

$$= \frac{12{,}000 \times 6}{33{,}000} \times \frac{2}{1}$$

$$= \frac{12{,}\cancel{000} \times 6 \times 2}{33{,}\cancel{000}}$$

$$= \frac{144}{33}$$

$$= 4.36 \text{ Horsepower}$$

Fig. 5-2 Finding the horsepower required to raise a 12,000-pound airplane six feet in one-half minute

You will notice that we can do a little mathematical juggling that simplifies the problem and makes it easier to work. We first set it up as 12,000 multiplied by 6, and divided by 33,000 multiplied by 1/2. We can divide by the 1/2 by inverting it and multiplying. Our problem has now become 12,000 × 6 × 2 divided by 33,000. We can cancel the three zeros in both the 12,000 and the 33,000 by dividing both numbers by 1000. We now have simply 12 × 6 × 2 divided by 33. This is 144/33, which gives us a horsepower of 4.36 required to raise the airplane six feet in one-half minute.

SECTION VI

Force

We have seen that work is the product of the amount of force applied to an object, times the distance through which the object is moved. In practical machines, we very often use a mechanical advantage to change the amount of force required to move an object, and some of the simplest mechanical advantage devices we have are the lever, the inclined plane, the pulley, and gears.

A. The Lever

In its most basic form, we can think of a lever as a board or a bar having no weight of its own but supported by a fulcrum, or a sharp line contact between two weights near its ends. In Fig. 6-1 the weight A has a tendency to rotate the board in a counterclockwise direction, while weight B tends to rotate it clockwise. Each weight produces a "moment," or a turning force that is equal to the product of the weight in pounds or grams, times the distance between the center of gravity of the weight and the fulcrum, measured in inches, feet, or meters, whichever is appropriate. The lever is balanced when the clockwise moments exactly balance the counterclockwise moments. When we arbitrarily assign signs to these directions, we can say that the lever is balanced when the algebraic sum of the moments is zero. If a ten-pound weight is located six feet to the left of the fulcrum, it has a moment of negative 60 foot-pounds. The moment of weight B and arm b is a positive 60 foot-pounds, and so the sum of the moments is zero, and the lever is balanced.

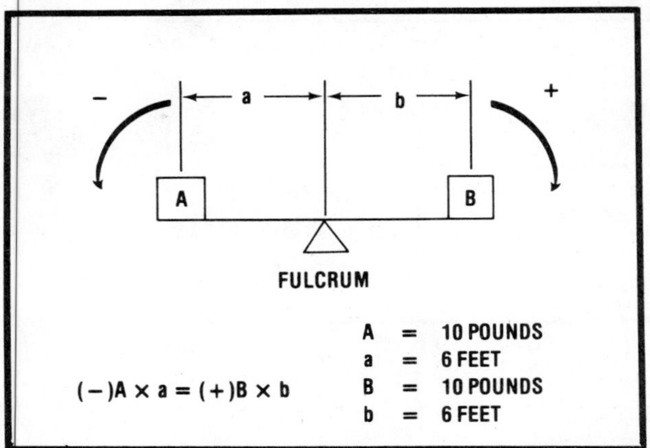

Fig. 6-1 In a balanced lever, the sum of the moments is zero.

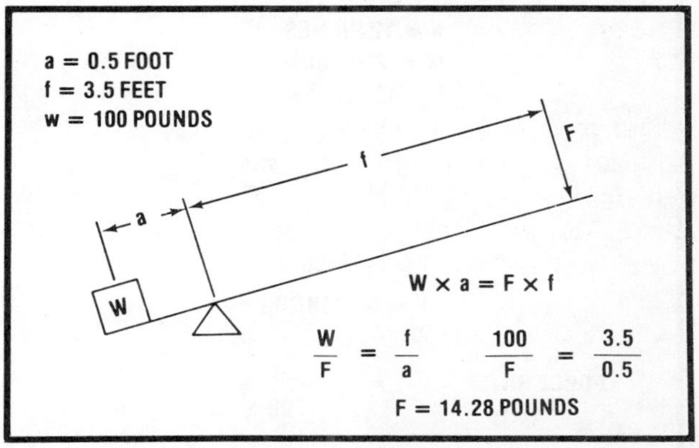

Fig. 6-2 A first-class lever

1. First-class lever

In Fig. 6-2 we have a practical application of a first-class lever. If we place the end of a four-foot bar under a 100-pound weight so the fulcrum is one-half foot from the center of gravity of the weight, we will have three and a half feet between the weight and the point at which we apply the force, so we need to use a force of only 14.28 pounds to move the 100-pound weight. We cannot gain any work with a lever, since work is the product of force and distance; so in order to lift the weight three inches, we must move the force through a distance of 21 inches. We can prove this by finding the moments:

$$(-)3 \times 100 = (+)21 \times 14.28$$

$$-300 = +300$$

2. Second-class lever

It is not always desirable to have the force acting in the direction opposite that which the weight is moved. For example, in a wheelbarrow, it is necessary to lift the weight upward, but we want to use the mechanical advantage of the lever to reduce the force required to carry the load.

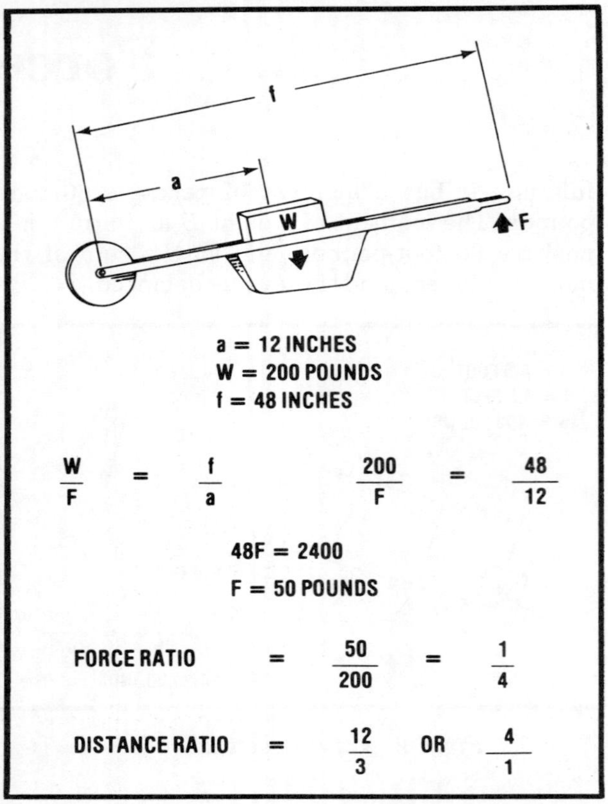

Fig. 6-3 A wheelbarrow is a form of second-class lever.

In the example of Fig. 6-3, we have a wheelbarrow in which the weight is concentrated 12 inches from the wheel axle, and the force is applied 48 inches from the axle. It requires a force of only 50 pounds to lift 200 pounds of weight, but the force must be moved through a distance of 12 inches to lift the support under the weight three inches off of the ground.

The mechanical advantage of a second-class lever is exactly the same as that of a first-class lever with the distances *a* and *f* the same in each case. The second-class lever is folded back on itself and the direction of the force application is opposite that of a first-class lever.

3. *Third-class lever*

A third-class lever is used to gain speed of movement at the cost of requiring a greater force to make the movement. In Fig. 6-5, we see a landing gear retracting mechanism. The force applied by the hydraulic cylinder is applied between the fulcrum and the weight. In order to raise a 200-pound wheel, the cylinder will have to apply a force of 1600 pounds, which is easy for a hydraulic cylinder. The piston need move only six inches to raise the wheel four feet, and it requires the same time for the wheel to move up the four feet as it does for the piston to move the six inches.

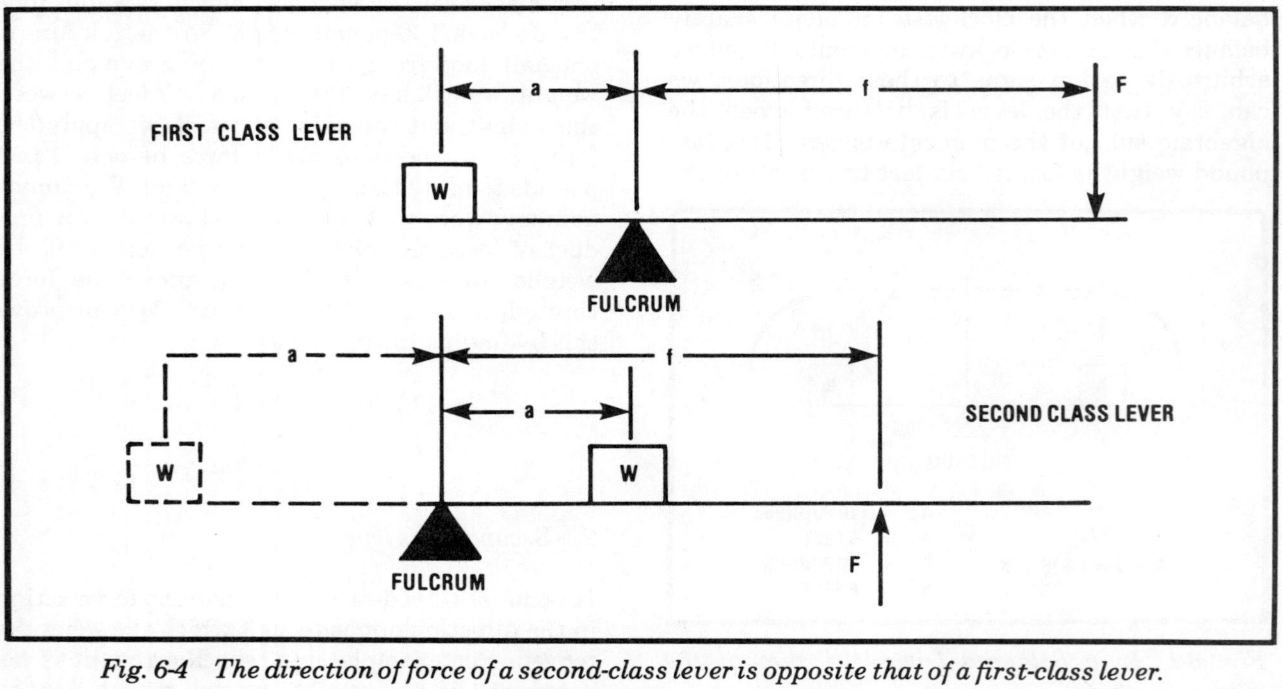

Fig. 6-4 The direction of force of a second-class lever is opposite that of a first-class lever.

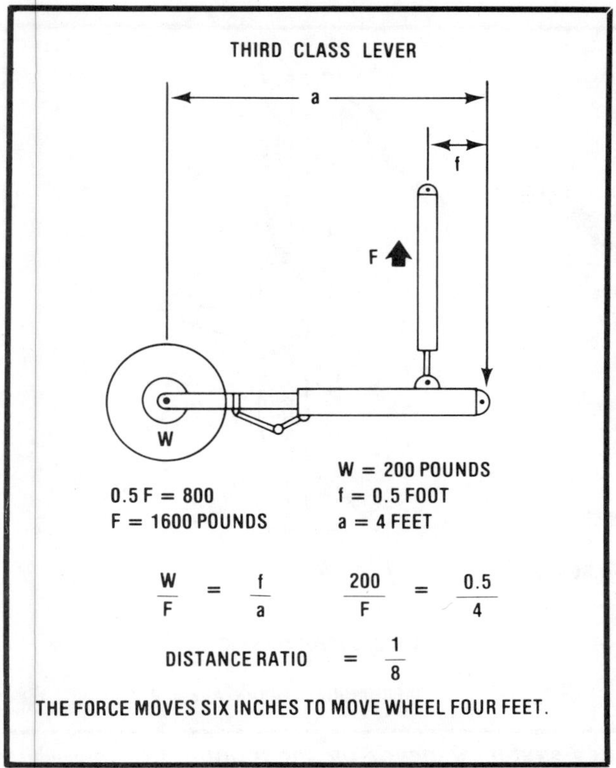

Fig. 6-5 A third-class lever

B. The Inclined Plane

The inclined plane is a form of simple machine that follows the law of the lever. If a 12-foot ramp is placed against a 4-foot loading dock, we can find the force needed to roll a 500-pound barrel up the slope.

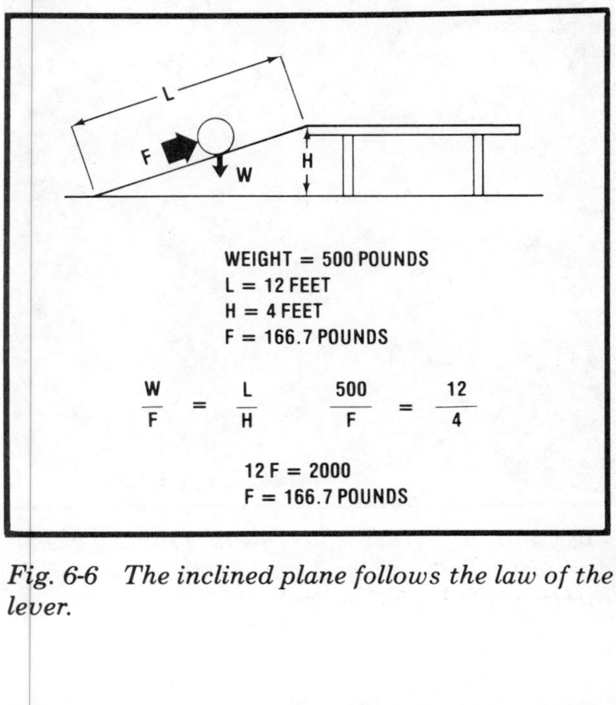

Fig. 6-6 The inclined plane follows the law of the lever.

Using the equation of Fig. 6-6, we find that it requires a force of 166.7 pounds to roll the barrel to the top. Lifting the barrel directly would require 500 × 4, or 2000, foot-pounds of work to be done, and since we are applying a force of 166.7 pounds for 12 feet, there is also 2000 foot-pounds of work being done.

Wedges and screw jacks are both examples of practical inclined plane applications used in our everyday life.

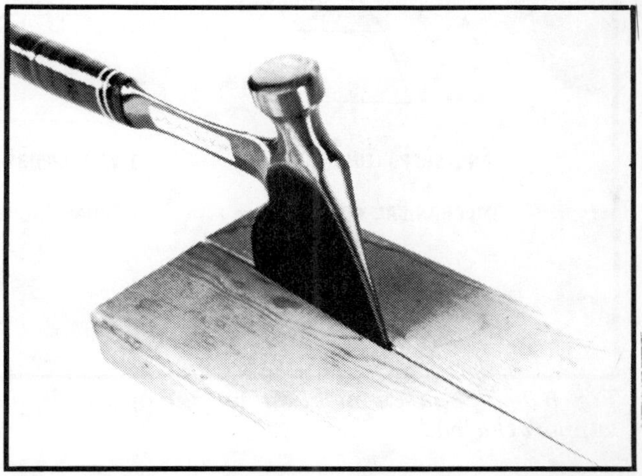

Fig. 6-7 A wedge is a form of inclined plane, and it also follows the law of the lever.

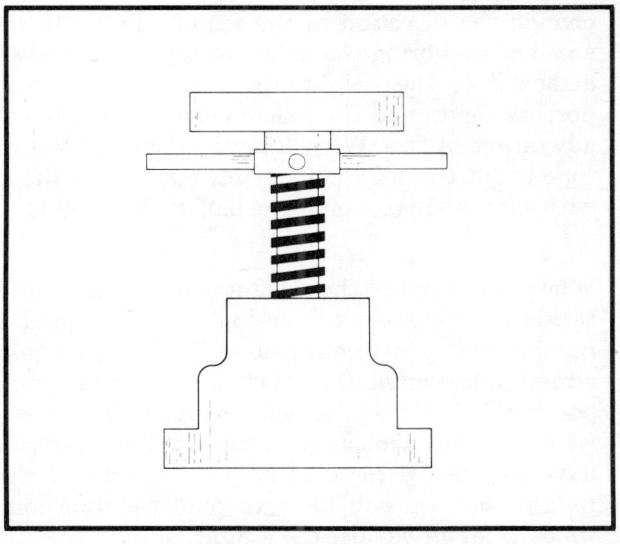

Fig. 6-8 A screw-type jack is a form of inclined plane.

C. The Pulley

The pulley is one of our basic machines in which we gain mechanical advantage by passing ropes around the circumference of a grooved wheel.

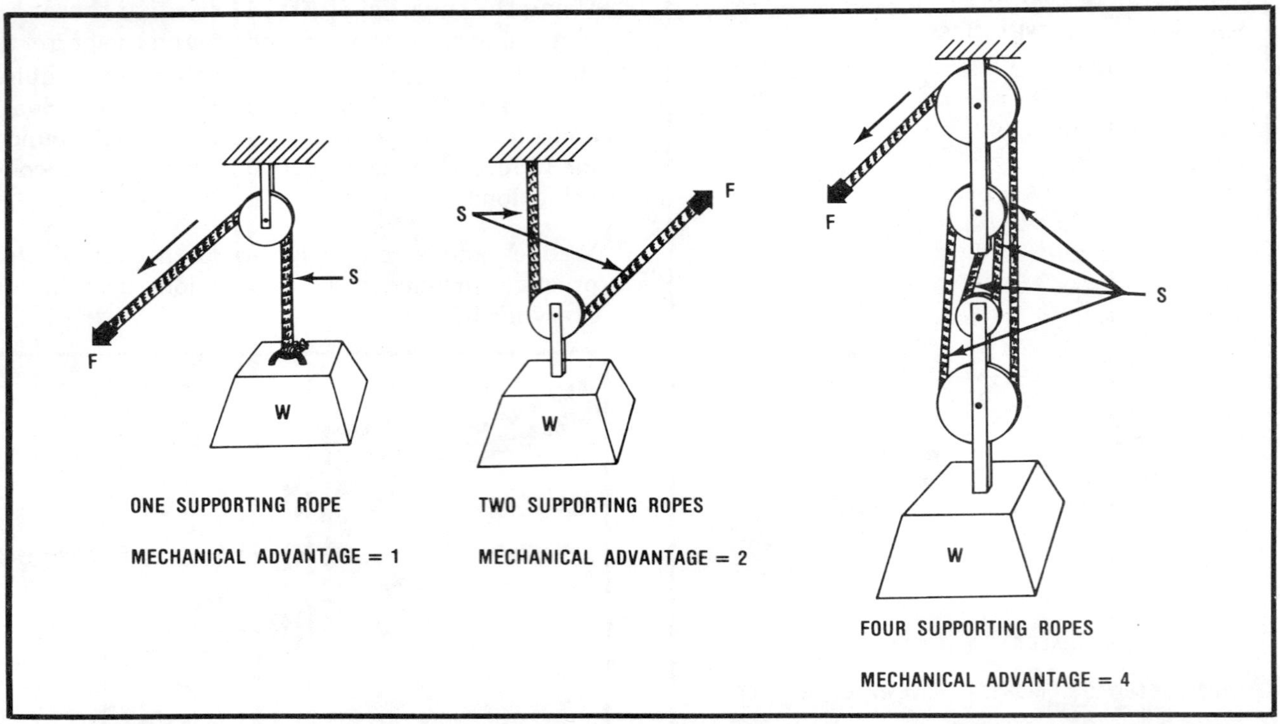

Fig. 6-9 The mechanical advantage of a rope-and-pulley system depends on the number of ropes that support the load.

When only one rope supports the load, the machine has a mechanical advantage of one, and in this instance the pulley is used simply to change the direction of the applied force. If, instead of anchoring the pulley to the structure, we attach it to the weight, we now have two supporting ropes, and the machine has a mechanical advantage of two. We will have to pull ten feet of rope to lift the weight five feet, but we can lift it with a force equal to only one-half of the weight.

When we support the weight with a block and tackle arrangement consisting of pulleys and a number of supporting ropes, we find that we gain a mechanical advantage each time we add a supporting rope. If we have four supporting ropes, we have a mechanical advantage of four. We will have to exert a force of only one-fourth of the weight, but we will have to pull the rope four times as far as we raise the weight.

D. Gears

There is probably no application of the basic machine that is used so much in our modern mechanical age as the gear, or toothed wheel. We use it in clocks and watches, in our automobiles and airplanes, and in just about every type of mechanical device. Gears may be used to gain mechanical advantage, or to change the direction of movement, and they do it smoothly and positively, with no slippage.

Fig. 6-10 External spur gears provide a mechanical advantage and reverse the direction of rotation of the drive shafts.

There are many forms and shapes of gears in use. Spur gears have their teeth cut straight across their circumference, and are used to connect two parallel shafts. The ratio of the speed of the two shafts is the same as the ratio of the number of teeth on the two gears. For example, if the drive gear has 50 teeth, and the driven gear, 100 teeth, the driven shaft will rotate at one-half the speed of the drive shaft, or the gear ratio will be 2:1. If both gears have external teeth, the shafts will turn in opposite directions, but if it is necessary for the shafts to turn in the same direction, one of the gears can have internal teeth like those in Fig. 6-11. Gear ratio is not affected by the direction of rotation of the shafts.

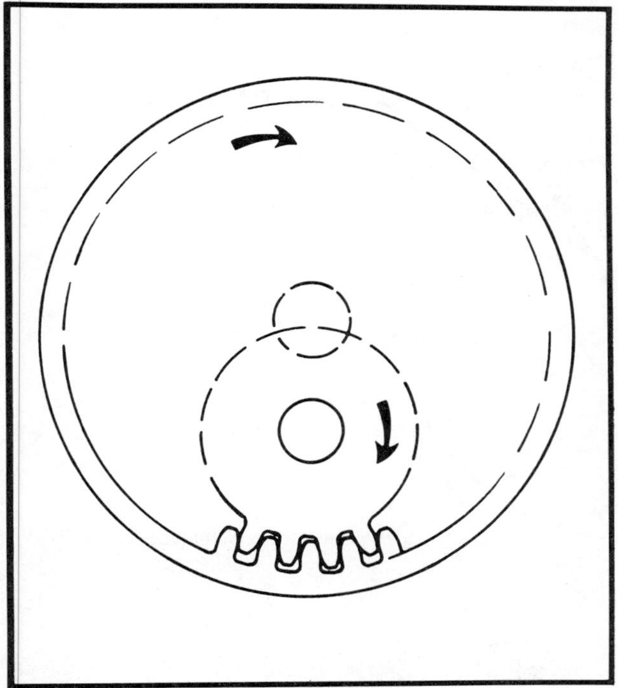

Fig. 6-11 A spur gear system with internal teeth on the driven gear provides a mechanical advantage without reversing the direction of rotation of the drive shafts.

When an extreme amount of mechanical advantage is needed, a worm gear may be used. This arrangement uses a spiral ridge around a shaft for the drive gear, and one complete rotation of the drave shaft moves the driven gear one tooth.

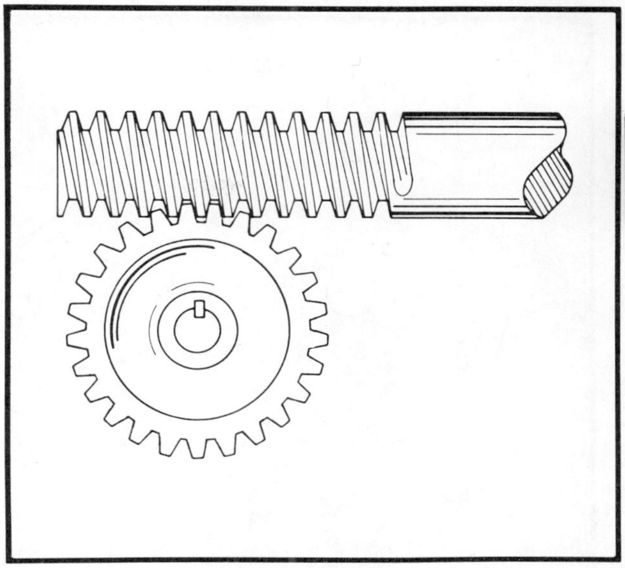

Fig. 6-12 A worm gear system provides a very high mechanical advantage.

The drive and driven gear shafts of a worm gear are usually at right angles to each other.

Planetary gear systems are used to reduce the propeller shaft speed on most of the more powerful aircraft engines, while allowing the engine to turn at a high RPM for power. The propeller mounts on a spider-like cage that holds the planetary gears and allows them to rotate around the fixed central sun gear and ride inside the driven ring gear. It is also possible for planetary reduction gear systems to drive the sun gear and hold the ring gear fixed in the nose section of the engine. The planetary gears are simple idlers in the system, and the number of their teeth does not enter into the gear ratio computations. The gear ratio of planetary gears may be found by dividing the sum of the number of teeth on the fixed and drive gear by the number of teeth on the drive gear. In Fig. 6-13 we see that if we have 100 teeth on the ring-type drive-gear and 50 teeth on the fixed sun-gear, the propeller shaft which is attached to the planetary gear spider will turn two revolutions when the crankshaft turns three.

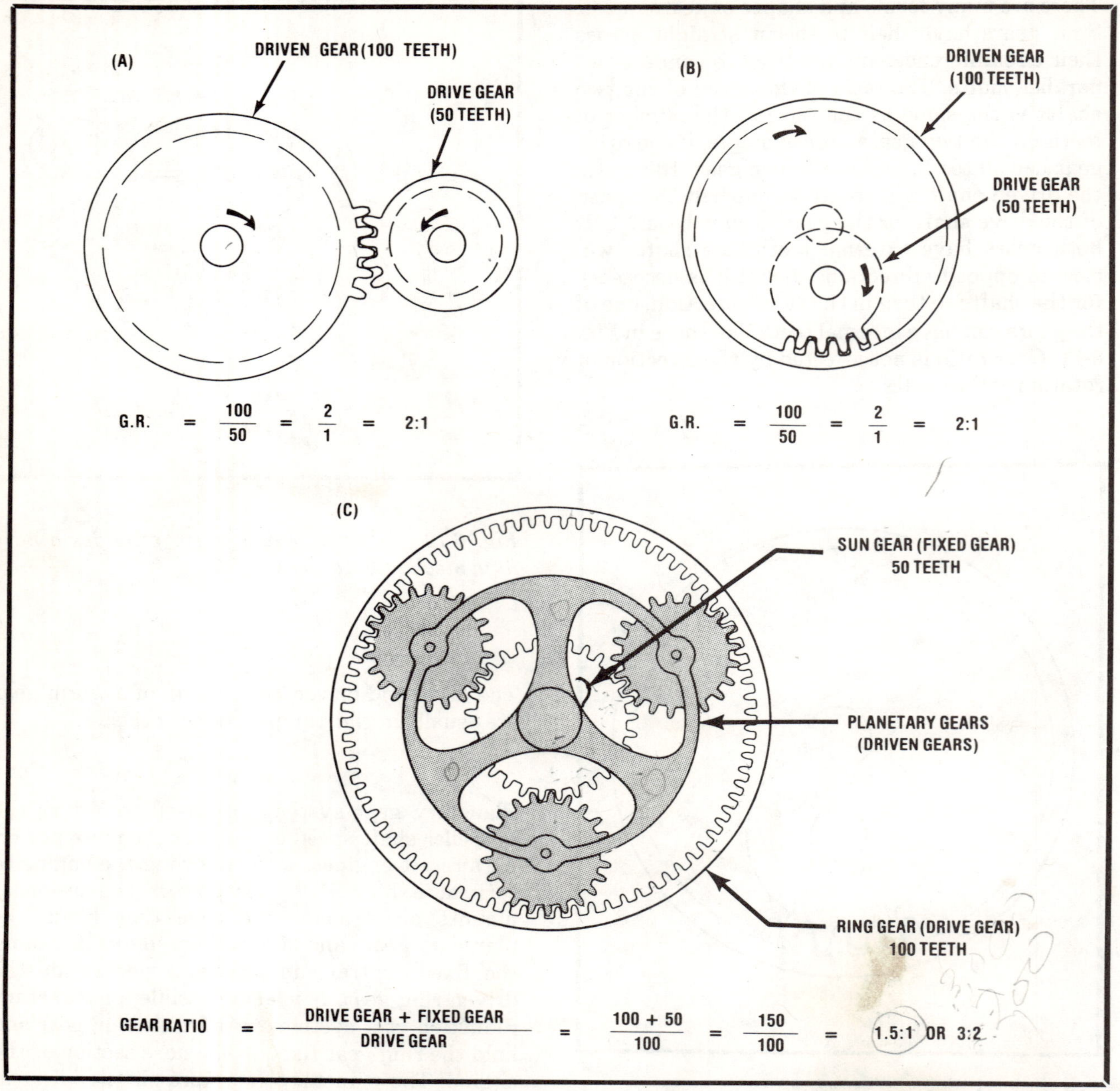

Fig. 6-13 A — Computing the gear ratio of an external spur gear system. B — Computing the gear ratio of an internal spur gear system. C — Computing the gear ratio of a planetary gear system.

SECTION VII

Stress

When an external force acts on a body, it is opposed by a force within the body. This force is called stress, and we can identify five different types of stress in mechanical bodies; tension, compression, torsion, bending, and shear. But, when we carefully analyze these stresses, we find that they can be broken down into *two* basic types, tension and compression. The other three types are special applications of these two.

A. Tension

Tension is one of the two fundamental stresses and is the one that tends to pull an object apart. The flexible steel cable used in aircraft control systems is an excellent example of a component that is designed to withstand tension loads only. It is easily bent, has little opposition to compression or torsion, and would be improperly used with shear loading. But it is excellent when subjected to a purely tension load, as it has great strength for its weight when loaded in this way.

Fig. 7-1 An aircraft control cable is designed to withstand tension stresses only.

B. Compression

Aircraft sheet metal skins are held together by rivets which are upset by applying a compressive force to their shank. The compression stresses in the rivet expand the shank as it is driven until it completely fills the hole and forms a head to hold the metal sheets together.

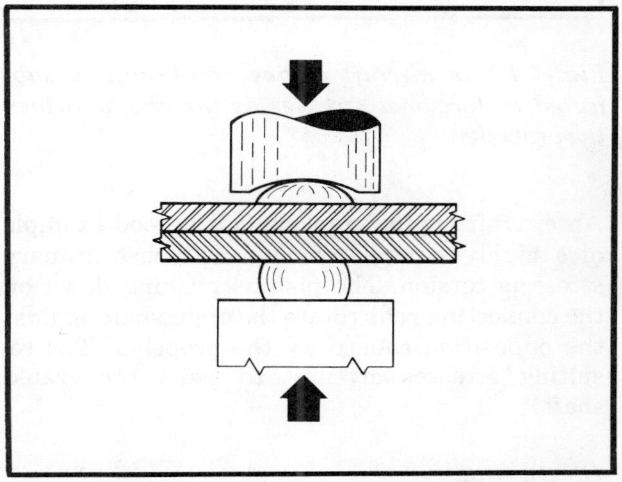

Fig. 7-2 An aircraft rivet is upset by the application of compression stresses.

C. Torsion

A torsional stress is one that is put into a material when it is twisted. When a bar is twisted, two stresses are created within it. One of these stresses is tension and operates diagonally across the bar. At right angles to the tension stress is a compressive stress. The combination of these two stresses is known as torsion.

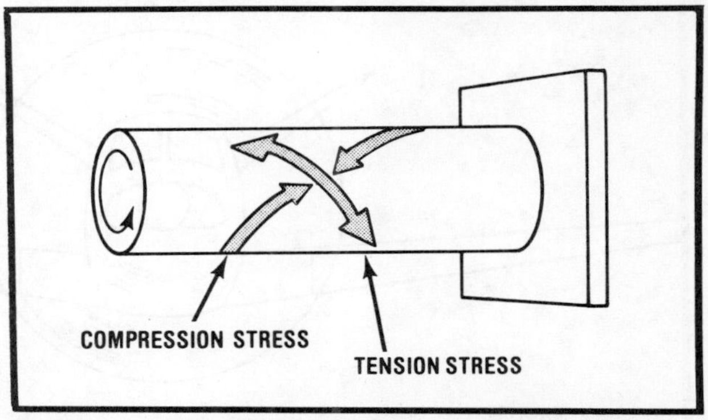

Fig. 7-3 Torsional stresses are a combination of tension and compression.

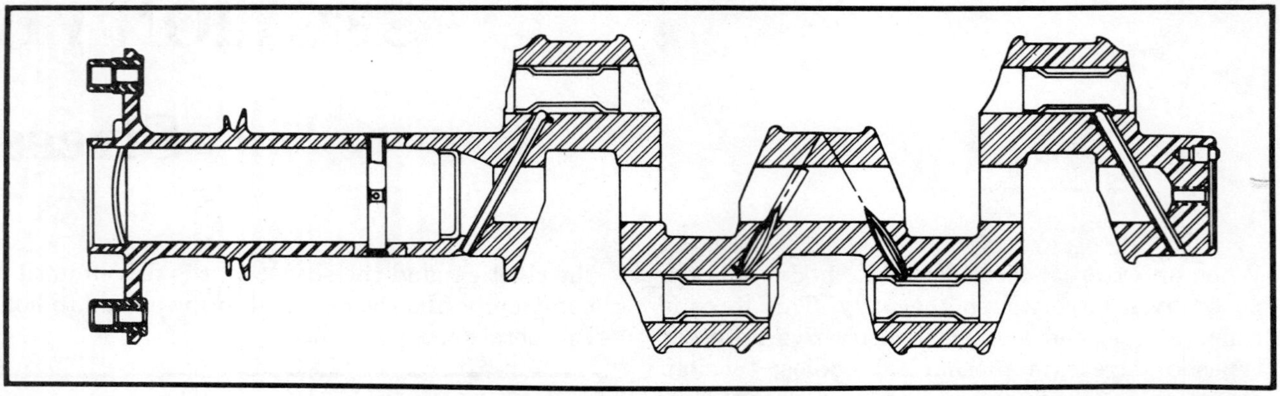

Fig. 7-4 An aircraft engine crankshaft is subjected to torsional stresses as the engine drives the propeller.

An aircraft engine crankshaft is a good example of a highly stressed component whose primary stress is torsion. The pistons pushing down on the connecting rods rotate the crankshaft against the opposition caused by the propeller. The resulting stresses attempt to twist the crankshaft.

D. Bending

When a wing is supporting an aircraft in flight, and the lift forces are attempting to bend the wing upward, the skin on top of the wing is subjected to a compression force, while that below the wing is being pulled by a tension force. When the airplane is on the ground and its weight is being supported by the landing gear, the stresses are reversed, and the top of the wing is under tension while the lower skin is under a compression stress.

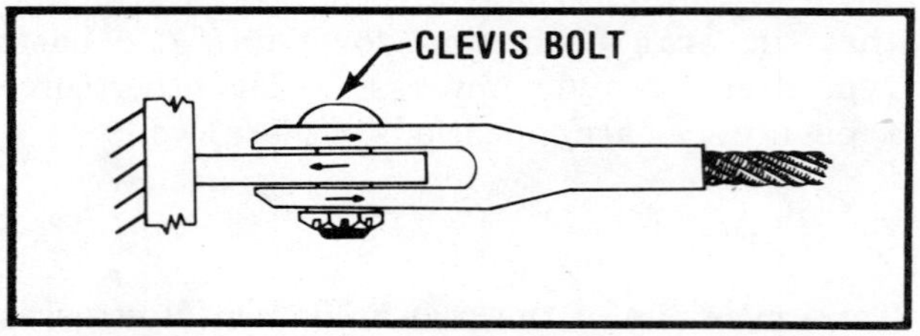

Fig. 7-6 A clevis bolt is subjected to shear stresses.

E. Shear

A shear stress is one that tries to slide a body apart. A clevis bolt in an aircraft control system is designed to take shear loads only. It is a high-strength steel bolt with a thin head and is fitted with a thin nut held on the threads by a cotter pin. The clevis bolt is free to turn in its hole. These bolts fasten the flexible steel cables to the controls and allow the cable to move with the surface without bending. The forces caused by the controls attempt to slide the bolt apart, or shear it.

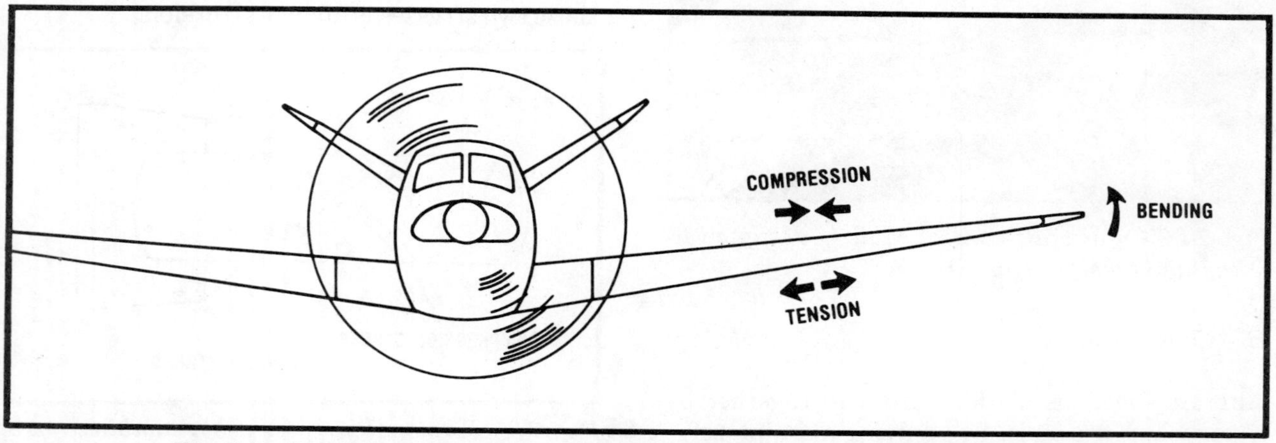

Fig. 7-5 Bending stresses are a combination of tension and compression.

SECTION VIII

Strain

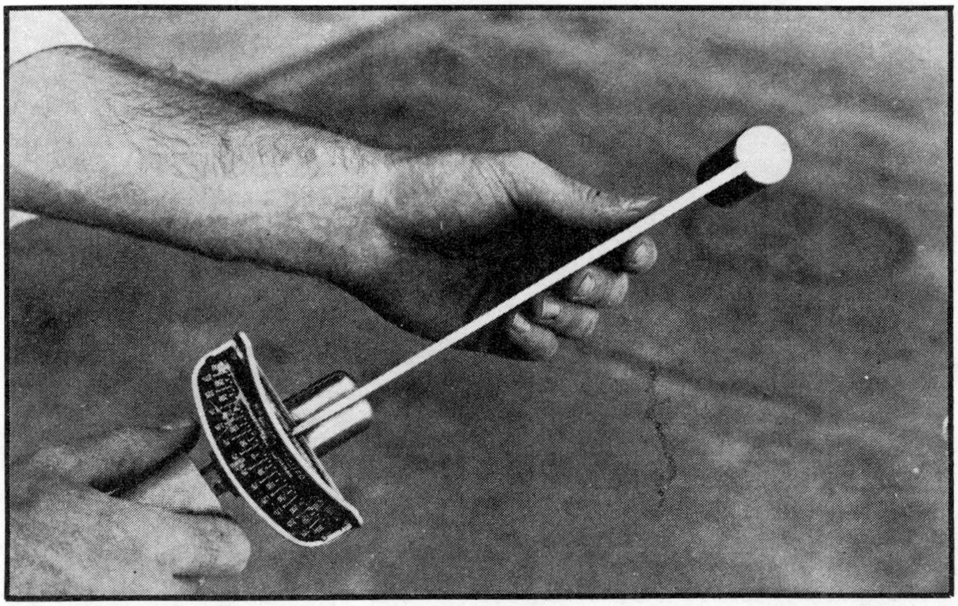

Fig. 8-1 The deflection, or strain, in the beam of this torque wrench is directly proportional to the stress applied to the bolt being tightened.

A stress is a force set up within a body to withstand an applied external force, but if the body is distorted by the stress, it is said to be strained.

If the strain does not exceed the elastic limit of the body, it will be in direct relationship to the applied stress, and this makes beams and springs useful as measuring devices. The hand torque wrench is an example of a flexible steel beam whose deformation, or bending, is directly proportional to the strain to which it is subjected, and because of this we can calibrate its deflection and use it as an indication of the amount of stress applied to the bolt. As soon as the stress is removed, the beam will return to its original condition.

SECTION IX

Motion

Fig. 9-1 When the airplane touches down on the runway, the wheels must accelerate from zero to the speed of the aircraft. The inertia of the wheel causes a great deal of rubber to be scraped off as the wheel comes up to speed.

A. Newton's Laws of Motion

The English physicist Sir Isaac Newton has given us three laws of motion that explain the effect of force on matter.

1. Newton's first law

Newton's first law of motion illustrates the effect of inertia on a body, as it states that a body at rest will tend to remain at rest, and a body in motion will tend to remain in uniform motion, in a straight line, unless it is acted on by some outside force. This law means, simply, that if an object were started in motion, it would continue indefinitely in a straight line, neither speeding up, slowing down, nor stopping. We know that this does not happen, and the reason is quite simple: friction is present. Every body in motion encounters friction and this is the outside force mentioned in the law.

2. Newton's second law

The acceleration produced in a mass by the addition of a given force is directly proportional to the force and is inversely proportional to the mass. When all of the forces acting on a body are in balance, there is no change in the velocity of the body, but if one of the forces exceeds the other, the body being acted on will change its velocity. That is, it will either speed up, accelerate, or will slow down, decelerate.

Acceleration, or the increase in velocity in time, is measured in the metric system in centimeters per second per second, or in the English system in feet per second per second, and it is one of the important considerations in the relationship between weight and mass. If a body is allowed to fall freely under the effect of gravity, it will accelerate uniformly 32.17 feet per second, or 980 centimeters per second every second it falls.

3. Newton's third law

For every force action, there is an equal and opposite reaction. This is one of the more familiar laws of physics, one for which we see daily applications. When we shoot a gun, the expanding gases forcing the bullet out of the barrel exert exactly the same amount of force back against our shoulders in the form of the familiar kick. The magnitudes of the two forces are exactly equal, but their directions are opposite.

It is an application of the third law of Newton that we observe under a hovering helicopter. In order for a heavier-than-air flying machine to remain in the air, it must force down a mass of air equal to its own weight. When it forces down more air than it weighs, it will rise, and when the mass is less than its weight, it will descend.

Fig. 9-2 When a helicopter hovers, it must force down a mass of air equal to its own weight.

B. Bodies in Motion

1. Speed and velocity

These two terms are often used interchangeably, but speed and velocity are actually quite different. Speed is simply a rate of motion, or the distance an object moves in a given time. It is usually expressed in such terms as miles per hours, feet per second, kilometers per hour, or knots, and it does not take into consideration any direction. Velocity, on the other hand, is the rate of motion in a given direction, and is expressed in such terms as five hundred feet per minute downward, or 300 knots eastward.

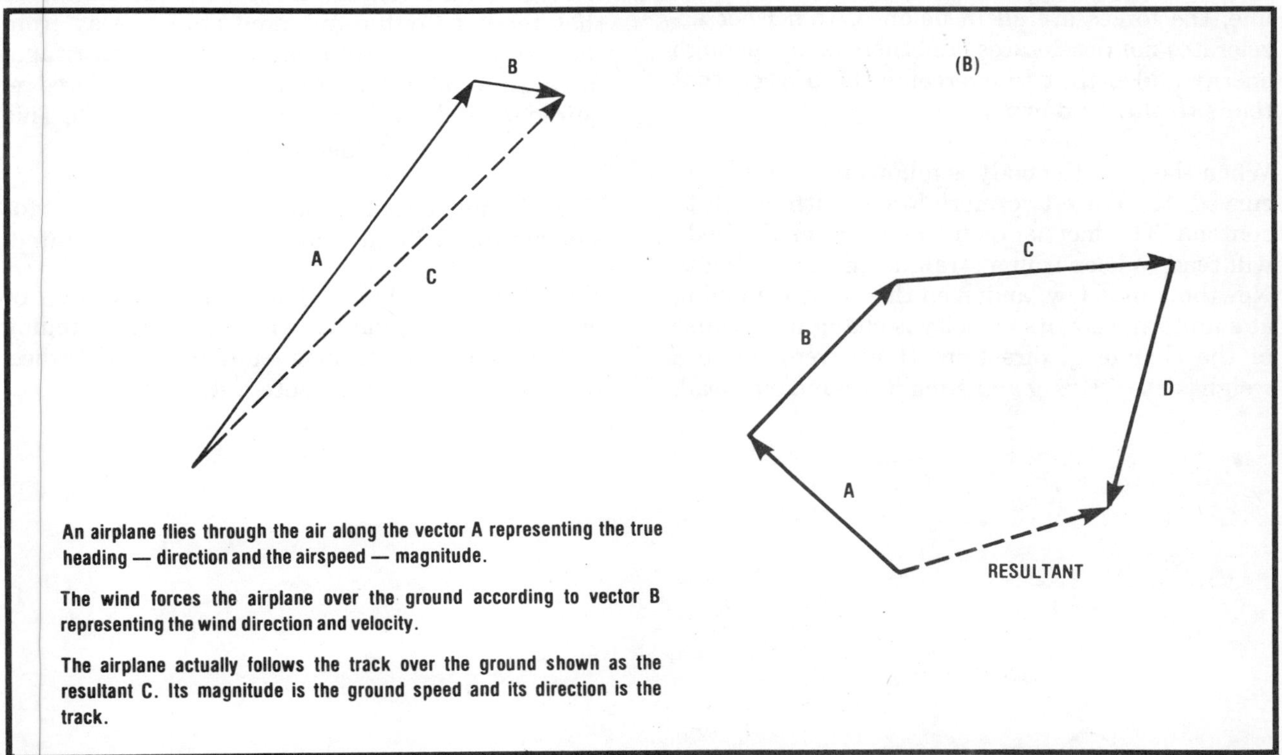

Fig. 9-3 A — An aerial navigation problem is a form of vector addition. B — The resultant of a series of vectors may be found by adding all of the vectors tail to head.

2. Vectors and vector addition

A vector quantity must have both magnitude and direction. Velocity is a vector quantity because it has both of these characteristics. Vector quantities may be added to each other and a resultant found that will produce the same velocity as the sum of all of the individual vectors.

One of the simplest ways to add vectors is to draw them to scale, using a direction measured clockwise from a reference line, and laid out to a given scale for length. In Fig. 9-3B we begin with vector *A*, and draw it out to the correct length and direction, then place the tail of vector *B* at the head of *A*. Continue until all of the vectors are laid out, and then draw the resultant from the starting point to the point of the last vector. This resultant will have the same velocity as the sum of all of the vectors you have laid out. When laying out the vectors, it makes no difference what sequence you use, the resultant will be the same.

3. Circular motion

When a body is in uniform motion in a straight line, the forces are all in balance. It neither accelerates nor decelerates, and there is just enough energy added to it to overcome the friction that tends to slow it down.

When the path the body is following is uniformly curved, there are two more forces with which to contend. The inertia, or momentum, of the body will tend to keep it in a straight line according to Newton's first law, and even though it is moving at a uniform rate, its velocity is changing because of the change in direction. If we were to tie a weight onto a string and sling it around our head, it would follow a circular path, as we see in Fig. 9-4. The force of inertia will continually try to pull the weight from a point such as *A* and cause it to go on a straight line to point *B*. The force that tends to pull a spinning object away from the center of its rotation is called centrifugal force. And there is an equal and opposite force required to hold the weight in its circular path. This force is called centripetal force.

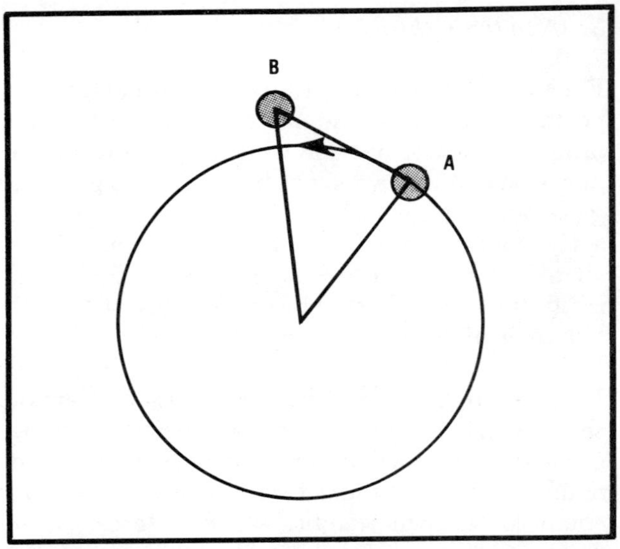

Fig. 9-4 A body in motion tends to remain in motion in a straight line. When the body is forced into a curved path, a centrifugal force is set up which tends to pull the body away from the center of the curve.

You will notice that grinding wheels are rated for a maximum RPM at which they may be operated.

The reason for this is that enough force can be built up in a spinning wheel that the binding materials are not strong enough to hold the wheel together, and it will disentegrate.

SECTION X

Heat

Heat is one of the most useful forms of energy because of the direct relationship between work and heat. For example, when a pilot wishes to slow an airplane after touchdown, he applies the brakes. What this actually does is to produce a great deal of friction between the rolling wheel and the aircraft, and the kinetic energy of the moving airplane is changed into energy of motion of the molecules within the brake disks. In high-energy stops, such as are required in aborted takeoffs, it is not uncommon for the brake disks to get red-hot.

In the English system of measurement, the relationship between heat and work is that one British Thermal Unit, or Btu, of heat energy will do 778 foot-pounds of work. In the metric system, we find that one calorie of heat energy is the equivalent of 4.186 joules of mechanical energy which will do 0.427 meter-kilograms, or 3.088 foot-pounds of work.

1 Btu = 778 Foot-Pounds or 107.59 Kilogram-Meters.

1 Btu = 252 Calories or 1055 Joules.

1 Calorie = 0.00397 Btu or 4.186 Joules.

1 Calorie = 3.088 Foot-Pounds or 0.426 Kilogram-Meters.

Fig. 10-1 Relationship between heat energy and work

A. Sensible Heat

Heat energy added to a body which causes a rise in its temperature, is called sensible heat. For example, when a pan of water is put on a stove, its temperature will go up until it begins to boil. Under standard sea level conditions, this will take place when the temperature reaches 212 degrees Fahrenheit or 100 degrees Celsius.

B. Latent Heat

If we leave the pan of water on the stove and continue to add heat energy, the temperature of the water will not rise any farther, but the water will change its state from a liquid to a vapor. The heat that causes the change of state is called the heat of vaporization, and is latent heat, which remains in the water as long as it is in its vapor form. We know that it requires 539.6 calories of heat energy to change one gram of liquid water into water vapor.

C. Dimensional Changes Caused by Heat

When heat energy is added to a body, the movement of the molecules is increased to the point that the material actually changes its physical dimensions, or is said to expand. This expansion is most noticeable in gases where the change is so great that we are able to use it in all forms of heat engines where it is the expanding gases which actually do the work.

Solids and liquids expand far less than gases, and the amount of their expansion can be calculated by using their coefficient of linear expansion, which is a constant for each material. In Fig. 10-2, we have a chart of these coefficients, and from this we see that brass expands somewhat less

Substance	Coefficient of linear expansion (per degree C.)
Aluminum	24×10^{-6}
Brass	19×10^{-6}
Copper	17×10^{-6}
Glass	4 to 9 $\times 10^{-6}$
Quartz	0.4×10^{-6}
Steel	11×10^{-6}
Zinc	26×10^{-6}

Fig. 10-2 Coefficient of linear expansion of various solids

than aluminum. This explains the reason we must shrink brass valve guides into an aluminum aircraft engine cylinder head. To do this, the valve guide is machined to a larger outside diameter than the hole into which it is to fit. The cylinder head is heated in an oven to increase the size of the hole and the valve guide is chilled with dry ice, to shrink it. The guide is pressed into the head, and when both the guide and the head expand from operational heat, the fit will still be tight enough to hold the guide in place.

D. Specific Heat

Different materials require different amounts of sensible heat energy to cause them to change their temperature. A comparative measure of the heat required to raise the temperature of one gram of the material one degree Celsius, with the amount of heat required to raise the temperature of one gram of water one degree Celsius is known as the specific heat of the material. In Fig. 10-3 we see the specific heat of several common materials. From this chart we see that it requires only about three percent as much heat energy to be added to lead to raise its temperature one degree Celsius as it does to change the temperature of an equal weight of water the same amount.

MATERIAL	SPECIFIC HEAT
Lead	0.031
Mercury	0.033
Brass	0.094
Copper	0.095
Iron or Steel	0.113
Glass	0.195
Alcohol	0.547
Aluminum	0.712
Water	1.000

Fig. 10-3 Specific heat of various materials

It is because of the high specific heat of water that our oceans and large lakes can serve as temperature stabilizers. Land surfaces have a much lower specific heat, and so they change their temperature within the course of a day, but the temperature of the large bodies of water remains very nearly constant.

E. Transfer of Heat

In order to use heat effectively, we must be able to move it from one place to another and, fortunately, there are three methods of doing this.

1. Conduction

When one body is in mechanical contact with another, heat will transfer from the body having the highest level of energy to the one having a lower level. This transfer is caused by the violent action of the molecules in the material affecting the molecules next to them. This action spreads until the heat is distributed throughout the material.

A good example of heat transfer by conduction is seen by the way we remove excess heat from an aircraft engine cylinder. When gasoline is burned inside the cylinder, it releases far more energy than is converted into useful work, and some of this heat must be removed, or it will damage the engine and cause the fuel-air mixture to explode or detonate. The heat from inside the cylinder passes by conduction to the outside of the cast-aluminum cylinder head and into the thin fins cast on the outside of the head. This heat is then conducted into the air flowing through the fins and is carried away from the engine.

Fig. 10-4 Heat is conducted away from an aircraft engine cylinder by conduction.

2. Convection

When heat is absorbed by a free-moving fluid which may be either a liquid or a gas, it will expand and its density will decrease. The fluid with the lower density will rise and drive the denser fluid down, so it can be heated. A pan of water on a stove is heated in this way. The water at the bottom of the pan is heated by conduction, and as it rises, it forces the water from the top down so it can be heated.

Hot-air furnaces heat a house by convection, but rather than using natural convection to move the cold air down through the furnace, a blower forces the warm air through the house and creates a low pressure that draws the cold air through the heater. This type of heating is known as forced convection.

Fig. 10-5 Water in the pan is heated by natural convection.

3. Radiation

The amount of heat transferred by either conduction or convection depends on the difference in the temperature between the hot and the cold bodies, and there must be some form of conductor between the two bodies. Heat can be transferred by radiation, even though there is a vacuum between the bodies, and the amount of transfer is proportional to the fourth power of the absolute temperatures of the two bodies. This means that transfer by this method is far more effective at high temperatures than it is when the temperature is low.

Heat from our primary source, the sun, reaches us by the process of radiation. Energy that is continually being released by the sun sets up a vibratory or wave motion in the ether which exists between the sun and the earth, and when this energy is received by the surface of the earth, it sets up a vibratory motion within the molecules it strikes. This increased molecular activity raises the temperature of the body. This form of heat transfer explains the reason we can be comfortably warm in front of a window through which the sun is shining, even though the temperature of the air through which the rays of the sun have passed is very low.

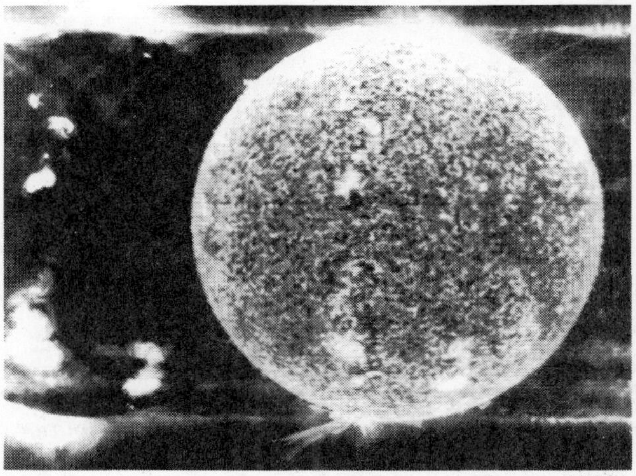

Courtesy of NASA.

Fig. 10-6 Energy is transmitted from the sun to the earth by radiation.

F. Temperature

Heat is a form of energy that causes molecular agitation within a material. And the amount of this agitation can be measured by a scale known as temperature.

In estalishing our temperature scales, two conditions which are easily duplicated have been chosen as the references. These conditions are the points under standard atmospheric conditions at which pure water turns from a liquid into a solid; that is, when it freezes; and when it changes from a liquid into a gas; when it boils. In the Fahrenheit system of measurement, the difference between these two points is divided into 180 equal divisions called "degrees." The temperature of a mixture of equal weights of snow and salt is used as zero, and it is found that this point is 32 degrees below the freezing point of water. On the Fahrenheit thermometer, then, we see that water freezes at 32 degrees and boils at 212 degrees.

A more convenient division was made in the Centigrade, or hundred graduations, scale when it was chosen to call the freezing point of water zero degrees and its boiling point 100 degrees. The

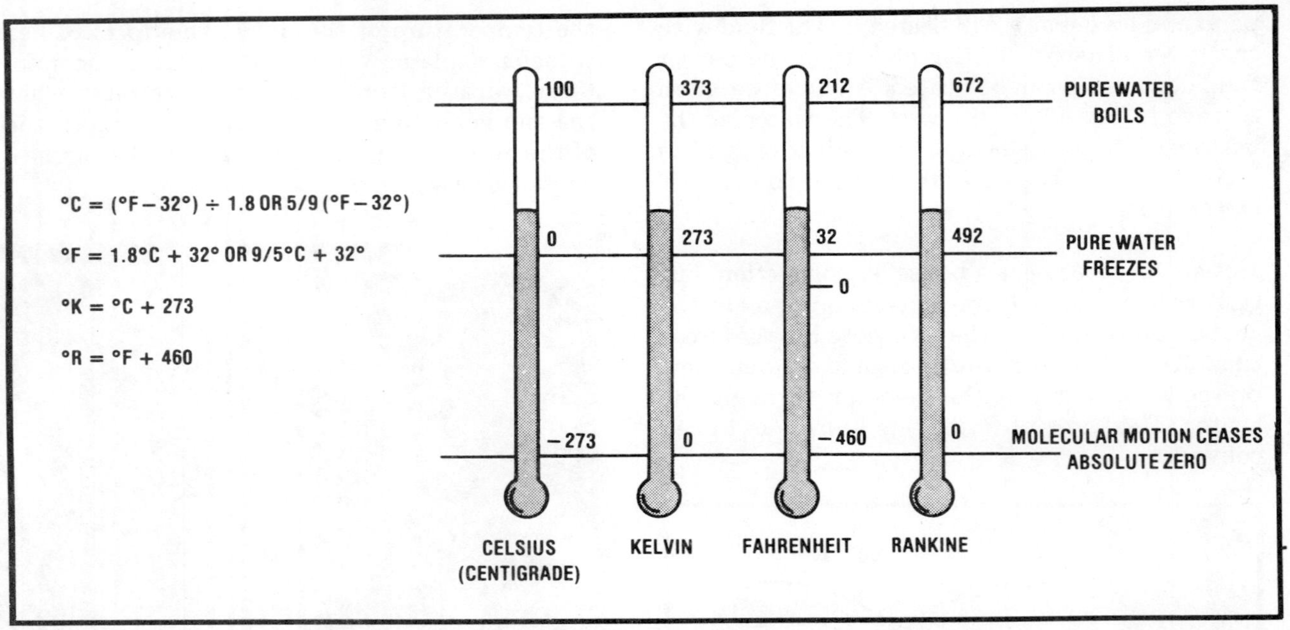

Fig. 10-7 The four temperature scales in common use.

Centigrade scale has been renamed the Celsius scale, in honor of the Swedish astronomer Anders Celsius, who in 1742 first described the centigrade scale.

In 1802, the French chemist and physicist Joseph Louis GayLussac found that when a gas was heated one degree Celsius, it expanded 1/273 of the volume it had at the temperature of freezing water. He reasoned that if it were cooled it would decrease its volume by the same amount, and by logic he concluded that if its temperature were decreased to 273 degrees below zero, the gas would decrease to zero volume, and there would be no more molecular activity. This point has been established as absolute zero. On the Fahrenheit scale, the same point is −460 degrees.

Many of our gas laws and laws relating to heat are based on or referenced from a condition of absolute zero, and to facilitate working with these terms, we have two absolute temperature scales. The Kelvin scale is based on the Celsius scale, with its one hundred graduations between the freezing and boiling points of pure water. Water freezes at 273 degrees Kelvin and boils at 373 degrees Kelvin.

The Rankine scale is based on the Fahrenheit scale with its 180 graduations between the two change-of-state conditions of water. Water freezes at 492° R. and boils at 672° R. Fig. 10-7 shows the relationship between the four temperature scales.

To convert Celsius temperature into Fahrenheit, we must remember that 100 Celsius degrees represents the same temperature difference as 180 Fahrenheit degrees, so we must multiply the Celsius temperature by 1.8. This is the same as the more familiar 9/5 used in many formulas. And since the Fahrenheit scale starts lower than the Celsius scale, we must add 32 degrees.

$$°F = 1.8\ °C + 32° \text{ or}$$

$$°F = 9/5\ °C + 32°$$

To convert Fahrenheit temperature into Celsius, we must first subtract the 32 degrees from the Fahrenheit temperature and then divide by 1.8 or, as it is more familiar, multiply by 5/9. The result is the same.

$$°C = (°F - 32°) \div 1.8 \text{ or}$$

$$°C = 5/9\,(°F - 32°)$$

When it is desired to convert Celsius degrees into Kelvin temperature, all we need to do is add 273 to the Celsius temperature.

$$°K = °C + 273$$

To find Rankine temperature when Fahrenheit is known, we need only add 460 to the Fahrenheit temperature.

$$°R = °F + 460$$

SECTION XI

Pressure

Pressure is a measurement of force per unit area and may be expressed in such units as pounds per square inch or grams per square centimeter. Atmospheric pressure is often expressed in inches or millimeters of mercury, which is the height of a column of mercury the pressure will support. Or it may be measured in atmospheres, in which one atmosphere is 29.92 inches of mercury, or 14.69 pounds per square inch.

In the metric system pressure may be expressed in millibars, which is one-thousandth of the force of a bar, or 1000 dynes per square centimeter. Standard sea level pressure in these units is 1013.2 millibars.

1 BAR	=	29.5306 INCHES OF MERCURY NORMAL SEA-LEVEL PRESSURE AT 45° NORTH LATITUDE AND A TEMPERATURE OF 32°F.
1 MILLIBAR	=	1/1000 BAR
	=	0.0295 INCHES OF MERCURY
1 ATMOSPHERE	=	29.92 INCHES OF MERCURY
	=	14.69 POUNDS PER SQUARE INCH
	=	1013.2 MILLIBARS NORMAL SEA-LEVEL PRESSURE AT 40° NORTH LATITUDE AND A TEMPERATURE OF 59°F (15°C.)

Fig. 11-1 The units of atmospheric pressure

There are three reference points from which pressure may be measured.

A. Absolute Pressure

If pressure is referenced from zero pressure, or a vacuum, it is known as absolute pressure. In a laboratory, absolute pressure is measured by a mercury barometer in which the pressure to be measured forces down on a container of mercury into which a closed-end tube filled with mercury is supported. There is a vacuum above the mercury in the tube, and so the pressure being measured is compared with or referenced from this vacuum.

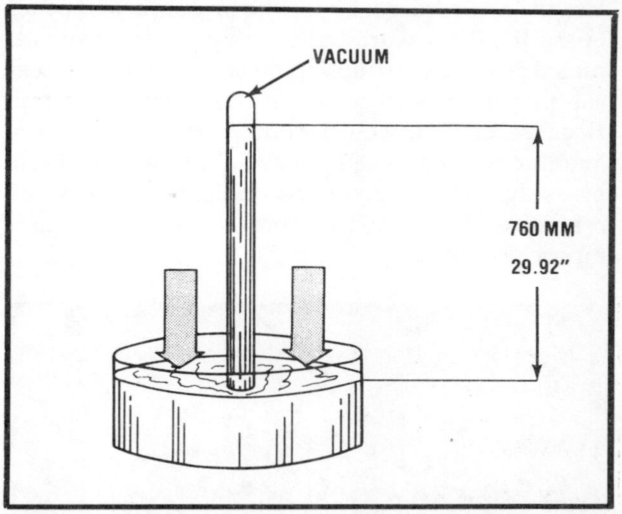

Fig. 11-2 Atmospheric pressure is measured by the height of a column of mercury the pressure will support. Under standard sea-level conditions the atmospheric pressure will support a column 29.92 inches, or 760 millimeters, high.

Fig. 11-3 An aircraft manifold pressure gage measures absolute pressure.

One of the power measuring instruments in an aircraft with a reciprocating engine is the manifold pressure gage which measures the absolute pressure in the induction system of the engine and expresses it in inches of mercury. You will notice that when the engine is not running, the gage reads the existing atmospheric pressure.

B. Gage Pressure

Engine oil pressure, hydraulic pressure, and other operational pressure built up by pumps have atmospheric pressure acting on the fluid as it enters the pump, as well as when it leaves the pump. Because of this, we are concerned only with the amount the pressure is raised above atmospheric pressure. Gage pressure indicators therefore reference the pressure from the existing barometric pressure.

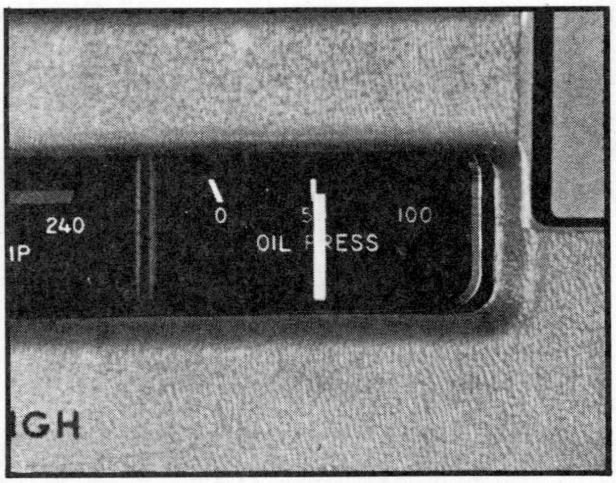

Fig. 11-4 An engine oil pressure gage measures gage pressure.

C. Differential Pressure

Since pressure is a measure of force acting on a given area, there must always be a reference from which to measure it. Absolute pressure is referenced from zero pressure, or a vacuum, and gage pressure is referenced from the existing barometric pressure. Differential pressure may be referenced from any pressure, and is actually nothing more than the difference between two pressures.

One of the most familiar differential pressure gages in an aircraft is the airspeed indicator. This instrument measures the difference between the ram pressure caused by movement through the air and the static, or still, pressure of the air through which the airplane is flying. This particular pressure is indicated to the pilot in terms of miles per hour, knots, or kilometers per hour.

Fig. 11-5 An airspeed indicator is actually a differential pressure gage, measuring the difference between impact air pressure and static air pressure.

SECTION XII

Gas Laws

Gases and liquids are both fluids, and both can be made to transmit forces. There is one great difference, however, between the two, and this difference has tremendous effects on the way the two are used. Gases are compressible, while liquids, for all practical purposes, are not.

The volume of a gas is determined by its temperature and its pressure, and the relationship between pressure, volume, and temperature is given to us in the gas laws of Boyle and of Charles.

A. Boyle's Law

In 1660, the British physicist Robert Boyle discovered that a change in volume of a confined gas, when held at a constant temperature, would create an opposite change in the pressure of the gas. In the form of an equation, this can be expressed as:

$$\frac{V_1}{V_2} = \frac{P_2}{P_1}$$

In the drawing of Fig. 12-1 we see that if the temperature of the gas in the container is held constant but the piston is moved down to decrease the volume of the gas, the pressure of the gas will increase.

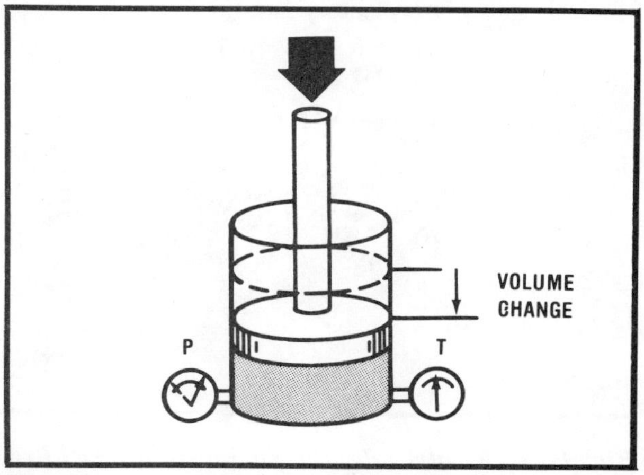

Fig. 12-1 If the temperature of a gas remains constant, the pressure of the gas will increase as its volume is decreased.

B. Charles's Law or Gay-Lussac's Law

In his investigation of the relationship between the temperature and volume of a gas, Joseph Louis Gay-Lussac found that if a gas is held at a constant pressure, it will expand 1/273 of the volume it had at zero degrees Celsius for every degree Celsius its temperature is raised. The statement of this fact is known as Gay-Lussac's law, or it sometimes goes by the name of Charles's law from J.A.C. Charles who, some fifteen years earlier, had made studies of the relationship between pressure and temperature of confined gases.

$$\frac{V_1}{V_2} = \frac{T_1}{T_2}$$

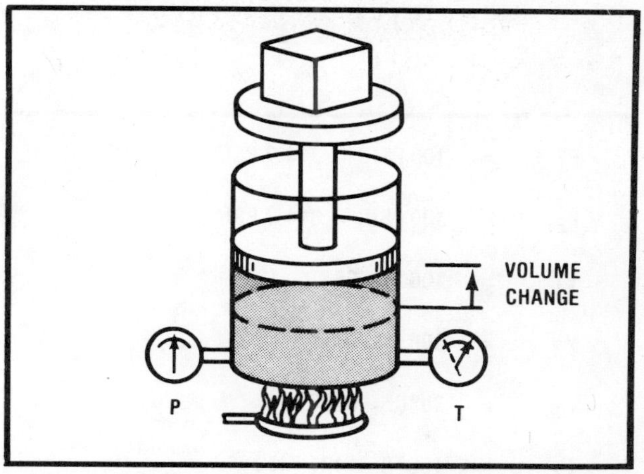

Fig. 12-2 When the pressure of a gas remains constant, the volume of the gas will increase as its temperature is increased.

Another way of looking at this same relationship is that if the volume of a confined gas is held constant, the pressure will vary in the same way as the absolute temperature of the gas.

$$\frac{P_1}{P_2} = \frac{T_1}{T_2}$$

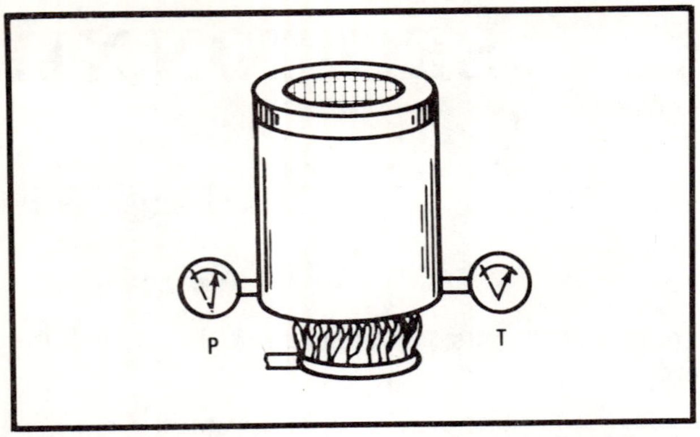

Fig. 12-3 When the volume of a gas is held constant, its pressure will increase as its temperature is increased.

C. General Gas Law

Boyle's law explains what happens to the pressure of a gas when its volume is changed, and Charles's laws help us to understand what happens to the volume and pressure of a gas as its temperature changes.

The general gas law combines these laws, and allows us to find the resulting pressure, volume, or temperature when two of the variables have been changed. The equation for this law is:

$$\frac{P_1V_1}{T_1} = \frac{P_2V_2}{T_2}$$

P_1 = 100 PSIG = 114.7 PSIA

P_2 = 300 PSIG = 314.7 PSIA

V_1 = 1000 LITERS

V_2 = 508.8 LITERS

T_1 = 30°C = 303° K

T_2 = 150°C = 423° K

$$\frac{P_1V_1}{T_1} = \frac{P_2V_2}{T_2}$$

$$\frac{114.7 \times 1000}{303} = \frac{314.7 \times \chi}{423}$$

$$95345.1\chi = 4851800$$

$$= 508.8 \text{ LITERS}$$

Fig. 12-4 An application of the general gas law

In the Fig. 12-4 example we have 1000 liters of a gas at a temperature of 30 °C, and a pressure of 100 psig. If we raise the pressure of the gas to 300 psig and its temperature to 150 °C, we can find its volume by using the equation of the general gas law. What we have done is to compress the gas, which decreases its volume and raises both its pressure and its temperature.

D. Dalton's Law

In 1801, the English chemist John Dalton announced his discovery that the total pressure exerted on the walls of a closed container by a mixture of gases is the same as the sum of the pressure exerted by each of the individual gases in the mixture.

The air we breathe is a physical mixture of gases of which about 78% is nitrogen and 21% is oxygen, and the remainder is made up of a mixture of other gases such as neon, argon, krypton, and xenon, and most air contains some water vapor.

If the pressure exerted by the atmosphere is 15 pounds per square inch, the oxygen, which accounts for 21% of the volume, would cause 21% of the pressure; or we can say that the partial pressure of the oxygen is 3.15 psi and the partial pressure of the nitrogen is 11.7 psi.

Partial pressure is important to us when we work with the environmental systems of an airplane. The mixture of the gases remains constant as the altitude increases, but by about 10,000 feet, the

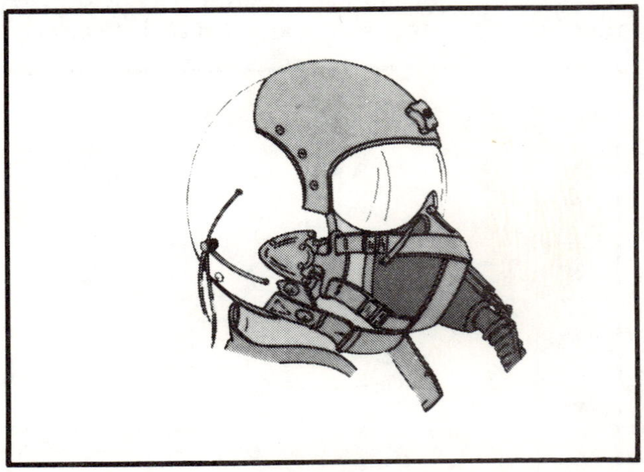

Fig. 12-5 At altitude, the partial pressure of the oxygen in the air becomes too low to force the needed oxygen into the blood, and supplemental oxygen is needed.

partial pressure of oxygen has dropped to such a level that it cannot force its way into the blood through the lungs. And so above this altitude either cabin pressurization or supplemental oxygen is needed to increase the amount of oxygen partial pressure so it can be forced into the blood.

E. Avagadro's Hypothesis

In the early 1800's, chemists came up with some unusual observations about gases. One volume of nitrogen and one volume of oxygen would produce two volumes of nitric oxide, but it requires one volume of oxygen and *two* volumes of hydrogen to produce two volumes of water vapor. And, even more strangely, it requires one volume of nitrogen and *three* volumes of hydrogen to produce two volumes of ammonia.

This strange behavior was explained in 1811 by the Italian physicist Amedo Avogadro when he stated that equal volumes of almost all gases under the same pressure and temperature will contain the same number of molecules. This explained the strange behavior of the gases. One N_2 + one O_2 → two NO, one O_2 + two H_2 → two H_2O, and one N_2 + three H_2 → two NH_3.

SECTION XIII
Fluid Mechanics

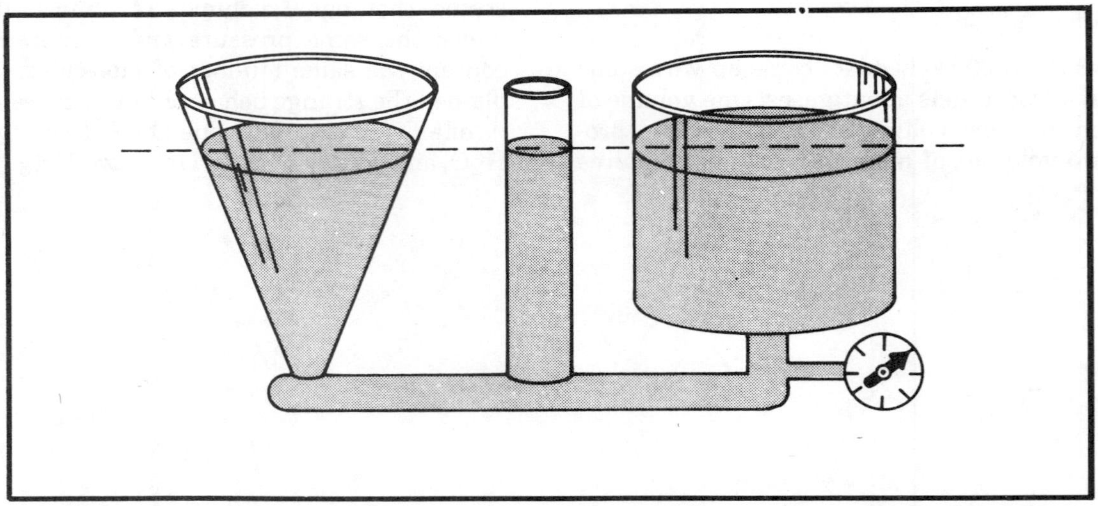

Fig. 13-1 The pressure exerted by a column of liquid is determined by the height of the column and is not affected by the volume of the liquid.

Much of the science of flight is based on the principles we find in the study of fluid mechanics. The air which supports our aircraft is a fluid, and as such it follows the laws we will discuss, and the liquid that flows in our hydraulic systems transmits forces by fluid mechanics.

A fluid is any substance which tends to flow or to conform to the outline of its container. Both liquids and gases are fluids and follow many of the same rules, but there is a difference. For all practical purposes, liquids are considered to be incompressible, while gases may be compressed.

A. *Pressure Exerted by a Fluid*

A column of fluid will exert a pressure at the bottom of the column that is determined by the density of the fluid and by the height of the column.

We know that pure water weighs 62.4 pounds per cubic foot or 0.0361 pounds per cubic inch. If we were to take 1000 cubic inches of water and stack them vertically, we would have a column with an area of one square inch at the base, and it would extend 1000 inches high and would weigh 36.1 pounds. There would be a pressure, or force per unit area, at the bottom of this column of 36.1 pounds per square inch.

Gasoline has a specific gravity of 0.72, and so a column of gasoline 1000 inches high would give us a pressure of only 26.0 pounds per square inch. If we wanted a pressure of one psi, the column would have to be 38.5 inches high.

The *amount* of fluid has no effect on the pressure at the bottom of the column. In Fig. 13-1, we have three different shaped containers connected together at the bottom, and we can see that the fluid level is the same in all of them. This shows us that the pressure anywhere along the connecting line is constant. From this we can derive a simple statement of fact that the pressure exerted by a column of fluid is determined by the height of the column and the density of the fluid, and it is independent of the volume of the fluid.

We can find a practical example of the use of this fact if we need to adjust the float level of a carburetor on a bench. Let's assume that the carburetor specifications require that we apply a pressure of three psi to the inlet of the carburetor. To do this, we first solve a proportion problem

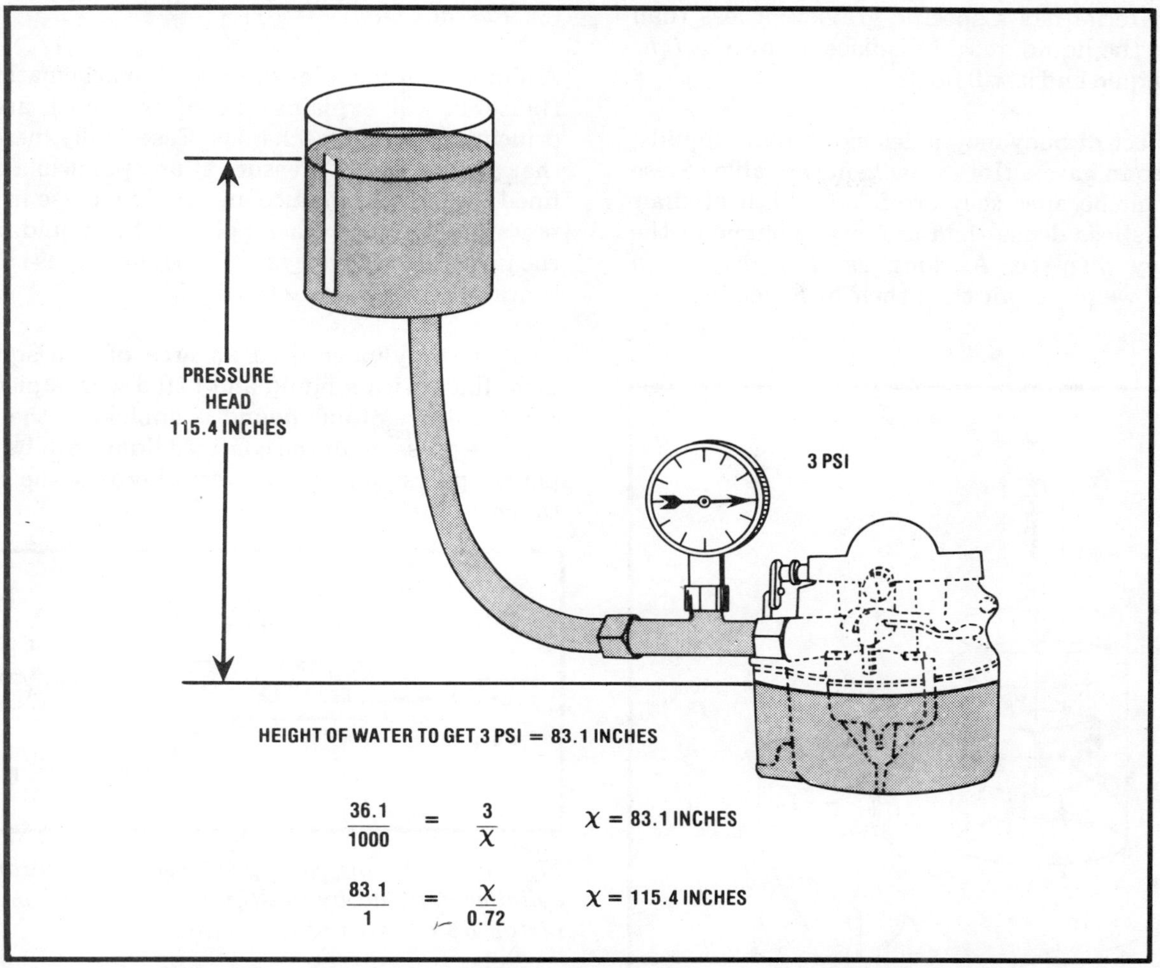

Fig. 13-2 A given pressure head of a liquid may be produced by raising a container of the liquid to a specified height.

and find that since 1000 inches of water give us a pressure of 36.1 psi, we will need a height of 83.1 inches of water to get three psi. We do not want to use water for our carburetor, but rather gasoline, which has a specific gravity of 0.72, so we divide the 83.1 by 0.72. When we do this, we find that our column of gasoline must be 115.4 inches high.

Neither the shape of the container nor the amount of fluid has an effect on the pressure, so all we need to do is to take a small container of gasoline, connect it to the carburetor by a piece of small-diameter tubing, and suspend the container so the level of the liquid is 115.4 inches above the inlet valve of the carburetor, and we have the correct pressure at the inlet.

B. *Buoyancy*

When a solid object is submerged in a liquid, it displaces a certain amount of the liquid. In the example of Fig. 13-3, we have a 100-cubic-inch block of aluminum which weighs 9.7 pounds. If this block is attached to a spring scale and lowered into a container of pure water, some of the water will be forced out the overflow as it is displaced by the block. The displaced water will weigh 3.6 pounds, and the spring scales, instead of indicating 9.7 pounds, will now read 6.1 pounds. This shows us that when a body is submerged in a liquid, it displaces its own *volume* of the liquid and is supported by a force equal to the *weight* of the liquid it has displaced. This force that caused the decrease in the scale reading is known as the buoyant force of the liquid.

If a material has a specific gravity of less than that of the liquid, it will displace its own *weight* of the liquid and it will float.

The effect of buoyancy is felt not only in liquids, but also in gases. Hot-air balloons are able to rise in the air because they are filled with heated air which is less dense than the large volume of the air they displace. As long as they displace a greater weight of air than their own weight, they will float.

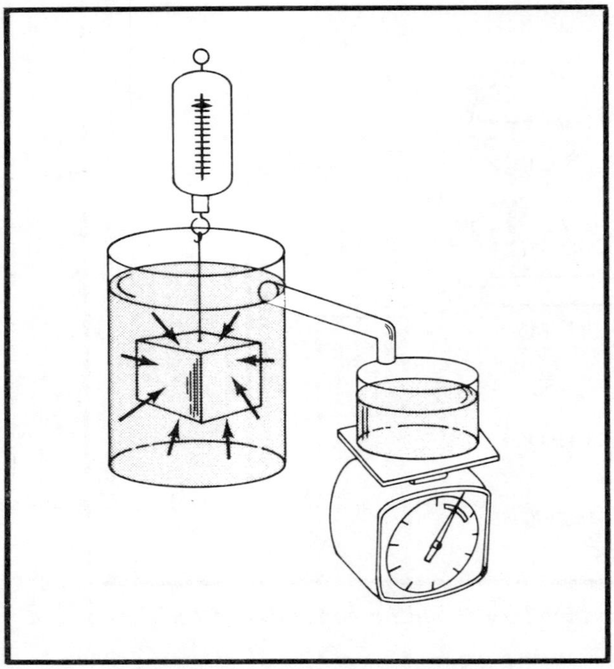

Fig. 13-3 A body immersed in a fluid is buoyed up by a force equal to the weight of the fluid it displaces.

C. Pascal's Law

A simple statement of the French mathematician Blaise Pascal explains one of the most useful principles of fluid mechanics. Essentially, he said that an increase in pressure at any point in a confined liquid will produce an equal increase in the pressure at every other point in the liquid, and the pressure will always act at right angles to the container that encloses the liquid.

Assume a cylinder with an area of one square inch, filled with a liquid and fitted with a piston. When a force of one pound is applied to the piston, the pressure of the confined liquid will be one pound per square inch everywhere in the container.

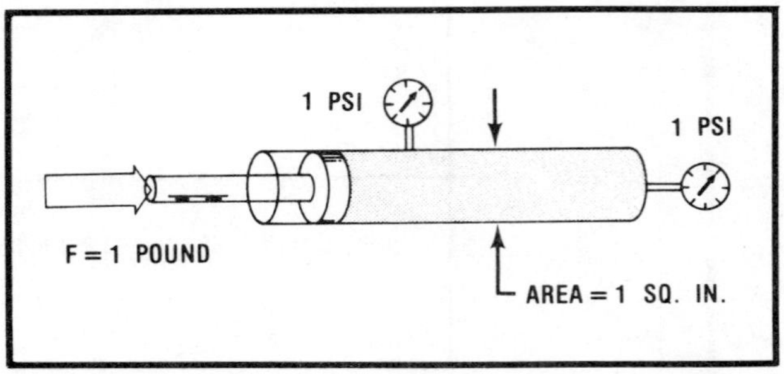

Fig. 13-4 The pressure produced in a hydraulic cylinder is found by multiplying the force on the piston by the area of the piston.

Since the shape of the container has no affect on the pressure in it, we can connect the piston through a small hose to a larger cylinder, one having, in our example of Fig. 13-5, an area of 10 square inches.

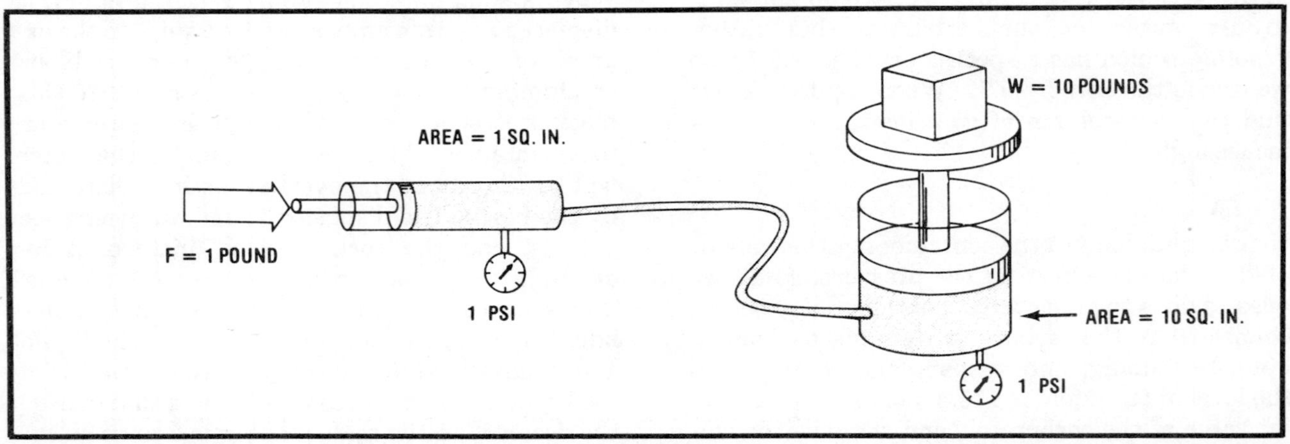

Fig. 13-5 A mechanical advantage may be obtained in a hydraulic system by using a piston with a small area to force fluid into a cylinder having a large area piston.

The pressure inside of both cylinders is one psi, but the piston in the larger cylinder has an area of 10 square inches, so there is a force of one pound acting on *every* square inch of the piston, and there is a total force of 10 pounds. This is the way we gain a mechanical advantage by the use of a fluid. When we apply a force of one pound to the small piston, we force upward against the large piston with a force of 10 pounds.

As with the lever, we do not gain any work. When we move the small piston inward five inches, we displace five cubic inches of fluid, and when this is spread out evenly over the 10 square inches of the larger cylinder, it will force the piston up only one-half inch. The same amount of work is done by both pistons, five inch-pounds of work.

The amount of force produced by a piston is determined by the pressure of the fluid and the area of the piston. In Fig. 13-6 we have a piston with an area of four square inches, but, one side of it has a one-square-inch rod connected to it which decreases its area by this amount. When a pressure of 1,000 psi is applied to the large side, it will produce a force of 4,000 pounds. But this same pressure applied to the side of the piston with the rod will produce a force of only 3000 pounds. We can find the amount of force produced by a hydraulic piston by multiplying the amount of pressure acting on the piston by the area of the piston.

We can find the area needed for a given amount of pressure to produce a required force by dividing the force by the pressure. In the same way, we can find the amount of pressure needed for a piston having a given area to produce the needed amount of force.

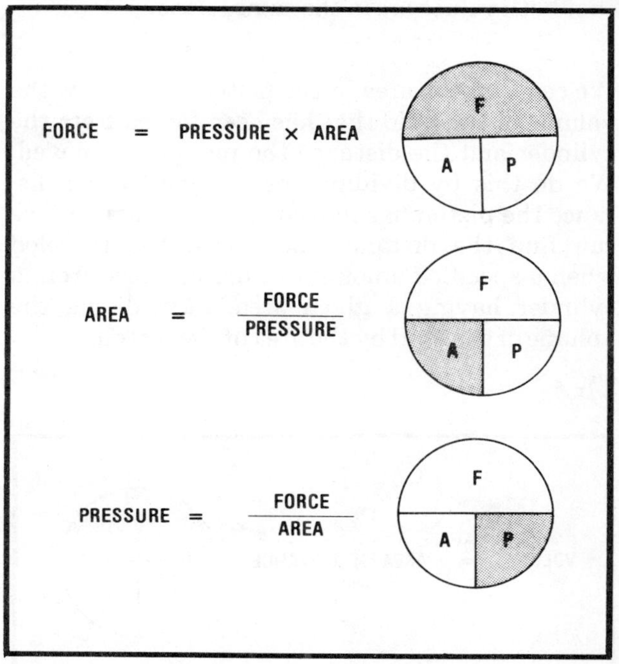

Fig. 13-7 The relationship between force, pressure, and area

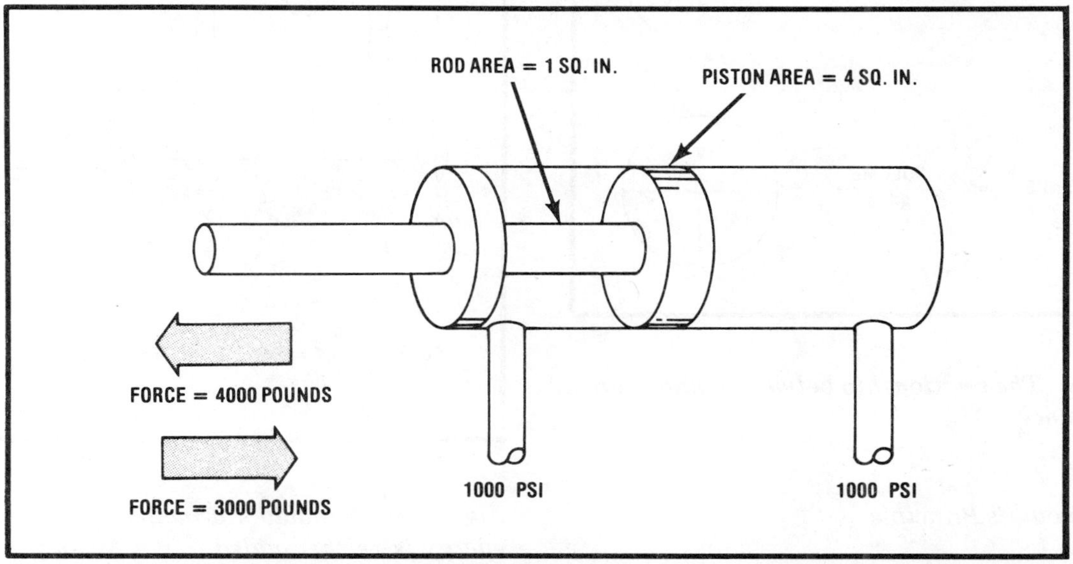

Fig. 13-6 A pressure of 1,000 psi will produce 4,000 pounds of force moving the piston outward, but because the area of the rod decreases the piston area, the same pressure will produce only 3,000 pounds of force moving the piston inward.

As a piston moves into the cylinder, it displaces a given amount of fluid, and a definite relationship exists between the area of the piston, the distance it moves, and the volume of the fluid it displaces. If we need to know the volume of the fluid displaced, we must multiply together the piston area and the distance the piston has traveled. Be sure that both values are in the same terms.

We can find the area of the piston if we know the volume of the fluid that has been forced from the cylinder and the distance the piston has moved. We do this by dividing the volume by the distance the piston has moved. In a like manner, we can find the distance the piston has traveled when a specified amount of fluid is forced from a cylinder having a given area. Just divide the volume of the fluid by the area of the piston.

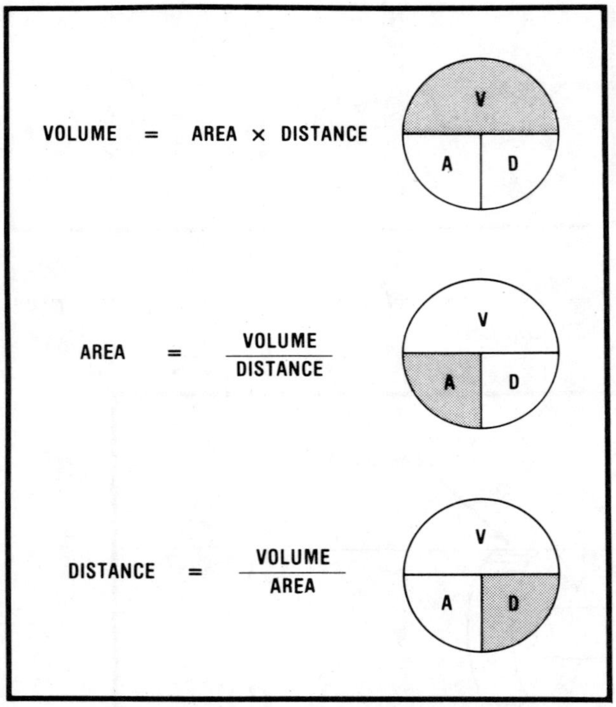

Fig. 13-8 The relationship between volume, area, and distance

D. Bernoulli's Principle

The Swiss mathematician and physicist Daniell Bernoulli has given us a principle bearing his name that explains the relationship between potential and kinetic energy in a fluid. And his is the explanation most frequently used to explain the production of lift by aerodynamic forces.

We must recognize that all matter contains energy in either its potential or kinetic form, or in both. The total energy in the fluid is the sum of the kinetic and potential energy. The potential energy can be thought of as that caused by the pressure of the fluid, and the kinetic energy as that caused by the velocity of movement of the fluid.

It is possible to exchange between kinetic and potential energy and if no energy is added to or taken from the fluid, the total energy will remain constant.

A venturi tube is a specially shaped tube smoothly narrowed down to a restrictor and then smoothly opening back up to its original size. When a fluid enters the tube, it has a given velocity and pressure. As it enters the restriction, it must speed up to get through, and in doing this it increases its kinetic energy. Now, since we have

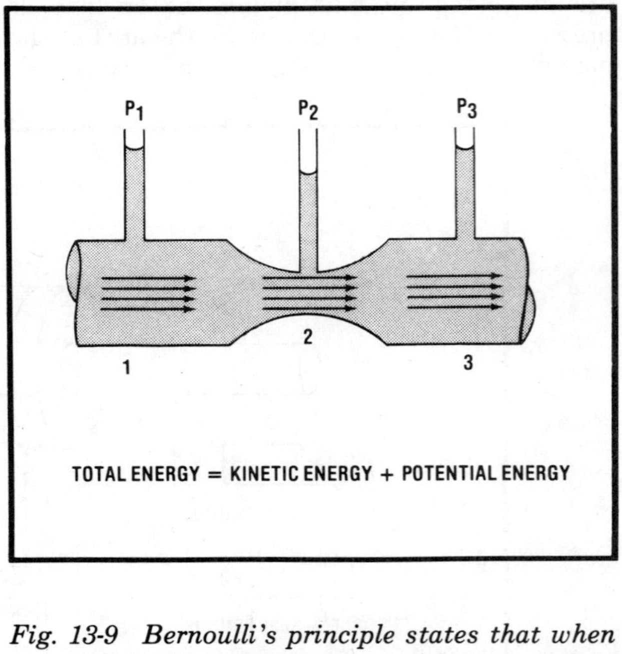

Fig. 13-9 Bernoulli's principle states that when energy is neither added to nor taken from a fluid in motion, the potential energy, or pressure, will decrease when the kinetic energy, or velocity, is increased.

not added any energy, nor have we taken any energy out of the fluid, when the kinetic energy is increased, the potential energy had to decrease. We can see this from the liquid levels in Fig. 13-9. At the point where the velocity is the greatest, the pressure is the lowest.

This principle is used to draw gasoline from the float bowl in a float-type aircraft carburetor. The discharge nozzle is in the narrowest part of the venturi, and as air is drawn into the cylinders, the velocity past the nozzle is increased to such a point that the pressure is lowered enough for the atmospheric pressure in the float bowl to force fuel out of the nozzle.

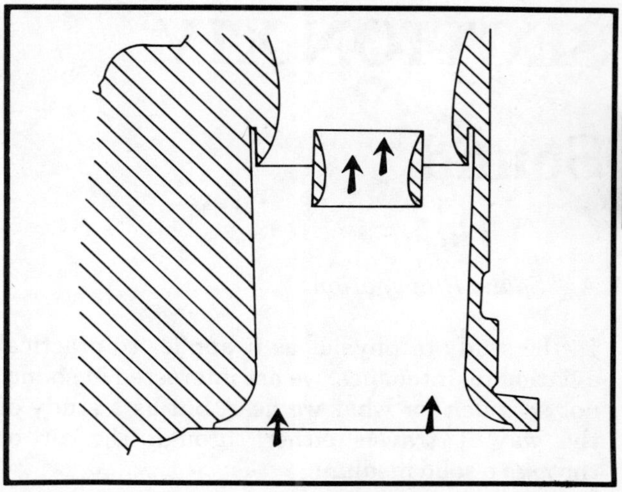

Fig. 13-10 The venturi in a carburetor is designed to speed up the flow of air entering the engine. This increase in velocity lowers the pressure, so fuel can be pulled out of the carburetor and into the engine.

SECTION XIV

Sound

A. Sound Propagation

In the study of physics as it applies to practical aviation maintenance, we are interested in sound, not so much for what we hear, but as a study of the way it travels either through the air or through a solid medium.

We can think of sound as any vibration of an elastic medium in the frequency range that affects our auditory senses. This range varies among different types of animals, and even among different humans, but it is generally considered to be between about 20 and 20,000 hertz.

Sound waves differ from water waves, as they are made up of a series of compressions and rarefactions. As the tine of the tuning fork moves away from the fork, the air in front of it is compressed, and as it moves back, the air is rarefied. These pressure changes, while extremely slight, are enough to vibrate the eardrum, and are strong enough that they can be measured by sensitive instruments.

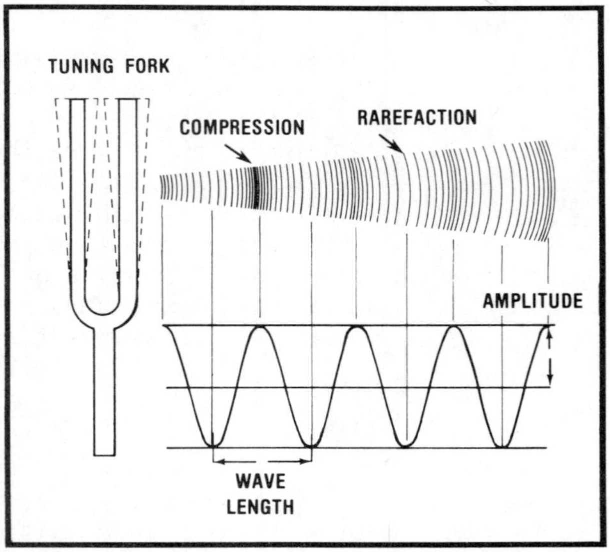

Fig. 14-1 Sound waves are made up of a series of compressions and rarefactions of the medium through which the sound is passing.

Sound travels through standard air at 661.2 knots, which is about 340.4 meters per second, and since water is so much denser than air, sound will travel much faster in water, at about 1435 meters per second, or 2789.6 knots.

B. Sound Intensity

Fortunately for us, the human ear does not register changes in sound in a uniform, or linear, scale, but rather we hear in a logarithmic fashion. We are able to detect very faint changes in a low level of sound, such as the ticking of a watch, but we cannot detect even tremendous differences in sound energy when the intensity is very high, as

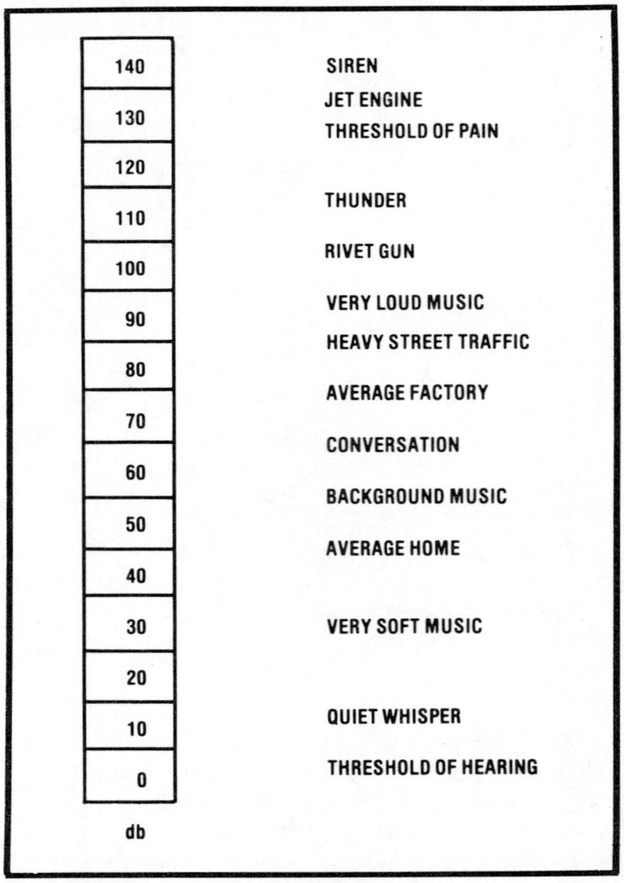

Fig. 14-2 Sound intensity is measured in decibels.

it is when the afterburner on a jet fighter is kicked in.

Sound intensity is measured in decibels, normally abbreviated db, and is the ratio of one sound to another. One db is the smallest change in sound intensity the human ear can detect. Around 60 db is the most comfortable level for conversation and for background music, while very loud music peaks out at more than 100 db. Up around 130 db, we are in the area that can cause pain.

C. *Doppler Effect*

When the source of sound is moving, the frequency ahead of it will be higher than it is behind it. This change in frequency is called the Doppler effect, and it accounts for the strange sound as an airplane passes over us. As the airplane approaches, the sound becomes both louder and more highly pitched. As it passes beyond us, the intensity drops off and the frequency of the sound decreases.

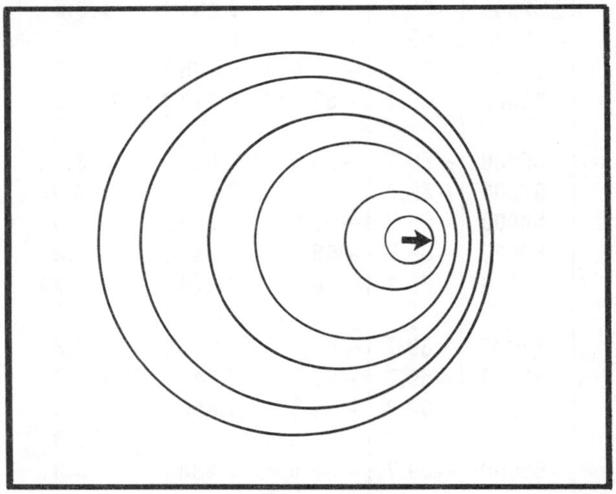

Fig. 14-3 When a body that is producing a sound is moving through the air, the sound waves progress out ahead of the body.

D. *Mach Number*

Vibrations cause waves which radiate out from their source at the speed of sound. If an airplane, which is a source of sound, is moving through the air more slowly than the speed of sound, the sound waves will move on out ahead of the airplane; but when the airplane moves as fast as the sound waves can travel, none of them can go faster than the airplane, and they will build up immediately ahead of the airplane in the form of a shock wave.

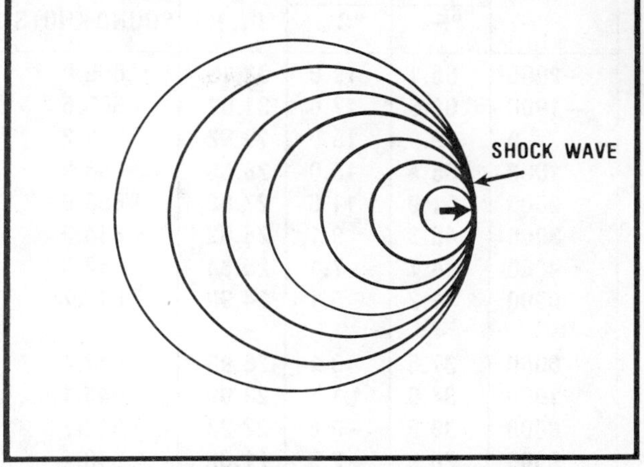

Fig. 14-4 When the body producing a sound is moving through the air at the speed of sound, the sound waves cannot get ahead of the body, and they pile up in the form of a shock wave that attaches to the body.

The Mach number is the ratio of the speed of the airplane to the speed of sound in the same atmospheric conditions. When the airplane is traveling at exactly the speed of sound, it is said to be traveling at a speed of Mach one. Two times the speed of sound is Mach two, and if the airplane is traveling at three-fourths the speed of sound, it is said to be traveling at a speed of Mach .75.

The speed of sound is dependent on the temperature of the air, and not on the air pressure. We can see in the table of Fig. 14-5 that the speed of sound decreases as the altitude increases and the temperature drops. This decrease continues until at slightly above 36,000 feet the temperature stabilizes out at −56.5 degrees Celsius, and the speed of sound remains at 575.3 knots until around 67,000 feet. Here the temperature begins to increase and the speed of sound begins to increase slightly.

E. *Resonance*

All bodies have a natural resonant frequency, which is the frequency at which an extremely small amount of excitation will cause a large amount of movement. It is possible for an aircraft structure to have a resonant frequency that is the same as certain vibrations from the engine or of

U.S. STANDARD ATMOSPHERE

ALT FT	TEMP °F	TEMP °C	PRESS IN. HG	SPEED OF SOUND KNOTS
−2000	66.1	19.0	32.15	666.0
−1000	62.6	17.0	31.02	663.6
0	59.0	15.0	29.92	661.2
1000	55.4	13.0	28.86	658.9
2000	51.9	11.0	27.82	656.6
3000	48.3	9.1	26.82	654.3
4000	44.7	7.1	25.84	652.0
5000	41.2	5.1	24.90	649.7
6000	37.6	3.1	23.98	647.4
7000	34.0	1.1	23.09	645.1
8000	30.5	−0.8	22.23	642.7
9000	26.9	−2.8	21.39	640.4
10000	23.3	−4.8	20.58	638.0
11000	19.8	−6.8	19.80	635.7
12000	16.2	−8.8	19.03	633.3
13000	12.7	−10.7	18.30	631.0
14000	9.1	−12.7	17.58	628.6
15000	5.5	−14.7	16.89	626.2
16000	2.0	−16.7	16.22	623.8
17000	−1.6	−18.7	15.58	621.4
18000	−5.1	−20.6	14.95	618.9
19000	−8.7	−22.6	14.35	616.5
20000	−12.2	−24.6	13.76	614.1
21000	−15.8	−26.6	13.20	611.6
22000	−19.4	−28.5	12.65	609.2
23000	−22.9	−30.5	12.12	606.7
24000	−26.5	−32.5	11.61	604.2
25000	−30.1	−34.5	11.12	601.7
26000	−33.6	−36.5	10.64	599.2
27000	−37.2	−38.4	10.18	596.7
28000	−40.7	−40.4	9.741	594.2
29000	−44.3	−42.4	9.314	591.7
30000	−47.8	−44.4	8.903	589.2
31000	−51.4	−46.3	8.506	586.6
32000	−54.9	−48.3	8.124	584.1
33000	−58.5	−50.3	7.756	581.5
34000	−62.1	−52.3	7.401	578.9
35000	−65.6	−54.2	7.060	576.3

U.S. STANDARD ATMOSPHERE

ALT FT	TEMP °F	TEMP °C	PRESS IN. HG.	SPEED OF SOUND KNOTS
36000	−69.2	−56.2	6.732	573.7
37000	−69.7	−56.5	6.417	573.3
38000	−69.7	−56.5	6.117	573.3
39000	−69.7	−56.5	5.831	573.3
40000	−69.7	−56.5	5.558	573.3
41000	−69.7	−56.5	5.299	573.3
42000	−69.7	−56.5	5.051	573.3
43000	−69.7	−56.5	4.815	573.3
44000	−69.7	−56.5	4.590	573.3
45000	−69.7	−56.5	4.375	573.3
46000	−69.7	−56.5	4.171	573.3
47000	−69.7	−56.5	3.976	573.3
48000	−69.7	−56.5	3.790	573.3
49000	−69.7	−56.5	3.613	573.3
50000	−69.7	−56.5	3.444	573.3
51000	−69.7	−56.5	3.284	573.3
52000	−69.7	−56.5	3.130	573.3
53000	−69.7	−56.5	2.984	573.3
54000	−69.7	−56.5	2.845	573.3
55000	−69.7	−56.5	2.712	573.3
56000	−69.7	−56.5	2.585	573.3
57000	−69.7	−56.5	2.465	573.3
58000	−69.7	−56.5	2.350	573.3
59000	−69.7	−56.5	2.240	573.3
60000	−69.7	−56.5	2.135	573.3
61000	−69.7	−56.5	2.036	573.3
62000	−69.7	−56.5	1.941	573.3
63000	−69.7	−56.5	1.850	573.3
64000	−69.7	−56.5	1.764	573.3
65000	−69.7	−56.5	1.682	573.3
66000	−69.6	−56.5	1.603	573.4
67000	−69.1	−56.1	1.528	573.8
68000	−68.5	−55.8	1.457	574.2
69000	−68.0	−55.5	1.390	574.6
70000	−67.4	−55.2	1.325	575.0

Fig. 14-5 The speed of sound in air varies as the temperature of the air.

Courtesy of Beech Aircraft Corp.

Fig. 14-6 Before an airplane is certificated, it must be subjected to an extensive series of vibration tests.

the air in the inlet duct. These vibrations could cause flutter or structural failure. For this reason, all aircraft must undergo extensive flight and vibration tests to be sure that they have no resonant frequency conditions that could become excited in flight and cause flutter.

AVIATION
TECHNICIAN
INTEGRATED
TRAINING
PROGRAM

EA-ITP-G2

Basic Electricity-DC

GENERAL
SECTION
2

EA-ITP-G2

TRAINING MANUAL
GENERAL SECTION
BOOK 2

BASIC ELECTRICITY — DC

by
Dale Crane

AVIATION MAINTENANCE PUBLISHERS, INC.

INTEGRATED TRAINING PROGRAM
FOR THE
AVIATION MAINTENANCE TECHNICAN

International Standard Book Number 0-89100-106-9

Printed in the United States of America

Table of Contents

PREFACE

Part 65 of the Federal Aviation Regulations requires that an applicant for a mechanic certificate with an Airframe or Powerplant rating have a general knowledge of the construction and maintenance of the aircraft covered by the rating he seeks. And each applicant must prove this knowledge by passing a written, an oral, and a practical examination, each given in five sections: General, Airframe Structures, Airframe Systems and Components, Powerplant Theory and Maintenance, and Powerplant Systems and Components.

Certificated Aviation Maintenance Technician Schools all operate under individual Federal Aviation Administration approvals, and all have in their approved curriculum enough information to satisfy the FAA that anyone completing their course will meet the knowledge requirements for certification as a mechanic.

In 1972, the Federal Aviation Administration published a series of three Advisory Circulars providing background information for mechanic certification applicants. These are entitled, *Airframe and Powerplant Mechanics General Handbook,* AC 65-9; *Airframe and Powerplant Mechanics Powerplant Handbook,* AC 65-12; and *Airframe and Powerplant Mechanics Airframe Handbook,* AC 65-15. These three handbooks were written to cover the area over which a mechanic applicant was to be tested, and they were an attempt by the FAA to coordinate the information available to the applicants with the tests that were given for certification. In 1976, the three handbooks were revised into their present A Revision format: AC 65-9A, AC 65-12A, and AC 65-15A.

Individuals attempting to certificate under FAR Part 65, as well as schools operating under FAR Part 147, have found that these handbooks actually do not follow the material used in the FAA examinations, and because of this, plus their lack of study helps, they fail to effectively provide the broad base of knowledge needed for certification and for a foundation on which to build in the aviation maintenance industry.

Realizing the need for updated information, presented in a format that is in keeping with the latest developments in educational technology, the Aviation Maintenance Publishers, Inc., is offering an Integrated Training Program (ITP) that presents the basic information needed for mechanic certification, in a form that will provide a good foundation in aviation maintenance knowledge. Developed in coordination with Crane Aeronautical Activities, Inc., the ITP training manuals, workbooks, study guides, and audio-visuals are designed to follow the FAA written examinations and are written to the level required for the FAR Part 147 schools.

All of this material can be used by an individual preparing for certification on his own, or by schools preparing groups of individuals for certification under FAR Part 65, but it has been prepared primarily for the Aviation Maintenance Technician Schools operating under FAR Part 147. Not only has every attempt been made to meet the requirements of Part 147, in producing this series, but the authors have followed many of the recommendations of the Aviation Mechanic Occupation Study conducted by the University of California, Los Angeles, under a contract from the U.S. Department of Health, Education, and Welfare.

One of the most basic premises of aviation maintenance is that all operations must be carried out according to FAA-approved data, and all information of a specific nature furnished by the manufacturer of an aircraft, powerplant, or component must be followed in detail. These training manuals are, of necessity, general in nature and cover a wide scope of aircraft, powerplants, and components. Because of this, it must be understood that they should *never* be allowed to take precedence over specific information furnished by the manufacturer.

The authors and publishers of this Integrated Training Program wish to express our appreciation to the various manufacturers, Aviation Maintenance Technician Schools, and FAA personnel who have aided our efforts to provide this coordinated training material to meet the challenge of better training for aviation maintenance technicians in the eighties.

SECTION I

Discovery of Electricity

One of the first recorded mentions of electricity was by the Greek philosopher Thales in about 600 B.C. He mentioned the fact that when substances such as amber and jet were rubbed, they would attract such light objects as feathers and bits of straw. Later in the eighteenth century, it was discovered that there were two kinds of forces, or charges, caused by rubbing certain kinds of materials. Charges of the same kind would repel each other while opposite charges would attract.

In about the middle of the eighteenth century, the practical mind of Benjamin Franklin found a way to prove that lightning was a form of electricity. In his famous kite experiment, he flew a kite into a thunderstorm and found that sparks would jump to the ground from a metal key attached to the wet string. Franklin made a logical assumption that whatever it was that came down the string was flowing from a high level of energy to a lower level. And so he assigned the term "positive" to the high energy, and "negative" to the lower level. It was not known what actually came down the string, but he used a term associated with the flow of water, and said that it was "current' that flowed down the string, from a positive charge to the negative.

This assumption of Franklin's was accepted until the discovery of the electron in 1897, and many textbooks on electricity are still in existence that speak of the flow of current from positive to negative. Not until it was discovered that it was actually electrons, or negative particles of electricity, that move through a circuit could we understand the true nature of electricity.

Fig. 1-1 Benjamin Franklin's practical mind found a way to prove that lightning was a form of electricity with his famous kite experiment.

SECTION II

Electron Theory

A. *Composition of Matter*

When a component such as a light bulb is connected to a source of electrical energy by solid conductors, or wires, there *appears* to be no movement within the conductor, even though it is obvious that something is happening.

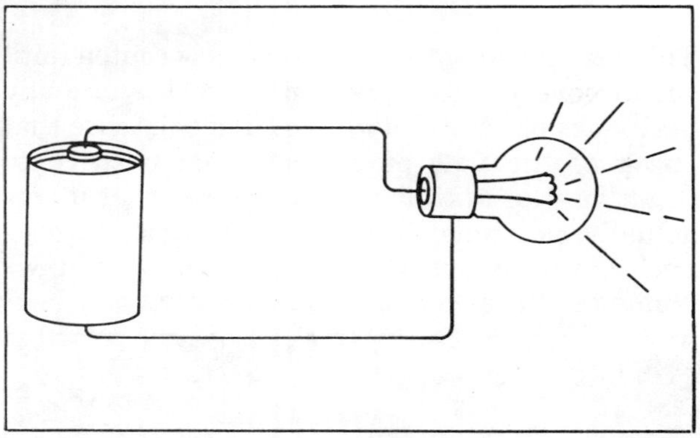

Fig. 2-1 There appears to be no movement within the conductor, but it is obvious that something is happening.

If we were able to see what is actually inside the wires that carry the electricity, we would see that they are not *really* solid at all, but actually contain far more empty space than matter.

1. *The atom*

All of the material from which our universe is made is composed of atoms, the smallest particles of an element which can exist, either alone or in combination with other atoms. Each atom consists of a nucleus, made up, usually, of positively charged protons, and neutrons, having no electrical charge but of the same weight as the protons. Surrounding this nucleus and traveling at incredible speed are electrons, each of which weighs only about 1/1845 as much as a proton and which has a negative charge.

The most simple atom is that of hydrogen, which consists of only one proton and no neutrons in the nucleus. Spinning around the nucleus is a single electron.

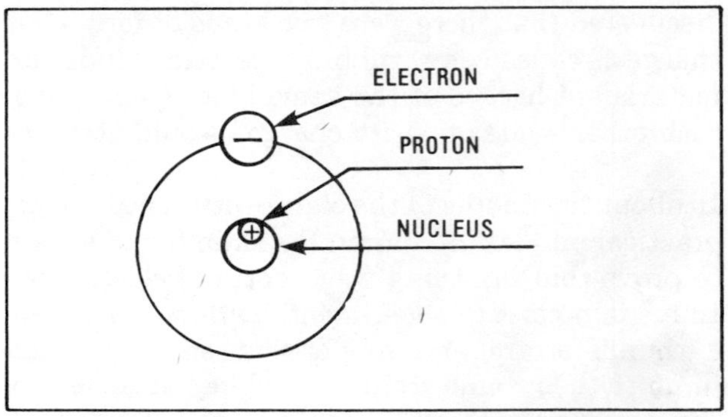

Fig. 2-2 The simplest atom, hydrogen, has one electron spinning around the nucleus consisting of one proton. There are no neutrons in the nucleus.

All electrons are alike, as are all protons and all neutrons, but the number and arrangement of these elementary building blocks in the atom determine the material the atoms make up. For example, copper has a nucleus made up of 29 protons and 36 neutrons. Surrounding the nucleus and spinning in four rings, or "shells," are 29 electrons. This combination gives us what is called a balanced atom, as there are exactly the same number of positive charges, protons, as there are negative charges, electrons. The neutrons have no electrical charge, and so they do not enter into the flow of electricity.

We are familiar with the diagrams of atoms in our textbooks and training manuals, but these cannot possibly be drawn to scale. For example, to have the same *mass* as a proton, the electrons would have to be about 1,845 times as large, and the distance between the electron and the nucleus would have to be about 100,000 times the diameter of the nucleus.

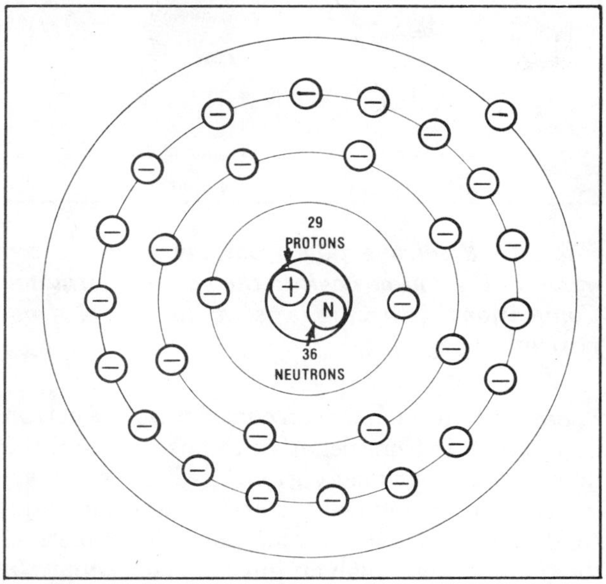

Fig. 2-3 The nucleus of an atom consists of positively charged protons and electrically neutral neutrons. Spinning around the nucleus are negatively charged electrons.

All matter contains energy, and energy in the atom causes the electrons to spin around the nucleus. As they spin, centrifugal force tends to pull them away from the nucleus. But there is an electrostatic field within the atom that produces a force which exactly balances this centrifugal force and holds the electrons a specific distance away from the nucleus.

The electrons spin around the nucleus in rings or shells, and when energy is added to an atom, such as is done when the material is heated, the orbiting radius, or the distance between the electrons and the proton, is increased, and the bond, or force of attraction, between the proton and the electron is decreased.

2. *Ions*

An electrical force outside the atom can attract electrons from the outer ring and leave the atom in an unbalanced condition. These unbalanced atoms are called ions. For example, copper has one electron in its outer ring, and if a positive force is applied to the atom, this outer ring electron, which is negative, will be drawn from the atom and leave it with more protons than electrons. It then becomes a positive ion and will attract an electron from a nearby balanced atom. Electrons constantly move about within a material from one atom to another, in a continuous but random fashion.

B. Electron flow

Let's see what happens when a conductor made of copper is connected across a source of electrons. An electron is attracted from an atom by the positive terminal of the source, and it leaves the conductor. The atom which lost the electron has now become a positive ion and pulls an electron away from the next atom. This exchange continues until the electron that left the conductor is replaced by one from the negative terminal of the source.

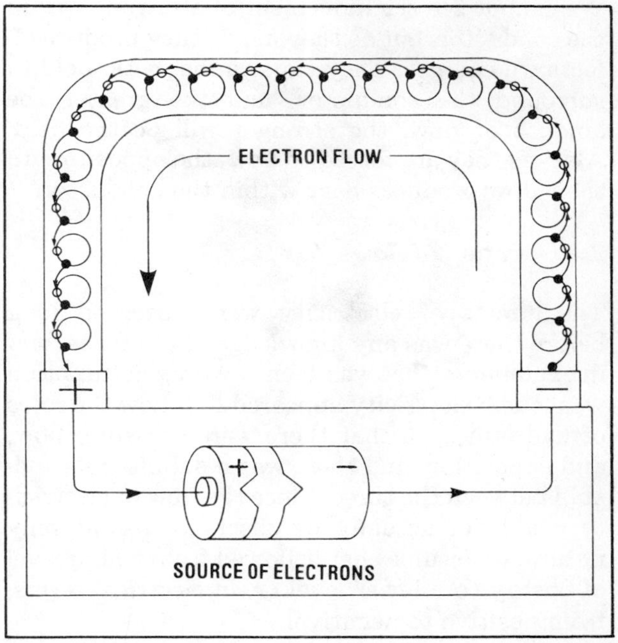

Fig. 2-4 When an electron is attracted from the conductor by the positive charge of the source, it leaves a positive ion. This ion attracts an electron from an adjoining atom. This exchange continues through the conductor until an electron is furnished by the negative terminal of the source to replace the one that was taken by the source.

It takes us a while to visualize what is happening inside the conductor, but electron movement takes place within the conductor at about the speed of light; that is, about 186,000 miles per second. This is not saying that a single electron moves through the conductor from one end to the other at this speed, but because of the domino effect, an electron entering one end of the conductor will almost immediately force an electron out of the other end.

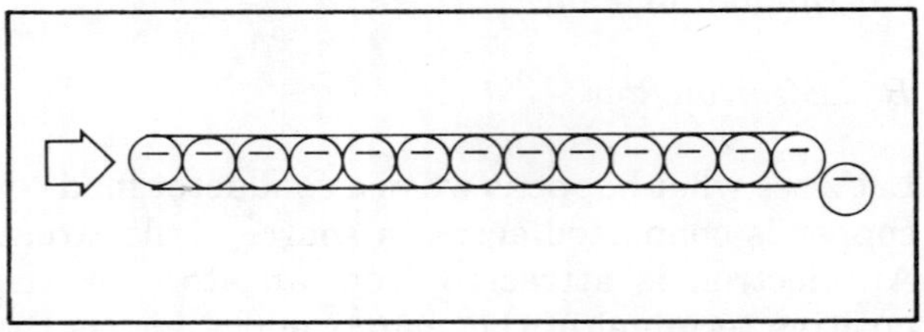

Fig. 2-5 When one electron is forced into the conductor, it immediately forces an electron out of the opposite end of the conductor.

1. Effects of electron flow

We cannot see the movement of electrons within the conductor, but as they move, they produce effects we can use. They cause a magnetic field to surround the conductor, and the greater the amount of flow, the stronger will be the field. Also, as they are forced to flow, the opposition to their flow produces heat within the conductor.

2. Direction of flow

The effects of electricity were observed long before there was any knowledge of electrons, and in explaining what was seen, a wrong assumption was made: electricity appeared to follow the rules of hydraulics, in that there were pressure, flow, and opposition, and there was a definite relationship between the three. Since the flow of electricity could not actually be observed, it was only natural to assume that it flowed from a high level of energy to a lower level or, in electrical terms, from positive to negative.

This theory worked well for years, and many texts have been written calling the flow of electrons "current flow," and assuming that whatever it was that flowed in the circuits moved from the positive terminal of the source to the negative.

As we have gained knowledge of the atom, it has become apparent that it was the electron with its *negative* charge that actually moved through the circuit, and the texts have had to be revised to explain electron flow as being from the *negative* terminal of the source through the load, back into the source of the *positive* terminal.

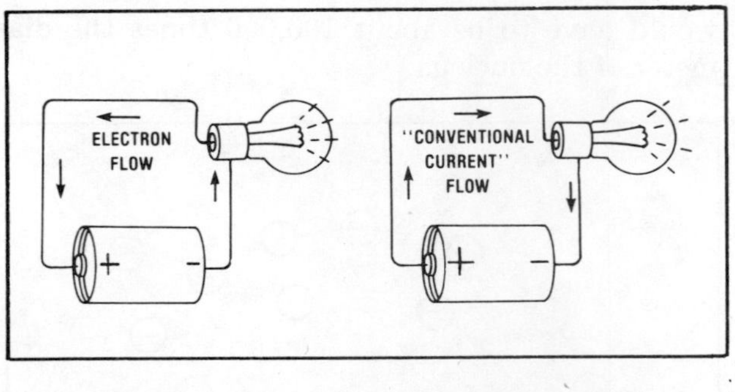

Fig. 2-6 Electrons flow from the negative terminal of the source back to the positive terminal. Conventional current is said to travel from positive to negative.

There are two ways we can consider flow: electron flow, which is from negative to positive, and the flow of "conventional current", which, while actually a myth, follows the arrows used on semiconductor symbols. You may use either method for tracing flow, but you must be consistent. *In this training manual, we will follow electron flow, from negative to positive, and we will use the terms electron flow and current interchangeably.* In later manuals, where we are dealing with semiconductor devices and using their symbols, we will follow the flow of "conventional current" from positive to negative, because the arrows used in semiconductor symbols point in the direction of conventional current flow.

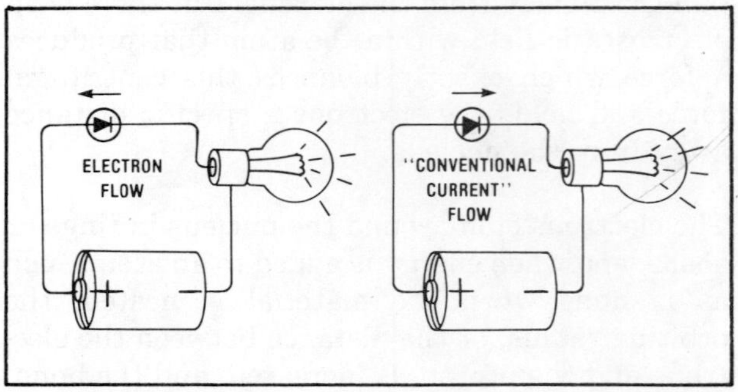

Fig. 2-7 Electron flow is opposite the arrows used in semiconductor symbols. It is easier to think in terms of conventional current, which, even though it is a myth, is easier to follow when semiconductor symbols are used.

C. Units of Electrical Measurement

1. Quantity

The electron is such an extremely small particle of electricity that an enormous number of them are required to have a measurable unit. The coulomb is the basic unit of electrical quantity and is equal to 6.28 billion, billion electrons. This is most generally written as 6.28×10^{18}, which means that the number of electrons is 6.28 followed by 16 zeros. The symbol for quantity is Q.

2. Flow

When one coulomb flows past a point in one second, there is a flow of one ampere, or one amp. It makes no difference whether we think of electron flow or conventional current, it is all generally called current, and its symbol is I.

3. Opposition

The ohm is the standard unit of resistance, or opposition to current flow, and is the resistance at a specific temperature of a column of mercury having a specified length and weight. More practically, it is the resistance through which a pressure of one volt can force a flow of one ampere. The symbol for resistance is R.

4. Pressure

The volt is the unit of electrical pressure and is the amount of pressure required to force one amp of flow through one ohm of resistance. There are a number of terms used to express electrical pressure. They are: voltage, voltage drop, potential, potential difference, EMF, and IR drop. These terms have slightly different shades of meaning, but are often used interchangeably. The symbol for electrical pressure is E.

VOLTAGE
VOLTAGE DROP
ELECTROMOTIVE FORCE (EMF)
POTENTIAL
POTENTIAL DIFFERENCE
IR DROP

Fig. 2-8 All of these terms may be used to express electrical pressure.

5. Power

The end result for practical electricity is power, and electrical power is expressed in watts. One watt is the amount of power dissipated when one amp of current flows under a pressure of one volt. The symbol for power is P.

D. Metric Prefixes and Powers of Ten

1. Prefixes

So many terms used in the study of electricity deal with numbers that are either extremely large or extremely small. Because of this, metric prefixes have been adapted to them.

Multiplier	Prefix	Symbol
1,000,000,000,000	tera	t
1,000,000,000	giga	g
1,000,000	mega	M
1,000	kilo	k
100	hecto	h
10	deka	dk
0.1	deci	d
0.01	centi	c
0.001	milli	m
0.000,001	micro	μ (mu)
0.000,000,001	nano	n
0.000,000,000,001	pico	p

Fig. 2-9 Metric prefixes.

The emergency frequency used for aircraft communications is 121,500,000 hertz, or cycles per second. This number is large and unwieldy, so we can divide it by one million and use the metric prefix mega-. The number becomes 121.5 megahertz.

These prefixes also help us in dealing with very small numbers. For example, the basic unit of capacitance, the farad, is much too large for practical use in aircraft electronics, and one of the commonly used capacitors has a capacity of 0.000,000,000,002 farad. A number such as this is awkward to work with, and its use encourages errors. A much more convenient way to express this same unit is to use the term two picofarads, or 2pf. This term, by the way, has previously been called a micromicro farad, and you may still see it referred to in this way. It may be written 2 mmf or 2μμf, with the Greek letter μ (mu) used.

2. *Powers of ten*

Multiplying and dividing very large and very small numbers is made easier by the use of powers of ten. In this method of handling numbers, convert every number into a number between one and ten by moving the decimal the proper number of places in the correct direction. For example, 0.000,000,002 can be converted into 2.0 by moving the decimal to the right nine places. Since the number is smaller than one, the number two will have to be multiplied by a *negative* power of ten. When 0.000,000,002 is converted into a power of ten number, it becomes 2×10^{-9}.

Numbers larger than one are converted in exactly the same way, except they are multiplied by a *positive* power of ten. One coulomb contains 6,280,000,000,000,000,000 electrons. This number is easier to work with when it is converted to 6.28×10^{18} by moving the decimal to the left 18 places and multiplying 6.28 by ten, 18 times.

Numbers that have been converted into powers of ten may be multiplied or divided by performing the required work on the numbers, and then adding the powers of ten (the exponents), to multiply, or subtracting the powers, to divide.

$$0.0025 \times 5{,}000 = 2.5 \times 10^{-3} \times 5 \times 10^{3} = 12.5$$

$$0.125 \times 0.5 = 1.25 \times 10^{-1} \times 5 \times 10^{-1} = 6.25 \times 10^{-2} = 0.0625$$

$$5{,}000{,}000 \div 250{,}000 = 5 \times 10^{6} \div 2.5 \times 10^{5} = 2 \times 10^{1} = 20$$

$$0.125 \div 0.5 = 1.25 \times 10^{-1} \div 5 \times 10^{-1} = 0.25 \times 10^{0} = 0.25$$

SECTION III

Static Electricity

A. Positive and Negative Charges

Electricity may be classified in two types, current and static. In current electricity, the electrons move through a circuit and perform work, either by the magnetic field created by their movement, or by the heat generated when they are forced through a resistance. Static electricity, on the other hand, normally serves little useful purpose, and is more often a nuisance rather than a useful form of electrical energy.

Static electricity is of real concern to us during the fueling operation of an aircraft. As the aircraft flies, friction between the air and the surface builds up a large static charge which cannot readily bleed off upon landing, because the rubber tires insulate the aircraft from the ground. Fuel trucks and fueling pits are grounded to the earth, so if the first contact with the aircraft is with the fuel nozzle in the open filler neck, a spark can jump in the explosive fumes and cause a serious fire. To prevent this, always connect the aircraft to the fuel truck and to the ground by the grounding cables which are provided on all fuel trucks for this purpose.

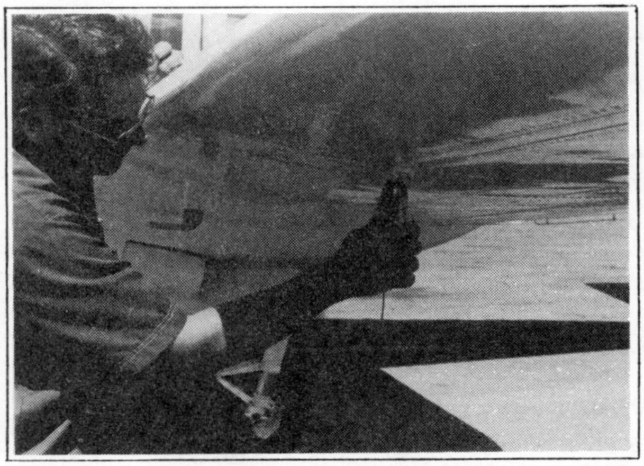

Fig. 3-1 Airplanes and fuel trucks should be grounded together to neutralize the charge of static electricity before the fueling nozzle is put into the tank.

We cannot see electricity, but the effects of both types are easy to observe. If we have a couple of balls of pith wood, we can suspend them by threads and observe the effects of static electricity as we charge them.

Rub a glass rod with a piece of wool or fur and the rod will pick up a lot of extra electrons, and will become negatively charged. When we hold the rod close to a ball which has a neutral charge, it will be attracted to the rod. But as soon as the ball touches the rod, electrons will flow to the ball and give it a negative charge. Now, it will be repelled by the rod.

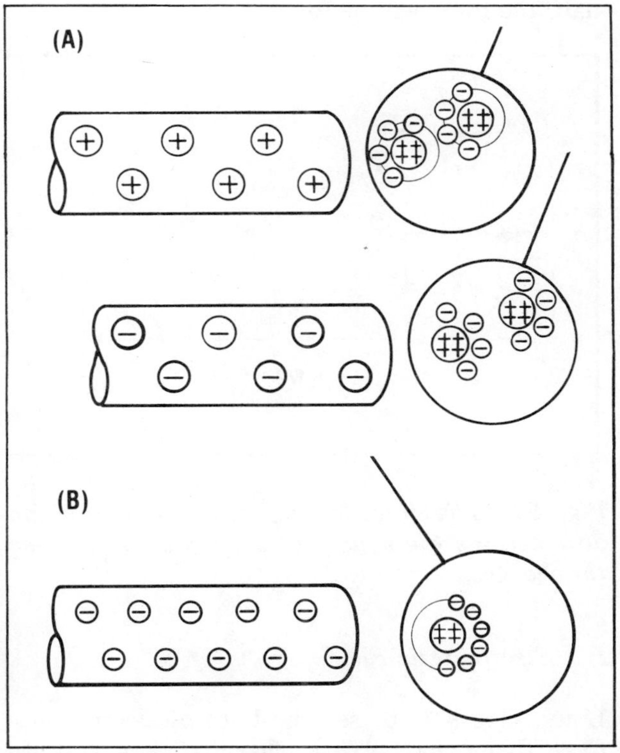

Fig. 3-2 A — An uncharged pith ball will be attracted to a rod that has either a positive or a negative charge. B — Once the charged rod has contacted the ball, the ball will assume the same charge as the rod and it will be repelled by the rod.

If we rub the glass rod with a piece of silk, the glass will give up electrons to the silk, and the rod will have a deficiency of electrons, or will become positively charged. And when we hold it near one of the balls, it, too, will attract the ball, and once it has touched it, the ball will be repelled. This is the same thing that happens with the first ball, but there is a difference. One of the balls is charged *negatively*, and the other is charged *positively*. When they are brought close together, they will be attracted to each other.

This demonstration proves that there are two types of charges and that like charges repel each other, while unlike charges attract. The strength of the repelling and attracting force varies as the square of the distance between the two charges. For example, if the distance between the two balls is doubled, the force of attraction will be reduced to one-fourth ($2^2 = 4$). If they are moved three times as far apart, the force will be only one-ninth of the original. On the other hand, if they are moved closer together, the force of attraction or of repulsion will *increase* as the square of the separation. If the distance is decreased to one-half, the force will be four times as great.

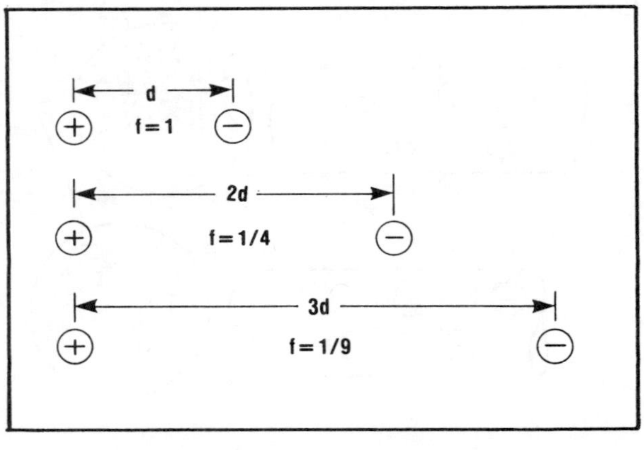

Fig. 3-3 The force of repulsion or attraction decreases as the square of the distance between the charges.

B. Electrostatic Fields

If we were able to see the lines of electrostatic force that exist between charges, we would see that the lines leave one charged object and, in our case, we will arbitrarily assume that they leave the one having the positive charge and enter the other, the one having the negative charge. If the charges are close together, all of the lines will link, and the two charges will form a neutral, or an uncharged, group. The lines of force from like charges repel each other and will tend to push the charges apart.

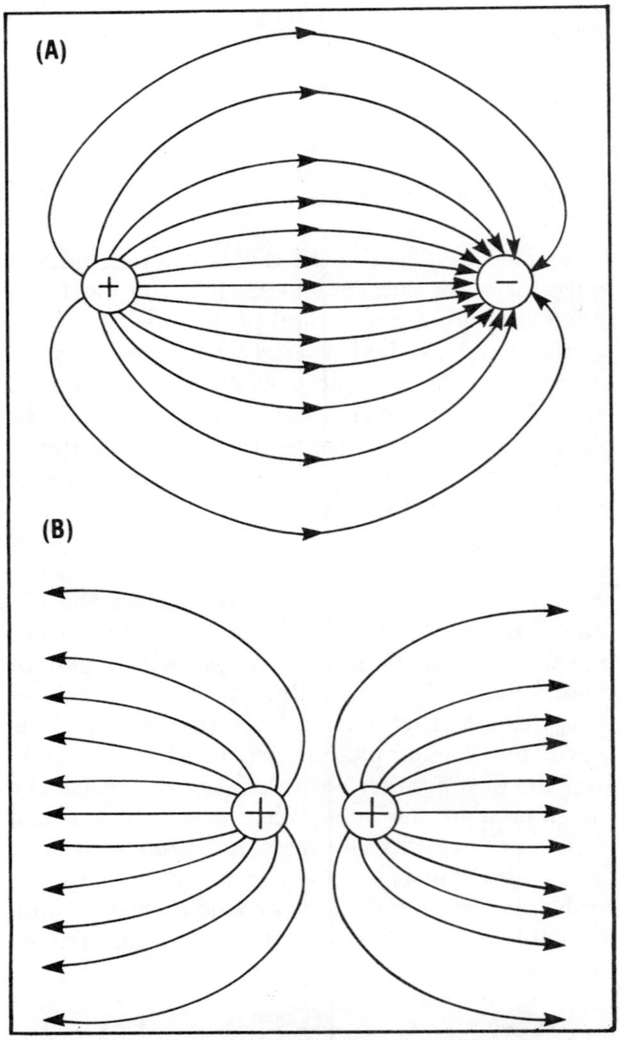

Fig. 3-4 A — Lines of force leave the charged body at right angles to its surface, and since they are polarized alike, the lines spread apart. They enter the oppositely charged body at right angles to its surface. B — Charged bodies reject lines of electrostatic force from other bodies having the same charge.

C. Distribution of Electrical Charges

When a body having a smooth or uniform surface is electrically charged, the charge will distribute evenly over the entire surface; but if the surface is irregular in shape, the charge will concentrate at the points or areas having the sharpest curvature.

This explains the action of static dischargers used on many aircraft control surfaces. As the airplane flies through the air, friction causes a large static charge to build up on the control surface. These surfaces are connected to the airframe structure by hinges which do not provide a particularly good conductive path, so the charge builds up in strength. Static dischargers are attached to the control surface to dissipate this charge. They have sharp points on which the static charges concentrate and will discharge into the air before they can build up on the smooth surface sufficiently high to jump across the hinges and cause radio interference.

As a further aid in preventing radio interference, the control surface is bonded to the structure. This means that a flexible metal braid is attached to both the control surface and the structure to act as a good conductor, so the charge can be neutralized as it forms.

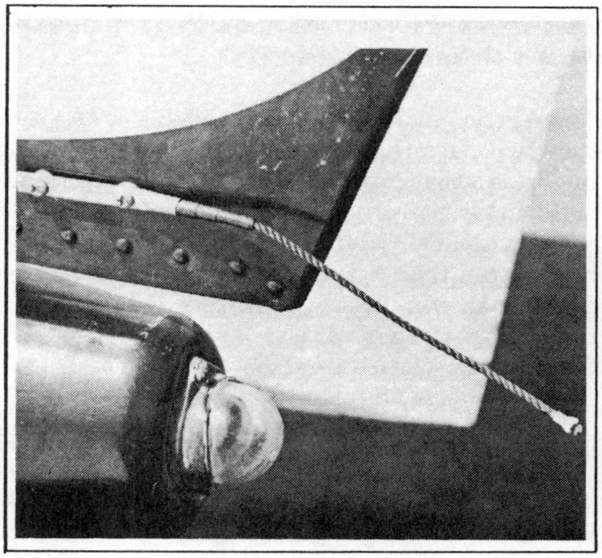

Fig. 3-5 Static dischargers provide sharp points from which static charges are dissipated into the air before they can build up to a high potential on the control surface.

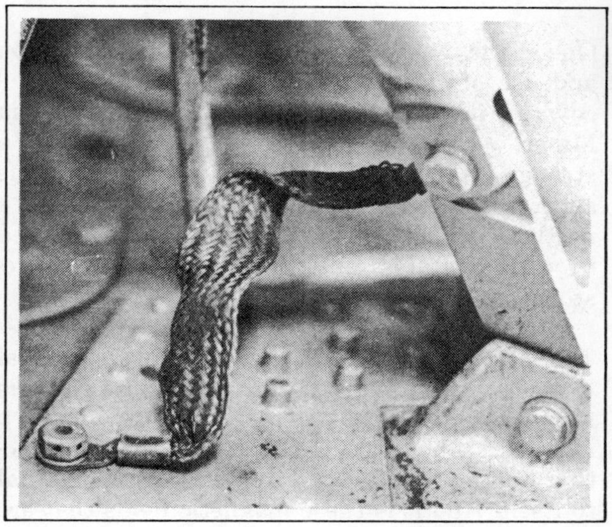

Fig. 3-6 Bonding straps provide a low-resistance path between the control surface and the aircraft structure to prevent the buildup of a static charge on the control surface.

SECTION IV

Magnetism

Magnetic Characteristics

One of the most useful devices in the production and use of electricity is the magnet. First discovered in the province of Magnesia in Asia Minor, a form of iron oxide demonstrated a strange property: when it was suspended in the air or floated on a chip, it would always turn in a northerly direction. This strange stone was used by early seafarers as a leading stone, or "lodestone" to aid in navigation.

The strange characteristics of these natural magnets have been studied, and now we are able to produce artificial magnets. By definition, a magnet is a body that has the property of attracting iron and producing magnetic fields external to itself. And it is these magnetic fields that are of interest to us.

Lines of magnetic force, or flux, are always complete loops that leave the magnet at right angles to its surface at the north pole, and since they are all polarized in the same direction, they repel each other and spread out. They draw closer together as they re-enter the magnet at the south pole, at right angles to its surface, and travel through the magnet to complete the loop.

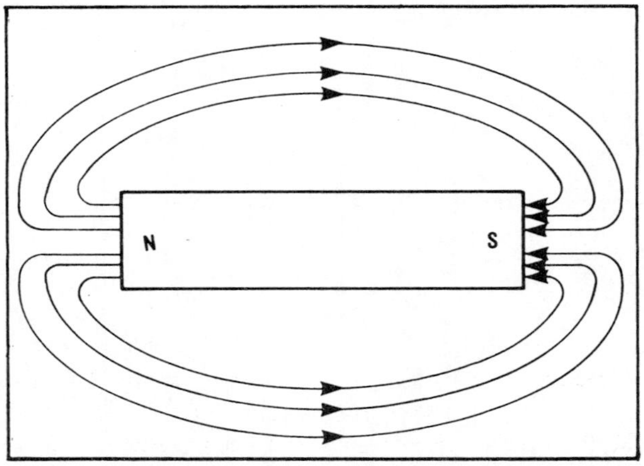

Fig. 4-1 Lines of magnetic force form complete loops, leaving the magnet at its north pole and returning at its south pole.

There is no insulation against the lines of flux, as they will pass through any material; but if it is important that a device be protected from magnetic fields, it can be entirely surrounded by a soft iron shield. Iron, as we will see, has a very high permeability and provides a much easier path for the flux than the air, and since the flux lines travel the path of least resistance, they will flow through the iron and leave an area inside the shield that has no magnetic field.

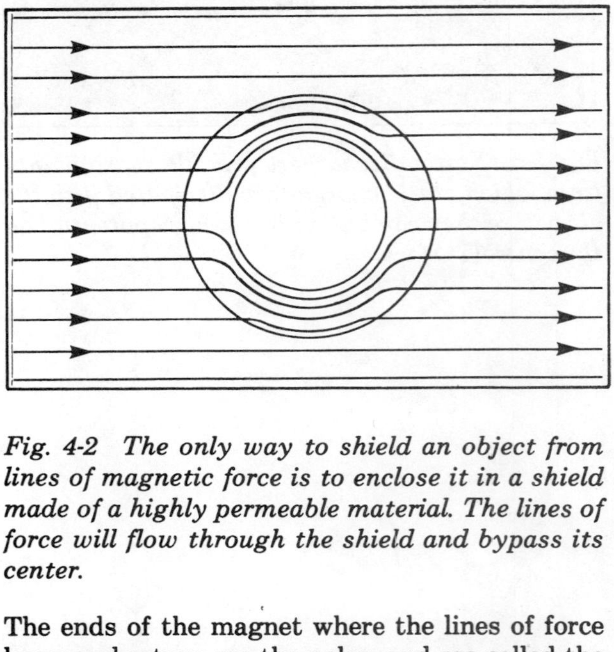

Fig. 4-2 The only way to shield an object from lines of magnetic force is to enclose it in a shield made of a highly permeable material. The lines of force will flow through the shield and bypass its center.

The ends of the magnet where the lines of force leave and return are the poles, and are called the north- and the south-seeking poles or, more commonly today, just the north and south poles.

If we were able to see inside a piece of unmagnetized iron, we would see that it contains an almost infinite number of magnetic fields oriented in such a random fashion that they cancel each other. Now, if this piece of iron were placed in a strong magnetic field, all of these little fields, or domains, as they are called, would align themselves with the strong field, and the iron would become a magnet, having a north and south pole, and lines of magnetic flux would encircle it.

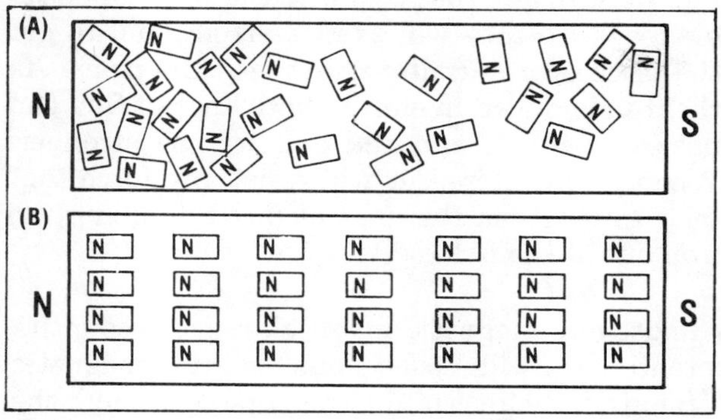

Fig. 4-3 A — In an unmagnetized material, all of the individual magnetic fields, or domains, are arranged in a random fashion and cancel each other. B — When the material is magnetized, all of the domains are aligned, and the material has a north and south pole, just like that of the individual domains.

The domain theory of magnetism is supported by the fact that each magnet has both a north and south pole, regardless of the size of the magnet. If we break a bar magnet in two, each half will demonstrate the characteristics of the original magnet, and if we break each of these halves in two, all of the small pieces will still be magnets having north and south poles, with lines of magnetic flux surrounding them.

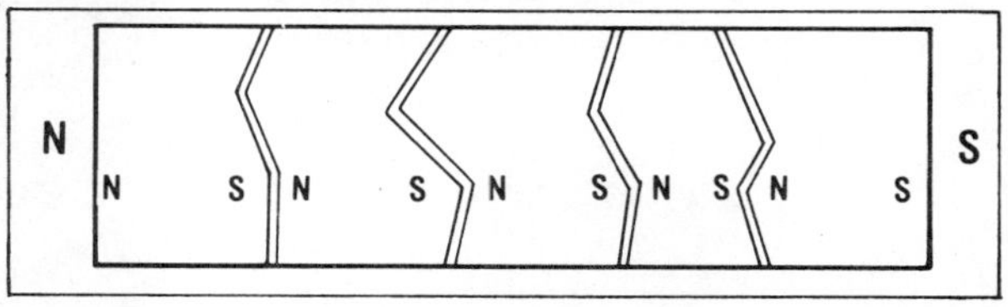

Fig. 4-4 Regardless of how small the pieces of a magnet may be broken, each piece will have a north and a south pole.

Soft iron has a very low retentivity, meaning that as soon as the magnetizing force is removed, the domains will lose their alignment, the fields will cancel each other, and the iron will no longer act as a magnet. Hard steel and some of the alloys of iron using aluminum, nickel, cobalt, and molybdenum have very high retentivities, and will retain the alignment of their domains long after the magnetizing force has been removed. It is materials of this type that are used for permanent magnets in aircraft magnetos, instruments, and radio speakers.

The number of lines of flux that loop through the magnet gives us an indication of its strength. One line of flux is called one maxwell. The flux density is the number of lines of flux for a unit area, and it is measured in gausses, with one gauss representing a density of one maxwell per square centimeter.

Lines of flux always follow the path of least resistance as they travel from the north pole to the south. They will even travel a longer distance if the traveling is easier. The measure of the ease with which the lines of flux can travel through a material or medium is measured in terms of permeability. Air is used as a reference and is given the permeability of one. Flux can travel through iron much more easily than through air, since it has a permeability of 7,000, and some of the extremely efficient permanent magnet alloys have permeability values as high as 1,000,000.

Lines of flux are invisible, but if we place a magnet under a piece of paper and sprinkle iron filings over it, they will form a definite pattern showing the lines they follow. They pass directly between the poles of a horseshoe magnet, but, if we place a piece of soft iron above the poles, the lines will enter the iron and flow through it to the south pole. This is because iron has so much greater permeability than air.

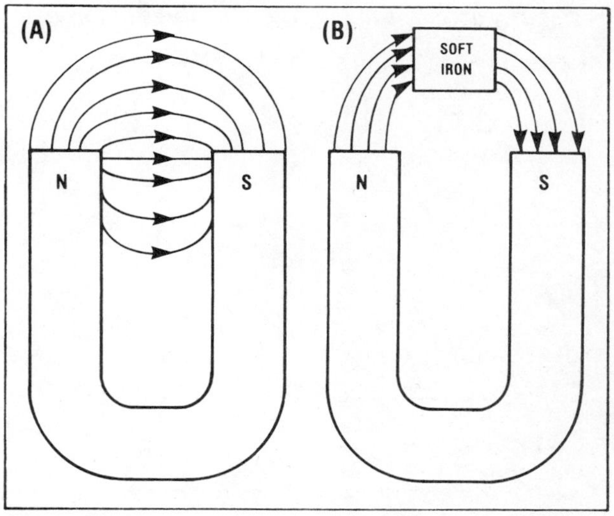

Fig. 4-5 A — Lines of magnetic flux are assumed to leave the north pole of a magnet at right angles to its surface and travel to the south pole, where they enter at right angles to its surface. B — The flux lines always seek the path of least resistance, even traveling longer distances if they can travel through a material with a high permeability.

It is the characteristic of the lines of flux to pass through a permeable material that explains the attraction of a piece of iron to a magnet. Remember that the lines of flux always seek the path of least resistance between the poles, and

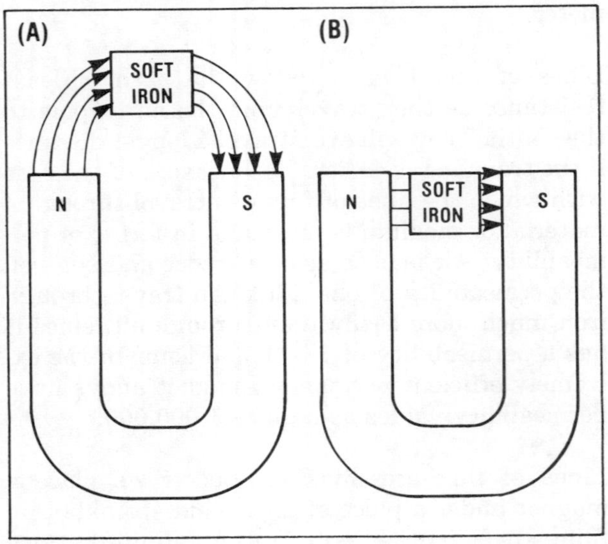

Fig. 4-6 A — The lines of flux pass through a material having a high permeability. In their effort to keep all of the force loops as short as possible, a force is exerted on the soft iron to pull it into the center of the magnetic field. B — When the soft iron is centered between the poles, it will resist any attempt to lengthen the lines of force.

since air has a very low permeability compared to iron, if a piece of iron gets within the field of a magnet, the lines of flux will travel through it rather than through the air around it. The lines of flux want to link the poles with the *shortest* possible loops, so they will exert a strong pull to get the piece of iron centered between the poles. As the iron is pulled in closer, more lines of flux can pass through it and the pull will be stronger. When it is centered, it will resist any force that tries to lengthen the lines of flux by pulling the iron out of the field.

Almost any magnetic material, regardless of its retentivity, will lose some of its magnetic strength if its lines of flux must pass through the air. Because of this, magnets whose strength is critical are stored with keepers of soft iron linking the poles to provide a highly permeable path for the flux.

Magnetism and lines of magnetic flux follow the same rules as charges of static electricity. Like poles repel each other, and the force of repulsion follows the inverse square law. This means that if the distance between the poles is doubled, the force of repulsion will be reduced to one-fourth. Unlike poles attract each other, and the force of attraction is squared as the distance is decreased. Halving the separation increases the force of attraction four times.

SECTION V

Electromagnetism

Though the effects of magnetism had been observed for centuries, it was not until 1819 that the relationship between electricity and magnetism was discovered. The Danish physicist Hans Christian Oersted discovered that the needle of a small compass would be deflected if it was held near a wire carrying electric current. This deflection was caused by an invisible magnetic field surrounding the wire.

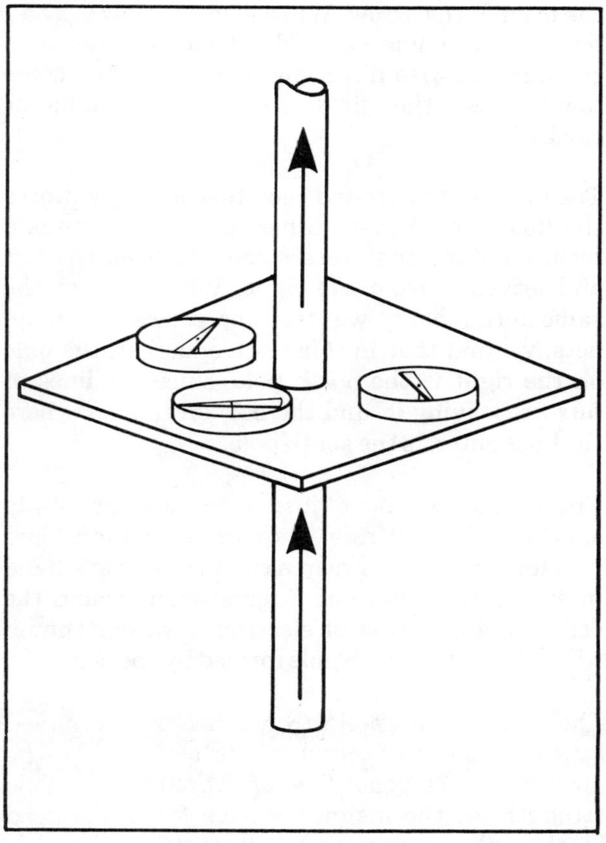

Fig. 5-1 The relationship between magnetism and electricity was discovered when it was found that the needle of a magnetic compass was deflected when it was placed near a current-carrying conductor.

We can see the effect of this field if we sprinkle iron filings on a plate that surrounds a current-carrying conductor. The filings will arrange themselves in a series of concentric circles around the conductor.

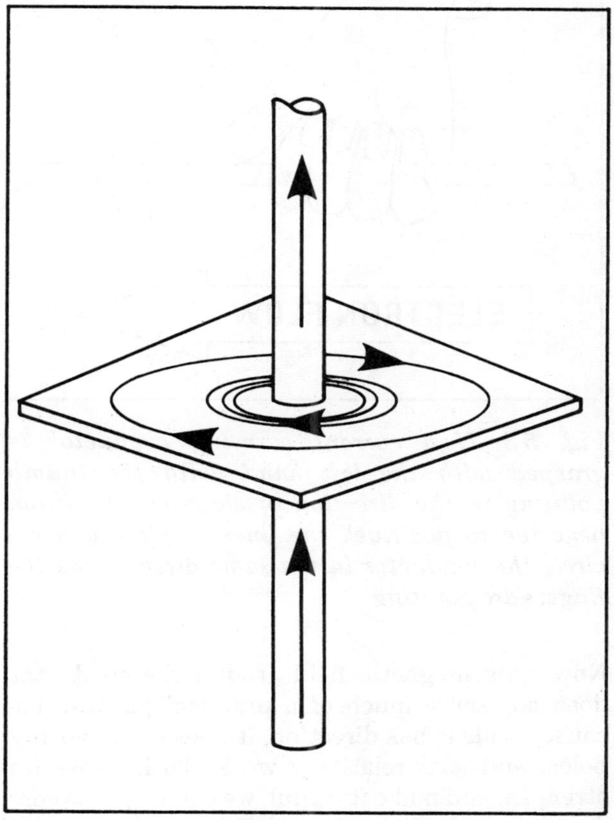

Fig. 5-2 Lines of magnetic force encircle a current-carrying conductor. These lines are relatively weak, and have no polarity.

By observing the effect of the field and remembering what we have learned about electron flow, we can understand a great deal about electromagnets, and about the way they serve us in such useful ways.

Electrons are negative charges of electricity that can be forced to flow through a conductor. As they travel, they produce the lines we have just observed. The greater the amount of flow, the stronger will be the magnetic field.

The left-hand rule gives us the direction of these lines of flux, and by knowing their direction, we can determine the direction of rotation of motors and the polarity of electromagnets. If we grasp the conductor in our left hand with our thumb pointing in the direction of electron flow, from negative to positive, our fingers will encircle the conductor in the direction of the lines of flux.

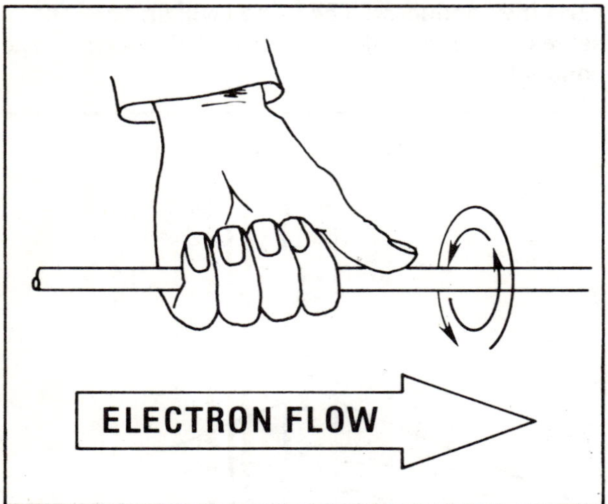

Fig. 5-3 If a current carrying conductor is grasped with the left hand, with the thumb pointing in the direction of electron flow (from negative to positive), the lines of force will encircle the conductor in the same direction as the fingers are pointing.

Now, this magnetic field around the conductor does not serve much of a practical purpose because, while it has direction, it does not have any poles, and it is relatively weak. To increase its strength, and make it useful, we can wind the conductor into the form of a coil. When we do this, we concentrate the lines of flux, and the coil attains the characteristics of a magnet.

In Fig. 5-4, the electrons flow into the coil from the right, and as the conductor passes over the top of the coil, the electrons are flowing away from us, as we see by the cross representing the tail of the arrow. Below the coil, the electrons flow toward us, as is seen from the dot representing

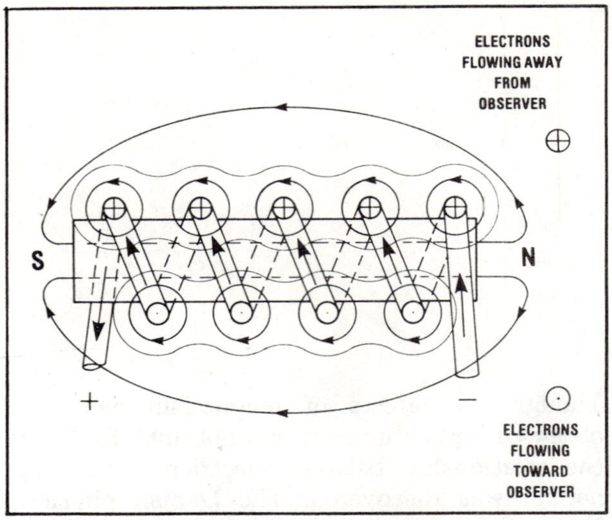

Fig. 5-4 By forming the conductor into a coil, the lines of flux can be concentrated, and the resulting coil will behave as a magnet and will have a north and a south pole.

the head of the arrow. When electron flow is away from us, the lines encircle the conductor in a counterclockwise direction, and when they come toward us, the field circles the conductor clockwise.

The lines of flux around each turn aid or reinforce the flux around every other turn, and there is a resultant field that enters the coil from the left and leaves it from the right. When we use the same terminology we used for permanent magnets, we find that in this electromagnet, the pole on the right is the north pole, since the lines of flux are leaving it, and the pole on the left where the lines enter, is the south pole.

From what we have just seen, we can apply another left-hand rule — this one, the left-hand rule for coils. If we grasp a coil with our left hand in such a way that our fingers wrap around the coil in the direction of electron flow, our thumb will point to the north pole formed by the coil.

The strength of an electromagnet is determined by the number of turns in the coil and by the amount of current flowing through it. This strength, or the magnetomotive force produced by the coil, is measured in gilberts. One gilbert is the amount of magnetomotive force produced by one maxwell (one line of flux) flowing through a magnetic circuit having one unit of "reluctance," the opposition of the circuit to the flow of magnetic flux. One gilbert is produced by a coil

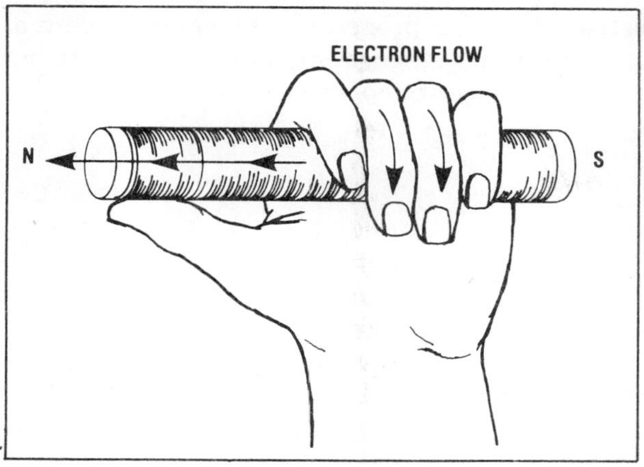

Fig. 5-5 If a coil is grasped with the left hand in such a way that the fingers encircle it in the same direction as the electron flow (from negative to positive), the thumb will point to the north pole of the electromagnet formed by the coil.

having $10/4\pi$ ampere-turns. More simply stated, an electromagnet having one ampere-turn will produce a magnetomotive force of 1.256 gilberts.

We can further increase the strength of an electromagnet by concentrating the lines of flux. This is accomplished by using some highly permeable material, such as soft iron, for the core.

There are two types of electromagnets used in practical applications, those having fixed cores, and those with movable cores. Fixed-core electromagnets are used in such devices as voltage regulators, where current proportional to the generator output voltage flows through the coil.

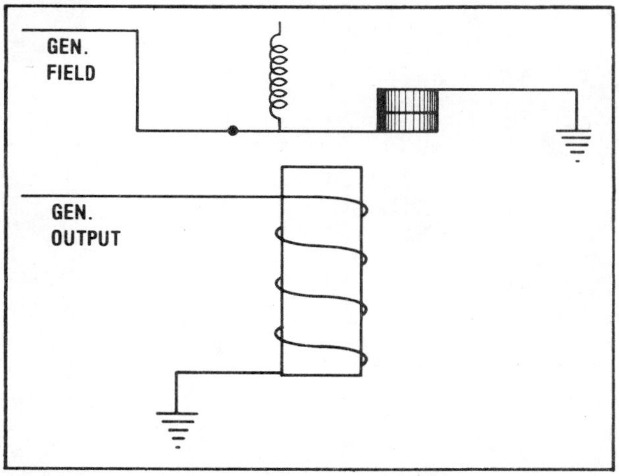

Fig. 5-6 An electromagnetically operated switch with a fixed core is called a relay.

When enough current flows, the magnet will become strong enough to pull the contacts open and lower the generator output.

Movable-core electromagnets are called solenoids. A soft iron core is held out of the center of the coil by a spring, but as soon as current flows in the coil, the magnetic field will pass through the core and, in its efforts to keep the loops of force as small as possible, will pull the core into the center of the coil.

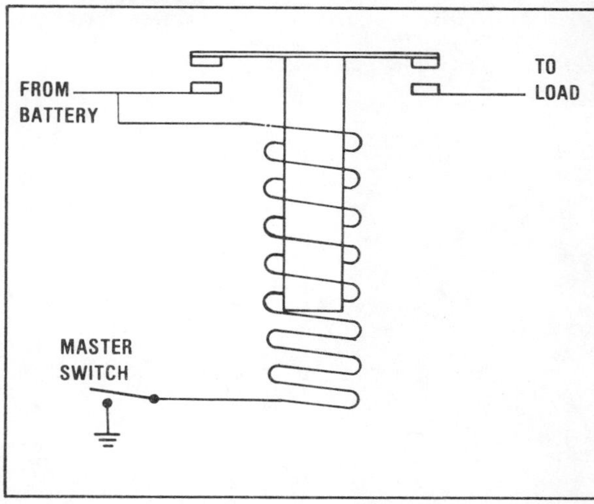

Fig. 5-7 An electromagnetically operated switch with a movable core is called a solenoid.

TERMS RELATING TO MAGNETISM AND ELECTROMAGNETISM:

ampere-turn The amount of magnetomotive force produced by an electromagnet of one turn when a current of one ampere flows through it. One ampere turn produces a magnetomotive force of 1.256 gilberts.

flux Lines of magnetic force.

gauss A measurement of flux density equal to one line of flux per square centimeter.

gilbert The amount of magnetomotive force required to maintain one maxwell in a magnetic circuit having one unit of reluctance. One gilbert = $10/4\pi$ ampere turns.

maxwell One line of magnetic flux.

permeability A measure of the ease with which lines of magnetic flux can pass through an object or body.

reluctance The opposition within a magnetic circuit to the flow of lines of magnetic flux.

residual magnetism The magnetism that remains in a material after the magnetizing force has been removed.

retentivity The property of a magnetic material to retain its magnetism after the magnetizing force has been removed.

SECTION VI

Sources of Electrical Energy

Man has not been given the prerogative of creating nor of destroying energy, but within rather wide guidelines, we are able to convert energy from one form into another. This ability to exchange forms makes electrical energy all the more valuable to us. As we will see shortly, we daily convert mechanical, thermal, chemical, and light energy into electricity, and fortunately, electricity can be converted back into all of these forms of energy.

A. Electromagnetic Induction

Lines of magnetic flux pass between the poles of a magnet, and if a conductor is moved through these lines of flux, they will transfer to the conductor and force electrons to flow through it. This is the principle used to generate most of our electricity today. Our aircraft carry generators or alternators to produce electricity by this method, and the atomic and hydro-electric powerplants produce power by the same procedure.

The amount of electricity that is generated depends on the *rate* at which the lines of flux are cut. This rate may be increased by increasing the number of lines of flux by making the magnet stronger, or by moving the conductor through the lines faster.

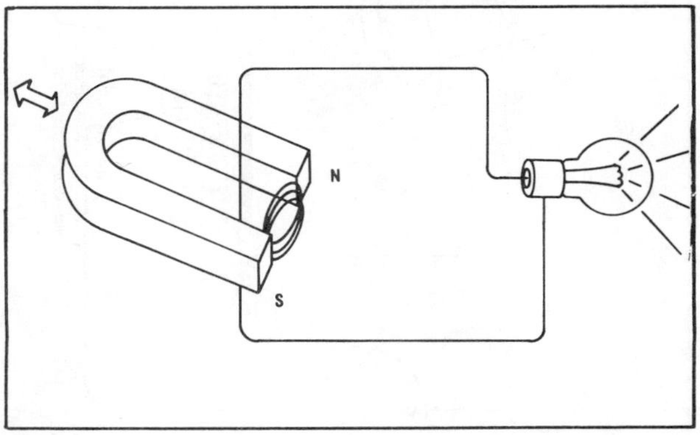

Fig. 6-1 The amount of electricity generated by electromagnetic induction is determined by the rate at which the conductor cuts through lines of magnetic flux.

B. Chemical

In the chemical make-up of materials, there can exist an imbalance of electrons and protons. If a material having an excess of electrons is connected by a conductor to a material having a deficiency, electrons will be forced through the conductor.

For example, if a piece of aluminum and a piece of copper are immersed in a solution of hydrochloric acid and water, and the two pieces of metal are connected by a piece of wire, electrons will leave the aluminum and flow to the copper. The electrons which leave the aluminum are replaced by the negative chlorine ions from the acid. When the chlorine combines with the aluminum, it eats away part of the metal and forms a gray powdery material on the surface. Positive hydrogen ions will be attracted to the copper where they are neutralized by the electrons that came from the aluminum, and bubble to the surface as free hydrogen gas.

C. Thermal (Heat)

When certain combinations of wire, such as iron and constantan or chromel and alumel, are joined into a loop with two junctions, a thermocouple is formed. An electrical current will flow through the wires when there is a *difference* in the temperature of the two junctions. A cylinder head temperature measuring system has one junction held tight against the engine cylinder head by a spark plug, while the other junction is in the relatively constant temperature of the instrument panel.

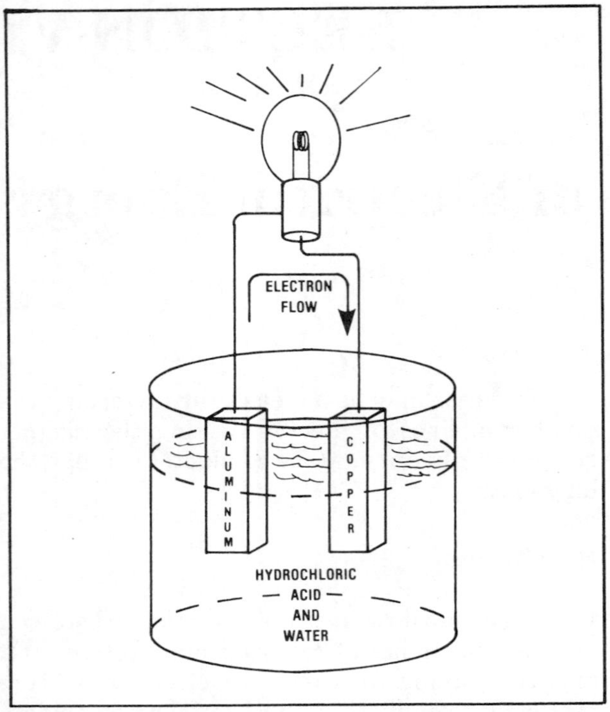

Fig. 6-2 Electrons will flow between two dissimilar metals when they are connected by a conductor and are both immersed in an electrolyte.

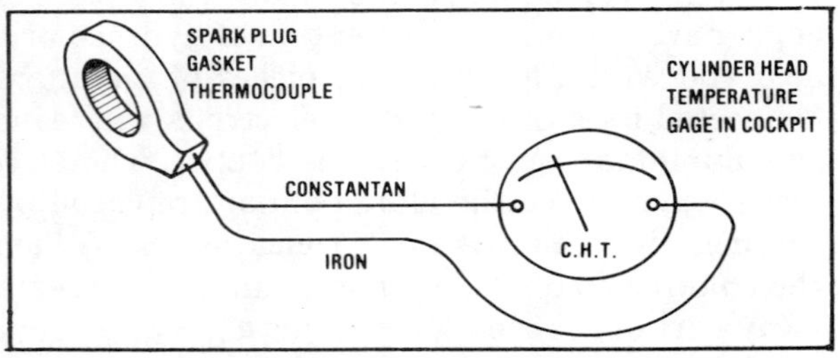

Fig. 6-3 Electrons will flow in a thermocouple made of certain dissimilar metals when there is a temperature difference between the two junctions.

D. Pressure

Crystalline material such as quartz has the characteristic that when it is bent or deformed by a mechanical force, an excess of electrons will accumulate on one surface, leaving the opposite surface with a deficiency. This feature is made use of in crystal microphones and phonograph pickups. And the fact that this interchange between mechanical and electrical energy is reversible makes crystals useful for producing alternating current for radio transmitters. A piece of crystal has only one natural frequency at which it will vibrate, and if it is excited by pulses of electrical energy, it will vibrate at this frequency. As it vibrates, it produces between its faces an alternating voltage having an accurate frequency.

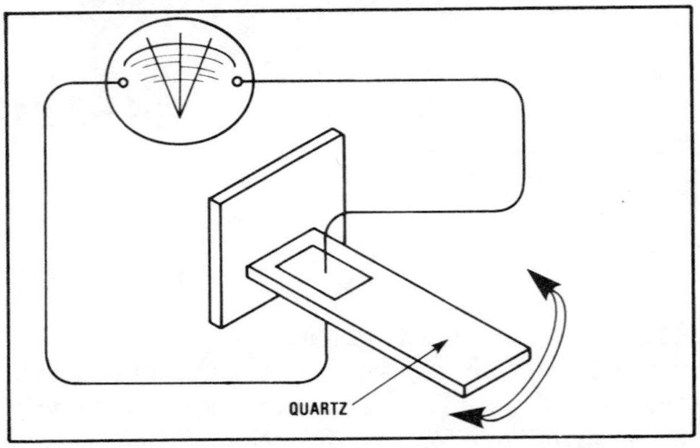

Fig. 6-4 An electrical potential difference will be built up across the faces of certain crystalline materials when they are bent or otherwise subjected to mechanical pressure.

E. Light

Light is a form of energy, and when it strikes certain materials that are photoemissive, electrons will be knocked off of it and forced to flow in a circuit and do work. Switches may be controlled by light-sensitive devices to turn airport lights on at dark and off at dawn, and sensitive light measuring meters are used in photography to determine the amount of light available, so that the proper exposure can be made.

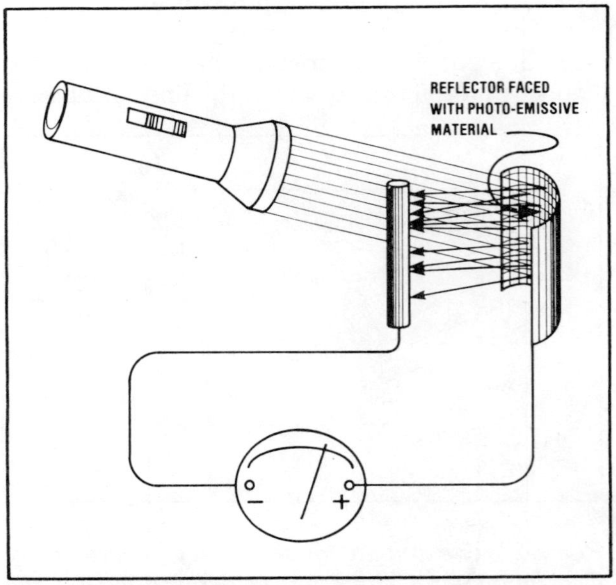

Fig. 6-5 A photoemissive material will emit electrons when it is struck by light.

SECTION VII

Current Electricity and Ohm's Law

A. Ohm's Law

In our study this far, we have seen that a concentration of electrons will produce an electrical pressure that will force electrons to flow through a circuit. And by assigning values to the pressure, flow, and opposition, we can understand the relationship that exists between them, and can accurately predict what will happen in a circuit under any given set of conditions.

It was the German scientist George Simon Ohm who proved the relationship between these values, and in 1826 he published his findings. Ohm's law is the basic statement which says in effect that the current that flows in a circuit is directly proportional to the voltage (pressure) that causes it, and inversely proportional to the resistance (opposition) in the circuit. The units we use make this relationship easy to see: one volt of pressure will cause one ampere of current to flow in a circuit whose resistance is one ohm.

For ease of handling these terms in formulas, voltage is represented by the letter E, current by the letter I, and resistance by the letter R. A statement of Ohm's law in the form of a formula is, therefore, $E = I \times R$. If we want to find the current, we use the formula $I = E/R$, and resistance may be found by the formula $R = E/I$.

Power in an electrical circuit is measured in watts, and one watt is the amount of power used in a circuit when one amp of current flows under a pressure of one volt.

The relationship between voltage, current, resistance, and power is such that any one value may be found when any two of the others are know. One easy way to find the correct formula is to use a series of divided circles representing the symbols in the formula. In Fig. 7-1, we see that the voltage in the circuit is equal to the product of the current and the resistance; $E = I \times R$. The top half of the circle is equal to the bottom half. We also know that the current may be found by dividing the voltage by the resistance, $I = E/R$, and the resistance is equal to the voltage divided by the current, $R = E/I$.

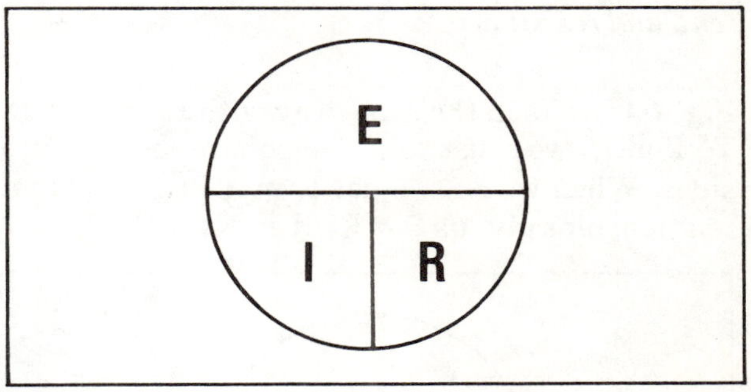

Fig. 7-1 The relationship between voltage, current, and resistance.

The same relationship may be found between power, current, and voltage. We know that power is equal to voltage times current, $P = I \times E$. Fig. 7-2 shows us an easy way to find current by dividing power by voltage, $I = P/E$, and to find the voltage by dividing power by the current, $E = P/I$.

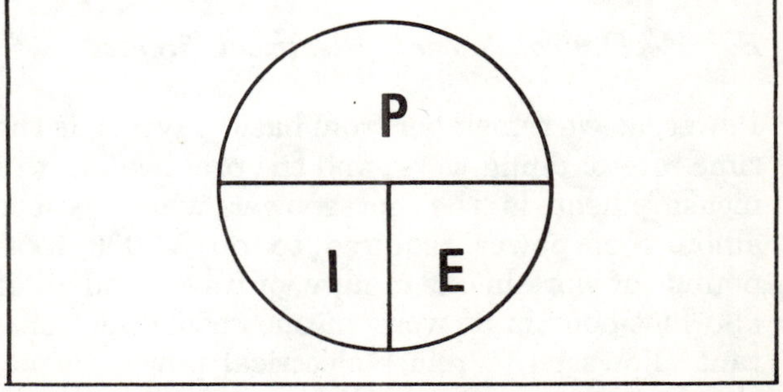

Fig. 7-2 The relationship between power, current, and voltage.

The other six relationships are just as easy to find when we use the circles of Figs. 7-3 and 7-4. In Fig. 7-3, we find that $P = I^2 \times R$, and $R = P/I^2$. Now we have one very small problem, $I^2 = P/R$, but we want I, not I^2, so we must take the square root of both sides of the equation. When we do this, we end up with $I = \sqrt{P/R}$.

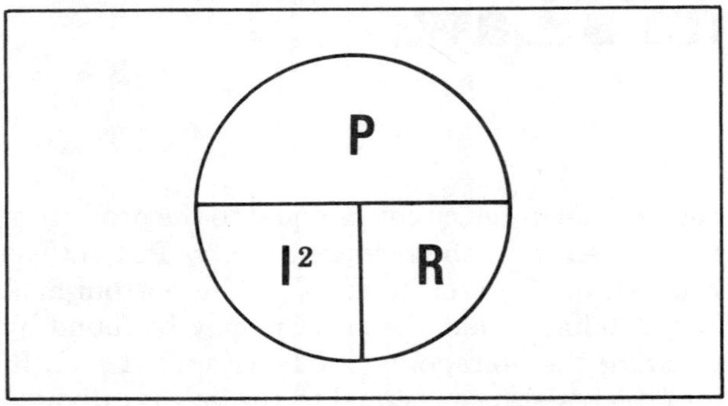

Fig. 7-3 The relationship between power, current, and resistance.

Fig. 7-4 works in the same way with $E^2 = P \times R$. To find E, we must take the square root of both sides. When we do, we get $E = \sqrt{PR}$. The other relationships give us $P = E^2/R$ and $R = E^2/P$.

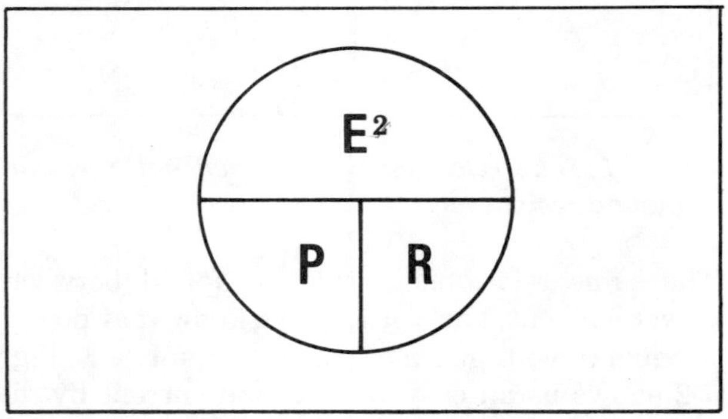

Fig. 7-4 The relationship between voltage, power, and resistance.

B. *Mechanical Power in Electrical Circuits*

Power, as we remember from basic physics, is the time-rate of doing work, and the practical unit of measurement is the horsepower which is the amount of power required to do 33,000 foot-pounds of work in one minute, or its equivalent of 550 foot-pounds of work in one second. A constant allows us to relate electrical power to mechanical power. Seven hundred and forty-six watts is the electrical equivalent of one horsepower.

If we have a 24-volt electric hoist, we can find the amount of current needed to raise a 1,000-pound load six feet in 30 seconds.

$$1{,}000 \times 6 = 6{,}000 \text{ foot-pounds}$$

$$\frac{6{,}000}{30} = 200 \text{ foot-pounds per second}$$

$$\frac{200}{550} = 0.364 \text{ horsepower}$$

$$746 \times 0.364 = 271.5 \text{ watts}$$

$$I = \frac{P}{E} = \frac{271.5}{24} = 11.31 \text{ amps}$$

C. *Heat in Electrical Circuits*

In circuits where mechanical work is not actually being done, power is still a very important consideration. For example, if you install a resistor in a light circuit to drop the voltage from 12 volts down to three, for a light bulb that requires 150 milliamps, you must find the resistance in ohms and the power in watts, this resistor must dissipate.

To solve this problem, we must first find the voltage to be dropped:

$$E = 12 - 3 = 9 \text{ volts}$$

Find the resistance required:

$$R = E/I = 9/0.15 = 60 \text{ ohms}$$

Find the power dissipated in the resistor:

$$P = I^2 \times R = 0.15^2 \times 60 = 1.35 \text{ watts}$$

The resistor will only have to dissipate 1.35 watts, but for practical purposes, you will most probably use a two-watt resistor.

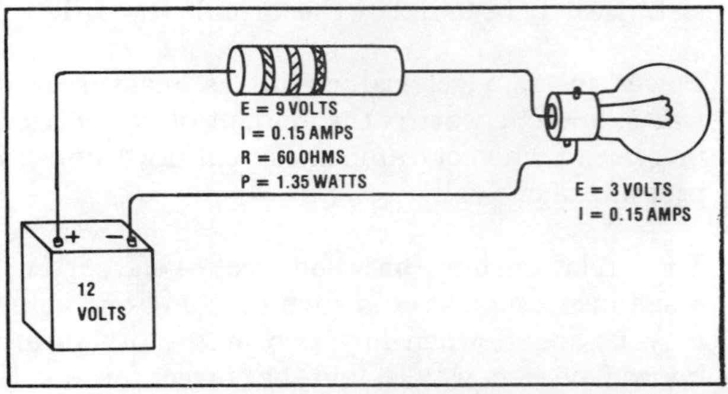

Fig. 7-5 Determining the characteristics of a resistor needed to drop voltage in a circuit.

SECTION VIII

Circuit Elements

All complete electrical circuits consist of three elements: a source of electrical energy, a load to use this energy, and conductors to join the source and the load.

The source may be a battery, a generator, or other apparatus that converts some form of energy into electricity, and these are discussed in the appropriate section of the text. The load may be anything that converts electrical energy into mechanical or chemical energy, or into heat or light. These devices, also, will be discussed in the appropriate section. Here, we are primarily concerned with the conductors used to join the source and the load, and components used to control the flow of current and protect the circuit from damage due to shorts.

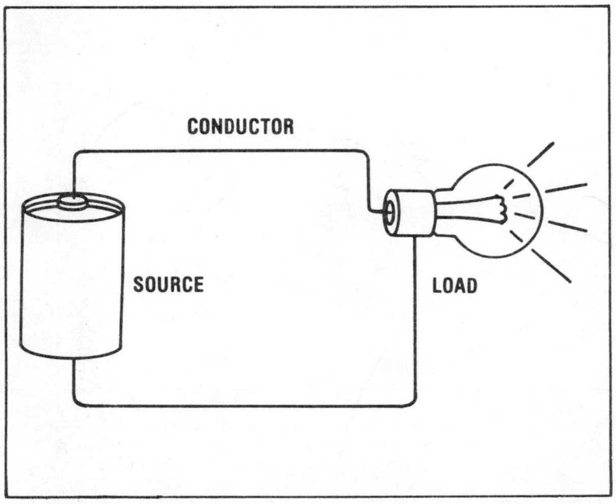

Fig. 8-1 All complete circuits must have a source of electrical energy, a load to use the energy, and conductors to join the source and the load.

A. Conductors

The purpose of a conductor is to provide a path for the electrons to flow from the source through the load, back to the source. It must do this with the minimum of resistance, but other factors must also be taken into consideration, so the choice of a conductor is often a compromise.

Most aircraft electrical systems are of the single-wire type, meaning that the aircraft structure itself provides the path through which the current returns from the load to the source. A great deal of weight is saved by using this type of system, but it is extremely important that a good connection capable of carrying all of the current is provided between the aircraft structure and the battery, generator, and all of the devices using the current.

The resistance of a conductor is affected by two things, its physical characteristics and its dimensions.

1. *Physical characteristics*

a. *Resistivity*

For most practical aircraft circuits, we use two types of conductors, copper and aluminum. Copper wire has only about two-thirds of the resistance of the equivalent gage of aluminum wire, and is the one most generally used. But for applications requiring a great deal of current, aluminum wire is often used. Its resistivity is higher than that of copper, and a larger conductor is needed, but since aluminum weighs so much less than copper, a great deal of weight may be saved by its use.

b. *Temperature*

Most metals have what is known as a positive temperature coefficient of resistance. This means that the resistance of the material will increase as its temperature increases. This characteristic is used in some temperature measuring instruments where the resistance change in a piece of wire is used to measure temperature. For practical purposes, however, both copper and aluminum have such a small change in resistance with temperature over the temperature range encountered in flight that it is normally not considered to be a problem.

2. *Dimensions*

a. *Length*

For all practical purposes, the resistance of a conductor varies *directly* as its length.

b. *Cross-sectional area*

The resistance of a conductor varies *inversely* as its cross-sectional area. Aircraft wire is measured by the American Wire Gage system, with the larger numbers representing the smaller wires. The smallest size wire normally used in aircraft wiring is the 22-gage size which has a diameter of about 0.025 inch. For carrying very large amounts of current, cables up to the 0000 size (spoken of as four aught size) are used. These large cables have a diameter of about 0.52 inch.

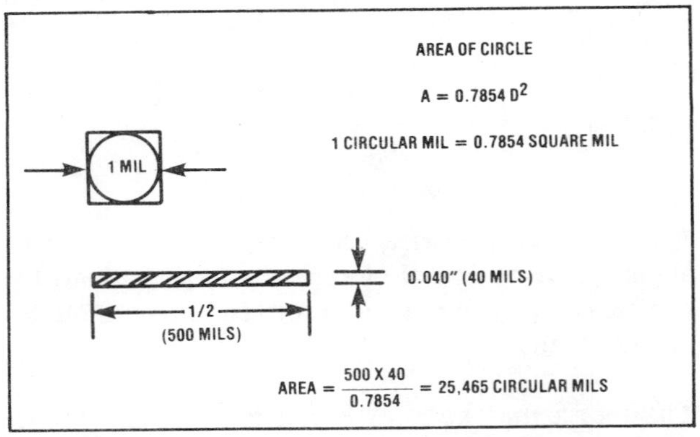

Fig. 8-2 Relationship between circular mils and square mils.

COPPER ELECTRIC WIRE CURRENT-CARRYING CAPACITY				
Wire size Specification MIL-W-5086	Single wire in free air—maximum amperes	Wire in conduit or bundled—maximum amperes	Maximum resistance—ohms/1,000 feet (20°C.)	Nominal conductor area—circular mills
AN-20	11	7.5	10.25	1,119
AN-18	16	10	6.44	1,779
AN-16	22	13	4.76	2,409
AN-14	32	17	2.99	3,830
AN-12	41	23	1.88	6,088
AN-10	55	33	1.10	10,443
AN-8	73	46	.70	16,864
AN-6	101	60	.436	26,813
AN-4	135	80	.274	42,613
AN-2	181	100	.179	66,832
AN-1	211	125	.146	81,807
AN-0	245	150	.114	104,118
AN-00	283	175	.090	133,665
AN-000	328	200	.072	167,332
AN-0000	380	225	.057	211,954
ALUMINUM ELECTRIC WIRE CURRENT-CARRYING CAPACITY				
Wire size—Specification MIL-W-7072	Single wire in free air—maximum amperes	Wire in conduit or bundled—maximum amperes	Maximum resistance—ohms/1,000 feet (20°C.)	Nominal conductor area—circular mills
AL-6	83	50	0.641	28,280
AL-4	108	66	.427	42,420
AL-2	152	90	.268	67,872
AL-0	202	123	.169	107,464
AL-00	235	145	.133	138,168
AL-000	266	162	.109	168,872
AL-0000	303	190	.085	214,928

Fig. 8-3 Characteristics of aircraft copper and aluminum wire.

Since it is the area rather than the diameter of a conductor that determines its current-carrying capability, and since almost all conductors are round rather than square, the circular mil measurement is handy to use. A circular mil is the standard measurement of cross-sectional area, and is the area of a circle whose diameter is one mil, or 1/1,000 of an inch. To find the cross-sectional area of a conductor in circular mils, you need only square its diameter in mils. For example, if a round wire has a diameter of 1/8 inch (0.125 inch or 125 mils) it would have an area of 15,625 circular mils.

Since the circular mil is smaller than the square mil, to find the area of a square or rectangular conductor in circular mils, divide the square mil area by 0.7854 (the relationship between the area of a circle and the square of its diameter). For example, to find the cross-sectional area of a strap of copper 0.040-inch thick and 1/2-inch wide, you would first find its area in square mils, which is 500 × 40 = 20,000. Now divide this by 0.7854. The area is 25,465 circular mils, which is about the same as a six-gage stranded copper wire.

Fig. 8-4 Most of the switches found in modern aircraft are either of the toggle or the rocker type.

B. Control Devices

1. Switches

a. Toggle or rocker switches

Switches used to control the flow of electrons in most aircraft circuits are of the enclosed toggle or rocker type. These switches are actuated by either moving the bat-shaped toggle or by pressing on one side of the rocker.

If the switch controls only one circuit and has only two connections and two positions, open and closed, it is called a single-pole, single-throw, or SPST switch. These may be used to control lights, either turning them on or off.

If the switch has three connections and either two or three positions, it is a single-pole, double-throw, SPDT switch. These switches are used to either raise or lower the landing gear, and may or may not have a neutral position in the middle.

Some switches are used to control more than one circuit and may be either the single- or double-throw type. The double-pole, single-throw, DPST switch could be used to control both the battery and the generator circuit so they would both be turned on and off at the same time. A double-pole, double-throw, DPDT switch controls two circuits and has either two or three positions.

Both toggle and rocker switches may have one or both of their positions spring-loaded so they will return to the Off position when your finger is removed.

Fig. 8-5 Wafer switches are used when it is necessary to select any of a large number of circuit conditions.

b. Wafer switches

When a switch is used to select one of a number of conditions, a wafer switch is often used. These switches may have several wafers stacked on the same shaft, and each wafer may have as many as twenty positions. Wafer switches are seldom used for carrying large amounts of current and are most generally open with the wires soldered to the terminals on the wafers.

c. Precision switches

Limit switches, uplock and downlock switches, and many other applications require the switch to be actuated by mechanical movement of some mechanism. In these applications it is usually important that the switch actuate when the mechanism reaches a *very definite and specific* location. These switches require an extremely small movement to actuate. Snap-acting switches of the type pioneered by Micro Switch division of Honeywell find a wide use in these applications. An extremely small movement of the plunger trips the spring so it will drive the contacts together. When the plunger is released, the spring snaps the contacts back open.

Fig. 8-6 Precision switches snap open or closed with an extremely small amount of movement of the operating control.

d. Relays and solenoids

It is often necessary to open or close a circuit carrying a large amount of current from a remote location and with a small switch. An example is the starter circuit for an aircraft engine. The starter motor requires a great deal of current and the cable between the battery and the starter must be as short as possible. But the starter switch must be located inside the cockpit and is often incorporated in the ignition switch. A solenoid, or contactor, is used for this application. A small amount of current is used to energize an electromagnet which closes the contacts in the circuit carrying the large amount of current. If the electromagnet has a fixed core that attracts a movable armature, the switch is called a relay, but if the core is movable, and is pulled into the hollow coil, it is called a solenoid.

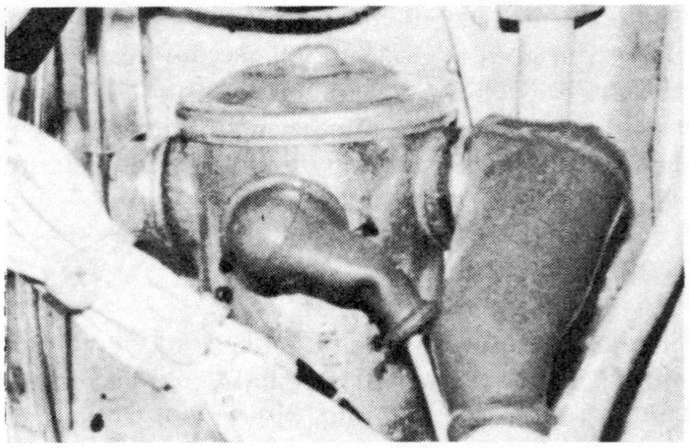

Fig. 8-7 Starter solenoid switches control large amounts of current, but they are operated by a very small current that flows through the coil.

C. Protective Devices

1. Fuses

A fusible link made of a low-melting-point alloy enclosed in a glass tube is used as a simple and reliable circuit protection device. The load current flows through the fuse, and if it becomes excessive, will melt the link and open the circuit. Some fuses are designed to withstand a momentary surge of current, but will melt if the current is sustained. These slow-blow fuses have a small spring attached to the link so when the sustained current softens the link, the spring will pull it in two and open the circuit.

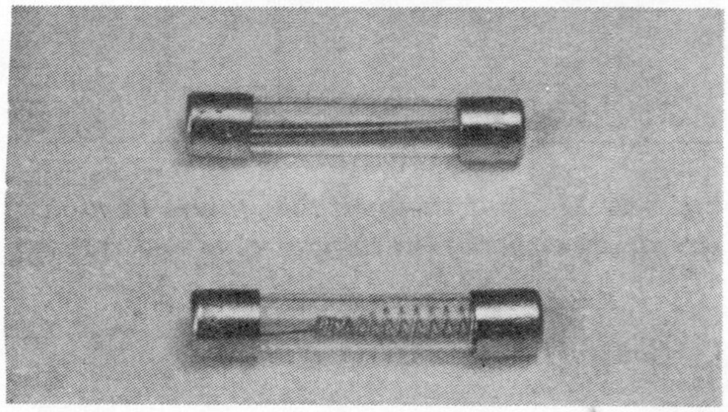

Fig. 8-8 The heat caused by excess current melts the fusible link and opens the circuit. Slow-blow fuses will stand a momentary excess of current, but sustained overload will soften the link and the spring will pull it apart.

2. *Circuit breakers*

It is often inconvenient to replace a fuse in flight, so most aircraft circuits are protected by circuit breakers that will automatically open the circuit if the current becomes excessive, but may be re-set by moving the operating control, which may be a toggle, a push-button, or a rocker. If the excess current was caused by a surge of voltage, or by some isolated and non-recurring problem, the circuit breaker will remain in and the circuit will operate normally. But if an actual fault such as a short circuit does exist, the breaker will trip again, and it should be left open.

Aircraft circuit breakers are of the trip-free type which means that they will open the circuit irrespective of the position of the operating control. With this type of breaker, it is impossible to hold the circuit closed, if an actual fault exists.

There are two operating principles of circuit breakers. Thermal breakers open the circuit when the excess current heats an element in the switch and it snaps the contacts open. The other type is a magnetic breaker which uses the strength of the magnetic field caused by the current to open the contacts.

Automatic re-set circuit breakers that open a circuit when excess current flows, but automatically close it again after a cooling-off period, are usually not used in aircraft circuits.

Fig. 8-9 Most circuit protection for modern aircraft is provided by circuit breakers that may be reset in flight.

D. *Resistors*

It is often necessary in electrical circuits to control current flow by varying voltage. We will discuss this in great detail in circuit arrangement, but here we want to discuss the physical characteristics of the devices we use to do this.

Resistors may be inserted into circuits to drop voltage by converting some of the electrical energy into heat. And these resistors may be classified as fixed or variable.

1. *Fixed resistors*

a. *Composition*

The great majority of resistors used to control small amounts of current are made of a mixture of carbon and an insulating material. The relative percentage of the two materials in the mix determines the amount of resistance a given amount of the material will have. Small amounts of material are used to dissipate small amounts of power and, for more power, more material is used. Composition resistors are normally available in sizes from 1/8 watt up to two watts. The larger the physical size of the resistor, the more power it will dissipate.

Most modern resistors are of the axial-lead type; that is, the leads come directly out of the ends of the resistor. The ohmic value of this type of resistor is indicated by three or four bands of color around one end.

The band nearest the end of the resistor is the first significant figure of the resistance, the second band represents the second figure, and the third band tells the number of zeros to add to the two numbers. For example, if the first band is green (5), the second is brown (1), and the third is yellow (0000), the resistor would have a resistance of 510,000 ohms.

If there is not a fourth color band, the resistor has a tolerance of plus or minus 20%. But if there is a fourth band of silver, the resistor is within plus or minus 10% of the indicated value. If the fourth band is gold, it is within plus or minus 5% of the value shown by the color code.

Resistors having gold or silver in the fourth band are rated according to the EIA (Electronic Industries Association) preferred values, and are those most commonly used. Certain precision resistors, however, have much closer tolerance than the 5%, 10%, and 20% of the preferred values, and these have a fourth band of one of the same colors as used in the first three bands. These resistors are marked according to the EIA alternative values and their tolerances are shown in the table of Fig. 8-10C.

(A)

(B)

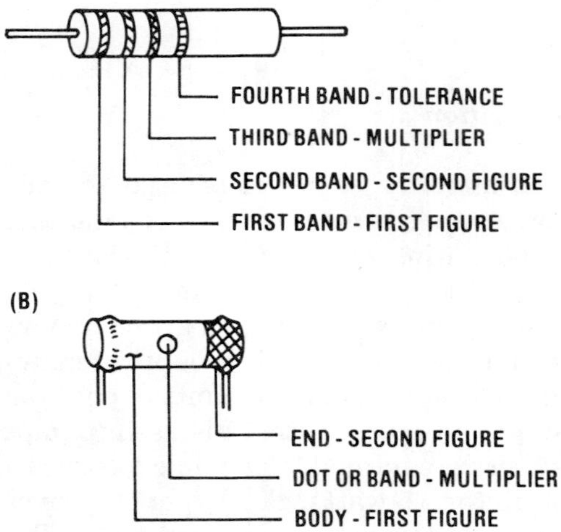

(C)

Color	Number	Multiplier	Tolerance EIA Preferred	Tolerance EIA Alternate
Black	0			
Brown	1	10		1%
Red	2	100		2%
Orange	3	1,000		3%
Yellow	4	10,000		4%
Green	5	100,000		5%
Blue	6	1,000,000		6%
Violet	7	10,000,000		7%
Gray	8	100,000,000		8%
White	9	1,000,000,000		9%
Gold		0.1	5%	
Silver		0.01	10%	
No Color			20%	

Fig. 8-10 A — Color code marking for axial-lead resistors. B — Color code markings for radial-lead resistors. C — Resistor color-code values.

Low-resistance composition resistors may have silver or gold as the third band. If the third band is silver, multiply the two significant figures by 0.01, and if it is gold, multiply them by 0.1. For example, a resistor having the first band of yellow, the second band of violet, and a third band of gold would have a resistance of 4.7 ohms. If the third band had been silver, the resistance would have been 0.47 ohms.

Some composition resistors have their leads coming off of the body radially instead of parallel to the resistor axis. These radial-lead resistors are color-coded with the same colors, but the color of the body is the first significant figure of the resistance, the color of the end is the second significant figure, and a dot or band of paint around the middle of the resistor gives us the number of zeros to add to the significant figures. If one end of the resistor is painted silver, the resistance is within plus or minus 10%. If it is gold, it is within plus or minus 5%, and if there is no gold or silver paint on the end, the tolerance is plus or minus 20% of the indicated resistance. As an example, if the body of the resistor is brown, one end is black, and the dot is green, and there is no silver or gold paint on the resistor, the resistance is 1,000,000 ohms or, as this is more commonly called, one megohm, plus or minus 20%.

b. Wire-wound resistors

When more power needs to be dissipated than can be handled by a composition resistor, special

Fig. 8-11 Wire-wound resistors are used when there is a great deal of power that must be dissipated.

resistors made of resistance wire wound over hollow ceramic tubes are used. Some of these resistors are tapped along the length of the wire to provide different values of resistance, and others have a portion of the wire left bare, so a metal band can be slid over the resistor, allowing it to be set to any desired resistance. When the screw is tightened, the band will not move from the selected resistance.

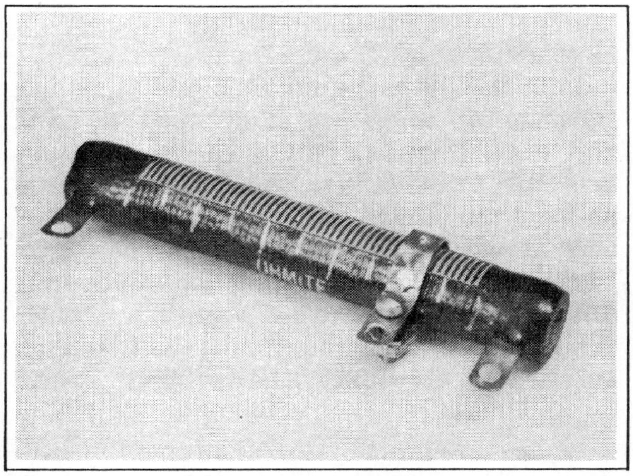

Fig. 8-12 Wire wound resistors may have a portion of the wire exposed and incorporate a movable tap.

2. *Variable resistors*

When it is necessary to change the amount of resistance in a circuit, variable resistors may be used. These may be of either the composition or the wire-wound type. In the composition resistor, the mix is bonded to an insulating disk, and a wiper, or sliding contact, is rotated by the shaft to vary the amount of material between the two terminals. We saw earlier in the discussion of resistance that the resistance varies with the length of the conductor, and the farther the sliding contact is from the fixed contact, the greater will be the resistance.

Resistance wire may be wound around a form which is shaped so that the sliding contact will touch the wire at the edge of the form. As the contact is rotated, the length of wire between the terminals varies and the resistance changes.

Variable resistors having only two terminals, one at the end of the resistance material, and the other, the sliding contact, are called rheostats and are used to vary the amount of resistance in a circuit.

Fig. 8-13 Variable resistors allow the amount of resistance in a circuit to be changed by rotating the shaft.

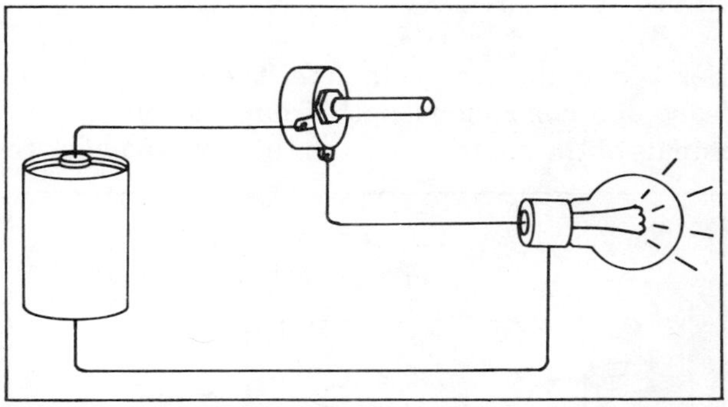

Fig. 8-14 Rheostats are used to vary the amount of resistance in a circuit.

If the resistor has three terminals, one for either end of the resistance material, and one for the slider, it is called a potentiometer, and may be used as a voltage divider, of which we will say more in the section on circuits.

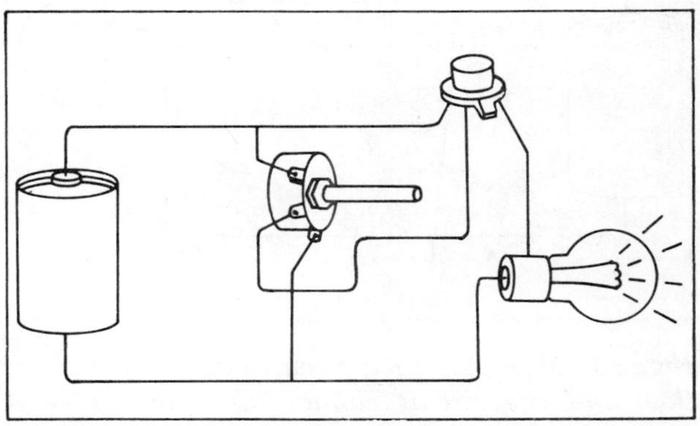

Fig. 8-15 Potentiometers are used as voltage dividers in a circuit.

SECTION IX

Circuit Arrangements

A. Circuit Considerations

All electrical circuits must contain three components: the source of electrical energy, a load device to use the energy, and conductors to connect the source and the load. Control devices such as switches and resistors and such protective devices as fuses or circuit breakers tailor the circuit to its purpose.

For a circuit to be complete, there must be at least one continuous path from one of the terminals of the source, through the load and back to the other terminal. If there is any interruption or break in this path, the circuit is said to be *open*, and there can be no flow of electrons. If, on the other hand, there is a path from one terminal of the source to the other without passing through the load, the circuit is *shorted*; and not only is there no work being done, but the lack of resistance in the circuit will allow excessive current to flow. In this case, unless a protective device such as a fuse opens the circuit, the wiring and even the source are likely to be damaged.

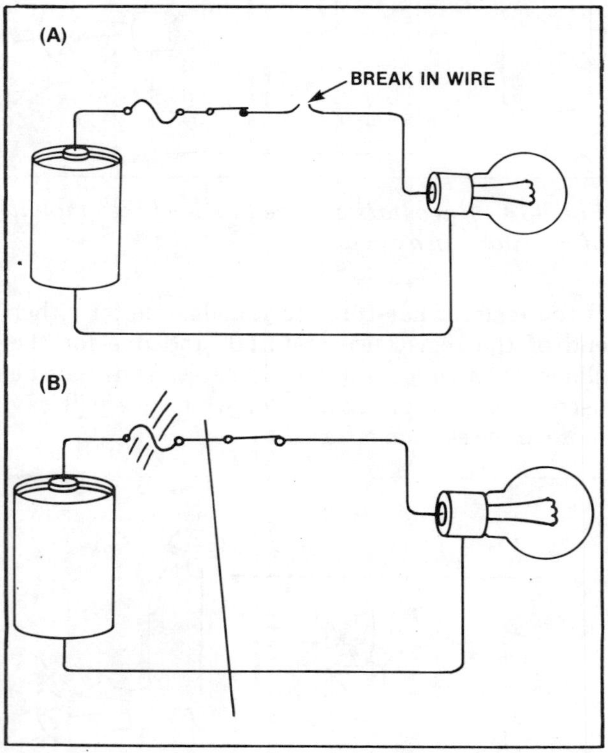

Fig. 9-1 A — An open circuit allows no current flow and the circuit cannot function. B — A short circuit causes excessive current flow, and can damage the circuit components or wiring unless a protecting device such as a fuse opens the circuit.

B. Types of Circuits

1. Series circuits

If there is only one path for electrons to flow, the circuit is called a series circuit. In Fig. 9-1 we have a typical circuit in which the battery is the source of power, and the lamp, which is our load, is in series with the rheostat, the switch, and the fuse. Since there is only one path for electron flow, if either the switch is open or the fuse is blown, the lamp cannot burn.

The rheostat is in series with the lamp, and when it is set for a minimum of resistance, the max-

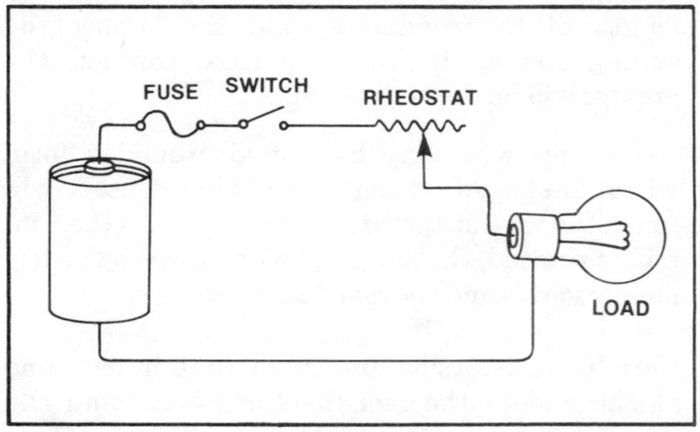

Fig. 9-2 A rheostat in series with the load can drop part of the voltage and control the brightness of the light.

imum amount of current will flow through the lamp, and it will burn with full brilliance. But when its resistance is increased part of the power from the battery is dissipated in the resistor in the form of heat. Less power is available for the lamp, and it will burn with less than full brilliance.

There are two laws propounded by the German physicist Gustav Robert Kirchhoff that go a long way in explaining the behavior of voltage and current in electrical circuits. His voltage law helps us understand the action in series circuits. Kirchhoff's voltage law states that "The algebraic sum of the applied voltage and the voltage drop around any closed circuit is equal to zero." What this means, simply, is that the voltage that is dropped across every part of the load must be exactly the same as the voltage supplied by the source. In Fig. 9-3 we have a good example of this. A six-volt battery is in series with a switch, three lamps, and a milliammeter. Each lamp is rated at two volts and requires 300 milliamps of current to burn at full brilliance. When the switch is closed, all of the lamps light up fully bright.

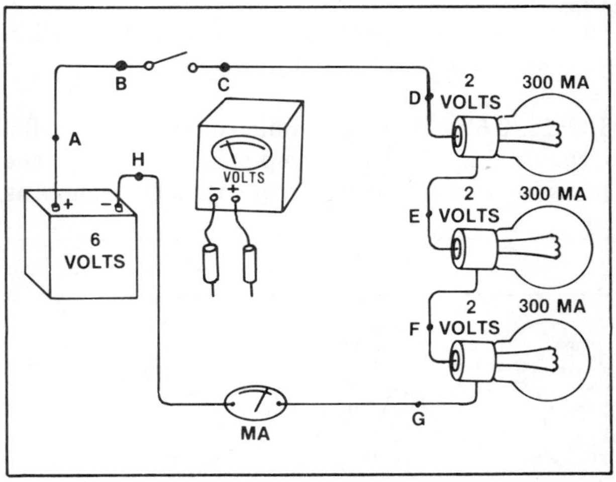

Fig. 9-3 This series circuit operates according to Kirchhoff's voltage law.

We can analyze our circuit and prove Kirchhoff's voltage law with a voltmeter that is capable of measuring up to six volts. Place the positive lead of the voltmeter at point A, and the negative lead at H. What do you read? You should read the full source voltage of six volts, because the meter is directly across the source.

Place the positive lead at position B and the negative lead at C. With the switch closed, it offers no resistance, and according to Ohm's law ($E = I \times R$) if the resistance is zero the voltage will also be zero, and so there is no voltage reading across the switch. But when the switch is opened, there will be no current flowing in the circuit, and so there will be no voltage dropped by the lamps, and the meter will read the full source voltage of six volts.

Now, with the switch closed and the lamps burning, measure between D and E. The meter should read two volts, indicating that the resistance of the lamp has dissipated enough power to drop this much pressure, or voltage.

Measure the voltage between points E and F. It also should be two volts, as should the voltage between points F and G. And the voltage between points D and G should be exactly the same as between A and H, proving that Kirchhoff's voltage law is true.

Ohm's law can also be applied to this circuit. Using a chart similar to that in Fig. 9-4, we can fill in the blanks. Since this is a series circuit with only one path for the electrons to flow, the current will be the same through each lamp as it is through the source, and in this case it is 300 milliamps.

The source voltage is six volts, and we have just found the voltage across each lamp to be two volts. The sum of the voltages dropped across the lamps does, in fact, equal the source, or applied voltage.

Now to find the resistance of the lamps. The operating resistance cannot be measured by an ohmmeter, because the resistance of a lamp is quite different when it is hot, from that when it is cold, so we find the resistance by Ohm's law.

$$R = E/I = 2/0.3 = 6.67 \text{ ohms}$$

There are three of these lamps in series, so we have a total circuit resistance of 20 ohms. To prove that this is correct, apply Ohm's law to the source:

$$E = I \times R = 0.3 \times 20 = 6 \text{ volts}$$

This is our source voltage.

The power dissipated by each lamp may be found by the formula:

$$P = I \times E = 0.3 \times 2 = 0.6 \text{ watt.}$$

The three lamps will dissipate 1.8 watts, which is exactly the same as we find when we multiply the source voltage by the total current.

Component	Voltage Volts	Current MA	Resistance Ohms	Power Watts
A	2	300	6.67	0.6
B	2	300	6.67	0.6
C	2	300	6.67	0.6
Source	6	300	20	1.8

Fig. 9-4 Circuit conditions in a series circuit.

To briefly summarize the characteristics of a series circuit:

1. There is only one path for the electrons to follow from the source, through the load back to the source.
2. The current is the same wherever it is measured in the circuit.
3. The sum of all of the voltage drops equals the source voltage.
4. The total resistance of the circuit is the sum of the individual load resistances.
5. The total power dissipated in the circuit is the sum of the power dissipated in each of the individual load resistances.

2. *Parallel circuits*

By far the most widely used circuit arrangement in use today is the parallel circuit. In this type of circuit, all of the load components are directly across the source, and if one should fail, it will have no effect on the others. You will remember, in a series circuit, if one of the lamps burned out, none of them could burn, but in a parallel circuit, if a lamp burns out, it will have no effect on the others.

The behavior of a parallel circuit is explained in part by Kirchhoff's current law which states that "The algebraic sum of the currents at any junction of conductors is zero." This means that all of the current that arrives at any junction within the circuit must leave that junction.

In Fig. 9-5 we have a simple three-lamp parallel circuit using the same components we had in the previous series circuit, *except* that the source is

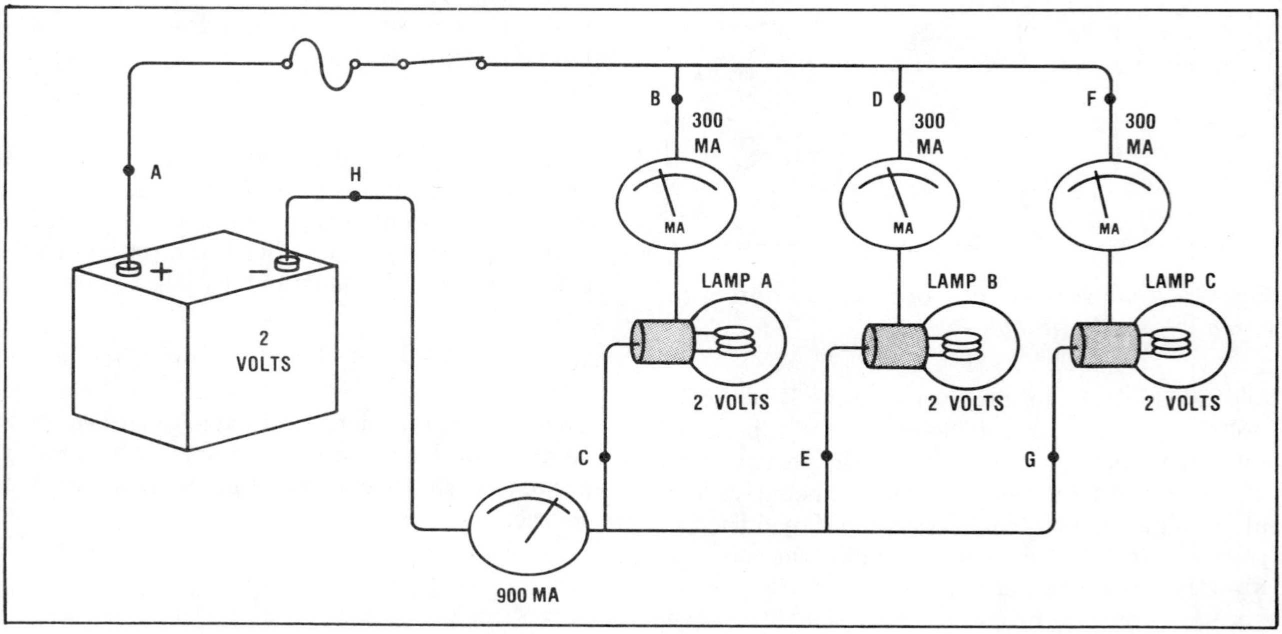

Fig. 9-5 This parallel circuit operates according to Kirchhoff's current law.

only two volts instead of six. All of the current flows through the fuse and the switch, and then it splits up, with some passing through each of the lamps. The amount that passes through each lamp is determined by its resistance, and in our example, we are assuming that all of the lamps are the same.

Notice that we have three separate paths from the source through part of the load and back to the source. And each route must obey Kirchhoff's voltage law. Since each path has only one load device, the voltage dropped must equal the source voltage, so each lamp has the full two volts across it. Measurement may be made with the voltmeter between A and H, B and C, D and E, and F and G, and all will read two volts. From this we see that in a parallel circuit, the voltage across each path is the same as the source voltage.

The milliammeters in each of the paths indicate that 300 milliamps flows through each lamp, and the total current is 900 milliamps, verifying Kirchhoff's current law, and showing that the total current is the sum of the current flowing in each of the paths.

We can find the resistance of each lamp by Ohm's law:

$$R = E/I = 2/0.3 = 6.67 \text{ ohms}$$

Now, since the resistance of each of the lamps is the same, the total resistance will be just one-third of the resistance of one lamp. This is because the voltage is the same across all of the lamps, but only one-third of the current flows through each of them. The total resistance is then 2.22 ohms. We can check this answer by using Ohm's law on the source and find that

$$R = E/I = 2/0.9 = 2.22 \text{ ohms.}$$

The power dissipated in the parallel circuit is computed the same way it was in the series circuit. Each of the lamps dissipates 0.6 watt (600 milliwatts). We find this by using the formula:

$$P = I \times E = 0.3 \times 2 = 0.6 \text{ watt.}$$

The total power found by adding the power dissipated in each lamp is 1.8 watts, and this may also be checked by using $P = I \times E$ for the source. $P = 2 \times 0.9 = 1.8$ watts.

Component	Voltage Volts	Current MA	Resistance Ohms	Power Watts
A	2	300	6.67	0.6
B	2	300	6.67	0.6
C	2	300	6.67	0.6
Source	2	900	2.22	1.8

Fig. 9-6 Circuit relationships in a parallel circuit.

We can summarize the characteristics of a parallel circuit:

1. There is more than one path for the electrons to follow from the source, through part of the load, back to the source.
2. The voltage is the same across any of the paths.
3. The current through each path is inversely proportional to the resistance of the path.
4. The total current is the sum of the current flowing through each of the individual paths.
5. The total resistance of the circuit is less than the resistance of any of the paths.
6. The total power dissipated in the circuit is the sum of the power dissipated in each of the individual load resistances.

Resistance in a parallel circuit

(1) All resistances have the same value

We saw in the previous problem that if the resistances are all the same in a parallel circuit, all you need to do to find the total resistance is to divide the value of a single resistor by the number of resistors. The formula looks like:

$$R_T = r/n$$

r = resistance of one resistor

n = number of resistances

(2) Two unlike resistances

If there are two unlike resistors in a parallel circuit, you may find the total resistance by dividing the product of the individual resistances by their sum:

$$R_T = \frac{R_1 \times R_2}{R_1 + R_2}$$

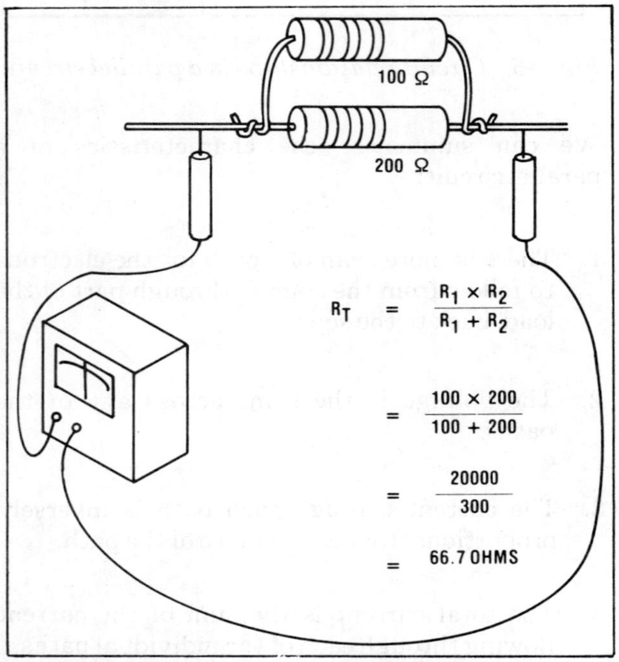

Fig. 9-7 Finding the equivalent resistance of two unlike resistors in parallel.

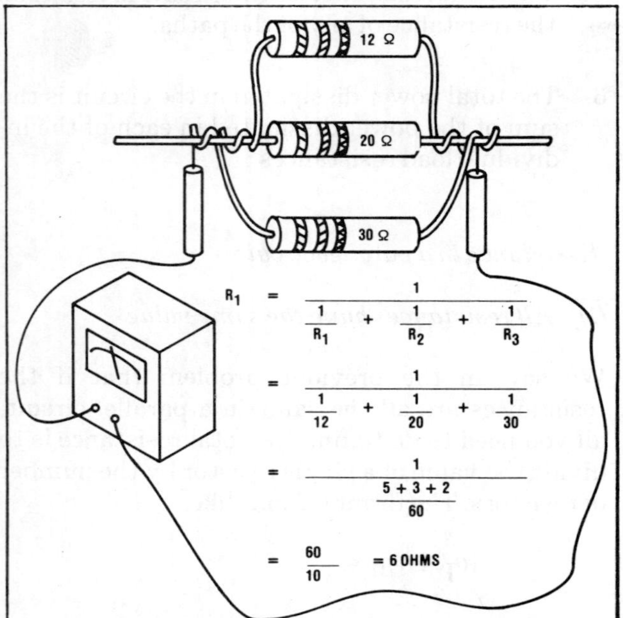

Fig. 9-8 Finding the equivalent resistance of more than two unlike reistors in parallel.

If we have a 100-ohm resistor in parallel with one having a resistance of 200 ohms, their total resistance found by this formula is 66.7 ohms. This, you will notice, is less than the resistance of the smaller resistor.

(3) More than two unlike resistances

When we have more than two unlike resistors in parallel, we can find the total resistance, by using a formula that gives us the reciprocal of the sum of the reciprocals of the individual resistances. What this means is that we use a formula such as the one in Fig. 9-8. We have a 12-ohm, a 20-ohm, and a 30-ohm resistor in parallel. When we use this formula, we find that these three resistors have a total resistance of six ohms. Again, this is less than the smallest of the three resistors.

3. Complex circuits

Many of our circuits are more complex than the simple series and parallel circuits we have just discussed, but by carefully analyzing them, we can usually combine the series and parallel components and find equivalent values that allow us to rearrange the entire circuit into a single equivalent resistance. Then, by applying Ohm's law, we can find all of the values we need to know. The arrangement in Fig. 9-10 is a good exercise in circuit analysis. We want to complete the chart in Fig. 9-9 with all of the values that are missing.

Item	Voltage Volts	Resistance Ohms	Current Amps	Power Watts
R_1		4		
R_2		20		
R_3		30		
R_4		12		
R_5		24		
R_6		40		
R_7		60		
Source	24			

Fig. 9-9 Find all of the values that are missing in this chart.

Our first step is to draw the circuit using the conventional symbols, Fig. 9-11. And, now, since we know all of the resistance values, our next step is to find the equivalent resistance of each of the parallel combinations. Starting with R_2 and R_3, we find their equivalent resistance to be 12 ohms.

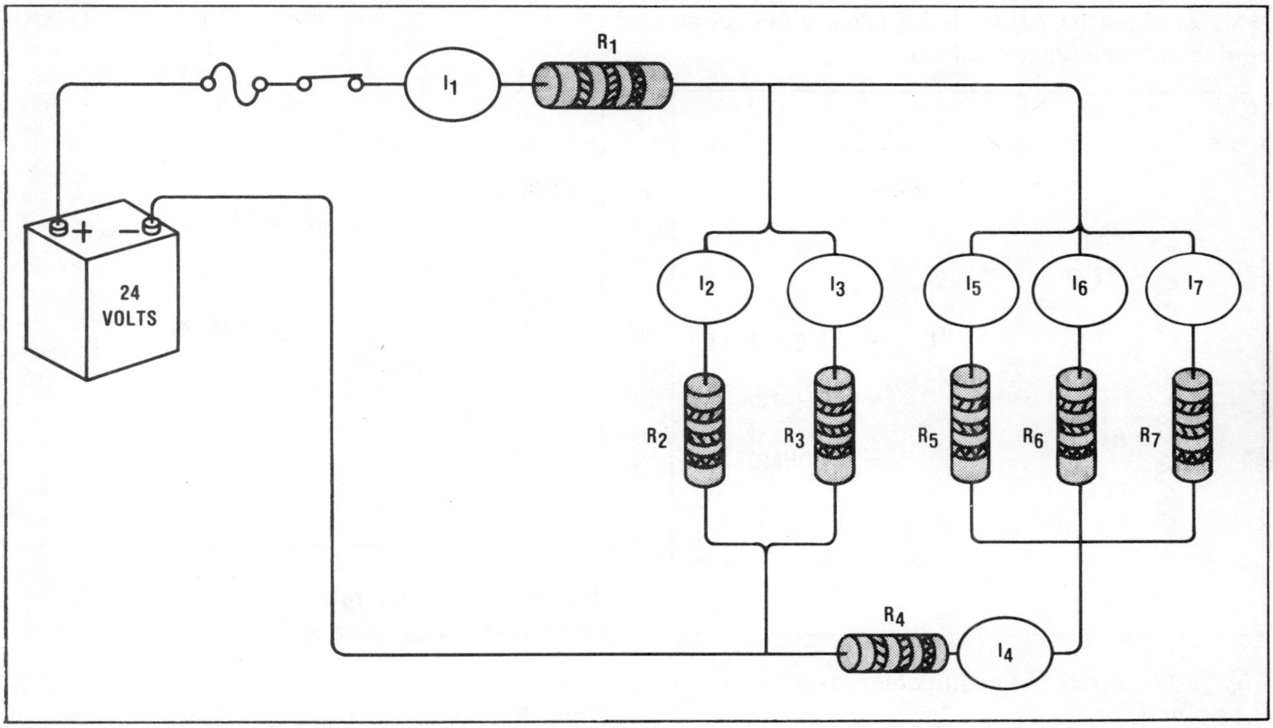

Fig. 9-10 A complex circuit having both series and parallel combinations of resistance.

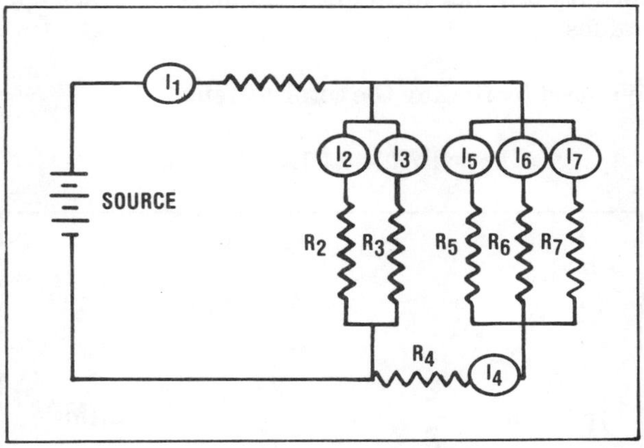

Fig. 9-11 The complex circuit of Fig. 9-10, using conventional symbols.

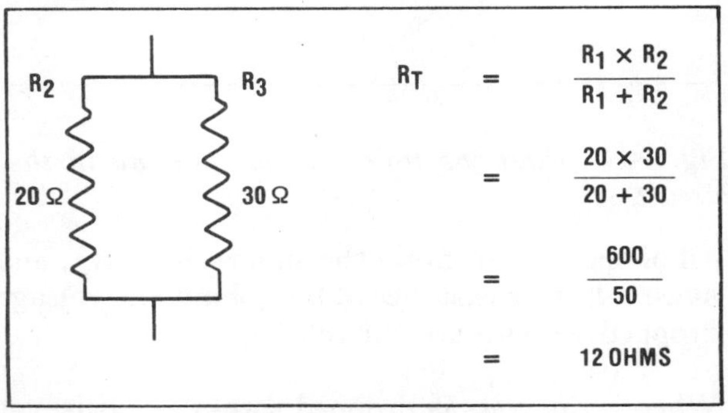

Fig. 9-12 Find the equivalent resistance of R_2 and R_3.

Now, our next step is to find the equivalent resistance of R_5, R_6, and R_7. This also happens to be 12 ohms.

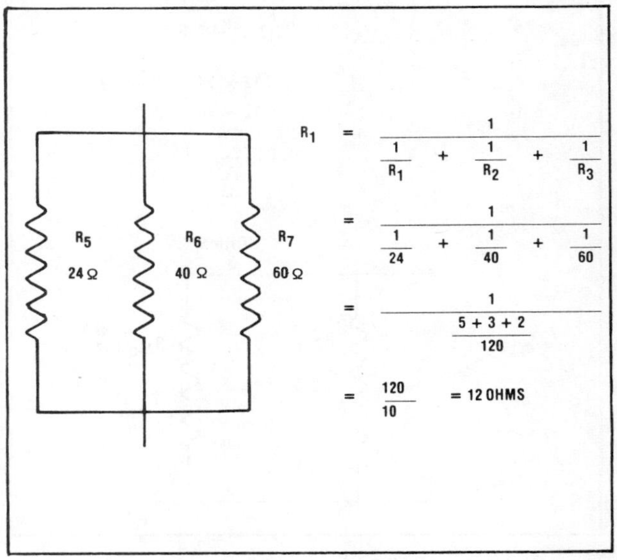

Fig. 9-13 Find the equivalent resistance of R_5, R_6, and R_7.

To make the problem easy to follow, re-draw the circuit, using the two equivalent resistances instead of the original combinations. When we do this, it becomes obvious to us that the next step is to combine $R_{5\text{-}6\text{-}7}$ with R_4. This gives us an equivalent resistance of 24 ohms.

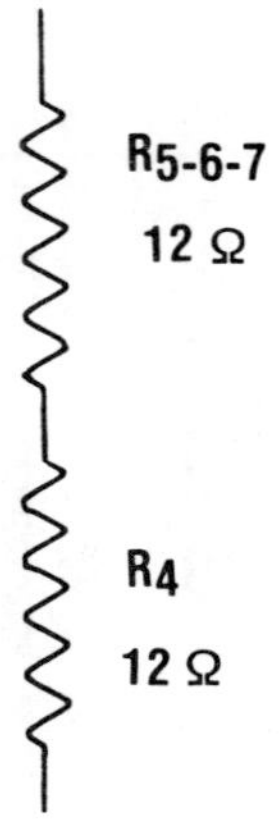

$$R_T = R_{5\text{-}6\text{-}7} + R_4 = 12 + 12 = 24 \text{ OHMS}$$

Fig. 9-14 Find the equivalent resistance of $R_{5\text{-}6\text{-}7}$, and R_4.

We now have a parallel combination of $R_{4\text{-}5\text{-}6\text{-}7}$ and $R_{2\text{-}3}$. The equivalent of these two combinations is eight ohms.

$$R_T = \frac{R_{2\text{-}3} \times R_{4\text{-}5\text{-}6\text{-}7}}{R_{2\text{-}3} + R_{4\text{-}5\text{-}6\text{-}7}} = \frac{12 \times 24}{12 + 24} = \frac{288}{36} = 8 \text{ OHMS}$$

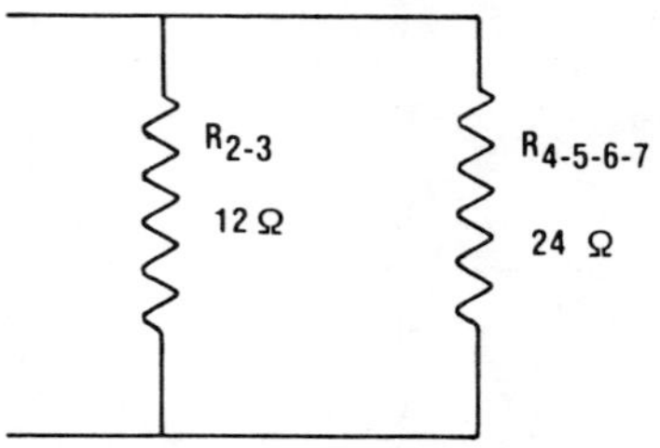

Fig. 9-15 Find the equivalent resistance of $R_{2\text{-}3}$, and $R_{4\text{-}5\text{-}6\text{-}7}$.

All we have to do now is to find the equivalent of the series combination of R_1 and $R_{2\text{-}3\text{-}4\text{-}5\text{-}6\text{-}7}$, which is 12 ohms. This is the equivalent resistance of the entire circuit.

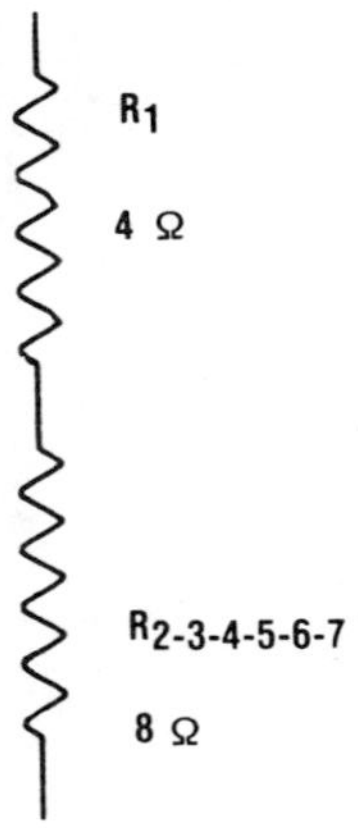

$$R_T = R_1 + R_{2\text{-}3\text{-}4\text{-}5\text{-}6\text{-}7} = 4 + 8 = 12 \text{ OHMS}$$

Fig. 9-16 Find the total equivalent resistance by adding R_1 to $R_{2\text{-}3\text{-}4\text{-}5\text{-}6\text{-}7}$.

Now that we know the equivalent resistance of the entire circuit, we can work the problem back into its original form and find all of the missing values.

We start by finding the total current.

$$I_T = E/R = 24/12 = 2 \text{ amps.}$$

$$I_T = \frac{E}{R} = \frac{24}{12} = 2 \text{ AMPS}$$

Fig. 9-17 Find the total current flowing in the circuit.

All of the current flows through resistor R_1, and since it has a resistance of four ohms, the voltage dropped across it is eight volts.

When eight volts is dropped across R_1, it leaves 16 volts across both the combinations $R_{2\text{-}3}$ and $R_{5\text{-}6\text{-}7}$.

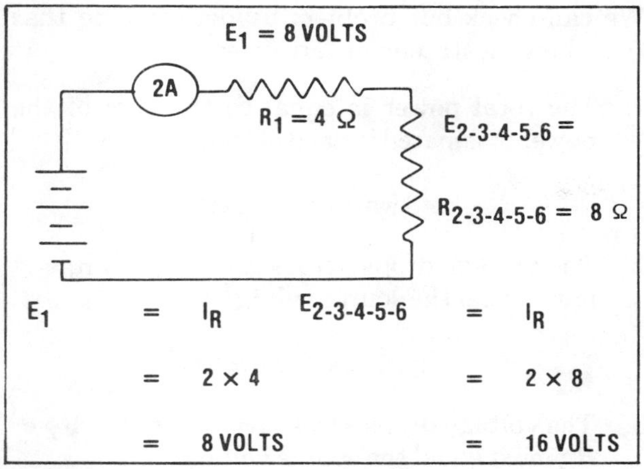

Fig. 9-18 Find the voltage dropped across R_1 and across $R_{2\text{-}3\text{-}4\text{-}5\text{-}6}$.

The current through the first combination is 1.33 amps, and that through the second is 0.67 amp.

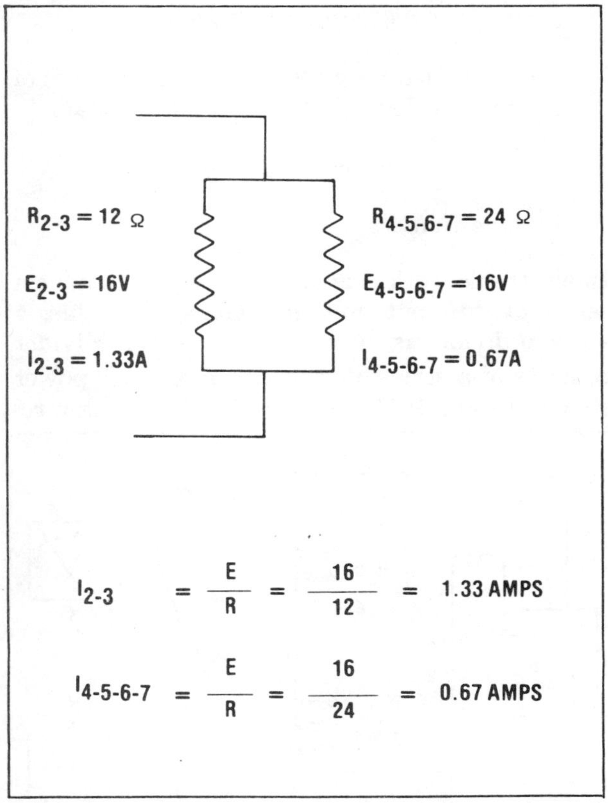

Fig. 9-19 Find the current flowing through $R_{2\text{-}3}$, and $R_{4\text{-}5\text{-}6\text{-}7}$.

The entire 0.67 amp flows through R_4, and the voltage drop across it is 8 volts. This same 0.67 amp flows through the combination $R_{5\text{-}6\text{-}7}$, and since its equivalent resistance is 12 ohms, there will also be eight volts dropped across it.

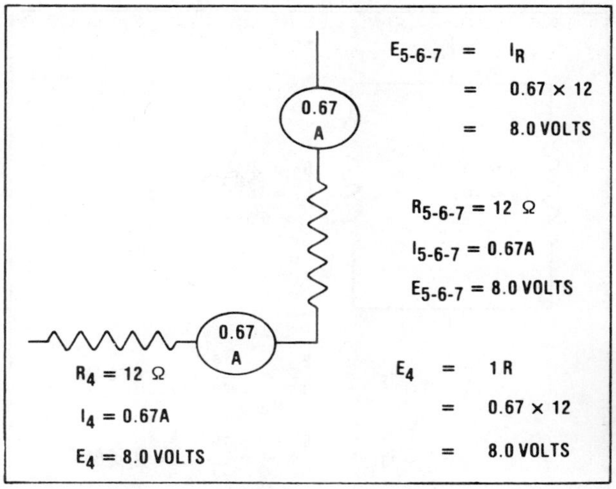

Fig. 9-20 Find the voltage dropped across R_4 and across $R_{5\text{-}6\text{-}7}$.

We can now find the current through resistors $R_{5, 6, 7}$.

$$I_5 = E/R = 8/24 = 0.33 \text{ amp}$$

$$I_6 = E/R = 8/40 = 0.2 \text{ amp}$$

$$I_7 = E/R = 8/60 = 0.14 \text{ amp}$$

The total current through these three resistors is the same 0.67 amp that flowed through R_4.

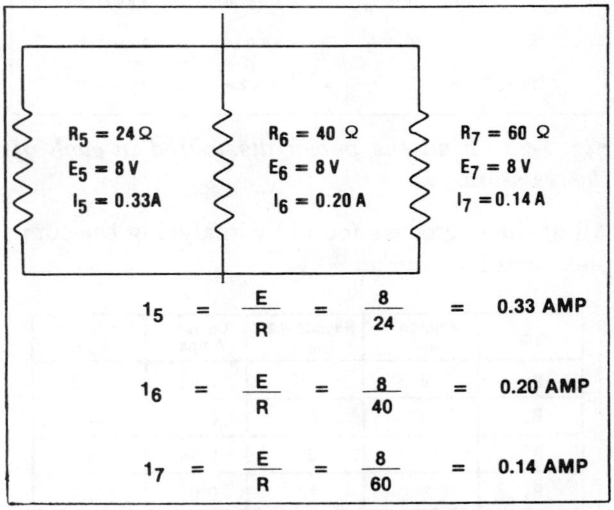

Fig. 9-21 Find the current through R_5, R_6, and R_7.

To find the current through R_2 and R_3, we use the formula:

$$I_2 = E/R = 16/20 = 0.80 \text{ amp}$$

$$I_3 = E/R = 16/30 = 0.54 \text{ amp}$$

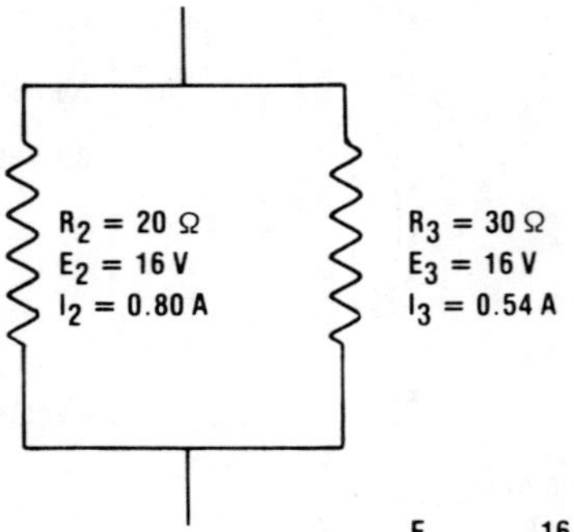

$$I_2 = \frac{E}{R} = \frac{16}{20} = 0.80 \text{ AMP}$$

$$I_3 = \frac{E}{R} = \frac{16}{30} = 0.54 \text{ AMP}$$

Fig. 9-22 Find the current through R_2 and R_3.

We can find the power dissipated in each resistor by simply multiplying the current through each of the resistors by the voltage drop across each resistor. This is done in Fig. 9-22.

P_1	=	$E_1 \times I_1$	=	8×2	=	16.0 WATTS
P_2	=	$E_2 \times I_2$	=	16×0.8	=	12.8 WATTS
P_3	=	$E_3 \times I_3$	=	16×0.54	=	8.6 WATTS
P_4	=	$E_4 \times I_4$	=	8×0.67	=	5.3 WATTS
P_5	=	$E_5 \times I_5$	=	8×0.33	=	2.6 WATTS
P_6	=	$E_6 \times I_6$	=	8×0.20	=	1.6 WATTS
P_7	=	$E_7 \times I_7$	=	8×0.14	=	1.1 WATTS
P_{TOTAL}	=	$E_T \times I_T$	=	24×2.0	=	48.0 WATTS

Fig. 9-23 Find the power dissipated in each of the resistors.

All of the values we found by analyzing the complex circuit are in Fig. 9-24.

Item	Voltage Volts	Resistance Ohms	Current Amps	Power Watts
R_1	8	4	2.00	16.0
R_2	16	20	0.80	12.8
R_3	16	30	0.54	8.6
R_4	8	12	0.67	5.3
R_5	8	24	0.33	2.6
R_6	8	40	0.20	1.6
R_7	8	60	0.14	1.1
Source	24	12	2.00	48.0

Fig. 9-24 Circuit relationships in a complex circuit.

We can check our problem by determining that the following statements are true:

1. The total power is equal to the sum of the power dissipated in each of the resistors.

   48 watts = 48 watts.

2. The voltage drops across R_1, R_4, and $R_{5\text{-}6\text{-}7}$ must equal the source voltage.

   24 volts = 24 volts.

3. The voltage drops across R_1, and either R_2 or R_3 must equal the source voltage.

   24 volts = 24 volts.

4. Current through R_4 is the same as the sum of the current through R_5, R_6, and R_7.

   0.67 amp = 0.67 amp.

5. The current through R_1 is equal to the sum of the current through R_2, R_3, R_5, R_6, and R_7.

   2 amps = 2 amps.

4. *Voltage dividers*

In electronic work it is often necessary to have a series of different voltages, and to get this, a voltage divider is often used. A voltage divider consists of a series of resistors across the power source. In Fig. 9-25, we have three 1,000-ohm re-

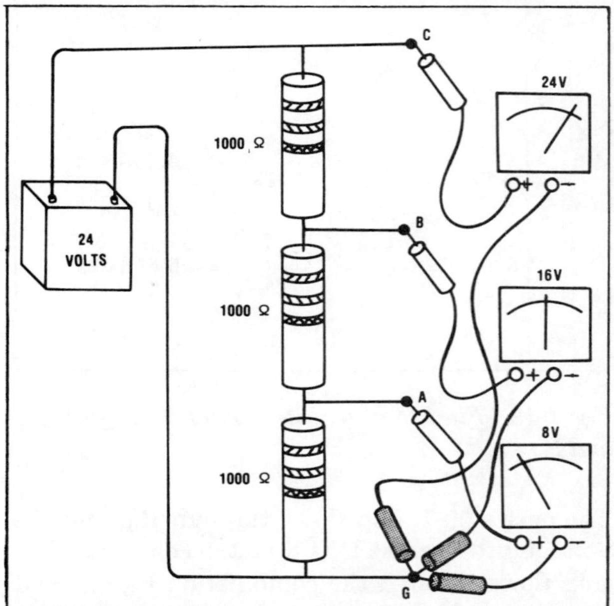

Fig. 9-25 An unloaded voltage divider.

sistors across a 24-volt battery. When there is no load attached to any of the terminals, there will be a current of eight milliamps (0.008 amp) flowing through the divider, and this current will produce an eight-volt drop across each of the resistors.

When a load is placed across any of the terminals of the divider, it will act in parallel with that portion of the resistance and lower the total resistance. This will increase the current through the circuit. If, for example, a load made up of a 1,000-ohm resistor is placed between the ground terminal and terminal A, the resistance between A and G will drop to 500 ohms, and the current will increase from eight milliamps to 9.6 ma. The voltage across R_3 and the load will be 4.8 volts instead of the eight volts that was between these same terminals without the load.

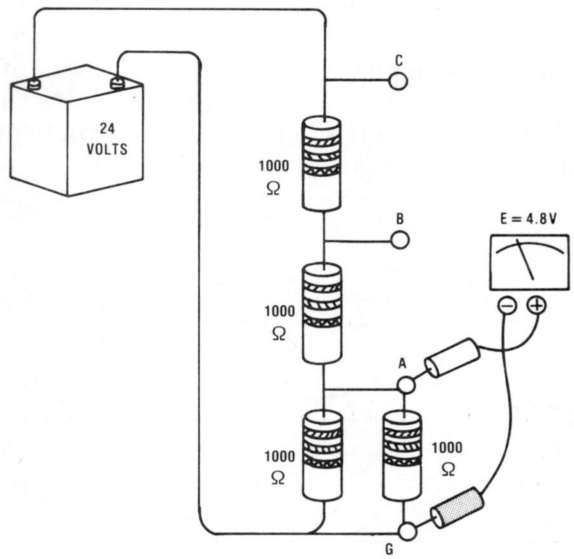

$$R_{A\text{-}G} = \frac{1000}{2} = 500\ \Omega$$

$$R_{A\text{-}B} = 1000\ \Omega$$

$$R_{B\text{-}C} = 1000\ \Omega$$

$$R_T = 2500\ \Omega$$

$$I = \frac{E}{R} = \frac{24}{2500} = 9.6\ \text{MA}$$

$$E_{A\text{-}G} = IR = 9.6 \times 500 = 4.8\ \text{VOLTS}$$

Fig. 9-26 A loaded voltage divider.

It is sometimes necessary in electronic circuits to have voltages that operate on either side of a reference value. For example, if we want voltages of -8 to $+16$ volts without a load, we can use a voltage divider made up of three 1,000-ohm resistors across our 24-volt source. Rather than using the negative terminal of the battery as our ground, or reference point, as we did previously, we can use the junction between R_2 and R_3.

The lower end of R_3 is eight volts negative with respect to the ground, the junction between R_2 and R_1 is eight volts positive with respect to ground, and the top of R_1 is 16 volts positive with respect to ground. These values will, of course, change when a load is applied, but the polarity with respect to ground will not change.

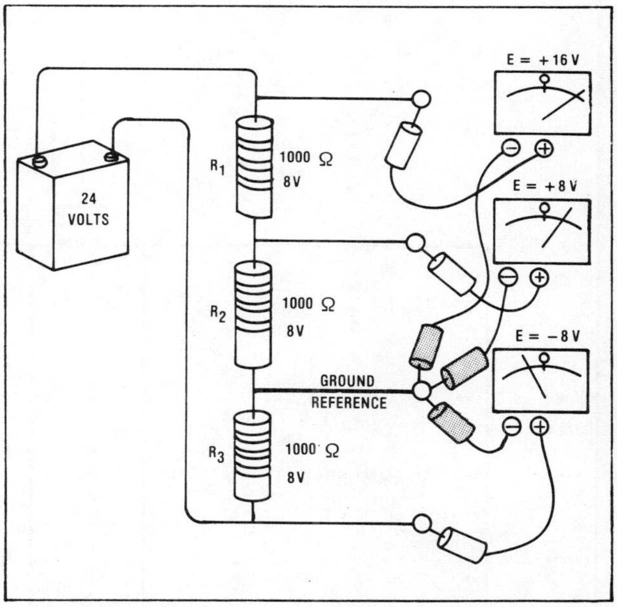

Fig. 9-27 A voltage divider that provides voltages on either side of ground, or the reference voltage.

SECTION X

Batteries

A battery is a device composed of two or more cells in which chemical energy is converted into electrical energy. The chemical nature of the battery components provides an excess of electrons at one terminal and a deficiency at the other. When the two terminals are joined by a conductor, electrons flow, and as they flow, the chemical composition of the active material changes. In a primary cell, the active elements become exhausted and cannot be restored, while in a secondary cell, electricity from an external source will restore the active material to its original, or charged, condition.

A. Primary Cells

1. Carbon-zinc cells

The most commonly used primary cell, or dry cell, as it is called, is of the carbon-zinc type. In this cell, a carbon rod is supported in a zinc container by a moist paste containing ammonium chloride,

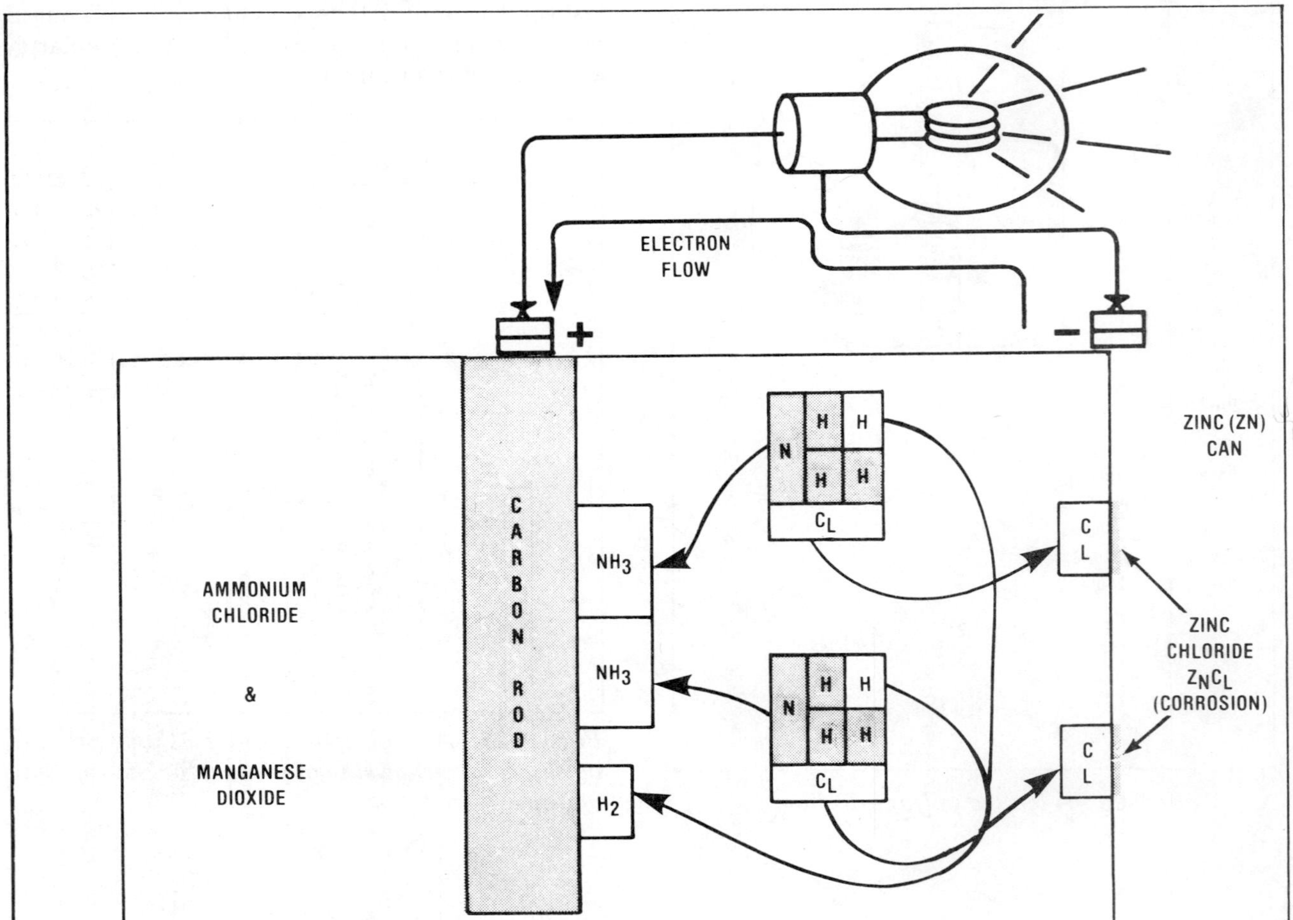

Fig. 10-1 Chemical changes that take place in a carbon-zinc primary cell when electrons flow.

manganese dioxide, and granulated carbon. When a conductor joins the zinc case and the carbon rod, electrons leave the zinc and flow to the carbon. The zinc is left with positive ions which attract negative chlorine ions from the ammonium chloride electrolyte and form zinc chloride on the inside of the can. Zinc chloride is actually corrosion, and it eats away the zinc, eventually causing the battery to leak.

The positive ammonium ions that are left in the electrolyte are attracted to the carbon rod, where they accept the electrons that have just arrived from the zinc. As they are neutralized, they break down into ammonia and hydrogen gasses which are absorbed into the moist manganese dioxide.

Carbon-zinc cells produce one and a half volts regardless of their size, but the size of the cell determines the amount of current it can supply.

Leakage, which at one time was a problem with carbon-zinc batteries, is minimized by the use of effective seals and by enclosing the zinc can inside of steel jacket.

2. Alkaline cells

When longer life is needed, alkaline cells are used, rather than the less expensive carbon-zinc variety. In the modern alkaline cell, the center electrode is a zinc rod supported in a carbon container by a moist electrolyte of potassium hydroxide. All of this is housed inside a steel case, with an insulating disc isolating the center electrode from the case.

Fig. 10-2 An alkaline cell uses a carbon container and a zinc center electrode with potassium hydroxide as an electrolyte.

Potassium hydroxide has a lower resistance than ammonium chloride, and these cells will produce more current than carbon-zinc cells and so are excellent for tape recorders and other devices which contain motors. In order for alkaline cells to be interchangeable with carbon-zinc cells, their polarity has to be reversed. This is done by mounting the cell in an insulated steel outer case. The negative center terminal of the cell bears against the outer case by a spring, and the case of the cell contacts only the center conductor of the outer shell. By doing this, both types of cells have negative outer cases and positive center terminals.

3. Mercury cells

Another type of alkaline cell is the mercury cell which is used in some hearing aids, cameras, and for other applications where a high capacity for small size overcomes their higher cost.

A pellet of mercuric oxide is placed inside a steel container, and a porous insulator or separator is placed over the pellet. A roll of extremely thin corrugated zinc is placed over the paper which is saturated by a solution of potassium hydroxide which serves as the electrolyte. The cell is then closed with a steel cap that contacts the zinc, and is insulated from the container. The container is the positive terminal, and the cap is the negative. In applications where it is necessary to have a negative case, the same procedure is used to reverse the polarity as was used by the other alkaline cell.

Fig. 10-3 A mercury cell is used when a high capacity is needed in a very small cell.

B. *Secondary cells*

It is possible to construct a cell which, instead of consuming or destroying one of its elements as it is used, will convert the element into a chemical that can be changed back into its original material by flowing electricity into it. Cells of this type are called secondary cells, and do not produce electrical energy, but merely store it. It is for this reason that batteries of these cells are called storage batteries.

1. *Lead-acid batteries*

The most commonly used storage battery for aircraft is the lead-acid battery.

a. *Construction*

The positive plate is made up of a grid of lead and antimony filled with lead peroxide. The negative plate uses a similar grid, but its open spaces are filled with spongy lead in which an expander material is used to prevent its compacting back into a dense form of lead, losing much of its surface area.

The cells are made up of a group of positive plates, joined together and interlaced between a stack of negative plates. Porous separators keep the plates apart and hold a supply of electrolyte in contact with the active material.

The battery container was formerly made of hard rubber, but is now made of a high-impact plastic and has individual compartments for the cells.

Fig. 10-4 A typical lead-acid aircraft battery.

Connector straps join the cells and provide the external terminals of the battery. A cover seals the cells in the case, and holes in the cover provide access to the cells for servicing them with water and for checking the condition of the electrolyte.

In aircraft batteries, these cell openings are closed with vented screw-in-type caps which have lead weights inside them to close the vent when the battery is tipped. This prevents the electrolyte spilling in unusual flight attitudes.

An electrolyte consisting of a mixture of sulfuric acid and water covers the plates and takes an active part in the charging and discharging of the cell.

b. *Chemical changes during discharge*

When a conductor connects the positive and negative terminals of the battery, electrons flow from the lead to the lead peroxide. When electrons leave the lead, it leaves behind positive ions which attract the negative sulfate radicals from the sulfuric acid in the electrolyte. This combination forms lead sulfate on the negative plate.

The electrons arriving at the positive plate drive the negative oxygen radicals from the lead peroxide. This oxygen joins up with the hydrogen in the electrolyte that had lost its sulfate radical, and this now becomes water, H_2O. The lead that was left on the positive plate attracts sulfate radicals from the electrolyte and becomes lead sulfate.

Now, with lead sulfate on both the positive and negative plates, and with the electrolyte diluted by the water that has formed in it, the battery is discharged and electrons no longer flow.

c. *Chemical changes during charge*

If a discharged battery is attached to a source of direct current having the proper voltage, and the positive plates of the battery connected to the positive terminal of the source, electrons will be drawn from the positive plate and forced into the negative plates.

Electrons arriving at the negative plates drive the negative sulfate radicals out of the lead sulfate back into the electrolyte, where they join with the hydrogen from the water to form sulfuric acid, H_2SO_4.

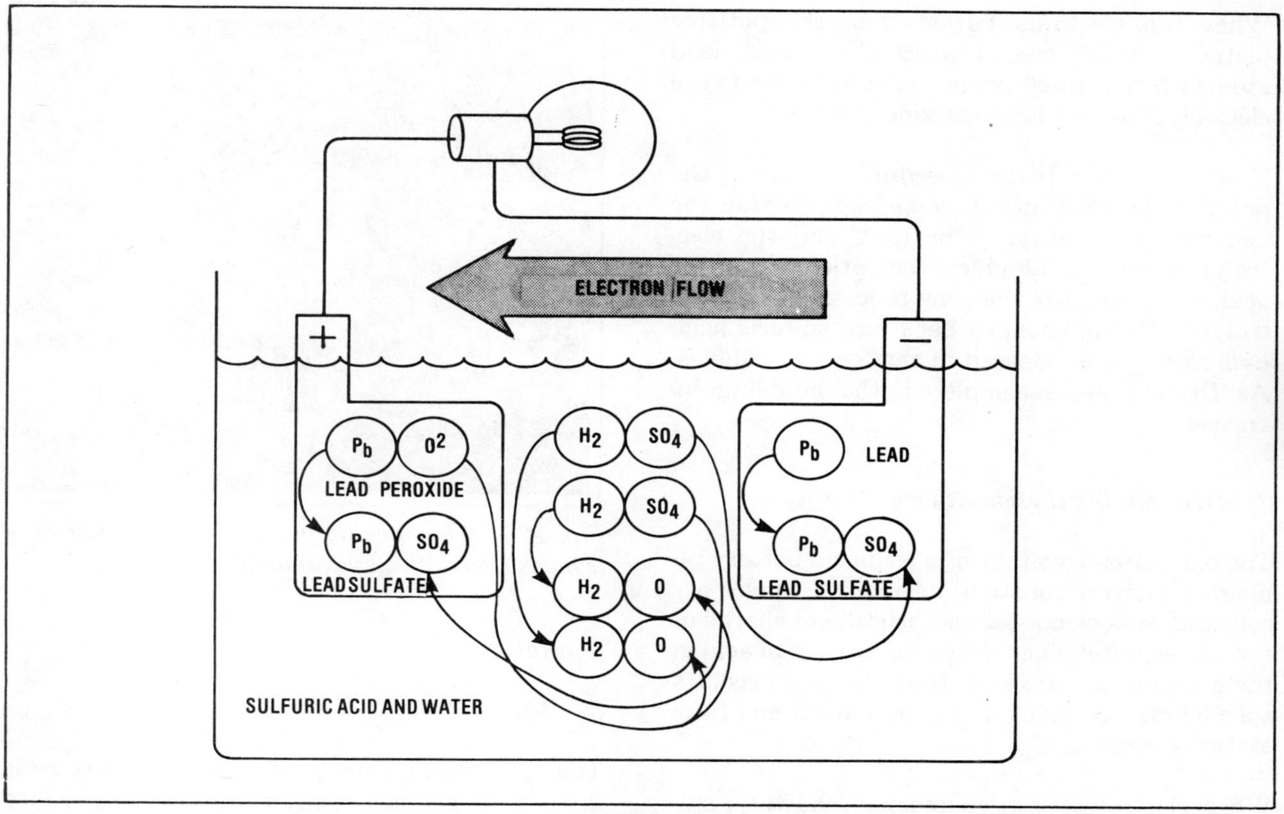

Fig. 10-5 Chemical changes that take place during discharge.

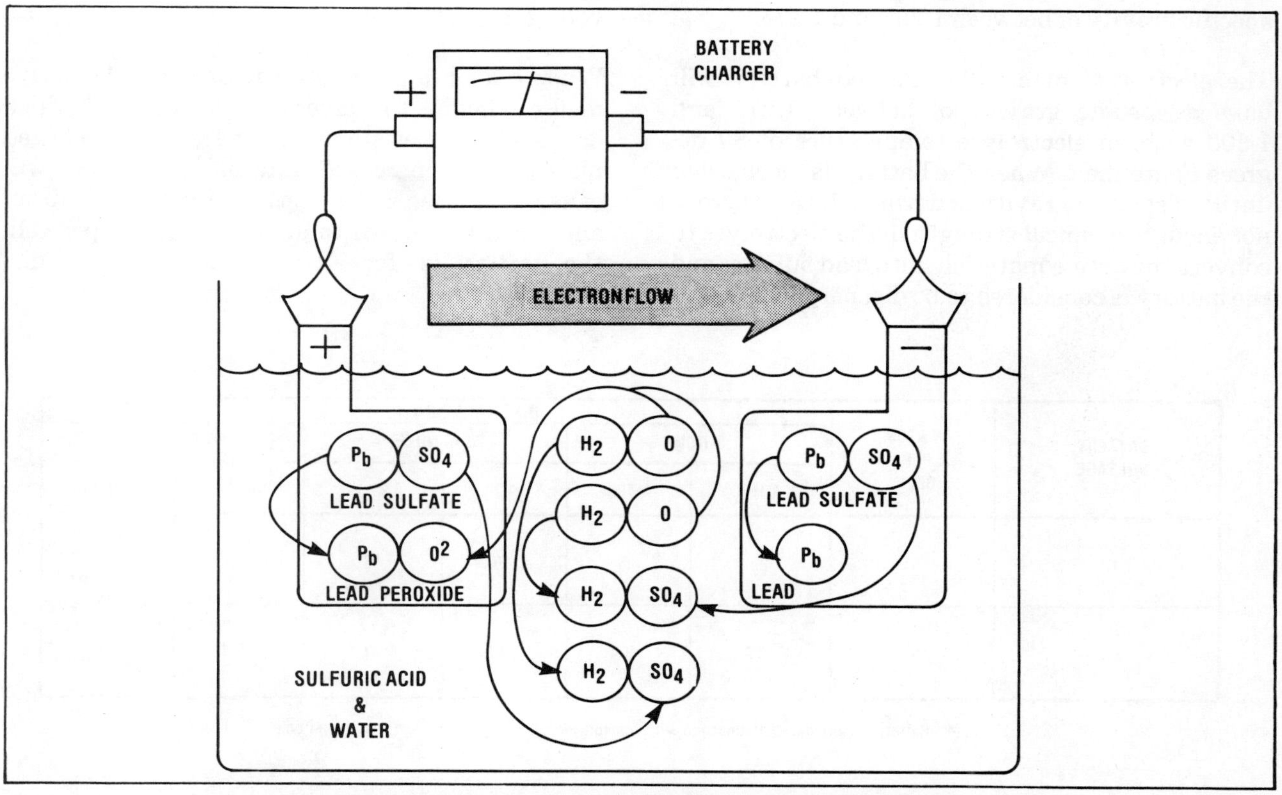

Fig. 10-6 Chemical changes that take place during charge.

When the electrons flowed from the positive plates, they left behind positively charged lead atoms which attract oxygen from the water in the electrolyte to form lead peroxide, PbO.

Now when the battery is fully charged, the positive plate has again become lead peroxide, the negative plate has become lead, and the electrolyte again has a high concentration of sulfuric acid. All during the charging process, as the electrolyte is being changed back into sulfuric acid, hydrogen gas is released in the form of bubbles. As the charge is completed, the bubbling increases.

d. Determining the condition of charge

The open-circuit voltage of a lead-acid battery remains relatively constant, at about 2.1 volts per cell, and so does not tell us the state of charge of the battery. But since the concentration of acid in the electrolyte changes as the battery is used, its specific gravity gives us a good indication of the state of charge.

When electrolyte is prepared for a battery, sulfuric acid having a specific gravity of 1.835 is mixed with pure water to get a mixture having a specific gravity of between 1.265 and 1.275.

The electrolyte of a fully charged battery will have a specific gravity of between 1.275 and 1.300 with an electrolyte temperature of 80 degrees Fahrenheit. When the battery is discharged until its specific gravity is down to 1.150, there is not enough chemical strength in the electrolyte to convert the active materials into lead sulfate, and the battery is considered to be discharged.

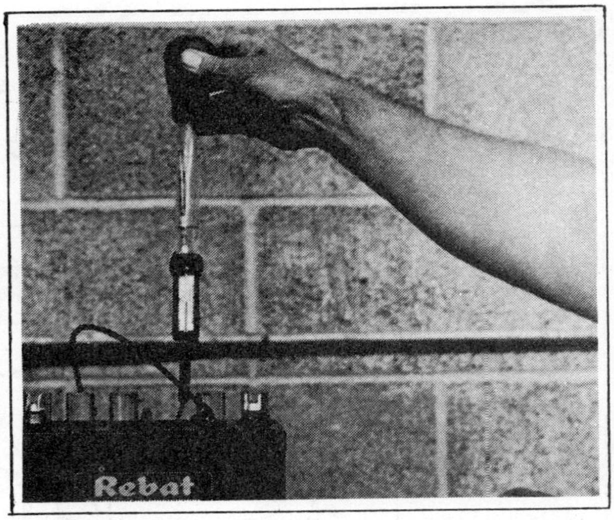

Fig. 10-7 The specific gravity of the electrolyte is measured with a hydrometer.

e. Ratings

(1) Voltage

The open-circuit voltage of a lead-acid battery is 2.10 volts per cell when the electrolyte has a specific gravity of 1.265. The physical size of the cell or the number of plates has no effect on this voltage.

When a load is placed on the battery, the active material begins to convert into lead sulfate which has a higher resistance than the fully charged plates. This increased internal resistance will cause the closed-circuit terminal voltage to drop, and when it is down to about 1.75 volts per cell, the battery is, for all practical purposes, discharged.

BATTERY VOLTAGE	PLATES PER CELL	DISCHARGE RATE *					
		5-HOUR		20-MINUTE		5-MINUTE	
		A.H.	AMPS	A.H.	AMPS	A.H.	AMPS
12	9	25	5	16	48	12	140
24	9	17	3.4	10.3	31	6.7	80

* Battery is considered discharged when closed-circuit voltage drops to 1.2 volts per cell.

Fig. 10-8 Relationship between ampere-hour capacity and discharge rate.

(2) Capacity

The capacity of a battery is its ability to produce a given amount of current for a specified time, and is expressed in ampere-hours. One ampere-hour of capacity is the amount of electricity that is put into or taken from a battery when a current of one ampere flows for one hour. Any combination of flow and time that moves the same amount of electricity is also one ampere-hour. For example, a flow of one-half amp for two hours or two amps for one-half hour is each one ampere-hour.

The amount of active material, the area of the plates, and the amount of electrolyte determine the capacity of a battery. But also of importance in rating a battery is the rate at which the current flows.

(a) Standard rate—five-hour discharge

The standard rating used to specify the capacity of a battery is the five-hour discharge rating. This is the number of ampere-hours of capacity of the battery when there is sufficient current flow to drop the voltage of a fully charged battery to 1.75 volts per cell at the end of five hours.

(b) High rate—twenty-minute discharge and five-minute discharge

Two ratings are used to specify the capacity of a battery when the current is taken out at a high rate. Both ratings will be lower than the five-hour rating. Fig. 10-8 shows the relationship between the capacity at the five-hour rating and that for the twenty-minute and the five-minute rating for typical 12- and 24-volt aircraft batteries.

f. Battery testing

(1) Specific gravity

As we have previously stated, the specific gravity of the electrolyte is one of the best indicators of the state of charge of a lead-acid battery.

(A)

Electrolyte Temperature °C	Electrolyte Temperature °F	Correction points Add or Subtract
60	140	+24
55	130	+20
49	120	+16
43	110	+12
38	100	+ 8
33	90	+ 4
27	80	0
23	70	− 4
15	60	− 8
10	50	−12
5	40	−16
− 2	30	−20
− 7	20	−24
−13	10	−28
−18	0	−32
−23	−10	−36
−28	−20	−40
−35	−30	−44

(B)

Specific Gravity	Freezing Point °C	Freezing Point °F
1.300	−70	−90
1.275	−62	−80
1.250	−52	−62
1.225	−37	−35
1.200	−26	−16
1.175	−20	− 4
1.150	−15	+ 5
1.125	−10	+13
1.100	− 8	+19

Fig. 10-9 A — specific gravity correction for non-standard temperature. B — Freezing point of lead-acid battery with various electrolyte specific gravities.

We determine the specific gravity with a hydrometer, an instrument which measures the depth to which the calibrated stem of a float sinks into the electrolyte. A sample of electrolyte is drawn from the cell by the syringe, and the float riding on the surface of the liquid is visible in the glass tube. The graduation on the stem which is even with the liquid shows the specific gravity of the electrolyte. The more dense the liquid, the less the float will sink, and the less dense, the deeper the float will be able to go.

The temperature of the electrolyte affects its specific gravity, and so a correction must be made if the temperature is not standard. Eighty degrees Fahrenheit is used as the reference, and if the electrolyte is warmer, it is less dense; so the float will sink deeper into it and will indicate that the state of charge is lower than it actually is. To correct for temperature effect, we use a correction chart similar to the one in Fig. 10-9.

If the specific gravity read by the hydrometer is 1.240, and the electrolyte temperature is 60 degrees, we know that the corrected specific gravity will be lower than is indicated because the lower temperature has made the electrolyte more dense. According to the chart, a correction of eight points should be made. The indication was 1.240, and after we subtract 0.008, we have a corrected specific gravity of 1.232.

(2) Cell test

If the battery construction is such that the voltage of the individual cells can be measured, we can get a good indication of the behavior of the cells under load.

We must first determine that the electrolyte level is proper, and if it is not correct, it should be adjusted.

Apply a heavy load to the battery for about three seconds by cranking the engine with the ignition switch off.

CAUTION: Be sure that the propeller is clear before engaging the starter.

Now, turn on the landing lights and taxi lights to draw about ten amps, and while the load current is flowing, measure the voltage of each cell.

A fully charged battery with all of the cells in good condition should have a voltage of 1.95 volts for each cell, and all of the cells should be within 0.05 volt of each other.

If some of the cells are below 1.95 volts, but all are within 0.05 volt of each other, the cells are in good condition but the battery is low. If any of the cells read higher than 1.95 volts and there is more than a 0.05-volt difference between any of them, there is a defective cell in the battery.

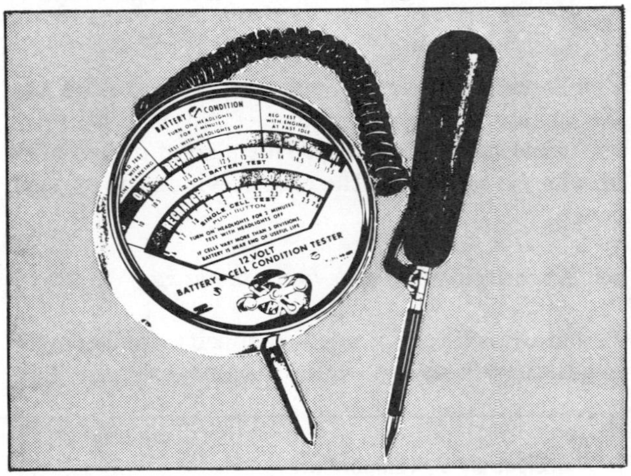

Fig. 10-10 A low-range voltmeter is needed to determine the condition of the cells under load.

g. Servicing and charging

(1) Inspection and servicing

One of the most important aspects of battery servicing is keeping the battery clean and all of the terminals tight and free of corrosion. Any corrosion on the battery terminals or the cables should be removed, and the battery box and the top of the battery should be scrubbed with a soft bristle brush and a solution of bicarbonate of soda (baking soda) and water. When washing the top of the battery, be sure that you do not get any soda in the cells, as it will neutralize the electrolyte. After the battery and box are perfectly clean, rinse them with clean water and dry them thoroughly; then coat the battery terminals with petroleum jelly or general purpose grease, and touch up any paint damage to the box or adjacent area with an acid-proof paint.

The electrolyte should just cover the plates, and most batteries have an indicator to show you the correct level. If it is low, add distilled or demineralized water to the level specified. You should never add acid to the battery unless it has been spilled, and then, follow the recommendations of the battery manufcturer in detail. The normal loss of liquid in a battery is the loss of water caused by its decomposition during charging.

Fig. 10-11 The electrolyte level should come up to the indicator inside the cell.

Most new batteries are received in a dry-charged state. These batteries have been fully charged when they were manufactured; then the electrolyte was dumped and the cells sealed. When putting one of these batteries into service, remove the cell seals, and pour in the electrolyte that is shipped with the battery. The battery is now ready for installation, but in order to have it fully charged, it can be given a slow freshening, or boost, charge. Allow the battery to sit for an hour or so after the boost charge and adjust the level of the electrolyte.

It is normally not necessary for you to mix electrolyte, as properly diluted acid is readily available. But if it should ever become necessary to dilute acid, *it is extremely important that the acid be poured into the water, and never the other way around.* If water is poured into the acid, the water, being less dense, will float on top of the acid, and a chemical action will take place along the surface where they meet. This action will generate enough heat to boil the water, and the boiling will splash acid out of the container and can cause serious injury if it gets on the skin or in the eyes.

If acid should get into your eyes, flush them with generous amounts of clean water, and get medical attention as soon as possible.

When acid is poured into the water, it will mix with the water and distribute the heat throughout the entire container, raising the temperature of the water, but not causing any violent reaction or boiling.

(2) Constant-current charging

The most effective way of charging a battery is by putting current back into it at a constant rate, and for the most efficient charging, this rate is usually no more than seven percent of the ampere-hour rating of the battery. For example, if you are charging a 35-ampere-hour battery, and do not have specific information from the battery manufacturer stating otherwise, you should charge it at a rate not exceeding two and a half amperes.

Constant-current chargers have a voltage adjustment so that the voltage can be increased as the charge progresses. In this way, the charger voltage can be kept just enough above the rising voltage of the battery to maintain a constant current of the correct rate.

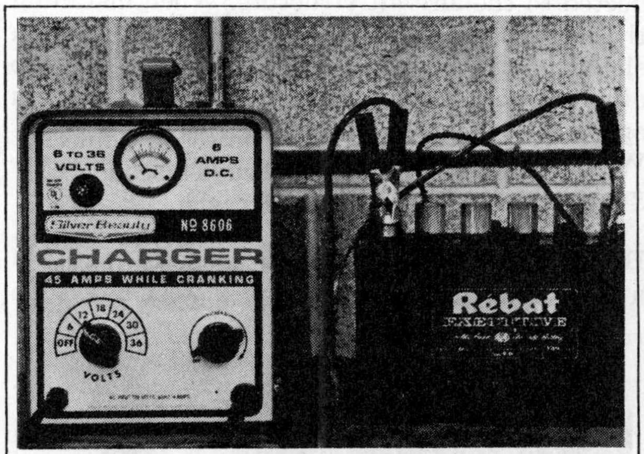

Fig. 10-12 The constant current charger is the most effective type of battery charger for shop use.

(3) Constant-voltage charging

The generating system in an aircraft charges the battery by the constant voltage method. A fixed voltage, slightly higher than the battery voltage, is connected across the battery.

The low voltage of a discharged battery will allow a large amount of current to flow into it when the charge is first begun, but as the charge continues, and the voltage rises, the current will decrease.

The voltage of the charger is sufficiently high that even when the battery is fully charged, about one ampere of current will flow into it.

Constant-voltage chargers are often used as shop chargers, but care should be exercised when using them that the battery is not overheated by the high charging rate when the charger is first connected across a fully discharged battery. The boost charge provided by a constant-voltage charger will not fully charge a battery, but it will usually supply enough charge to start the engine and allow the aircraft generating system to complete the charge.

h. Battery installation

Be sure, before installing any battery in an aircraft, that you know that the battery is correct for the aircraft, that the voltage and ampere-hour ratings are as specified, and that the battery fits the battery box properly.

Most aircraft use a single-wire electrical system, with the negative terminal of the battery connected to the aircraft structure. When installing the battery, connect the "hot" lead first. If you should short-circuit between the battery and the aircraft with your wrench, you will not cause a spark if the ground lead has not been connected. When removing a battery, always disconnect the ground lead first for the same reason.

Be sure that the battery box is properly vented, if a vent is required, and that the battery box drain extends through the aircraft skin. Some batteries are of the manifold type, which do not require a separate battery box, but there is a cover over the cells, and the area above the cells is vented to the outside of the aircraft structure.

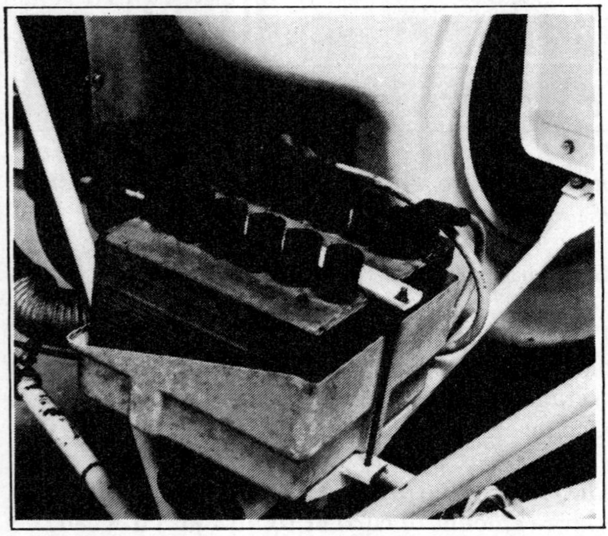

Fig. 10-13 A typical lead-acid battery installation in an aircraft.

The fumes emitted from storage batteries as they are charged are highly corrosive to the metals of which aircraft are made, and they must be neutralized before they are released into the atmosphere. Many battery installations have vent sump jars containing absorbent pads moistened with a solution of bicarbonate of soda and water.

No battery installation is complete until you know that the battery will supply enough current to crank the engine, and that the aircraft generating system will keep the battery charged.

2. Nickel-cadmium batteries

Turbine engines with their demand for extremely high current for starting have encouraged the development of an alkaline battery for aircraft use.

One of the disadvantages of the use of a lightweight lead-acid battery for high current requirements, is its buildup of internal resistance and subsequent voltage drop during high-rate discharges. The nickel-cadmium battery has a very distinct advantage in this regard. Its internal resistance is very low, and so its voltage remains constant until it is almost totally discharged. This low resistance is also an advantage in recharging, as it will accept high charging rates without damage.

a. Construction

Most nickel-cadmium batteries are made up of individual removable cells. The positive plates are made of powdered nickel called plaque, fused, or sintered, to a nickel wire-mesh screen. This porous material is impregnated with nickel hydroxide.

The negative plates are made of the same type of plaque as was used on the positive plates, but the negative plates are impregnated with cadmium hydroxide.

Separators of nylon and cellophane prevent the plates touching each other, and the stack-up of plates and separators is assembled into a polystyrene or nylon cell case, and the case is sealed.

A thirty-percent, by weight, solution of potassium hydroxide and water is used as the electrolyte. The specific gravity of this liquid is between 1.24 and 1.30 at 80 °F. and since it acts only as a conductor during charging and discharging, its specific gravity is no indication of the state of charge of the battery.

The individual cell produces an open circuit voltage of between 1.55 and 1.80 volts, and batteries for 12-volt aircraft may use either nine or ten cells, while batteries for 24-volt aircraft are made up of 19 or 20 individual cells.

Fig. 10-14 Nickel-cadmium batteries are made up of individual cells in transparent cases.

b. Chemical changes during charge

When charging current flows into the battery, oxygen is driven from the cadmium oxide on the negative plate and leaves metallic cadmium. The nickel hydroxide on the positive plate accepts some of the released oxygen and becomes more highly oxidized. The charge continues until all of the oxygen is removed from the negative plate, and then gassing occurs in the cell as the electrolyte is decomposed by electrolysis. When the battery is fully charged, the electrolyte is driven from the plates, and immediately after the charge is completed, will be at its highest level.

c. Chemical changes during discharge

As electrons flow from the negative terminal of the nickel-cadmium battery, oxygen is driven from the positive plate and is recovered by the negative plate.

d. Servicing nickel-cadmium batteries

The electrolytes used by nickel-cadmium and lead-acid batteries are chemically opposite, and either type of battery can be contaminated by fumes from the charging of the other. For this reason it is extremely important that separate facilities be used for servicing nickel-cadmium batteries, completely away from the area used for lead-acid batteries.

The alkaline electrolyte used in nickel-cadmium batteries is corrosive, and it can burn your skin or cause severe injury if it gets into your eyes. Be careful when handling this liquid, and if any of it is spilled, neutralize it with vinegar or boric acid, and flush the area with clean water.

Most nickel-cadmium batteries will get an accumulation of potassium carbonate on top of the cells. This white powder forms when electrolyte spewed from the battery combines with carbon dioxide. Scrub it off with a nylon or other type of *non-metallic* bristle brush, and dry the battery thoroughly with a soft flow of compressed air.

Check for electrical leakage between the cells and the steel case by using a milliammeter between the positive terminal of the battery and the case. If there is more than about 100 milliamps of leakage, the battery should be disassembled and thoroughly cleaned.

Fig. 10-15 Cell-to-case leakage should be measured with a milliammeter.

The only way of actually determining the condition of a nickel-cadmium battery is to fully charge it, and then discharge it at a specified rate and measure its ampere-hour capacity. When charging, use the five-hour rate, and charge it until the cell voltage is that specified by the battery manufacturer. When it is fully charged, and immediately after it is taken off of the charger, measure the level of the electrolyte. If it is low, adjust it by adding distilled water. If the level is not checked immediately after the charge is completed, the level will drop so the battery should be allowed to sit for about three hours and then the electrolyte adjusted to a lower specified level.

When the battery is fully charged and the electrolyte adjusted, it must be discharged at a specified rate and its ampere-hour capacity measured. If the capacity is less than it should be, it is an indication that some of the cells are unbalanced, and they must be equalized by a process known as deep-cycling.

Fig. 10-16 Short all of the cells to assure that they will all be exactly the same before recharging the battery during a deep-cycle operation.

To deep-cycle the battery, continue to discharge it at a rate somewhat lower than that used for the capacity test. When the cell voltage is down to around 0.2 volt per cell, short across each cell with a shorting strap. Leave the strap across the cells for three to eight hours, to be sure that all of the cells are *completely* discharged. Now, remove the shorting straps and put the battery on a constant-current charger and charge it at the five-hour rate for seven hours. This will give it 140% of its rated current.

If the battery has been properly serviced and is in good condition, each cell should have a voltage of between 1.55 and 1.80 at a temperature between 70 and 80 degrees Fahrenheit. If the voltage of any cell is not within this range, the cell must be replaced.

Fig. 10-17 Complete the deep-cycle operation by charging the battery to 140% of its ampere-hour capacity.

AVIATION
TECHNICIAN
INTEGRATED
TRAINING
PROGRAM

EA-ITP-G3

Basic Electricity-AC

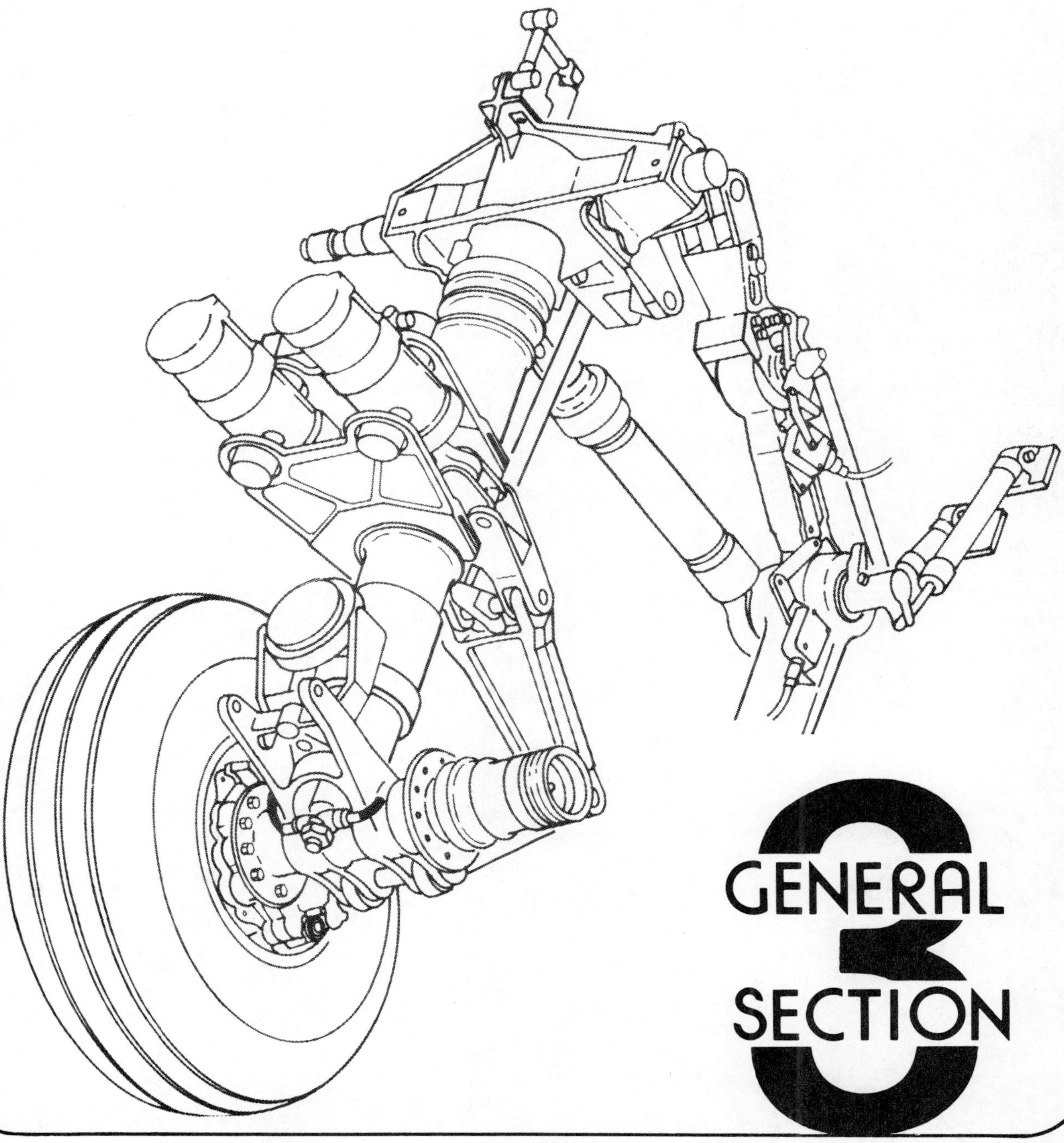

GENERAL
SECTION
3

EA-ITP-G3

TRAINING MANUAL
GENERAL SECTION
BOOK 3

BASIC ELECTRICITY — AC

by
Dale Crane

AVIATION MAINTENANCE PUBLISHERS, INC.

INTEGRATED TRAINING PROGRAM
FOR THE
AVIATION MAINTENANCE TECHNICIAN

International Standard Book Number 0-89100-154-9

Printed in the United States of America

Table of Contents

PREFACE

Part 65 of the Federal Aviation Regulations requires that an applicant for a Mechanic Certificate with an Airframe or Powerplant rating have a general knowledge of the construction and maintenance of the aircraft covered by the rating he seeks. And each applicant must prove this knowledge by passing a written, an oral and a practical examination, each given in five sections: General, Airframe Structures, Airframe Systems and Components, Powerplant Theory and Maintenance, and Powerplant Systems and Components.

Certificated Aviation Maintenance Technician Schools all operate under individual Federal Aviation Administration approvals, and all have in their approved curriculum enough information to satisfy the FAA that anyone completing their course will meet the knowledge requirements for certification as a mechanic.

In 1972, the Federal Aviation Administration published a series of three Advisory Circulars providing background information for mechanic certification applicants. These are entitled: *Airframe and Powerplant Mechanics General Handbook, AC 65-9; Airframe and Powerplant Mechanics Powerplant Handbook, AC 65-12; and Airframe and Powerplant Mechanics Airframe Handbook, AC 65-15.* These three handbooks were written to cover the area over which a mechanic applicant was to be tested, and they are an attempt by the FAA to coordinate the information available to the applicants with the tests that were given for certification. In 1976, the three handbooks were revised into their present A Revision format: AC 65-9A, AC 65-12A, and AC 65-15A.

Individuals attempting to certificate under FAR Part 65, as well as schools operating under FAR Part 147, have found that these handbooks actually do not follow the material used in the FAA examinations, and because of this, plus their lack of study helps, they fail to effectively provide the broad base of knowlege needed for certification and for a foundation on which to build in the aviation maintenance industry.

Realizing the need for updated information, presented in a format that is in keeping with the latest developments in educational technology, the Aviation Maintenance Publishers, Inc., is offering an Integrated Training Program (ITP) that presents the basic information needed for mechanic certification, in a form that will provide a good foundation in aviation maintenance knowledge. Developed in coordination with Crane Aeronautical Activities, Inc., the ITP training manuals, workbooks, study guides, and audio-visuals are designed to follow the FAA written examinations and are written to the level required for the FAR Part 147 schools.

All of this material can be used by an individual preparing for certification on his own, or by schools preparing groups of individuals for certification under FAR Part 65, but it has been prepared primarily for the Aviation Maintenance Technician Schools operating under FAR Part 147. Not only has every attempt been made to meet the requirements of Part 147 in producing this series, but the authors have followed many of the recommendations of the Aviation Mechanic Occupation Study conducted by the University of California, Los Angeles, under a contract for the U.S. Department of Health, Education, and Welfare.

One of the most basic premises of aviation maintenance is that all operations must be carried out according to FAA-approved data, and all information of a specific nature furnished by the manufacturer of an aircraft, powerplant, or component must be followed in detail. These training manuals are, of necessity, general in nature and cover a wide scope of aircraft, powerplants, and components. Because of this, it must be understood that they should *never* be allowed to take precedence over specific information furnished by the manufacturer.

The authors and publishers of this Integrated Training Program wish to express our appreciation to the various manufacturers, Aviation Maintenance Technician Schools, and FAA personnel who have aided our efforts to provide this coordinated training material to meet the challenge of better training for aviation maintenance technicians in the eighties.

SECTION I:

Advantage of Alternating Current Over Direct Current

Fig. 1-1 Transformers are used to change the values of alternating current and voltage.

Alternating current which continually changes its value and periodically reverses its direction of flow has many advantages over direct current, in which the electrons flow in one direction only. AC is much easier to generate in the large quantities needed for our homes and industries and for our large transport aircraft.

More important, though, is the ease with which we can change the values of current and voltage to get the most effective use of our electrical energy. For example, in our homes and shops, most of our AC electricity has a pressure of about 115 volts, and if we need a kilowatt of power, almost nine amps of current must flow. The current flowing in a conductor determines the amount of heat generated, and therefore the size of conductor needed. So, if we can get the same amount of power with less current, we can use smaller conductors which will save both money and weight.

In order to use the smallest conductor possible for cross-country transportation of electrical power, the voltage of the electricity carried in our transmission lines is boosted up to several thousand volts. For example, at 15,000 volts, we will need to transmit only 0.067 amp for one kilowatt of electrical power. Before the electricity is brought into our homes or shops, it is transformed down to a usable value of around 115 volts, so it will be safer and more convenient for us to handle. Between the generation of alternating current and its final use, the voltage and current will be changed many times, and the transformers that do this changing are quite efficient and very little energy is lost.

SECTION II:

Generation of Alternating Current Electricity

We have seen in our study of direct current electricity that there is a close relationship between magnetism and electricity. Any time electrons flow in a conductor, a magnetic field surrounds the conductor, and the strength of this field is determined by the amount of electron flow. We have also found that when a magnetic field is moved across a conductor, electrons are forced to flow in it, and the amount of this flow is determined by the *rate* at which the lines of magnetic flux are cut by the conductor. Increasing the number of lines of flux by making the magnet stronger or increasing the speed of movement between the conductor and the magnet will increase the amount of electron flow.

If a conductor, wound in the form of a coil, is attached to an electrical measuring instrument, and a permanent magnet is moved back and forth through the coil, the meter will deflect from side to side. This indicates that the electrons flow in one direction when the magnet is moved into the coil and reverse and flow in the opposite direction when the magnet is withdrawn. This is alternating current, AC.

The AC electricity with which we are most familiar has been generated by a rotary generator in which the conductor in the form of a coil is rotated inside a magnetic field. If we were to watch on an oscilloscope the changing values of the voltage as the coils are rotated, we would see that they start at zero, rise to a peak, and then drop back off to zero. As the coils continue to rotate, the voltage builds up in the opposite direction to a peak and then back to zero. There is one

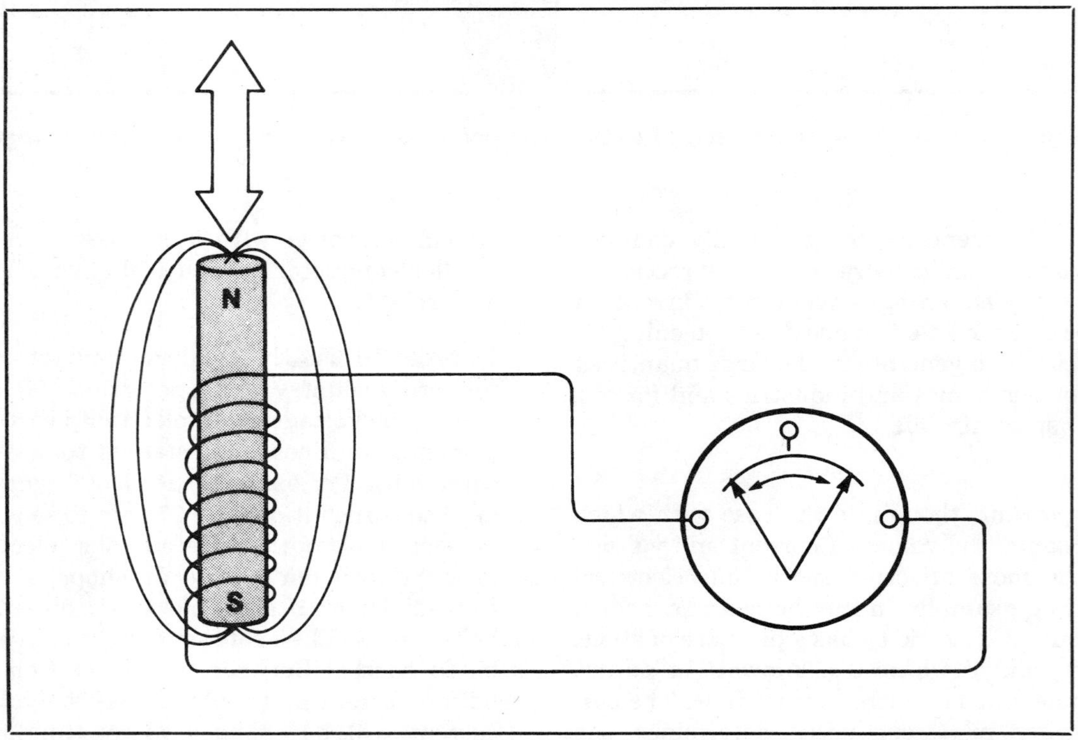

Fig. 2-1 Current will flow in a conductor when lines of magnetic flux from a permanent magnet cut across it.

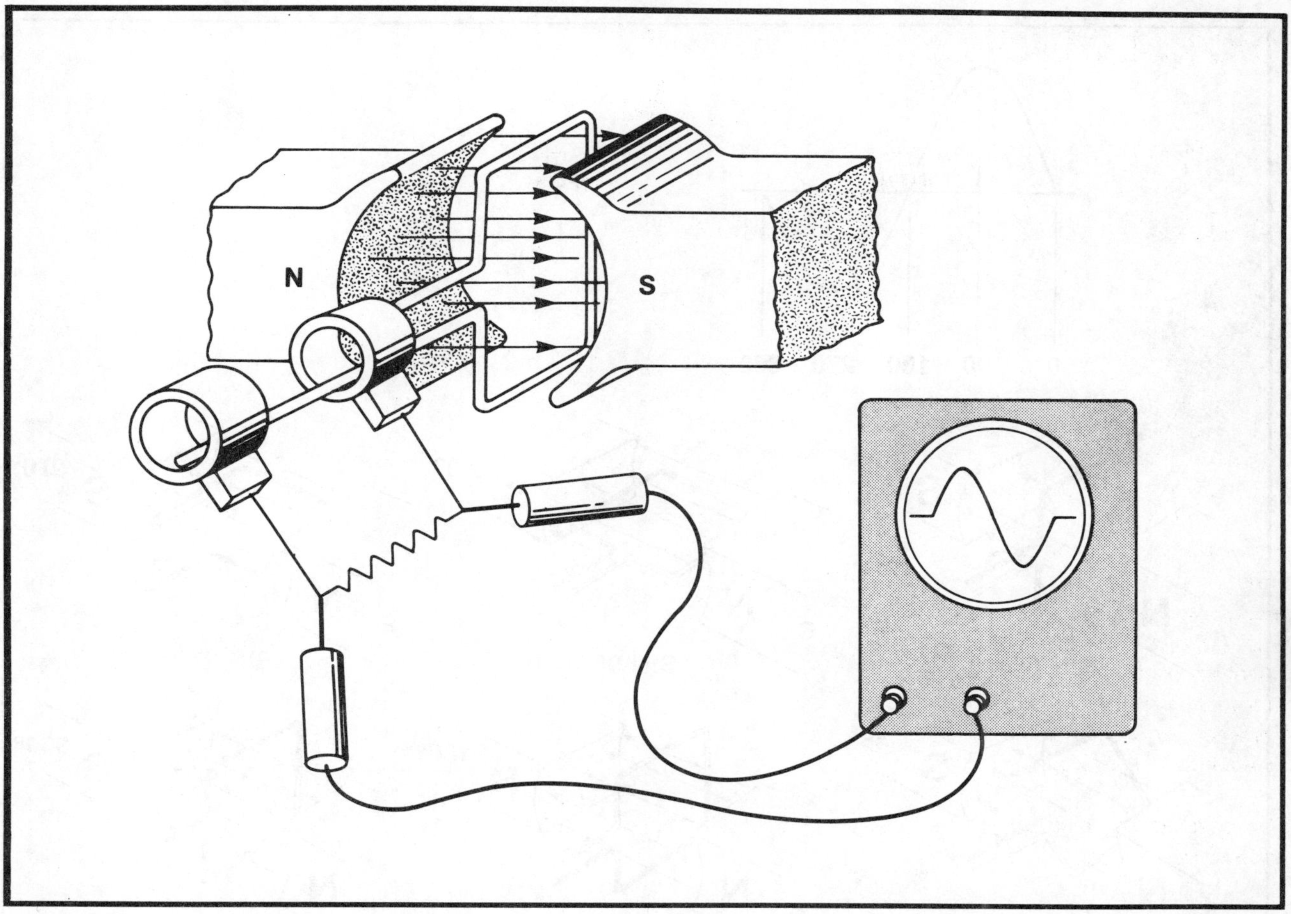

Fig. 2-2 Alternating current is produced in a conductor when it is rotated within a magnetic field.

complete cycle of voltage changes for each complete revolution of the coil.

The wave form of alternating current or voltage produced by a rotary generator is called a sine wave, because the voltage or current changes according to the sine of the angle through which the generator has rotated.

The sine of an angle is the relationship between the length of the hypotenuse of a right triangle to the length of the side opposite the angle we are considering. When the coil is in position *A* of Fig. 2-3, it does not cut any lines of flux from the magnet as it moves, so no voltage is generated. This is our starting point, or the zero-degree angle.

By the time the coil rotates 45 degrees (point *B*), it cuts across some of the lines of flux, and the voltage generated at this point is 0.707 times the peak amount. As the coil continues to rotate, it reaches point *C* where it cuts across a maximum number of lines of flux for each degree it rotates, and here it produces the peak voltage. Further rotation decreases the number of lines of flux cut for each degree of rotation until, at point *D*, the coil is cutting no flux, and the output is again zero.

Rotation beyond this point brings the opposite side of the coil down through the flux near the south pole of the magnet, and the voltage builds up in the opposite direction. The change in voltage is continuous and smooth, and its instantaneous value may be found by multiplying its peak value by the sine of the angle through which the generator has rotated since its zero-voltage position.

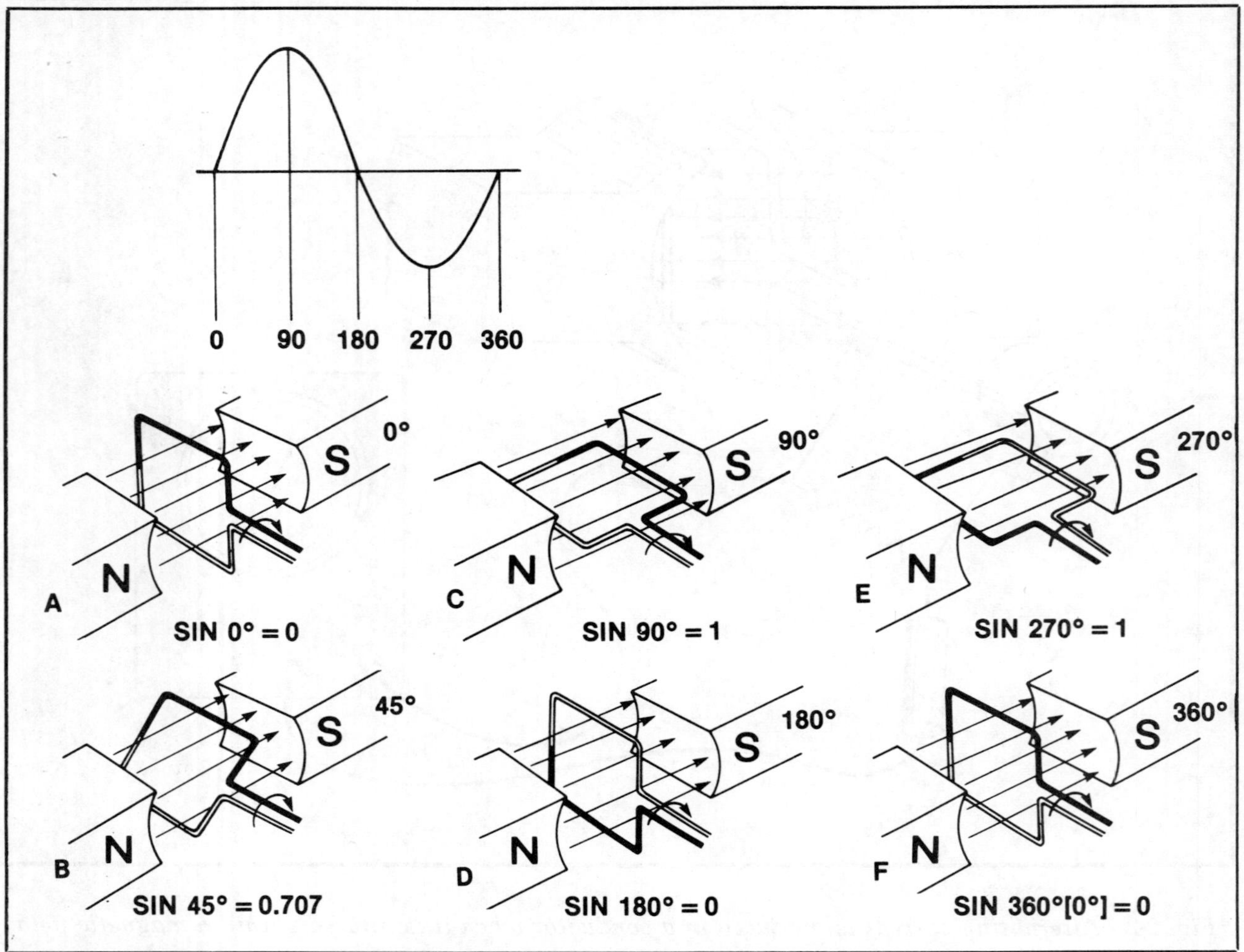

Fig. 2-3 Production of sine wave alternating current

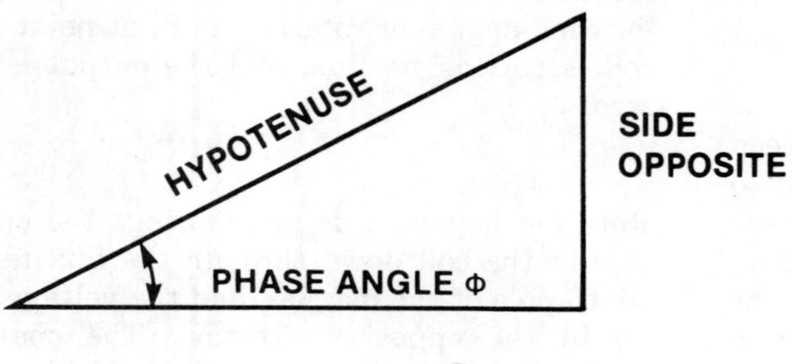

$$\text{SIN } \Phi = \frac{\text{SIDE OPPOSITE}}{\text{HYPOTENUSE}}$$

ANGLE	SINE
0	0
45	+ 0.707
90	+ 1.000
135	+ 0.707
180	0
225	− 0.707
270	− 1.000
315	− 0.707
360	0

Fig. 2-4 The sine of the phase angle is the ratio of the length of the side opposite the angle to the length of the hypotenuse.

SECTION III:

Alternating Current Terms and Values

A. *Cycle*

A cycle is one complete sequence of voltage or current changes from zero through a positive peak to zero, then through a negative peak, back to zero where it can start over and repeat the sequence.

B. *Alternation*

An alternation is one-half of an AC cycle in which the voltage or current rises from zero to a peak and back to zero.

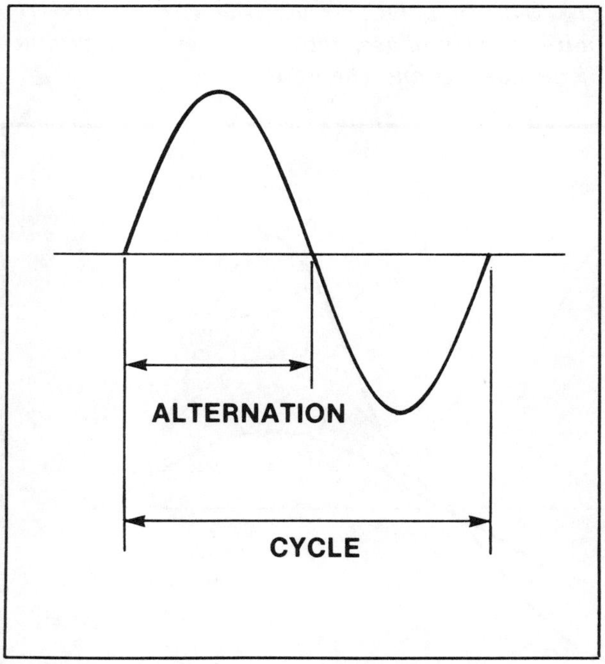

Fig. 3-1 Sine wave values

C. *Period*

The time required for one cycle of events to occur is known as the period of the alternating current or voltage.

D. *Frequency*

The number of complete cycles per second is the frequency of the AC, and it is express in hertz. One hertz is one cycle per second. The frequency of the alternating current produced by a generator is determined by the number of *pairs* of magnetic poles in the generator and the speed in revolutions per minute of the rotating coils. Frequency may be found by the formula:

$$\text{Frequency (Hz)} = \frac{\text{Poles}}{2} \times \frac{\text{RPM}}{60}$$

The frequency of commercial alternating current in the USA is 60 hertz, while in some foreign countries it is 50 hertz. The AC power used in most aircraft is 400 hertz, for reasons we will discover later.

E. *Phase*

In a DC circuit, a change in the voltage will cause an immediate change in the current, and so the voltage and current are said to be in phase.

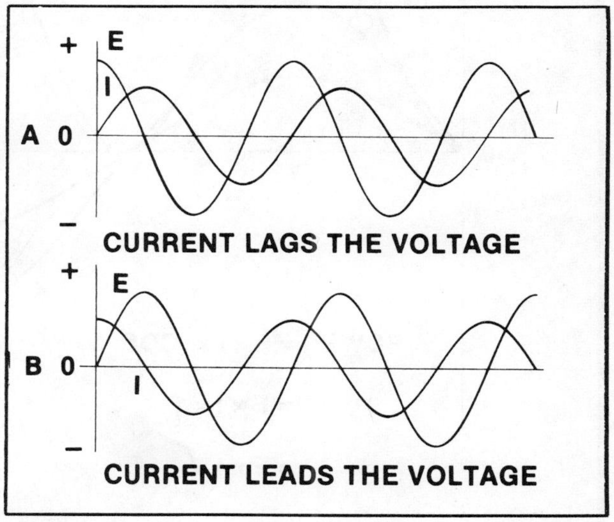

Fig. 3-2 A — The current lags behind the voltage. B — The current leads the voltage.

In alternating current where the values are constantly changing, certain circuit components cause a phase shift between the voltage and the current. Some components cause the current to change before the voltage changes, and the current is said to *lead* the voltage, while other components cause the voltage to change before the current, and the current is then said to *lag* the voltage.

F. Power

1. Apparent power

You will remember in our study of direct current electricity, we saw that electrical power is the product of the voltage and the current and it is measured in watts. In alternating current this is not true, because the current is not necessarily in phase with the voltage, and the product of the voltage and the current is the *apparent* power and is expressed in volt-amps rather than in watts.

If the current is in phase with the voltage, we have the condition we see in Fig. 3-4B. The power developed at any instant is the product of the voltage and the current, and as long as the voltage and the current are in phase, the power will be positive; that is, the generator is supplying power to the load. We will remember from our study of multiplication of signed numbers that when we multiply signed numbers having like signs, whether they are positive or negative, the product will be positive.

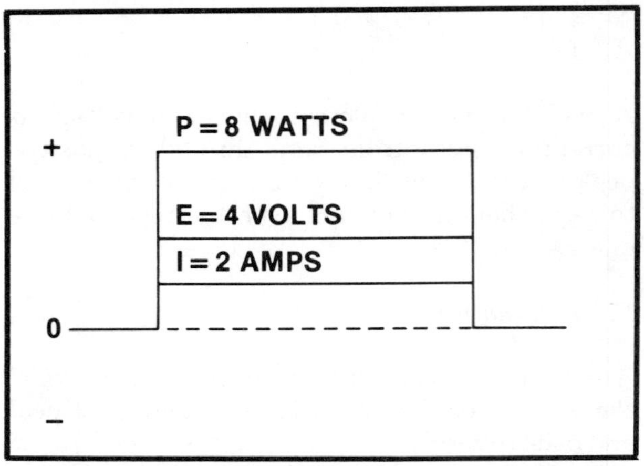

Fig. 3-3 In a DC circuit, the current and the voltage are in phase, and the power is the product of the current and the voltage.

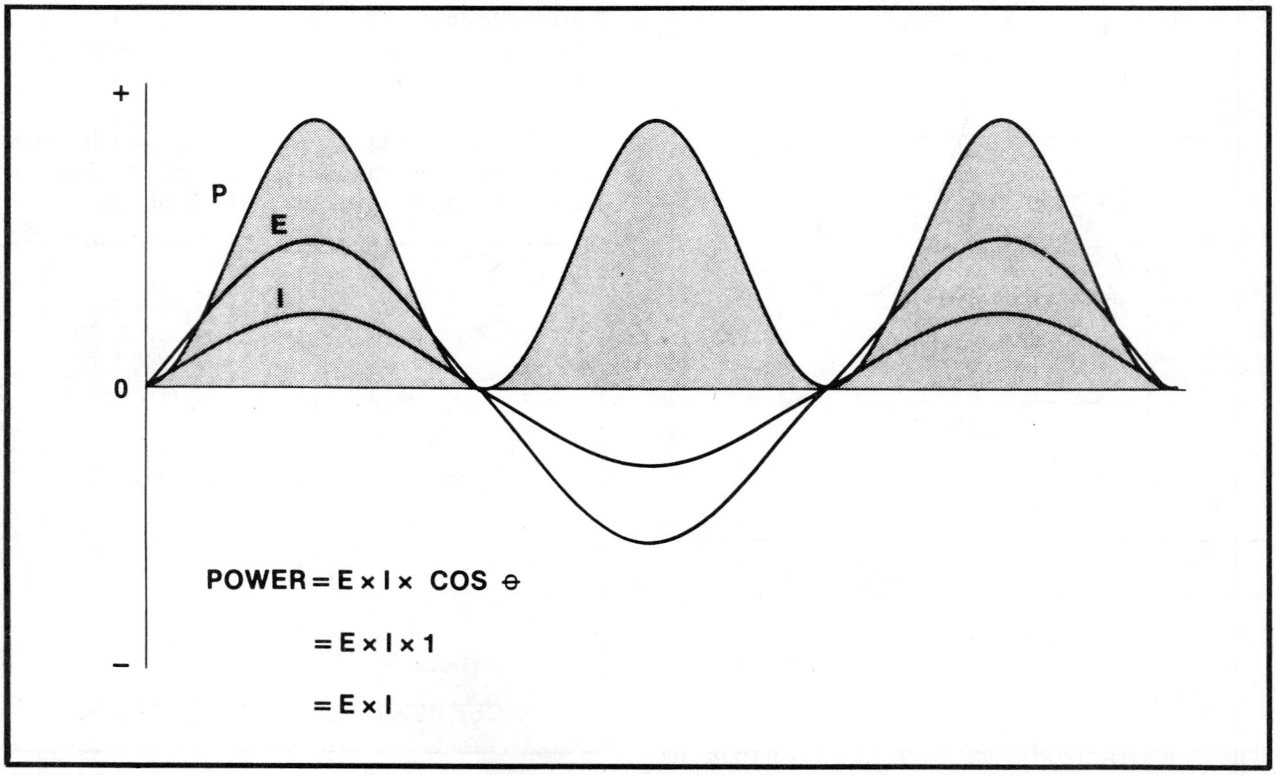

Fig. 3-4 In a purely resistive AC circuit, the current and the voltage are in phase, and the power is the product of the current and the voltage.

2. True power

If the current and the voltage are not continually in phase — that is, if the current either leads or lags the voltage — there will be at least part of the cycle in which the voltage is positive while the current is negative. The product of unlike signed numbers is always negative and the power that is produced during this portion of the cycle is negative power; that is, the load will be forcing power back into the generator.

True power is expressed in watts and is the product of the voltage and only that portion of the current that is in phase with the voltage.

3. Power factor

"Power factor" is a term used to indicate the amount of the current that is in phase with the voltage, and it may be found as the ratio between the true power and the apparent power. If the power factor is 0.5, only 50% of the current is in phase with the voltage. If all of the current is in phase with the voltage, as it is in a circuit having no opposition other than resistance, the power factor will be 1.0, meaning that all of the current is in phase with the voltage.

If the amount of phase shift between the voltage and the current is known, we can find the power factor, as it is the cosine of the phase angle. In Fig. 3-5, we see that in a right triangle, the ratio of the length of the side adjacent the phase angle to the length of the hypotenuse is the cosine of the angle. When the current and the voltage are in phase, or the phase angle is zero, the side adjacent the zero degree-angle and the hypotenuse are the same length, and so the ratio is 1.00. The real power is equal to the apparent power. For a 45° phase angle, the length of the side adjacent (the true power) will be only 0.707 as long as the hypotenuse (the apparent power). When the current and the voltage are 60° out of phase, the true power is only 0.50 as much as the apparent power. And if the current and voltage are 90° out of phase, the adjacent side will have a length of zero, and the power factor will also be zero. There can be no real power produced in the circuit, even though voltage is present and current is flowing.

The true power in a circuit is found by multiplying the product of the voltage and the current by the power factor.

$$\text{Power (watts)} = E \times I \times \text{power factor}$$

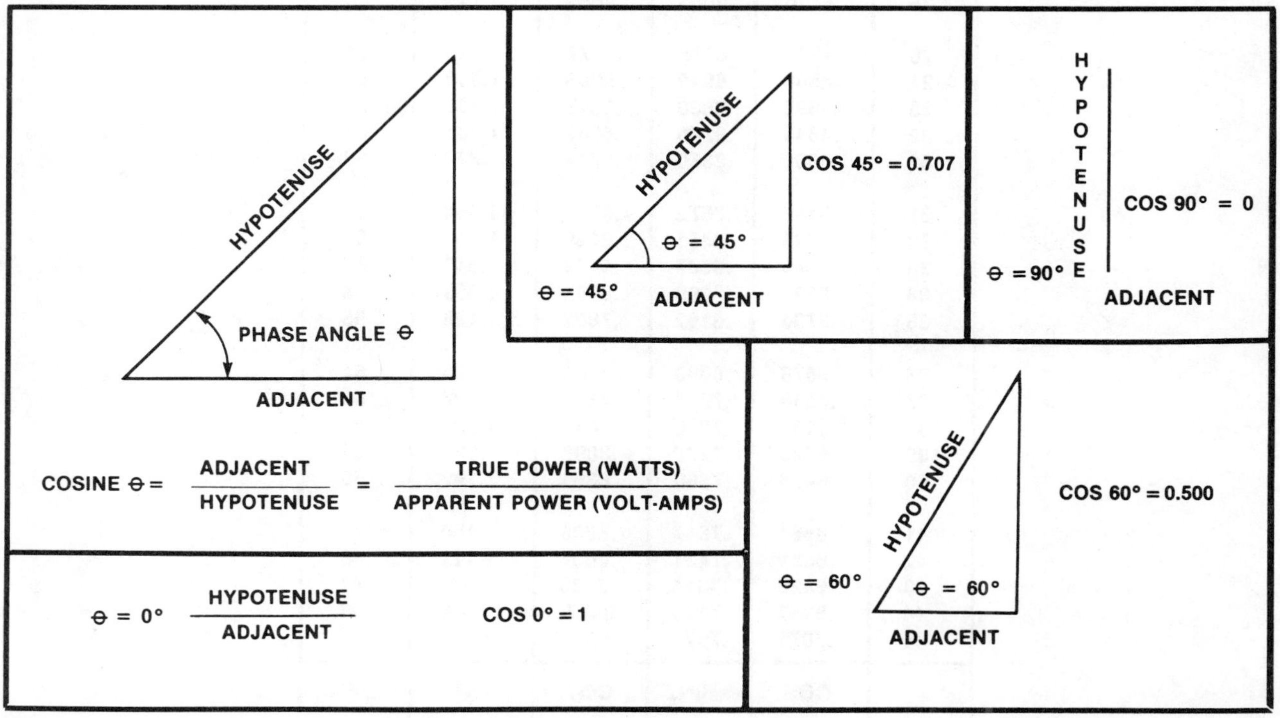

Fig. 3-5 The power factor of an AC circuit is the cosine of the phase angle, which is the ratio of the length of the side adjacent the angle to the length of the hypotenuse.

DEG.	SIN.	COS.	TAN.	COT.	
0	.0	1.000	.0	+ ∞	90
1	.0175	.9999	.0175	57.29	89
2	.0349	.9994	.349	28.64	88
3	.0523	.9986	.0524	19.08	87
4	.0698	.9976	.0699	14.30	86
5	.0872	.9962	.0875	11.43	85
6	.1045	.9945	.1051	9.514	84
7	.1219	.9926	.1228	8.144	83
8	.1392	.9903	.1405	7.115	82
9	.1564	.9877	.1584	6.314	81
10	.1737	.9848	.1763	5.671	80
11	.1908	.9816	.1944	5.145	79
12	.2079	.9744	.2309	4.331	78
13	.2250	.9744	.2309	4.331	77
14	.2419	.9703	.2493	4.011	76
15	.2588	.9659	.2680	3.732	75
16	.2756	.9613	.2868	3.487	74
17	.2924	.9563	.3057	3.271	73
18	.3090	.9511	.3249	3.078	72
19	.3256	.9455	.3443	2.904	71
20	.3420	.9397	.3640	2.747	70
21	.3584	.9336	.3839	2.605	69
22	.3746	.9272	.4040	2.475	68
23	.3907	.9205	.4245	2.356	67
24	.4067	.9136	.4452	2.246	66
25	.4426	.9063	.4663	2.145	65
26	.4384	8988	.4877	2.050	64
27	.4540	.8910	.5095	1.963	63
28	.4695	.8830	.5317	1.881	62
29	.4848	.8746	.5543	1.804	61
30	.5000	.8660	.5774	1.732	60
31	.5150	.8572	.6009	1.664	59
32	.5299	.8481	.6249	1.600	58
33	.5446	.8387	.6494	1.540	57
34	.5592	.8290	.6745	1.483	56
35	.5736	.8192	.7002	1.428	55
36	.5878	.8090	.7265	1.376	54
37	.6018	.7986	.7536	1.327	53
38	.6157	.7880	.7813	1.280	52
39	.6293	.7772	.8098	1.235	51
40	.6428	.7660	.8391	1.192	50
41	.6561	.7547	.8693	1.150	49
42	.6691	.7431	.9004	1.111	48
43	.6820	.7314	.9325	1.072	47
44	.6947	.7193	9657	1.036	46
45	.7071	.7071	1.0000	1.000	45
	COS.	SIN.	COT.	TAN.	DEG.

Fig. 3-6 Trigonometric function table

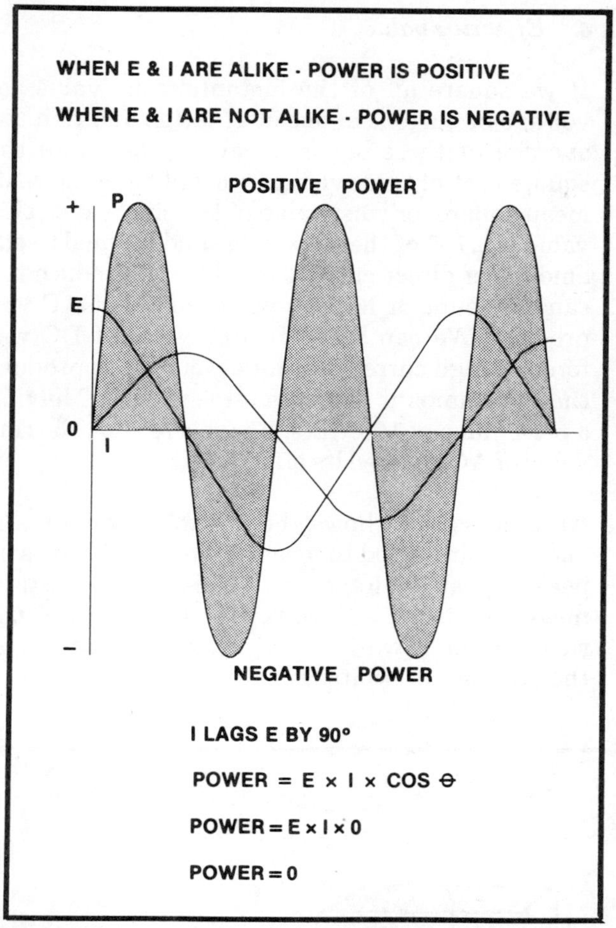

Fig. 3-7 The negative power equals the positive power when the current and the voltage are 90 degrees out of phase.

As we will see very soon, circuits that have inductive or capacitive reactance as the chief opposition to the current flow have very low power factors and while a great deal of current flows, very little true power is produced.

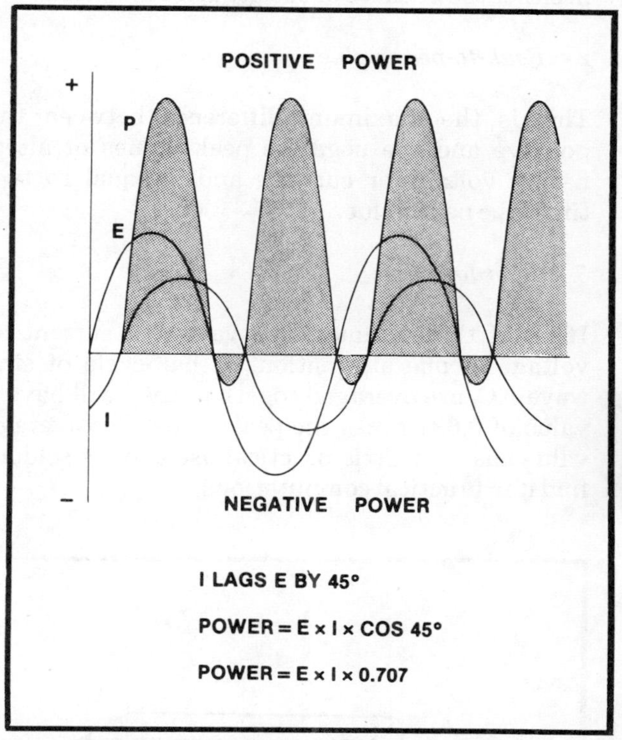

Fig. 3-8 When the voltage and current are 45 degrees out of phase, the true power is 0.707 times the apparent power.

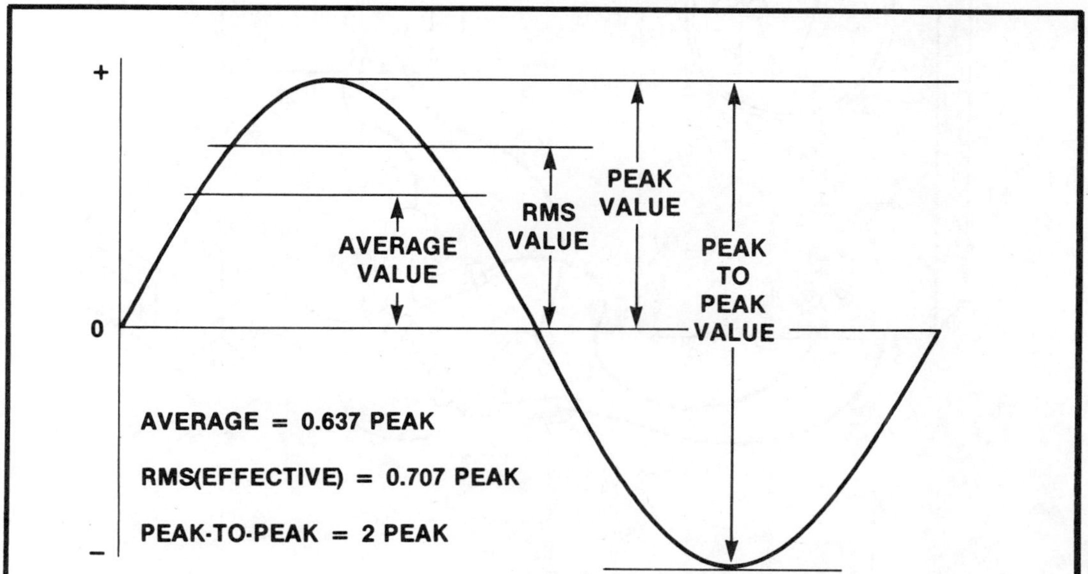

Fig. 3-9 Values of sine wave alternating current

G. *Sine Wave Values*

1. *Peak value*

This is the maximum value of voltage or current in either the positive or the negative direction.

2. *Peak-to-peak value*

This is the maximum difference between the positive and the negative peak values of alternating voltage or current, and is equal to two times the peak value.

3. *Average value*

If all of the instantaneous values of current or voltage in one alternation, or half-cycle of sine wave AC, are averaged together, they will have a value of 0.637 times the peak value. The average value has very little practical use and we seldom find it in practical computations.

4. *Effective value*

If we square all of the instantaneous values of voltage or current in one alternation and find the average of these squared values, then take the square root of this average, we will have the root-mean-square, or rms, value of the sine wave. This value is 0.707 of the peak value and is equal to the amount of direct current required to produce the same amount of heat the peak value of AC will produce. We can see that 70.7 volts of DC will force enough current through a circuit to produce the same amount of heat as sine wave AC having a peak value of 100 volts. For this reason, the rms value of AC is called its effective value.

An oscilloscope shows the full waveform of AC, and when it is used to measure voltage, it gives us peak-to-peak voltage, but this same voltage measured on an AC voltmeter will indicate the rms voltage. Two hundred peak-to-peak volts is the same as 70.7 volts rms.

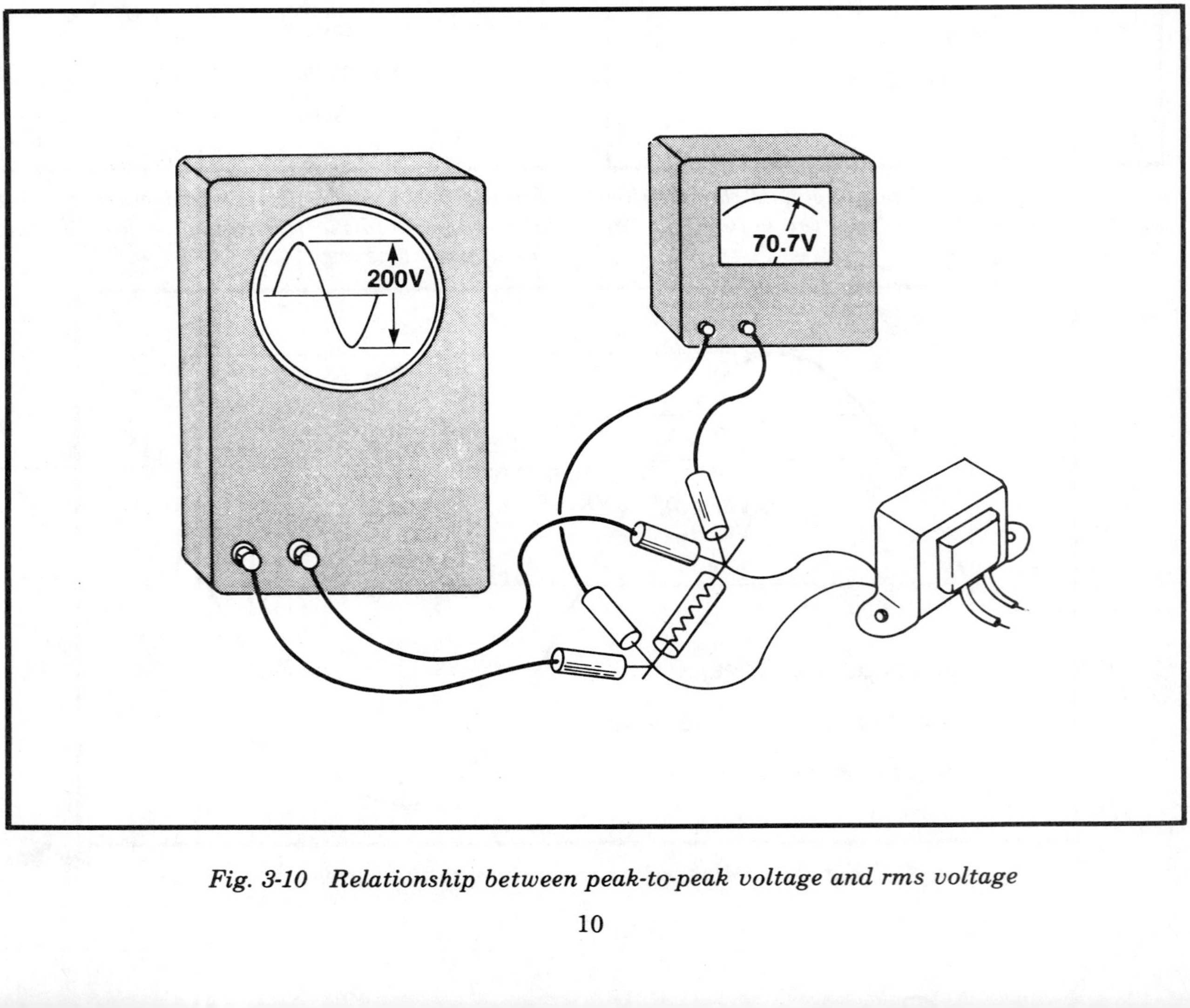

Fig. 3-10 Relationship between peak-to-peak voltage and rms voltage

SECTION IV:

Alternating Current in a Circuit Containing Only Resistance

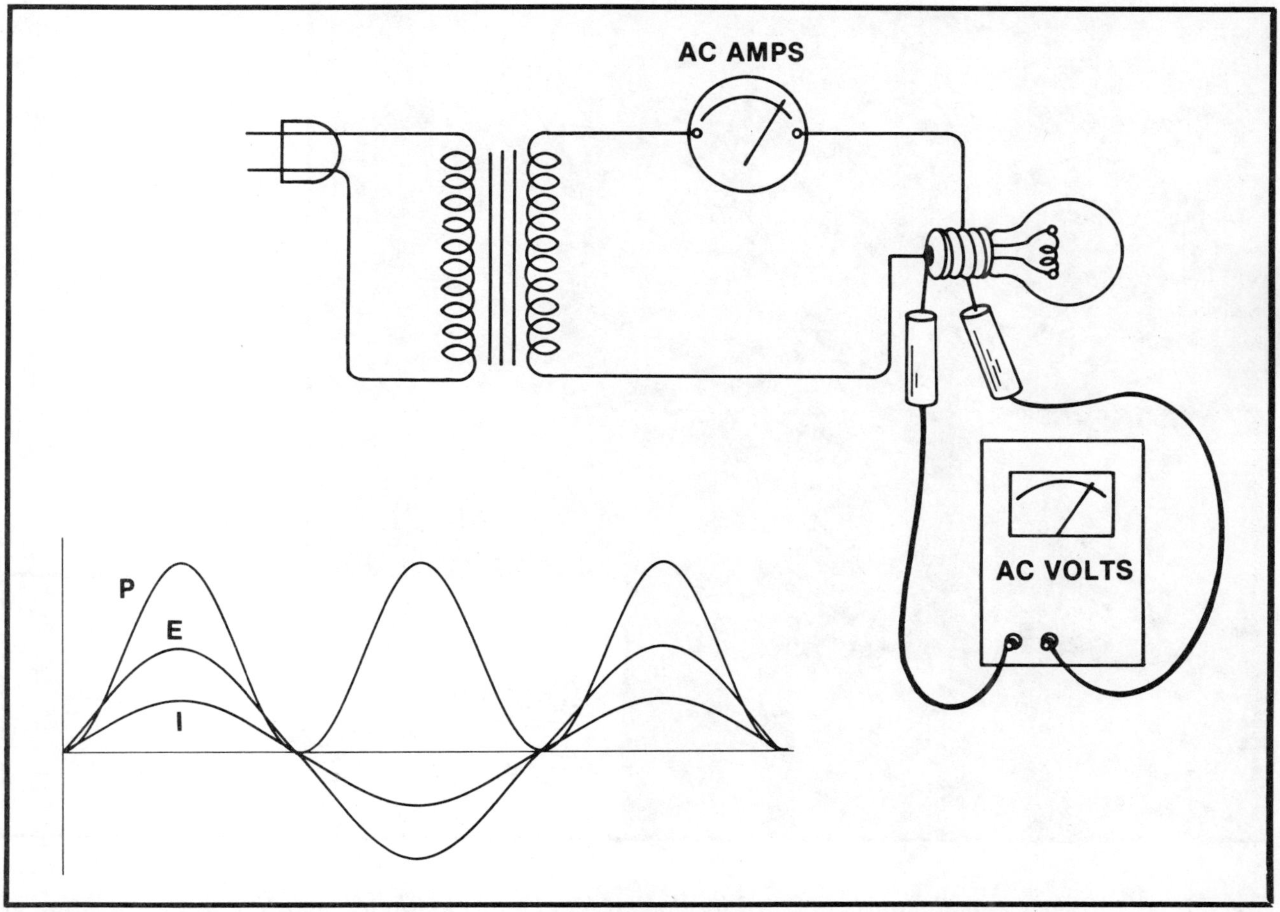

Fig. 4-1 In a purely resistive AC circuit, the current and voltage are in phase.

Circuit components such as light bulbs, heaters, and composition resistors do not normally cause a phase shift in an AC circuit. Circuits containing only this type of devices are called resistive circuits, and the current and voltage will be in phase; that is, they will both pass through zero at the same time, going in the same direction.

In a purely resistive circuit, the power factor is one, and so the apparent power and the true power are the same. The power in watts is the product of the effective value of the voltage times the effective value of the current.

SECTION V:

Inductance

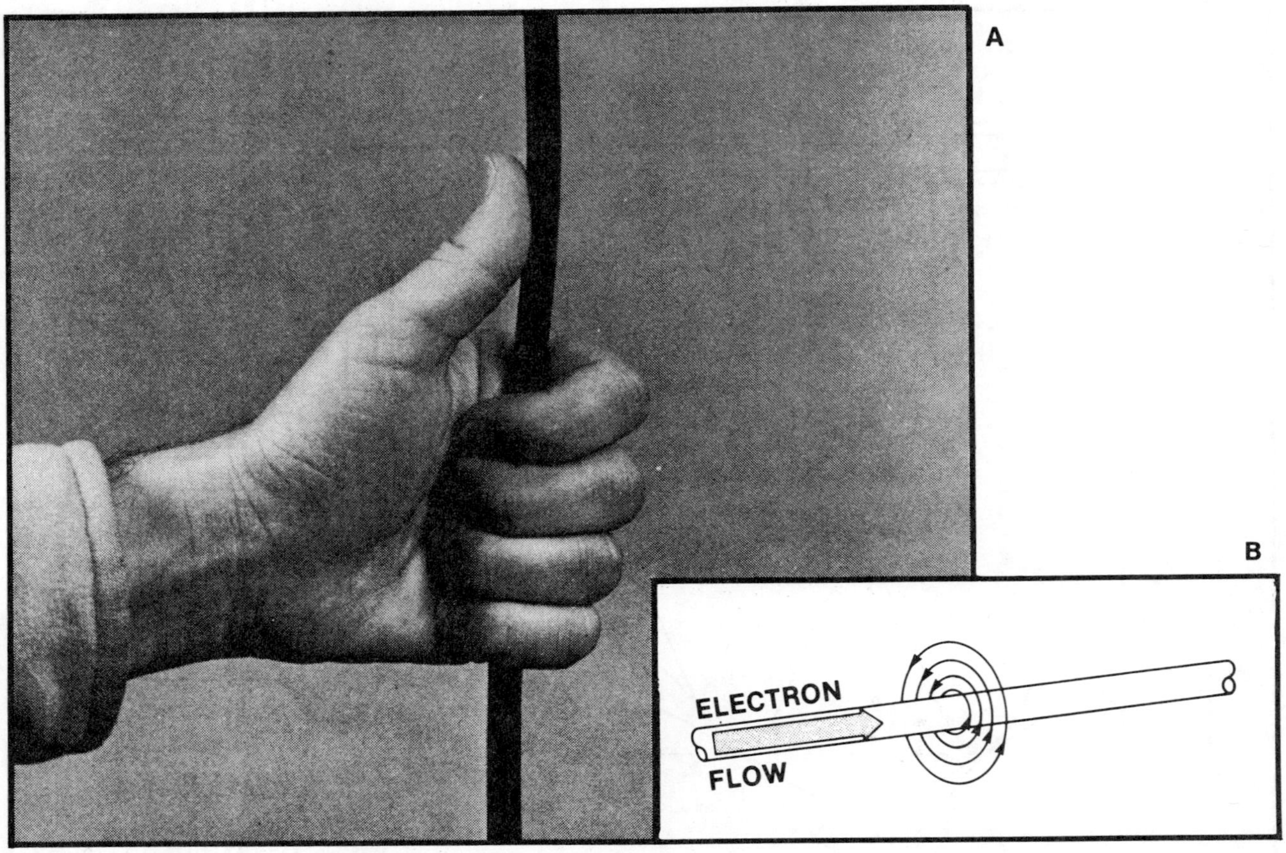

Fig. 5-1 A — When a conductor is grasped by the left hand with the thumb pointing in the direction of electron flow, from negative to positive, the fingers will encircle the conductor in the same direction as the lines of magnetic flux. B — The relationship between the direction of electron flow and the lines of magnetic flux

Anytime current flows in a conductor, a magnetic field surrounds it and the strength of this field is determined by the amount of current flow. The direction of the lines of flux around the conductor may be found by the left-hand rule for generators, which states that if the conductor is held in the left hand so that the thumb points in the direction of electron flow (from the negative to the positive terminal of the source), the fingers will encircle the conductor in the direction of the lines of flux.

As the amount of current flow changes, the magnetic field expands or contracts, and as it does, the flux cuts across the conductor and induces a voltage into it. According to Lenz's law, the voltage that is induced into the conductor is of such a polarity that it opposes the change that caused it. For example, as the voltage begins to rise and the current increases, the expanding lines of flux cut across the conductor and induce a voltage into it that opposes, or slows down, the rise. When the current flow in the conductor is

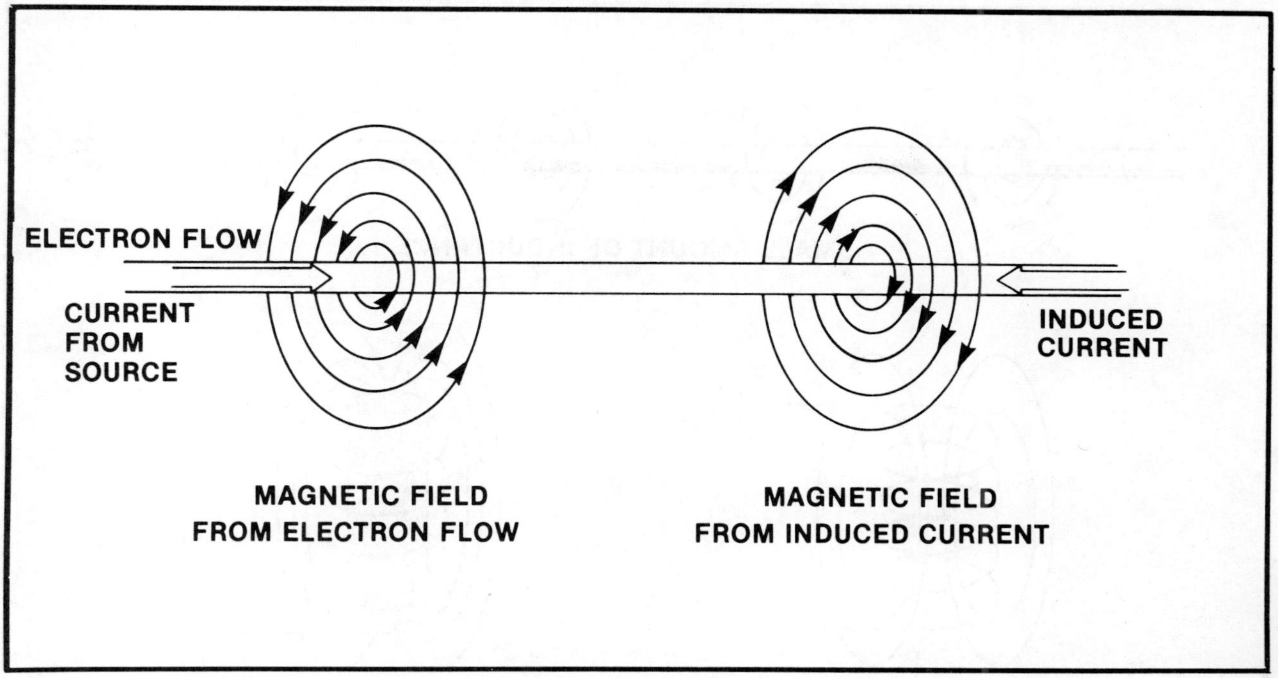

Fig. 5-2 The magnetic field from the induced current encircles the conductor in the direction opposite that of the magnetic field produced by the source current.

steady, lines of flux surround it, and since there is no change in the amount of current, these lines do not cut across the conductor, and so there is no voltage induced into it. When the current decreases, the lines of flux cut across the conductor as they collapse, and they induce into it a voltage that opposes the decrease.

When a conductor carries alternating current, both the amount and the direction of the current continually change, and so an opposing voltage is constantly induced into the conductor. This induced voltage acts as an opposition to the flow of current and we will discuss it in detail under its proper name, inductive reactance.

A. Factors Affecting Inductance

Inductance opposes a change in current by the generation of a back voltage, and all conductors have the characteristic of inductance, since they all generate back voltage any time the current flowing in them changes. The amount of inductance is increased by anything that concentrates the lines of flux, or causes more of the flux to cut across the conductor. If the conductor is formed into a coil, the lines of flux surrounding any one of the turns cut not only across the conductor itself, but also across each of its turns, and so it generates a much greater induced current to oppose the source current.

If a soft iron core is inserted into the coil, it will further concentrate the lines of flux and cause a still higher induced current, which allows less source current to flow.

Inductance is measured in henries, and one henry will generate one volt of induced voltage when the current changes at the rate of one ampere per second.

The inductance of a coil is determined by the number of turns in the coil; by the spacing between the turns; by the number of layers of winding; and by the wire size. The ratio of the diameter of the coil to its length and the type of material used in the core also affects the amount of inductance in the coil. Since all of these factors are variable, there is no simple formula we can use to find the inductance of a coil.

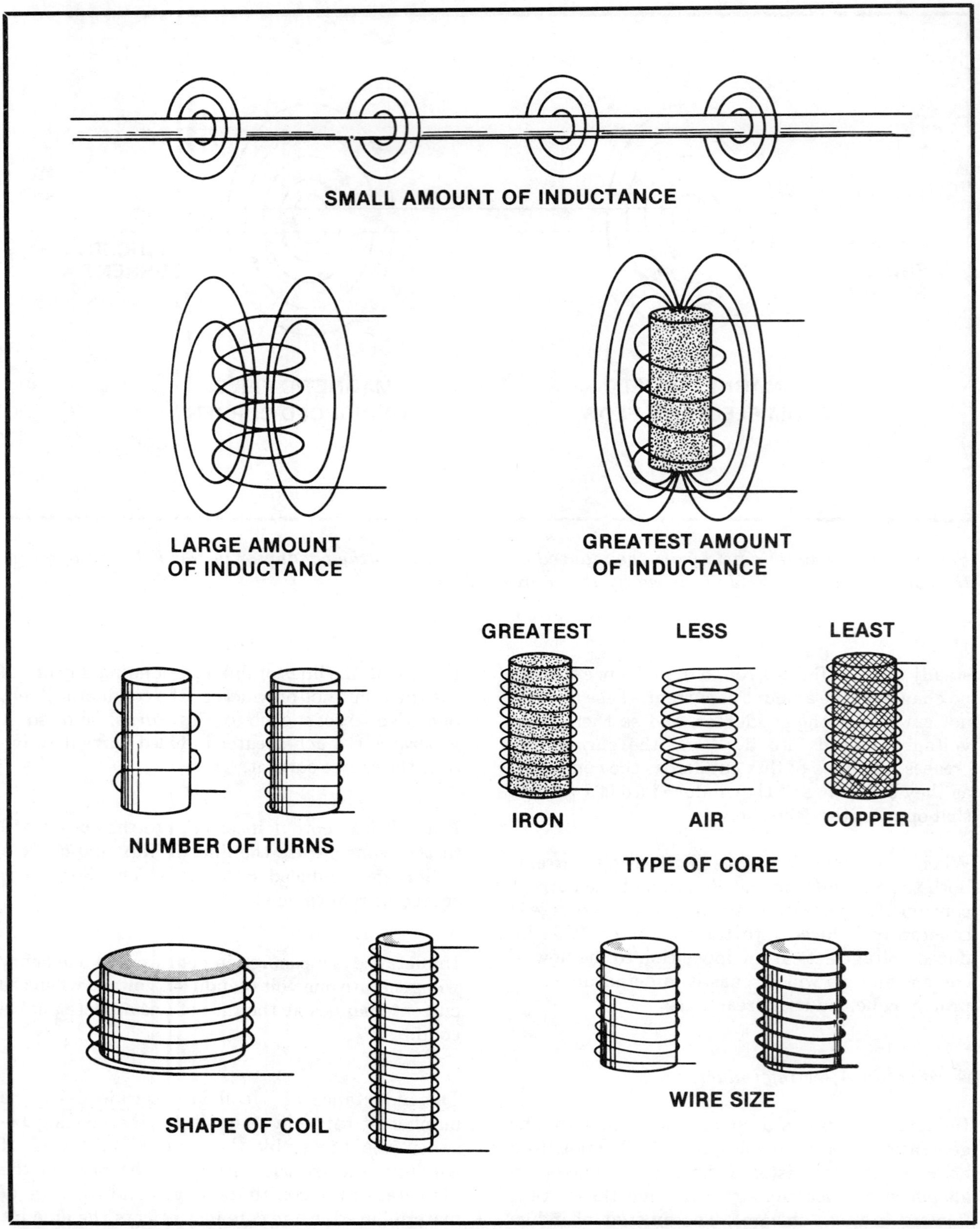

Fig. 5-3 The amount of inductance of a conductor may be increased by forming it into a coil, and the inductance of the coil is determined by the number of turns, the shape of the coil, the wire size, and the type of core.

B. *Series and Parallel Inductors*

It is sometimes necessary to use more than one inductor, or coil, in a circuit. If they are connected in series in such a way that the changing magnetic field of one does not affect the others, the total inductance will be equal to the sum of their individual inductances.

$$L_T = L_1 + L_2 + L_3 \ldots$$

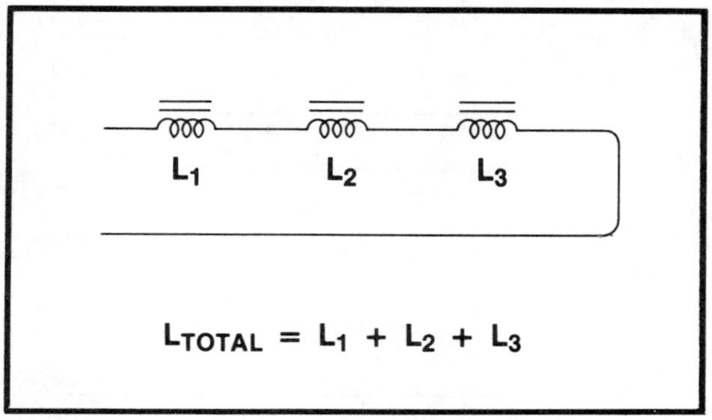

Fig. 5-4 Inductances in series

When inductors are connected in parallel, the total inductance will be less than that of any of the individual inductors. The formulas used for finding the total inductance of parallel connected inductors are the same as those used for finding the total resistance of parallel connected resistors. If the inductors are all the same, the total inductance may be found by dividing the inductance, in henries, of one inductor by the number of inductors:

$$L_T = \frac{L}{n}$$

If there are only two inductors, the total inductance may be found by dividing the product of the two inductors by their sum:

$$L_T = \frac{L_1 \times L_2}{L_1 + L_2}$$

When more than two inductors having different amounts of inductance are connected in parallel, the formula used is the same as that for finding the total resistance of several unlike resistors connected in parallel; the reciprocal of the sum of the reciprocal of the inductances:

$$L_T = \frac{1}{\frac{1}{L_1} + \frac{1}{L_2} + \frac{1}{L_3} \ldots}$$

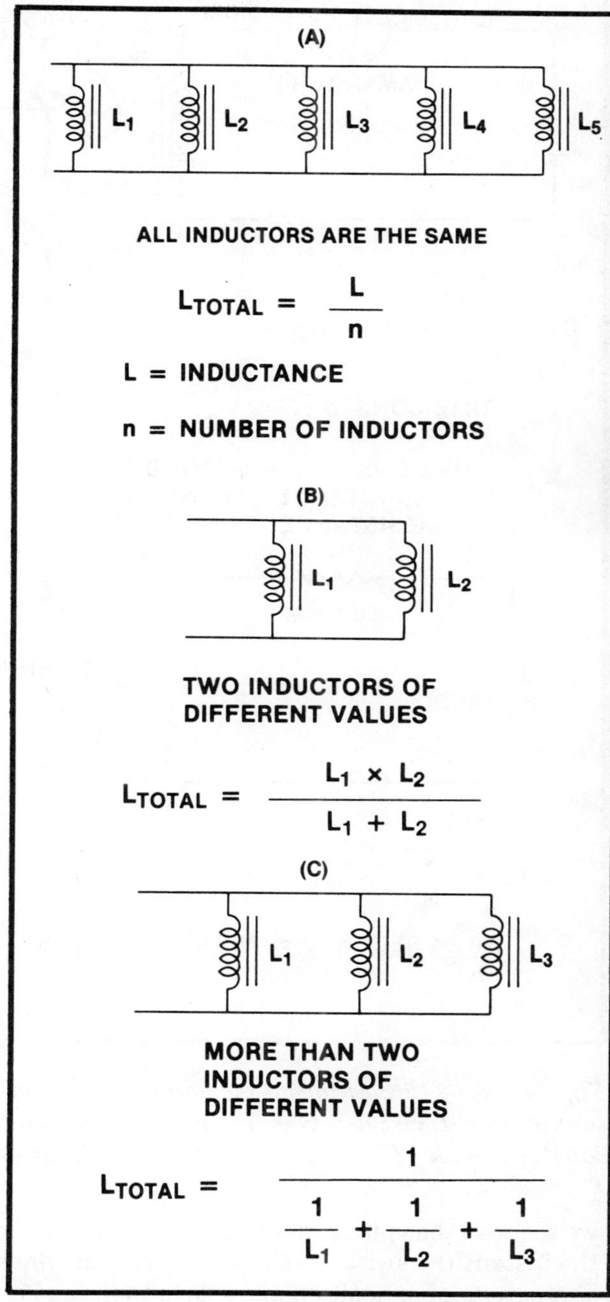

Fig. 5-5 Inductances in parallel

C. *Time Constant of Inductors*

If a source of direct current is placed across an inductor and controlled by a switch, S_1, Fig. 5-6A,

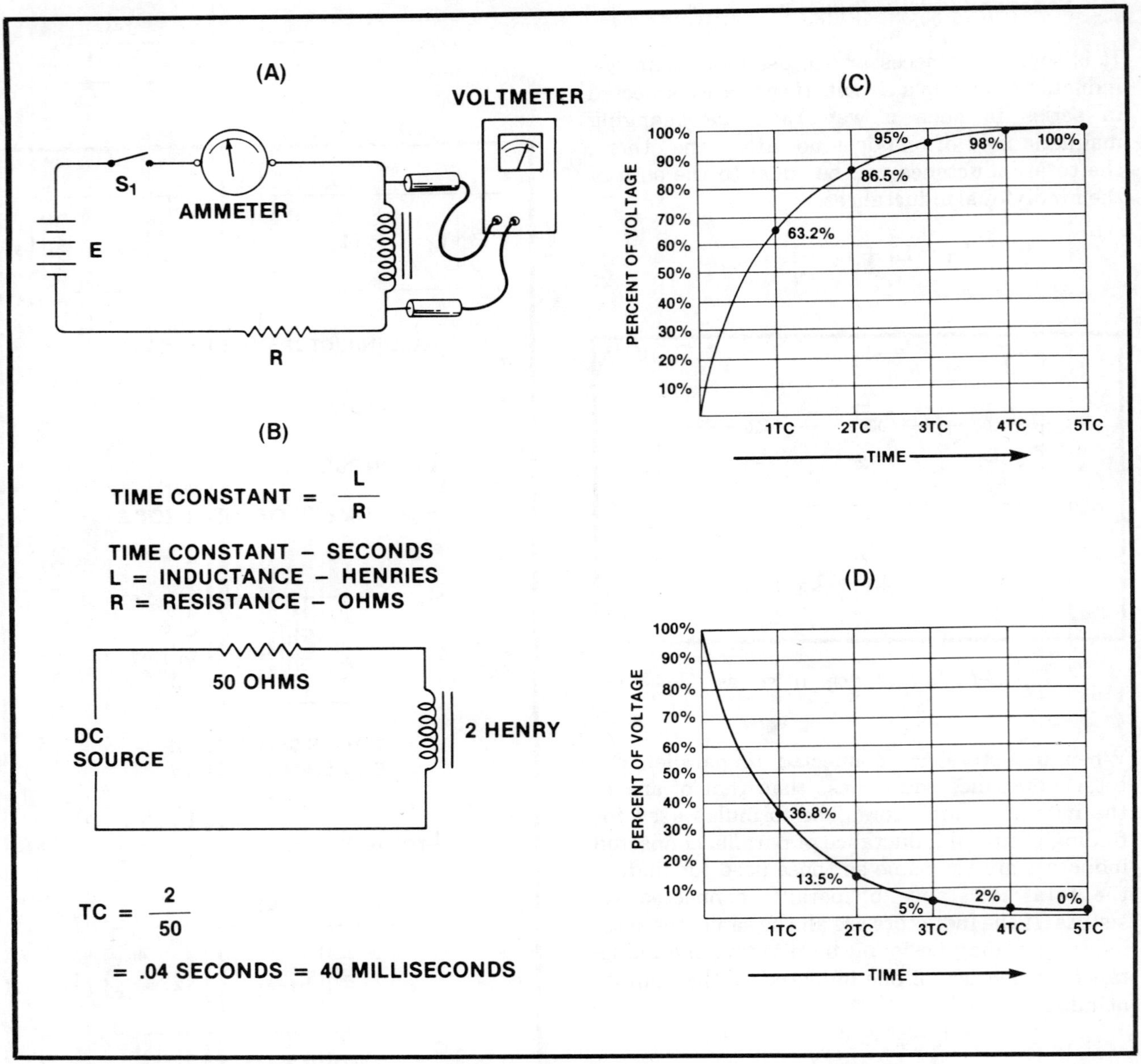

Fig. 5-6 A — Circuit for measuring the time constant of an R-L circuit B — Computing the time constant of an R-L circuit C — Time constant curve of an R-L circuit as the voltage builds up D — Time constant curve of an R-L circuit as the voltage drops off

we see that the voltage does not rise instantly. At the instant the switch is closed, the current finds a minimum of opposition and starts to flow at its greatest rate. But this great change in the rate of current flow from zero to maximum induces a maximum back voltage that opposes the current and causes the voltage rise to decrease. The time required for the voltage to rise 63.2% of its peak value is known as the time constant of the circuit and is determined by the value of inductance and resistance in the circuit.

$$\text{Time Constant} = \frac{\text{L (henries)}}{\text{R (ohms)}}$$

For example, a circuit containing 2.0 henries of inductance and 50 ohms of resistance would have a time constant of 0.04 second, which would normally be spoken of as 40 milliseconds.

The voltage will rise to 63.2% of its peak value in a period of time equal to one time constant (0.04

second). In two time constants, or 0.08 second, it will rise to 86.5% of its peak value. In three time constants (0.13 seconds), it will rise to 95%, in four (0.16 seconds) to 98%, and it will take five time constants or 0.2 second for the voltage to rise to the full value of the source.

When the switch is opened, the current *starts* to drop off immediately, but the changing current induces a voltage that opposes the change and so drop-off is slowed down. In a period of time equal to one time constant, it will drop off to 36.8% of its peak value. In two time constants, it will drop to 13.5%; in three, to 5%; in four time constants, to 2%; and it will take five time constants for it to drop off completely to zero.

D. *Inductive Reactance*

The opposition to the flow of alternating current caused by the generation of a back voltage as the magnetic flux cuts across the conductor is called inductive reactance, and its symbol is X_L. It is measured in ohms, and it varies directly as the frequency of the AC that produces it. The formula for inductive reactance is:

$$X_L = 2\pi fL$$

Two pi is the constant 6.28, the frequency is measured in hertz, and the inductance, in henries.

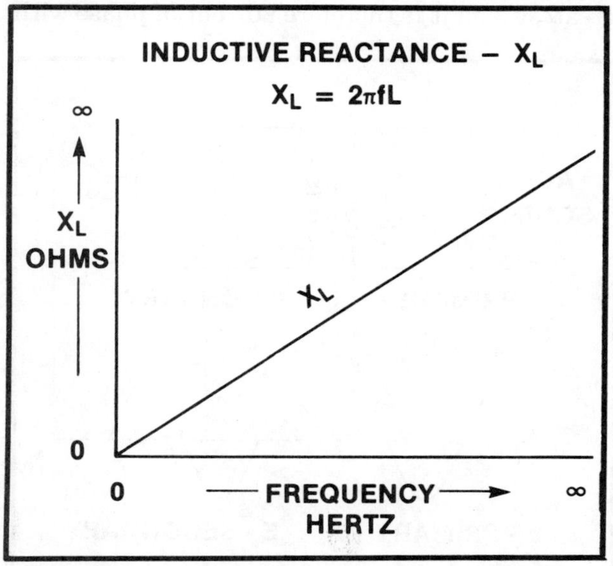

Fig. 5-7 Inductive reactance is a function of the amount of inductance and the frequency of the circuit. X_L increases with an increase in both frequency and inductance.

From this formula, we see that for zero frequency, as we have in direct current, there is no inductive reactance. This is because there is no *changing* magnetic field. But as the frequency increases, the amount of change in the magnetic field increases, and the inductive reactance increases. An infinite frequency would produce an infinite amount of inductive reactance.

Remember that the opposition is caused by the generation of a counter, or back, electromotive force; and unlike the opposition caused by resistance, no heat is generated in the circuit by this opposition, and therefore no power is dissipated.

If the circuit is purely inductive — that is, there is no resistance present — the current will not begin to flow until the voltage has risen to its peak value. The amount of current is determined by the *rate* of change of the voltage, and we can see in Fig. 5-8 that when the voltage is at its peak, there is no change, and so the current is zero. As the voltage begins to drop off, the current rises until — at the point the voltage passes through zero — its *rate* of change is the greatest and the current is the maximum. From this, we can see that in a purely inductive circuit, the change in the current lags the change in voltage by 90 degrees, and with a 90-degree phase angle the power factor is zero.

There can be no true power in a purely inductive circuit, because the negative power equals the positive power. The load returns as much power as it receives from the source.

If an inductor of the proper size is placed in series with a light bulb, the inductive reactance will cause most of the source voltage to be dropped across the inductor, and the bulb will burn very dimly, if at all.

Most aircraft use 400-hertz alternating current, because the inductive reactance at this frequency is high enough to allow smaller transformers and motors to be used. If a transformer designed for 400-hertz AC is used in a 60-hertz circuit, the lower inductive reactance caused by the lower frequency will allow enough current to flow that the transformer will be burned out. But is a 60-hertz transformer is used in a 400-hertz circuit, there will be so much inductive reactance that the efficiency of the transformer will be too low for practical use.

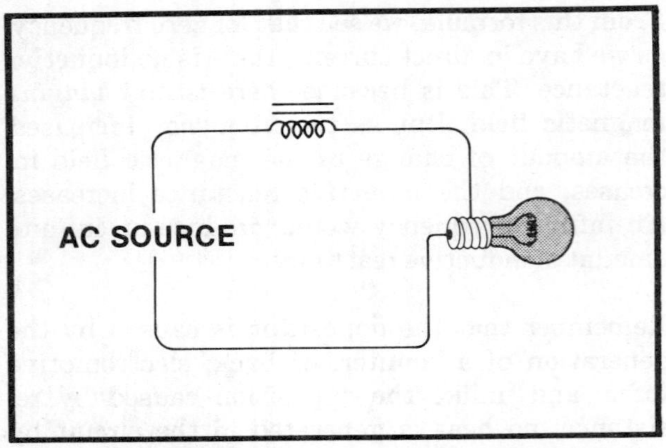

Fig. 5-8 An inductor in series with the load in an AC circuit can prevent current reaching the load.

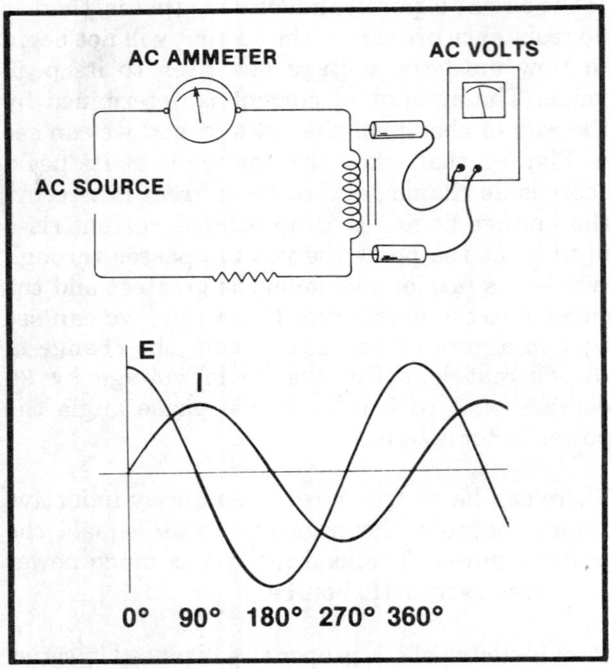

Fig. 5-9 In an inductive circuit, the current lags behind the voltage.

E. Mutual Inductance

When alternating current flows in a conductor, the changing lines of flux radiate out and cut across any other conductor that is nearby, and anytime they cut across a conductor, they generate a voltage in it even though there is no electrical connection between the two. This voltage is said to be generated by mutual inductance, and is the basis for transformer action that is so important to us in our use of alternating current, as it allows us to change the values of AC voltage and current in our circuits.

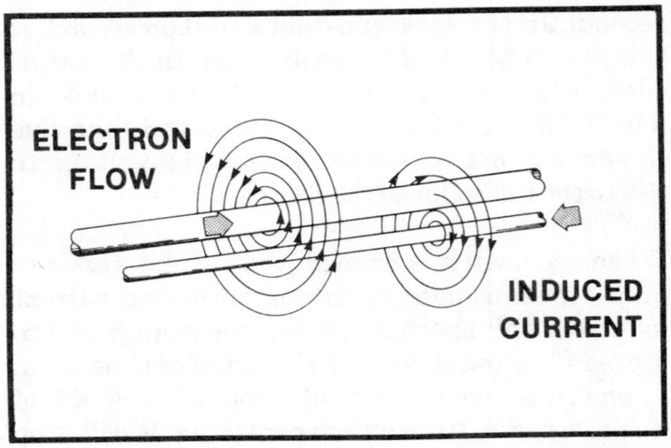

Fig. 5-10 Mutual induction causes a voltage to be induced into a conductor not electrically connected to the conductor through which the source current flows.

Consider in Figure 5-11 two coils of wire, a primary and a secondary, wound around a common core, but not connected electrically.

When an alternating current flows in the primary, a voltage will be induced into the secondary, and current will flow in it.

Since we can consider these windings to be purely inductive, we see that the current in the primary winding lags the source voltage by 90°. The voltage induced into the secondary winding will be greatest when the change in current is the greatest, and it is therefore 90° out of phase with

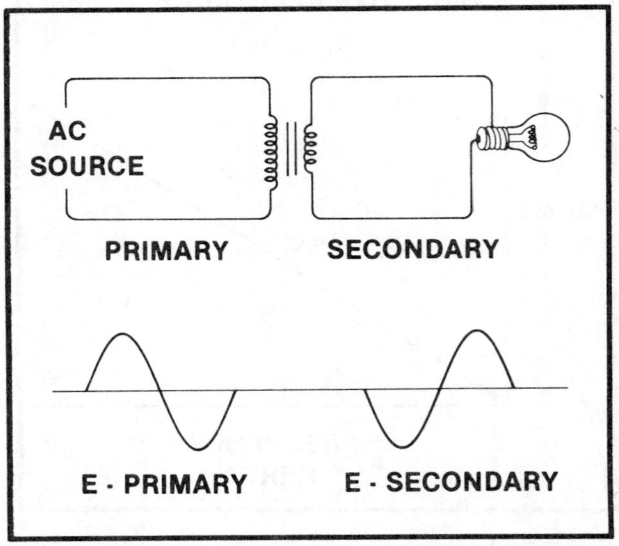

Fig. 5-11 The voltage induced into the secondary of a transformer by mutual induction is 180° out of phase with the source voltage.

the current in the primary. So when the two phase shifts are added, the voltage in the secondary winding will be 180° out of phase with the voltage in the primary.

The amount of voltage generated in the secondary winding of a transformer is equal to the voltage in the primary, times the turns ratio between the primary and the secondary windings. For example, if there are 100 turns in the primary winding and 1,000 turns in the secondary, we have a turns ratio of 1:10, and if there are 115 volts across the primary, there will be 1,150 volts across the secondary. A transformer does not generate any power, so the product of the voltage and the current in the secondary must be the same as that in the primary. Because of this, there must be a flow of one ampere in the primary winding to produce a flow of 100 milliamps in the secondary.

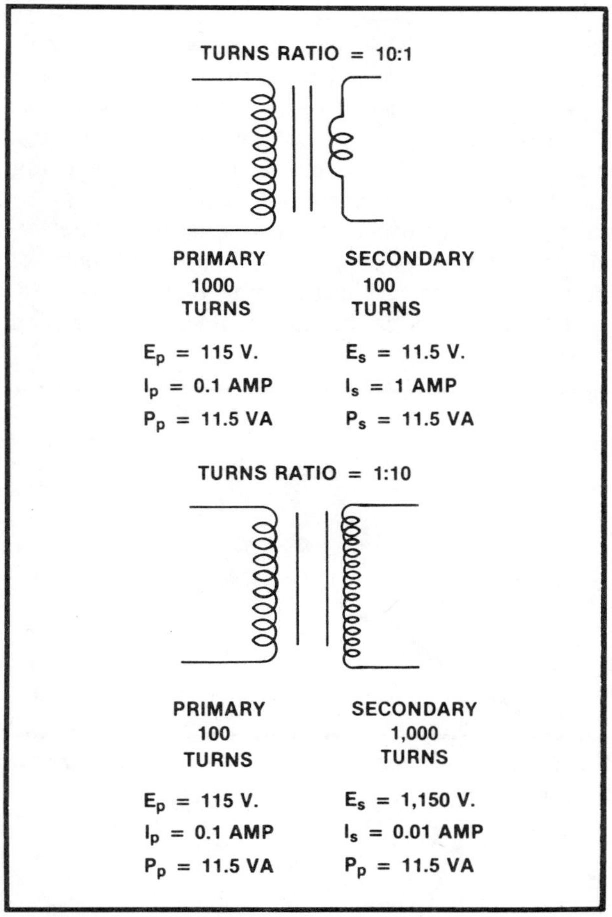

Fig. 5-12 Voltage increases directly as the turns ratio between the primary and secondary windings of a transformer, while the current decreases in proportion to the turns ratio.

A transformer may have its primary winding connected directly across the AC power line, and as long as there is an open circuit in the secondary winding, the back voltage produced in the coil will block the source voltage enough that there will be almost no current flowing through the primary winding. But, as we see in Fig. 5-13, when the push button in the secondary circuit is pressed to complete the circuit for the light, secondary current will flow and its flux will oppose that which created the back voltage so source current will flow in the primary. Only when the switch in the secondary circuit is pressed, will there be any current flow indicated on the AC ammeter in the primary circuit.

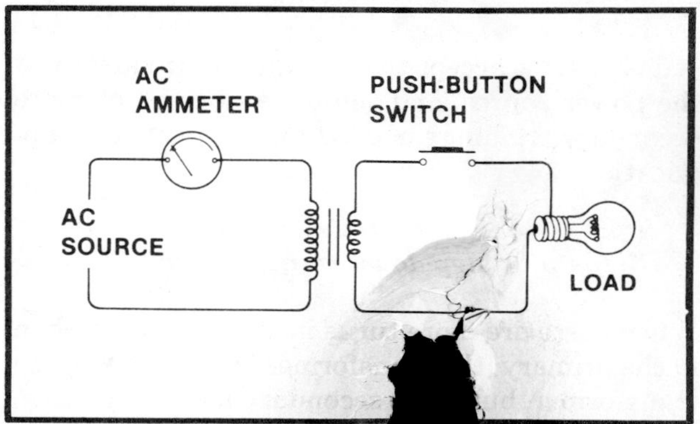

Fig. 5-13 Primary current flows only when the secondary circuit is completed.

F. Types of Practical Inductors

1. Chokes

It is often necessary to install an inductor, or a choke, in a circuit to impede the flow of alternating current of a particular frequency while not affecting the flow of AC below that frequency.

When alternating current in the power frequencies of 50 or 60 hertz is rectified, or changed into direct current, the output is in the form of pulsating direct current, and if an inductor is placed in series with the load, the changing current will induce a back voltage that will tend to smooth out the pulsations, or ripples. Chokes of this type have laminated iron cores and often have an inductance of more than one henry.

2. Transformers

We can get almost any voltage of alternating current by using a transformer. The primary winding

Fig. 5-14 Forms of commonly used inductors

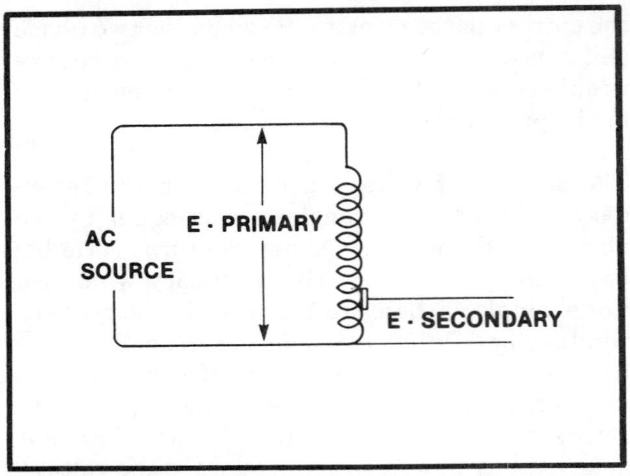

Fig. 5-15 An autotransformer uses all of the winding for the primary and a portion of it for the secondary.

is designed to accept the voltage and frequency of the power source, and there may be one or more secondary windings needed for the particular application.

a. Step-up or step-down transformers

When there are more turns in the secondary than in the primary, the transformer is called a step-up transformer, but if the secondary has fewer turns, it is called a step-down transformer. Step-down transformers are often used to get the high current necessary for operating some motors.

b. Autotransformers

An autotransformer is a form of variable transformer such as we have in Fig. 5-15. There is only one winding, the primary, which is connected across the source. One lead of the secondary is common with the primary, and the other secondary lead is connected to a brush which is movable and makes contact with a bare spot in the primary winding. The amount of secondary voltage is determined by the position of the brush. The greater the number of turns between the two secondary leads, the greater will be the secondary voltage.

c. Current transformers

It is possible to measure the amount of AC current flowing in a line by the use of a clamp-on-type ammeter which uses a current transformer. The probe consists of a coil that can be opened up, and it is clamped over one of the wires carrying the current to be measured. The flow of alternating current induces a much smaller current into the coil that is porportional to the amount of load current flow. The meter that reads this induced current is calibrated in terms of the load current.

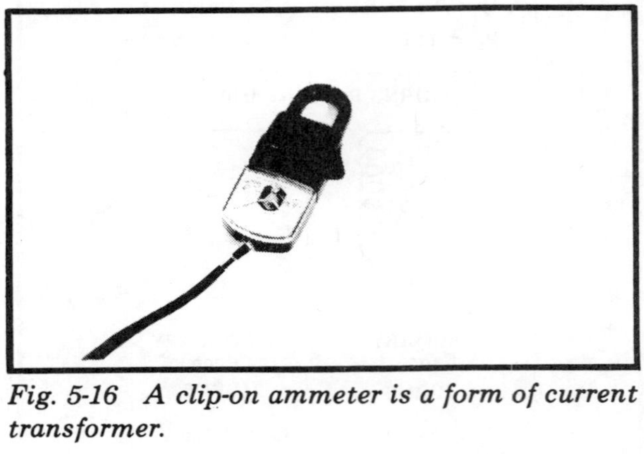

Fig. 5-16 A clip-on ammeter is a form of current transformer.

SECTION VI
Capacitance

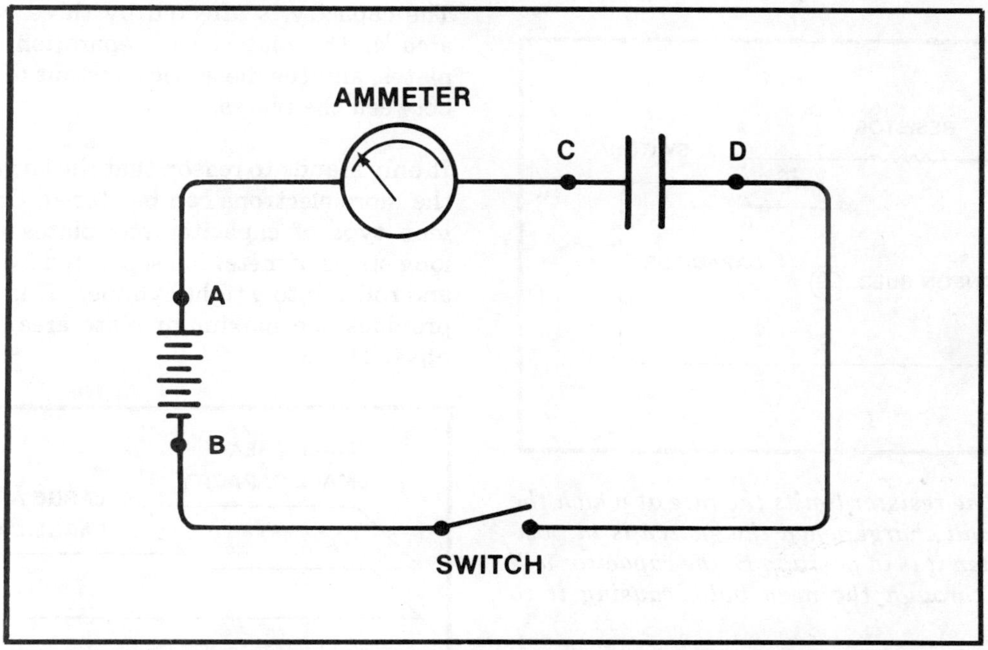

Fig. 6-1 Current flows only when the capacitor is charging or discharging.

A. Energy Stored in Electrostatic Fields

Electrical energy, as we have seen, may be stored in the magnetic field which surrounds a conductor through which electrons are moving. It may also be stored in *electrostatic* fields caused by an accumulation of electrical charges that are not moving, but are static. The electromagnetic field strength is determined by the amount of *current* flowing in the conductor, but the strength of the electrostatic field is determined by the amount of pressure, or voltage, on the static charges.

A capacitor, sometimes called a condenser, is a device that stores electrical energy in the electrostatic fields that exist between two conductors that are separated by an insulator, or a dielectric. Let's consider the circuit in Fig. 6-1 where we have two flat metal plates arranged so they face each other, but are separated by an insulator. One of the plates is attached to the positive terminal of the power source and the other to the negative terminal. When the switch is closed, electrons will be drawn from the plate attached to the positive terminal and will flow to the plate attached to the negative. There can be no flow across the insulator, but the plates will become charged. If the voltmeter reading were taken across points *C* and *D*, it would be found to be exactly the same as that taken across points *A* and *B*. Current flow would be indicated by the ammeter during the time the plates are being charged, but when they become fully charged, no more current will flow.

In the circuit of Fig. 6-2, we have a power source connected to a capacitor through a resistor, which limits the amount of current that initially flows into the capacitor, but does not prevent the voltage across the capacitor rising to that of the source.

When the voltage across the capacitor rises to the voltage of the source, no more current will flow. Now the switch can be moved to position *B* to complete the circuit across the neon light, and the capacitor will immediately discharge through the light and cause it to flash. If the switch is placed in its neutral position, when the capacitor is charged, the capacitor will remain charged until the electrons eventually leak off through the dielectric.

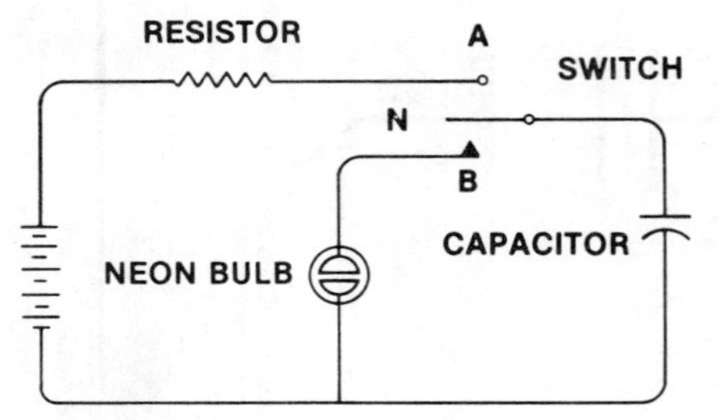

Fig. 6-2 The resistor limits the rate at which the capacitor will charge when the switch is in position A. When it is in position B, the capacitor will discharge through the neon bulb, causing it to flash.

B. Factor Affecting Capacitance

A capacitor is a device to store an electrical charge, and its capacity is measured in farads, with one farad the amount of capacity that will hold one coulomb of electricity (6.28×10^{18} electrons) under a pressure of one volt.

$$C = \frac{Q}{E}$$

C = capacity in farads

Q = charge in coulombs

E = voltage in volts

The farad is such a large unit that it is seldom used in pracitical circuits. Instead, most capacitors are measured in microfarads which are millionths of a farad, or in picofarads which are millionths of millionths of a farad. Picofarads have formerly been called micro-micro farads and may still be referred to in this way in some texts. The Greek letter mu (μ) is used to represent the prefix micro.

1 microfarad (μf) $= 1 \times 10^{-6}$ farad

1 picofarad (Pf or $\mu\mu$f) $= 1 \times 10^{-12}$ farad

The capacity is affected by three variables; the area of the plates, the separation between the plates, and the dielectric constant of the material between the plates.

It only stands to reason that the larger the plates, the more electrons can be stored. One very common type of capacitor has plates made of two long strips of metal foil separated by waxed paper and rolled into a tight cylinder. This construction provides the maximum plate area for its small physical size.

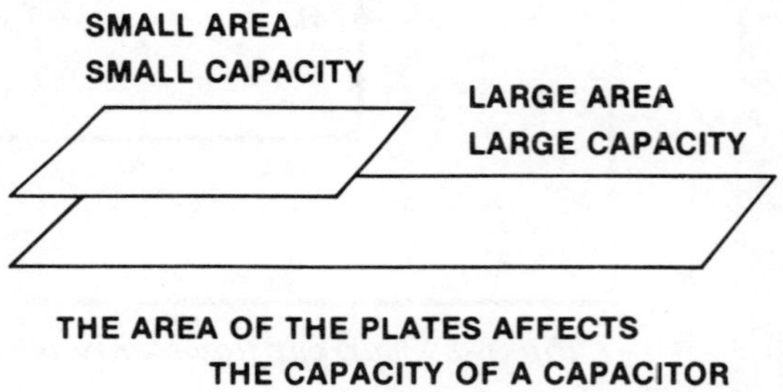

Fig. 6-3 The greater the area of the plates, the greater will be the capacity of the capacitor.

The distance the plates are apart determines the strength of the electrostatic field between them, and this affects the capacity. If the plates are widely separated, the field will be weak, and it will not pull very many electrons onto the negative plate. If, on the other hand, they are very close together, the attraction caused by the unlike charges will produce a very strong field in the dielectric, and many electrons will be held on the negative plate. The strength of the electrostatic field increases inversely as the separation between the plates. When the space between the plates is cut in half, the strength of the electrostatic field will double.

One problem with getting the plates too close together is the possibility of the field becoming so strong that electrons will be pulled across the insulator and actually flow to the positive plate. When this happens, in most capacitors, the dielectric will be damaged and a conductive path set up, shorting the capacitor and making it useless. For this reason, all capacitors are rated with regard to their working voltage which is a DC measurement indicating the strength of the dielectric.

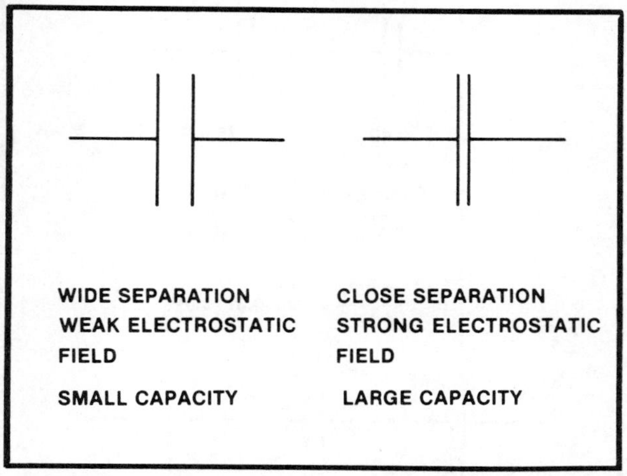

Fig. 6-4 The closer the plates of the capacitor, the greater will be the capacity.

The third factor which affects the capacity of a capacitor is the material of the dielectric. More specifically, it is the dielectric constant of the insulating material. Energy is stored not only in the stress across the dielectric, but by the distortion of the orbits of the electrons in the material of which the dielectric is made. Air is used as the reference for measuring the dielectric constant and it is given a value of one. If glass, which has a dielectric constant of eight, is substituted for air as an insulator, the capacity will increase eight times because of the energy stored in the distortion of the electron orbits within the glass.

C. *Series and Parallel Connection of Capacitors*

1. *Capacitors in series*

It is often necessary to connect more than one capacitor into a circuit, and when they are installed in series, the effect is similar to increasing the separation between the plates, and so the total capacity will be less than that of any of the capacitors. The formulas for finding the total series capacitance are the same as those used for finding the total resistance of resistors connected in *parallel*. When more than one similar capacitor

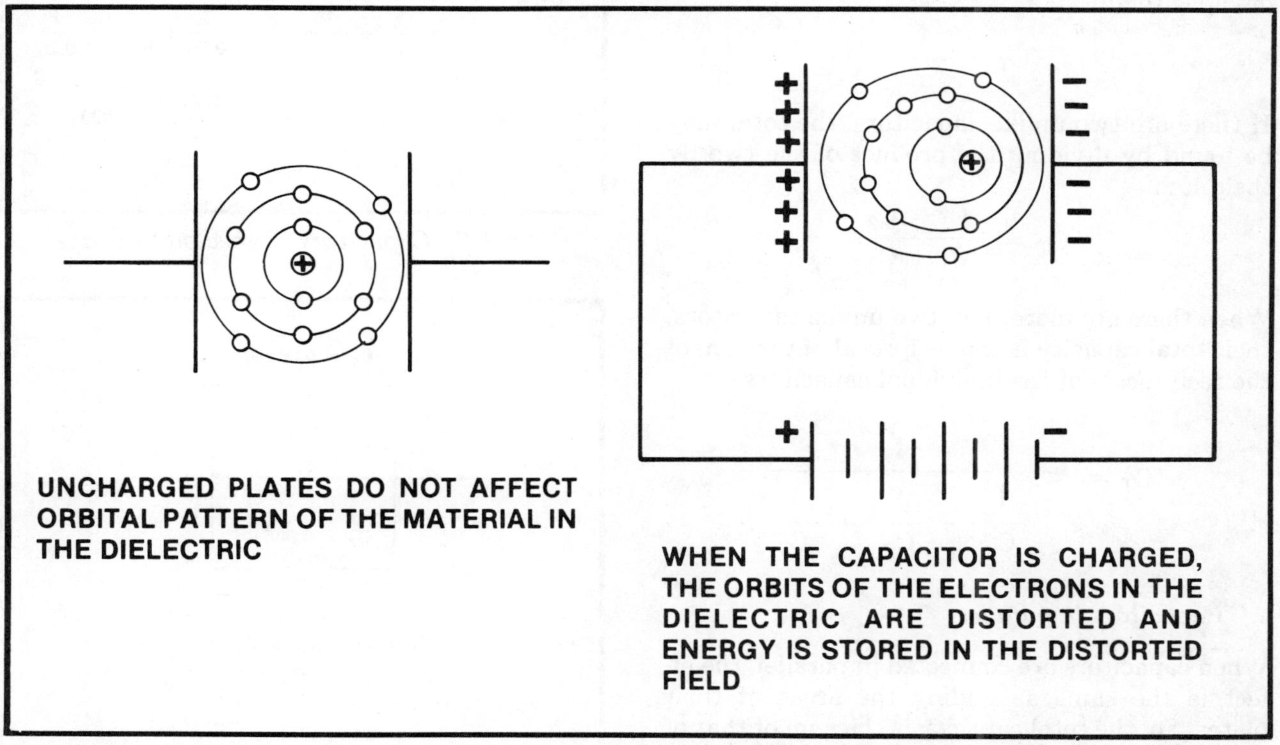

Fig. 6-5 The dielectric constant of the insulation between the plates of a capacitor determines its capacity. Energy is stored in the stress of the dielectric.

Kind of Dielectric	Approximate* K Value
Air [at atmospheric pressure]	1.0
Bakelite	5.0
Beeswax	3.0
Cambric [varnished]	4.0
Fibre [red]	5.0
Glass [window or flint]	8.0
Gutta Percha	4.0
Mica	6.0
Paraffin [solid]	2.5
Paraffin Coated Paper	3.5
Porcelain	6.0
Pyrex	4.5
Quartz	5.0
Rubber	3.0
Slate	7.0
Wood [very dry]	5.0

***These values are approximate, since true values depend upon quality or grade of material used, as well as moisture content, temperature, and frequency characteristics of each.**

Fig. 6-6 Table of dielectric constants

are connected in series, the total capacitance is the value of one capacitor divided by the number of capacitors:

$$C_T = \frac{C}{n}$$

If there are two unlike capacitors, the total may be found by dividing the product of the two by their sum.

$$C_T = \frac{C_1 \times C_2}{C_1 + C_2}$$

When there are more than two unlike capacitors, their total capacity is the reciprocal of the sum of the reciprocals of the individual capacitors.

$$C_T = \frac{1}{\frac{1}{C_2} + \frac{1}{C_2} + \frac{1}{C_3} \ldots}$$

2. *Capacitors in parallel*

When capacitors are connected in parallel, the effect is the same as adding the areas of their plates. So, the total capacity is the sum of that of the individual capacitors.

$$C_T = C_1 + C_2 + C_3 \ldots$$

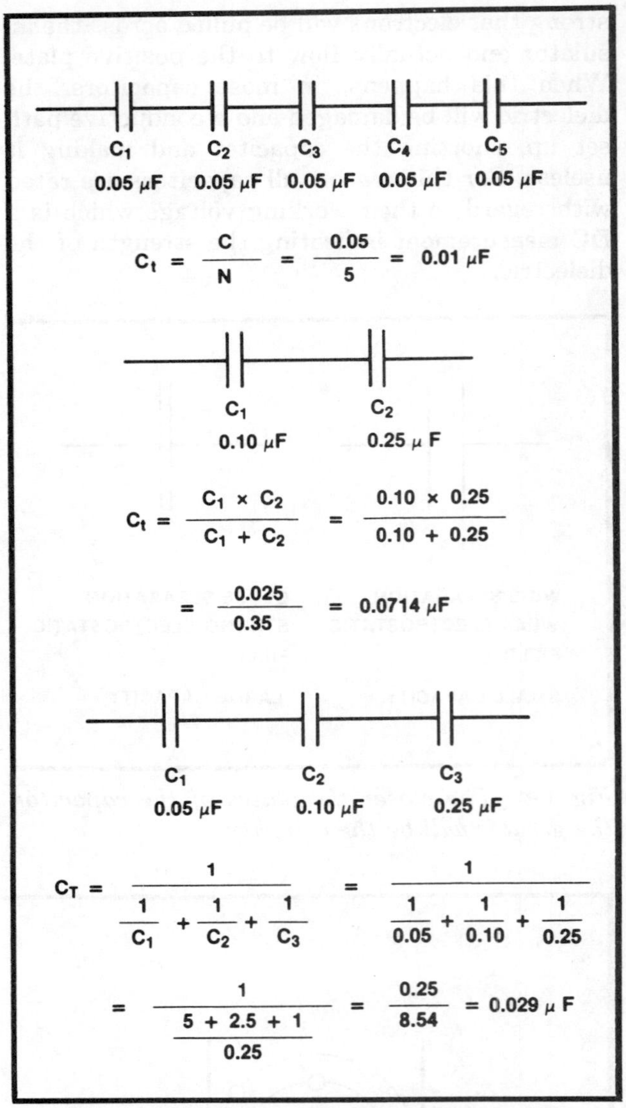

Fig. 6-7 Capacitors connected in series

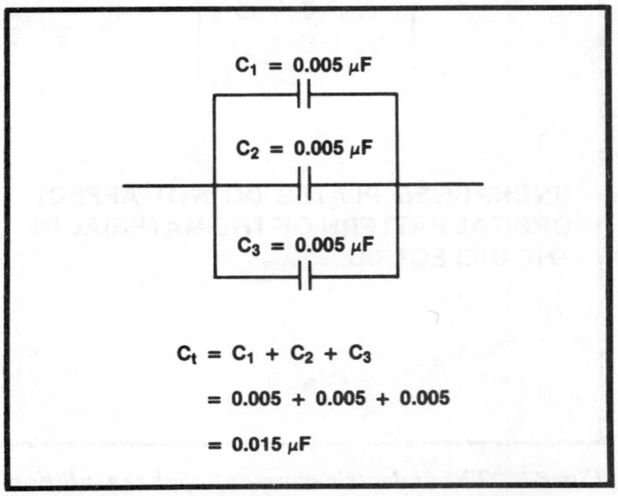

Fig. 6-8 Capacitors connected in parallel

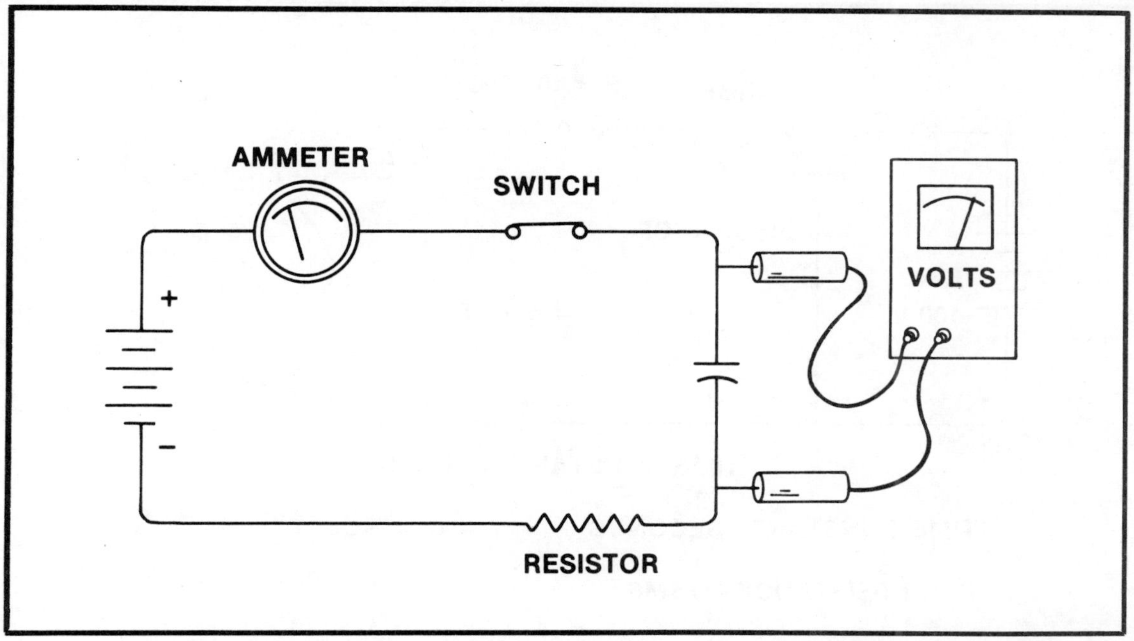

Fig. 6-9 Circuit for measuring the time constant of a R-C circuit.

D. Time Constant of Capacitors

We saw in our study of inductors that when a voltage source was placed across the inductor, the inductance slowed the rise of current in the circuit and, as a result, the changes in current lagged behind the changes in the voltage.

In a capacitive circuit, such as we have in Fig. 6-9, the results are the opposite. Changes in the current lead the changes in the voltage. When the switch is closed, current immediately begins to flow as electrons are attracted away from the positive plate and forced onto the negative plate. The voltage across the plate does not jump up immediately, but it rises as the plates become charged.

The time constant of a capacitive circuit is the time, in seconds, required for the voltage across the capacitor to reach 63.2% of the source voltage and is determined by both the capacitance and the resistance of the circuit.

$$TC = R \times C$$

TC = Time constant in seconds

R = Resistance in ohms

C = Capacitance in farads

Timing circuits are often made using a capacitor and a resistor in series. In Fig. 6-11 we have a 100,000-ohm resistor in series with a 100-microfarad capacitor, and both of these across a 100-volt power source. When the switch is closed, current will begin to flow, but it will be limited in its rate by the opposition caused by the resistor. In 10 seconds, the voltage will rise to 63.2 volts; in 20 seconds, it will be up to 86.5 volts; in 30 seconds, to 95 volts; in 40 seconds to 98 volts; and in 50 seconds the voltage will be equal to the source of 100 volts, and no more current will flow.

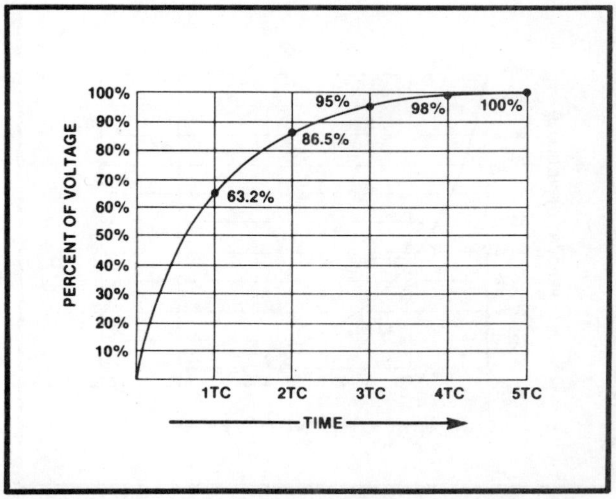

Fig. 6-10 Time constant curve of a capacitor during charge

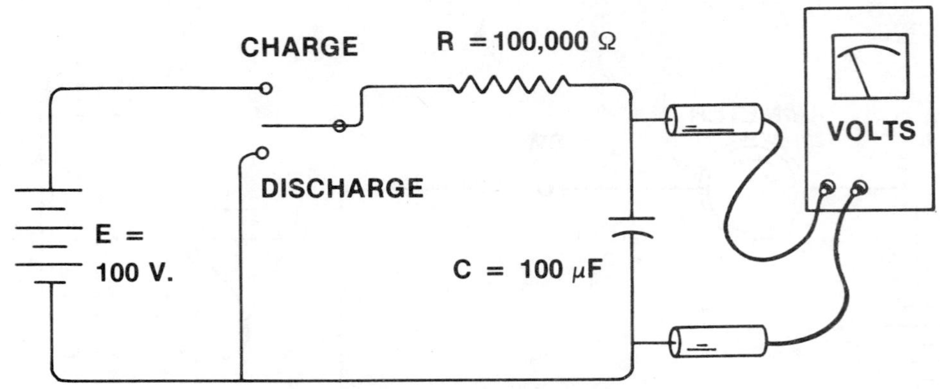

TIME CONSTANT = R × C

TIME CONSTANT - SECONDS

R = RESISTANCE - OHMS

C = CAPACITANCE - FARADS

$$TC = RC$$
$$= 1 \times 10_5 \times 1 \times 10^{-4}$$
$$= 1 \times 10^1$$
$$= 10 \text{ SECONDS}$$

Fig. 6-11 Measurement of the time constant of a R-C circuit

When the switch in Fig. 6-11 is placed in the discharge position, the voltage will begin to drop as the capacitor starts to discharge. It will follow the curve of Fig. 6-12, and in 10 seconds it will be down to 36.8 volts, and then its drop will slow down. In 20 seconds, it will be down to 13.5 volts; in 30 seconds, to 5 volts; in 40 seconds to 2 volts; and it will not be down to zero volts until the current has flowed through the resistor for 50 seconds, or five time constants.

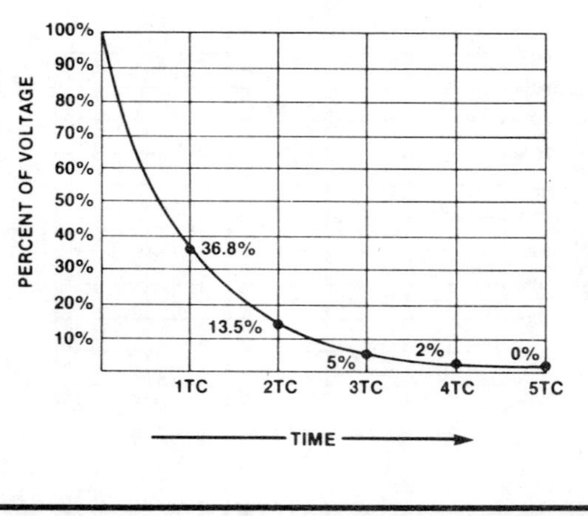

Fig. 6-12 Time constant curve of a capacitor being discharged

E. Capacitive Reactance

1. Factors affecting capacitive reactance

Capacitive Reactance, abbreviated X_C, is the opposition to the flow of alternating current caused by the capacitance in a circuit. For direct current (frequency = zero) the opposition is infinite because *no* current can flow *through* a capacitor.

As the frequency increases from zero, current flows into and out of the capacitor, and, as you have noticed in the time constant curve, the greatest rate of flow occurs when the current flows into a discharged capacitor, or out of a fully charged capacitor.

The amount of current flow in a capacitive circuit increases with either an increase in the size of the capacitor or with an increase in the frequency of the AC. And since capacitive reactance is the *opposition* to the flow of current, the larger the capacitor, or the higher the frequency, the *lower* will be the capacitive reactance. The formula for capacitive reactance shows us this relationship:

$$X_C = \frac{1}{2\pi f C}$$

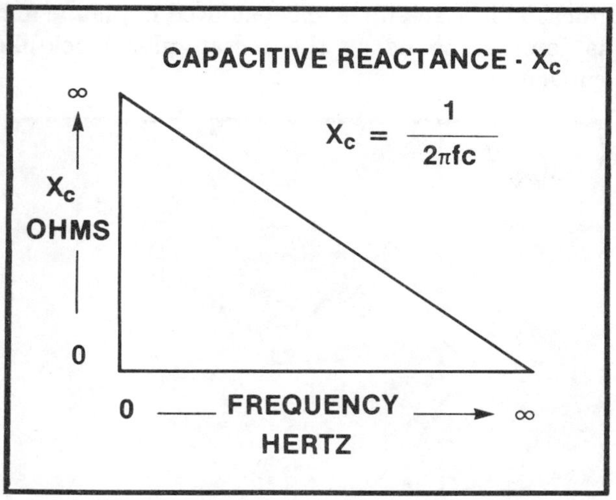

Fig. 6-13 Capacitive reactance is a function of frequency and the capacity of the circuit. Capacitive reactance decreases as either the frequency or the capacity increases.

2. Phase shift in a capacitive circuit

From Fig. 6-14, we can see that in a purely capacitive circuit, the changes in current *lead* the changes in voltage by 90°. The current must flow into the capacitor before the voltage across it can rise. When the capacitor is fully charged, the voltage will be maximum and the current will be

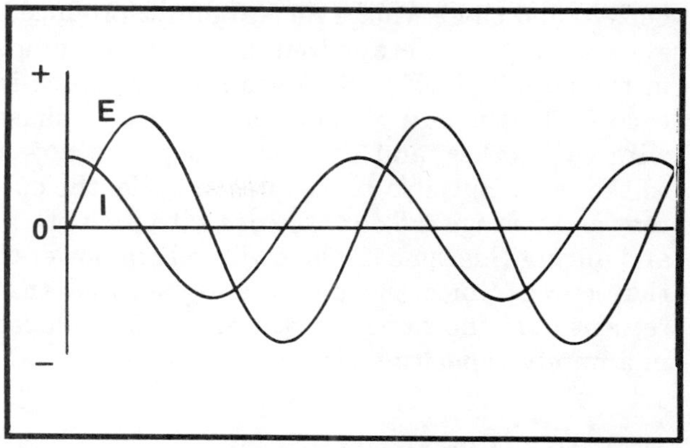

Fig. 6-14 The current leads the voltage in a capacitive circuit.

zero. Now, as the capacitor begins to discharge, the current begins to flow out of it and the voltage across it starts to drop. The current flow will be the greatest as the voltage passes through zero, and as the voltage begins to build up in the opposite direction, the current flow starts to drop off until the capcitor is fully charged and the current will again be zero.

3. Power in a capacitive circuit

Since the current leads the voltage by 90° in a purely capacitive circuit, the power factor will

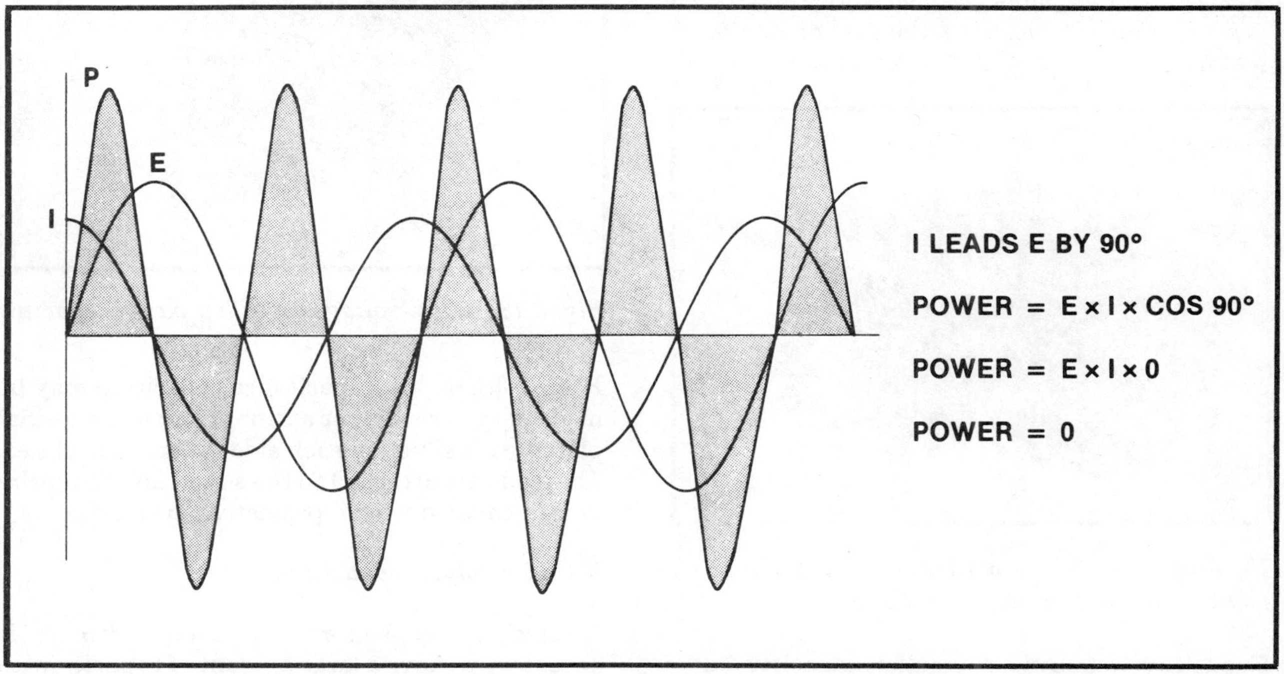

Fig. 6-15 In a purely capacitive circuit, the negative power equals the positive power, and no true power is produced.

be zero and there will be no real power produced, even though there is apparent power or volt-amps in the circuit. In Fig. 6-15, we see that twice in each cycle, the current and voltage are in phase with each other, and the source supplies power to the load. But also twice in each cycle, the current and voltage are *out of phase* with each other, and during this time the load will return power to the source. Since the power supplied and that returned are the same, no real power is produced in a purely capacitive circuit.

F. Types of Practical Capacitors

1. Fixed capacitors

Capacitors may be divided basically into two types, fixed and variable, and the fixed capacitors may be further divided into electrolytic and non-electrolytic types.

a. Non-electrolytic capacitors

When relatively low values of capacitance are needed, non-electrolytic capacitors are used, and one of the most common types of non-electrolytic capacitors is the paper capacitor. The plates are made of two strips of very thin metal foil and are separated by a strip of waxed paper. The sandwich is coiled into a tight roll, and wire leads are attached to the plates. The assembly may be encapsulated in plastic, or, in the case of an aircraft magneto capacitor, sealed in a metal can.

Fig. 6-16 A magneto capacitor is a paper capacitor sealed in a metal container.

Capacitors requiring a smaller capacity, but a higher working voltage, are made using stacks of thin metal foil sandwiched between thin sheets of mica. This stack-up is encapsulated in plastic and is familiar to us as the rectangular block-like capacitor.

Fig. 6-17 Block-type mica capacitor

For high-voltage applications paper capacitors may be enclosed in a metal container filled with an insulating oil. If a voltage surge should break through the insulator, the oil will flow in and restore its insulating characteristics. These are called self-healing capacitors.

Fig. 6-18 High-voltage, oil-filled paper capacitor

High-voltage, low-capacitance capacitors may be made of either a disc or a tube of ceramic material plated with silver on each side to form the plates. The leads are attached to the silver, and the entire unit is covered with a protective insulation.

b. Electrolytic capacitors

Electrolytic capacitors are used when it is necessary to have a large amount of capacity with a relatively low working voltage. These capacitors are polarized, meaning that they can be

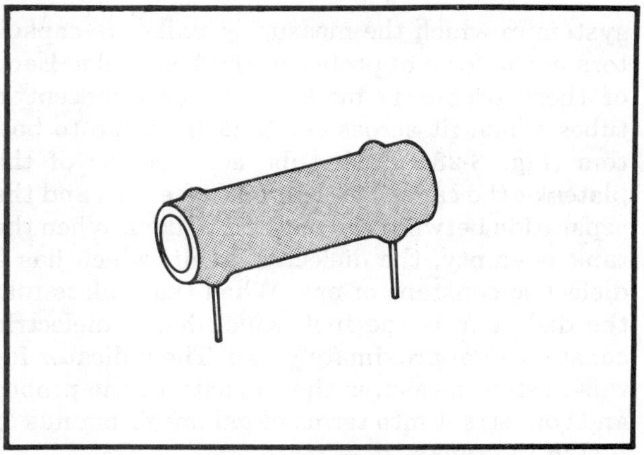

Fig. 6-19 *Tubular ceramic capacitor*

used only in DC circuits because they act as capacitors only when they are properly connected into the circuit. If they are installed with the wrong polarity, current will flow through them, causing them to overheat and actually explode.

The reason electrolytic capacitors have such a high capacity for their small physical size is their extremely thin dielectric. The positive plate is made of aluminum foil and has electrolytically deposited on its surface an extremely thin oxide film which serves as the dielectric. A liquid or paste electrolyte which will conduct current saturates a gause, which is held in contact with both the positive plate and the negative can in which the capacitor is sealed. The moist electrolyte forms the second plate of the capacitor, and this combination of metal plate, oxide film, and conductive liquid or paste makes a capacitor which has the maximum capacity for its size.

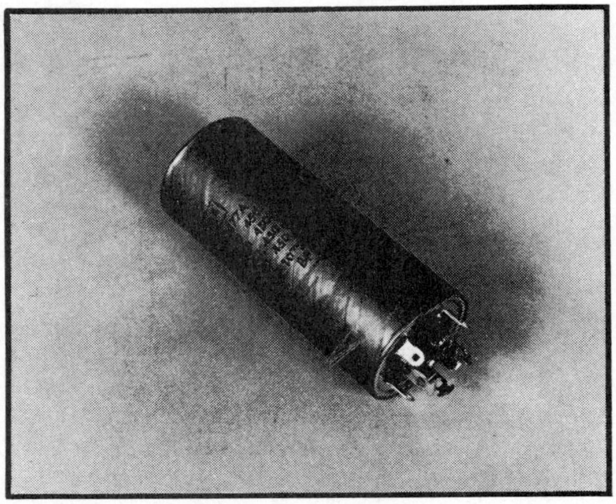

Fig. 6-20 *Electrolytic capacitor*

2. Variable capacitors

The capacity of a capacitor is determined by three things; the area of the plates, the separation of the plates, and the type of dielectric. If we are able to change any of these three factors, we can change the capacity.

a. Changing the area of the plates

Most radios have variable capacitors in which we change the area of the plates. The plates are made of thin sheets of aluminum and are meshed together with one group of plates fixed and serving as the stators. The other plates, called the rotors, are mounted on a shaft and mesh with the fixed plates, but do not touch them. The air between the plates serves as the dielectric. When the plates are fully meshed, the capacity is maximum, and as the shaft is rotated, the area of the plates that are meshed decreases and the capacity becomes less.

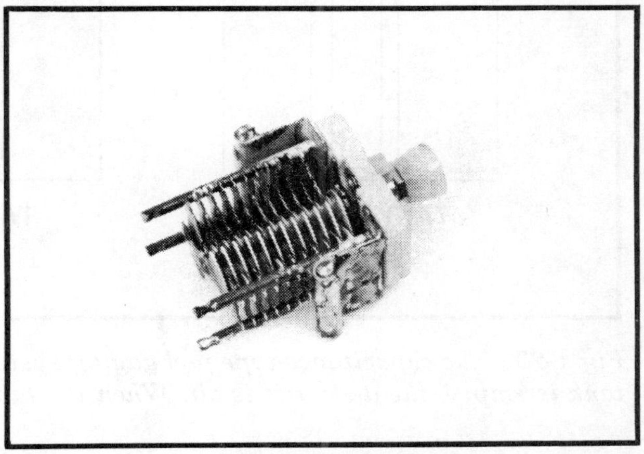

Fig. 6-21 *Variable-plate-area tuning capacitor, using air as the dielectric*

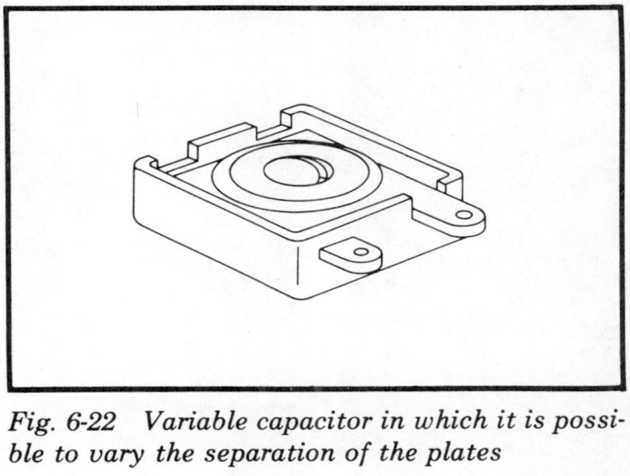

Fig. 6-22 *Variable capacitor in which it is possible to vary the separation of the plates*

b. Changing the spacing between the plates

While variable-area capacitors are used for the main tuning capacitor for a radio, small trimmer and padder capacitors are used for fine tuning. These small capacitors are made up of a stack of metal foil plates separated by thin sheets of mica. A screw adjustment allows the plates to be squeezed tightly together to increase the capacitance, or relaxed to decrease the capacitance.

c. Changing the dielectric constant

The most popular fuel quantity measuring system used in our modern aircraft is the capacitance system in which the measuring units are capacitors in the form of probes in the fuel tanks. Each of these probes is made up of two concentric tubes which fit across the tank from top to bottom (Fig. 6-23). Each tube acts as one of the plates of the capacitor, and both the area and the separation between the plates are fixed. When the tank is empty, the dielectric is air, which has a dielectric constant of one. When the tank is full, the dielectric is the fuel, which has a dielectric constant of approximately two. The indicator for this system measures the capacity of the probes and converts it into terms of gallons or pounds of fuel in the tanks.

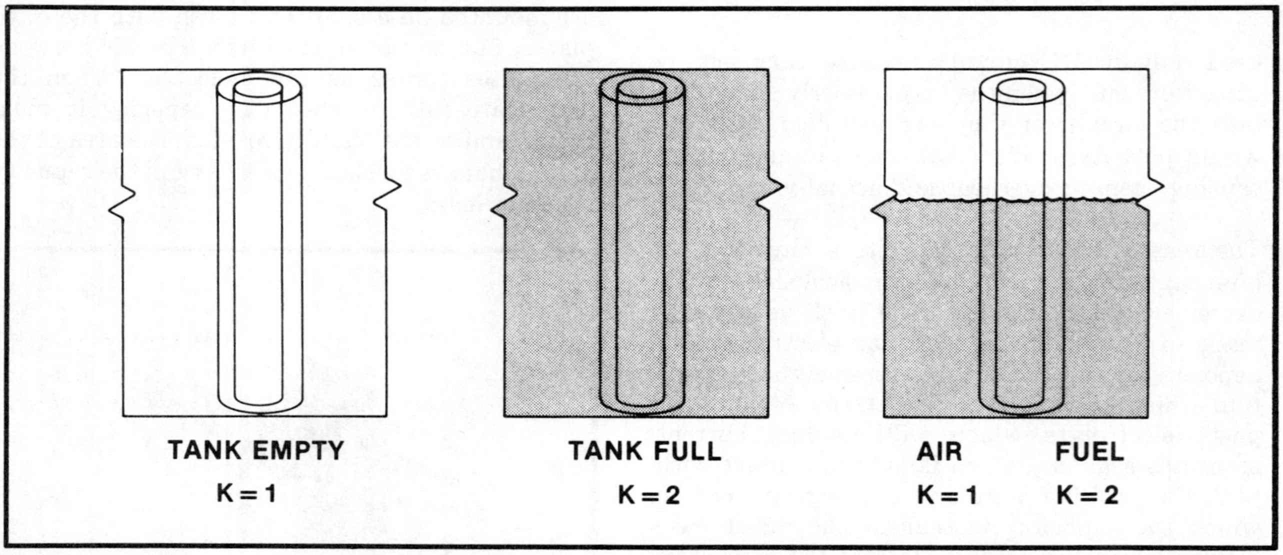

Fig. 6-23 The capacitance-type fuel gaging system uses capacitors as probes in the fuel tank. When the tank is empty, the dielectric is air. When the tank is full, fuel is the dielectric.

SECTION VII

Series Alternating Current Circuits

A. *Impedence*

The flow of current in an AC circuit is opposed by three things: by resistance, which converts the electrical energy into heat, and by inductive and capacitive reactance which oppose the flow but do not produce heat. All circuits have some resistance, some inductance, and some capacitance, but the total opposition is not just the arithmetic sum of the three individual oppositions. Since both the inductance and the capacitance cause phase shifts between the voltage and the current, the total opposition will be the *vector* sum of the three oppositions.

Let's begin our study of Ohm's law in an AC circuit by examining a 400-hertz circuit in which the capacitance is negligible, but one which has a total resistance of 100 ohms and an inductance of 20 millihenries.

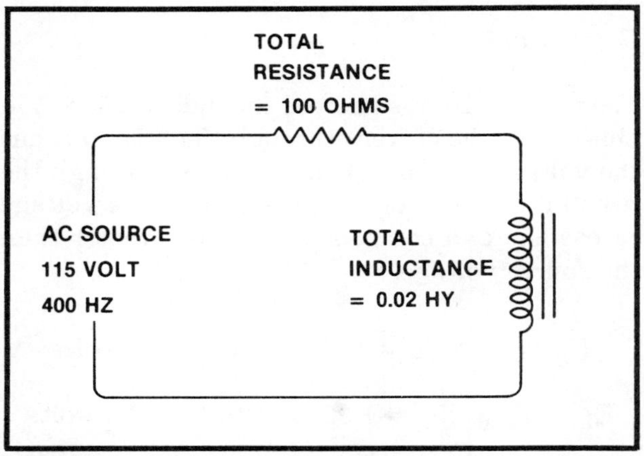

Fig. 7-1 A series R-L circuits

A vector is a quantity that has both magnitude and direction, and for solving AC problems, we will plot the opposition caused by the resistance of the circuit on a horizontal line extending to the right of the zero point on our graph, Fig. 7-2. Draw this to scale with a length representing 100 ohms of resistance. This is the total resistance of the circuit and includes not only any resistors in the circuit, but the resistance of the inductor as well.

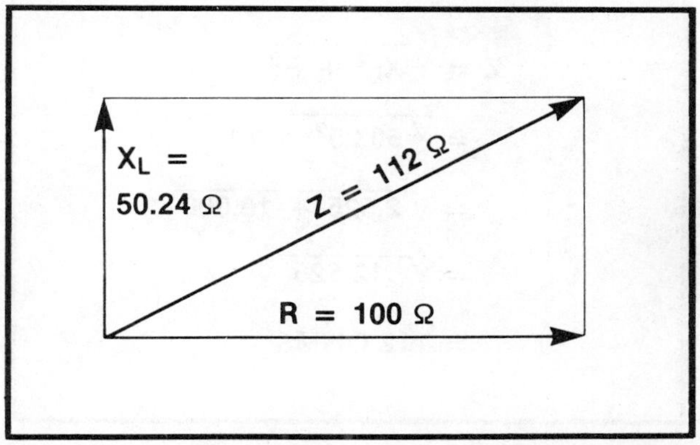

Fig. 7-2 The total opposition, or the impedance, of an R-L circuit is the vector sum of the inductive reactance and the resistance.

We must find the inductive reactance of the 20-millihenry inductor, and to do this, we use the formula:

$$X_L = 2\pi fL$$

When we multiply the constant 2π (6.28) by the frequency of 400 hertz, and the inductance of two millihenries, we get an inductive reactance of 50.24 ohms, Fig. 7-3. We plot the vector for the inductive reactance vertically upward from the zero point.

$$X_L = 2\pi fL$$

$$= 6.28 \times 4 \times 10^2 \times 2 \times 10^{-2}$$

$$= 50.24 \text{ OHMS}$$

Fig. 7-3 The formula for finding the inductive reactance

We now have the vector quantities for resistance and inductive reactance at right angles to each other, and their vector sum is the circuit impedance (Z) which is the square root of the sum of their squares.

$$Z = \sqrt{X^2 + R^2}$$

By using the formula shown in Fig. 7-4, we find that the total circuit opposition, or the impedance, is 112 ohms.

$$\begin{aligned} Z &= \sqrt{X_L^2 + R^2} \\ &= \sqrt{50.25^2 + 100^2} \\ &= \sqrt{2{,}525 + 10{,}000} \\ &= \sqrt{12{,}525} \\ &= 112 \text{ OHMS} \end{aligned}$$

Fig. 7-4 The formula for finding impedance

B. Power Factor

Not all of the current and voltage in this circuit are in phase, and power is produced only by that portion of the voltage and current that are in phase, so we must find this percentage which is the power factor. We saw earlier in this study that the power factor was the ratio of the true power to the apparent power. It is also the ratio of the resistance of the circuit to the impedance, Fig. 7-5. In this problem, the power factor is 0.89, or 89% of the current is in phase with the voltage.

$$\begin{aligned} \text{POWER FACTOR} &= \frac{\text{TRUE POWER}}{\text{APPARENT POWER}} = \frac{R}{Z} \\ &= \frac{100}{112} \\ &= 0.893 \end{aligned}$$

Fig. 7-5 The formula for finding the power factor

C. Phase Angle

The phase angle is the angle whose cosine is the power factor, and by looking at the trigonometric function chart of Fig. 3-6, we see that 0.89 is the cosine of 27 degrees. In this circuit, the current lags behind the voltage by 27 degrees.

D. Current

Rather than using $E = I \times R$ as an expression of Ohm's law in an AC circuit, this law must be stated as $E = I \times Z$ because we must consider the *total* opposition, the impedance, rather than just the resistance. In our circuit, we found the impedance to be 112 ohms, and we know the source voltage is 115 volts, so we can find the total current through the series circuit to be $I = E/Z = 115/112 = 1.03$ amps.

E. True Power

The true power developed in this circuit is found by the formula $P = E \times I \times$ power factor, and is equal to $115 \times 1.03 \times 0.89 = 105.4$ watts.

F. Apparent Power

This is the product of the source voltage time the total current and is 115×1.03, or 118.5 volt-amps.

G. Voltage

The voltage dropped across the inductor is not in phase with the current through the inductor, but the voltage and current *are* in phase through the resistor. Because of this, the sum of the voltage across the two components will be greater than the source voltage.

$$E_R = I_T \times R = 1.03 \times 100 = 1.03 \text{ volts}$$

$$E_L = I_T \times X_L = 1.03 \times 50.24 = 51.7 \text{ volts}$$

H. Circuits Having Capacitance and Resistance

Anytime we analyze a series circuit, we should make a chart similar to the one in Fig. 7-6 to record all of the information about the circuit. Now, let's look at a circuit in which there is negligible inductance, but there is a measurable capacitance and resistance, and we will complete the chart of Fig. 7-8.

E = 115 VOLTS

F = 400 HERTZ

R = 100 OHMS

L = 0.02 HENRY

X_L = 50.24 OHMS

Z = 112 OHMS

I_T = 1.03 OHMS

I_R = 1.03 OHMS

I_L = 1.03 OHMS

E_R = 103 VOLTS

E_L = 51.7 VOLTS

P_{TRUE} = 105.4 WATTS

$P_{APPARENT}$ = 118.5 VOLT-AMPS

POWER FACTOR = 0.89 [89%]

PHASE ANGLE = 27° LAGGING

Fig. 7-6 Values for a series R-L circuit

In the circuit of Fig. 7-7, we have a 115-volt, 400-hertz power source and a total circuit resistance of 100 ohms and a capacitance of 5 microfarads.

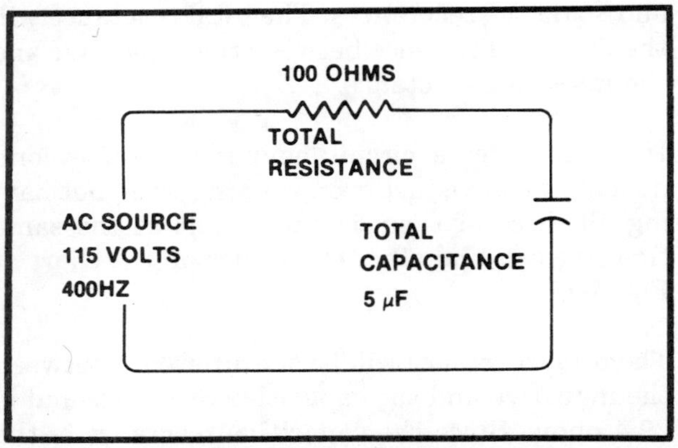

Fig. 7-7 A series R-C circuit

E = 115 VOLTS

F = 400 HERTZ

R = 100 OHMS

C = 5 μF

X_C = 79.6 OHMS

Z = 128 OHMS

I_T = 0.9 AMPS

I_R = 0.9 AMPS

I_C = 0.9 AMPS

E_R = 90 VOLTS

E_C = 71.6 VOLTS

POWER FACTOR = 0.78 (78%)

P_{TRUE} = 80.73 WATTS

$P_{APPARENT}$ = 103.5 VOLT · AMPS

PHASE ANGLE = 39° LEADING

Fig. 7-8 Values for a series R-C circuit

Using the formula

$$X_C = \frac{1}{2\pi fC}$$

we find the the capacitive reactance is 79.6 ohms.

The impedance is found by the formula

$$Z = \sqrt{X^2 + R^2}$$

to be 128 ohms.

The total circuit current is found by the formula

$$I = \frac{E}{Z}$$

to be 0.9 amp, and since this is a series circuit, the same current will flow through the resistor and the capacitor.

The voltage across the resistor is found by

$$E = I \times R$$

to be 90 volts, and since this is considered to be a pure resistance, the voltage and the current are in phase.

$$Z = \sqrt{X_C^2 + R^2}$$
$$= \sqrt{79.6^2 + 100^2}$$
$$= \sqrt{6{,}336 + 10{,}000}$$
$$= \sqrt{16{,}336}$$
$$= 128 \text{ OHMS}$$

$$I = \frac{E}{Z} = \frac{115}{128} = 0.9 \text{ AMPS}$$

$$X_C = \frac{1}{2\pi fC}$$
$$= \frac{1}{6.28 \times 4 \times 10^2 \times 5 \times 10^{-6}}$$
$$= \frac{1}{6.28 \times 4 \times 5 \times 10^{-4}}$$
$$= \frac{10{,}000}{125.6}$$
$$= 79.6\ \Omega$$

$$E_R = IR = 0.9 \times 100 = 90 \text{ VOLTS}$$

$$E_{X_C} = IX_C = 0.9 \times 79.6 = 71.6 \text{ VOLTS}$$

$$PF = \frac{R}{Z} = \frac{100}{128} = 0.78 = 78\%$$

PHASE ANGLE = ANGLE WHOSE COSINE IS 0.78
= 39° CURRENT IS LAGGING

$$P_{APPARENT} = E \times I$$
$$= 115 \times 0.9 = 103.5 \text{ VOLT-AMPS}$$

$$P_{TRUE} = E \times I \quad PF = 115 \times 0.9 \times 0.78$$
$$= 80.73 \text{ WATTS}$$

Fig. 7-9 Computation of the values for the series R-C circuit

The voltage dropped across the capacitor, found by the formula $E = I \times X_C$, is 71.6 volts. And since the current changes must occur before there can be any voltage change, the current leads the voltage.

The power factor is found by dividing the circuit resistance by the impedence and is 0.78 or 78%.

By referring to the table of trigonometric functions of Fig. 3-6, we see that the angle whose cosine is nearest to 0.78 is 39 degrees, and in this circuit, the current is leading the voltage.

The apparent power is the product of the source voltage and the total current and is 103.5 volt-amps, while the true power, which is the product of the apparent power and the power factor, is 80.73 watts.

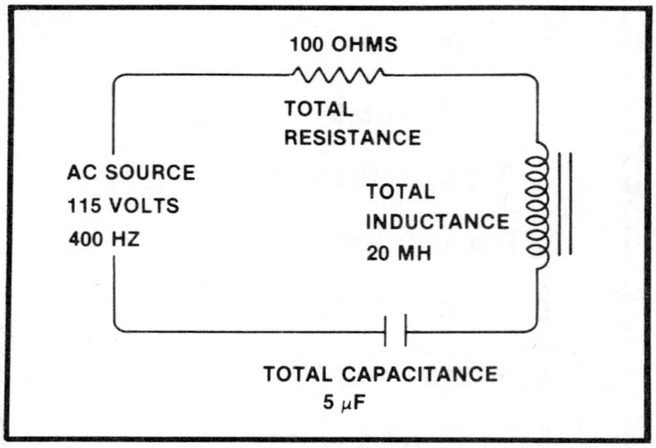

Fig. 7-10 A series R-L-C circuit

I. Circuits Having Inductance, Capacitance, and Resistance

If all three of the variables are measurable in an AC circuit, we can analyze it in the same way we have just done. The only difference is that we must find the total reactance and use it to find the impedance.

In a capacitive circuit, the current leads the voltage by 90 degrees, and in an inductive circuit, it lags by 90°, so the current in the two components is 180° out of phase, and as far as the circuit is concerned, the currents cancel each other, and so do the reactances. The total reactance will then be the difference between the capacitive and the inductive reactance.

If we consider a circuit having the same components we have just been working with, but having all three of them in the circuit at the same time, we can find all of the values on the chart of Fig. 7-12.

The total reactance will be the difference between the inductive and the capacitive reactance and is 29.4 ohms. Since the capacitive reactance is the greater, this total reactance is capacitive.

E = 115 VOLTS	I_R = 1.1 AMP
F = 400 HERTZ	I_L = 1.1 AMP
R = 100 OHMS	I_C = 1.1 AMP
L = 0.02 HENRY	E_R = 110 VOLTS
C = 5 MICROFARAD	E_C = 87.6 VOLTS
X_L = 50.2 OHMS	E_L = 55.2 VOLTS
X_C = 79.6 OHMS	P_F = 0.96 (96%)
X = 29.4 OHMS - CAPACITIVE	$P_{APPARENT}$ = 126.5 VOLT-AMPS
Z = 104 OHMS	P_{TRUE} = 121.4 WATTS
I_T = 1.1 AMP	PHASE ANGLE = 16° LEADING

Fig. 7-11 Values for the series R-L-C circuit

The impedance is found by using the total reactance, and it is found to be 104 ohms.

The total current is 1.1 amps, and since this is a series circuit, the current through all of the components is the same. The voltage dropped across each of the components is found by multiplying the total current by the resistance and the inductive and the capacitive reactance.

The power factor, phase angle, and the true and apparent powers are all found in the same way they were found in the previous problems.

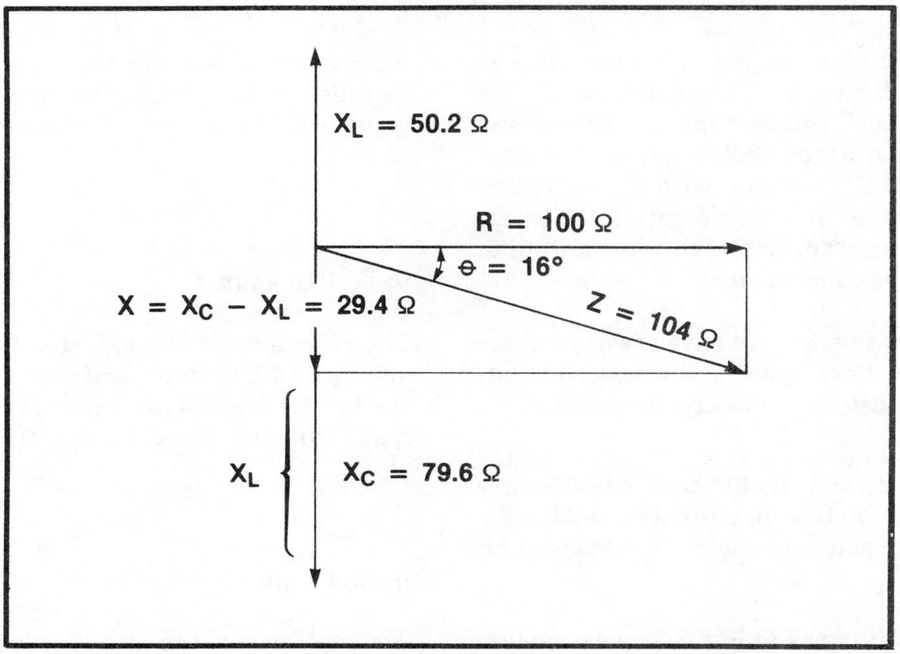

Fig. 7-12 Graphic determination of impedance in a series R-L-C circuit

SECTION VIII:

Parallel Alternating Current Circuits

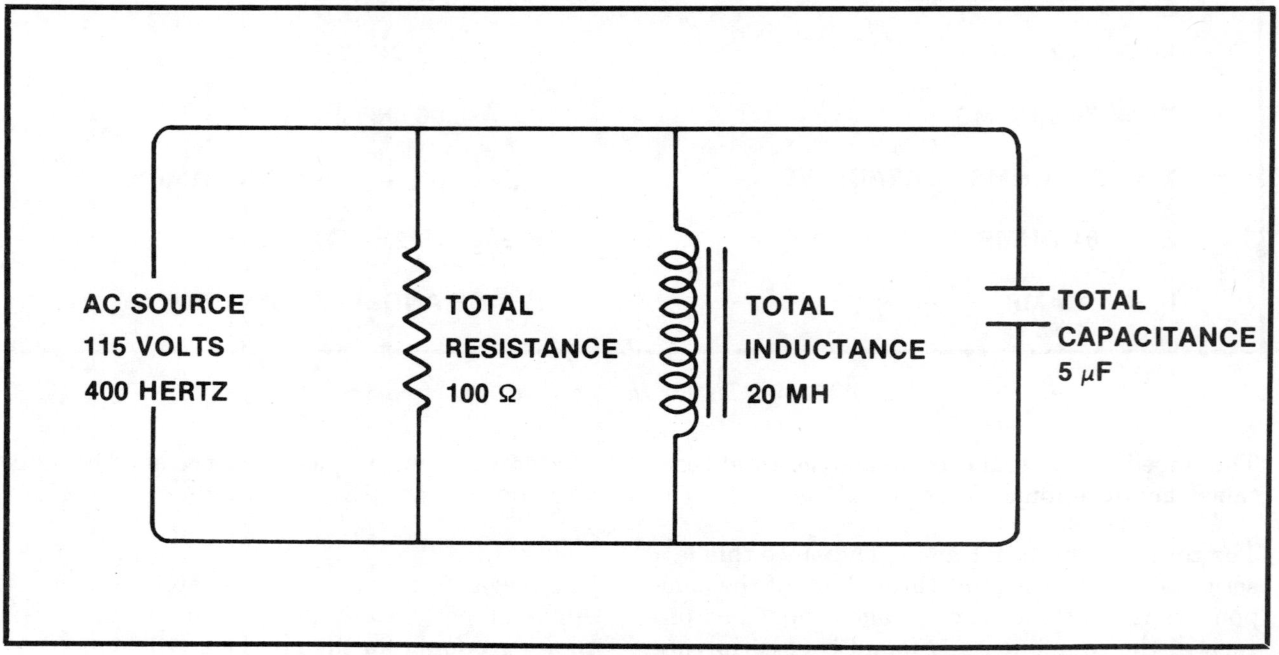

Fig. 8-1 A parallel R-L-C circuit

Almost all of the alternating current circuits in our shops and homes have the components connected in parallel rather than in series, and parallel AC circuits are handled in much the same way as parallel DC circuits, with the exception that we must take into consideration the phase shifts that occur between the current flowing in each of the three components.

In the circuit of Fig. 8-1, we have the same three components we have used in our series circuit, only this time they are connected in parallel.

The circuit has a 115-volt, 400-hertz power source and consists of a 100-ohm resistor, a 20-millihenry inductor, and a five-microfarad capacitor, all in parallel.

We can follow the work in Fig. 8-3 as we analyze the circuit, starting with the resistor. The voltage across the resistor is the same as the source voltage, or 115 volts. The current through the resistor is found by the formula

$$I = \frac{E}{R}$$

to be 1.15 amps.

The reactance of the inductor for the 400-hertz voltage is 50.24 ohms, and since the inductor has the full source voltage applied across it, the current through it found by the formula

$$I = \frac{E}{X_L}$$

is 2.29 amps.

The reactance of the capacitor is found to be 79.6 ohms, and the current through it is 1.44 amps.

E = 115 VOLTS	I_R = 1.15 AMPS
f = 400 HERTZ	I_L = 2.29 AMPS
R = 100 OHMS	I_C = 1.44 AMPS
L = 0.02 HENRY	I_X = 0.85 AMPS
C = 5 μF	I_T = 1.43 AMPS
X_L = 50.2 OHMS	Z = 80.42 OHMS
X_C = 79.6 OHMS	P_F = 0.80 (80%)
E_R = 115 VOLTS	PHASE ANGLE = 37° LAGGING
E_L = 115 VOLTS	$P_{APPARENT}$ = 164.45 VOLT-AMPS
E_C = 115 VOLTS	P_{TRUE} = 131.56 WATTS

Fig. 8-2 Values for a parallel R-L-C circuit

Before we can find the circuit impedance, we must find the total current which is the vector sum of the current flowing through the three components. The current through the inductor is 180 degrees out of phase with that flowing through the capacitor, so the total reactive current is the difference between the two, which is 0.85 amps, and since the inductive reactance is greater than the capacitive reactance, the total reactance is inductive.

The total current is found as the square root of the square of the reactive current plus the square of the resistive current. This is 1.43 amps.

The impedance of the circuit is now found by dividing the total voltage by the total current, and this is 80.42 ohms.

The power factor is the ratio of the current through the resistor to the total circuit current, and it is 0.80. Eighty percent of the voltage and current are in phase with each other.

The phase angle is found by referring to the table of trigonometric functions (Fig. 3-6). The angle whose cosine is nearest 0.80 is 37 degrees, and since the circuit is more inductive than capacitive, the current lags behind the voltage.

The apparent power is the product of the total voltage and the total current, and is 164.45 volt-amps. And the true power, which is the product of the apparent power and the power factor, is 131.56 watts.

$$X_L = 2\pi fL = 6.28 \times 4 \times 10^2 \times 2 \times 10^{-2} = 50.2 \text{ OHMS}$$

$$X_C = \frac{1}{2\pi fC} = \frac{1}{6.28 \times 4 \times 10^2 \times 5 \times 10^{-6}} = \frac{10{,}000}{125.6} = 79.6 \text{ OHMS}$$

$$I_T = \sqrt{I_R^2 + (I_L - I_C)^2} = \sqrt{1.15^2 + 0.85^2} = \sqrt{1.32 + 0.72} = \sqrt{2.04} = 1.43 \text{ AMP}$$

$$I_R = \frac{E_T}{R} = \frac{115}{100} = 1.15 \text{ AMPS}$$

$$I_L = \frac{E_T}{X_L} = \frac{115}{50.2} = 2.29 \text{ AMPS}$$

$$I_C = \frac{E_T}{X_C} = \frac{115}{79.6} = 1.44 \text{ AMPS}$$

$$I_X = I_{XL} - I_{XC} = 2.29 - 1.44 = 0.85 \text{ AMPS}$$

$$Z = \frac{E_T}{I_T} = \frac{115}{1.43} = 80.42 \text{ OHMS}$$

$$PF = \frac{I_R}{I_T} = \frac{1.15}{1.43} = 0.80 = 80\%$$

PHASE ANGLE =
ANGLE WHOSE COSINE IS 0.80 = 37°
THE CURRENT LAGS THE VOLTAGE

$$P_{APPARENT} = E_T \times I_T = 115 \times 1.43 = 164.45 \text{ VOLT-AMPS}$$

$$P_{TRUE} = E_T \times I_T \times PF = 115 \times 1.43 \times 0.80 = 131.56 \text{ WATTS}$$

Fig. 8-3 Computation of the values for a parallel R-L-C circuit

SECTION IX

Resonance in An AC Circuit

Inductive reactance in a coil is zero when the frequency is zero, and it increases smoothly until at an infinite frequency, it is infinity; that is, the back voltage generated in the inductor is equal to the source voltage, and no current can flow.

The reactance in a capacitor changes in exactly the opposite way as that in an inductor. At zero frequency, no current can flow through the capacitor, so the reactance is infinite. But as the frequency increases, the capacitive reactance decreases, until at an infinite frequency, there is no capacitive reactance. Fig. 9-1 shows the way the two types of reactance vary with the frequency.

The lines representing the two reactances cross at the resonant frequency of the circuit. Here the inductive and the capacitive reactances are equal.

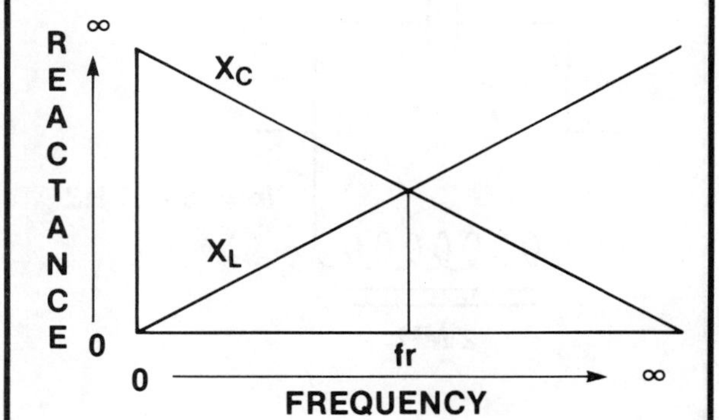

Fig. 9-1 The resonant frequency of a circuit is that frequency at which the inductive and capacitive reactances are equal.

It makes no difference whether the capacitor and inductor are connected in series or in parallel, the resonant frequency is the same, and it may be found by the formula:

$$F_R = \frac{1}{2\pi\sqrt{LC}}$$

The resonant frequency in hertz is found by dividing one by the constant two pi times the square root of the product of the inductance in henries and the capacitance in farads.

A. Series Resonant Circuit

At the resonant frequency of a series R-L-C circuit, the current flowing in the coil and the capacitor are equal, but they are 180 degrees out of phase with each other. The inductive and capacitive reactances are also exactly the same, but, because of the phase difference, they cancel each other, and the total reactance is zero. The total opposition the circuit offers to the flow of AC is that of the resistance, and so the circuit impedance is minimum at resonance and is equal to the circuit resistance.

The voltage drop across the entire circuit is very low, or is zero, and the current flow is maximum. But the voltage across either of the reactances is quite high and can even be higher than the source voltage. Remember, the voltage across the inductor and the voltage across the capacitor are out of phase with each other so the sum of the individual voltages does not equal the source voltage, as it does in a DC circuit.

A series resonant circuit acts as a pure resistance, and the source voltage and current are in phase, so the power factor of the circuit is one.

B. Parallel Resonant Circuit

In a parallel R-L-C circuit at its resonant frequency, there is a large amount of current flowing between the capacitor and the inductor, storing energy first in an electrostatic field in the capacitor and then in an electromagnetic field

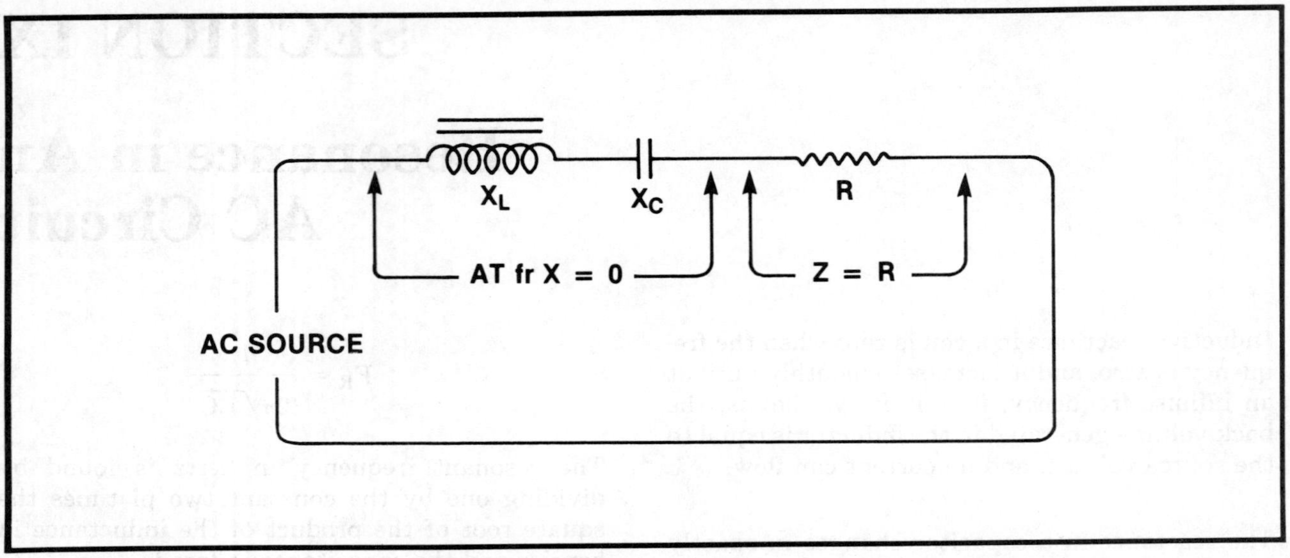

Fig. 9-2 In a series R-L-C circuit at its resonant frequency, the inductive and capacitive reactances cancel each other.

around the inductor. If there were no resistance in the circuit, once the exchange of energy between the two types of fields had started, the circulating current, as this current is called, would continue to flow back and forth indefinitely. But in practice, all circuits have some resistance which causes this current to die down unless extra energy is added from the source.

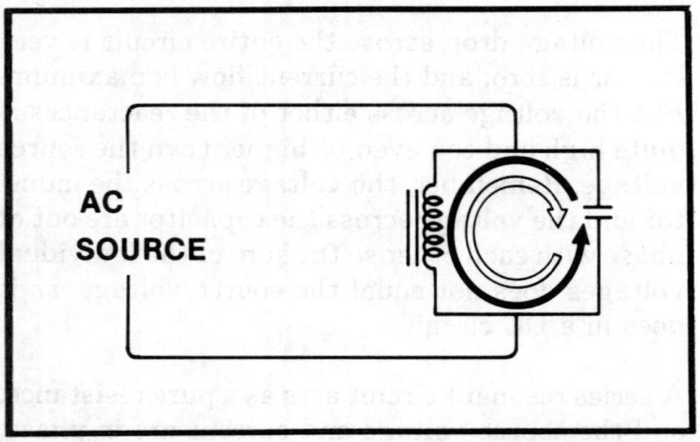

Fig. 9-3 In a parallel L-C circuit at its resonant frequency, the circulating current is maximum, but the source current is minimum.

At the resonant frequency, the circulating current in the inductor and capacitor is high, but there is almost no current supplied from the source, so the source sees the parallel circuit as having a high impedance. The reactances cancel each other, and so the opposition is purely resistive, and the power factor of the circuit is one.

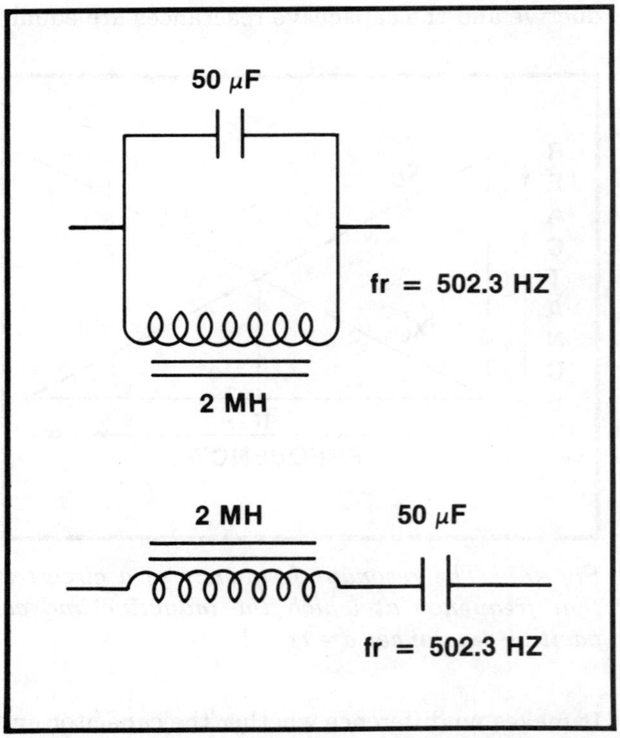

Fig. 9-4 The resonant frequency is the same for a series or a parallel L-C circuit.

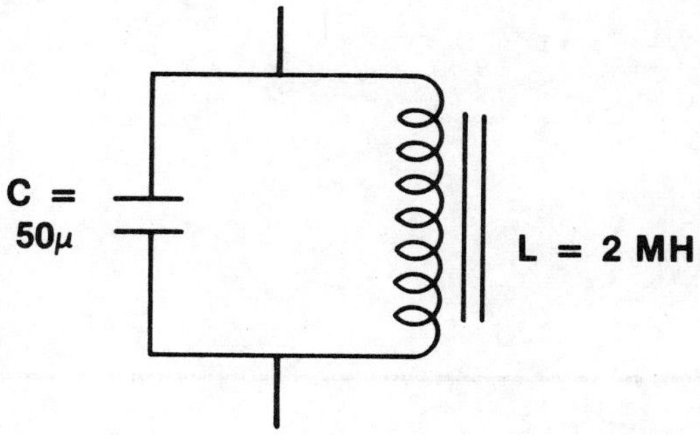

$$fr = \frac{1}{2\pi\sqrt{LC}}$$

$$= \frac{1}{6.28\sqrt{5 \times 10^{-5} \times 2 \times 10^{-3}}}$$

$$= \frac{1}{6.28\sqrt{10 \times 10^{-8}}}$$

$$= \frac{1}{6.28 \times 3.17 \times 10^{-4}}$$

$$= \frac{1 \times 10^{4}}{6.28 \times 3.17}$$

$$= 502.3 \text{ HZ}$$

$$\text{RESONANT FREQUENCY} = \frac{1}{2\pi\sqrt{LC}}$$

RESONANT FREQUENCY – HERTZ
2π – CONSTANT = 6.28
L = INDUCTANCE – HENRIES
C = CAPACITANCE – FARADS

Fig. 9-5 Computation of the resonant frequency of a L-C circuit

SECTION X

Three-Phase Alternating Current

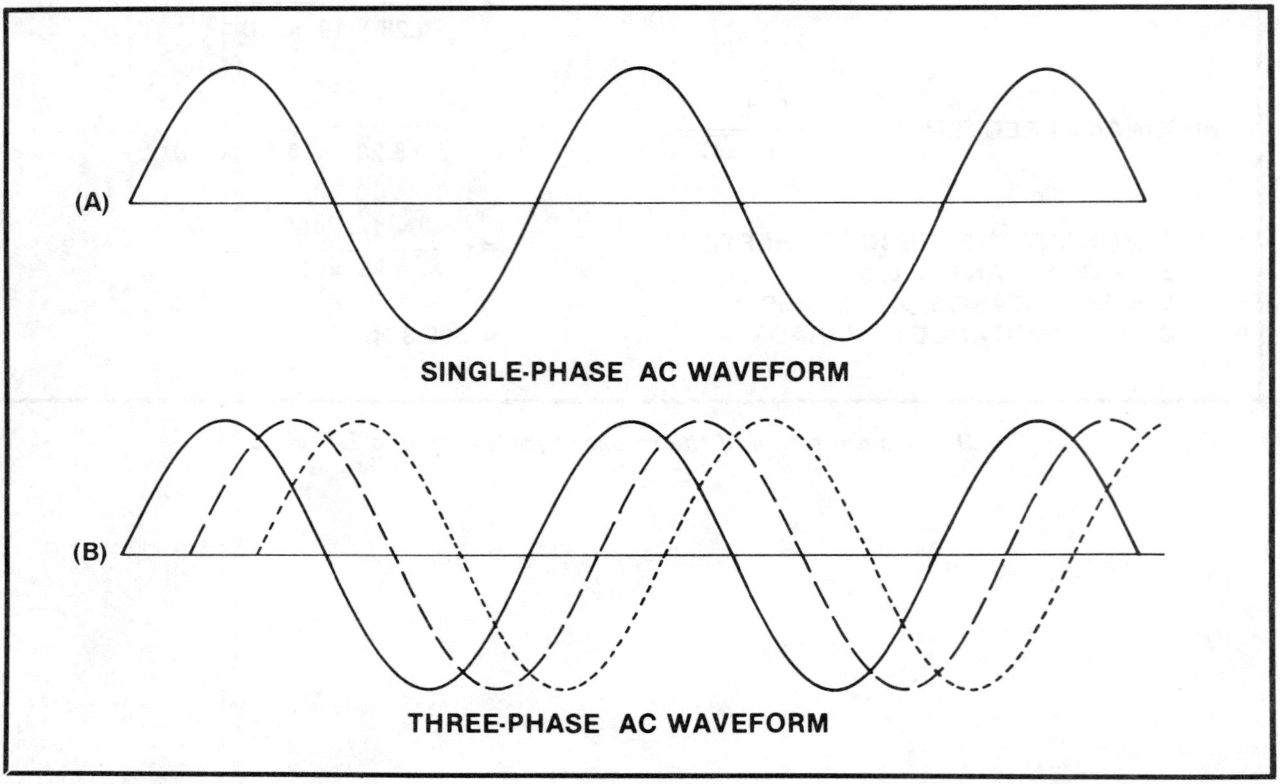

Fig. 10-1 A — Single-phase sine wave AC waveform B — Three-phase sine wave AC waveform

When it is necessary to get the maximum amount of power from an alternating current circuit, three-phase generators, or alternators, as they are more properly called, are used. These alternators have three sets of output windings excited by a single rotating field, and the voltage in each winding is 120 degrees out of phase with that in the other windings. There are several advantages of using three-phase AC, and one of the main advantages lies in the fact that when the AC is rectified, or changed into direct current, there are three times as many pulses of rectified current as there are in single-phase AC, and these pulses overlap so that the current never drops to an instantaneous value of zero. The higher frequency of these pulses also makes it easier to filter, or smooth out the ripples.

A. Y-connected Alternators

There are two ways of connecting the three output windings of an alternator. We will first discuss the Y-connected windings, in which one end of each of the three windings is connected together at a common, or neutral point. The other ends of the three windings are brought out of the alternator as the output leads, each of which is across two of the windings in series. Since the voltage across each winding is 120 electrical degrees out of phase with that in the other windings, the output voltage will never be twice that of one of the phase windings. It will, instead, be 1.73 times that of a single-phase winding. Many three-phase alternators produce 120 volts across each phase, and the voltage between any two of the output leads is 208 volts.

Since the windings are in series between the output leads, the output current is the same as the phase current.

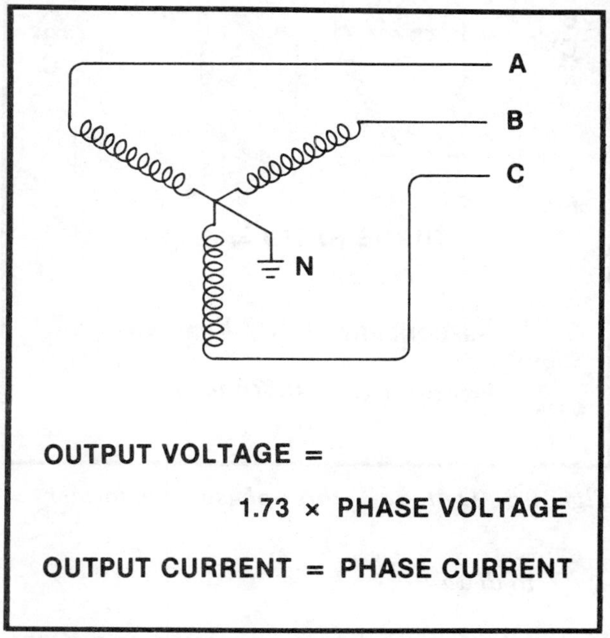

Fig. 10-2 Y-connected, three-phase alternator

B. Delta-connected Alternators

Both ends of each phase winding may be connected to the ends of the other windings to form a loop, or a delta connection. (So-called because of the triangular symbol formed by the coils resembling the Greek letter delta [Δ].) An output lead is brought from each junction so that the output voltage will always be the same as the phase voltage. But, as you see in Fig. 10-3, there are two coils in series across, or in parallel, with each of the phase windings, and since the current in each of these windings is 120 degrees out of phase with that in the other windings, the output current is 1.73 times that of the current in the phase winding.

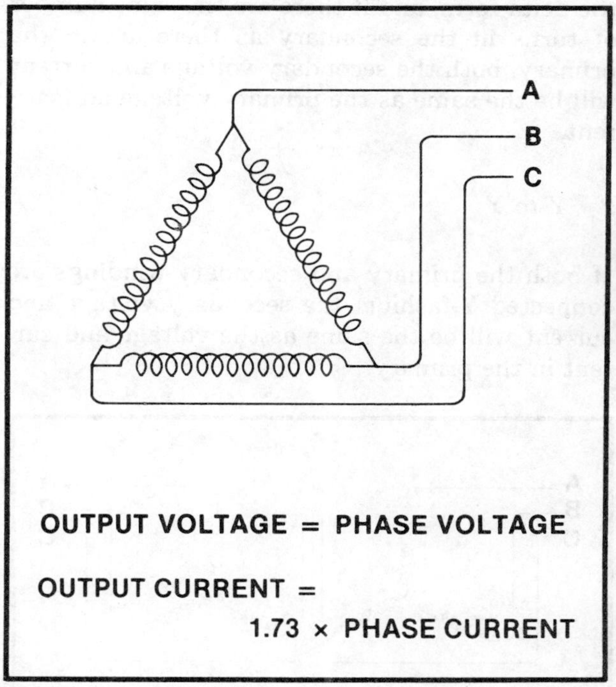

Fig. 10-3 Delta-connected, three-phase alternator

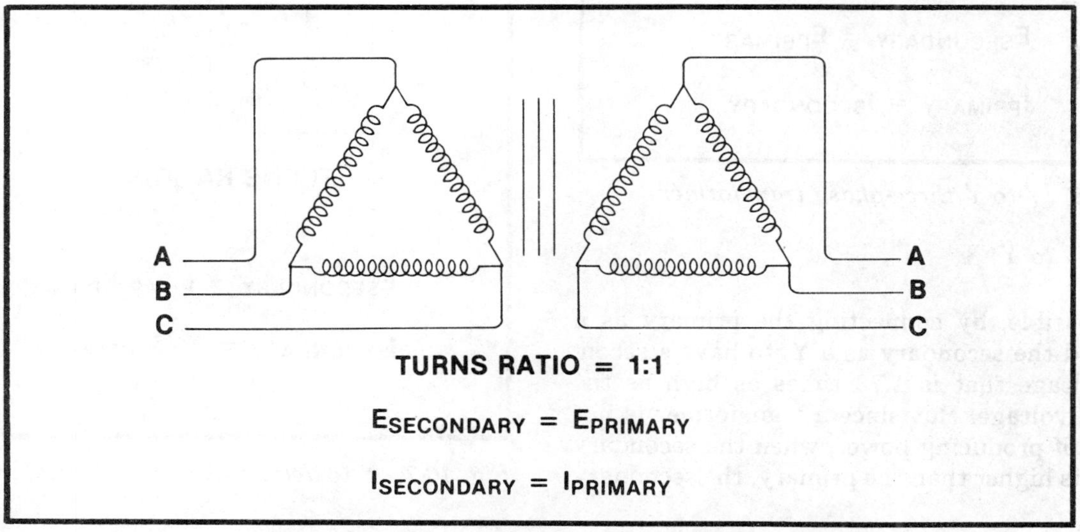

Fig. 10-4 Delta-to-delta three-phase transformer

C. *Three-phase Transformers*

Three-phase transformers may have their primary and secondary windings connected in either a Y or a delta to provide the needed output.

1. *Delta to delta*

In this type of connection, both the primary and secondary windings have their leads connected in the delta form, and if there are the same number of turns in the secondary as there are in the primary, both the secondary voltage and current will be the same as the primary voltage and current.

2. *Y to Y*

If both the primary and secondary windings are connected Y-fashion, the secondary voltage and current will be the same as the voltage and current in the primary.

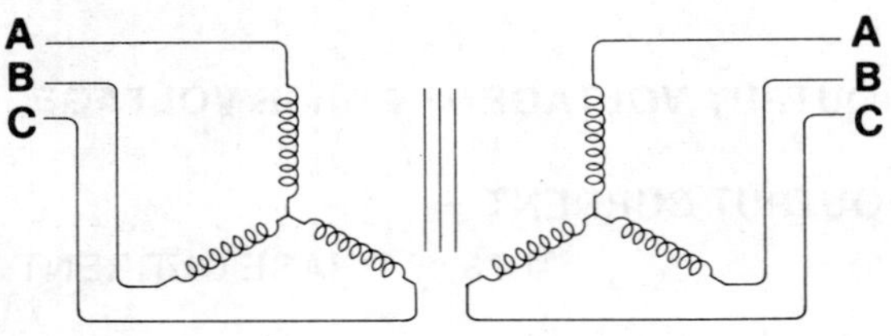

TURNS RATIO = 1:1

$E_{SECONDARY} = E_{PRIMARY}$

$I_{PRIMARY} = I_{SECONDARY}$

Fig. 10-5 Y-to-Y three-phase transformer

3. *Delta to Y*

It is possible, by connecting the primary as a delta and the secondary as a Y, to have a secondary voltage that is 1.73 times as high as the primary voltage. But since a transformer is not capable of producing power, when the secondary voltage is higher than the primary, the secondary current must be lower. The current in the secondary will, with this form of connection, be only 0.578 times the primary current.

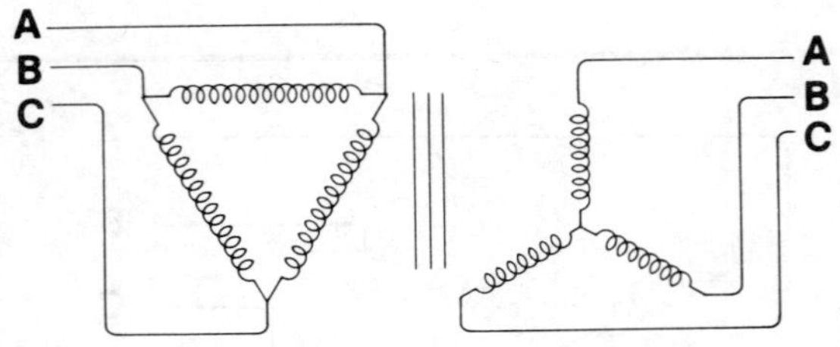

TURNS RATIO = 1:1

$E_{SECONDARY} = 1.73\ E_{PRIMARY}$

$I_{SECONDARY} = 0.578\ I_{PRIMARY}$

Fig. 10-6 Delta-to-Y three-phase transformer

4. *Y to delta*

If the primary winding is connected as a Y and the secondary as a delta, the secondary voltage will be only 0.578 of the primary voltage, but the secondary current will be 1.73 times the primary current.

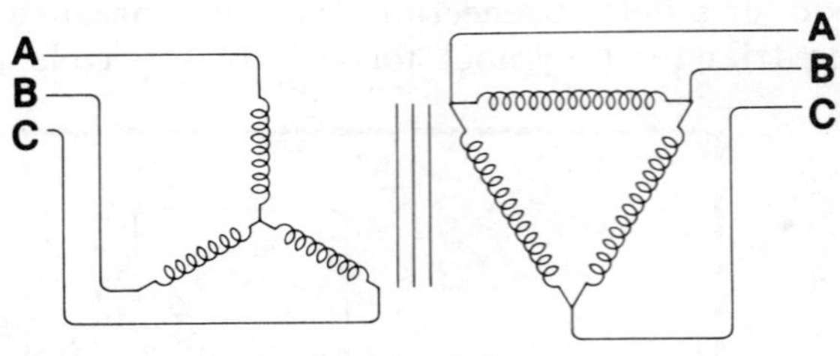

TURNS RATIO = 1:1

$E_{SECONDARY} = 0.578\ E_{PRIMARY}$

$I_{SECONDARY} = 1.73\ I_{PRIMARY}$

Fig. 10-7 Y-to-delta three-phase transformer

SECTION XI

Converting Alternating Current into Direct Current

It is often necessary to convert alternating current into direct current, and this is accomplished very simply by a rectifier circuit.

A. *The Diode Rectifier*

There have been a number of different types of rectifiers used to change AC into DC. All of these devices are simply electron check-valves that allow electrons to pass in one direction, but block them when they attempt to flow in the opposite direction. We will discuss these devices in detail later in this text, but here, we want to recognize some of the more common types.

1. *The vacuum tube*

Inside the glass envelope of a vacuum tube is an electrode called the cathode. When it is heated with a small electric heater, the electrons in its material loosen their bond and are attracted to the plate during the half-cycle of the AC when the plate is positive. During this half-cycle the current can flow, but during the half-cycle when the plate is negative, the electrons are forced back to the cathode and no current flows. This type of rectifier has been used for many years, but its large size and the need for an electric heater have caused it to be replaced by the more modern solid-state diode.

2. *The dry-disc rectifier*

Another type of rectifier that has been replaced by the solid-state diode is the dry-disc rectifier. Small discs of copper have an oxide film deposited on one side, and a lead disc is pressed against the oxide to form a conductor. When this

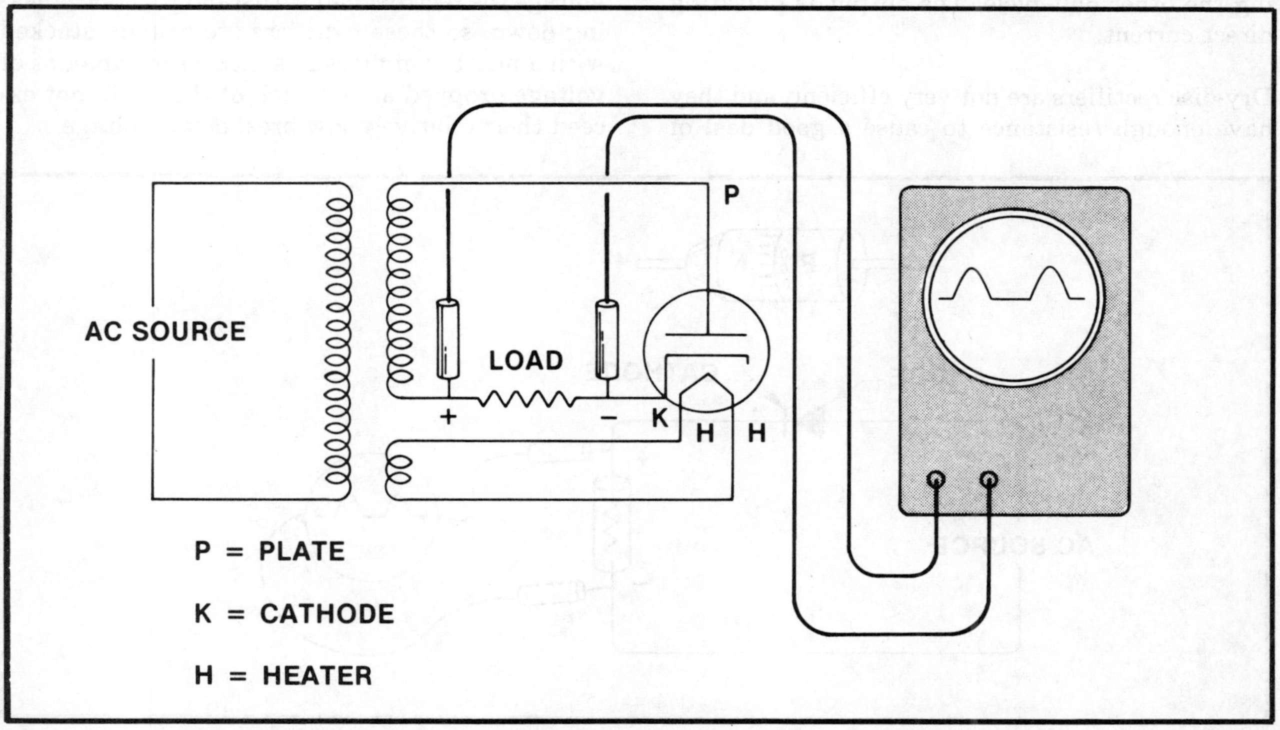

Fig. 11-1 A diode vacuum tube used as a half-wave rectifier

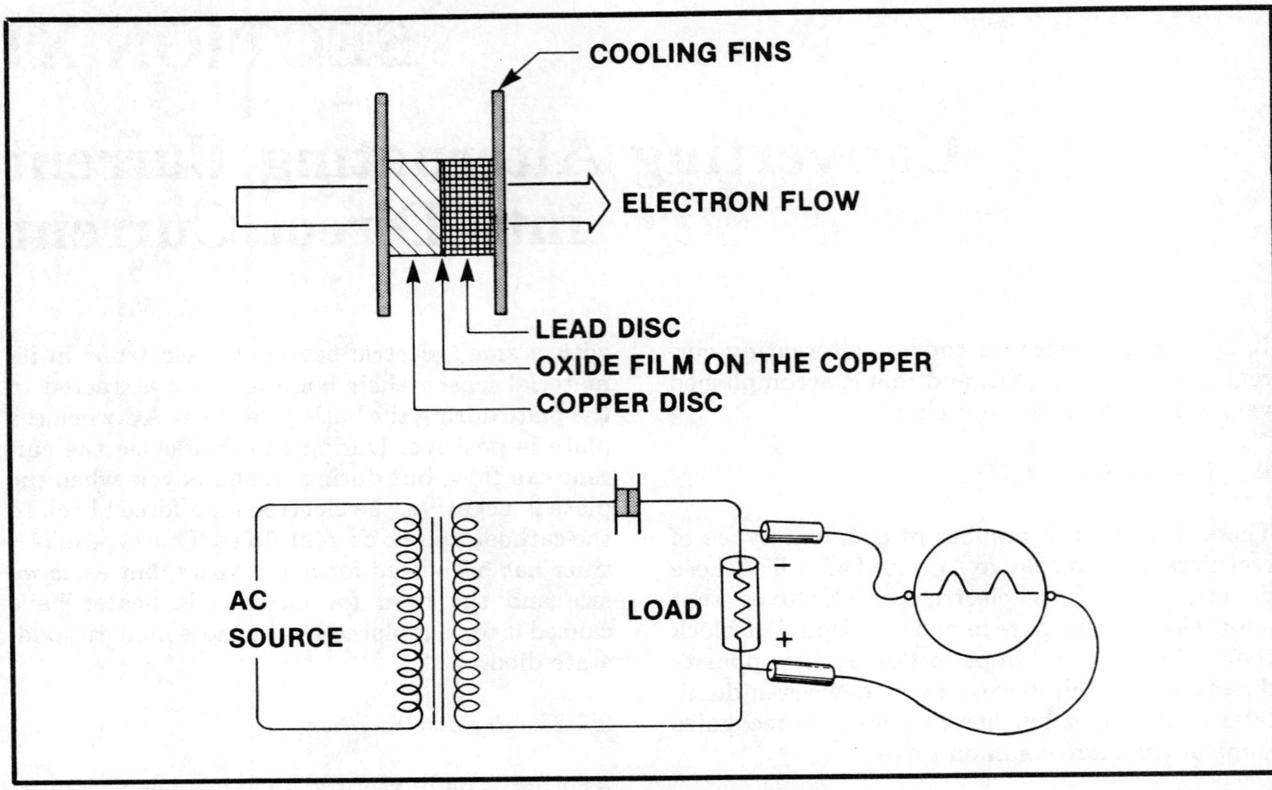

Fig. 11-2 A dry-disc rectifier

diode is placed in an AC circuit such as we have in Fig. 11-2, electrons will flow during the half-cycle when the copper is negative, but will not flow during the other half-cycle. The output is pulsating direct current.

Dry-disc rectifiers are not very efficient, and they have enough resistance to cause a good deal of heat to be generated; cooling plates are required to dissipate the heat so the rectifier won't be damaged. Another problem is the relatively low voltage the rectifier can withstand without breaking down, so these rectifiers are usually stacked with a number of discs in series so the amount of voltage dropped across each of them will not exceed their relatively low breakdown voltage.

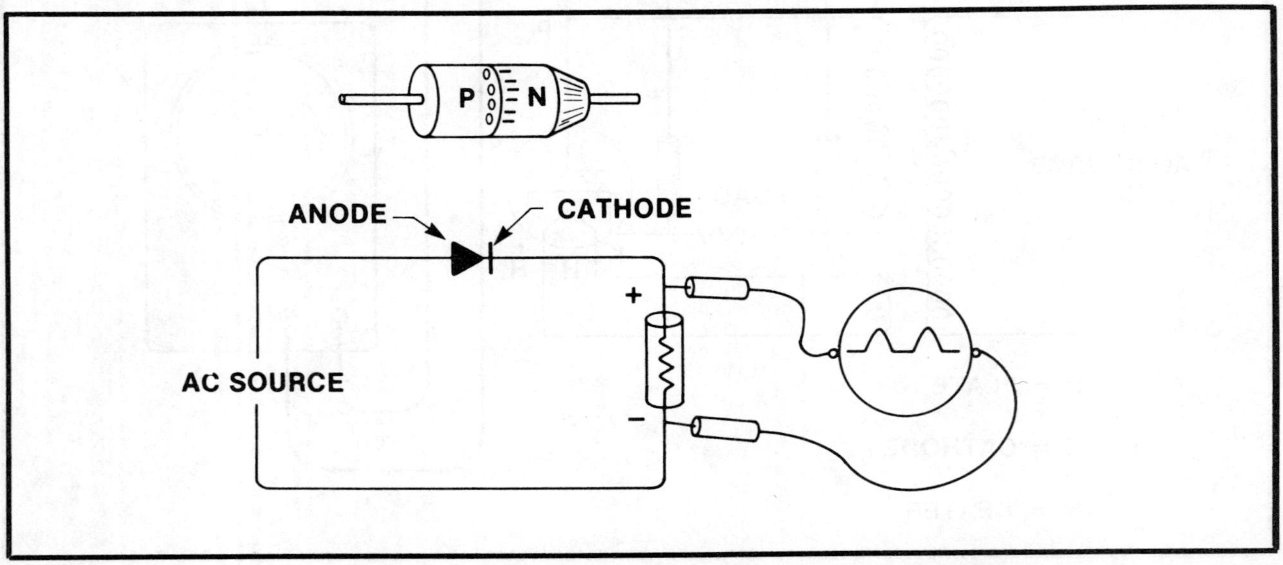

Fig. 11-3 a semiconductor diode used as a rectifier

3. The solid-state diode

Solid-state electronics ushered in a revolution in communications, control, and in actually almost all aspects of electricity. One small piece of an insulating material such as germanium or silicon is "doped" with an element having an excess of electrons and is called an N-material. Another piece of the same material is doped with an element having a deficiency of electrons, and this is called the P-material. These two pieces are fused together and they act as a check valve. Electrons can easily pass from the N-material to the P-, but they cannot flow from the P- to the N-.

When a solid-state diode is placed in the circuit such as that in Fig. 11-3, the alternating current is changed into pulsating direct current.

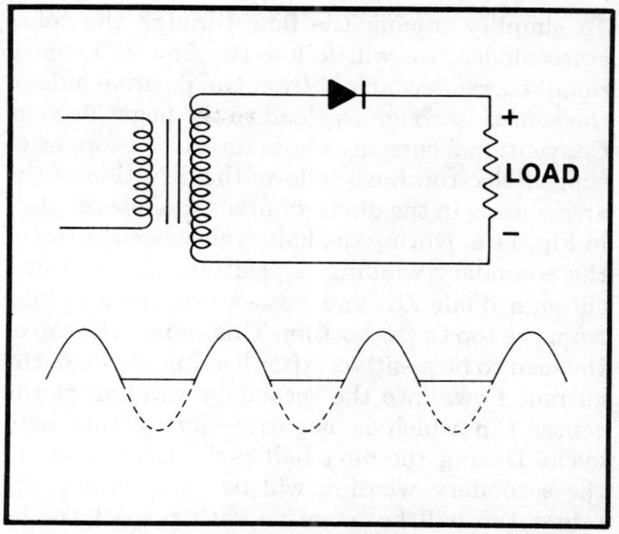

Fig. 11-4 A semiconductor diode half-wave rectifier with its output waveform

B. Half-wave Rectifier

A half-wave rectifier circuit uses a single diode in series with the voltage source, which is the secondary of the transformer in Fig. 11-4, and the load. Electrons can flow only during the half-cycle when the cathode, represented by the bar across the arrowhead, is negative. The output waveform of this type of rectifier is one-half of the alternating-current wave, and because of this, it is a very inefficient type of circuit.

C. Full-wave Rectifier

In order to change both halves of the AC cycle into DC, we can use a transformer with a center-tapped secondary winding and two diodes.

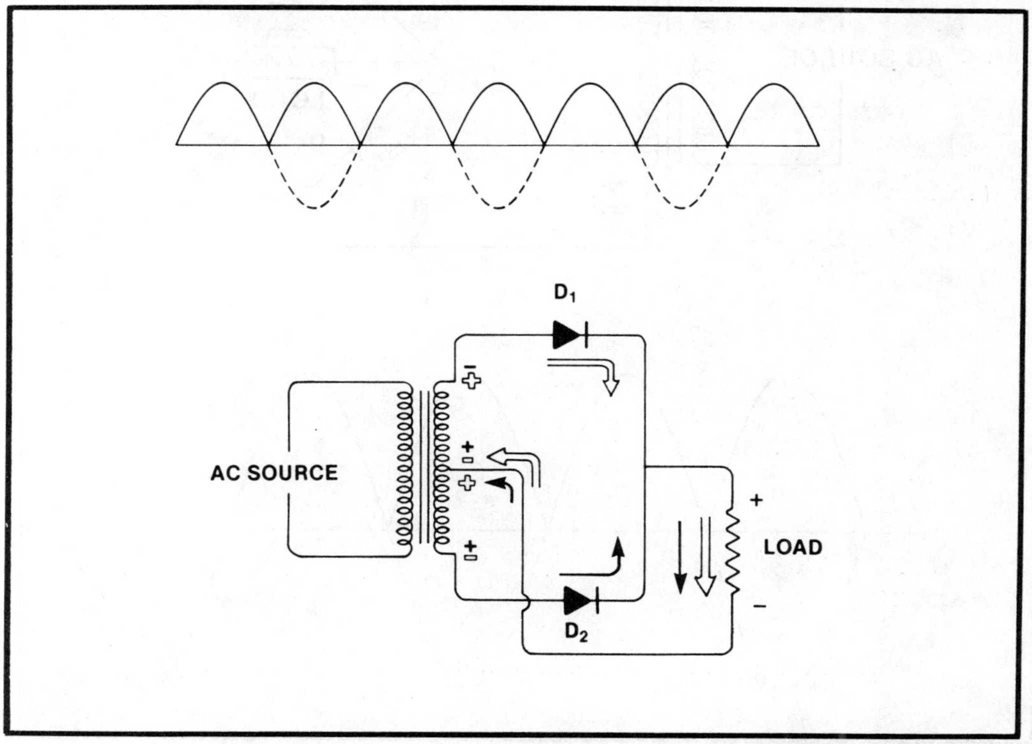

Fig. 11-5 A two-diode, full-wave rectifier with its output waveform

To simplify tracing the flow through the solid-state diodes, we will follow the flow of *conventional* current, which is from the positive side of the source, through the load to the negative side. Conventional current, whose direction is opposite that of electron flow, follows the direction of the arrowheads in the diode symbols. Follow the flow in Fig. 11-5. During the half-cycle when the top of the secondary winding is positive, current flows through diode D_1, and passes through the load from the top to the bottom. This causes the top of the load to be positive. After leaving the load, the current flows into the secondary winding at the center tap which is negative during this half-cycle. During the next half-cycle, the bottom of the secondary winding will be positive and the center tap will be negative with respect to the bottom. Current will flow through diode D_2 and through the load resistor in the same direction it passed during the first half-cycle. The output waveform is pulsating direct current, whose frequency is twice that of the pulsating DC produced by a half-wave rectifier.

D. *Bridge-type Full-wave Rectifier*

The two-diode full-wave rectifier requires a transformer that will give the desired output voltage across just one-half of the secondary winding. This is inefficient, and this inefficiency may be overcome by the bridge-type full-wave rectifier, in which we can use the entire secondary winding. To do this, we must use four diodes instead of two.

We can trace the flow through the load in Fig. 11-6. During the half-cycle when the top of the transformer secondary is positive, the current flows through the diode D_1 and through the load resistor from the right to the left, and then down through diode D_2 and back to the bottom of the secondary winding, which is negative. During the next half-cycle, the polarity of the secondary has reversed, and current flows through diode D_3, through the load in the same direction as before, and up through diode D_4 to the top side of the transformer, which at this time is negative. The

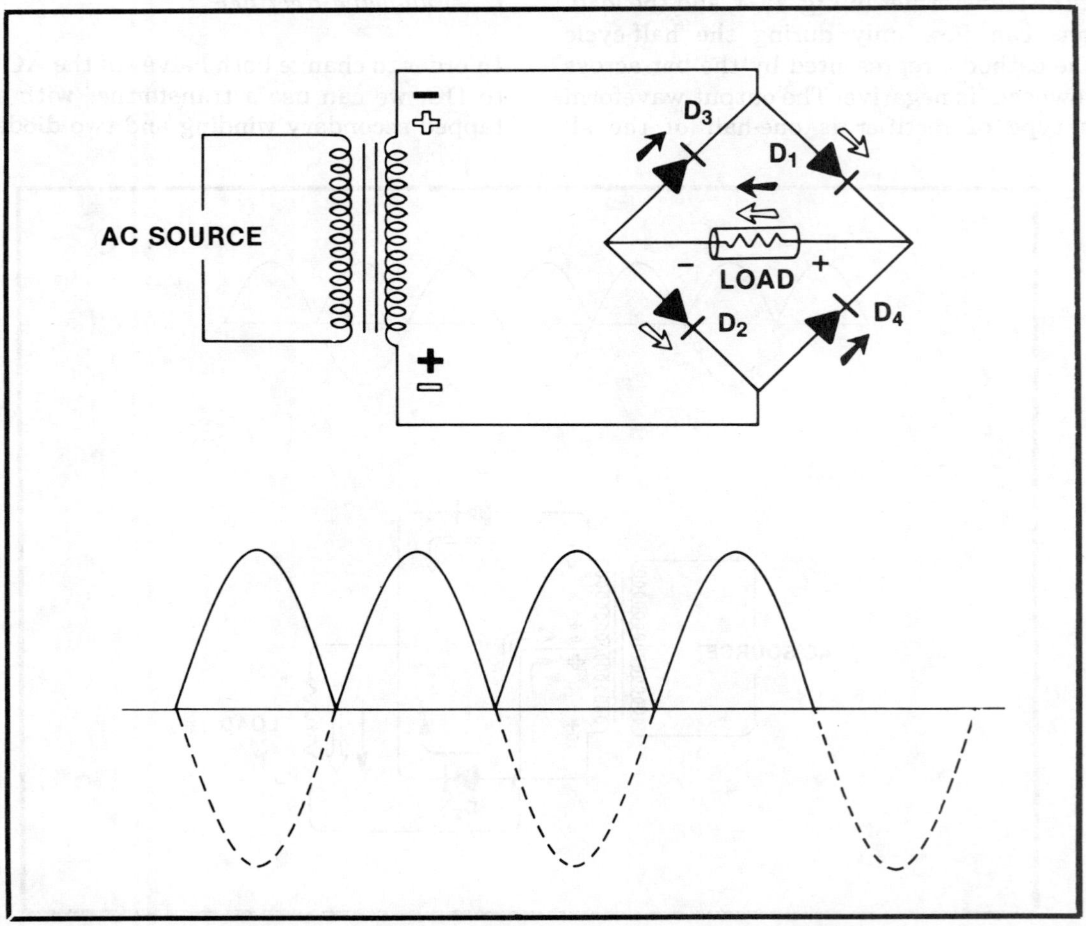

Fig. 11-6 A four-diode, bridge-type, full-wave rectifier with its output waveform

output waveform is similar to that produced by the two-diode full-wave rectifier, but the voltage is quite a bit higher because with this circuit the entire secondary winding is used.

E. Full-wave Three-phase Rectifier

The most familiar six-diode three-phase full-wave rectifier is the one found in the DC alternator, which is rapidly replacing the generator as a device for producing direct-current electricity for our modern aircraft. This alternator uses a three-phase stator and six silicon diodes arranged as those in Fig. 11-8.

Fig. 11-7 Most aircraft direct current is produced by a three-phase alternator.

Let's examine the current flow through the load resistor for one complete cycle of all three phases. Remember that we are tracing conventional current which is opposite the electron flow, but which follows the direction indicated by the arrowheads in the diode symbols. In that portion of the cycle when the output end of phase *A* is positive, current leaves it and flows through diode D_1 and down through the load, making its top end positive. After leaving the load, the current flows through diode D_2 and coil *C*, whose output lead is negative. Now, as the alternator field rotates, it causes the output end of the coil *B*to become positive and the output of coil *A* to be negative. Current flows out of *B*, through diode D_3, the load, diode D_4, and back through coil *A*. Continued rotation causes the output of coil *C* to be positive and *B* to be negative, so the current leaves *C*, passess through diode D_5, the load and diode D_6 and back into coil *B*.

The output waveform of this rectifier gives us a very steady direct current as the current from the

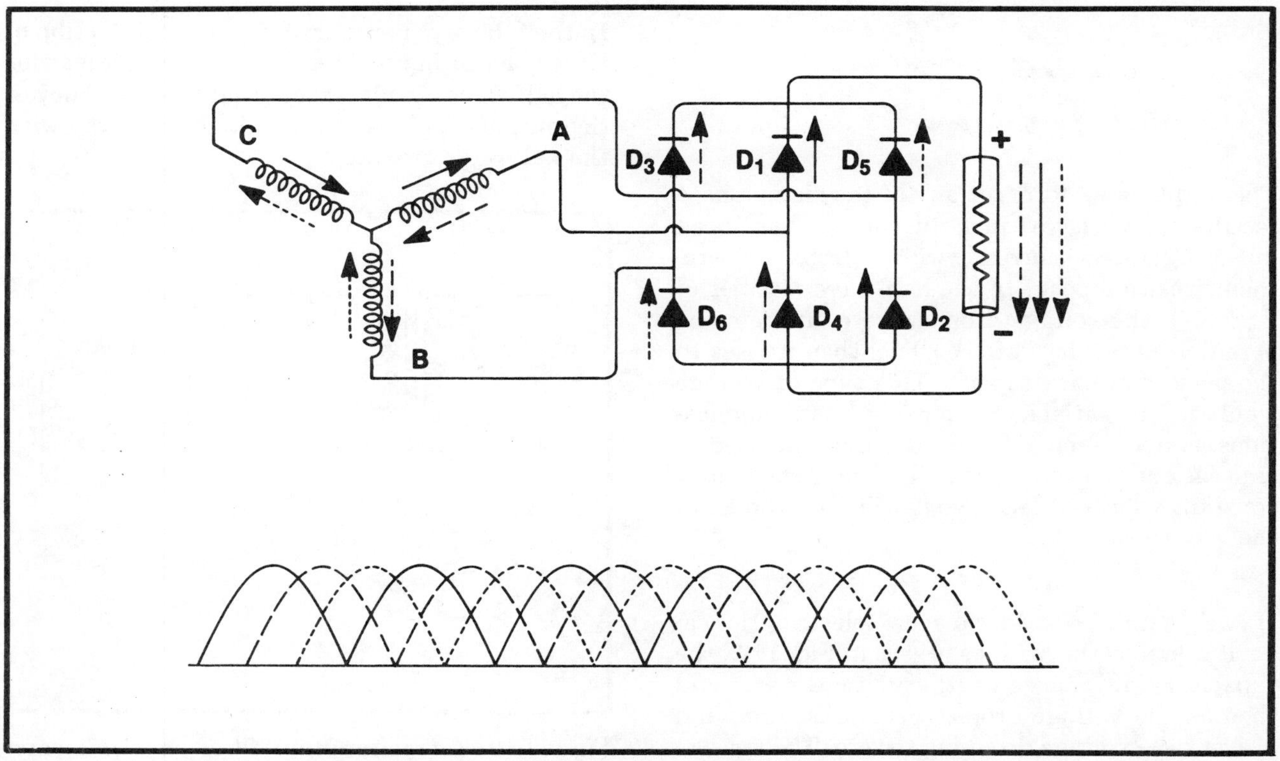

Fig. 11-8 Three-phase, six-diode, full-wave rectifier and its output waveform

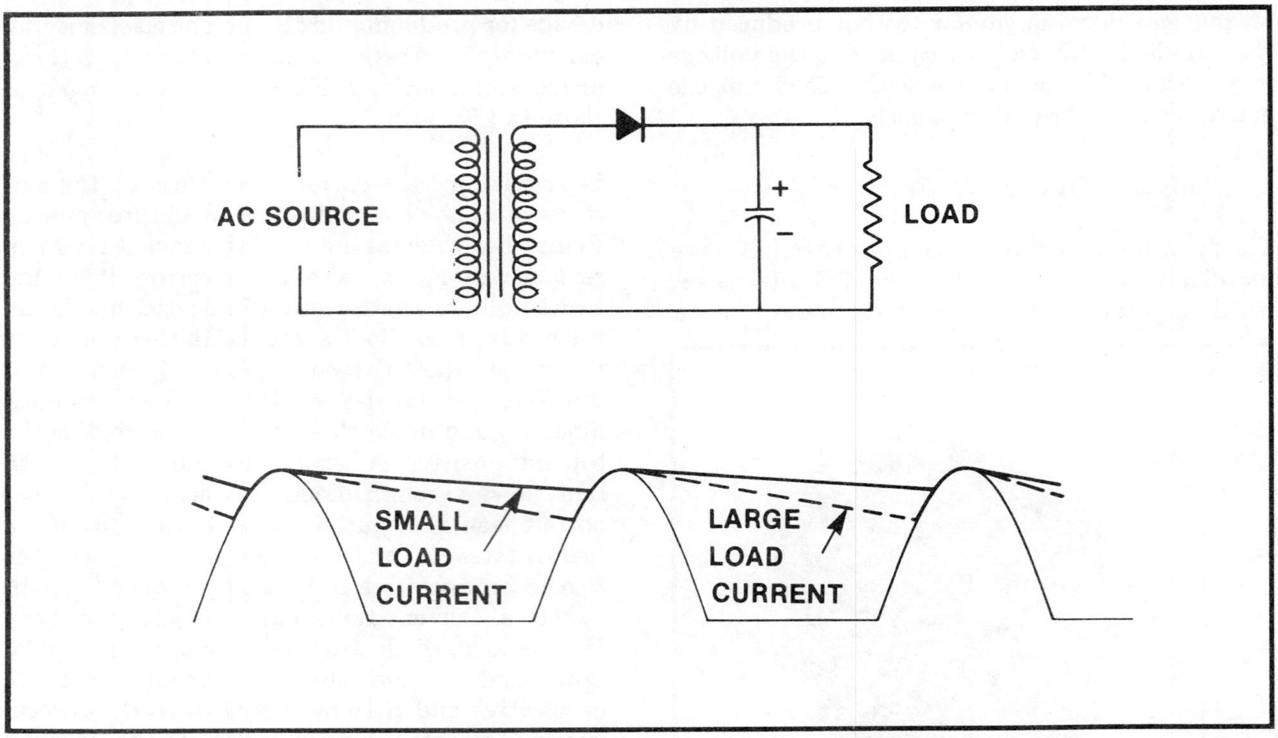

Fig. 11-9 Capacitor-input filter on a half-wave rectifier

three phases overlaps and there is never a time when the current drops to zero.

F. Filters

1. Capacitor-input filters

The output waveform for any of the single-phase rectifiers we have just discussed has been pulsating direct current, which drops to zero, then rises to a peak. In the half-wave rectifier of Fig. 11-4, the voltage drops to zero and remains here during one-half of the cycle; then it rises to the peak, and back to zero. This type of voltage would not be satisfactory for very many applications, and so some form of filter must be used to smooth out the DC so it will remain at a fairly constant value—at least, so it will not drop all of the way to zero.

If we install a large-capacity electrolytic capacitor across the load, as we see in Fig. 11-9, the capacitor will charge as the voltage rises, and then as the voltage drops, it will discharge and prevent the voltage dropping to zero. The voltage will follow the curve shown in Fig. 11-9. A large flow of load current will cause the voltage to drop more than a small current.

In the full-wave rectifier of Fig. 11-5, the action of the capacitor input filter is the same as it was for the half-wave circuit, except that the frequency of the output ripple is twice as high as it was with the half-wave circuit.

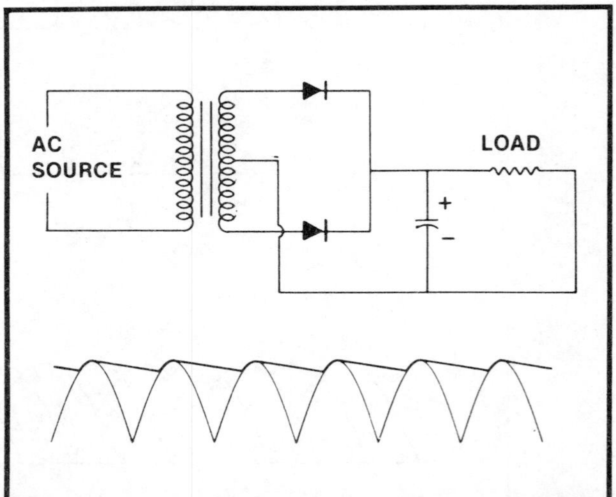

Fig. 11-10 Capacitor-input filter on a full-wave rectifier

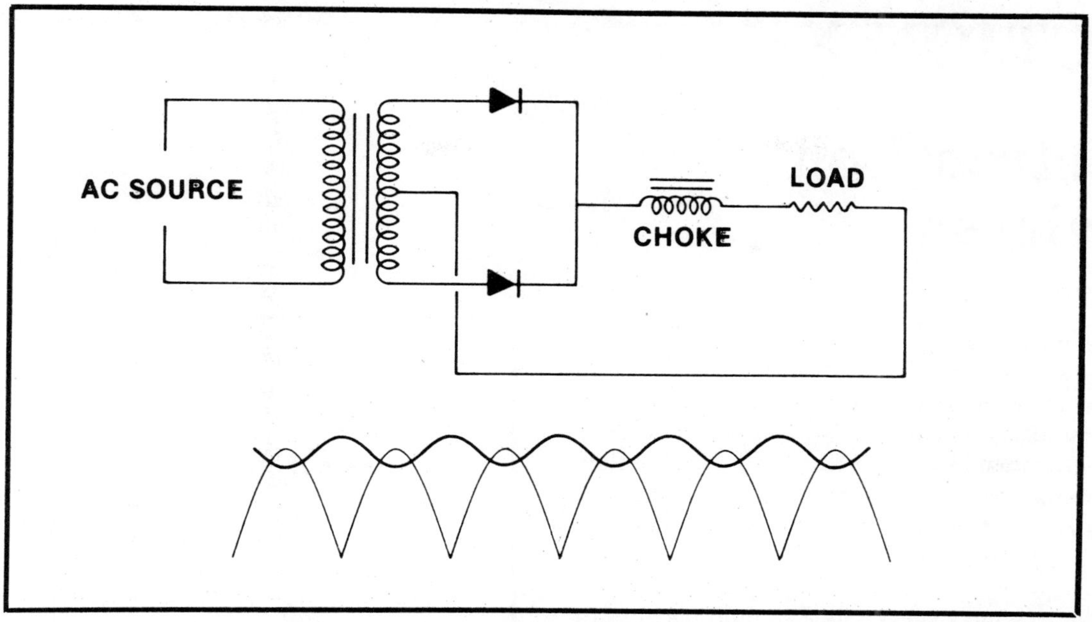

Fig. 11-11 Choke-input filter on a full-wave rectifier

2. *Choke-input filter*

If a choke (inductor) is placed in series with the load, the effect of the inductance will oppose any change in the load current. During the part of the cycle in which the current is increasing, the magnetic field in the inductor will induce a current that opposes the rise. When the source current starts to drop off, the inductor will produce a current that opposes the drop-off. The resulting waveform is a low alternating voltage whose frequency is twice that of the source.

3. *Pi filter*

There are a number of arrangements of capacitors and inductors for filtering the ripple from rectified AC, but one of the most popular is that of a capacitor input filter whose circuit diagram resembles the Greek letter pi (π). In Fig. 11-12, we see a pi filter connected to a full-wave rectifier. The two capacitors are high-capacity electrolytics which have a minimum reactance to the ripple frequency of the rectified AC, but which will not allow any flow of DC to ground. The inductor has a high opposition to the AC caused by the ripple, but offers very little opposition to the flow of the DC. With this arrangement, the ripple frequency AC is passed to ground and leaves an almost pure DC output.

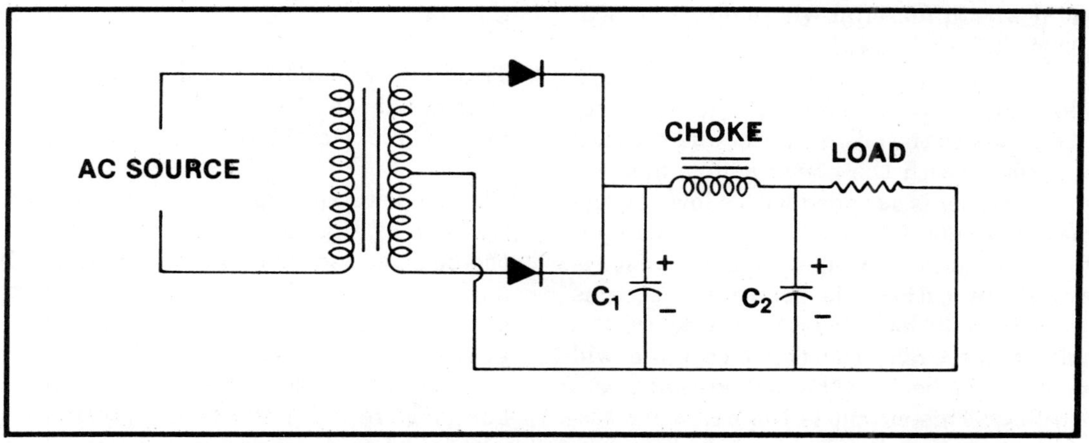

Fig. 11-12 Pi-filter on a full-wave rectifier

SECTION XII:

Electron Control Devices

A. *Vacuum Tubes*

The age of electronics was ushered in with the discovery of the vacuum tube, or the Audion, by Dr. Lee DeForest in 1907. With this device the flow of electrons could be controlled by electrical charges rather than by mechanical switches and power dissipating resistors.

These devices have served us well, and because of them we have radio communication, television, and the electronic computer. Because of their large physical size, however, the great amount of power they require, and the heat they generate, they have been replaced almost entirely by the more modern solid-state devices and by the still more revolutionary integrated circuit (IC) chips.

We do need to understand the working of the vacuum tube, as its principle gives us a background for these more efficient and even exotic electron control devices.

1. *Diodes*

The simplest vacuum tube is the diode, or two-element tube. In a glass or metal container called an envelope, we have two active elements and a small electric heater. The cathode is a small metal tube coated with a material whose electrons will weaken their bond when it is heated.

The diode is installed in the circuit with its plate attached directly to the power source and the load resistor in series with the power source and the cathode. The heater is supplied with a low voltage which may be either AC or DC. During the half-cycle of the AC source voltage, when the plate is positive and the cathode is negative, electrons will flow through the load resistor. But during the other half of the cycle, the negative plate will repel the electrons back to the cathode and there will be no flow. This circuit is the half-wave rectifier we have previously discussed, and its output is pulsating direct current.

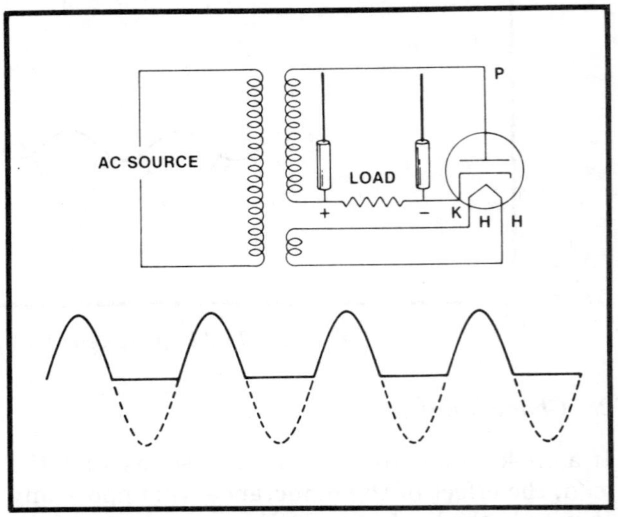

Fig. 12-1 Vacuum tube diode used as a half-wave rectifier

2. *Triodes*

The diode acts as an electron check-valve and allows electrons to flow only when the plate is positive with respect to the cathode. But the triode, or three-element tube, is a *control* valve rather than a check valve, and by varying the voltage on the grid of the tube we can control the amount of electron flow between the cathode and the plate.

Triodes have a plate, a cathode, and a heater just as the diode has, but between the cathode and the plate, there is a spiral of wire called the grid.

The triode tube is connected into a circuit such as we have in Fig. 12-2, with the plate connected to the positive terminal of a high-voltage DC power supply through the load resistor. The cathode is connected to the negative terminal, or the ground, of the power source. When the heater is warmed up, electrons are emitted from the cathode and are attracted to the positively charged plate, and there is electron flow within the tube. Now, if a negative voltage is placed on the grid, it will repel the electrons back to the cathode so they cannot be influenced by the positive plate. A high enough negative potential on the grid will completely stop the flow of electrons. But, if a positive potential is placed on the grid, electrons will be accelerated on their way to the plate, and the flow will increase beyond the amount of flow with no charge on the grid. The grid is so small relative to the plate, and the positive potential on the grid is so much lower than that on the plate, that only an extremely few electrons will be attracted to the grid, and the grid current is negligible compared to the flow between the cathode and the plate.

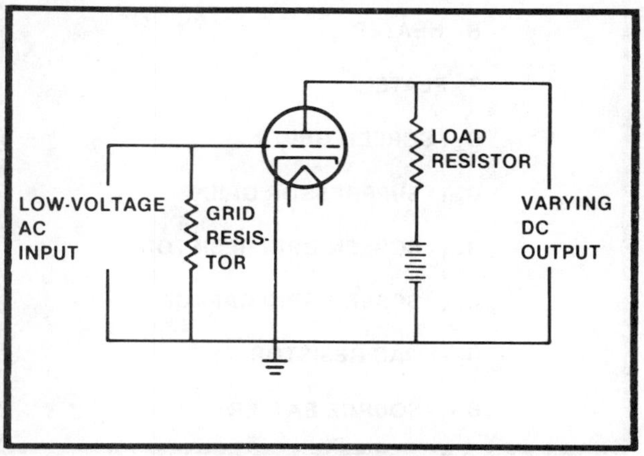

Fig. 12-2 Triode vacuum tube connected as an amplifier

If a low-voltage AC signal is placed across the grid resistor between the grid and the ground, the amount of *current* through the tube will vary in the same way as the *voltage* on the grid. In Fig. 12-3, we see the way an AC input causes a varying DC output. When there is no signal voltage on the grid, the output is a steady flow of direct current. When the grid has a positive voltage on it, the electron flow through the tube increases, and the voltage drop across the load resistor increases. The polarity of the voltage across the load resistor is opposite that of the power source and, as a result, as the grid becomes *more* positive, the output becomes *less* positive. When a negative voltage on the grid makes it so negative that all flow through the tube is shut off, there is no voltage drop across the load resistor, and the output voltage rises to the value of the DC source. In this circuit, the triode tube acts as an amplifier, which changes the value of the input signal and, as we see, it also reverses the phase of the amplified signal.

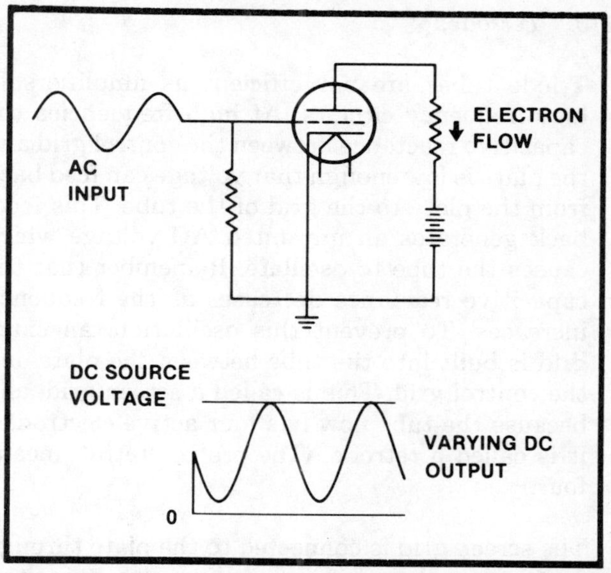

Fig. 12-3 A triode vacuum tube amplifier increases the voltage of the input signal and inverts its phase.

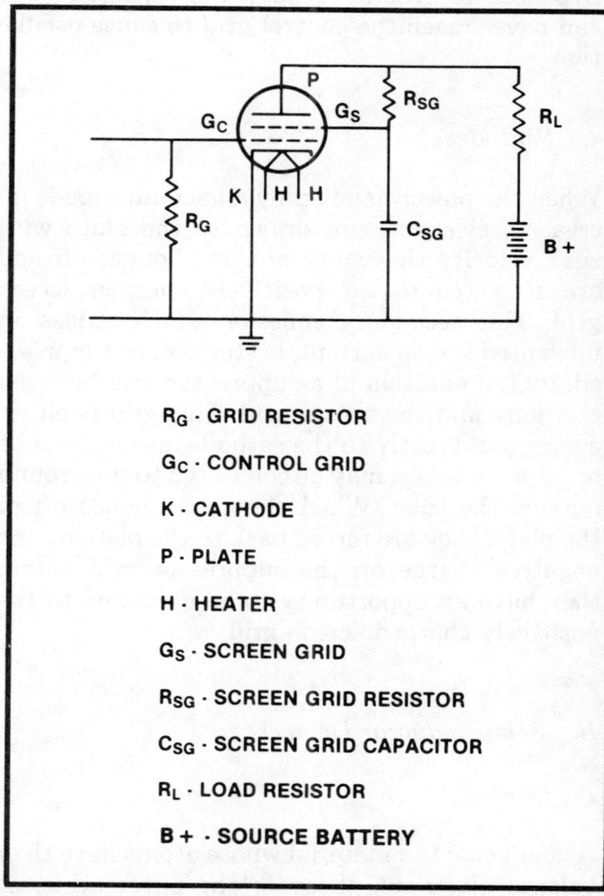

Fig. 12-4 A tetrode vacuum tube uses a screen grid to minimize the interelectrode capacitance between the plate and the control grid.

3. *Tetrodes*

Triode tubes are not efficient as amplifiers in high-frequency circuits. At high frequencies the capacitive reactance between the control grid and the plate is low enough that voltage can feed back from the plate to the grid of the tube. This feedback generates an unwanted AC voltage which causes the tube to oscillate. Remember that the capacitive reactance decreases as the frequency increases. To prevent this oscillation, an extra grid is built into the tube between the plate and the control grid. This is called a screen grid, and because the tube now has four active electrodes, it is called a tetrode. (The prefix "tetra" means four.)

The screen grid is connected to the plate through a resistor and to the ground through a capacitor. Since the screen grid is between the control grid and the plate, any AC signal fed back from the plate will be intercepted by the screen grid and bypassed to ground through the capacitor so it can never reach the control grid to cause oscillation.

4. *Pentodes*

When the power handled by a vacuum tube is increased, electrons are drawn to the plate with such velocity that some of them bounce off and are attracted to the positively charged screen grid. This secondary emission which causes an unwanted screen current, is stopped, or suppressed, by the addition of a suppressor grid between the plate and the screen grid. This grid is either connected directly to the cathode inside the tube or, in some tubes, may be connected to the ground outside the tube. When electrons bounce off of the plate, they are forced back to the plate by the negative charge on the suppressor grid before they have an opportunity to be attracted to the positively charged screen grid.

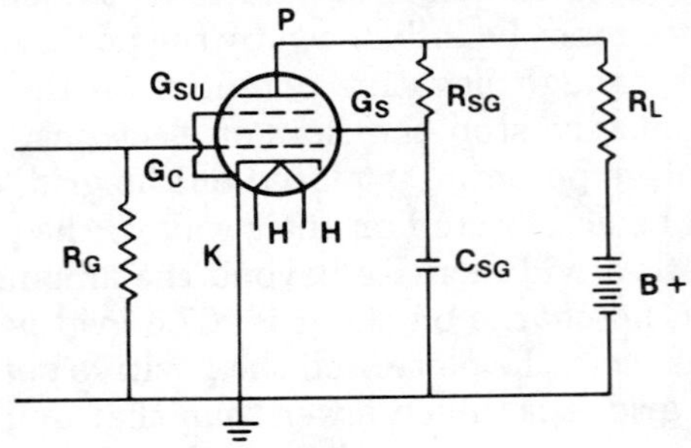

R_G - GRID RESISTOR

G_C - CONTROL GRID

K - CATHODE

H - HEATER

P - PLATE

G_S - SCREEN GRID

G_{SU} - SUPPRESSOR GRID

R_{SG} - SCREEN GRID RESISTOR

C_{SG} - SCREEN GRID CAPACITOR

R_L - LOAD RESISTOR

B + - SOURCE BATTERY

Fig. 12-5 A pentode vacuum tube uses a suppressor grid to suppress secondary emission back into the plate.

B. *Semiconductor Devices*

1. *Diodes*

A conductor is a material whose atoms have their valence electrons, those in the outer orbit, so loosely bound to the nucleus that a relatively small force can cause them to move out and be replaced by electrons dislodged from an adjoining atom. An insulator, on the other hand, has its valence electrons so tightly bound with covalent bonds to those in the adjacent atoms that a high voltage is needed to dislodge and move them.

Germanium and silicon both have four electrons in their outer shell and are good examples of insulators. The valence electrons of one of their atoms are held tightly in covalent bonds with those of the adjoining atoms, and there are no loose electrons to give up, nor are there any vacancies to accept free electrons.

If the silicon or germanium are doped—that is, if a few parts per million of an element such as arsenic, bismuth, or antimony, which have five electrons in their outer orbit, is mixed in—there will be extra, or free, electrons in the material after all of the covalent bonds are formed. These electrons are free to move, and so the material is called a donor, or an N-type material.

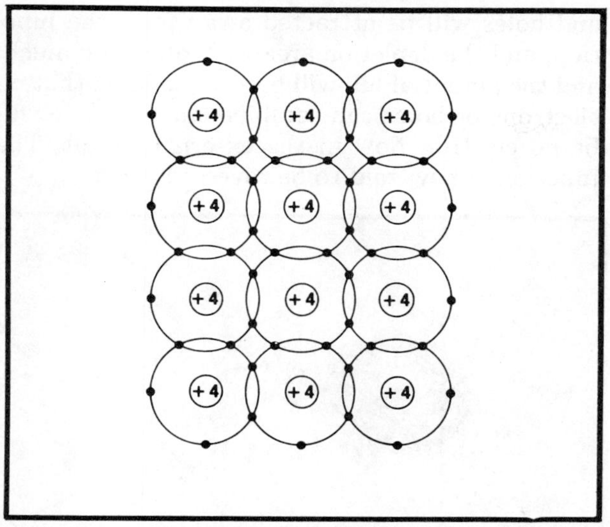

Fig. 12-6 All of the valence electrons in an insulator are tightly bound with covalent bonds, and the material will not accept electrons from an outside source.

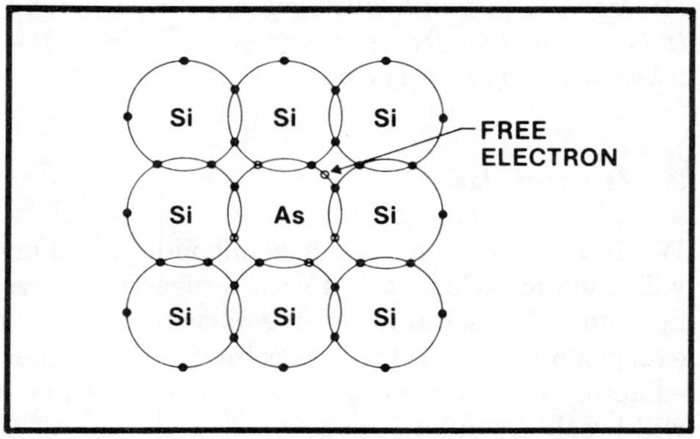

Fig. 12-7 When an insulator is doped with an impurity having five valence electrons, it produces an N-type material, one with an excess of electrons.

Elements such as boron, indium, and gallium have only three valence electrons, and so when a few parts per million is alloyed with silicon or germanium there will be areas where covalent bonds have not formed. These areas are called holes and will accept electrons from an outside source. Material doped in this way is called an acceptor, or a P-type material.

N- and P-type silicon or germanium may be joined either by a junction or a point contact, to form a semiconductor diode. To understand the way this diode works, let's consider a junction formed by a piece of P-type and a piece of N-type silicon. The

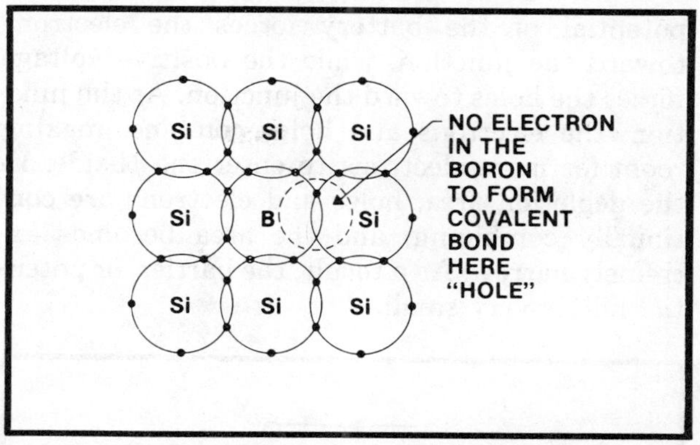

Fig. 12-8 When an insulator is doped with an impurity having only three valence electrons, it produces a P-type material which will accept free electrons.

holes in the P-type material attract the electrons in the N-type and, at the junction, they combine, leaving a depletion area where there are no more free electrons or holes. When the electrons leave the N-material, its side of the junction assumes a slight positive charge because of the lack of electrons. As the holes leave the edge of the P-material, its side assumes a negative charge. These charges constitute a barrier, or a potential hill, whose intensity is proportional to the width of the depletion area.

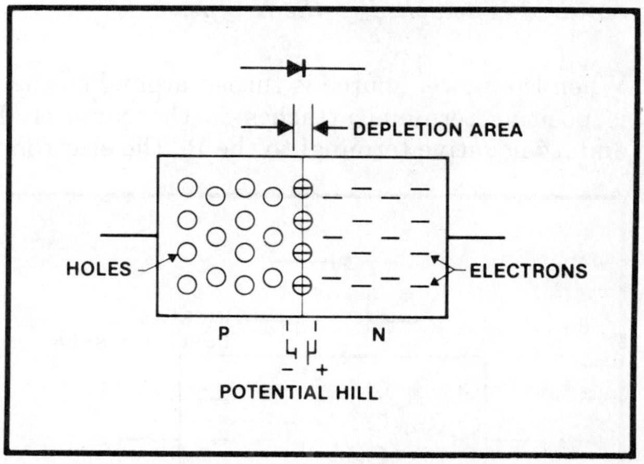

Fig. 12-9 When a piece of P-type and N-type silicon are joined, the electrons and holes combine along the junction and form a depletion area, or a potential hill.

If a voltage source is attached to a semiconductor diode with its positive terminal connected to the P-material and its negative terminal to the N-material, the diode is said to be forward-biased, and electrons can flow through it. The negative potential of the battery forces the electrons toward the junction, while the positive voltage forces the holes toward the junction. At the junction, the electrons and holes combine, making room for more electrons to enter the N-side. In the depletion area, holes and electrons are continually combining, and the area becomes extremely narrow. As a result, the barrier, or potential hill, is very small.

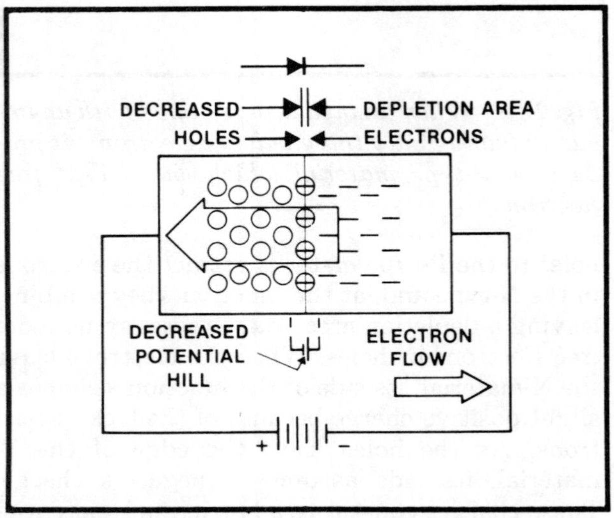

Fig. 12-10 A semiconductor is forward-biased when the positive terminal of the power source is attached to the P-type material and the negative terminal is attached to the N-type.

When the power source is turned around so that is positive terminal attaches to the N-material and its negative terminal to the P-, the electrons

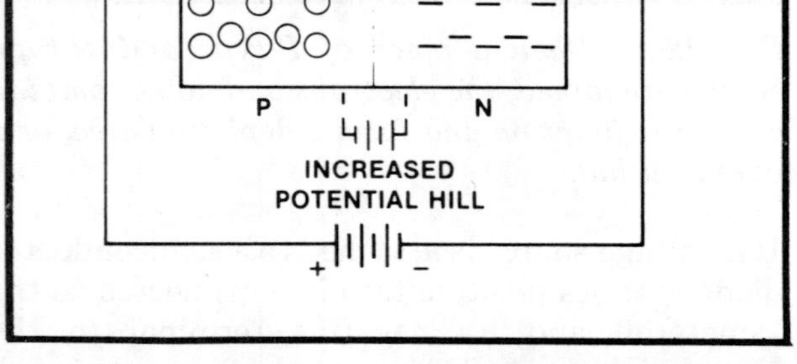

Fig. 12-11 When a semiconductor is reverse-biased, no current can flow.

and holes will be attracted away from the junction; and the depletion area will enlarge so much, and the potential hill will become so high, that no electrons or holes can combine, and so there will be no electron flow in the external circuit. The junction is now said to be reverse-biased.

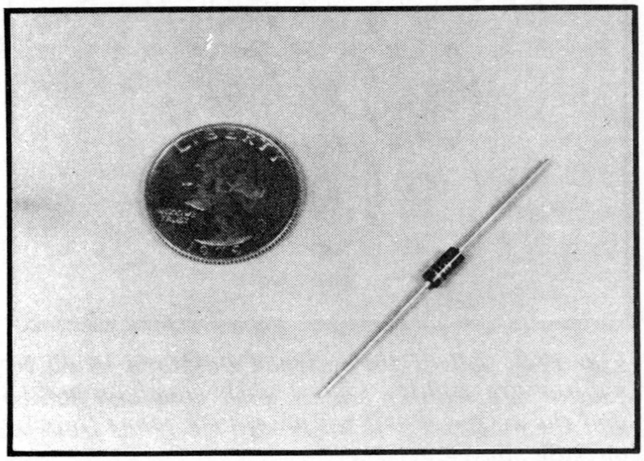

Fig. 12-12 A semiconductor diode acts as an electron check valve, allowing current flow when it is forward-biased, but not allowing any flow when it is reverse-biased.

2. *Zener diodes*

We have just seen that a semiconductor diode will conduct when it is forward-biased and will not conduct when it is reverse-biased. There is an exception to this. When a specific voltage is placed across the diode in its reverse-bias direction, it will break down and conduct.

In Fig. 12-13 we have a curve showing the way current through a diode varies with the voltage applied between its two ends. When it is forward-biased, the current increases almost uniformly as the voltage is increased, until at a certain voltage there will be enough current through the diode that it will burn out. When the diode is reverse-biased, there will be only an extremely small amount of current flow until a voltage known as the zener voltage of the diode is reached. At this voltage, the diode will break down and will act as an extremely good conductor. Unless a limiting resistor is installed in the circuit to limit the current, the diode can be burned out.

Zener diodes are used in circuits to protect against spikes of induced voltage and are familiar to all of us as the voltage-sensing component in solid-state voltage regulators.

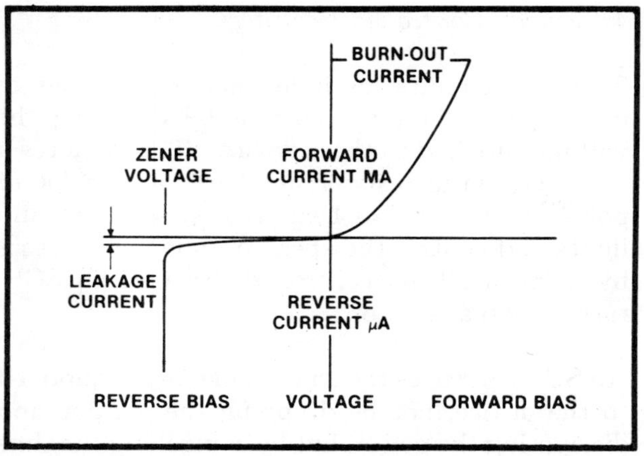

Fig. 12-13 Zener diode conduction characteristics

3. *Transistors*

A solid-state diode acts as a check valve, but a three-element device can be used as a control valve. In much the same way that the triode vacuum tube opened up the field of electronics, the triode semiconductor, or the transistor, has made possible the miniaturization of electronic devices.

***Fig. 12-14** The transistor serves as an electron control valve in much the same way as a vacuum tube, but it is much smaller and does not need power for a heater.*

A transistor is essentially a sandwich of N-type silicon or germanium between two pieces of P-type, or of a piece of P-type material between two pieces of N-type. The center piece of material is extremely thin and is called the base and is the control element of the transistor. One end piece is the emitter and the other is the collector.

Transistors having the emitter and collector of N-type material are known as NPN transistors and are identified in circuit diagrams by the arrowhead on the emitter pointing *out of* the transistor. PNP transistors have the arrowhead on the emitter pointing *into* the symbol. The operation of these two types of transistors is the same, the only difference being the polarity of the power sources required to provide the bias for their operation.

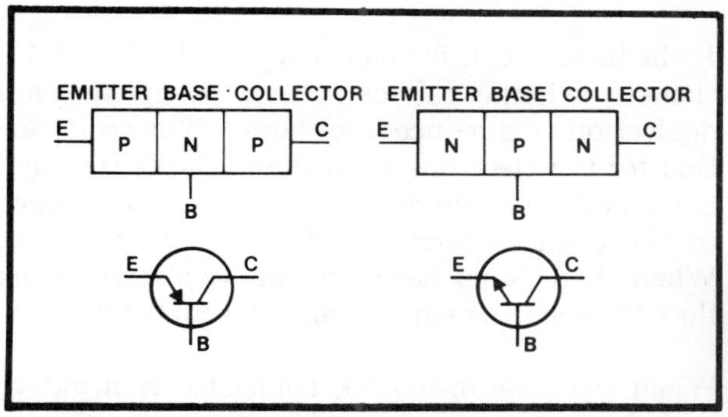

Fig. 12-15 Commonly used transistor symbols

For a transistor to conduct current between its emitter and its collector, the emitter-base junction must be forward-biased, while the collector-base junction is reverse-biased. A small amount of current flows between the emitter and the base, and this causes an extremely narrow depletion area. The relatively large reverse bias between

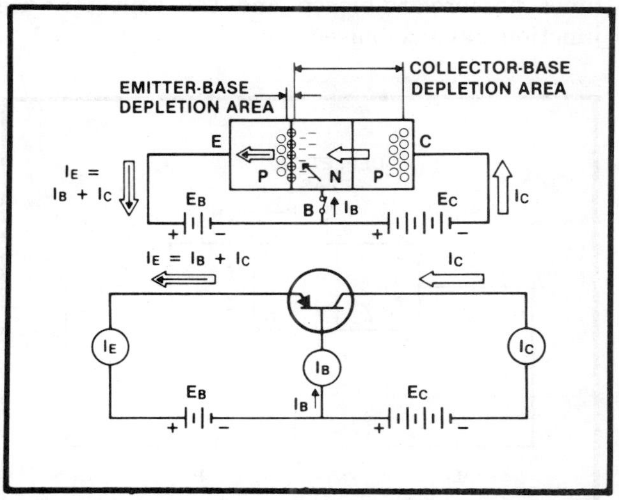

Fig. 12-16 When the emitter-base junction of a transistor is forward-biased, current will flow. The large emitter-collector current is controlled by varying the much smaller emitter-base current.

the collector and the base forces electrons to the emitter-base junction where they are attracted to the positive voltage source attached to the emitter. Since the base is so thin, the electrons which leave the collector, rather than returning to the positive terminal of the collector source, return to the positive terminal of the emitter source. A very small base current is all that is needed to keep the emitter-base depletion area small enough that a much larger current can flow between the collector and the emitter.

If the base circuit is opened, as it is in Fig. 12-17, there is no longer a force to keep the emitter-base deplection area reduced; so there will be no attraction for the electrons from the negative terminal of the collector source across both depletion areas to the positive terminal of the emitter source. When there is no base current, there can be no flow between the emitter and the collector.

When the base of a PNP transistor is negative with respect to its emitter, base current will flow, and the transistor conducts between its emitter and collector. It is sometimes necessary in a circuit for a positive voltage to cause a transistor to conduct. For this reason, there are NPN transistors. NPN transistors are similar in almost every way to the PNP, except for the arrangement of the doped areas. When biasing an NPN transistor, the voltage polarities are exactly opposite those of the PNP. For maximum conduction with either type, the emitter-base junction must be forward-biased and the collector-base junction, reverse-biased.

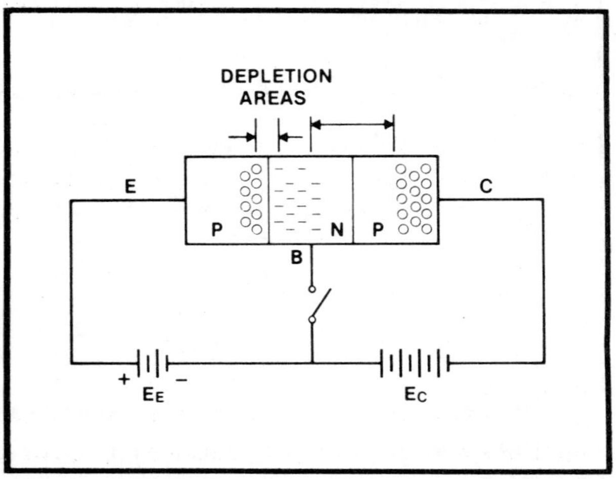

Fig. 12-17 When the base circuit is open, no base current flows, and there will be no emitter-collector current.

4. Silicon Controlled Rectifiers

Electrical circuits for lights and some types of motors may limit the current by dropping the voltage supplied to these devices. This requires a relatively large resistor and dissipates a lot of power in the form of heat. It is possible to dim lights and control the speed of universal motors by using a silicon controlled rectifier, an SCR, rather than a resistor.

An SCR decreases the amount of current supplied to the device, not be dropping the voltage and dissipating the power, but by controlling the time in the cycle of AC that the current is allowed to flow.

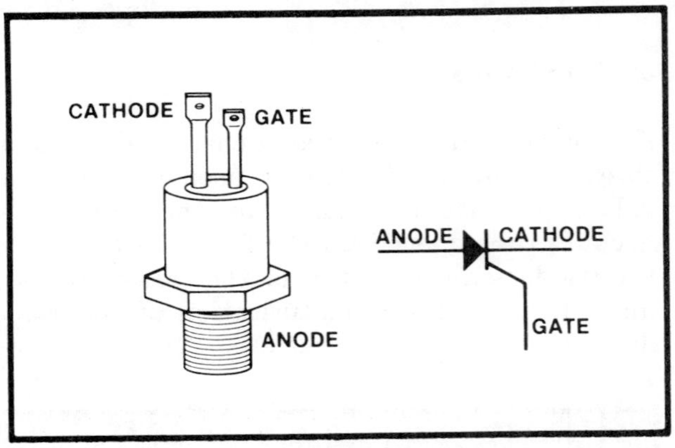

Fig. 12-18 A typical silicon controlled rectifier and its symbol

A silicon controlled rectifier is similar to a silicon diode in its outward appearance, except for its extra terminal, its gate. Another difference is that the case of a stud-type SCR is its anode, while on a regular diode, the case is the cathode.

An SCR is a special type of diode having three junctions. Two of these junctions are forward-biased, and one is reverse-biased so no electrons can flow through the load. If the gate is *momentarily* connected to the positive voltage at the anode, the reverse-biased junction will become forward-biased, and electrons will flow through the SCR. Once this flow has started, it will maintain the forward bias and flow will continue until the voltage across the SCR is removed. No further voltage need be applied to the gate.

The SCR, as its name implies, is a rectifier, and it can be used to produce DC from AC. It not only produces DC, but it is selective in the amount that is produces.

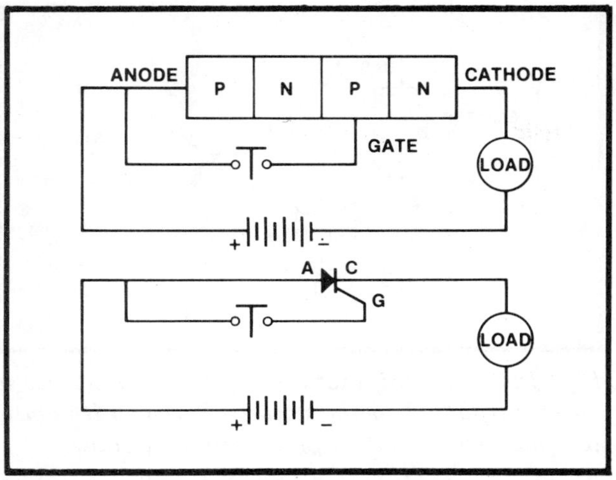

Fig. 12-19 A pulse of positive voltage on the gate of a silicon controlled rectifier will cause it to conduct, and it will continue to conduct until the voltage across the anode and the cathode is removed.

The diode D_1, Fig. 12-20, and the voltage divider R_1, R_2, and R_3, provide an adjustable direct current to trigger the SCR. Looking at the waveform of the output controlled by the SCR, we see that there is no flow at all in one-half of the cycle, that half when the anode is negative. Neither is there any flow during the half-cycle when the anode is positive *until* the voltage across the divider has risen sufficiently to charge the capacitor C_1

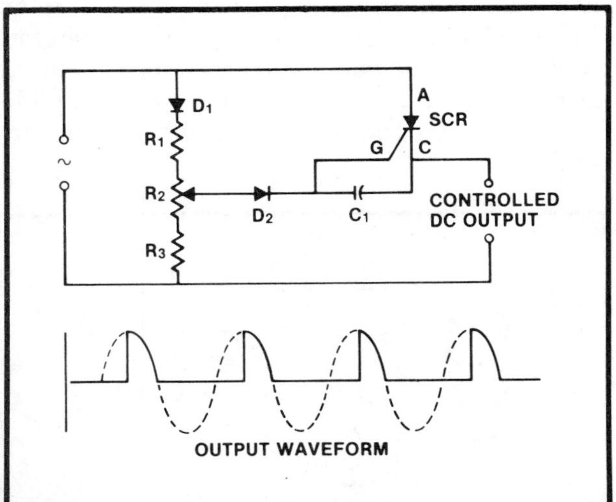

Fig. 12-20 The position of the control **R_2** *determines the time in the cycle the silicon controlled rectifier will conduct. Current will flow only during the positive half-cycle after the SCR has been triggered.*

enough to trigger the gate of the SCR. When the voltage rises high enough, the gate is triggered and the charged capacitor provides the needed pulse to start the SCR conducting. Once it starts conducting, it will continue until the supply voltage reverses, as it does in the next half-cycle.

5. *Triacs*

One of the limitations of a silicon-controlled rectifier is that is controls only one-half of the cycle of AC. A triac, on the other hand, overcomes this by acting as though it were two SCR's connected side by side, in opposite directions. An SCR requires a positive pulse to trigger its gate, but a triac can be triggered by a pulse of either polarity, and the output waveform will appear as that in Fig. 12-21. For full power, the triac is triggered at the beginning of the cycle and all of the current flows; but if it is triggered later in the cycle as seen here, only about one-half of the current flows.

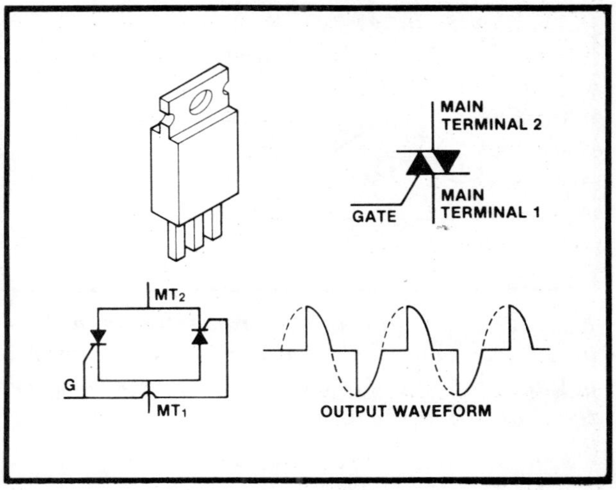

Fig. 12-21 A triac may be triggered with a pulse of either polarity, and it will conduct during the remainder of the cycle after it has been triggered.

6. *Field-effect transistors*

A transistor is a *low-impedance* device, which depends on the control of the *current* flow into or out of the base to control the flow of current between the emitter and the collector. A vacuum tube, on the other hand, is a *high-impedance* device and variations in *voltage* on the grid will control the flow of current between the cathode and the plate.

Sometimes we need a solid-state device that will control the load current flow by the control of the voltage rather than current, and the field-effect transistor, the FET, has been developed to do just this. An FET is constructed of a channel of either N-type or P-type silicon with the gate sitting in the channel and acting as a valve. One end of the channel is called the source and the other end the drain. An N-channel FET has a P-type gate, so when a positive voltage is applied to the gate, the FET is forward-biased, and there will be a greater flow of electrons between the source and the drain. If a negative voltage is applied to the gate, the FET will be reverse-biased, and the flow between the source and the drain will be pinched off.

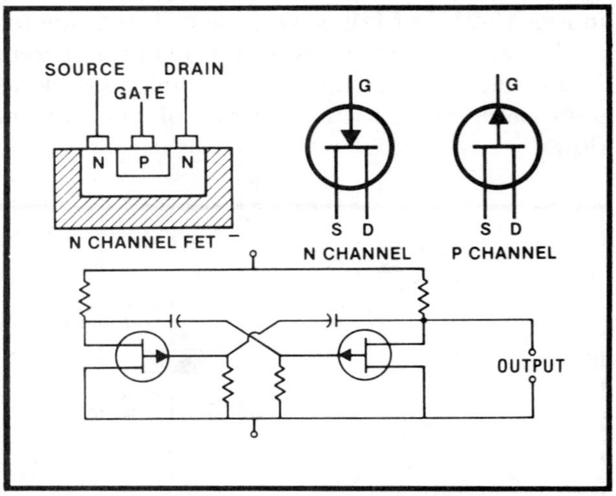

Fig. 12-22 A field-effect transistor is a high-impedance control device that uses a change in voltage on its gate to control the current between the source and the drain.

7. Unijunction transistors

A unijunction transistor, or UJT, is sometimes called a double-base diode, and it is made up of a single crystal of uniformly doped N-type silicon, and has contacts at each of its ends, and a small P-type emitter is located near its middle. A UJT acts like a good insulator until the voltage at the emitter becomes sufficiently high to trigger it, at which time it conducts with a minimum of resistance. UJT's are used in such circuits as relaxation oscillators, where it is necessary to provide short, high-intensity current pulses when the control voltage rises to a given value. They may also be used to provide the gate pulses for silicon controlled rectifiers or triac circuits.

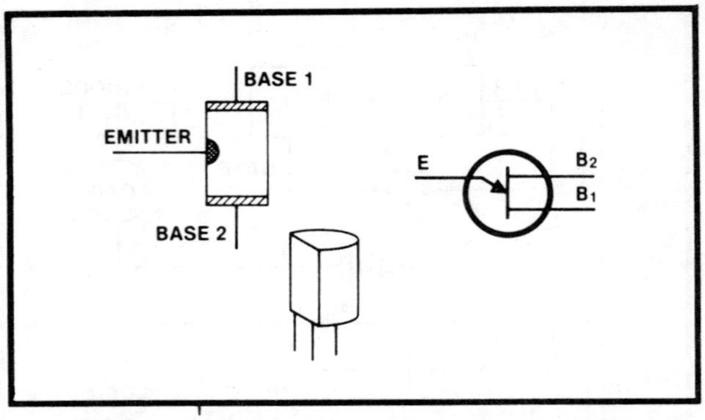

Fig. 12-23 A unijunction transistor produces a pulse output when the voltage between its emitter and base rises to a predetermined value.

C. Magnetic Amplifiers

Vacuum tubes and most of the semiconductor devices are used to control a relatively small amount of alternating current output, and for the control of large amounts of current, the magnetic amplifier, or saturable reactor is often employed.

This "magamp", as it is commonly called, is a special form of transformer with one of its windings supplied with a flow of *direct current* that is used to change the permeability of the core and in this way control the amount of alternating current that is allowed to flow.

To understand the operation of a magamp, let's consider a coil of wire in series with an electric light bulb in an AC circuit. If the coil has an air core, it will have very little inductance; therefore there will be a small amount of back voltage generated, and most of the source voltage will be dropped across the light bulb, and it will burn brightly.

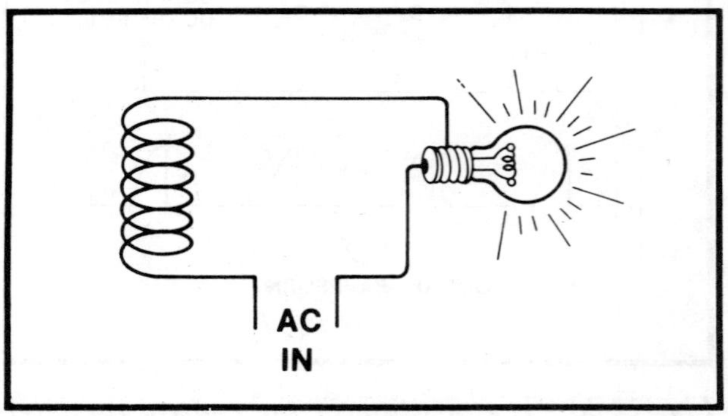

Fig. 12-24 The inductance of an air-core coil may be low enough that there will not be enough back voltage generated to oppose the flow of alternating current.

Now, if a highly permeable core is inserted into the coil to concentrate the lines of flux, an appreciable back voltage will be generated in the coil, and the bulb will receive much less voltage, so it will burn very dimly, if at all.

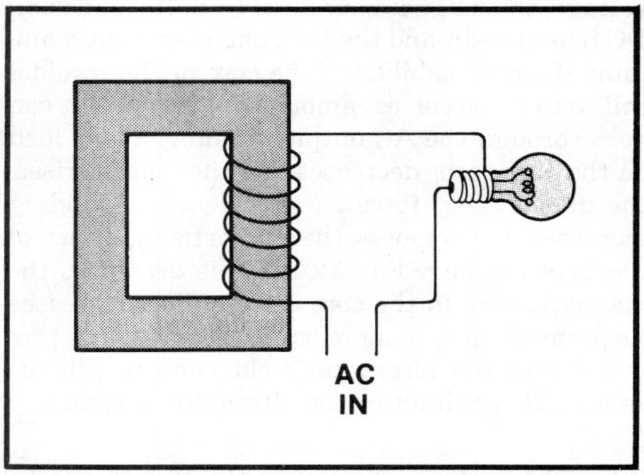

Fig. 12-25 A highly permeable core in the coil will concentrate the flux and produce enough back voltage to prevent the light illuminating.

From these two conditions, we can see that the amount of inductance in series with the electrical load determines the amount of current that can flow through it.

In Fig. 12-26, we have a second winding on the core of the coil, and by controlling the small amount of direct current that flows in this winding, we can control the large amount of alternating current that flows in the output winding. When there is very little or no flux from the DC, the iron core will accept lines of flux from the AC and concentrate them to generate the back voltage that opposes the source voltage, and there will be very little current flow through the load.

When the DC control current is increased, its magnetic field will saturate the core so it will no longer accept the changing lines of magnetic flux from the AC coil, and the core now appears to have a low permeability, approaching that of air, so there will be very little back voltage generated and the bulb will glow brightly. By controlling a small amount of DC, a large amount of AC can be controlled. This is the reason this saturable reactor is called a magnetic *amplifier*.

Magnetic amplifiers are often wound on a ring-shaped soft iron core to form a toroidal coil. This type of core concentrates the lines of magnetic flux and provides the maximum inductance with the least loss of the lines of flux. Toroidal coils can be mounted quite close together without there being any magnetic reaction or coupling between them.

A practical example of a magnetic amplifier circuit in an aircraft electrical system is the voltage

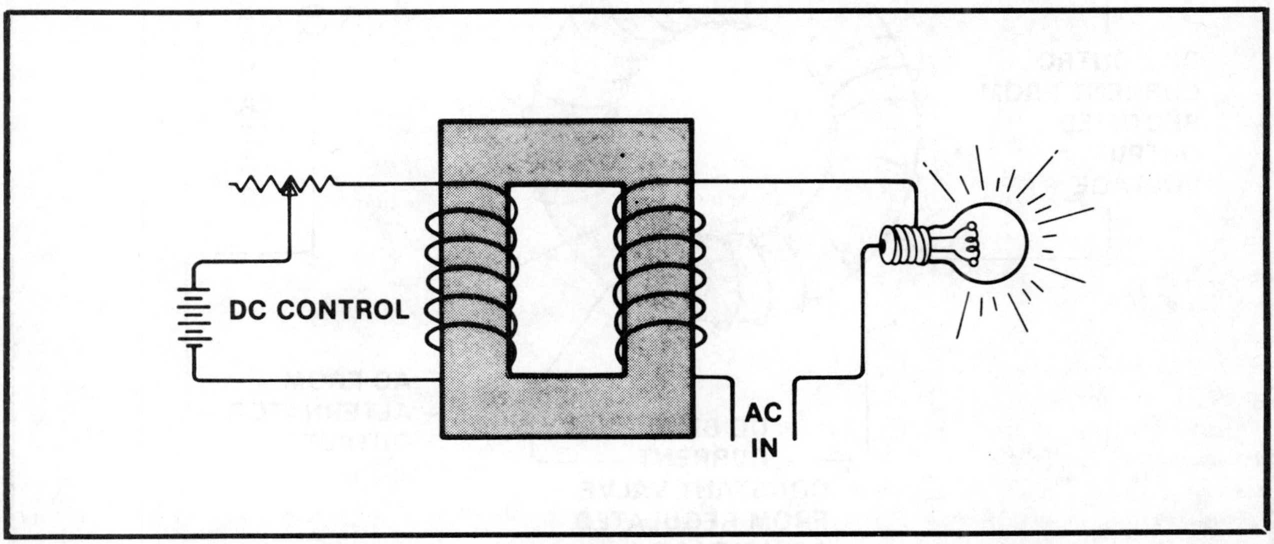

Fig. 12-26 Direct current flowing in the control coil can magnetically saturate the core and decrease its permeability so that there will not be enough inductive reactance to restrict the flow of alternating current.

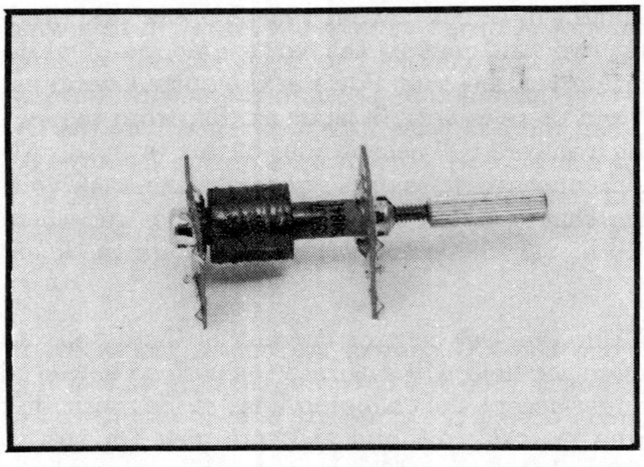

Fig. 12-27 A toroidal coil concentrates the lines of magnetic flux and prevents any stray fields interfering with other coils.

sensing circuit in some of the AC voltage regulators. In Fig. 12-28, we have the principle of this type of circuit. A toroidal coil is wound with three windings, one a DC bias-voltage coil that is supplied with a constant amount of current from a regulated DC source. Another winding is supplied with direct current that varies as the alternator output voltage. This is the DC control winding. The polarity of these two windings is such that they tend to cancel the effect of each other. The excitation current for the alternator field flows through the third winding.

When the alternator is producing the proper voltage, the magnetizing force from the opposing DC bias current and the DC control current maintains the permeability of the core of the toroidal coil so that the correct amount of field current can pass through the AC output winding. If the load on the alternator decreases and the voltage rises, the magnetizing force from the control winding increases and opposes the magnetizing effect of the bias winding, Fig. 12-29. This decreases the magnetization in the core and in effect increases its permeability, so more back voltage will be produced and the alternator field current will decrease. This will lower the alternator output.

When the load is increased on the alternator, the output voltage will drop, and the magnetizing effect of the control winding will decrease. This will allow the bias winding to more nearly saturate the core so the permeability will decrease and there will be less back voltage generated. The

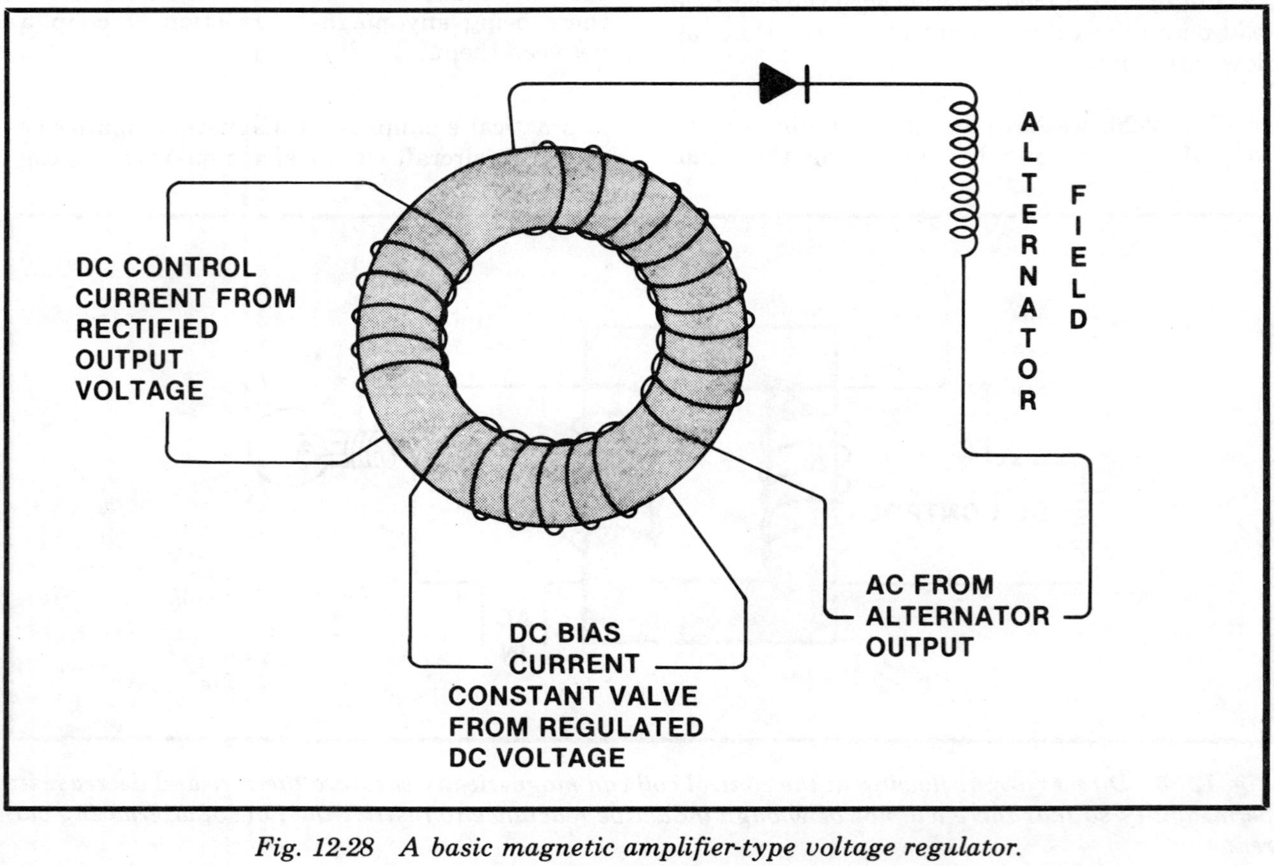

Fig. 12-28 A basic magnetic amplifier-type voltage regulator.

alternator field current will increase, bringing the output voltage back to the desired value.

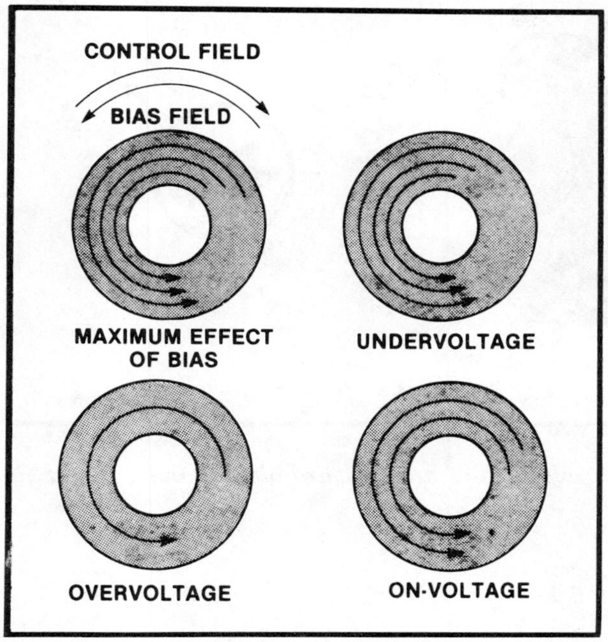

Fig. 12-29 The flux from the control field cancels the magnetic effect of the bias field in a magnetic amplifier voltage regulator.

D. Optoelectric Devices

1. Photodiodes

As the science of electronics becomes more sophisticated, we are able to use the energy found in light for our devices and circuits. Light energy is electromagnetic in nature, and one of its characteristics is that it will increase the reverse, or leakage, current in a semiconductor device. A photodiode is a special diode, made in such a way that light shining through an aperature in its case will release free electrons into the depletion area and cause it to conduct.

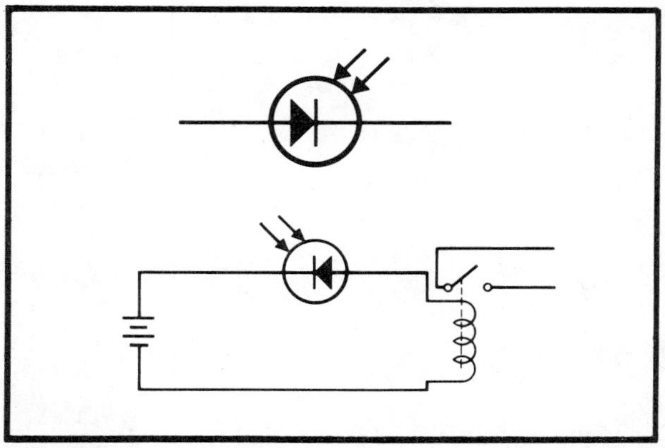

Fig. 12-30 A photodiode will conduct when sufficient light strikes its junction to cancel the effect of the reverse bias.

A practical circuit using a photodiode is seen in Fig. 12-30, where we have a photodiode in the coil circuit of a relay. When the photodiode is dark, no current can flow and the relay is open; but when light shines through the aperture, the diode will break down so current from the battery can flow through the relay coil and close the contacts.

2. Phototransistor

A photodiode is limited in the amount of current it can conduct, and so there are phototransistors that use the energy released by light to forward-bias the emitter-base junction and allow current to flow through the emitter-collector circuit.

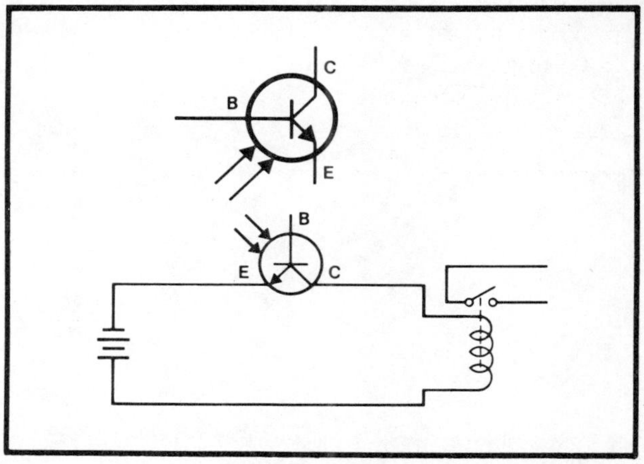

Fig. 12-31 A phototransistor will conduct when sufficient light strikes its emitter-base junction to provide a forward bias.

3. Light-emitting diodes

One of the most familiar optoelectronic devices is the light-emitting diode, or the LED. Tiny bars or dots of photo-conductive material can be made to light up when current flows through them. When these LED's are reverse-biased, no current can flow and the diode is dark; but when it is forward-biased, current can flow through the diode and as it does, the diode emits light.

The most important application of LED's is that of numerical displays in our watches, calculators, and many different types of aircraft instruments.

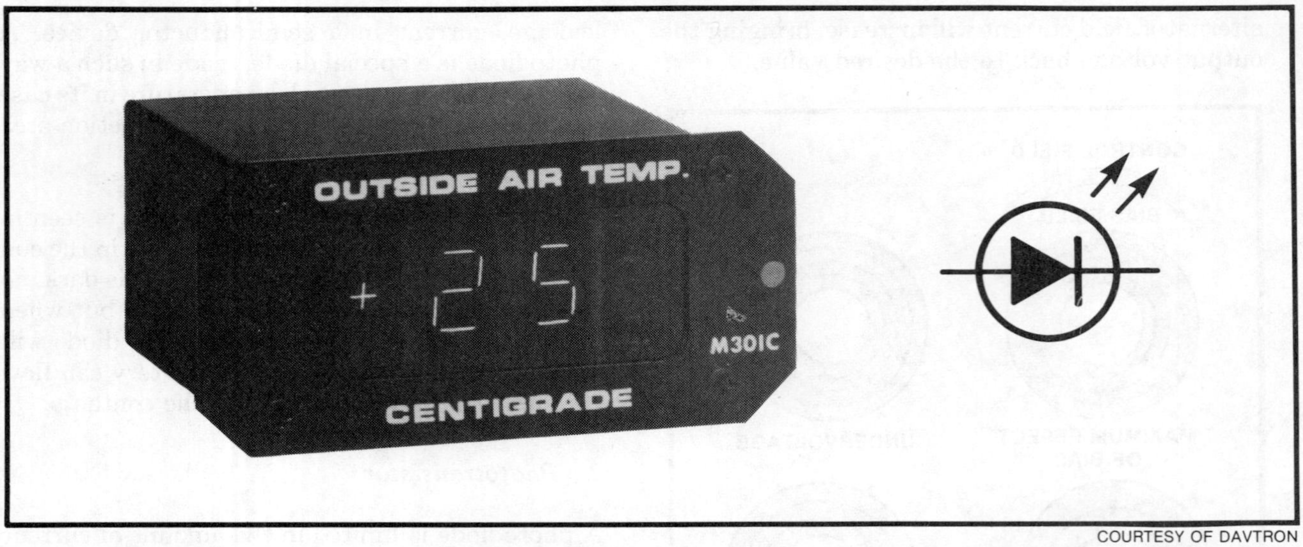

COURTESY OF DAVTRON

Fig. 12-32 Light emitting diodes (LED's) emit light when they are forward-biased and current flows through them.

SECTION XIII

Basic Semiconductor Circuits

There are only three basic types of circuits in most of our electronic equipment. Rectifiers change alternating current into direct current. Amplifiers change the amplitude of either the voltage or the current, and oscillators produce alternating current from direct current.

We have already discussed rectifiers in Section XI of this book, and here we will discuss basic amplifier and oscillator circuits.

A. Amplifiers

An amplifier uses, as an input, a very small change in either voltage or current to produce a large change in the output voltage or current. A transistor is a low-impedance device, which means that in a transistor amplifier, we will use a small change in current in the emitter-base circuit to cause a large change in the current in the emitter-collector circuit.

In the common-emitter amplifier, the transistor is installed in the circuit of Fig. 13-1 in such a way that the forward bias for the emitter-base junction is provided by the voltage drop across resistor R_B, and the reverse bias for the collector-base junction is provided by the voltage drop across resistors R_A and R_C. When there is no signal voltage across R_B, there will be a steady current flowing through the transistor and resistor R_C, whose voltage drop is opposite in polarity to that of the battery. The result is that the steady output current is negative, but not as negative as that of the battery.

When a negative signal is put on the input between the base and the emitter, across R_B, the forward bias across the emitter-base is increased. There will be more current flow through the transistor and R_C, and the voltage drop across R_C will increase, causing the output to become less negative; or, if you would like to think of it that way, it would go positive.

During the half-cycle of the input signal when it is positive, the forward bias of the emitter-base junction decreases and the transistor conducts

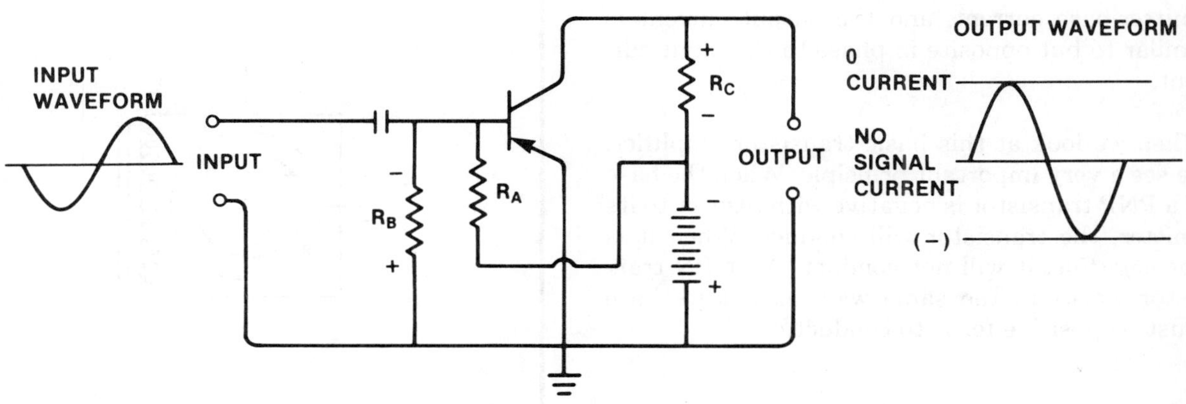

Fig. 13-1 A small voltage applied to the input of a transistor amplifier will cause a large change in the output current, and the output voltage will be 180 degrees out of phase with the input voltage.

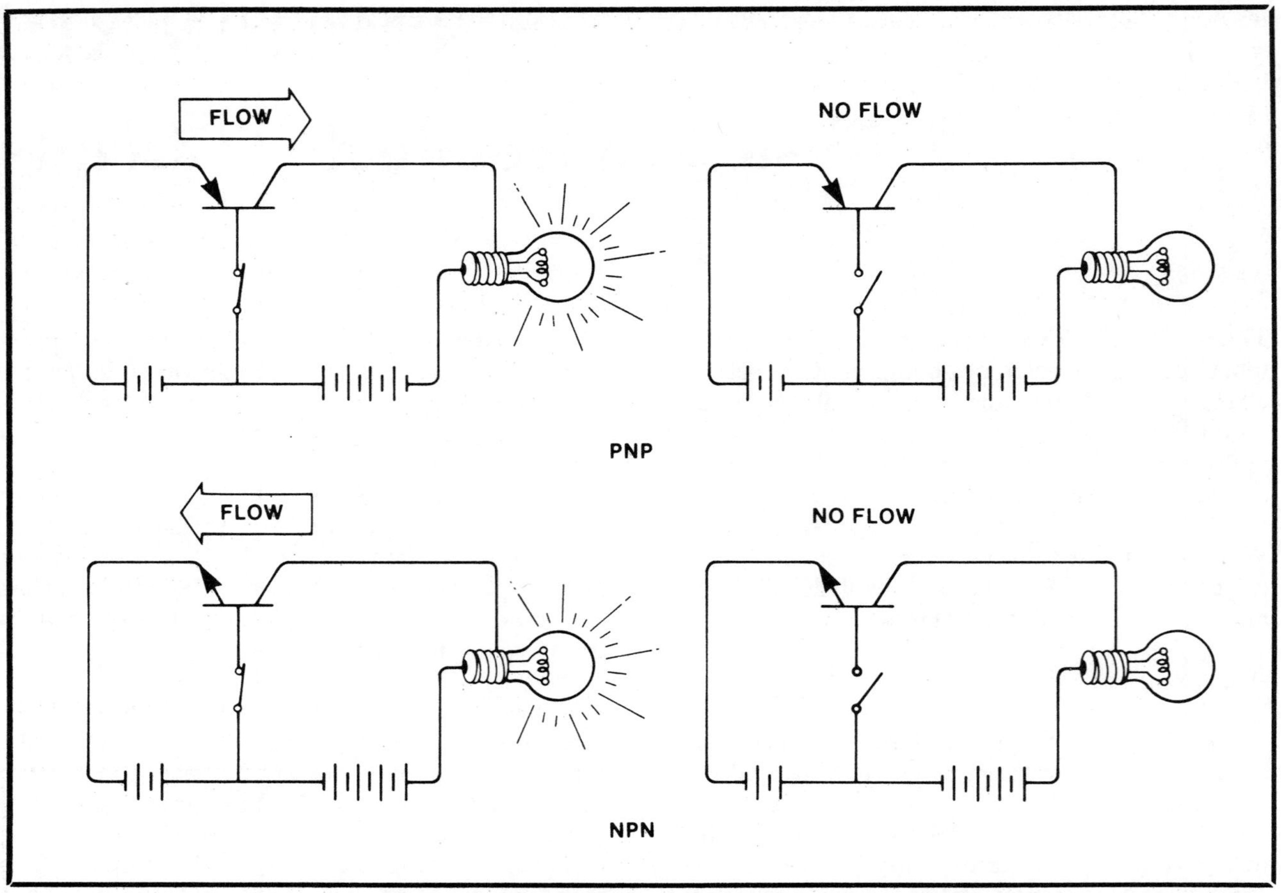

Fig. 13-2 Current will flow in the emitter-collector circuit of a transistor amplifier when current flows in the emitter-base circuit.

less, decreasing the voltage drop across the resistor R_C so the output will go more negative.

In a common-emitter transistor amplifier, the emitter-collector current is much greater than the emitter-base current, and the output current is similar to but opposite in phase to the input current.

When we look at this basic transistor amplifier, we see a very important principle. When the base of a PNP transistor is negative with respect to its emitter, the transistor will conduct. When it is *not negative*, it will not conduct. An NPN transistor works in the same way, except its base must be positive for it to conduct.

B. Oscillators

In order to best understand electronic oscillation, let's briefly review what we have already learned about resonance in a parallel L-C circuit which is called a tank circuit. We saw earlier that when the value of a capacitor and an inductor are chosen so that their reactances at a particular frequency are the same, the energy stored in the electro-

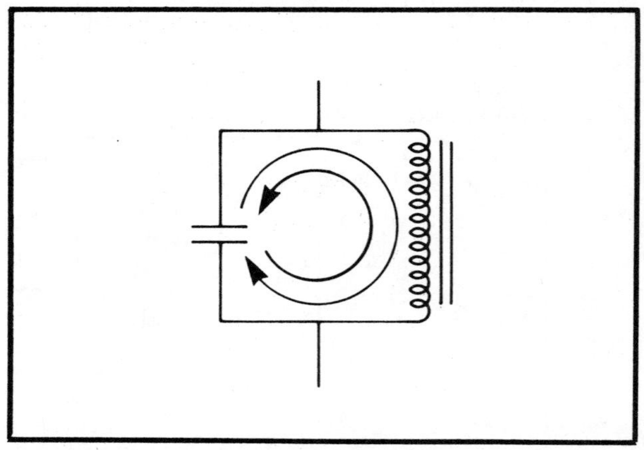

Fig. 13-3 Selection of the values of inductance and capacitance determine the frequency of oscillation of an L-C oscillator. An inductor and capacitor in parallel is called a tank circuit.

magnetic field around the inductor is exactly the same as the energy stored in the electrostatic field across the plates of the capacitor. There is a large amount of circulating current flowing back and forth between the inductor and the capacitor, but very little current flows through the circuit from the source. The current would continue to flow back and forth, or oscillate, indefinitely if there were no resistance in the circuit. But in any practical circuit, there is always some resistance, and so for an electronic oscillator to function, there must be amplification and feedback from the output back into the input with the proper phase to replace the energy that is lost in the resistance.

When the switch *S* in Fig. 13-4, is closed, current begins to flow in the circuit, biasing the transistor through the voltage divider, R_B and R_A, and flowing through $2L_1$, the lower half of the tapped coil. As the current increases in this coil, it induces a voltage into the upper half of the coil that charges capacitor C_2 in such a direction that it increases the forward bias on the transistor, further increasing the current flow.

By the time the circuit is saturated—that is, all of the current is flowing that can flow—there will be no more voltage induced into L_2 from L_1, and all of the energy stored in the tank circuit is in the electrostatic fields in the capacitor C_1. Since there is no more force to push electrons into capacitor C_1, it will discharge, and the energy lost will be made up by that stored in capacitor C_2 as it discharges. As C_2 discharges, the forward bias on the emitter-base junction of the transistor decreases, and begins to conduct less, decreasing the current in L_1. This decrease in the current induces a voltage into L_2, which causes it to charge

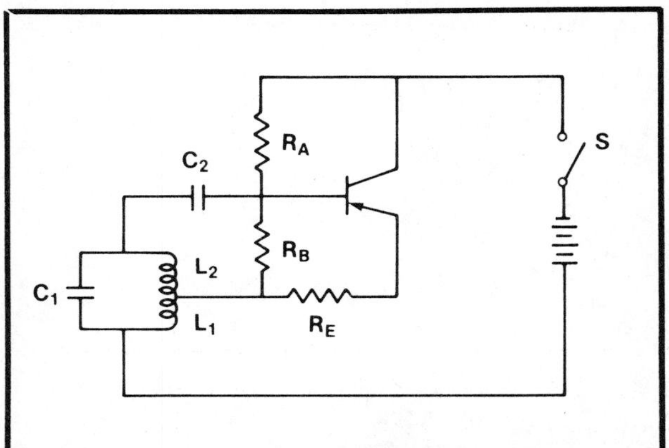

Fig. 13-4 A Hartley oscillator using a transistor

capacitor C_2 in such a direction that it *decreases* the forward bias and eventually reverse-biases the emitter-base junction of the transistor, driving it all the way to cut-off.

When the current flow stops, C_1 and C_2 are both fully charged and the cycle can repeat itself. The frequency of the oscillation is determined by the values of the capacitor, and the inductor and by varying either one, the frequency of the alternating current produced by the oscillator can be changed.

C. Voltage Control Circuits

1. Full-wave voltage doubler

A circuit consisting of three capacitors and two diodes can be used to provide direct current with a voltage almost twice the peak voltage of the input AC.

In Fig. 13-5A, during the half of the cycle when the top of the power source is positive, current

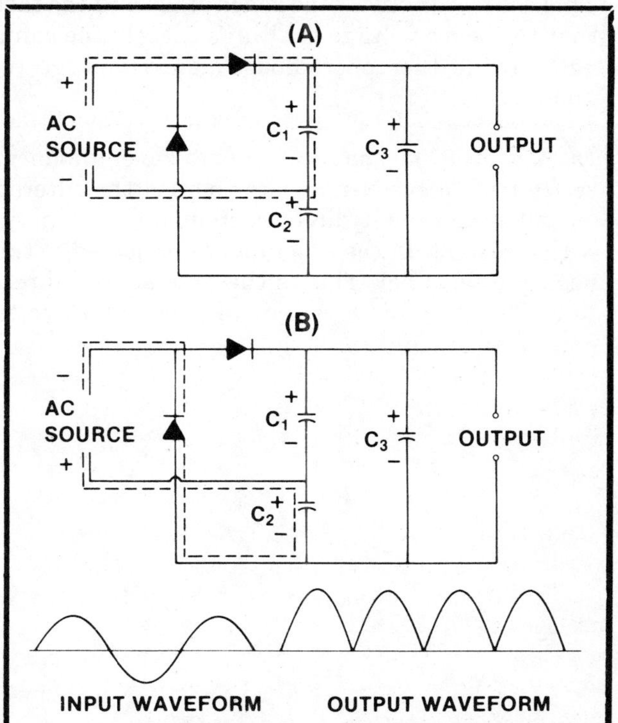

Fig. 13-5 A — Capacitor C_1 of this full-wave voltage doubler is charging to full line voltage. B — During this half-cycle, capacitor C_2 charges to full line voltage, and since C_1 and C_2 are in series across C_3, it will charge to double the line voltage.

flows through diode D_1 to charge capacitor C_1 to full line voltage. During the next half-cycle, when the bottom terminal of the power source is positive, current flows through D_2 and charges capacitor C_2 to the full line voltage. Capacitors C_1 and C_2 are in series across capacitor C_3, which will charge to *twice* the voltage of the input. Actually the DC output voltage will depend to a great extent on the amount of load current that flows. With low-output current, the DC voltage will be approximately twice the peak value of the source AC.

2. Voltage regulator

The zener diode, as you will remember, is a special form of diode with a specifically controlled reverse voltage. Below this voltage, the diode acts as a check valve, allowing flow when it is forward-biased, but preventing flow when it is reverse-biased. When the reverse-bias voltage exceeds the zener voltage of the diode, it will break down and conduct in its reverse direction.

This characteristic of the zener diode and the fact that the zener voltage is always exactly the same value make the zener diode a good voltage regulator.

In Fig. 13-6, we have a 15-volt zener diode in series with a bleeder resistor across the filtered output of a 24-volt direct current power supply, with the anode of the zener diode connected to the negative terminal. This is the reverse-bias direction. The zener diode will break down at 15 volts

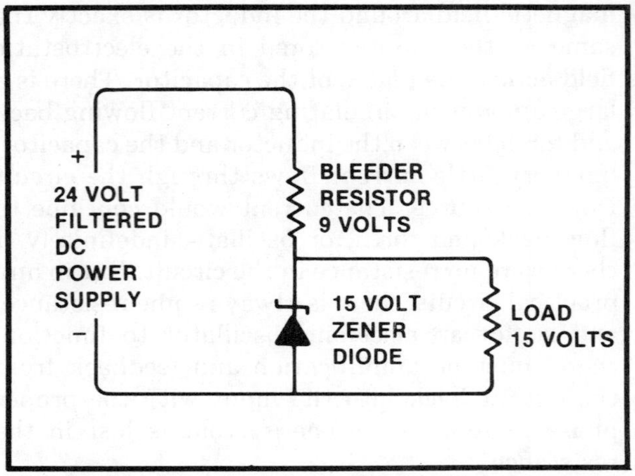

Fig. 13-6 A zener diode voltage regulator circuit

and allow enough current to flow through the bleeder resistor to maintain a nine-volt drop across it.

The load is connected across the zener diode, and since it will maintain a 15-volt drop across it, there will always be exactly 15 volts across the load. When the load current increases, the current through the zener diode will decrease enough to maintain the nine-volt drop across the bleeder resistor, and when the load current decreases, rather than allowing the voltage to rise, the current through the zener diode will increase.

The current through the zener diode will vary so that the total current through the load and the zener diode will be the correct amount to maintain the nine-volt drop across the bleeder resistor.

SECTION XIV

Electrical Measuring Instruments

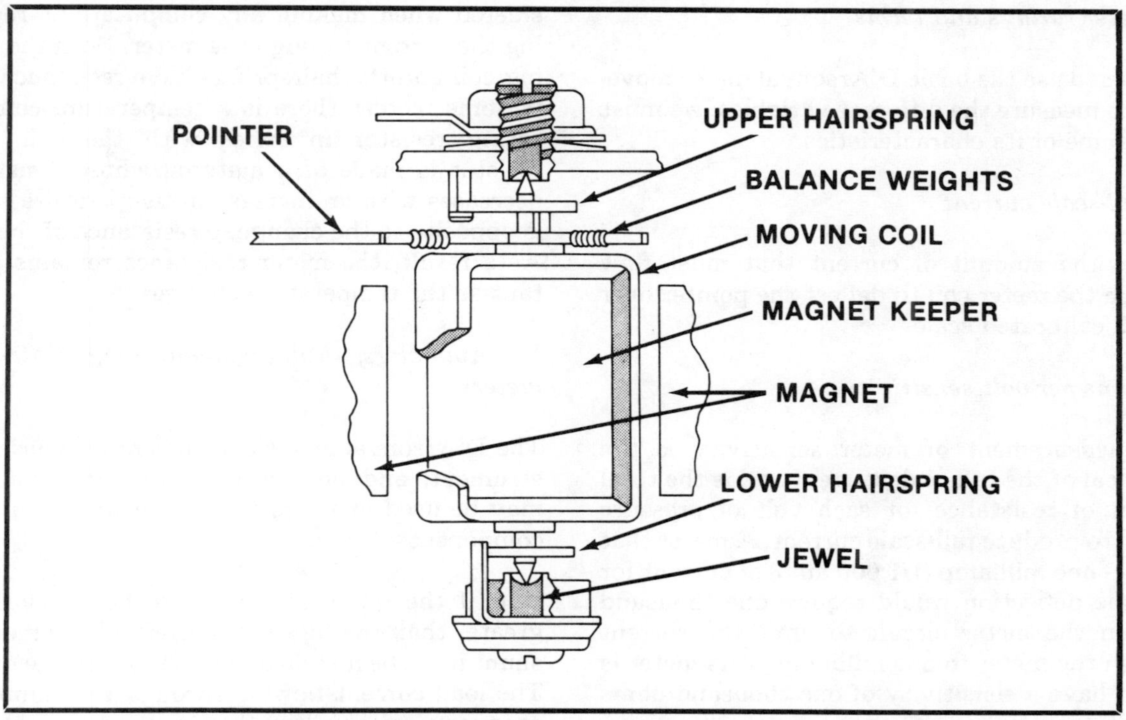

Fig. 14-1 The moving coil of a D'Arsonval-type meter

A. *The D'Arsonval Meter*

In order to get the results we need from electrical circuits and components, we must be able to measure each of the four variables: current, voltage, resistance, and power.

There have been a number of principles used for these measurements, but by far the one most commonly found is that which uses electromagnetism, and is based on two fundamental assumptions:

1. The strength of an electromagnetic field is proportional to the amount of current that flows in the coil.
2. Voltage, resistance, and power all relate to a flow of current, and if the amount of current is known, the other values may be found.

The most widely used meter movement is the D'Arsonval movement whose pointer deflects an amount proportional to the current flowing through its moving coil. A reference magnetic field is created by a horseshoe-shaped permanent magnet, and its field is concentrated by a cylindrical keeper in the center of the open end.

Surrounding the keeper and supported by hardened steel pivots riding in smooth glass jewels, is a coil through which the current to be measured flows. The current enters and leaves the coil through calibrated hairsprings, one surrounding each of the pivots. Current flowing through this coil creates a magnetic field whose polarity is the same as that of the permanent magnet, and thus the two fields oppose each other. The opposing force rotates the coil on its pivots until the force of the hairspring exactly balances the force caused by the magnetic fields.

Oscillation of the pointer is minimized by electromagnetic damping. The moving coil is wound around a thin aluminum bobbin, or frame, and as this frame moves back and forth in the concentrated magnetic field, eddy currents are generated within the bobbin that produce their own fields which oppose the movement.

1. Meter ratings and terms

In order to use the basic D'Arsonval meter movement to measure the different variables, we must know some of its characteristics.

a. Full-scale current

This is the amount of current that must flow through the meter coil to deflect the pointer over the full calibrated scale.

b. Ohms-per-volt sensitivity

This measurement of meter sensitivity is the reciprocal of the full-scale current and is the total amount of resistance for each volt of pressure needed to produce full-scale current. A meter that requires one milliamp (1/1,000 amp) of current for full-scale deflection would require one thousand ohms in the meter circuit to limit the current through the meter to one milliamp. This meter is said to have a sensitivity of one thousand ohms per volt.

Many multimeters have a sensitivity of 20,000 ohms per volt and these meters require 1/20,000 amp, or 50 microamps, of current to move the pointer full scale.

c. Meter resistance

The total resistance of the meter must be considered when making any computations regarding the current through the meter. Both the moving coil and the hairsprings have resistance, and in some meters there is a temperature compensating resistor in series with the coil. This resistor is made of a material whose resistance decreases with an increase in temperature which is opposite to the change in resistance of the coil. As a result, the meter resistance remains constant as the temperature changes.

B. Ammeters, Milliammeters, and Microammeters

The D'Arsonval meter is a current measuring instrument, and one with the proper sensitivity may be used in a circuit without any additional components.

But, if the range of current to be measured is greater than the full-scale current of the meter, a shunt must be installed in parallel with the meter. The load current flowing through the shunt will produce a voltage drop that is proportional to the

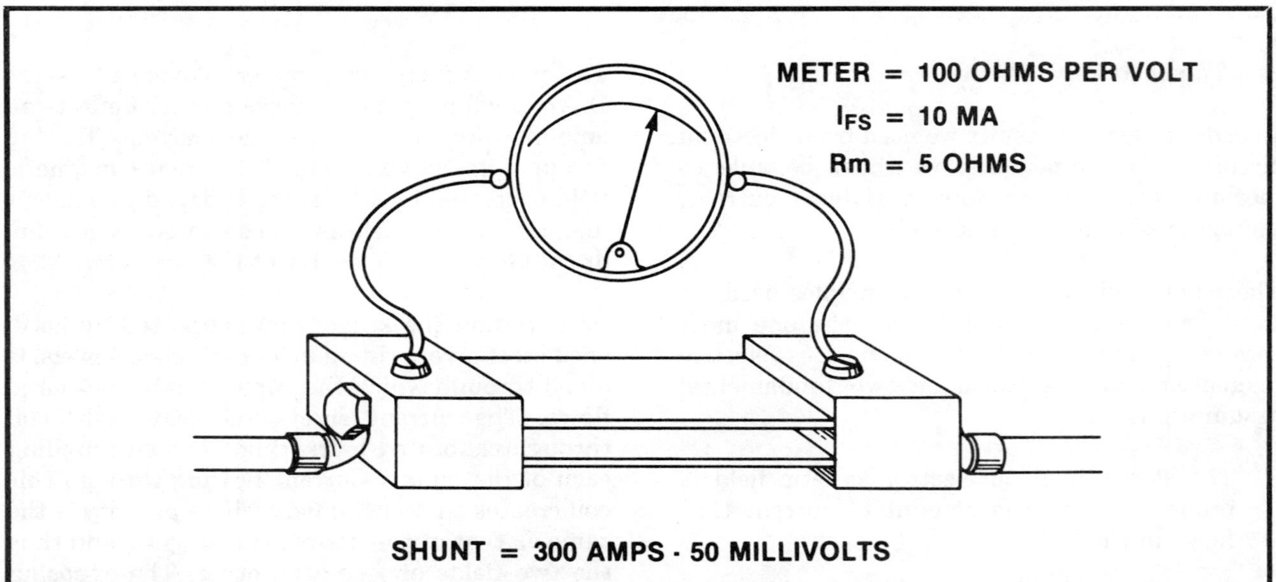

Fig. 14-2 The voltage drop across an ammeter shunt is proportional to the amount of current flowing through it.

amount of current, and the meter will read the voltage drop across the shunt and display this voltage in terms of amps, milliamps, or microamps on its scale. The standard aircraft shunt produces a voltage drop of fifty millivolts when its rated current flows through it.

It is sometimes necessary to extend the range of an ammeter by using a precision resistor as a shunt, and we can determine the resistance needed by the simple use of Ohm's law. Let's assume that we want our meter to deflect full scale when ten milliamps flows through the meter and shunt combination. If the meter requires one milliamp for full-scale deflection and has an internal resistance of 50 ohms, it will deflect full scale when it is connected across a voltage of 50 millivolts. The shunt, therefore, must produce a voltage drop of 50 millivolts when 9 milliamps flows through it. Remember that one milliamp flows through the meter when they are connected in parallel.

$$R_{shunt} = E/I = 0.050/0.009 = 5.55 \text{ ohms}$$

According to the Ohm's law formula, R = E/I, we find that if the meter is connected in parallel with a 5.55-ohm shunt, one milliamp will flow through the meter when nine milliamps flow through the shunt. In other words, the meter will deflect full scale when ten milliamps flow through the circuit.

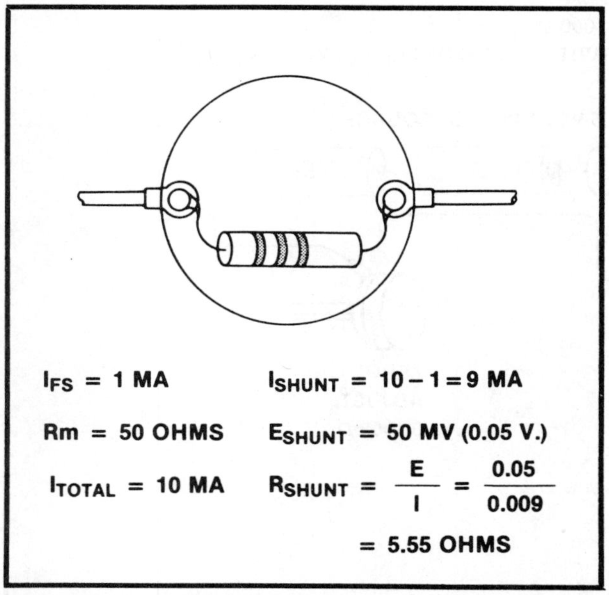

Fig. 14-3 Computation of the resistance of a milliammeter shunt

C. Voltmeters

A D'Arsonval meter can be used to measure voltage by using enough resistance in series with the meter movement to limit the current to the value for which the meter will give full-scale deflection. The meter sensitivity is rated in ohms per volt, which is the number of ohms of resistance we must have in the circuit for each volt we wish to measure. If, for example, our meter has a sensitivity of 1,000 ohms per volt and a resistance of 1,000 ohms, we would need a resistor of 500 ohms in series with the meter to limit the current to one milliampere when the meter is placed across a 1.5-volt battery.

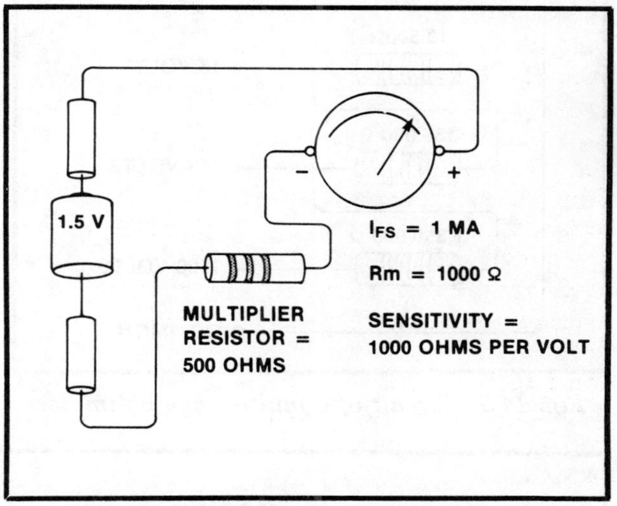

Fig. 14-4 Circuit of a voltmeter

The resistors that are placed in series with the meter movement are called multiplier resistors, or simply multipliers, because they are used to multiply the range of the basic meter.

Multi-range voltmeters use one meter movement with several different multipliers. These multipliers are usually arranged so the current for each succeedingly higher range flows through the multipliers for all of the lower ranges. When the meter in Fig. 14-5 is used to measure 1.5 volts, the test leads are connected to the common terminal and the 1.5-volt terminal. Current then flows through the meter and the 500-ohm resistor. To measure 15 volts, connect to the common and the 15-volt terminals and the current will flow through the meter, the 500-ohm, and the 13,500-ohm multiplier resistors. The total resistance in the circuit is 15,000 ohms. To measure voltages as high as 1,500, the current must flow through all of the resistors in the voltmeter circuit.

Most multi-range meters use a selector switch rather than separate terminals. When using this type of meter, the range is set for a voltage higher than that anticipated. After the meter is connected and the needle is deflected, reduce the range with the selector switch to get a needle deflection in the center third of the scale, where the meter is most accurate.

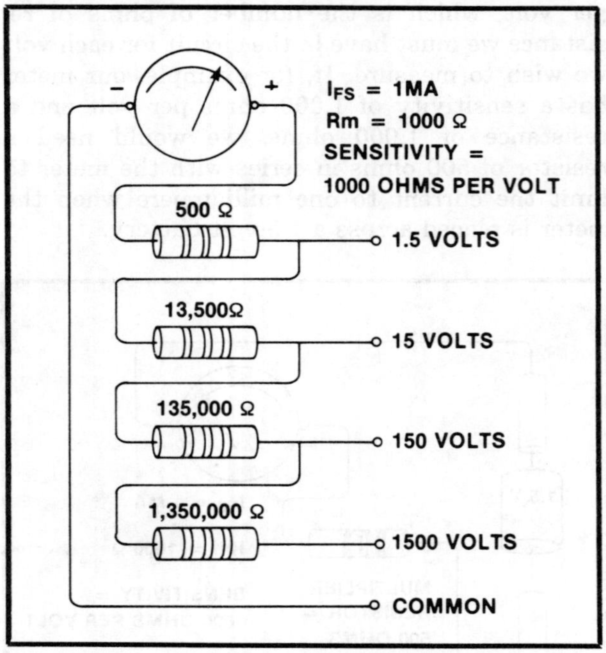

Fig. 14-5 Circuit of a multi-range voltmeter

D. Ohmmeter

Resistance is most easily determined by measuring the current through the unknown resistor when a known voltage is placed across it.

1. Series ohmmeter

The series ohmmeter uses small flashlight or penlight batteries in series with a fixed and an adjustable resistor and the meter.

For the meter to deflect full-scale, the total resistance must equal 4,500 ohms, since we are using a 4.5-volt battery, and the meter has a sensitivity of 1,000 ohms per volt. Because the battery voltage changes with use, a variable resistor is included in the circuit to "zero", or standardize, the meter before each use. To set the meter up for use, short the test leads together and turn the zero adjusting knob until the meter reads full-scale. Here, the scale is marked zero ohms, indicating that there is no resistance between the test leads. When the leads are separated, the nee-

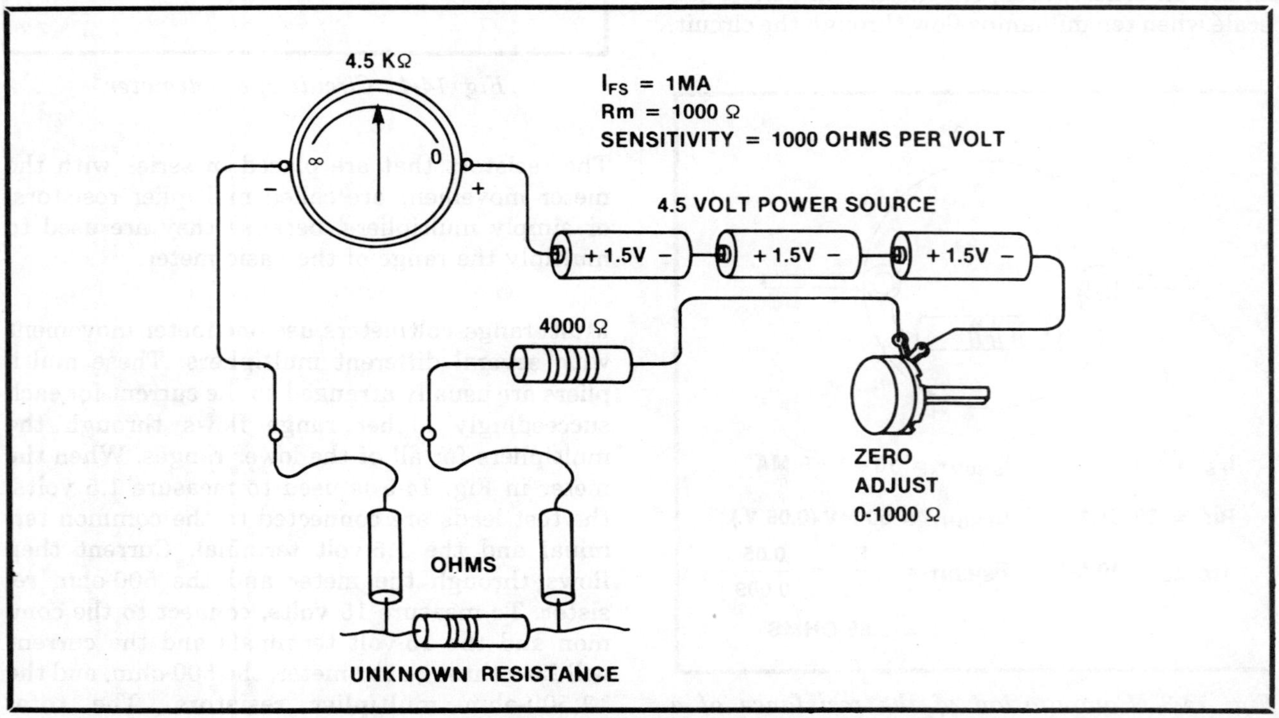

Fig. 14-6 Circuit of a series-type ohmmeter

dle will drop back to the left side of the scale where the dial is marked ∞. This is the symbol for infinity, and indicates that there is an infinite resistance between the test leads, and no current is flowing.

In this particular circuit, if a resistance of 4,500 ohms, the same as the total circuit resistance, is placed between the test leads, the current will drop to one-half of its original value. At this point, the dial is marked 4.5 k ohms. The scale on a series ohmmeter is highly non-linear, meaning that there is no uniform distance between the graduations. The numbers are fairly widely separated at the low-range end, on the right side of the dial, but are crowded very tightly at the high end, the left side. For most accurate measurement of resistance, you should use a scale that will give a pointer deflection in the center third of the dial.

Different resistance ranges are selected by using different values of battery voltage and of the fixed resistor.

2. Potentiometer-type ohmmeter

The simple series-type ohmmeter has a problem of seriously crowding the resistances on the high end of the scale, and this problem is solved to some extent by the potentiometer-type ohmmeter. The scale is still non-linear, but it is not crowded nearly so badly.

A low-resistance standard resistor is in series with the battery and the resistance to be measured. This sets up a voltage divider circuit, and the voltage across the standard resistor is proportional to the current through the unknown resistance.

A sensitive meter having a high internal resistance is used to measure the voltage drop across the standard resistor; hence the name potentiometer-type ohmmeter. Rather than being calibrated in volts or millivolts, the meter dial is calibrated in ohms. When the test leads in Fig. 14-7 are shorted together, all of the voltage of the 1.5-volt battery is dropped across the standard

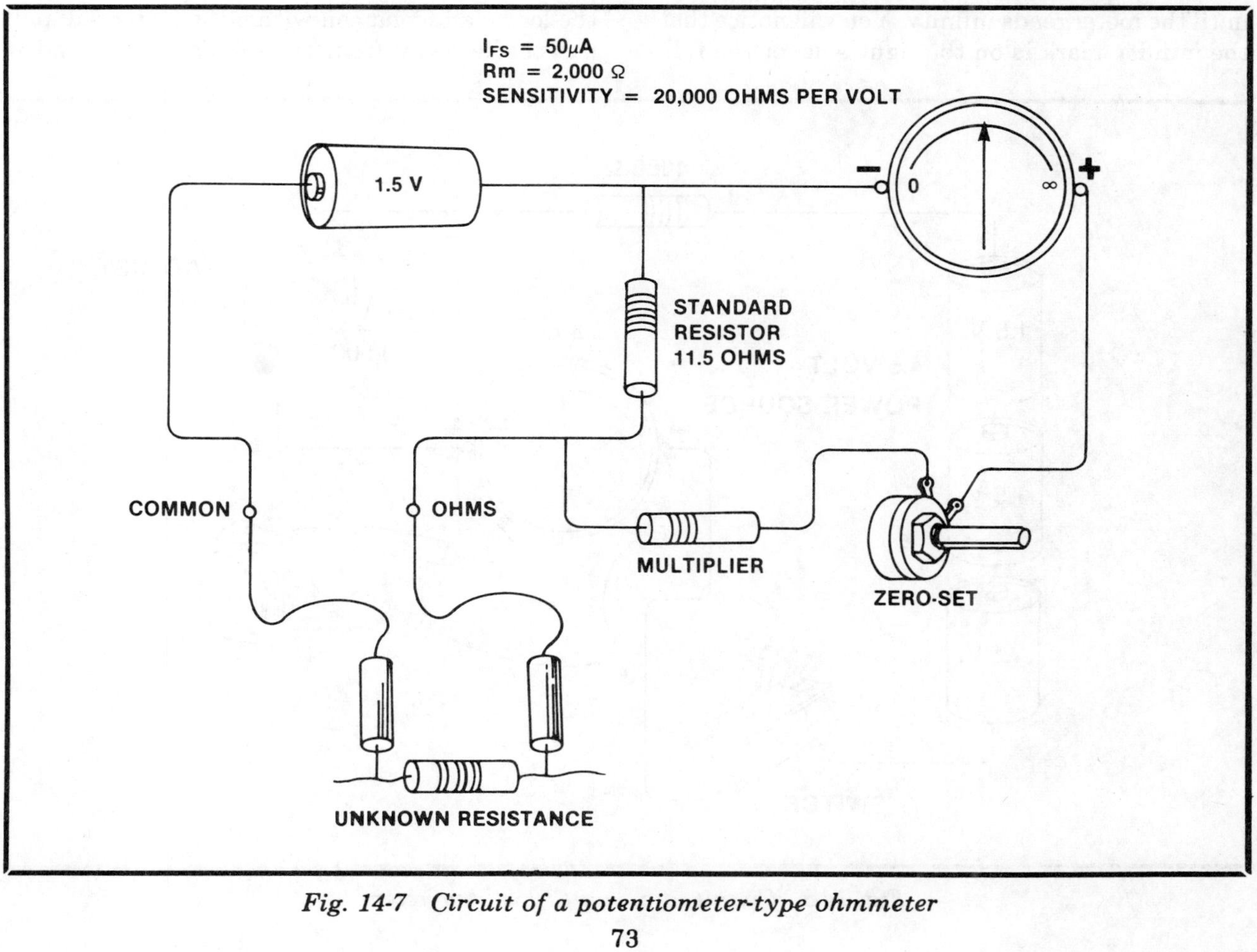

Fig. 14-7 Circuit of a potentiometer-type ohmmeter

resistor, and the meter can be adjusted with the zero-set variable resistor to give full-scale deflection. At this point, the dial is marked zero, since there is no resistance between the leads. When the leads are separated, the battery circuit is open and no current can flow. The point at which the needle rests is marked ∞, indicating that there is an infinite resistance between the leads.

3. Shunt-type ohmmeter

It is sometimes necessary to measure very low resistances; for example, the resistance of the primary winding of a magneto coil. Neither a series nor a potentiometer-type ohmmeter is suitable for this kind of measurement, but a shunt-type ohmmeter may be used.

The shunt-type ohmmeter in Fig. 14-8 uses a meter having a very low internal resistance of four ohms in series with a switch, a 4,000-ohm fixed resistor, and a 600-ohm variable resistor. All of this is in series with three 1.5-volt batteries. When the switch is closed and the test leads are open, the variable resistor is adjusted until the meter reads infinity. You will notice that the infinity mark is on the right side, or the full-scale side, of the dial which is opposite that of the other types of ohmmeters.

When the test leads are shorted, all of the current will flow through them and bypass the meter, so the meter will read zero. If a resistance of four ohms is placed between the leads, half of the current will flow through the meter and half through the resistance. The meter will then read half-scale.

4. Megohmmeter

We sometimes need to measure very high resistance values that require a voltage in excess of that provided in any of the standard ohmmeters. When we have a requirement of this nature, the megohmmeter, or "Megger" made by the Charles Biddle Company, is used. A hand-cranked generator with a slip clutch allows the operator to produce a voltage of several hundred volts. When the leads are separated and the crank is turned, the pointer will deflect fully to the left, indicating that there is an infinite resistance between the leads. When a high resistance is placed between the leads, a second coil within the meter will pull the needle away from the infinity mark, and it

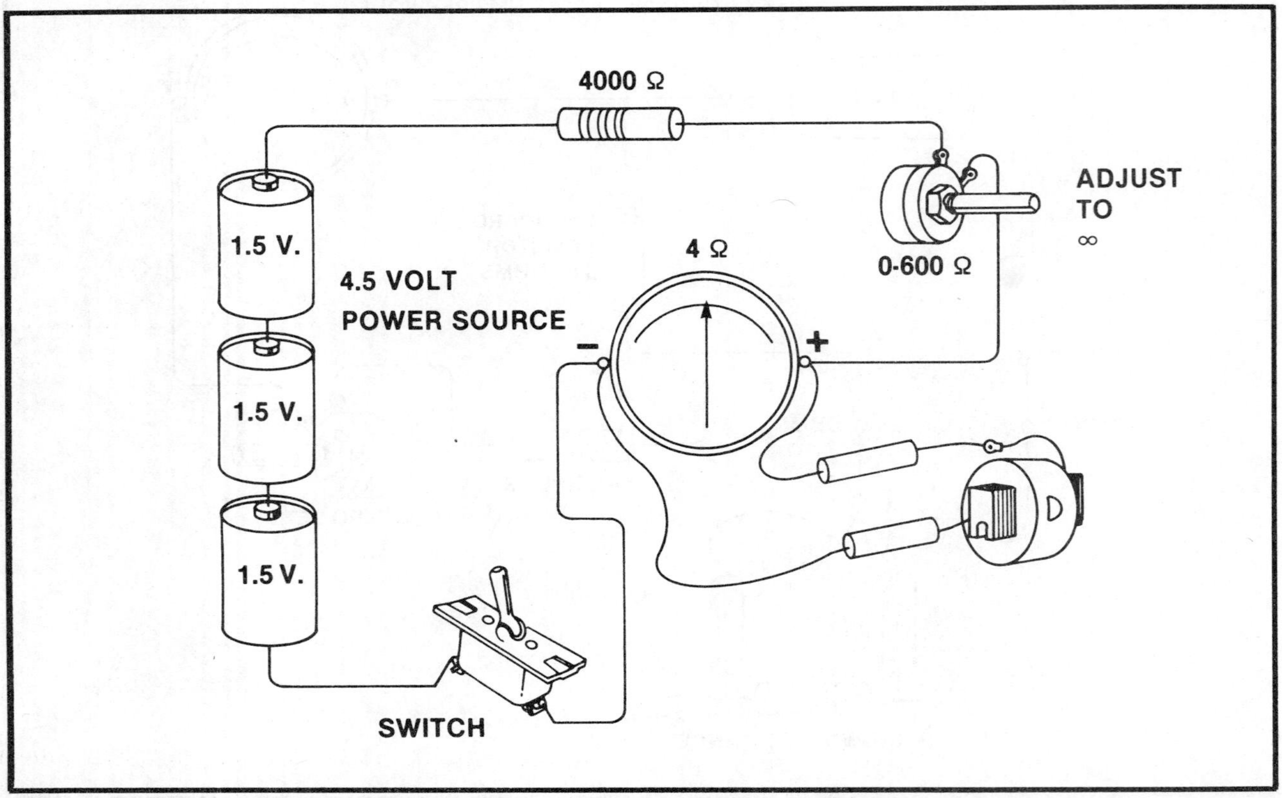

Fig. 14-8 Circuit of a shunt-type ohmmeter

will come to rest at the proper resistance measurement. Meggers are most often used for measuring insulation resistance in ignition systems and other high-voltage circuits.

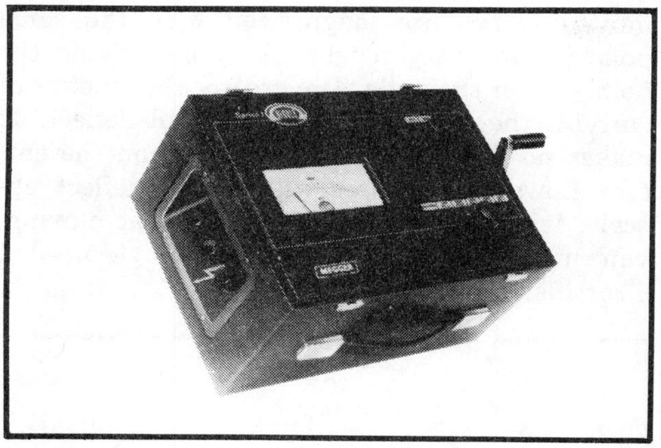

Fig. 14-9 A megohmmeter measures high resistances by applying a high voltage across the resistance being measured.

E. Multimeters

The most versatile electrical measuring instrument used by the aircraft technician is the multimeter. This handy tool has a single meter movement and a selector switch which may be used to select any of a number of circuits.

The Simpson Model 260 is one of the standard meters found in many aircraft maintenance shops. It has voltage ranges from 0 to 2.5, 10, 50, 250, 1,000, and 5,000 volts for both AC and DC. It can measure direct current in ranges of 100 microamperes, 10, 100, and 500 milliamps, and 10 amps. Resistance can be measured in ranges from 0-2,000 ohms, 0-200,000 ohms and 0-20 megohms. The meter has a sensitivity of 20,000 ohms per volt for measuring DC, but because of the rectifier circuit, its sensitivity for AC is 1,000 ohms per volt.

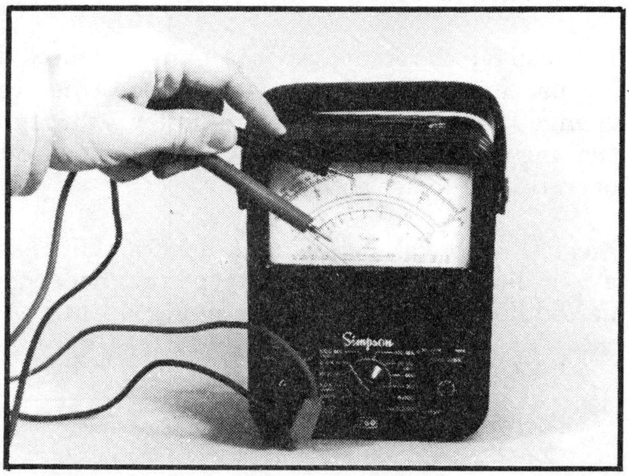

Fig. 14-10 The multimeter is one of the most useful electrical measuring instruments used by the A&P technician.

F. Electrodynamometer

1. Electrodynamometer wattmeter

An electrodynamometer operates in a manner similar to a D'Arsonval meter, except that an electromagnet is used instead of the permanent magnet for the fixed field.

A large coil having a few turns of heavy wire is connected in series with the load, and the strength of the magnetic field is proportional to the amount of current flowing through the load. The movable voltage coil is connected across the load, and its magnetic strength is proportional to the amount of voltage dropped across the load. The magnetic fields caused by the current and the voltage react with each other to move the pointer an amount that is proportional to the power dissipated by the load.
Electrodynamometer wattmeters may be used in either DC or AC circuits. In an AC circuit, they measure the true power, because if the current and voltage are out of phase in the circuit, they will also be out of phase with each other in the coils of the instrument, and the resultant field will cause the pointer to deflect an amount proportional to the true power rather than the apparent power.

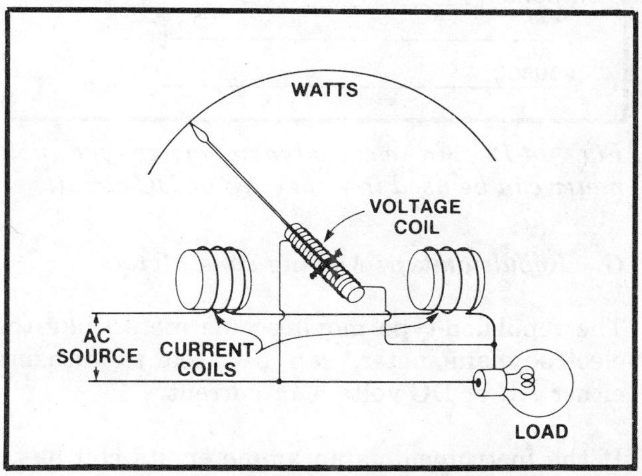

Fig. 14-11 An electrodynamometer-type wattmeter

The apparent power in an AC circuit may be found by measuring the current with an AC ammeter and the voltage with an AC voltmeter. The product of these two values is the apparent power.

Apparent power (volt-amps) = volts × amps

The power factor of the circuit can be found as the quotient of the true power divided by the apparent power.

$$\text{Power factor} = \frac{\text{True power (watts)}}{\text{Apparent power (volt-amps)}}$$

2. *Electrodynamometer voltmeters and ammeters*

Electrodynamometers may be used for voltmeters and ammeters to measure AC as well as DC values, because the polarity of both the fixed and movable fields reverse at the same time. The sensitivity of this type of meter is considerably lower than that of the D'Arsonval-type meter.

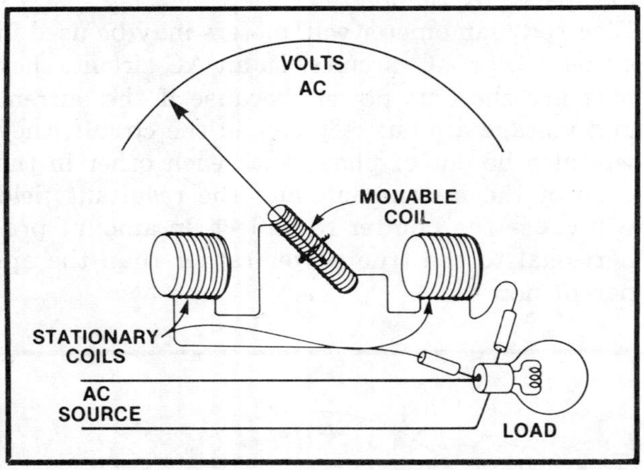

Fig. 14-12 An electrodynamometer-type voltmeter can be used in either AC or DC circuits.

G. Repulsion-type Moving-vane Meters

The repulsion-type moving-vane meter, like the electrodynamometer, can be used to measure either AC or DC voltage or current.

If the instrument is an ammeter, its coil has a relatively few turns of heavy wire, but if it is designed as a voltmeter, the coil will have many turns of fine wire. Inside the coil there are two vanes, one fixed and the other attached to the pointer staff and free to move beside the fixed vane, but restrained by a calibrated hairspring. When current flows in the coil, both the fixed and moving vanes are magnetized with the same polarity, and they repel each other, driving the pointer over the scale. The greater the amount of current, the farther the pointer will deflect. It makes no difference in which direction the current flows, the pointer will always deflect upscale. It is because of this feature that moving-vane meters are usable on AC without the need of a rectifier.

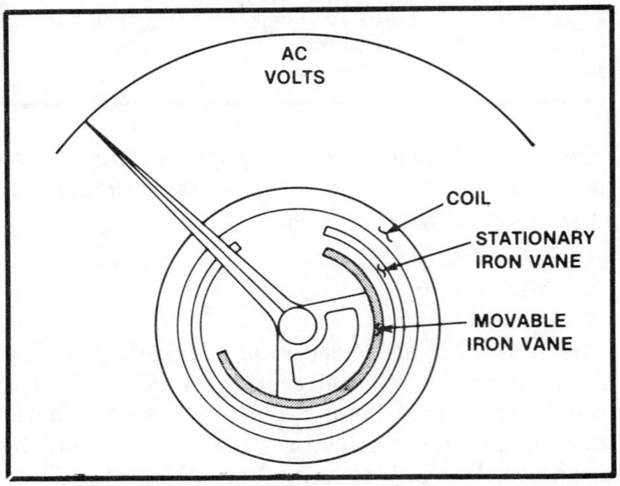

Fig. 14-13 A repulsion-type moving-vane meter can be used in either AC or DC circuits, and this principle may be used for ammeters or voltmeters.

H. D'Arsonval Meters with Rectifiers

D'Arsonval meter movements may be adapted for use in AC circuits by using a rectifier to change the AC into DC before it flows through the meter coil. After the AC is rectified, it operates the meter as DC.

Most D'Arsonval meters used in AC circuits use a four-diode full-wave bridge-type rectifier, connected in the circuit as we see in Fig. 14-14.

I. Thermocouple-type Ammeters

Low-frequency alternating current can be measured with an electrodynamometer or by a repulsion-type moving-vane meter, but when the

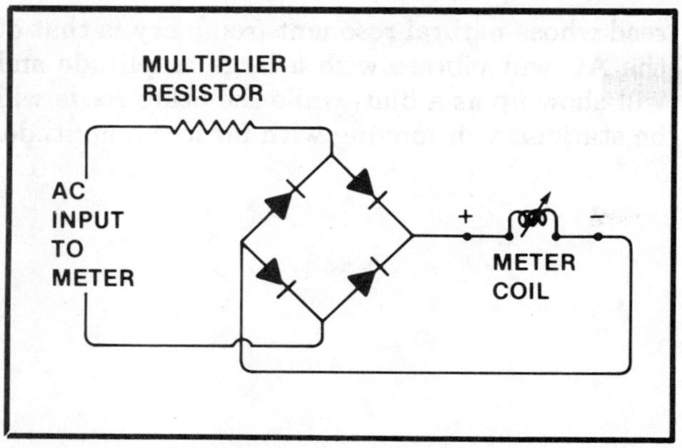

Fig. 14-14 D'Arsonval meters must use a rectifier such as this four-diode, bridge-type, full-wave rectifier when they are used in an AC circuit.

frequency is in the range of kilohertz or megahertz, this type of meter is not usable, and the thermocouple-type indicator is used.

The alternating current to be measured passes through a small piece of resistance wire inside the meter case. The greater the amount of current, the more the wire will be heated. A thermocouple, made of two dissimilar metals welded together, is welded to the resistance wire. The other ends of the thermocouple are connected to the moving coil of a D'Arsonval-type meter movement. A voltage is generated in a thermocouple that is proportional to the *difference* in the temperature between the two junctions. And since the junction at the meter movement has a relatively constant temperature, the amount of voltage and therefore the current through the meter is proportional to the temperature of the resistance wire. This temperature is proportional to the amount of the current. The meter scale is calibrated in amperes, and since the amount of heat produced in the wire is a function of the *square* of the current ($P = I^2 \times R$), the scale is not uniform, but the numbers are bunched up on the low end of the scale and spread out as the current increases. When the current doubles, there is four times as much deflection.

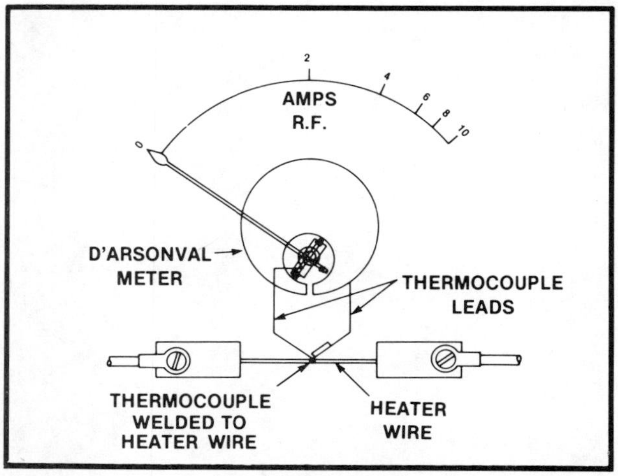

Fig. 14-15 Thermocouple-type ammeters are useful for measuring high-frequency alternating current.

Thermocouple-type ammeters are usable for direct current or audio-frequency or radio-frequency alternating current.

J. *Vibrating-reed Frequency Meters*

For precise frequency measurement, integrated circuit chips having clock circuits are used to actually count the cycles in a given time period and display the frequency as a digital display. But a much simpler type of frequency meter is used for determining the frequency of the AC produced by aircraft alternators.

These frequency meters use a series of metal reeds of different lengths. The center reed has a resonant frequency of exactly 400 hertz, and the reeds on one side are progressively longer, while those on the other side are progressively shorter.

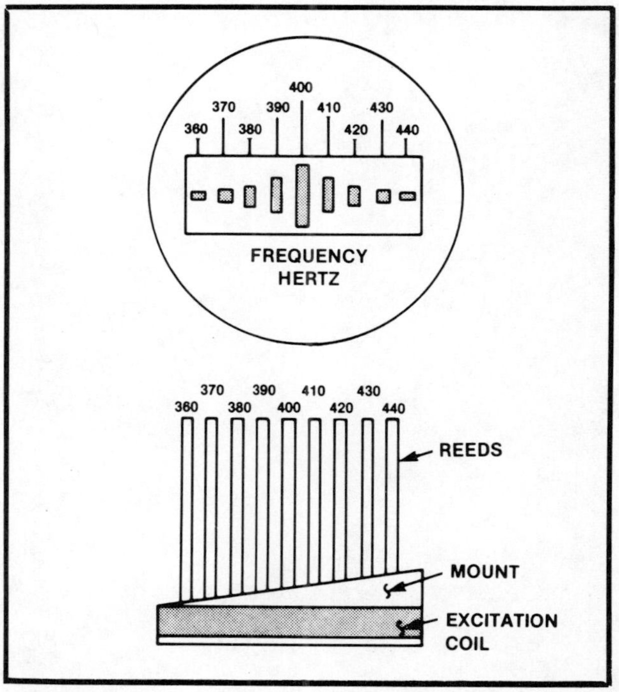

Fig. 14-16 A vibrating-reed-type frequency meter

A coil through which alternating current flows is wound around the fixture that holds these reeds, and the magnetic fields from the AC cause the fixture to vibrate at the frequency of the AC. The reed whose natural resonant frequency is that of the AC will vibrate with a large amplitude and will show up as a blur, while the other reeds will be stationary or moving with far less amplitude.

SECTION XV

Circuit Analysis and Troubleshooting

A. Principle of Troubleshooting

One of the most important jobs of an aviation maintenance technician is the analysis of faults or, as it is more commonly called, troubleshooting. Systematic troubleshooting allows us to find the cause of a problem and intelligently choose the correct solution.

Electrical troubleshooting is actually little different from that involving any other system. When there is a malfunction or a problem, *we must first find out exactly what should be happening.* Only when we have clearly in mind what *should be happening,* are we ready to find out exactly what *is* happening. The difference between what is actually happening and what should happen is a clear statement of our problem, and more often than not, when our problem is clearly stated in our minds, its cause and solution are not too far behind.

Once we clearly understand the problem, we can determine its cause, and then the solution is an easy matter of logic.

B. System Troubleshooting

Let's take, for an example, a problem with the navigation lights on an airplane. The trouble has been reported that the navigation lights do not burn. When we get a problem such as this, the first thing we should do is to review exactly what should be happening. Let's list the conditions that describe the situation as it should be:

1. The battery should be in the airplane and connected.
2. The master switch should be on.
3. The navigation light circuit breaker should be in.
4. The navigation light switch should be on.
5. All three navigation lights should be burning.

Our next step is then to go to the airplane and see what is actually happening:

1. Turn the master switch on. If the battery is in the airplane and is properly connected, you should hear the master contactor close, and some of the instruments and cockpit lights will give an indication that power is actually being supplied to the aircraft power bus. If you do not get any indication of power, the problem is between the battery and the main bus. But, let's assume that you do have power to the bus. This takes care of our items one and two.
2. Check the navigation light circuit breaker to be sure that it is not tripped out. If it is out, you must find the reason; but, again, let's assume that the circuit breaker is in as it should be. This clears item three.
3. Be sure the navigation light switch is on, and then go out and actually look at the three navigation lights. If we go to the left wing tip and see that the red light is not burning, but the green light on the right wing and the white light on the tail are burning, we know that our problem is isolated in the left wing.

The most logical condition, if only one light of the three does not burn, is a burned-out light bulb. The filament does not appear to be broken, so we must look further for the problem. Remove the bulb and check the continuity of the filament with a ohmmeter. With the meter set on the R × 1 scale, there should be continuity between the base of the bulb and the contact at its end. If there is continuity, the trouble is in the aircraft wiring, not in the bulb.

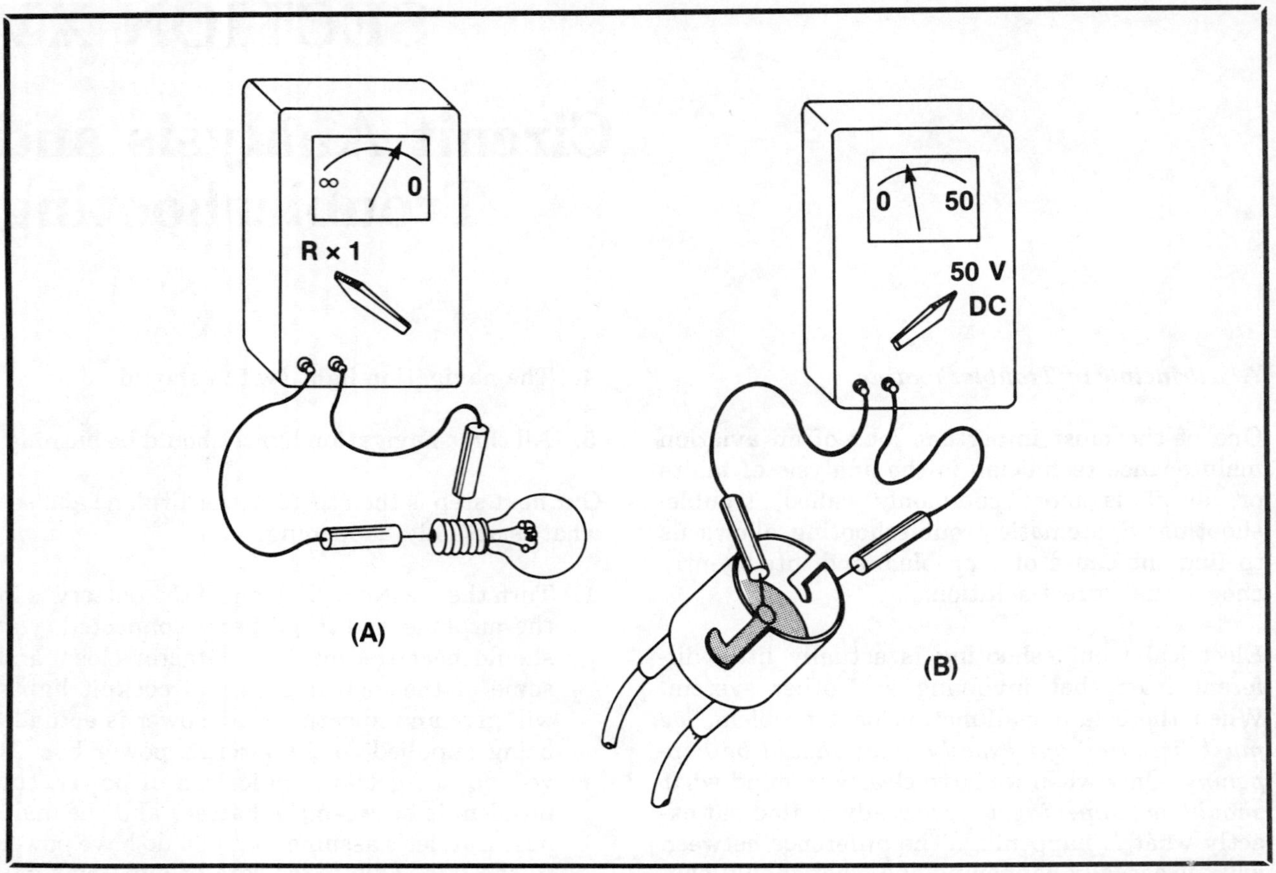

Fig. 15-1 A — Measuring continuity of a light bulb to check it for a burned out filament. B — Measuring voltage at the lamp socket.

Set the multimeter on a DC voltage scale, higher than the aircraft system voltage, and place the negative test lead against the socket into which the lamp fits and the positive lead against the center conductor. Let's assume that there is no voltage here.

Look at Fig. 15-2 to find what we have. We know that there is voltage at point *A,* because the green and white lights are burning, but there is no voltage between points *B* and *C.* There are two possibilities. The wire connecting the lamp socket to the airframe may not be making a good ground connection, or there is no connection between point *A* and *B.* The easiest place to check is between point *C,* the case of the lamp socket, and the airframe. Set the multimeter to the R × 1 ohmmeter scale and hold one test lead on the aircraft structure while you touch the lamp socket with the other lead. If the ground connection is good, the ohmmeter should give an indication of zero resistance. If you find that the lamp fixture

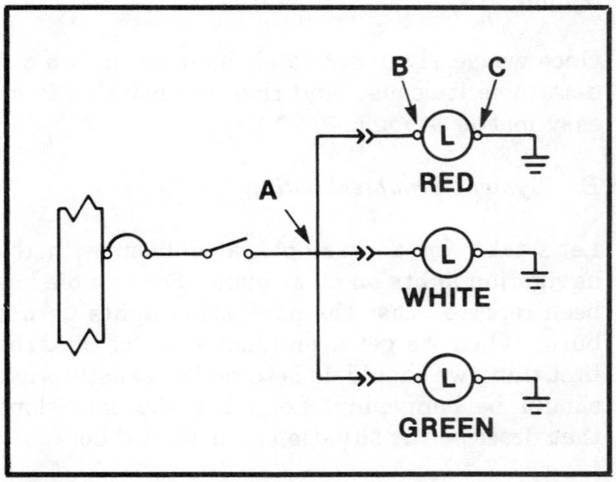

Fig. 15-2 A typical aircraft navigation light circuit

is properly grounded, the trouble is most probably in the quick-disconnect fitting in the wing root that allows the electrical circuits in the wing to be disconnected when the wing is removed.

C. *Component Troubleshooting*

The two basic instruments used to troubleshoot components in an electrical system are the voltmeter and the ohmmeter. It is extremely important that you understand that the voltmeter is used to find circuit problems with the aircraft power on the circuit, and the ohmmeter is used *only when there is no power on the circuit.* If an ohmmeter is used on a circuit that is energized, the meter will probably be damaged.

1. *Relays and contactors*

Many of the main battery contactors are connected into the circuit as we have in Fig. 15-3. To energize the contactor, the ground circuit for the coil is completed through the aircraft master switch.

If, when the master switch is closed, there is no voltage on the main bus through the ammeter, you can easily check the contactor with a voltmeter. Since the contactor is connected directly to the battery, there should be battery voltage at terminal *B,* and when the master switch is OFF, there should also be battery voltage at terminal *SW.* Since the master switch connects this end of the coil to ground, there should be *no* voltage at the terminal *SW* when the master switch is ON. If there is voltage at *SW* with the master switch ON, there is a problem in the master switch circuit. There should be no voltage between terminal *A* and ground when the master switch is OFF, but when it is ON, there should be battery voltage between *A* and ground.

If there is no voltage at the *SW* terminal when the master switch is off, there is a possibility of an open coil. To find out if this is the problem, disconnect the wire from the *SW* terminal, and with the *battery disconnected,* measure the continuity between terminals *B* and *SW* with an ohmmeter set on the R × 1 scale. There should be a low resistance, indicating that the coil is not open. If there is an infinite reading, the coil is open, and the contactor will have to be replaced.

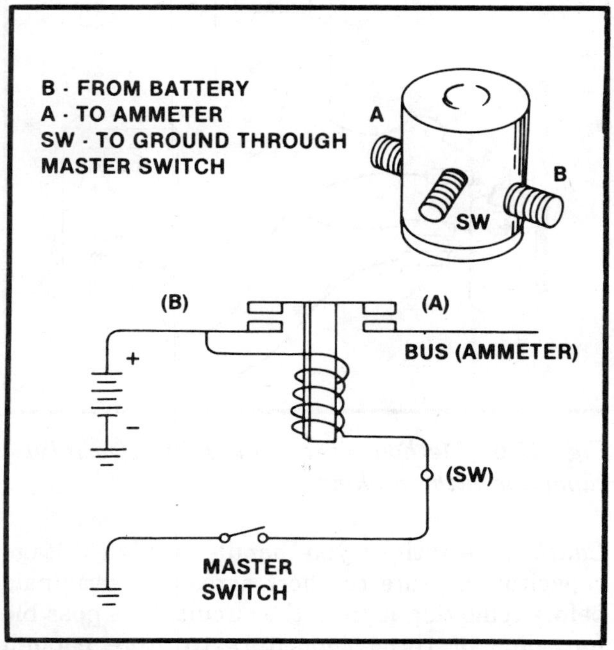

Fig. 15-3 An aircraft master battery contactor circuit

2. *Semiconductor diodes*

A semiconductor diode is an electrical check valve and, as such, it allows electron flow in one direction but not in the other. To check for a shorted or an open diode, disconnect it from the circuit and measure its resistance with an ohmmeter set on the R × 1 scale. With the leads across the diode in one direction, there should be a *low* resistance, and when the leads are reversed, there should be a *high* resistance. If the readings are both low, the diode is shorted, and if they are both high, the diode is open.

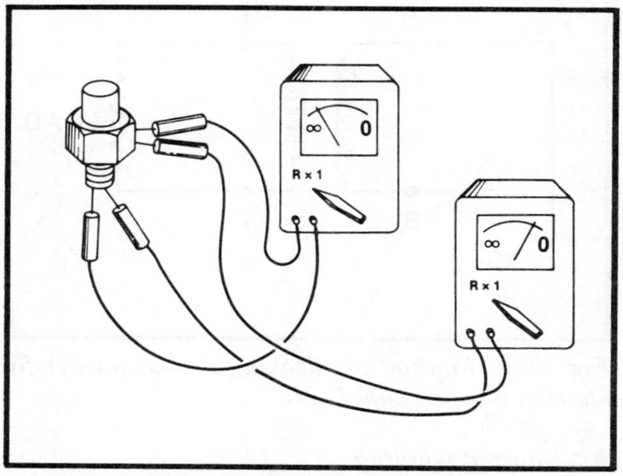

Fig. 15-4 Measuring the front-to-back resistance ratio of a semi-conductor diode

3. *Transformers*

There are two problems that could cause a transformer to not function as it should; one of the windings could be open, or one of them could be shorted.

a. Open winding

If either winding of a transformer is open, there will be no voltage in the secondary. If there is an open primary winding, there will be full source voltage between points *A* and *B*, but if the source is disconnected at either *A* or *B*, there will be an infinite resistance between them.

An open secondary winding will prevent there being voltage across points *C* and *D*. If the load is disconnected at either point, there should be some resistance indicated across these points if the transformer is good. If the transformer is a step-up transformer, the secondary will usually have a higher resistance than the primary, but if it is a step-down transformer, the secondary winding will usually have a very low resistance.

It is important when measuring the resistance of either of the windings that the transformer be disconnected from the circuit, because the shunting effect of the rest of the circuit will make the resistance reading inaccurate. For example, in Fig. 15-5, if you attempt to measure the resistance of the secondary winding without disconnecting the load, the ohmmeter will read the resistance of the load as well as that of the winding.

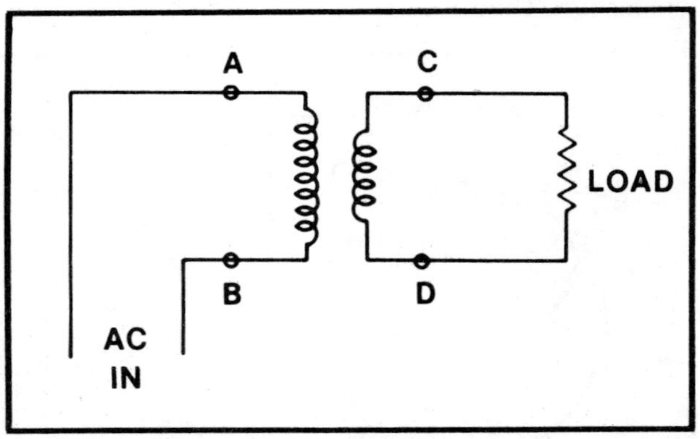

Fig. 15-5 Method of checking a transformer for shorted or open windings

b. Shorted winding

It is often difficult to find a shorted transformer winding by measuring its resistance unless you have an extremely accurate ohmmeter and know exactly what the resistance should be. It is easier to determine whether or not any coils are shorted by feeling the transformer after it has been operating for a while. If any of the windings are shorted, the transformer will be hot.

Another way to check for a shorted winding is to measure the output voltage. If the input voltage is the proper value, and the output voltage is low, it is a good indication that some of the windings are short-circuited.

4. Capacitors

It is difficult to check low-capacity capacitors for open circuits without a special tester, because even good capacitors will show up as an open circuit with an ohmmeter. If they do not indicate infinite resistance, they are probably short-circuited.

An electrolytic capacitor, however, can be tested by using an ohmmeter on its R × 1 scale. When you measure its resistance, the ohmmeter needle should deflect slightly up-scale as the capacitor charges and then come to rest at a high value of resistance. Now, reverse the leads. The needle should read *far* up-scale (low resistance) for a short time and then come back to a high-resistance reading. This large and temporary deflection of the pointer is caused by the capacitor discharging before it recharges in the opposite direction.

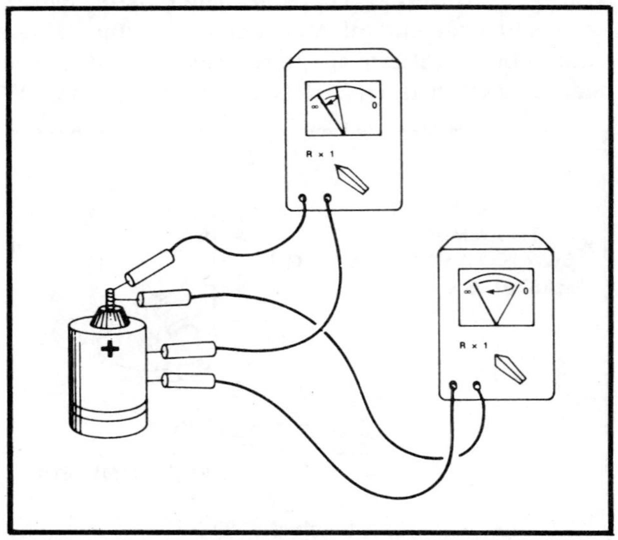

Fig. 15-6 Method of checking an electrolytic capacitor with an ohmmeter

Caution: Anytime you handle a high-voltage capacitor, be sure to short across its terminals before removing it from the circuit. It is possible for some of these capacitors to store enough charge that you can be hurt if it discharges through your body.

5. Electrical instruments

A word of warning is needed here about checking circuits that include an electrical measuring instrument such as a voltmeter or an ammeter. Many of these meters use external multiplier resistors or shunts, and if an attempt is made to check continuity through the meter itself, it is possible for the ohmmeter to put out enough current when set on its R × 1 range to damage the meter being examined. If it is necessary to measure across a meter, be sure that the ohmmeter is on its *highest* range. In this range, the minimum current flows through the test leads, and most meters can be checked for an open circuit without damaging them.

AVIATION
TECHNICIAN
INTEGRATED
TRAINING
PROGRAM

EA-ITP-G4

Aircraft Drawings
Weight and Balance
Fluid Lines and Fittings

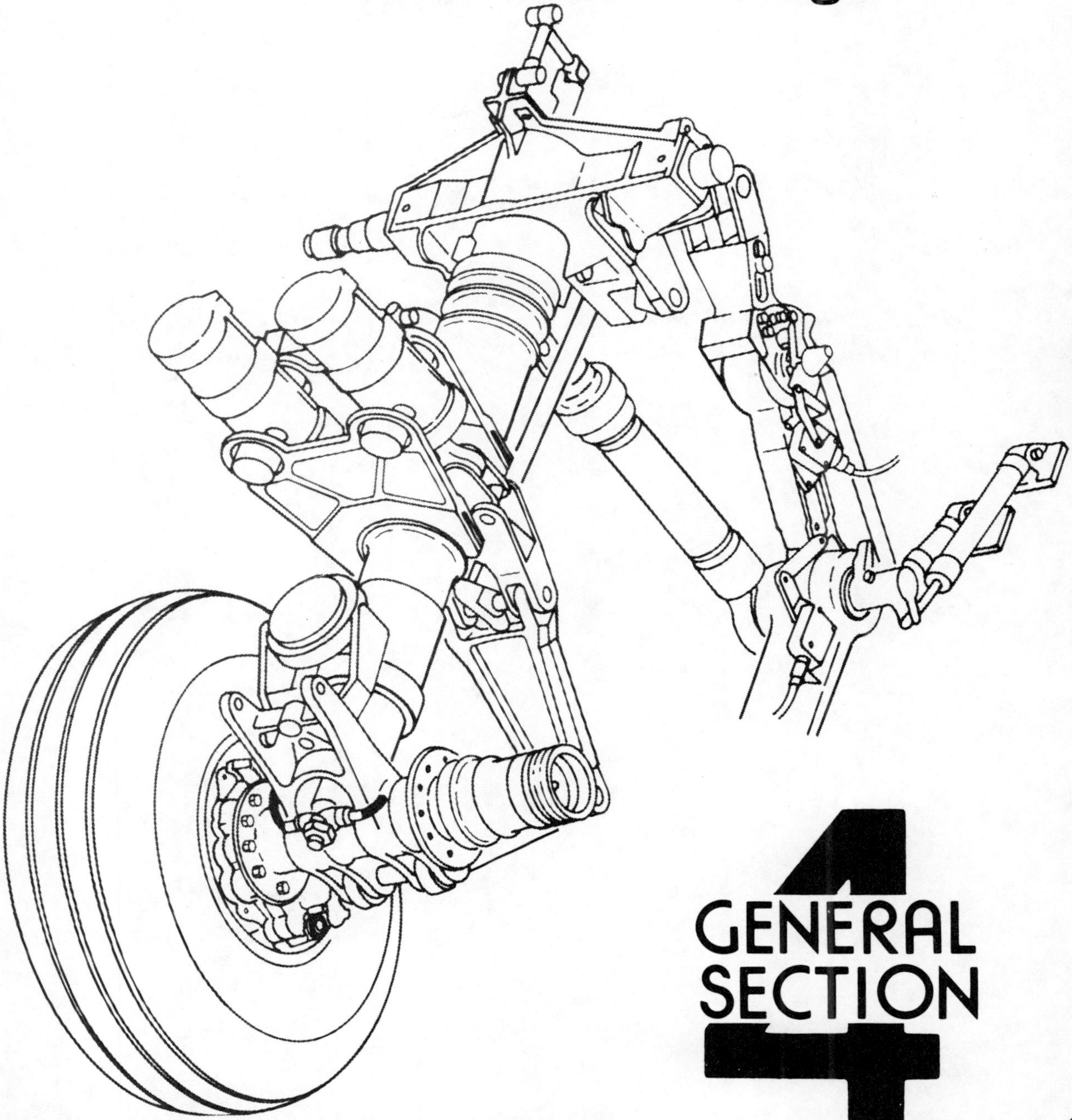

GENERAL
SECTION
4

EA-ITP-G4

TRAINING MANUAL
GENERAL SECTION
BOOK 4

AIRCRAFT DRAWINGS
WEIGHT AND BALANCE
FLUID LINES AND FITTINGS

by
Dale Crane

AVIATION MAINTENANCE PUBLISHERS, INC.

INTEGRATED TRAINING PROGRAM
FOR THE
AVIATION MAINTENANCE TECHNICIAN

International Standard Book Number 0-89100-126-3

Printed in the United States of America

TABLE OF CONTENTS

Preface

Part 65 of the Federal Aviation Regulations require that an applicant for a Mechanic Certificate with an Airframe or Powerplant rating have a general knowledge of the construction and maintenance of the aircraft covered by the rating he seeks. And each applicant must prove this knowledge by passing a written, an oral and a practical examination, each given in five sections: General, Airframe Structures, Airframe Systems and Components, Powerplant Theory and Maintenance, and Powerplant systems and Components.

Certificated Aviation Maintenance Technician Schools all operate under individual Federal Aviation Administration approvals, and all have in their approved curriculum enough information to satisfy the FAA that anyone completing their course will meet the knowledge requirements for certification as a mechanic.

In 1972, the Federal Aviation Administration published a series of three Advisory Circulars providing background information for mechanic certification applicants. These are entitled: *Airframe and Powerplant Mechanics General Handbook, AC 65-9; Airframe and Powerplant Mechanics Powerplant Handbook, AC 65-12;* and *Airframe and Powerplant Mechanics Airframe Handbook, AC 65-15.* These three handbooks were written to cover the area over which a mechanic applicant was to be tested, and they were an attempt by the FAA to coordinate the information available to the applicants with the tests that were given for certification. In 1976, the three handbooks were revised into their present A revision format: AC 65-9A, AC 65-12A, and AC 65-15A.

Individuals attempting to certificate under FAR Part 65, as well as schools operating under FAR Part 147, have found that these handbooks actually do not follow the material used in the FAA examinations, and because of this, plus their lack of study helps, they fail to effectively provide the broad base of knowledge needed for certification and for a foundation on which to build in the aviation maintenance industry.

Realizing the need for updated information, presented in a format that is in keeping with the latest developments in educational technology, the Aviation Maintenance Publishers, Inc., is offering an Integrated Training Program (ITP) that presents the basic information needed for mechanic certification, in a form that will provide a good foundation in aviation maintenance knowledge. Developed in coordination with Crane Aeronautical Activities, Inc., the ITP training manuals, workbooks, study guides, and audio-visuals are designed to follow the FAA written examinations and are written to the level required for the FAR Part 147 schools.

All of the material can be used by an individual preparing for certification on his own, or by schools preparing groups of individuals for certification under FAR Part 65, but it has been prepared primarily for the Aviation Maintenance Technician Schools operating under FAR Part 147. Not only has every attempt been made to meet the requirements of Part 147 in producing this series, but the authors have followed many of the recommendations of the Aviation Mechanic Occupation Study conducted by the University of California, Los Angeles, under a contract from the U.S. Department of Health, Education, and Welfare.

One of the most basic premises of aviation maintenance is that all operations must be carried out according to FAA-approved data, and all information of a specific nature furnished by the manufacturer of an aircraft, powerplant, or component must be followed in detail. These training manuals are, of necessity, general in nature and cover a wide scope of aircraft, powerplants, and components. Because of this, it must be understood that they should *never* be allowed to take precedence over specific information furnished by the manufacturer.

The authors and publishers of this Integrated Training Program wish to express our appreciation to the various manufacturers, Aviation Maintenance Technician Schools, and FAA personnel who have aided our efforts to provide this coordinated training material to meet the challenge of better training for aviation maintenance technicians in the eighties.

SECTION I

Aircraft Drawings

I. Purpose and Function of Aircraft Drawings

When an airplane is conceived in the minds of the engineers and designers, it is nothing but an idea until it is put on paper so others can add their expertise and ideas, and these thoughts can be translated into aluminum and steel.

Written descriptions are often difficult to follow, and this fact has been recognized since ancient times when some wise Chinese said that a picture was worth a thousand words.

Drawings take many forms, with some used to illustrate principles, and others to create impressions and moods. But in the aviation maintenance industry, we are concerned only with drawings that give us information on the construction of our aircraft and details of the proper way for us to maintain them.

When an airplane is designed, drawings are made of every single part. These are called detail drawings, and where detail parts must be assembled, assembly drawings are made to give the production workers in the aircraft factory all of the information necessary.

Every aircraft drawing is numbered and a careful record is maintained, so if any changes are required, the information will be passed on to the persons affected so that all of the changes can be incorporated into the parts as they are built.

The actual drawings are made on translucent vellum or Mylar by the draftsman, and after they have been checked for accuracy, they are released and sent to the reproduction department. Here, as many copies as are needed are printed by a process in which the black lines made by the draftsman come out on the paper as blue lines. The prints are then recorded and distributed throughout the factory as they are needed.

The most popular type of reproduction for machine drawings in the past has been the blueprint, in which the black lines made by the draftsman reproduce as white lines on a blue paper. But blueline drawings have been adopted by the aircraft manufacturing industry generally, because the pencil lines used to make the drawing do not need to be traced in ink before they are reproduced, as is required for blueprints. This allows a real saving in time and labor and reduces the cost of preparing the drawings.

All aircraft factories have a drawing room manual that details all of the lines, symbols, and conventions used by that company for their drawings. And rather than the drawings produced by each factory being different, there have, throughout the years, been trends to standardize the drawings. In this text we will show typical applications of aircraft drawings as they are most generally used.

II. Types of Aircraft Drawings

A. Detail Drawings

When a part is designed, the draftsman must make a drawing of it with such detail that the part can be fabricated in the shop. All dimensions are included, and the material and finish of the part is specified.

A drawing number is assigned to each drawing, and this becomes the part number that this particular part will carry with it through its fabrication and on to its assembly and installation into the finished unit, and as it is carried in the service parts inventory.

When a detail drawing is made, it is carefully and accurately drawn to a scale, and is dimensioned. But when the print is made from the drawing, the paper shrinks or stretches, and the part shown on the print is seldom the exact size as it was on the original drawing. For this reason, you should never scale a part from a print. Always use the dimensions that are given.

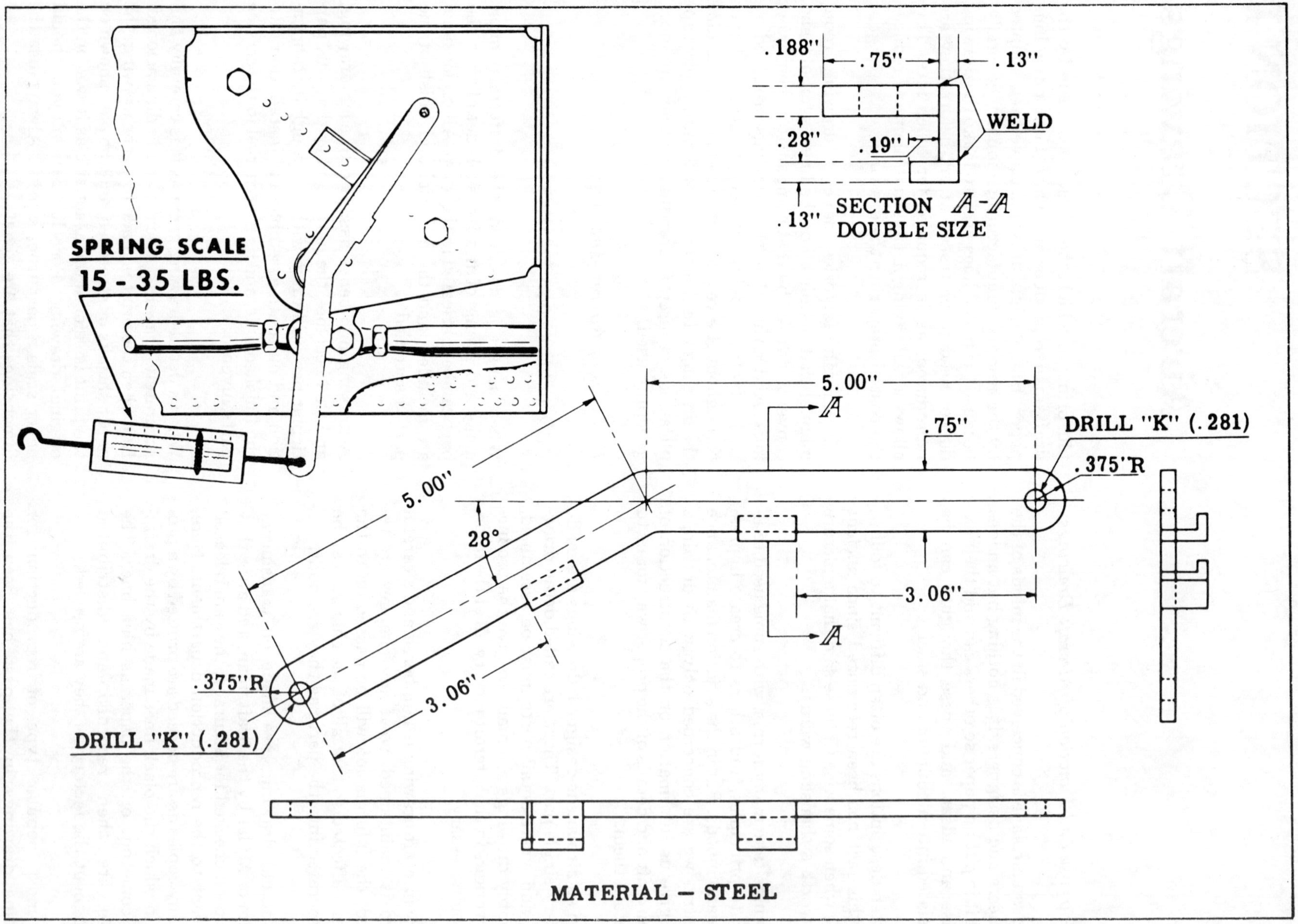

Fig. 1-1 A detail drawing must include enough information that the individual part can be fabricated.

B. *Assembly Drawing*

After all of the detail parts have been fabricated, they are assembled into the various subassemblies according to the information on assembly drawings. These drawings reference the detail parts by their part number and specify the type and number of fasteners to be used to join them. Only such dimensions as are needed for the assembly of the parts are included, and since there are already detail drawings for every component, no materials are specified on an assembly drawing.

NOTE

Do NOT subsitute automotive gaskets for any gaskets used in this assembly. Use only approved gaskets listed in the Parts Catalogs.

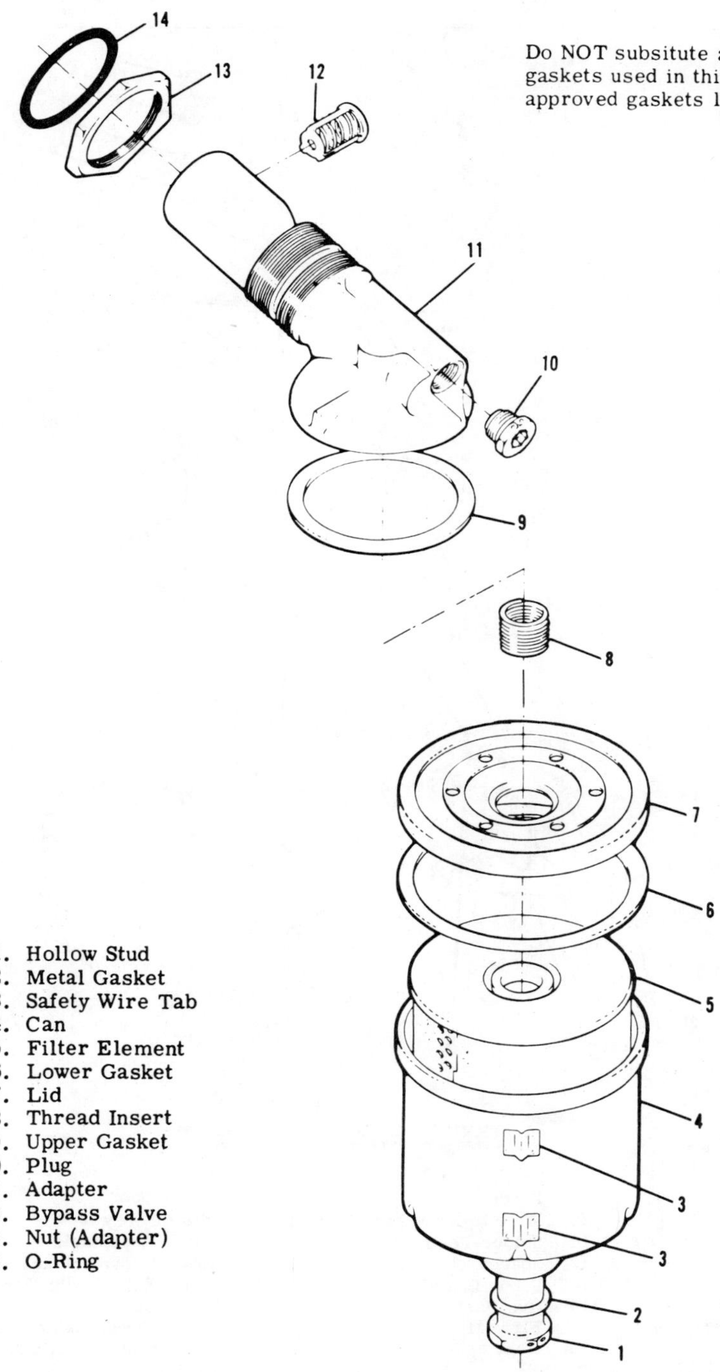

1. Hollow Stud
2. Metal Gasket
3. Safety Wire Tab
4. Can
5. Filter Element
6. Lower Gasket
7. Lid
8. Thread Insert
9. Upper Gasket
10. Plug
11. Adapter
12. Bypass Valve
13. Nut (Adapter)
14. O-Ring

Fig. 1-2 An assembly drawing shows the way two or more parts are assembled into a complete unit.

C. Installation Drawings

All of the subassemblies are brought together in an installation drawing. The bill of materials on these drawings lists every fastener needed, and notes on the face of the drawing furnish any information required for the installation. Dimensions are given for any adjustments that must be made for the part to function.

Often, portions of the aircraft that are not actually involved in the installation are shown, to help you locate the part being installed, and these are drawn a with phantom lines. The part is identified with its name, and the word "Ref." is noted beside it to indicate that the part is there only as a reference.

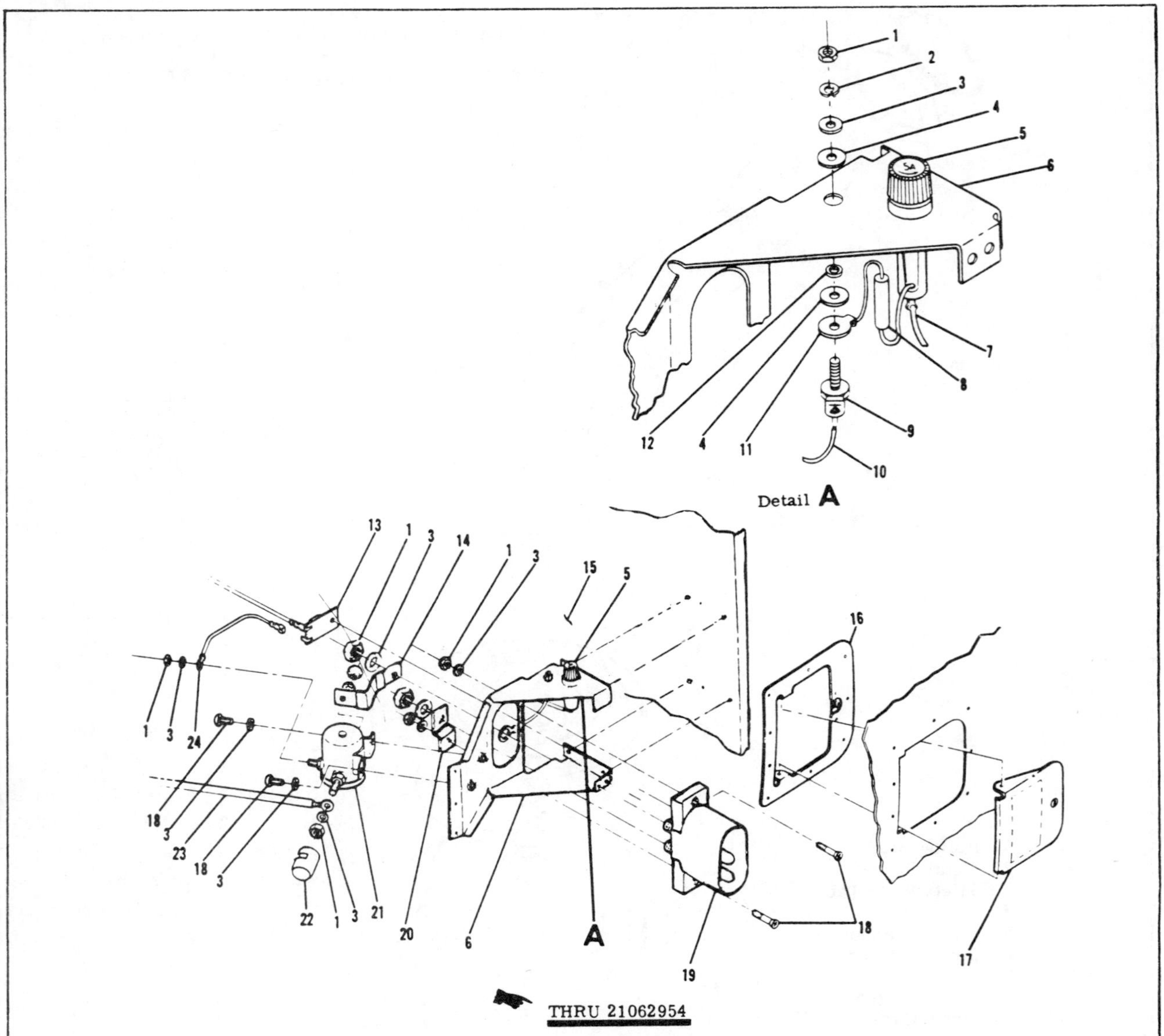

1. Nut
2. Lockwasher
3. Washer
4. Insulating Washer
5. Fuse
6. Bracket
7. Wire to Battery Contactor
8. Resistor-.75 ohm
9. Diode
10. Wire to Ext. Pwr. Recept.
11. Solder Terminal
12. Spacer
13. Diode Board
14. Bus Bar
15. Firewall
16. Doubler
17. Door
18. Screw
19. Receptacle
20. Ground Strap
21. Ext. Power Contactor
22. Cover
23. Power Cable
24. Contactor Control Wire

Fig. 1-3 An installation drawing shows the way parts are installed on the aircraft.

D. Sectional Drawings

It is often necessary to show the internal construction or shape of a part, and this is most generally done with the use of a sectional drawing.

If only the shape of the cross section of a part needs to be shown, this can be done with either a revolved or a removed section. In Fig. 1-4A, we have a removed section. The I-beam in this case has been broken in two places with long break lines, and the cross section is shown between the breaks.

In Fig. 1-4B we see a bar cut with a cutting plane line and the section removed and identified with the letters used at the end of the cutting plane arrows.

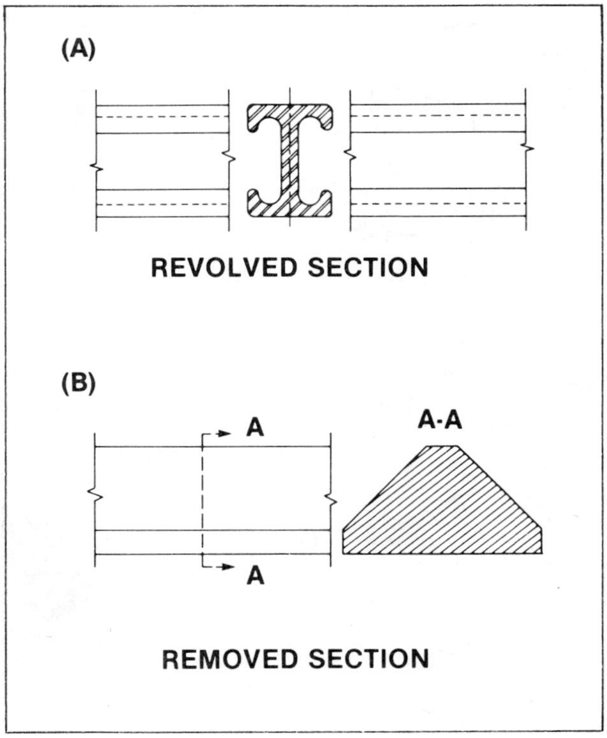

Fig. 1-4 A — Section drawing using a revolved section B — Section drawing using a removed section

Complex assemblies such as the driveshaft (Fig. 1-5) are shown in complete sectional views. Because of the proper use of section lines, it is easy for us to identify the shaft, the housing, the bearings, and the pulley, as well as the seal and the seal-retaining nut.

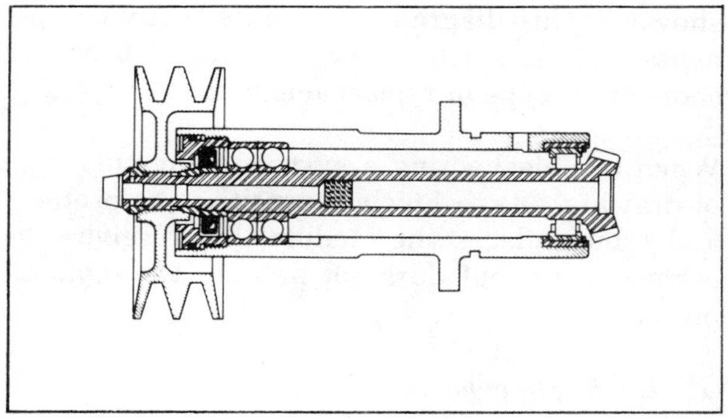

Fig. 1-5 Complete sectional view

If it would be helpful to see the outside of the part as well as the inside, a half-sectional view can be made, as we have in Fig. 1-6. The upper half of this drawing shows the internal construction of the assembly, while the lower half shows the entire assembly as it appears from the outside.

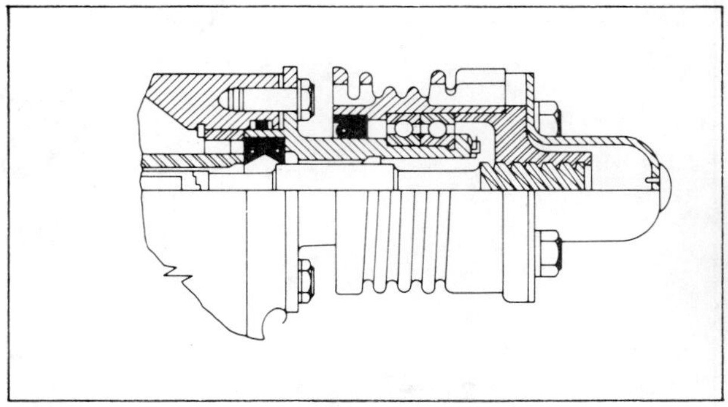

Fig. 1-6 Half-sectional view

E. Exploded-view Drawing

Illustrated parts lists often make use of exploded-view drawings to show every part that is in an assembly. All of the parts are in their relative positions, but are expanded outward, so that each part can be identified both by its physical appearance and by its name or by a reference number that is coded to the parts list.

F. Block Diagrams

With electrical and electronic systems and components becoming more complex all of the time, methods and procedures have been developed to aid the technician in locating troubles. And since much of the maintenance of these systems consists of identifying a malfunctioning subassembly and replacing it, block diagrams such as the one shown in Fig. 1-8 are useful. Each block shown in this diagram represents many components and is usually a printed-circuit board or some other type of replaceable module.

When troubleshooting a system using this type of drawing, the technician identifies the problem and then replaces the module that receives the correct input, but does not produce the required output.

G. Logic Flowcharts

As a further aid to troubleshooting, the logic flowchart has been developed and is shown in Fig. 1-9. When troubleshooting a system using this aid, the technician starts at the oblong *START* symbol and follows the arrows. Rectangular boxes tell what to do, and the diamonds are decision points. Match the condition that exists after the instructions in the rectangular boxes have been followed. Each diamond has one input and only two outputs so, by using logic, the chart can be followed to the oblong *END OF TEST* symbol.

This type of chart identifies the probable faults a system can develop and specifies the fix for each one. And by using this chart, troubleshooting time is reduced to a minimum.

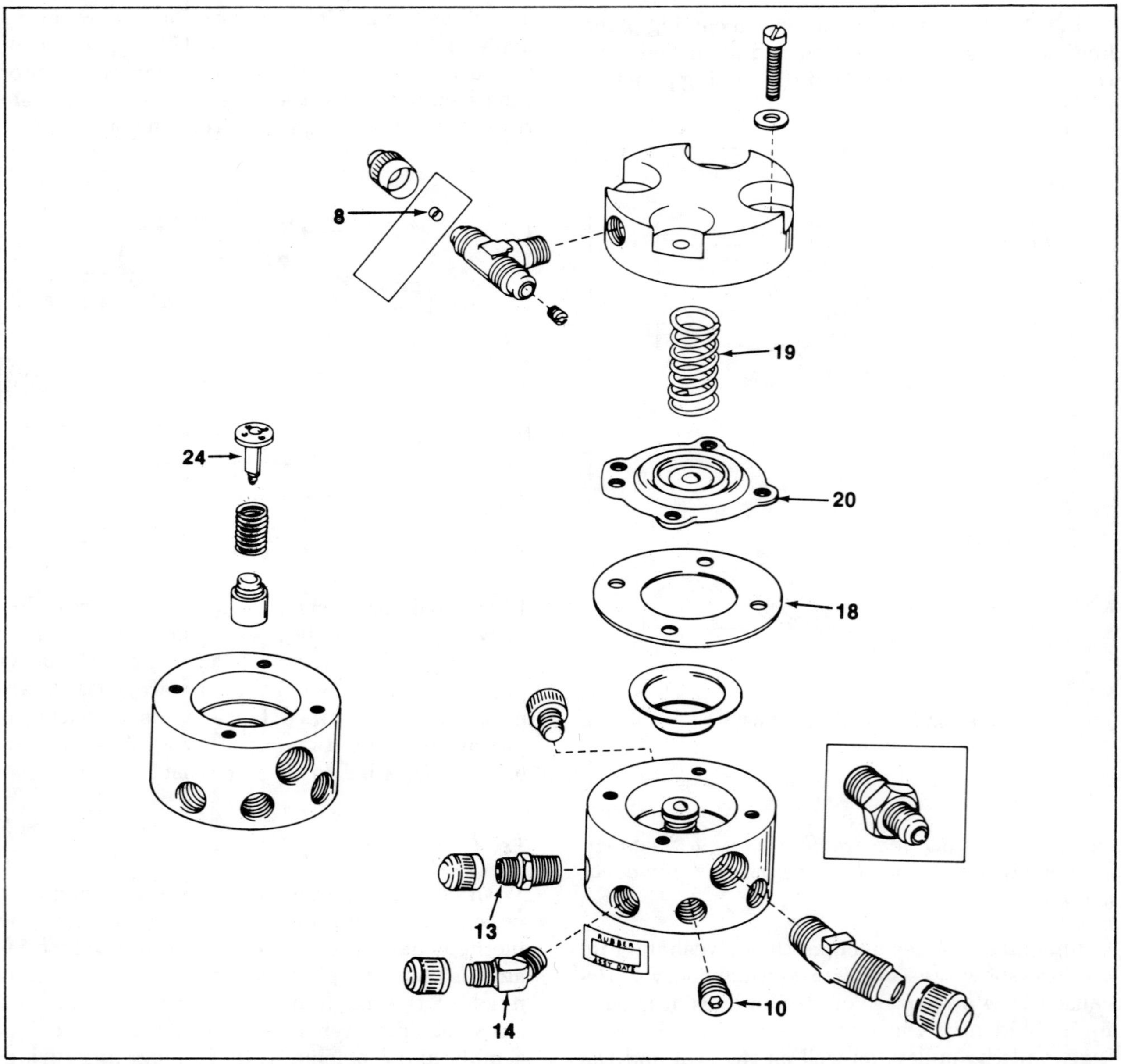

Fig. 1-7 An exploded view of a part shows the relationship of all of the components.

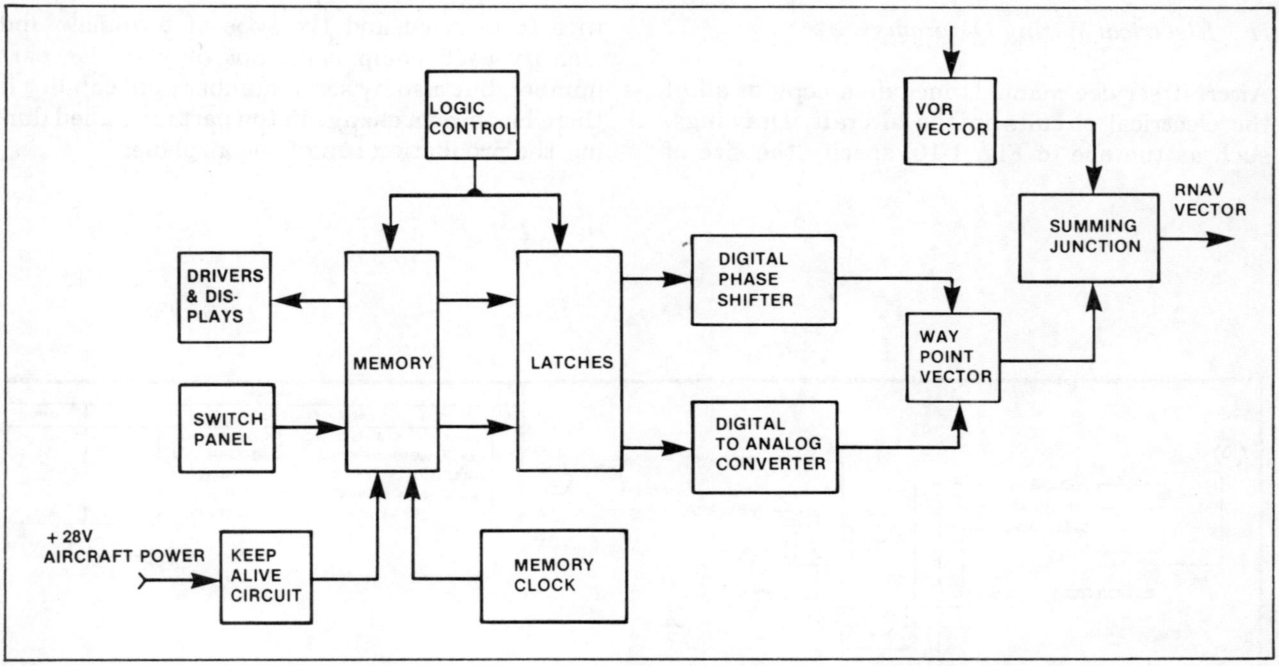

Fig. 1-8 A block diagram shows the various operations rather than the components.

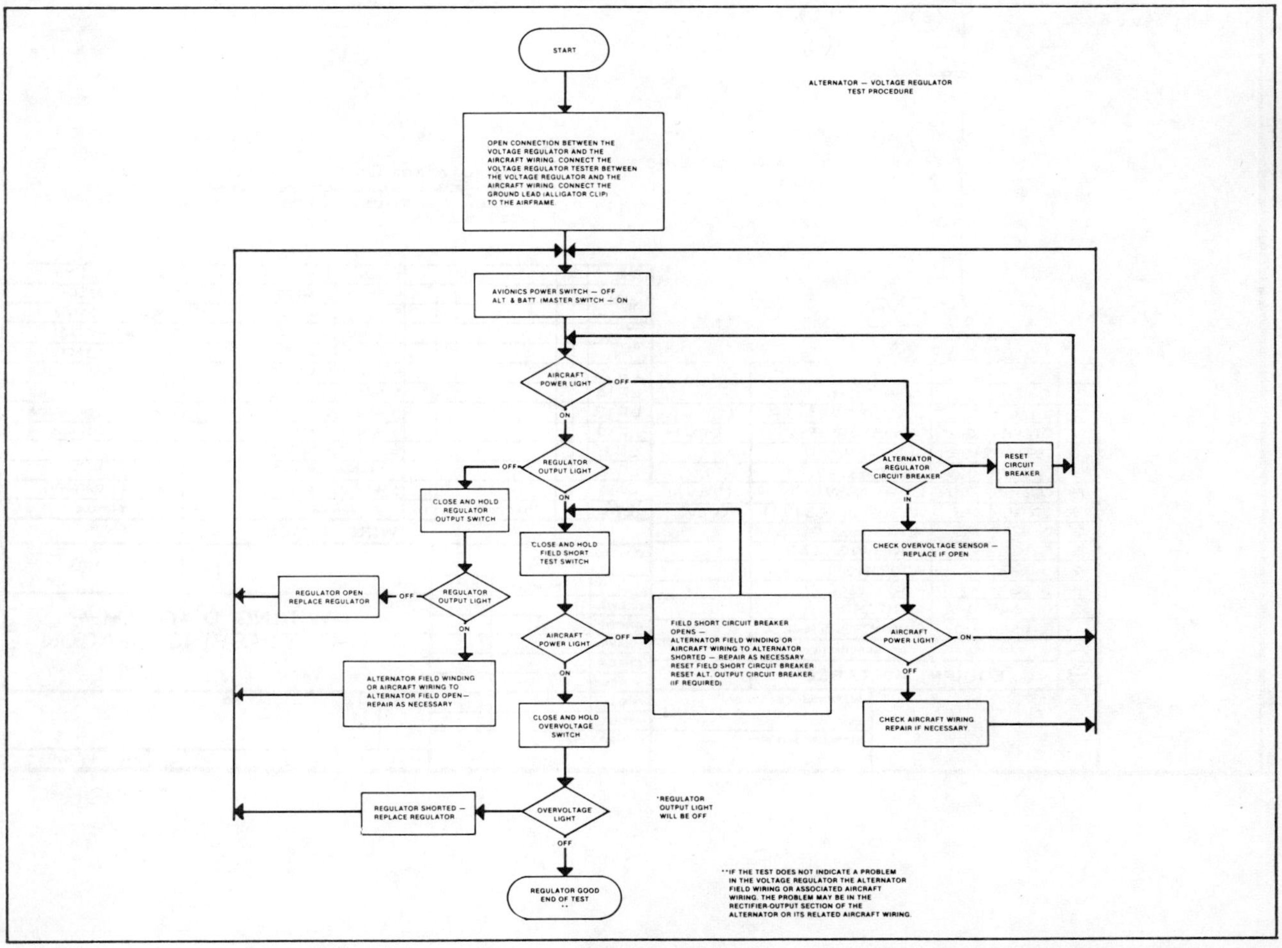

Fig. 1-9 A logic flowchart directs troubleshooting steps in a logical step-by-step manner.

H. Electrical Wiring Diagrams

Aircraft service manuals include a copy of all of the electrical circuits in the aircraft. Drawings, such as the one in Fig. 1-10, specify the size of wire to be used and the type of terminals and identify each component, not only by its part number, but also by serial number applicability if there has been a change in the parts installed during the production run of the airplane.

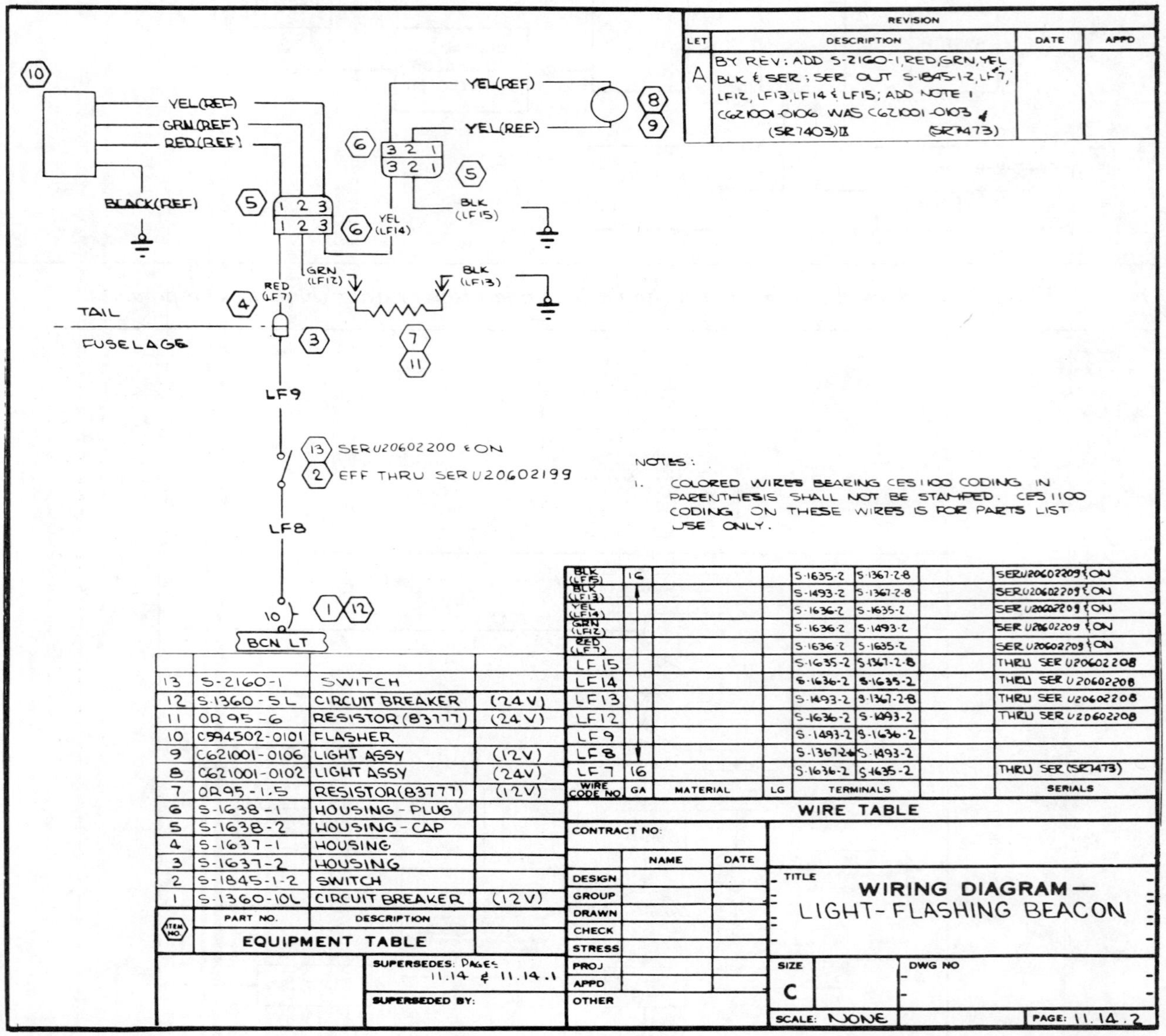

Fig. 1-10 An electrical wiring diagram allows a technician to trace all of the wires in a circuit.

I. Pictorial Electrical Diagrams

Pilot's handbooks and some training manuals often use pictorial diagrams of electrical and hydraulic systems. Line drawings represent the components, rather than the more conventional symbols used in the schematic diagrams. Diagrams such as that shown in Fig. 1-11 help visualize the operation of an electrical power system of an airplane.

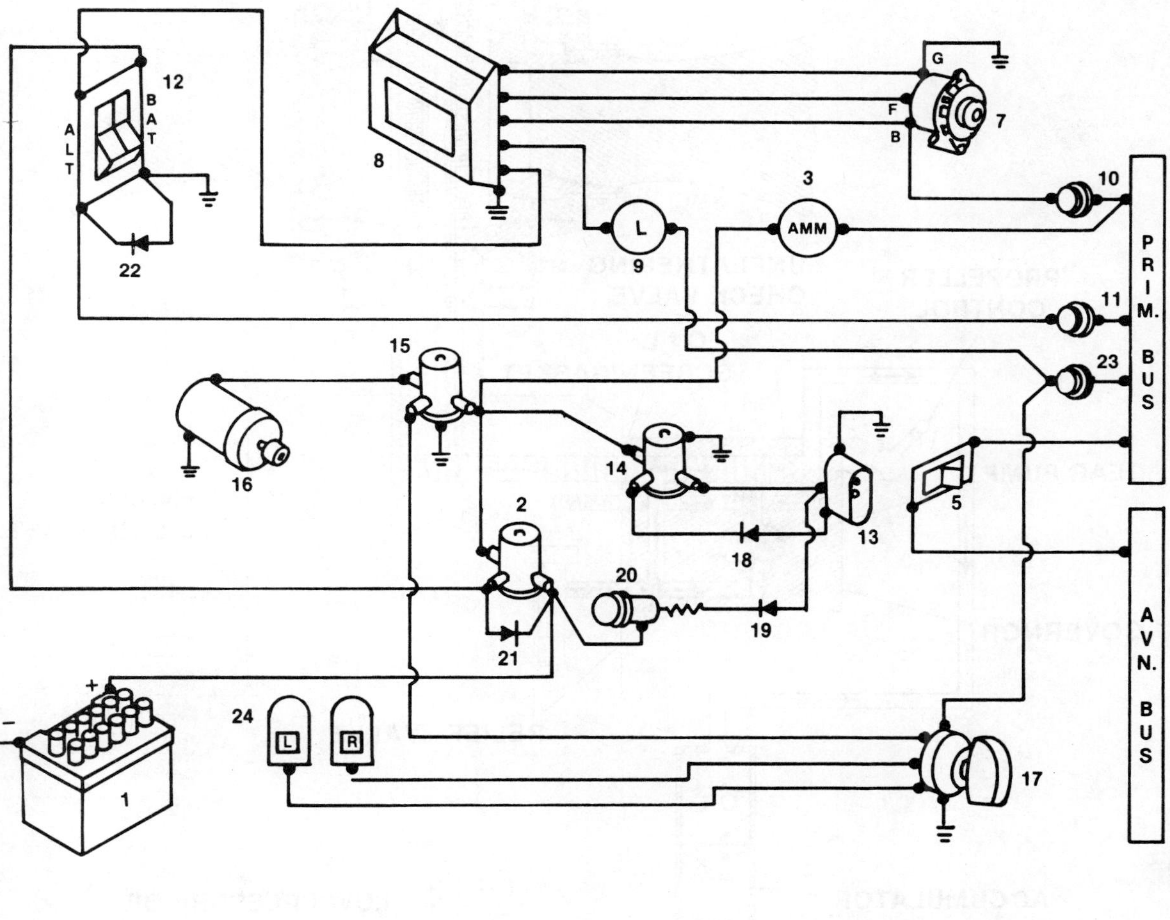

Fig. 1-11 A pictorial electrical diagram makes it easy for a technician to visualize the components involved in an electrical circuit.

J. Schematic Diagrams

A schematic diagram is used to explain a principle of operation, rather than to show the parts as they acually appear, or as they actually function. Fig. 1-12 helps us understand the operation of a constant-speed propeller, without showing any of the details of its construction or of its installation.

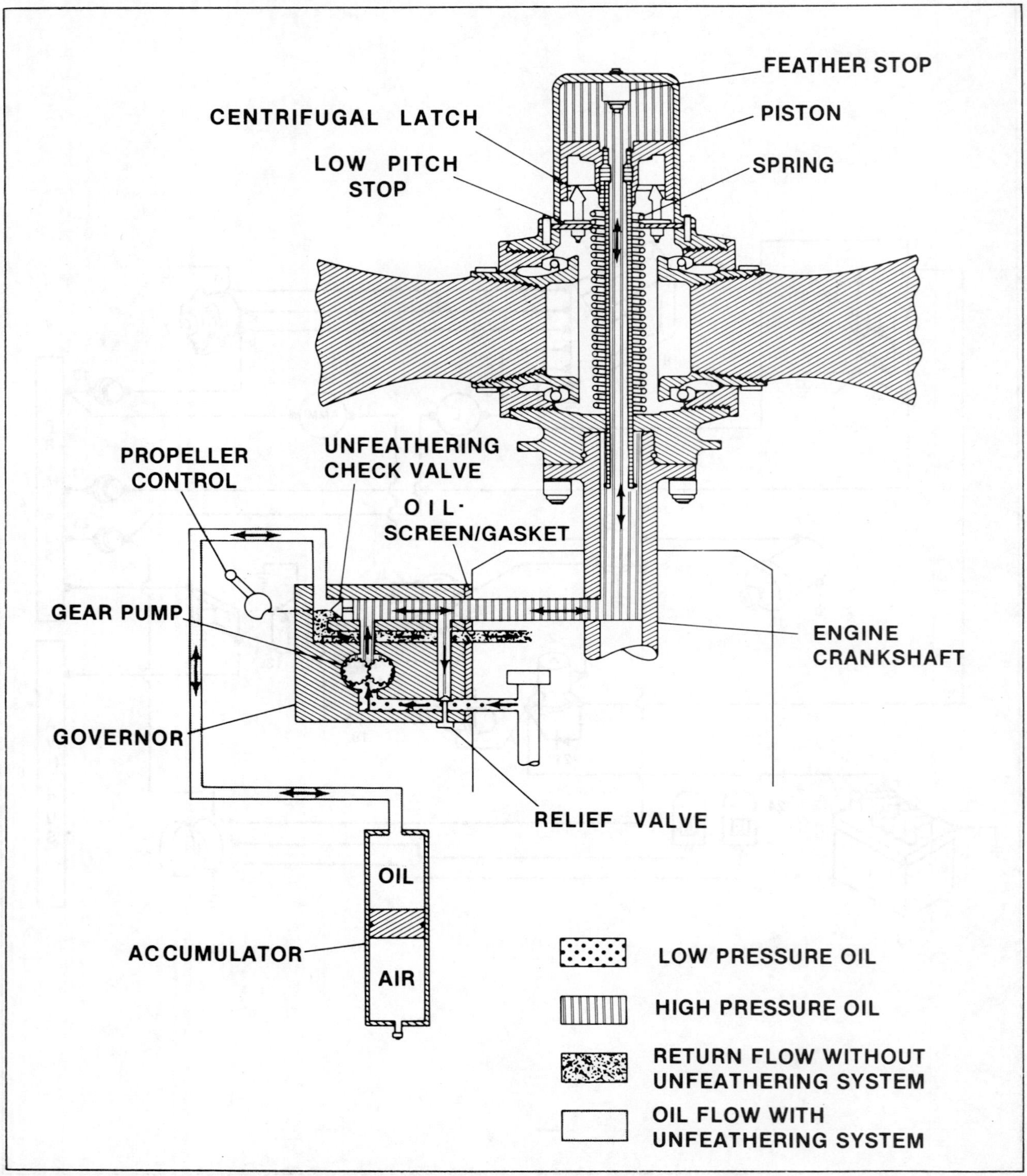

Fig. 1-12 A schematic diagram explains a principle of operation rather than showing the parts as they actually are.

III. Methods of Illustration

A. Orthographic Drawings

Almost all drawings used for the construction of a detailed part of an aircraft are made by the orthographic projection method. The part is shown in one, two, or three views, each drawn on a flat plane as it would appear if it were drawn on the surface of a transparent box. All of the rays extending from the object are parallel and perpendicular to the plane.

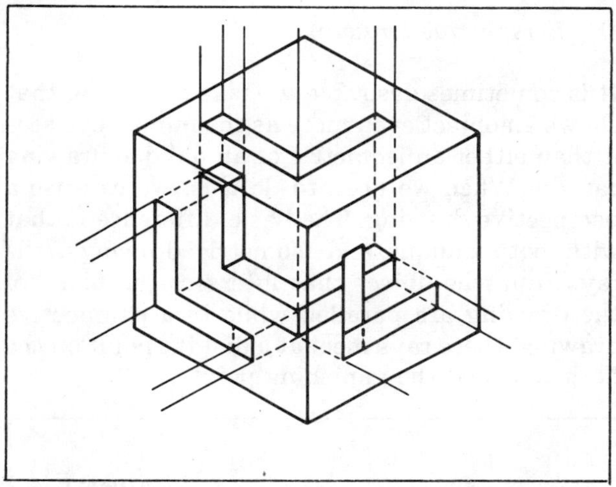

Fig. 1-13 An orthographic drawing shows six views of an object. These views are: front and rear, left and right, top and bottom.

Most of the drawings used in aviation maintenance show three views of the part: a front view, a side view, and a top view. These are the same as the views projected on the surfaces of our transparent box and are positioned on the paper in the same relationship they would have if the sides of the transparent box were opened out.

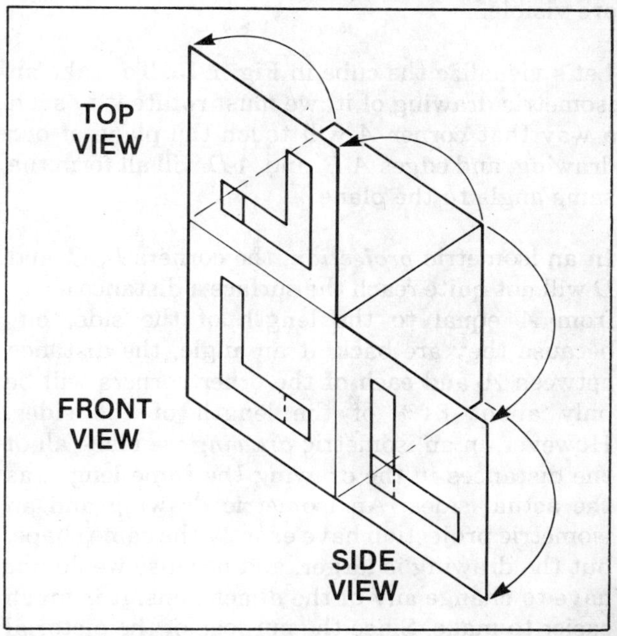

Fig. 1-14 The orthographic views are laid out on a flat drawing in the way they would appear if the sides of the box enclosing the object were opened out.

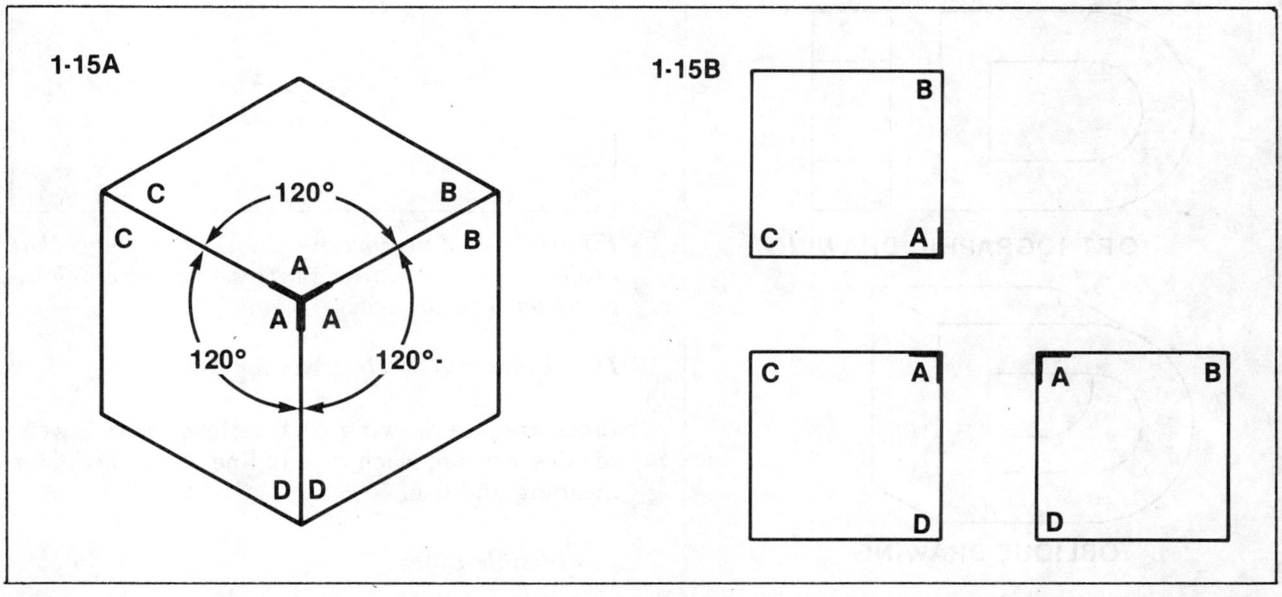

Fig. 1-15 A — An isometric drawing is a projection of three sides of an object on a flat plane. The angles formed by the three sides of the object are equal. B — The corner of this cube is the only part of the object that touches the plane of the drawing.

B. *Isometric Drawings*

The form of pictorial presentation that is most used for aircraft drawing and sketching is the isometric drawing. This is a projection of a three-dimensioned object on a flat plane, with the object rotated in such a way that three of its faces are visible.

Let's visualize the cube in Fig. 1-15. To make an isometric drawing of it, we must rotate it in such a way that corner *A* will touch the plane of our drawing, and edges *A-B,* and *A-D* will all form the same angle to the plane.

In an isometric *projection*, the corners *B, C,* and *D* will not quite reach the surface a distance away from *A* equal to the length of the side, but because they are back at an angle, the distance between *A* and each of the other corners will be only about 81% of the length of the sides. However, in an isometric *drawing* we make all of the distances in the drawing the same length as the actual sides. An isometric drawing and an isometric projection have exactly the same shape, but the drawing is larger, and because we do not have to change any of the dimensions, it is much easier to make. Since the purpose of the pictorial drawing is to allow a person to visualize a part, an isometric drawing serves our purpose well.

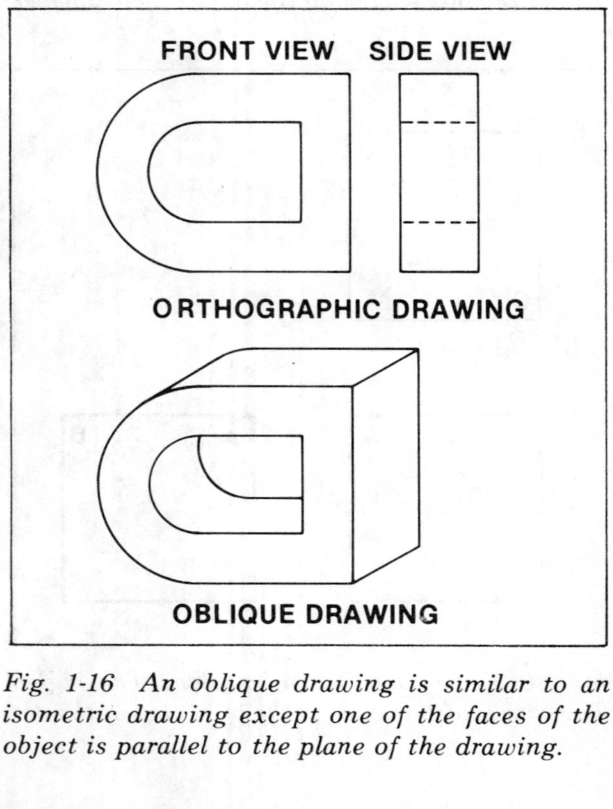

Fig. 1-16 An oblique drawing is similar to an isometric drawing except one of the faces of the object is parallel to the plane of the drawing.

C. *Oblique Drawings*

An oblique drawing is similar to an isometric drawing, except that one of the faces of the object is made parallel to the plane of the drawing. Because of this, two of the axes are perpendicular to each other, and the front face of the drawing is identical to the front view of an orthographic drawing. The depth axis of an oblique drawing may be angled up at any convenient angle, and for most oblique drawings it is about 30 degrees.

D. *Perspective Drawings*

It is sometimes desirable to make a drawing that shows an object even more as the human eye sees it than either an isometric or an oblique drawing can do. When we need to do this, we can use a perspective drawing. The basic difference is that with both oblique and isometric drawings, the rays from the object that intersect the plane of the drawing are parallel, while in a perspective drawing, these rays meet at a distant point on the horizon called the vanishing point.

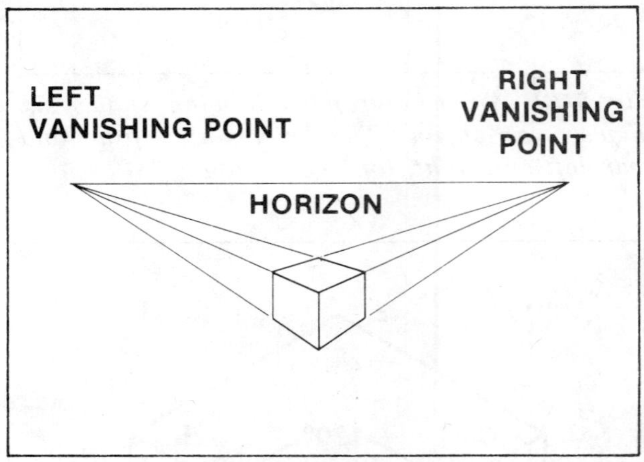

Fig. 1-17 In a perspective drawing, the rays that project from the drawing intersect at a vanishing point on the horizon.

IV. *Lines and Their Meaning*

Lines are to a drawing what letters are to a written description. Each type of line has a particular meaning and use.

A. *Visible Lines*

Any edge of a part that is visible in the view represented by the drawing is shown with a medium-weight solid line.

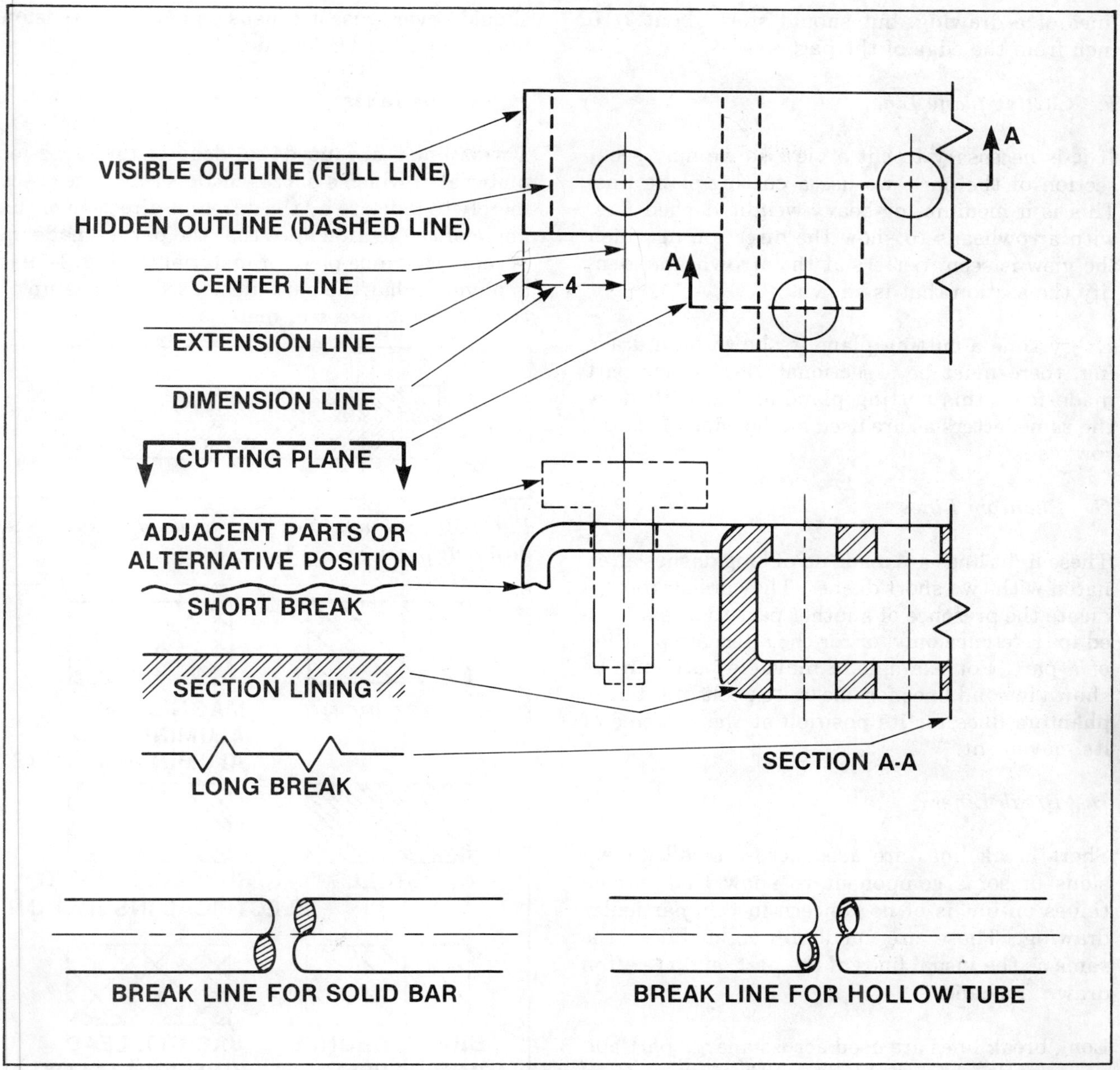

Fig. 1-18 The meaning of lines

B. Hidden Lines

If an edge of a part is hidden from view, it is represented on the drawing by a dashed line of the same weight as the visible outlines.

C. Center Lines

The centers of holes, and lines used to show the middle of a symmetrical part, are drawn as light lines, made up of alternate long and short dashes. The exact center of a hole is marked by the intersection of two of the short dashes.

D. Dimension Lines

These lines are light and solid, but broken in their center for the dimension to be inserted. An arrowhead is placed at each end, touching an extension line from the part, and shows the exact location from which the dimension is made.

E. Extension Lines

These are the light lines that extend from the point at which a measurement is made. The extension lines should not actually touch the visible lines of a drawing, but should start about 1/16 inch from the edge of the part.

F. Cutting-plane Line

If it is necessary to cut a view so we may see a section of the part, we use a cutting-plane line. This is a medium- or heavy-weight dashed line, with arrowheads to show the direction in which the view is seen. Letters at the arrowheads identify the section that is shown.

Every time a cutting plane is shown on a drawing, there must be a sectional view of the part made from this cutting plane and identified by the same letters as are used at the ends of the arrowheads.

G. Phantom Lines

These light lines are made up of long dashes alternated with two short dashes. They are used to indicate the presence of another part that is included for reference only, or for the alternate position of a part. For example, a movable part may be shown in solid lines for one of its positions, but in phantom lines for its position at the extreme of its movement.

H. Break Lines

Short break lines are used across small dimensions of some component to show that it continues on but is of no concern in this particular drawing. These are medium-weight lines, the same as the visual lines of the part, and are often drawn freehand.

Long break lines are used across a large part and are made up of a light line with a series of irregular breaks to show that the part continues on. Long break lines usually extend beyond the solid lines which indicate the edges of the part.

Bars, tubes, and rods have break lines drawn in such a way that they show a partial sectional view of the part.

I. Leader Lines

These are light lines with arrowheads that extend from a note to a part of the drawing to which the note applies. They extend from the center of the beginning of the first line of the note, or the center of the end of the last line. Leader lines should never cross a dimension line, an extension line, or another leader line.

J. Section Lines

These lines are used to denote the type of material of which a part is made, or in some cases merely to indicate by the different direction of the angle of the section lines that the part is made up of more than one piece of material. In Fig. 1-19 it is obvious that the part illustrated is made up of two different pieces of material.

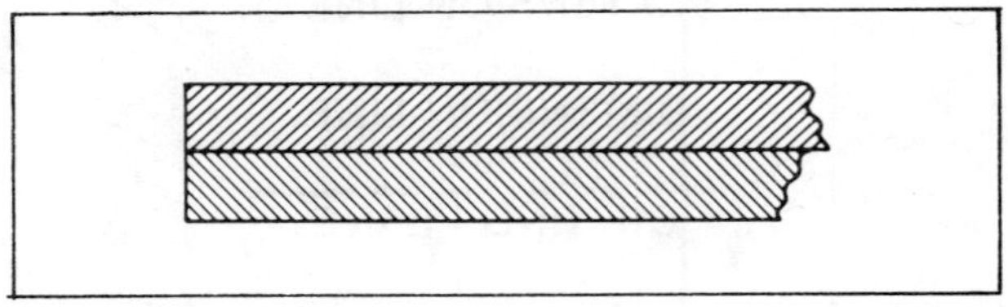

Fig. 1-19 Section lines may be used to show the different parts of an object.

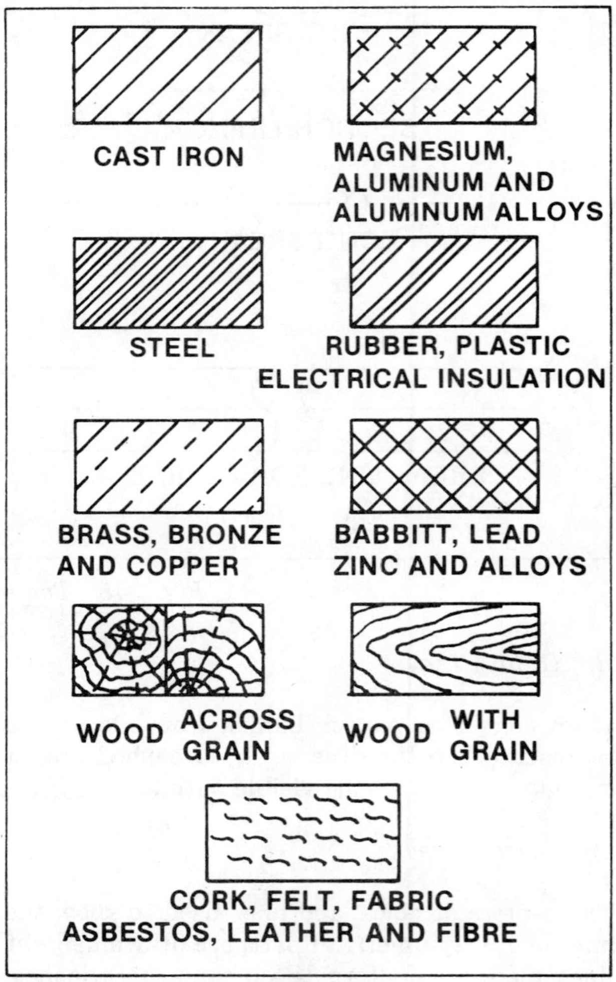

Fig. 1-20 Section lines are used to indicate the type of material of which a part is made.

K. Lettering

The most important consideration for an aircraft drawing is that it accurately portray the information the draftsman wants to give. But also of importance is the economy with which this information is presented.

For legibility and speed of doing the job, all lettering on aircraft drawings is done freehand, using single-stroke gothic upper-case letters, such as those seen in Fig. 1-21.

Very light guide lines are drawn, and the letters are spaced so there is approximately the same amount of white space between them. This spacing does not put the letters the same distance apart, but they *appear* uniform, and it is the appearance that makes the lettering attractive and easy to read. Words should be separated by the amount of space required for the letter *O* with no space on each side of it.

Fig. 1-21 Single-stroke gothic letters are used on almost all aircraft drawings.

V. Dimensioning of a Drawing

A drawing, to be meaningful, must not only show us the shape of the part, but it must accurately give us all of the needed dimensions.

A. Tolerance

The precision with which a part is to be made is indicated by the way the dimensions are written. In aircraft drawings, any dimension given in a common fraction normally assumes a dimensional tolerance of plus or minus 1/64 inch. If a dimension is given in one decimal, such as 2.5 inches, the tolerance is plus or minus 0.1 inch. If the dimension appears as 2.50 inches, the tolerance is plus or minus 0.04 inch.

More precise measurements are made using three decimal places, such as 2.500, which denotes a tolerance of plus or minus 0.010 inch. If still more exact measurement is needed, the dimensions may specify the limits such as:

2.500 + 0005 − 0000.

This dimension requires that the part be no larger than a half-thousandth inch over the base dimension and that it be no smaller than the given dimension.

B. Placement of Dimensions

Most aircraft drawings are dimensioned using a datum edge or surface as a reference from which all of the dimensions are made. In Fig. 1-22, we show three holes drilled in a piece of material. These holes are all located with reference to one corner of the material. The center of each hole is easy to locate, and there will be no cumulative errors, as there could be if one hole were measured from the center of the adjacent hole.

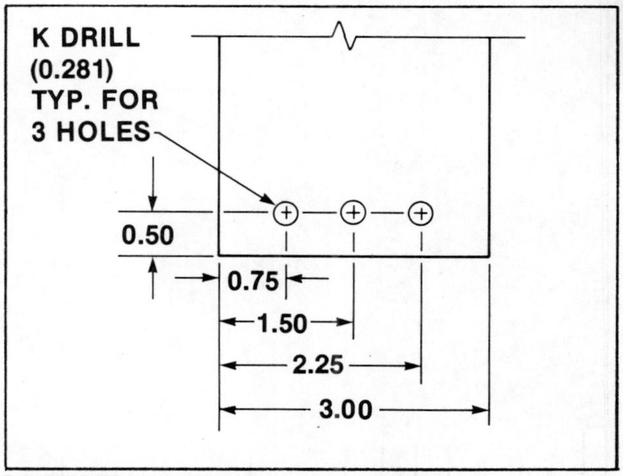

Fig. 1-22 Typical dimensioning

There are two ways of placing the dimensions on the drawing. One is to write all of the dimensions perpendicular to the dimension lines. With this type of dimensioning, the numbers will be parallel to the right edge of the drawing for vertical dimension lines and at various angles across the drawings for parts that have angled surfaces. The convention used for aircraft drawings, however, is to have all of the dimensions written so they are parallel to the bottom of the drawing.

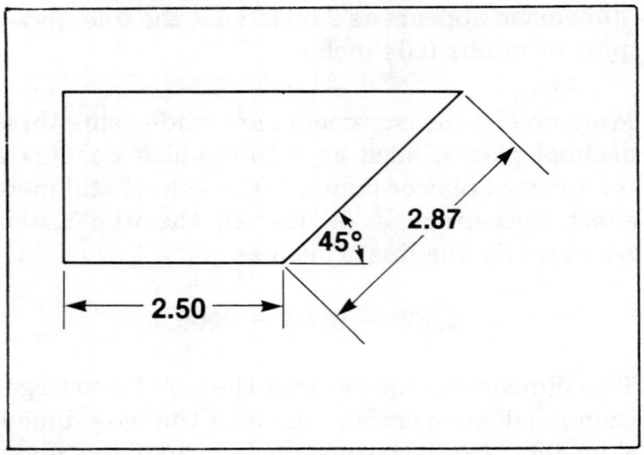

Fig. 1-23 All of the dimensions on an aircraft drawing are parallel to the bottom of the drawing.

VI. Aircraft Production Drawings

Every aircraft factory has its own drawing room manual that gives the standards to be used in making all of the drawings produced by that company. The information given in our manual is typical, and while it may differ from some aircraft drawings, you will find that this information applies to most of them.

A. Title Block

A title block is printed in the lower right-hand corner of every aircraft drawing, and it contains information the manufacturer feels is necessary for his product. Special or additional information may be specified to the left of the title block. In Fig. 1-24, we have a typical title block, and we will examine the information that is to be used in these blocks.

1. Title

In this space the name of the part is given, and in most companies, the noun is used first. For exam-

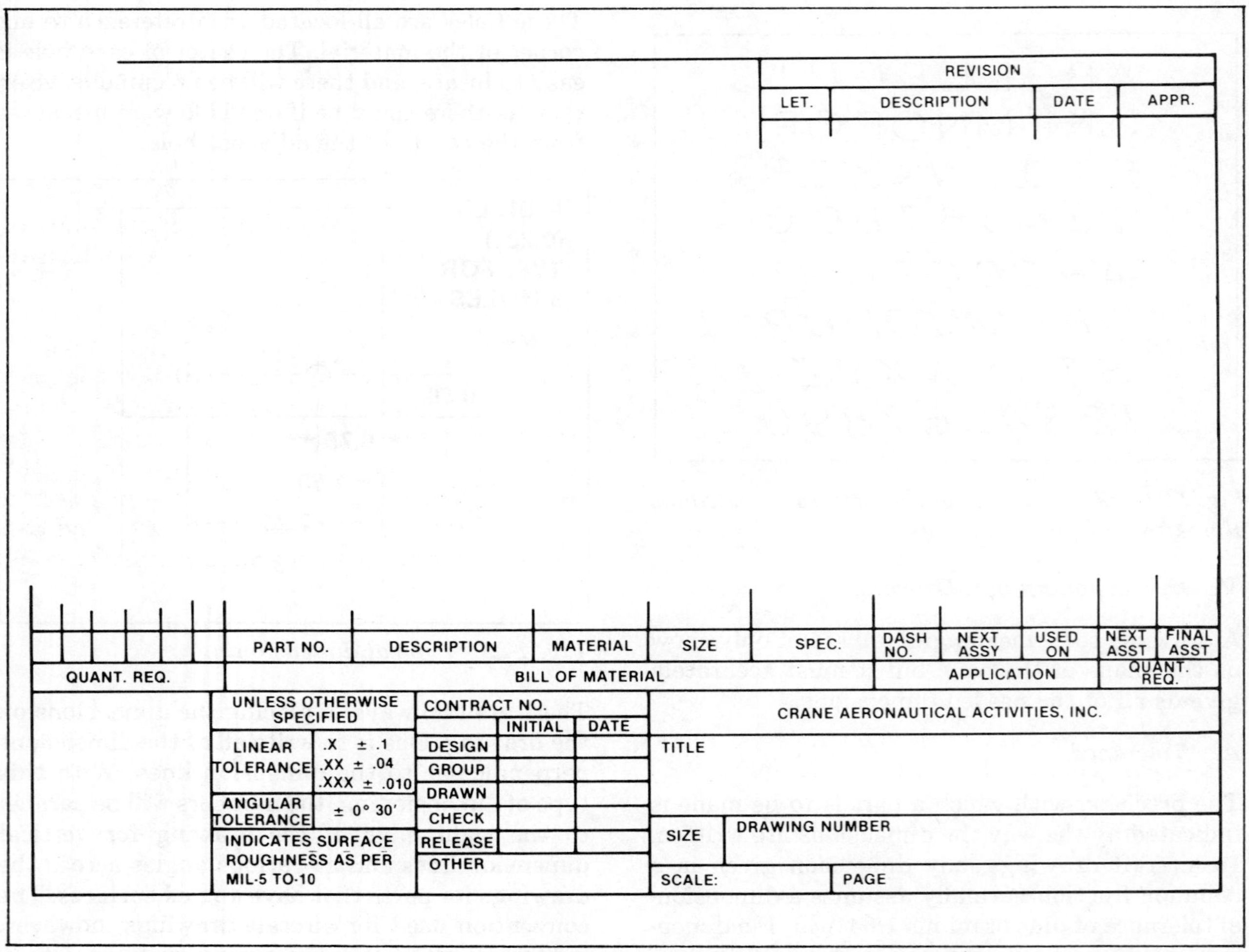

Fig. 1-24 Title block of typical aircraft drawing

ple, if this drawing were for an autopilot servo bracket, its title would be "Bracket, Autopilot Servo."

2. Size

This is the size of the drawing and is specified by letters representing the standard sizes. An *A*-size drawing is 8-1/2 × 11 inches, *B*-size is 11 × 17 inches, *C*-size is 17 × 22 inches, and a *D*-size drawing is 22 × 34 inches. Larger drawings are made on rolls of paper 36 to 42 inches wide and are specified as *R*-size drawings.

3. Drawing number

The number assigned to the drawing is the number the part will carry with it throughout its entire life. If an aircraft has two similar parts, one right-hand and one left-hand, they will normally have the same part number; but the left-hand part will have a -1 following the part number, and the right-hand part will be a -2 part.

4. Scale

This indicates the scale to which the drawing is made, and is noted with such indications as FULL, HALF, OR 1 INCH = 1 FOOT. If the drawing is essentially full-scale, but there are some details drawn to another scale, the notation will be FULL EXCEPT AS NOTED, and a note must be placed by each detail that is not drawn full-scale.

If the drawing does not include any parts that are drawn to scale, the word NONE will be used in this blank.

5. Page

If these drawings are assembled into a book, as electrical wiring diagrams often are, this is the page number of the book.

6. Responsibility

The column to the left of the title includes a space for the initials of the person who made the drawing, with the date the drawing was finished. The drawing is checked for accuracy and completeness, and is initialed by the checker. The design engineer and the group engineer, as well as the person who releases the drawing, all initial it in this column with the appropriate date.

7. Standards

In the column entitled UNLESS OTHERWISE SPECIFIED are the standards of manufacturing tolerance used by this company. This particular title block specifies the linear and the angular dimensional tolerance and a note that the MIL-SPEC-10 tolerance for surface roughness is used.

8. Bill of materials

Starting just above the title block and extending upward as far as necessary is the space for a list of every piece of material that is used in this part. The quantity required, the part number, and the description, size, and specification number can all be written in the space provided.

9. Application

This column indicates where this part is to be used. It shows the model of the aircraft on which it is used, the part number of the next assembly, and the quantity of these parts that are needed for both the next assembly and for the final assembly of the complete aircraft.

B. Revision Block

In the upper right-hand corner of the drawing, there is a block in which all of the revisions made to the drawing are recorded.

If revision *A* is issued by the engineering department, the letter *A* will be written in the revision block when it is made, along with the description of the revision, the date the revision was released, and the initials of the person approving the revision. A person can check with the engineering department to find the latest revision, and then by looking in this column can tell whether or not this drawing is up to date.

C. Zones

Aircraft drawings of the roll size are usually 36 to 42 inches wide and may be several feet long. Many detail views are included on some of these large drawings, and a system of zones is used to make location of details easy. The edges of the drawing are marked every 12 inches, vertically and horizontally. The vertical marks are identified by letters, with *A* being for the bottom 12 inches, *B* for the next 12 inches, and so on, up to the top of the drawing.

Starting at the right side and moving to the left, the zones are marked with numbers every 12 inches. Number one is the extreme right side, and the numbers increase to the left.

If a revision is noted as a dimension change in zone *C-7*, we would know that this change is made somewhere in the portion of the drawing three feet up from the bottom and seven feet from the right edge.

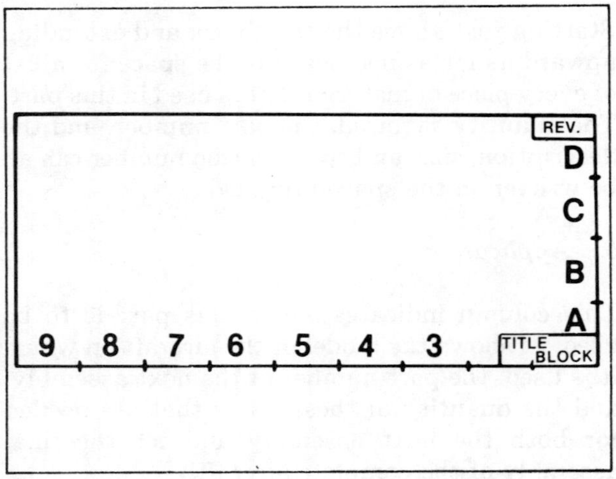

Fig. 1-25 Details on a large roll-size drawing are made easy to locate by the use of zone numbers and letters.

D. Station Numbers

The location of a part on an aircraft is specified by station numbers, water lines, and buttock lines.

1. Fuselage stations

These are locations fore and aft along the fuselage, measured from a datum which is called station zero. This datum may be either on the fuselage or ahead of it. In the example of Fig. 1-26, the datum is six inches ahead of the nose of the fuselage, and the center line of the main spar, which is an easy to identify position, is station 137. All fuselage frames and bulkheads are identified by their fuselage station numbers.

2. Water lines

Vertical measurements are made along water lines and are measured from a reference line called water line zero. We see in Fig. 1-26 that the floor of the main cabin is at water line -16, which means that it is 16 inches below water line zero.

3. Buttock lines

Distances to the right or the left of the centerline of the fuselage are measured by buttock lines. In Fig. 1-26, the distances out along the horizontal stabilizer and elevator are measured in buttock lines, with the tip of the horizontal surface located at buttock line 108.88. This means that it is 108.88 inches from the centerline of the fuselage.

4. Wing stations

Wing stations are measured from the centerline of the fuselage, or buttock line zero, and indicate the distance out along the wing toward the tip, in inches.

VII. Applied Geometry

In the construction of aircraft drawings we must make use of geometry, and here we see the way some of the more common layout problems are solved by the use of applied geometry.

A. Find the Center of a Line

We can find the center of line *A-B* by using a pencil compass. Open the compass to a distance slightly longer than halfway across the line and draw arcs A_1 - A_2 and B_1 - B_2. Connect the intersection of these two arcs with a straight line which will divide line *A-B* into two equal parts, and the line we have just drawn will be perpendicular to line *A-B* and is called a perpendicular bisector.

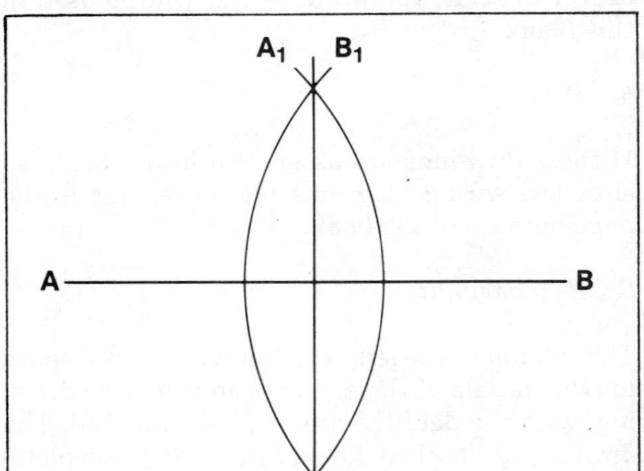

Fig. 1-27 Finding the center of a line

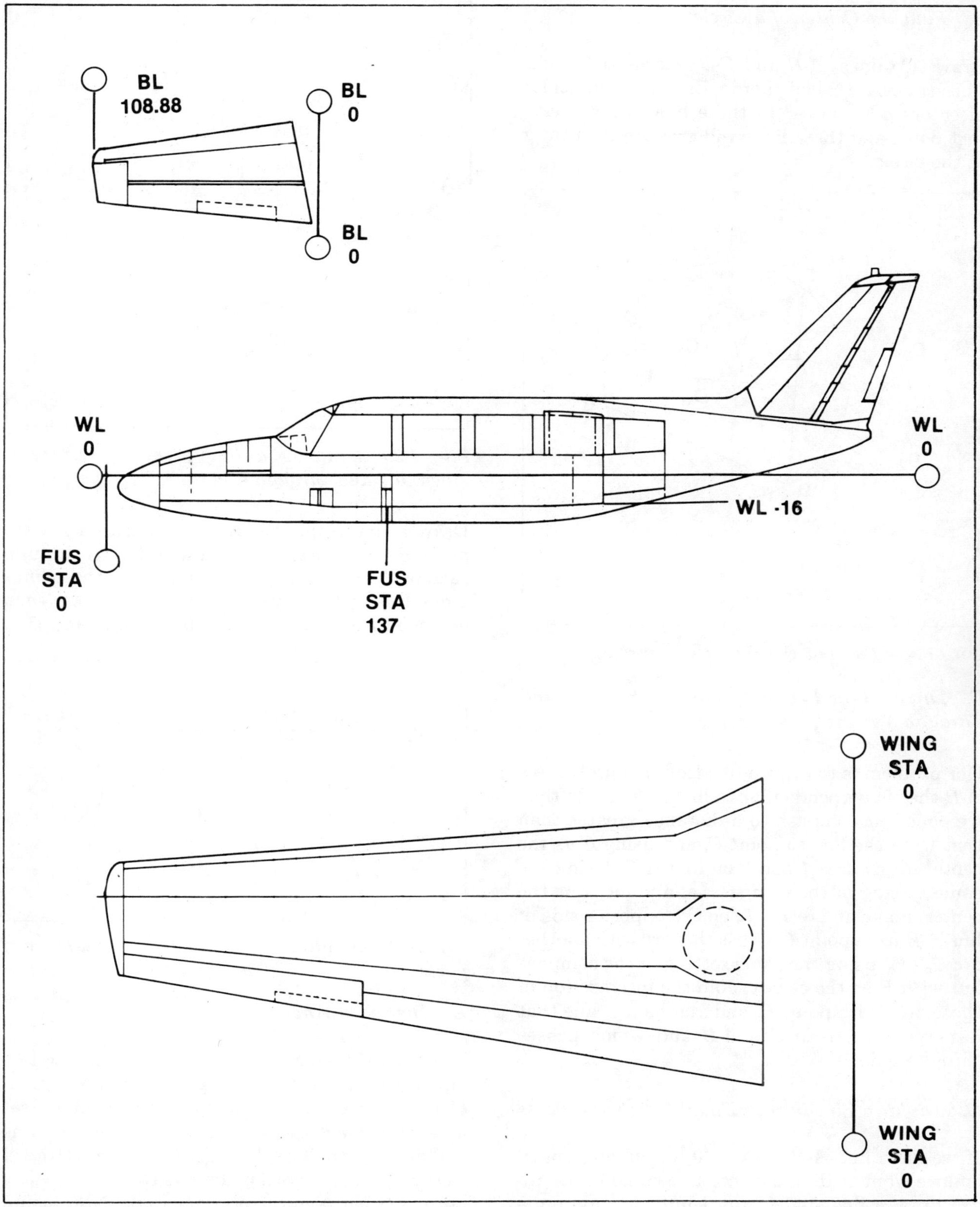

Fig. 1-26 Fuselage stations are used for locations fore and aft along the fuselage. Water lines locate positions vertically on the fuselage. Buttock lines locate points to the right or left of the centerline of the fuselage. Wing stations are measurements along the span of a wing with wing station zero the centerline of the fuselage.

B. Find the Center of a Circle

Draw two lines, *A-B* and *C-D*, across the circle. These lines are called chords. Erect perpendicular bisectors *a-b* and *c-d* to these lines, as we have just done, and these lines will cross in the center of the circle.

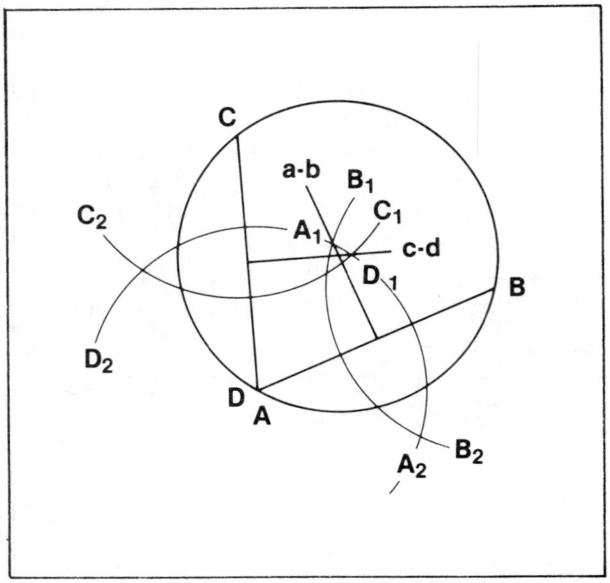

Fig. 1-28 Finding the center of a circle

C. Draw a Line Perpendicular to a Baseline and Through a Point

Our problem is to draw a line from point *C* to line *A-B* that is perpendicular to line *A-B*. To do this, we open our compass to a distance greater than that from the line to point *C*, and using *C* as the center, draw arcs *1* and *2* on line *A-B*. Using the same setting of the compass, and point *A* as the center, make an are 1_a - 1_b on the opposite side of line *A-B* from point *C*. Cross this arc with another arc, 2_a - 2_b, using the same setting of the compass but with *B* as the center. Join the intersection of these arcs with point *C*, and you have a line that is perpendicular to line *A-B* and which passes through point *C*.

D. Divide a Line into an Equal Number of Parts

If we have line *A-B*, Fig. 1-30, of an odd length and we want to divide it into five equal parts, we can draw a line *A-C* of any length at any angle through point *A* and divide it into five equal parts. In this case, we have made line *A-C* more than five inches long and have marked it every inch.

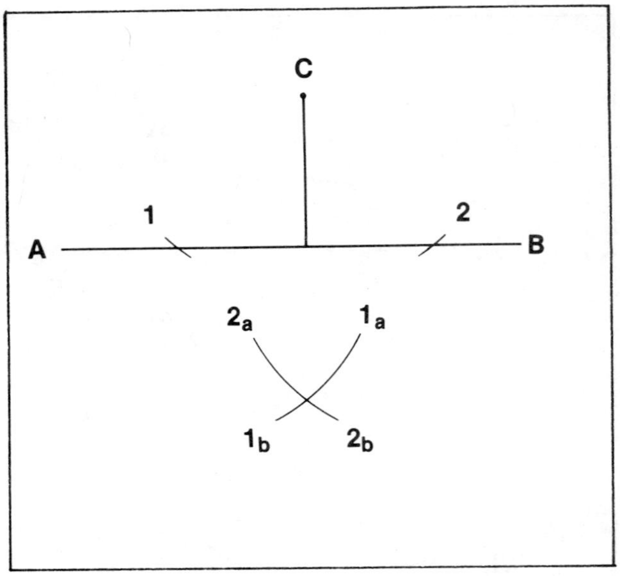

Fig. 1-29 Drawing a line perpendicular to a given line and through a point

Draw a line from the five-inch mark on *A-C* to point *B*. Now, draw lines parallel to *5-B* through each of the one-inch marks, and where these lines cross line *A-B*, they form division marks, each one one-fifth of the distance between *A* and *B*.

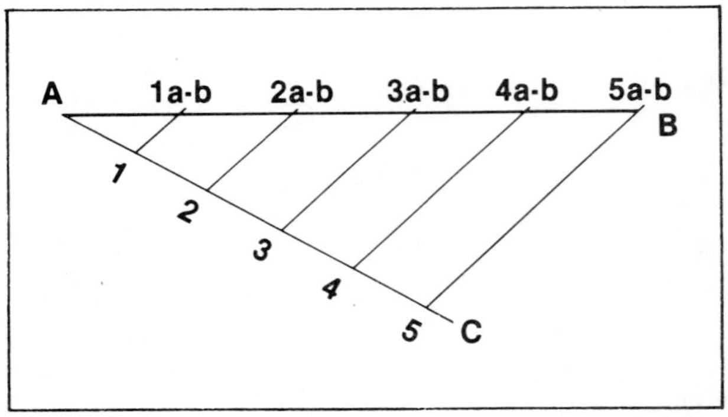

Fig. 1-30 Dividing a line into an equal number of parts

E. Bisect an Angle

If we have angle *B-A-C*, we can divide it into two angles so that angle *B-A-D* is equal to angle *D-A-C*. Use *A* as the center and with the compass set at a convenient distance, make an arc *1* on line *A-C* and arc *2* on line *A-D*. Use the same setting of the compass and with point *1* as the center, make arc 1_a - 1_b between the original angle. Now, using the same setting, and point *2* as the center, cross the first arc with arc 2_a - 2_b. Draw a line from the intersection of the two arcs to point *A*, and this line will bisect angle *B-A-C*.

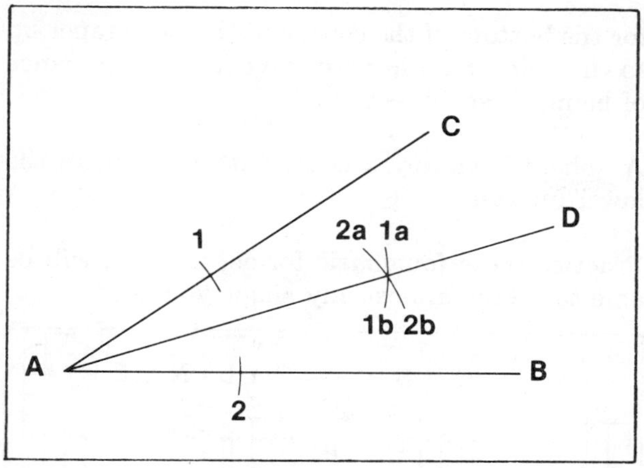

Fig. 1-31 Bisecting an angle

VIII. Sketching

A. Basic Sketching

Few aircraft technicians are professional drafts-men, and this depth of skill is not required for aircraft maintenace. But it is essential that a technician be able to graphically express himself well enough to describe a repair, or to be able to make a simple drawing of a part that must be made.

Sketching is the form of graphics most used by the technician and one that is easy to master to the level that allows us to express ourselves. The most important prerequisite for sketching is that we understand the principles of the basic forms of drawing that are used and be able to break our view down into its component shapes.

Almost all sketches we need to make for aircraft maintenance are either orthographic or isometric. If a simple front, end, and top view will do, it is easy to look at the part and make a sketch showing the shape of the object, and then we can place the extension lines and dimensions where they are needed.

Rather than make a three-view drawing of the object, we may find that an isometric sketch will make the part easier to visualize.

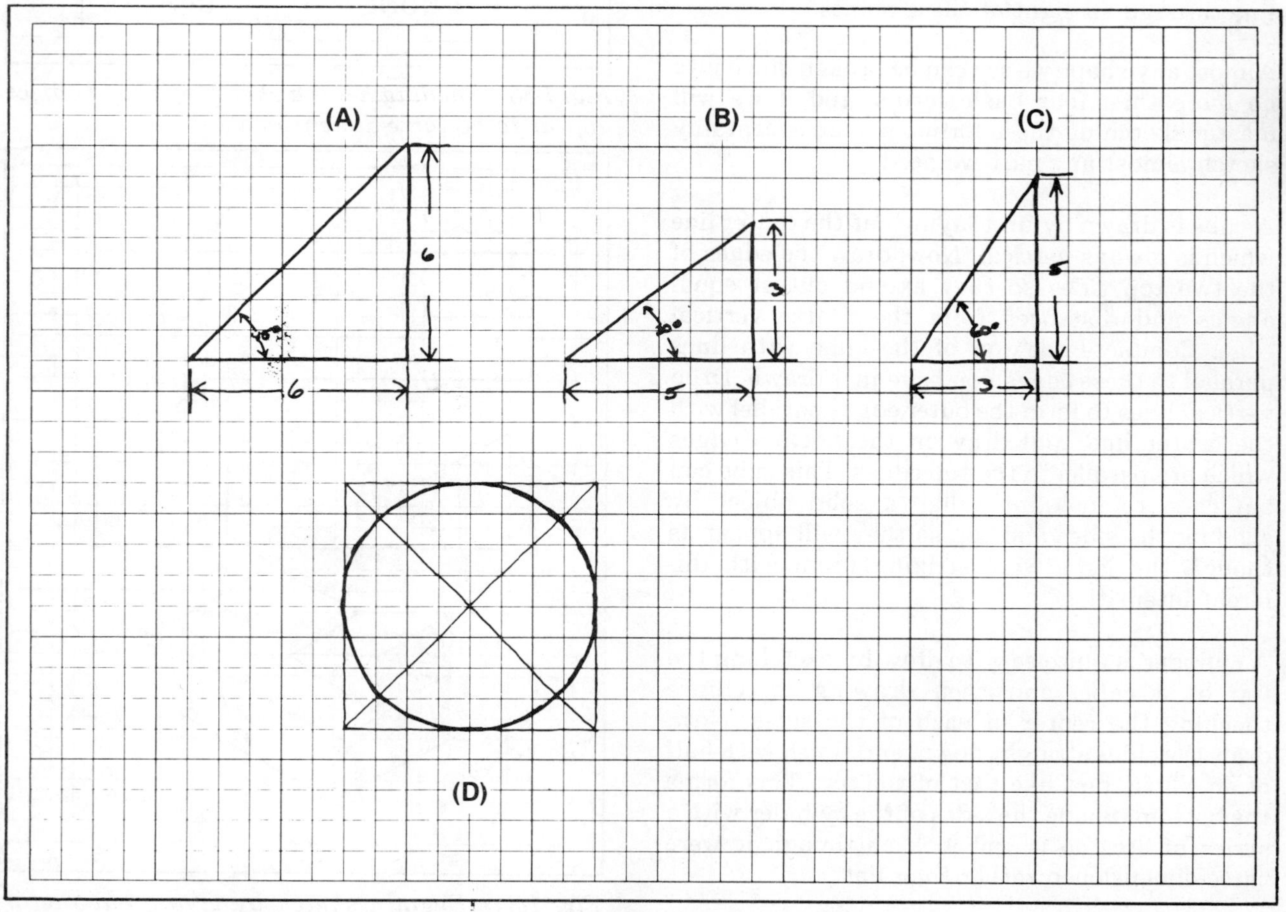

Fig. 1-32 A — Sketching a 45-degree triangle B — Sketching a 30-degree triangle C — Sketching a 60-degree triangle D — Sketching a circle

By their very nature, sketches are made without the use of drafting instruments, but one of the greatest aids to sketching is a pad of paper that is ruled in squares. This type of paper is available with either four or five squares to the inch.

Angles are easily estimated on square-ruled paper by making triangles as we see them in Fig. 1-32. A 45-degree angle is formed in a triangle having two equal sides. A 30-degree angle is formed when the long side of the triangle is five units long and the short side three. The other angle of this triangle is 60 degrees. A 15-degree angle can be approximated by making the long side five units and the short side only one and one-half units.

A circle is easily sketched by marking a point as the center and then making four marks, each an equal distance from the center on the horizontal and vertical lines; then, between each of these marks, the same distance from the center, make four more marks. Now, lightly sketch a curved line through all eight of these points.

Almost any shape we see can be broken down into no more than four basic forms, and if we will master sketching these forms, we can acceptably sketch almost any part we need.

A cube is drawn by first laying out the center line which is always vertical. Now, draw the edges of the two top sides so they extend out at equal angles and distances from the center vertical edge. Complete the top of the cube with lines parallel to these edges you have just drawn. Drop vertical lines to form the outer edges parallel with the center line, and draw in the bottom edges which are parallel to the top edges. This cube can be made to look more like a solid object by shading the sides and top so they will appear as though the light were striking them with different intensities.

A cylinder is quite easy to draw by sketching the top of a cube, and then drawing an ellipse touching the center of each of the sides. Now, draw the sides straight down, and finish with half of an ellipse just like that of the top. This forms the bottom. Shade the sides of the cylinder with a series of arcs so it will look as though it were three-dimensioned rather than flat.

A cone is made in much the same way as the cylinder, except that only part of the ellipse is used for the bottom of the cone, and the sides taper up to the point. Shading arcs give it the appearance of being three-dimensioned.

A sphere is simply a circle with shading in the lower quarter.

Practice these four basic forms, and you will be able to sketch almost any shape you need.

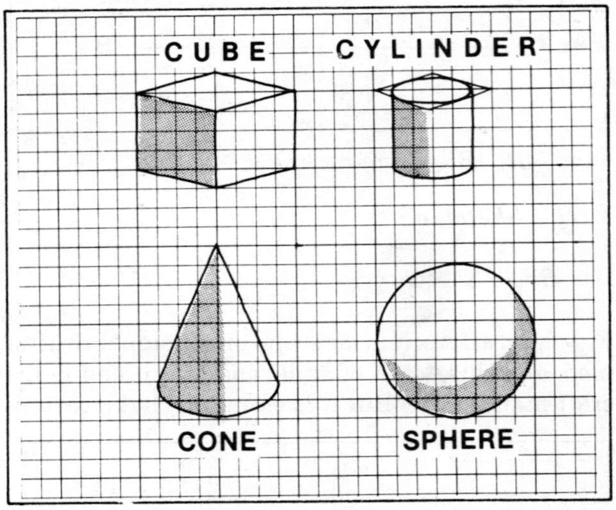

Fig. 1-33 Shading helps a sketch of a solid object appear to be three-dimensional.

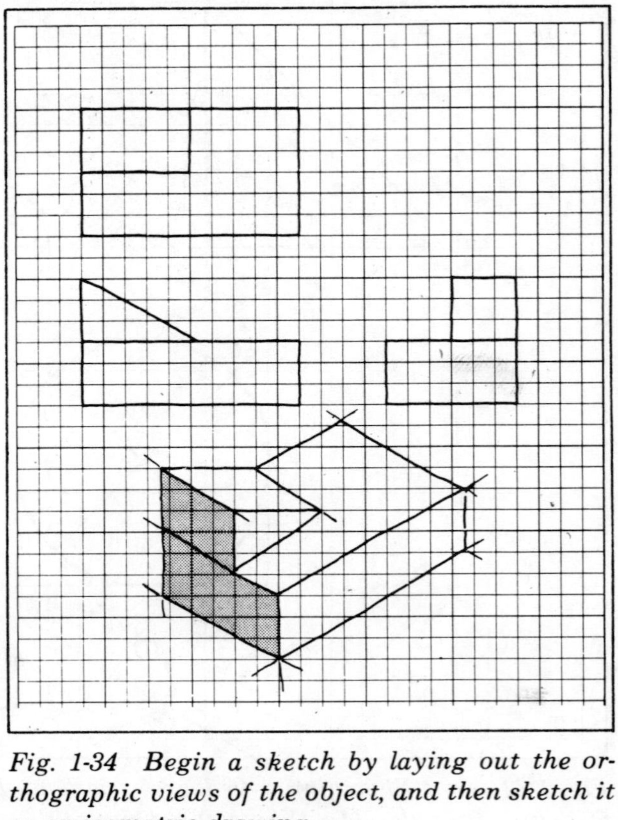

Fig. 1-34 Begin a sketch by laying out the orthographic views of the object, and then sketch it as an isometric drawing.

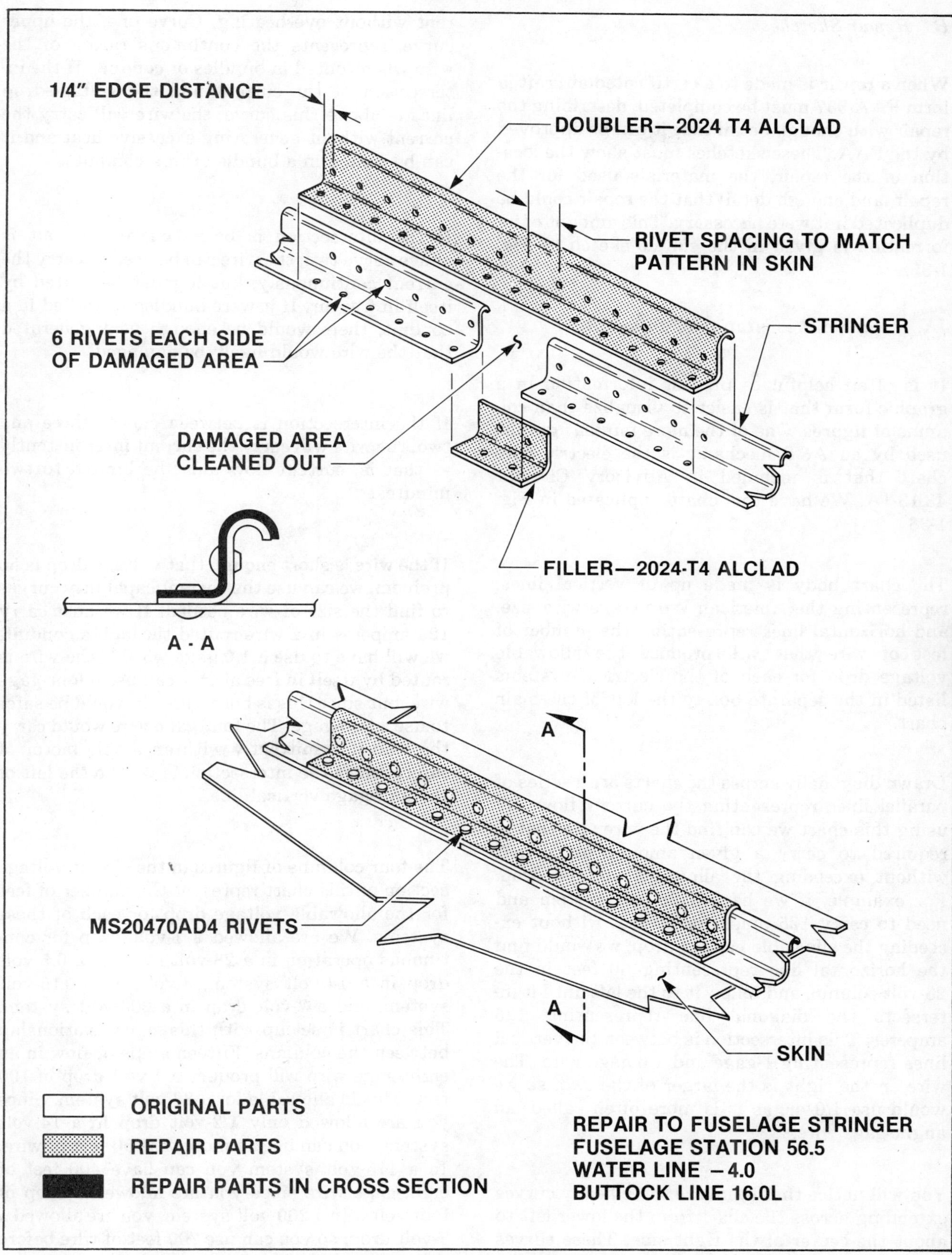

Fig. 1-35 A repair sketch

B. Repair Sketches

When a repair is made to a certificated aircraft, a form FAA-337 must be completed, describing the repair with enough detail that it can be approved by the FAA. These sketches must show the location of the repair, the materials used for the repair, and enough detail that the repair could be duplicated if it were necessary. This amount of information is given in the simple sketch of Fig. 1-35.

IX. Graphic Presentation of Information

It is often helpful to present information in a graphic form that is easier to visualize than columns of figures. One of the more familiar graphs used by an A&P mechanic is the electric wire chart that is included in Advisory Circular 43.13-1A. We have this chart duplicated in Fig. 1-36

The chart body is made up of vertical lines, representing the American Wire Gage wire size, and horizontal lines representing the number of feet of wire that will produce the allowable voltage drop for each of the electrical systems listed in the separate box to the left of the main chart.

Drawn diagonally across the charts are a series of parallel lines representing the current flow. By using this chart we can find the wire size that is required to carry a given amount of current without exceeding the allowable voltage drop. For example, if we have a 28-volt system and need to carry 125 amps for 60 feet without exceeding the allowable voltage drop, we would find the horizontal line representing 60 feet in the 28-volt column, and follow it to the left until it intersects the diagonal line representing 125 amperes. This intersection is between the vertical lines representing 1-gage and 1/0-gage wire. The wire on the right is the larger of the two, so we would use 1/0 or, as it is more often called, an aught-gage wire.

You will notice that there are three heavy curves extending across the chart from the lower left to about the center of the right side. These curves represent the ability of the wire to carry the current without overheating. Curve one, the upper curve, represents the continuous rating of the wire when routed in bundles or conduit. If the intersection of the current and the voltage-drop lines is above this curve, the wire will carry the current without generating excessive heat and it can be routed in a bundle or in a conduit.

If the intersection is below curve one, but is above curve two, the wire can be used to carry the current continuously, but it must be routed by itself in free air. If it were bundled or routed in a conduit, there would be enough heat generated that the wire would likely be damaged.

If the intersection is between curves three and two, the wire will carry the current intermittently — that is, current flow must be limited to two minutes.

If the wire is short enough that voltage drop is no problem, we can use the current capability curves to find the size of wire needed. If we nust carry 125 amperes in a wire routed through a conduit, we will have to use a 1/0-gage wire. If the wire is routed by itself in free air, we can use a four-gage wire, but since this is borderline, it would be safer to use a two-gage. The four-gage wire would carry the load intermittently with an ample factor of safety, since the intersection is well to the left of the four-gage vertical line.

The four columns of figures in the circuit voltage section of this chart represent the number of feet for the allowable voltage drop for each of these systems. We are allowed a 1-volt drop for continuous operation in a 28-volt system, a 0.5-volt drop in a 14-volt system, 4-volts in a 115-volt system, and a 7-volt drop in a 200-volt system. This chart is set up with this same relationship between the columns. Fifteen amps of flow in an eight-gage wire will produce a 1-volt drop in 100 feet. This is allowable for a 28-volt system. Since you are allowed only 1/2-volt drop in a 14-volt system, you can use no more that 50 feet of wire. In a 115-volt system you can have 400 feet of eight-gage wire since you are allowed a drop of four volts. In a 200-volt system, you are allowed a 7-volt drop, so you can use 700 feet of wire before the voltage drop becomes excessive.

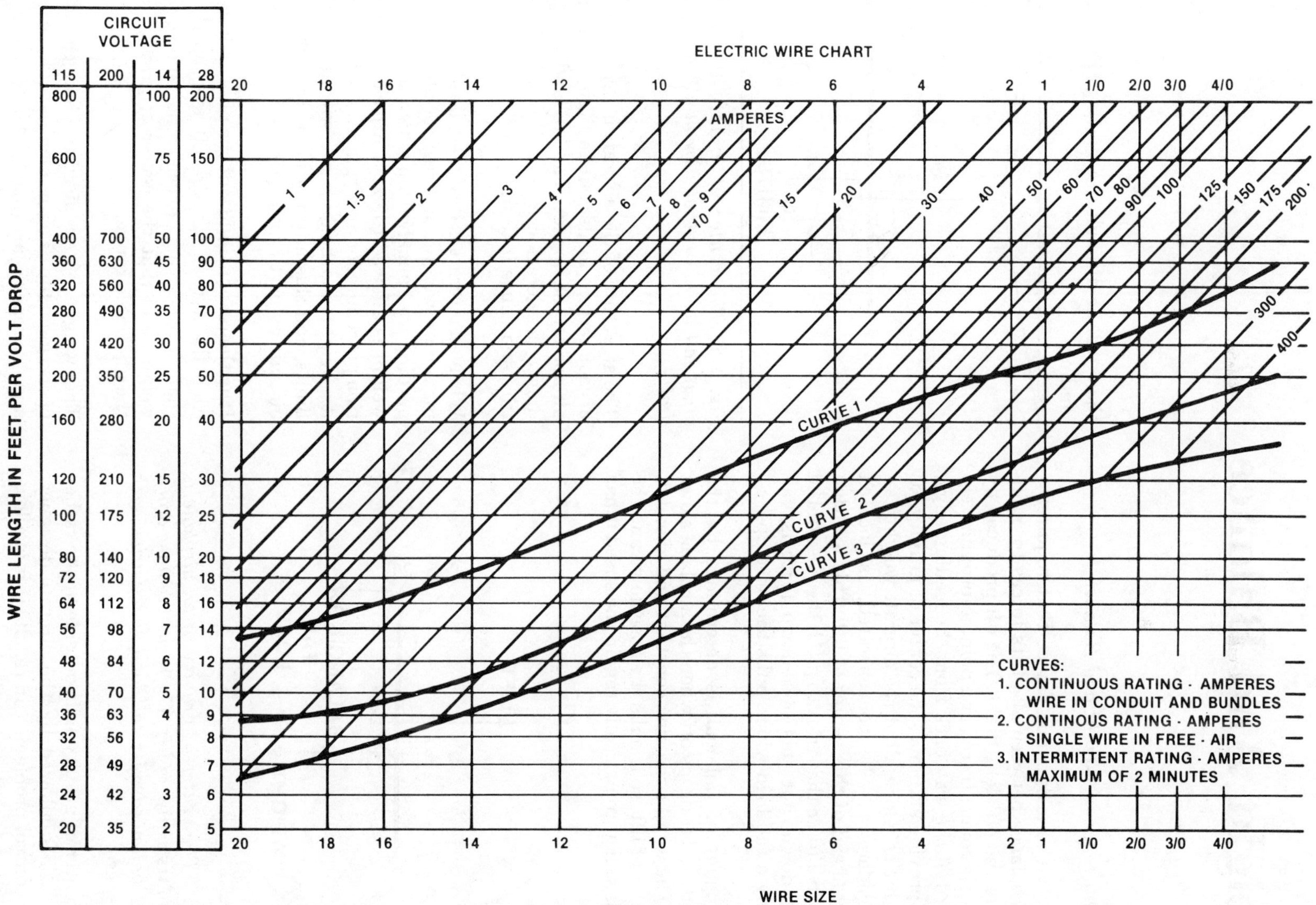

Fig. 1-36 Electric wire chart

SECTION II

Weight And Balance

I. Importance of Weight and Balance

An aircraft is a dynamic device that requires a careful balance between all of its forces to maintain safe and efficient flight. The lift produced by the wing is concentrated at a point approximately one-third of the way back from the leading edge, and to provide stability, the center of gravity, or that point at which all of the aircraft weight can be considered to be concentrated, is located slightly ahead of this center of lift. This location results in a force that tries to pitch the nose of the aircraft down, but this nose-down force is balanced by a tail load which pulls the nose up.

The amount of tail load is determined by the airspeed, and it drops off when the airplane slows down. The weight remains constant and ahead of the center of lift, so it pulls the nose down and the airplane will automatically regain the speed it has lost.

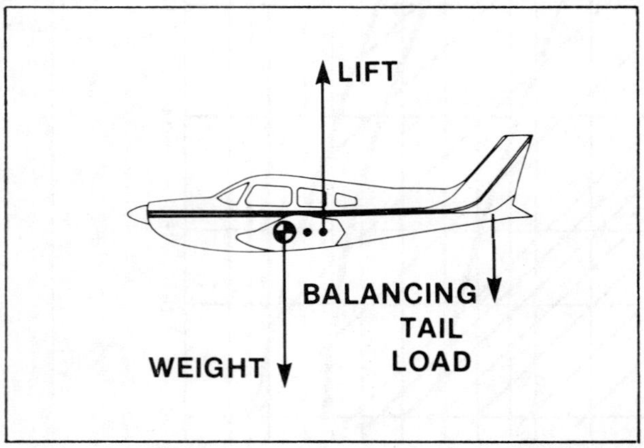

Fig. 2-1 The center of gravity of an airplane is ahead of its center of lift. The resulting nose-down pitch is balanced by a tail load that varies with the airspeed.

If the center of gravity falls outside of the rather narrow limits allowed by the aircraft designer, serious control problems can result. If it is allowed to get too far aft, the stall characteristics will be adversely affected, and if it is too far forward, there will be difficulty in slowing the airplane for landing.

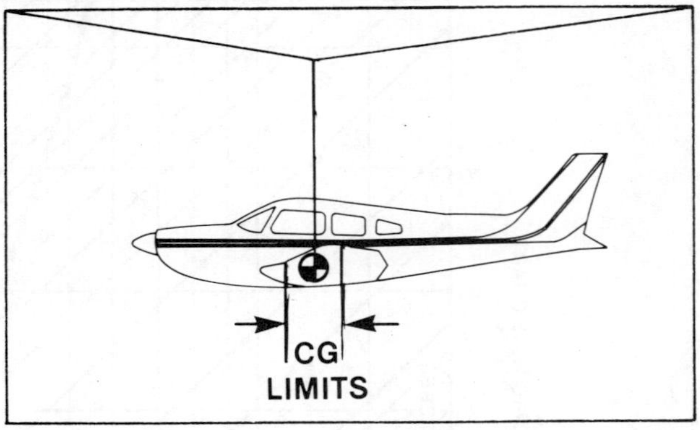

Fig. 2-2 The center of gravity of an airplane must fall within rather narrow center of gravity limits.

The structure of the aircraft is designed to safely accept certain loads, but in flight through rough air and on the impact of a hard landing, the forces due to acceleration may well overload the structure and cause it to fail.

When an aircraft is designed, limits are put on its maximum weight, and restrictions are set up regarding the range within which the center of this weight is allowed to vary. A part of the certification procedure for an airplane is to determine that its weight and balance are within the allowable limits, and this information is furnished with the aircraft as part of its operations manual.

It is the responsibility of the pilot to know before each flight that his aircraft is properly loaded, that it does not exceed the allowable gross weight, and that the center of gravity is within the allowable range.

The weight of an aircraft changes during its operational life as equipment is added or removed and as repairs are made. All of these changes must be monitored and the weight and balance information used by the pilot must be kept up to

Fig. 2-3 The seats, baggage compartment, and fuel tanks of this small aircraft are located quite near the aircraft center of gravity.

date. This is the responsibility of the aviation maintenance technician.

Very close track must be kept of the weight and balance of aircraft used to carry passengers or cargo for hire, and they must be reweighed periodically and have their center of gravity recomputed.

The fuel tanks of some of the smaller training aircraft are located in the wing, and the seats and baggage compartment are located directly below the wing, so it is not likely that these aircraft will be loaded in such a way that the center of gravity will fall outside of the allowable center of gravity range. Most of the larger aircraft, however, have several rows of seats, some of which are ahead of the center of gravity range and some behind it. and there are often both forward and aft baggage compartments. This wide range of loading possibilities makes the use of charts or other aids to loading a necessity for the pilot to be sure that

Fig. 2-4 The great distance between the rows of seats and the baggage compartments of this type of aircraft makes careful loading important to keep the center of gravity within the allowable limits.

the center of gravity is within the allowable range.

II. Principles of Weight and Balance

The actual principle involved in finding the center of gravity of an aircraft is quite simple, and it is easy to visualize when we consider the playground teeter-totter, or seesaw, which is a practical example of weight and balance.

Fig. 2-5 The principle of weight and balance is as simple as the balance of these children on the seesaw.

When a large child and a small child get on a seesaw, the large child must slide up close to the support, or the fulcrum, to balance the small child who is farther away from the fulcrum. The distance from the fulcrum to the center of gravity of the weight is called the "arm" of the weight, and it may be measured in such units as feet, inches, or meters. The amount of force — in this case, the weight of the child — is measured in pounds, grams, or kilograms, and is considered to be concentrated at a point called its center of gravity.

The product of the weight and the arm is the moment of the force and is expressed in pound-feet, pound-inches, or in gram- or kilogram-meters. And since a moment is a force that causes rotation, we must specify the direction, either clockwise or counterclockwise, in which the force causes the weight to rotate.

To best understand the principles of weight and balance, let's consider that the board we use does not have any weight of its own, and that all of the weight is concentrated at the center of gravity of the weights themselves.

In the illustration of Fig. 2-7, we have a board on which two weights are located. The weight on the left is 25 pounds, and the one on the right is 50 pounds, and there are 12 feet between the centers of gravity of the two weights. We want to find the location of the fulcrum about which the two weights will balance.

We begin by choosing the location of a datum, or a reference line, from which we will make all of our measurements. This line can be anywhere, as

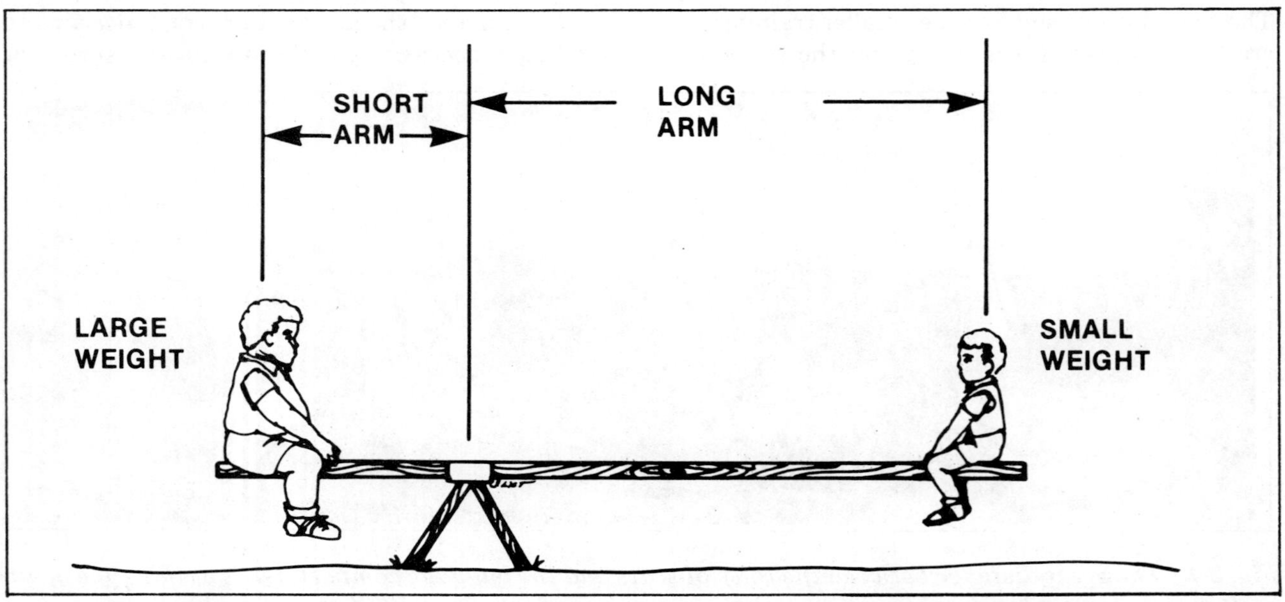

Fig. 2-6 The small child on the long arm balances the large child on the short arm.

we will soon see, but for this initial explanation, let's assume it to be located at the center of gravity of one of the weights. In our case we will choose the weight *A*, the one on the left side.

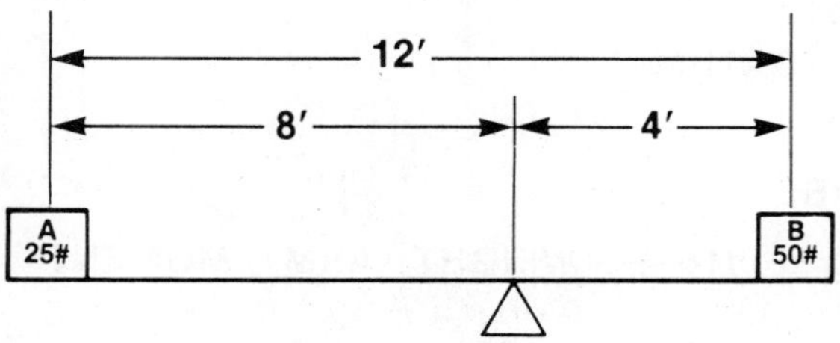

Fig. 2-7 A simple lever problem, using weight **A** *as the datum.*

To visualize the computations more clearly, let's make a chart such as the one we have in Fig. 2-8. Since weight *A* is on the datum, its arm is zero and when we multiply any number by zero, the product, or the moment, is also zero. The arm of weight *B* is 12 feet, and its moment is 12 × 50, or 600 pound-feet, and its direction of rotation is clockwise.

ITEM	WEIGHT	ARM	MOMENT
A	25	Φ	Φ
B	50	12′	600
TOTAL	75	8′	600

Fig. 2-8 Chart for the lever problem of Fig. 2-7.

To find the balance point, we must divide the total moment by the total weight. The total moment is 600 pound-feet, and the total weight is 75 pounds; this places the balance point eight feet to the right of the datum. In this example, there are no counterclockwise moments and so the total moment is clockwise.

To check our work and prove that the board is really balanced about the point we have just discovered, we can make a chart similar to that in Fig. 2-9. Here we have moved the datum from the center of gravity of weight *A* to the fulcrum, and we compute all of our moments from this new location. Any counterclockwise moment is considered to be negative, and a moment that causes a clockwise rotation is positive.

Weight *A* has an arm of negative eight feet, and its moment is −200 pound-feet. The arm of weight *B* is positive four feet, and when this is multiplied by its weight of 50 pounds, it gives us a moment of +200 pound-feet. The sum of the moments is zero, which means that the board does actually balance about the fulcrum.

(A)

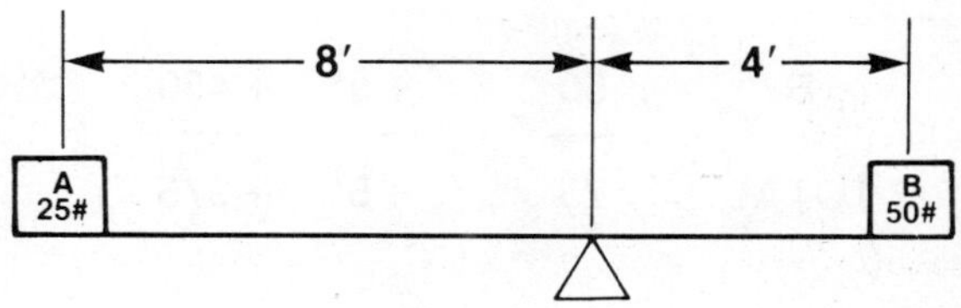

(B)

ITEM	WEIGHT	ARM	MOMENT
A	25	−8	−200
B	50	+4	+200
TOTAL	75	Φ	Φ

*Fig. 2-9 A — Simple lever problem, using the fulcrum as the datum B — Chart for the problem of Fig. 2-9***A**

We can easily show that the datum can be placed anywhere by working this same problem, using two different locations for the datum. In Fig. 2-10, we place the datum between the two weights, three feet to the right of weight *A*. The arm of *A* is now negative three feet, and its moment is −75 pound-feet. The arm of weight *B* is positive nine feet and its moment is +450 pound-feet.

The total moment is +375 and the weight is 75 pounds, so the balance point is five feet to the right of the datum, which places it in exactly the same location we previously found, eight feet to the right of *A*

(A)

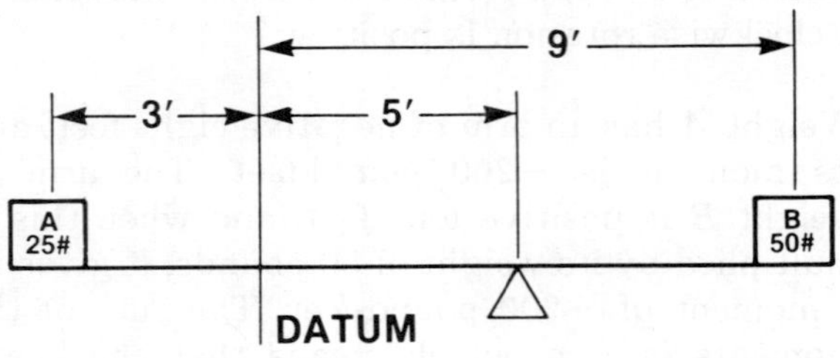

(B)

ITEM	WEIGHT	ARM	MOMENT
A	25	−3′	− 75
B	50	+9′	+450
TOTAL	75	+5′	+375

Fig. 2-10 A — Simple lever problem, placing the datum between weight **A** *and the fulcrum B — Chart for the problem of Fig. 2-10***A**

(A)

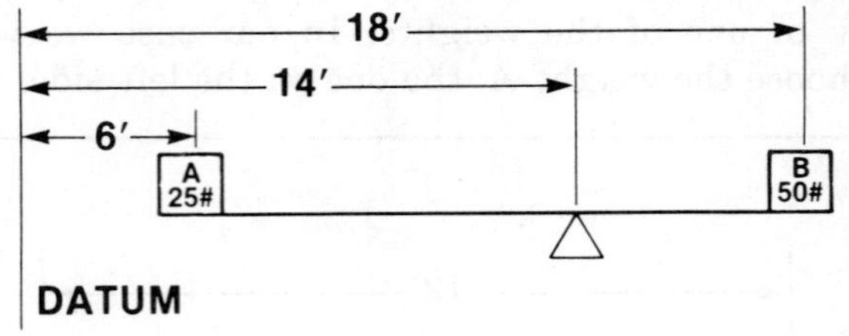

(B)

ITEM	WEIGHT	ARM	MOMENT
A	25	+ 6	+ 150
B	50	+18	+ 900
TOTAL	75	+14	+1050

*Fig. 2-11 A — Simple lever problem, placing the datum to the left of the lever B — Chart for the problem of Fig. 2-11***A**

Some aircraft manufacturers place the datum a given distance ahead of the aircraft so all of the moments will be positive, and we can see by the example of Fig. 2-11 that this does not change the answer. The datum in this example is located six feet to the left of weight *A* and the moment of *A* is +150 pound-feet. Weight *B* is 18 feet from the datum, and its moment is +900 pound-feet. The total moment is +1,050 pound-feet and when this is divided by the total weight of 75 pounds, the balance is found to be 14 feet to the right of the datum. This again is the same location as we have found it in the previous two computations, eight feet to the right of weight *A*.

We can continue with our teeter-totter explanation to find where we would place a third weight to balance the board. In Fig. 2-12, we have a board on which weights *A* and *B* are placed. Since the moment of weight *B* is greater than that of *A*, there will be a net force tending to rotate the board in a clockwise direction. The moment of this force is +550 pound-inches.

We want to balance the board by placing 50-pound weight *C* the proper distance from the fulcrum. Weight *C* must have a moment of −550 pound-inches, because for a board to balance about a point, the sum of the moments about that point must equal zero. The moment of *C* is −550 pound-inches and its weight is 50 pounds, so its arm must be −11 inches, or the center of weight *C* must be 11 inches to the left of the fulcrum.

(A)

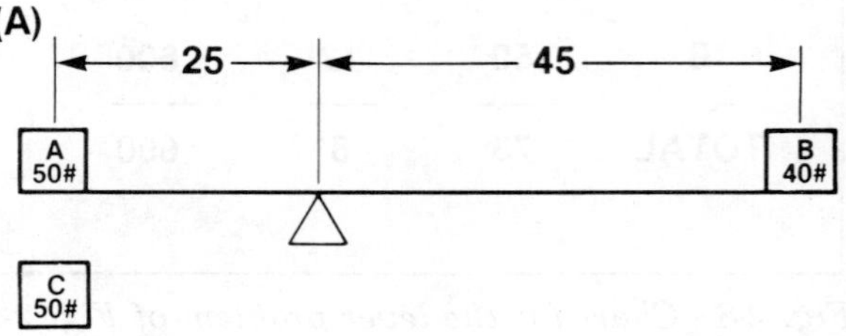

(B)

ITEM	WEIGHT	ARM	MOMENT
A	50	−25	−1250
B	40	+45	+1800
TOTAL			+ 550
C	50	−11	− 550

*Fig. 2-12 A — Determining the placement of a third weight to balance lever B — Chart for the problem of Fig. 2-12***A**

In Fig. 2-13, we see this balance proven. The sum of the two negative moments is − 1,800 pound-inches, and the positive moment is + 1,800 pound-inches, so the board balances.

(A)

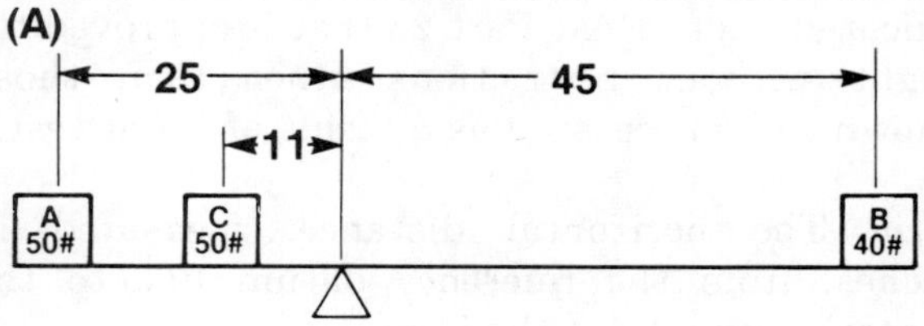

(B)

ITEM	WEIGHT	ARM	MOMENT	
A	50	−25	−1250	
B	40	+45		+1800
C	50	−11	− 550	
TOTAL	140		−1800	+1800

TOTAL MOMENT IS ZERO

*Fig. 2-13 A — Proof of the balance found in the problem of Fig. 2-12*A *B — Chart for the problem of Fig. 2-13*A

It is necessary for an aircraft center of gravity to fall within a given range, and we sometimes need to add ballast to the aircraft to move the empty-weight center of gravity into the allowable range. We can again use our board to see the way we do this.

Let's assume that our board in Fig. 2-14 balances at a point 37.5 inches from item *A*, but we want to balance at a point 42.5 inches away. Weight *B* is 180 inches from weight *A*, and there is a location 170 inches from *A* at which we can place our ballast weight. Our problem is to find the amount of weight we will have to add 170 inches from weight *A* in order to move the point of balance five inches to the right.

The formula we will use is:

$$\text{Ballast} = \frac{\text{Total weight} \times \text{distance needed to shift balance point}}{\text{Arm of ballast--arm of desired balance point}}$$

The total weight is 480 pounds, and we need to shift the balance point five inches. The arm of the ballast is 170 inches, and the arm of the new balance point is 42.5 inches. When we work the problem, we find that we must add 18.82 pounds of ballast at 170 inches to move the balance point five inches to the right.

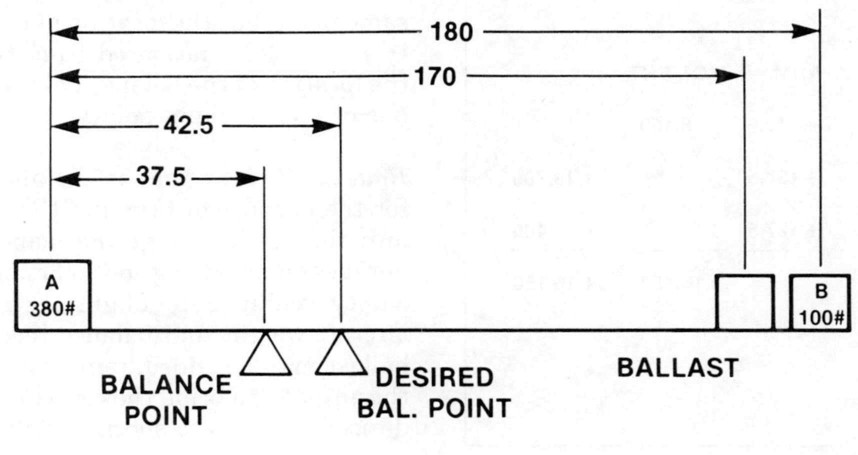

$$\text{BALLAST} = \frac{\text{TOTAL WEIGHT} \times \text{DISTANCE NEEDED TO SHIFT BALANCE POINT}}{\text{ARM OF BALLAST} - \text{ARM OF NEW BALANCE POINT}} = \frac{480 \times 5}{170 - 42.5} = \frac{2400}{127.5} = 18.82 \text{ POUNDS}$$

Fig. 2-14 Determining the amount of ballast required to move the balance point a given distance

We can check our computations in Fig. 2-15. If the sum of all of the moments about the new balance point is equal to zero, we have added the correct amount of ballast.

Item *A* has a weight of 380 pounds, and it is located at an arm of −42.5 inches. This gives it a moment of −16,150 pound-inches. Item *B* has a weight of 100 pounds and an arm of +137.5 inches, giving it a moment of +13,750 pound-inches. The ballast weighs 18.82 pounds and is located at +127.5, and so it has a moment of +2,400 pound-inches. The total positive moment is +16,150 pound-inches and the total negative moment is −16,150 pound-inches. The sum of the moments about the new balance point is zero, so our ballast was correct.

(A)

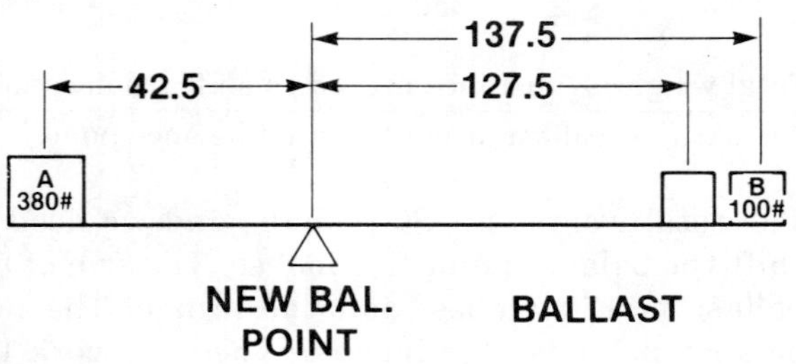

(B)

ITEM	WEIGHT	ARM	MOMENT	
A	380	− 42.5	− 16,150	
B	100	+ 137.5		+ 13,750
Ballast	18.82	+ 127.5		+ 2,400
Total			− 16,150	+ 16,150

Fig. 2-15 A — Proof of the accuracy of the ballast determined in the problem of Fig. 2-14. B — Chart for the problem in Fig. 2-15A

III. Terms Used in Weight and Balance

In any branch of technology, there are certain technical terms with which we must be familiar before we can really understand the subject, so let's look at some of the more common terms used in weight and balance computations.

Acrobatic category airplane An airplane certificated under FAR Part 23 that is approved for flight without restrictions other than those shown to be necessary as a result of flight test.

Arm The horizontal distance, measured in inches, from the reference datum line to the center of gravity of the item.

Algebraic sign The plus and minus convention assigned to numbers used in weight and balance computations. The arm of a weight ahead of the datum is a negative (−) arm, and that behind the datum is a positive (+) arm. When weight is added to the aircraft, it is a positive (+) weight, but when weight is removed, it is negative (−). A moment that causes the airplane to nose down is a negative (−) moment, while one that causes a nose-up pitch is a positive (+) moment. When the signs of the arm and the weight are alike, both either plus or minus, the moment will be positive, but if the signs are different, the moment will be negative.

Balance point The point about which a body balances and the point about which the algebraic sum of all of the moments is zero. The balance point and the center of gravity are physically the same point, but the location of the center of gravity is normally measured from the datum, while the location of the balance point is measured from one of the weighing points.

Ballast Weight that is installed in an aircraft for the purpose of bringing the center of gravity into the desired range. Permanent ballast must not be removed without changing the aircraft empty-weight center of gravity as recorded in the aircraft weight and balance records. Temporary ballast may be added, removed, or moved within the aircraft, to bring the center of gravity into the desired range for a specific flight condition.

Center of Gravity (CG) This is the point about which an aircraft would balance if it were possible to suspend it from that point. Its location is normally specified as being a given distance from the datum.

Center of gravity limits The maximum forward and aft location allowable for the center of gravity. These limits are established by the designer of the aircraft and are approved by the Federal Aviation Administration. They are furnished to the aviation maintenance technician and to the pilot, and it is the responsibility of the pilot to see that these limits are never exceeded in flight.

Center of gravity range The distance between the forward and the aft center of gravity limits of an aircraft. The Federal Aviation Administration prohibits an aircraft being operated if its center of gravity limits do not fall within this range.

Datum An imaginary vertical line or plane from which all measurements are taken. The location of the datum is arbitrarily chosen by the aircraft designer, and it is identified in the Aircraft Specifications or the Type Certificate Data Sheets, and the location of all equipment and the center of gravity limits are specified from this location.

Delta The Greek letter Δ used in weight and balance computations to indicate a change in values. For example, Δ CG indicates a change in the center of gravity.

Empty weight The weight of an aircraft and engine, with all of the items of operating equipment which have a fixed location installed. It includes all of the optional and special equipment and full reservoirs of hydraulic fluid and such liquids as anti-icing fluid. Until recently the empty weight included only the unusable fuel and the undrainable oil, but on March 1, 1978, Change 8 of Federal Aviation Regulation Part 23 was released which changed this definition for aircraft certificated under FAR Part 23. The empty weight for all aircraft certificated under FAR Part 23 includes fixed ballast, unusable fuel, full operating fluids *including oil*, hydraulic fluid, and other fluids required for normal operation of airplane systems except potable water, lavatory precharge water, and water intended for injection in the engines.

Empty-weight center of gravity range The most forward and most rearward empty-weight center of gravity locations that will not allow the legally loaded center of gravity of an aircraft to fall outside of the loaded center of gravity range.

Fulcrum The support about which a lever rotates.

Fuel load The expendable part of the fuel carried in an aircraft. It includes only the usable fuel — not the fuel in the lines or that part of the fuel which remains in the tank sumps.

Leveling means The method specified by the aircraft designer by which the technician can determine that the aircraft is properly leveled for weighing. This information is included in the Aircraft Specifications or in the Type Certificate Data Sheets.

LEMAC The leading edge of the mean aerodynamic chord.

Maximum landing weight The maximum allowable weight at which an aircraft may be landed. It is usually less than the maximum takeoff weight, because the stresses during a landing are greater than those during takeoff.

Maximum ramp weight The maximum allowable weight of an aircraft while it is on the ramp. It differs from the allowable takeoff weight by the weight of the fuel that will be consumed in taxiing to the point of takeoff.

Maximum takeoff weight The maximum allowable weight for an aircraft at the beginning of the takeoff roll.

Mean aerodynamic chord (MAC) The cord of an imaginary airfoil that has the same aerodynamic characteristics as the actual wing.

Minimum fuel The minimum amount of fuel considered to be in the tanks when computing an adverse-loaded center of gravity condition. This is no more that the quantity of fuel necessary for one-half hour of operation at rated maximum continuous power. An approximation of this value may be found by dividing the maximum-except-takeoff (METO) horesepower of the engines by 12 to get the number of gallons required. Then, multiply the gallons by six to convert them into pounds. We can go directly to pounds by simply dividing the METO horsepower by two. This method of finding the weight of the minimum fuel is sufficiently accurate for computing an adverse-loaded center of gravity condition.

For turbine aircraft, the minimum fuel for these computations is specified by the aircraft manufacturer.

Moment The product of the weight of an item times its distance from the datum or from a weighing point. Moments are forces that tend to produce rotation; for aircraft weight and balance purposes they are specified in pound-inches.

Moment index A moment divided by a constant such as 100, 1,000, or 10,000. It gives a much smaller and therefore handier number for use in weight and balance computations.

Net weight The scale reading, less the tare weight.

Normal category The category of aircraft certificated under FAR Part 23, which is limited to airplanes intended for nonacrobatic operation.

Payload That portion of the useful load of an aircraft from which revenue may be derived. It includes passengers and baggage.

Reduction factor The number that the moment is divided by to get the moment index.

Standard weights For general weight and balance purposes the following weights are considered standard:

Gasoline: 6 pounds per gallon

Turbine fuel: 6.7 pounds per gallon

Lubricating oil: 7.5 pounds per gallon

Water: 8.35 pounds per gallon

Crew & passengers: 170 pounds per person

FAR 135, which covers air taxi operators and commercial operators of small aircraft, has added a few more standard weights:

Adults (summer): 160 pounds per person

Adults (winter): 165 pounds per person

Children: 80 pounds per person

Crew, other than flight attendants: 170 pounds per person

Female flight attendants: 130 pounds per person

Male flight attendants 150 pounds per person

Checked-in baggage: 23.5 pounds per item

Carry-on baggage: 5 pounds per item

Station A location in an aircraft which is identified by a number denoting its distance in inches from the datum.

Tare The weight of all items such as chocks or blocks that are used to hold the aircraft on the scales while it is being weighed. The tare must be subtracted from the scale reading to find the net weight of the aircraft.

Unusable fuel That portion of the fuel in an aircraft fuel system that will not flow to the carburetor or fuel metering system in normal flight attitudes. The amount and location of this fuel is specified in the Specification Sheets or in the Type Certificate Data Sheets.

Undrainable oil That portion of the oil in an aircraft lubricating system that will not drain out when the drain plug is removed or when the drain valve is opened with the aircraft in its normal ground attitude.

Useful load The weight of the pilot, copilot, passengers, baggage, usable fuel, and drainable oil. It is actually the empty weight of the aircraft subtracted from the maximum allowable gross takeoff weight. For Part 23 aircraft, this does not include any oil.

Utility category That category of aircraft, certificated under Federal Aviation Regulation Part 23, which are intended for limited acrobatic operation.

Weighing points Specified points on an aircraft where the scales are placed for weighing the aircraft.

Zero-fuel weight The operational weight of the aircraft including the payload, but excluding the fuel load.

IV. Weighing Procedure

A. Preparation

Because of the large surface area of an airplane, it is possible for it to accumulate enough dirt to give it an inaccurate weight. When the aircraft is prepared for weighing, it should be washed, and all of the accumulation of dirt and debris should be cleaned out of the inside.

The equipment list should be checked to be sure that all of the required equipment is actually installed, and that there is nothing installed that is not on the equipment list.

The hydraulic reservoir should be full, and if the aircraft is equipped with an anti-icing system, the anti-icer fluid tank should also be full.

The fuel tanks should be drained, and only the unusable fuel — that is, the fuel that cannot flow to the engines in the normal flight attitude — should be in the aircraft when it is weighed. This unusable fuel is part of the aircraft empty weight.

If it is impractical to drain the fuel tanks, they may be filled completely full. The temperature of the fuel is taken when the aircraft is weighed, and a correction is applied to compute the actual weight of the fuel on board. This total fuel weight, less the weight of the usable fuel as is specified in the Type Certificate Data Sheets, is deducted from the as-weighed weight of the aircraft.

The oil tanks or sumps for aircraft certificated under FAR Part 23 should be filled to the full mark. Aircraft that are not certificated under this part should have the tanks or sumps drained and the drain plugs left out or the drain valves left open. Only the undrainable oil, that oil which is trapped in the engine and will not drain out, is included in the empty weight of these aircraft. Again, if it is not practical to drain the oil when the aircraft is weighed, the tanks or sumps should be filled and the weight of the known amount of oil should be deducted from the as-weighed weight of the aircraft. Since the amount of oil is quite small compared with the amount of fuel carried on board, it is not necessary to calculate the weight change of the oil caused by temperature changes.

B. Equipment

There are two types of equipment that may be used to weigh an aircraft. Platform scales are normally used for the smaller aircraft, and electronic load cells are used to weigh larger aircraft.

The smaller aircraft are lifted off the floor with jacks and are lowered onto platform scales with a scale under each wheel. These scales are specially built with low beams so they will fit under the wing of low-wing airplanes.

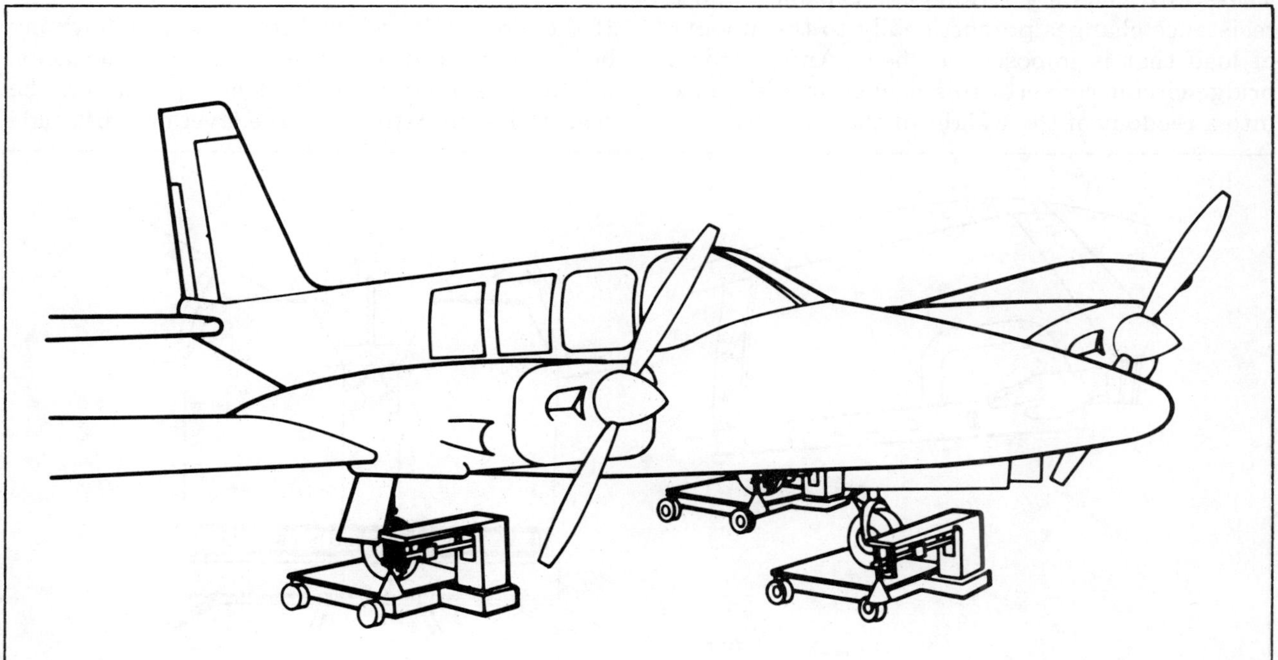

Fig. 2-16 Special low-beam scales are required for weighing a low-wing airplane.

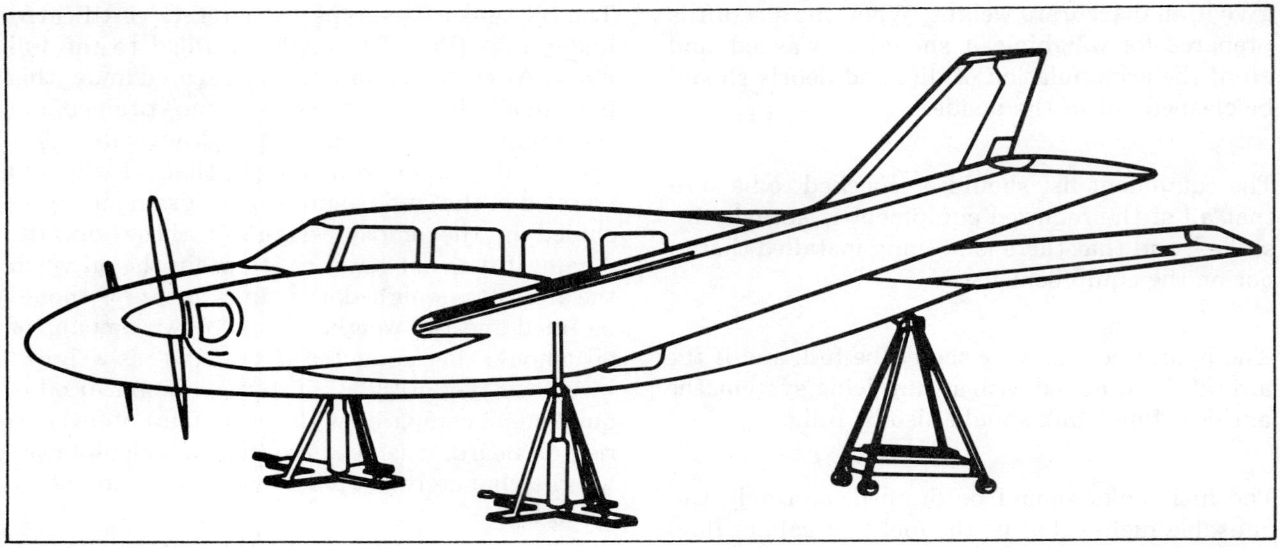

Fig. 2-17 Electronic load cells may be placed between the jacks and the jack points to weigh an airplane.

Chocks are used on the scale platform to prevent the aircraft rolling, as the brakes must be released to prevent an uneven application of force to the platform. The weight of the chocks is called the tare weight, and it must be subtracted from the scale reading to get the net weight of the aircraft.

Larger aircraft are weighed by placing load cells between the jack and the jack pad on the aircraft. These load cells are strain-gage capsules whose resistance changes proportionally to the amount of load that is imposed on them. An electronic bridge circuit converts this change in resistance into a readout of the weight of the aircraft.

The aircraft is leveled according to the leveling instructions in the Type Certificate Data Sheets. The level of some aircraft is determined by placing a spirit level across two screws on the side of the fuselage, while others are leveled by dropping a plumb bob from a specified screw, usually located in the cabin door frame. Longitudinal level is the most important condition, but the aircraft should be laterally level as well.

If the aircraft is weighed from the jacks, it may be leveled by adjusting the extension of the jacks, but if it is resting on its landing gear on the scales, fine adjustment of the level may be made

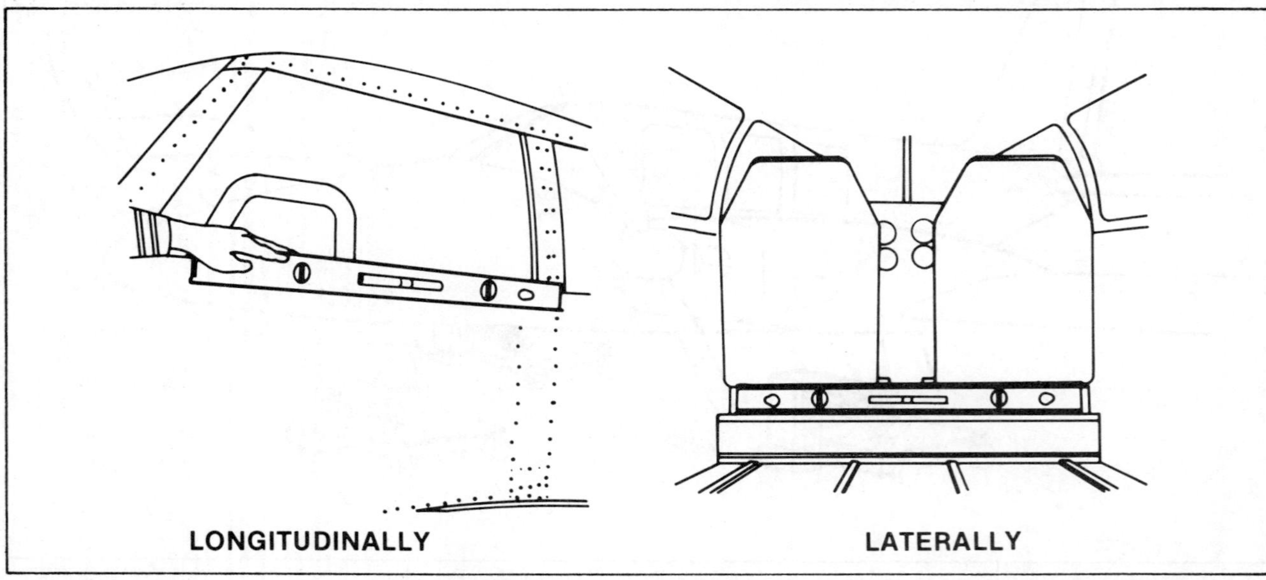

Fig. 2-18 The leveling means of an airplane are specified in the Type Certificate Data Sheets.

by changine the amount of inflation of the tires or of the shock struts.

It is extremely important when jacking or weighing an aircraft that the recommendations of the aircraft manufacturer be followed in detail, as the procedures required for each aircraft differ.

When the aircraft is on the scales and is level, plumb bobs may be dropped from the datum and from the weighing points so that lines may be chalked on the floor from which to make accurate measurements to determine the center of gravity.

C. *Recording the Data*

A weighing form such as the one in Fig. 2-19 is useful to systematically record all of the information and to find the empty-weight center of gravity.

The aircraft make and model with its serial number and N-number are recorded, as well as the location of the datum as found in the Type Certificate Data Sheet. Quite often the datum is located ahead of the aircraft, and when it is, it is specified as being so many inches ahead of a specific location on the aircraft. In Fig. 2-20 we see that the datum is 100 inches ahead of the forward side of the front pressure bulkhead. A plumb bob is dropped from the forward side of this bulkhead, and the hangar floor is marked at this point as station 100.

The weighing conditions of the aircraft specify whether or not any fuel or oil were on board when it was weighed. The amount of fuel and its temperature should be recorded, as well as the amount of oil. The location of the weighing points should be specified in inches forward or aft of the datum.

WEIGHING FORM

MAKE...................... MODEL..................... SERIAL...................... N.......................

DATUM LOCATION:

WEIGHING CONDITIONS:

Main weighing point is located (—............" forward) (+.........." aft) of datum.

Tail or nose weighing point is located (—........" forward) (+............" aft) of datum.

Weighing Point	Scale Reading	− Tare	= Net Weight	× Arm	= Moment
Left Main					
Right Main					
Nose or Tail					
Total as weighed					

Space for listing of Items when Aircraft is not Weighed Empty.

Item	Net Weight	Arm	Moment
Aircraft as Weighed			
Fuel Gal. Temp. °F #/gal			
Aircraft Empty Weight & c.g.			

Maximum Allowable Gross Weight________________ **Computed by:**________________

A&P Number:________________

Useful Load________________ **Date**________________

Fig. 2-19 Typical weighing form

DEPARTMENT OF TRANSPORTATION
FEDERAL AVIATION ADMINISTRATION

A12CE
Revision 11
BEECH
60
A60

May 7, 1973

TYPE CERTIFICATE DATA SHEET NO. A12CE

This data sheet which is part of type certificate No. A12CE prescribes conditions and limitations under which the product for which the type certificate was issued meets the airworthiness requirements of the Federal Aviation Regulations.

Type Certificate Holder	Beech Aircraft Corporation Wichita, Kansas 67201

I - Model 60, 4 or 6 PCLM (Normal Category), Approved February 1, 1968
Model A60, 4 or 6 PCLM (Normal Category), Approved January 30, 1970

Engines	Lycoming TIO-541-E1A4 or TIO-541-E1C4 (2 of either or 1 of each)	
Fuel	100/130 minimum grade aviation gasoline	
Engine limits	For all operations, 2900 r.p.m. (380 b. hp.)	
Propeller and propeller limits	(a) Two (in any combination) Hartzell three-blade propellers Diameter: 74 in., (Normal) Minimum allowable for repair 73 1/2 in. (No further reduction permitted) Pitch settings at 30 in. sta.: low 14°, high 81.7° HC-F 3 YR-2/C7479-2R or HC-F-3 YR-2/C7497B-2R or HC-F 3 YR-2F/FC 7479-2R or HC-F 3 YR-2F/FC 7479B-2R (b) Beech 60-389000-3 governor	
Airspeed limits	Never exceed	235 knots
	Maximum structural cruising	208 knots
	Maneuvering	161 knots
	Maximum flap extension speed	
	Approach position 15°	175 knots
	Full down position 30°	135 knots
	Landing gear extended	174 knots
	Landing gear operating	174 knots
C.G. range (landing gear extended)	(+134.2) to (+139.2) at 6725 lb. (+128.0) to (+139.2) at 51[illegible] lb. or less Straight line variation between points given Moment change due to retracting landing gear (+857 in.-lb.)	

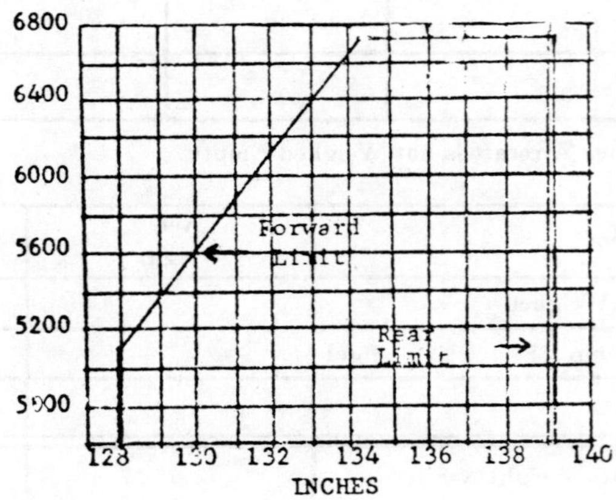

Empty wt. C.G. range	None	
Maximum weight	Takeoff and landing	6725 lb.
Ramp weight		6819 lb.

Fig. 2-20 Typical Type Certificate Data Sheet.

No. of seats	4 (2 at +141, 2 at +173) (add 2 at +205)
Maximum baggage (structural limit)	500 lb. at +75 (nose compartment 655 lb. at +212 (aft area of cabin)
Fuel capacity	142 gal. (+138) comprising two interconnected cells in each wing or 204 gal. (+139) comprising three cells in each wing and one cell in each nacelle (four cells interconnected) See NOTE 1 for data on system fuel
Oil capacity (wet sump)	26 qt. (+88)
Max. operating limit	30,000 ft. pressure altitude

Control surface movements

Wing flaps			Maximum	30°
Aileron	Up	25°	Down	15°
Aileron tab (L.H. only)	Up	10°	Down	10°
Aileron tab anti-servo	Up	12°	Down	7°
Elevator	Up	17°	Down	15°
Elevator tab (L.H. only)	Up	10°	Down	30°
Elevator tab servo	Up	6°	Down	7°
Rudder	Right	33°	Left	28°
Rudder tab	Right	26°	Left	26°

Serial Nos. eligible	Model 60: P-3 thru P-126 (except P-123) Model A60: P-123, P-127 and up (see NOTE 3)
Datum	Located 100 in. forward of front pressure bulkhead
Leveling means	Drop plumb line between leveling screws in cabin door frame rear edge
Certification basis	Part 23 of the Federal Aviation Regulations effective February 1, 1965 as amended by 1, 2, 3, and 12; and Special Conditions dated May 16, 1967, forwarded with FAA letter dated June 1, 1967; approved for flight into known icing conditions when equipped as specified in the approved airplane flight manual. Application for Type Certificate dated December 22, 1965. Type Certificate No. A12CE issued February 1, 1968, obtained by the manufacturer under delegation option procedures.
Production basis	Production Certificate No. 8 issued and Delegation Option Manufacturer No. CE-2 authorized to issue airworthiness certificates under delegation option provisions of Part 21 of the Federal Aviation Regulations.

Equipment

The basic required equipment as prescribed in applicable airworthiness regulations (see Certification basis) must be installed in the aircraft for certification. This equipment must include, for all operations, Airplane Flight Manual P/N 60-590000-5D dated January 15, 1971, amended July 1, 1971, or later issue.

In addition:

1. For flights into known icing conditions, these flight manual supplements and the equipment noted therein:

    60-590001-17 Flight Into Known Icing Conditions.

    60-590001-11 Continuous Pressure Operated Surface Deice System.

    60-590001-13 Goodrich Electrothermal Propeller Deice System.

2. For all other operations:

    Pre-stall warning indicator P/N 151-6, 151-7, or 190-2 (Safe Flight Corporation).

Fig. 2-20 Typical Type Certificate Data Sheet. (continued)

NOTE 1. Current weight and balance data including list of equipment included in certificated empty weight and loading instructions when necessary must be provided for each aircraft at the time of original certification.

The certificated empty weight and corresponding center of gravity locations must include unusable fuel of 24 lb. at (+135).

NOTE 2. The following placard must be displayed in front of and in clear view of the pilot:

"This airplane must be operated in the normal category in compliance with the operation limitations stated in the form of placards, markings and manuals."

NOTE 3. Fuselage pressure vessel structural life limit - refer to the latest revision of the Airplane Flight Manual for mandatory retirement time.

NOTE 4. Model 60 (S/N P-3 thru P-126 except P-123) when modified to Beech dwg. 60-5008 and Model A60 (S/N P-123, P-127 and up) eligible for a maximum weight of 6775 lb.

NOTE 5. A landing weight of 6435 lb. must be observed if 10 PR tires are installed on aircraft not equipped with 60-810012-15 (LH) or 60-810012-16 (RH) shock struts.

. . . END . . .

Fig. 2-20 Typical Type Certificate Data Sheet (Continued)

The scale reading is taken and recorded for both the left and right main weighing points and for the nose or tail point. And the weight of any tare used to hold the aircraft on the scales is subtracted from the scale reading to get the net weight at each of the three weighing points. The distance as measured along the floor between the datum and the weighing points is recorded in the arm column. If the weighing point is ahead of the datum, the arm is negative, but if it is behind the datum, the arm is positive. The moment is the product of the arm in inches times the net weight in pounds, and it is expressed in pound-inches.

The algebraic sum of the moments is recorded in the bottom line of the moments column, and the total net weight is the sum of the three scale readings, less the tare weight. The total moment divided by the total weight gives us the arm of the center of gravity.

If the aircraft was not weighed empty, the weight and moments of the fuel must be computed out. The as-weighed weight, arm, and moment of the aircraft are recorded in the lower chart, and the weight, arm, and moment of the oil and the fuel are recorded. Since these weights are to be subtracted from the weight of the aircraft, they are negative weights, and if their arm is negative, the moment will be positive; but if the arm is positive, the moment will be negative.

When the algebraic sum of the moments is divided by the net weight, less the weight of the fuel and oil, the arm will be the empty-weight center of gravity of the aircraft.

The weighing form is completed by noting the maximum allowable gross weight, as is found in the Type Certificate Data Sheet. The difference between the maximum allowable gross weight and the empty weight is the useful load.

The information is made valid by the A&P mechanic signing the weighing form and including his certificate number and the date the aircraft was weighed.

V. Locating the Balance Point

It is sometimes convenient to locate the balance point of an aircraft from the centerline of the main wheels. After this balance point is located, it is then an easy matter to relate this location to the datum.

A chart is made, similar to the one in Fig. 2-21, and since the main wheels are used as the datum, their arm and their moment will both be zero. The weight and arm of the nose wheel are recorded, and the balance point is found by dividing the total moment by the total weight. In our illustration, the balance point is found to be 14.5 inches ahead of the centerline of the main wheels.

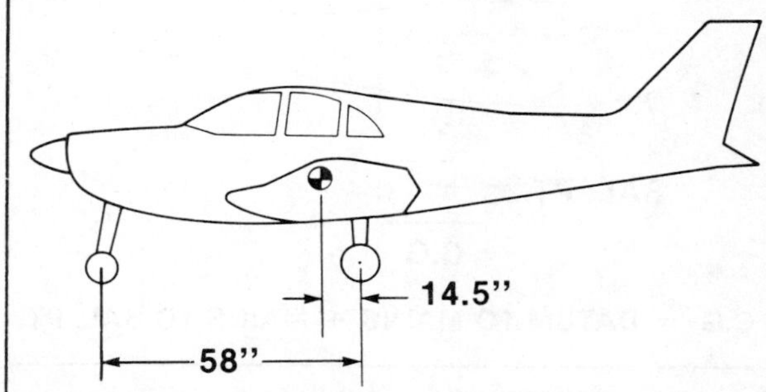

ITEM	WEIGHT	ARM	MOMENT
R. MAIN	406	Φ	Φ
L. MAIN	404	Φ	Φ
NOSE WHEEL	271	– 58	– 15,718
TOTAL	1081	– 14.5	– 15,718

Fig. 2-21 Finding the balance point of an airplane from the weighing points

If the datum is located ahead of the main wheels, the center of gravity from the datum may be found by subtracting the distance between the main wheels and the balance point from the distance between the datum and the main wheels.

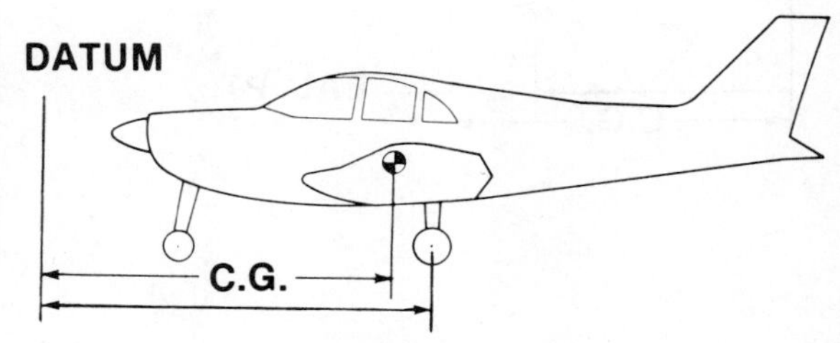

Fig. 2-22 Finding the center of gravity when the balance point is known. The datum is ahead of the airplane. (Tricycle landing gear.)

If the datum is located behind the main wheels, the center of gravity from the datum is found by adding the distance between the main wheels and the balance point to the distance between the datum and the main wheels.

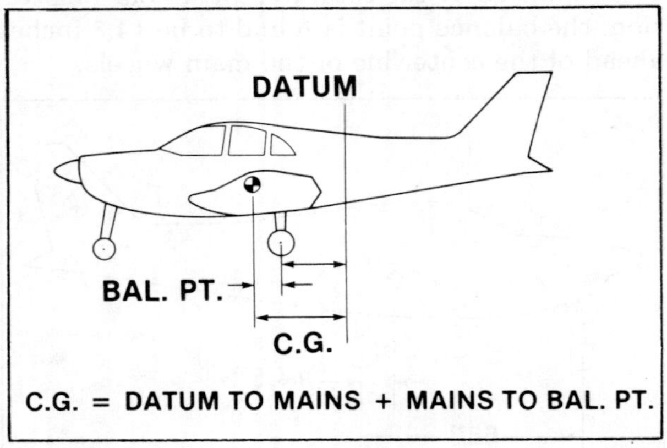

Fig. 2-23 Finding the center of gravity when the balance point is known. The datum is behind the main wheels. (Tricycle landing gear.)

If the airplane has a tail wheel, the balance point will be behind the main wheels, and if the datum is located ahead of the main wheels, the center of gravity may be found by adding the distance between the balance point and the centerline of the main wheels to the distance from the datum to the main wheels.

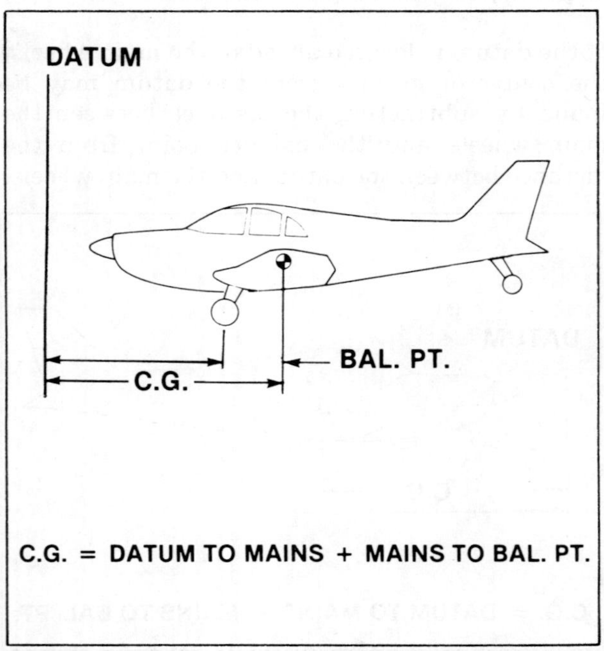

Fig. 2-24 Finding the center of gravity when the balance point is known. The datum is ahead of the airplane. (Tailwheel airplane.)

If the datum is located behind the main wheels of a tail wheel airplane, the center of gravity from the datum is found by subtracting the distance between the main wheels and the balance point from the distance between the main wheels and the datum.

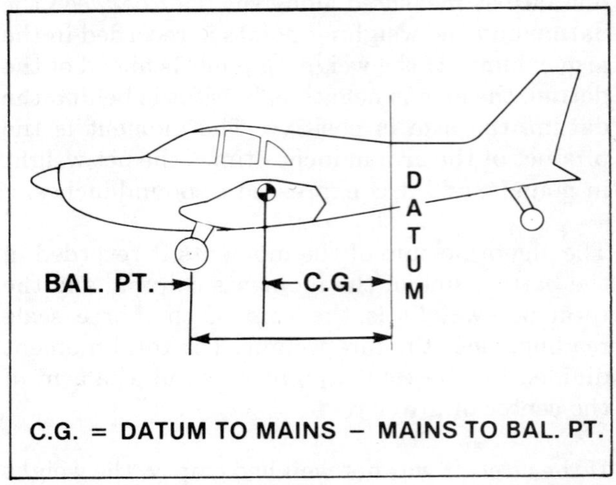

Fig. 2-25 Finding the center of gravity when the balance point is known. The datum is behind the center of gravity. (Tailwheel airplane.)

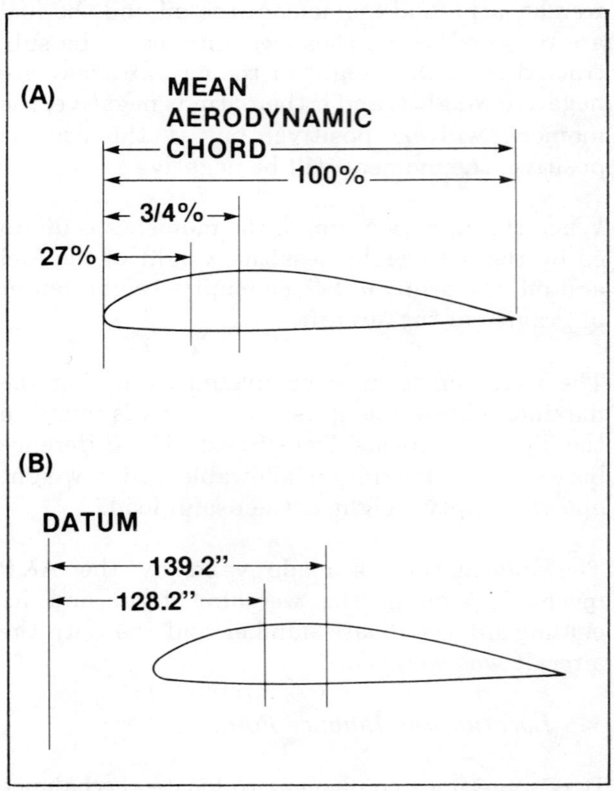

Fig. 2-26 A — The center of gravity range expressed in percent of the mean aerodynamic chord. B — Center of gravity range expressed as inches from the datum.

VI. Center of Gravity Range

When an aircraft is designed, the engineers determine the most forward and the most rearward location for the center of gravity that will allow the aircraft to have all of the required flight characteristics. These locations are measured as percentages of the mean aerodynamic chord, or, as it is spoken of in engineering terms, in % MAC. In order for this information to be usable to the technician, it is given to him in terms of inches from the datum.

A. Empty-Weight Center of Gravity Range

Some of the smaller aircraft, whose fuel tanks are located in the wing and whose two seats are side-by-side with a small baggage compartment immediately behind the seats, have such a limited movement of the center of gravity that the manufacturer includes an empty-weight center of gravity range in the Specifications Sheets. In Fig. 2-27 we see that the empty-weight center of gravity range is inside the loaded center of gravity range. If the empty-weight center of gravity of the aircraft falls within the empty-weight center of gravity range, the aircraft cannot be legally loaded in such a way that its center of gravity in flight will fall outside of the loaded center of gravity range. If the empty-weight center of gravity does not fall within the allowable empty-weight range, the technician must compute an adverse-loaded center of gravity condition, and if the aircraft can be made to fall outside of the loaded center of gravity range, it must be placarded to prevent the pilot loading the aircraft improperly. In extreme cases, fixed ballast may be installed in the aircraft to prevent its falling outside of the empty-weight center of gravity range.

B. Loaded Center of Gravity Range

Larger aircraft having several rows of seats and with both forward and aft baggage compartments, do not have an empty-weight center of gravity range, but on their Type Certificate Data Sheets they have a center of gravity graph, such as the one in Fig. 2-28. Above the graph are the figures from which the graph was made.

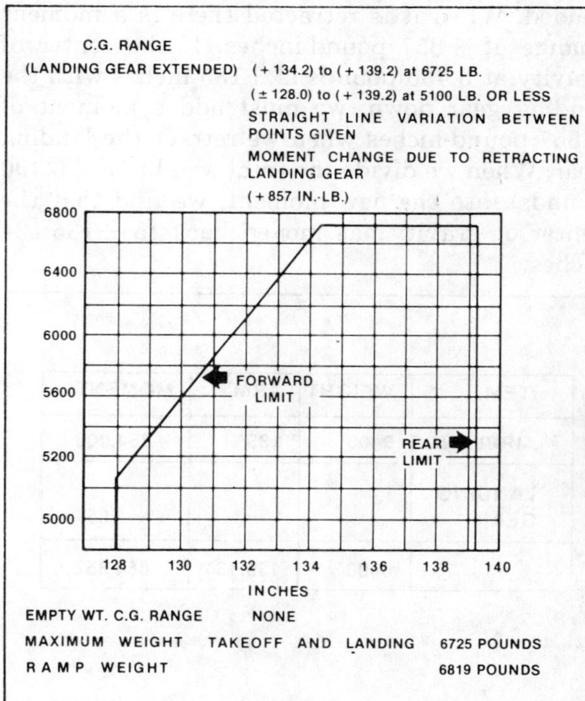

Fig. 2-28 Center of gravity graph

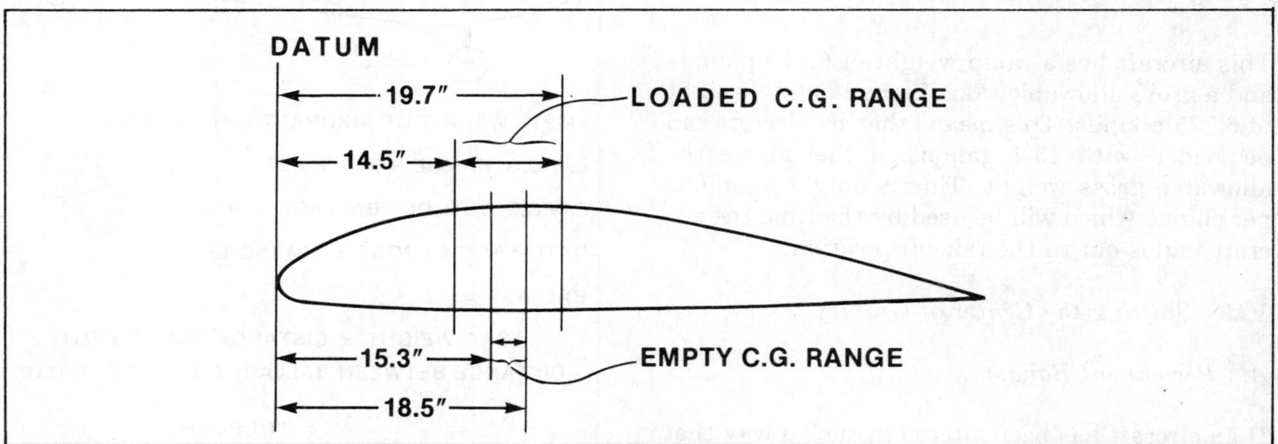

Fig. 2-27 If the empty-weight center of gravity falls within the empty-weight center of gravity range, the airplane cannot be legally loaded so its operational center of gravity will fall outside the operational range.

Up to a gross weight of 5,150 pounds, the forward center of gravity limit is +128 inches, and the aft limit is +139.2 inches. At the gross weight of 6,725 pounds, the forward limit is +134.2 inches and the aft limit is still +139.2. The forward limit moves back in a straight line between the two weights. Using the graph, we see, for example, that at a gross weight of 6,400 pounds the forward limit is +133.0 inches, and the aft limit is +139.2 inches.

You will notice in the information that this graph applies to the aircraft with the landing gear extended. When it is retracted there is a moment change of +857 pound-inches. If the center of gravity at 6,400 pounds is +135 inches with the landing gear down, we must add a moment of +857 pound-inches when we retract the landing gear. When we divide the total weight (still 6,400 pounds) into the new moment, we find that the center of gravity has moved back to +135.134 inches.

ITEM	WEIGHT	ARM	MOMENT
AIRPLANE	6400	135	+ 864,000
LANDING GEAR			+ 857
	6400	135.134	864,857

Fig. 2-29 Determining the amount the center of gravity shifts when the landing gear is retracted

This aircraft has a ramp weight of 6,819 pounds and a gross allowable takeoff-and-landing weight of 6,725 pounds. This means that he aircraft can be loaded with 15.7 gallons of fuel above the allowable gross weight. This is only 7.8 gallons per engine which will be used by the time the aircraft taxies out to the takeoff position.

VII. Shifting the Center of Gravity

A. Permanent Ballast

If an aircraft has been altered in such a way that its center of gravity is outside of its allowable range, it may be brought back into range by adding permanent ballast. If ballast is needed, it should be installed with as long an arm as possible so the weight will be minimum. The aircraft structure to which the ballast is attached must be strong enough to support the weight under all flight conditions, and the ballast is normally painted red and marked "Permanent Ballast — Do Not Remove."

In Fig. 2-30 we have an example of a small two-place airplane that has had a powerplant alteration performed which moved its center of gravity two inches ahead of the forward empty-weight center of gravity limit. In order to bring the center of gravity back within limits, a bar of lead may be attached to the tailpost. To determine the amount of lead needed, we can divide the out-of-range moment by the arm of the ballast from the desired center of gravity position.

$$\text{Ballast weight} = \frac{\text{Aircraft weight} \times \text{distance out of limits}}{\text{Distance between ballast and desired CG}}$$

In order to move the center of gravity from ^22 inches to +24 inches, we must install a weight of 11.1 pounds on the tailpost.

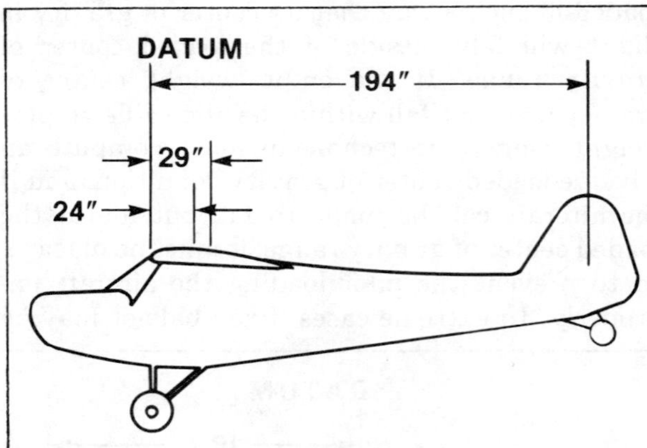

EMPTY WEIGHT OF AIRCRAFT = 945 POUNDS

E.W.C.G. RANGE +24 TO +29

E.W.C.G. WITHOUT BALLAST = +22

DATUM TO TAILPOST = 194 INCHES

$$\text{BALLAST} = \frac{\text{ACFT. WEIGHT} \times \text{DISTANCE OUT OF LIMIT}}{\text{DISTANCE BETWEEN BALLAST AND DESIRED C.G.}}$$

$$= \frac{945 \times 2}{170} = 11.1 \text{ POUNDS}$$

Fig. 2-30 Addition of permanent ballast to move the empty-weight center of gravity

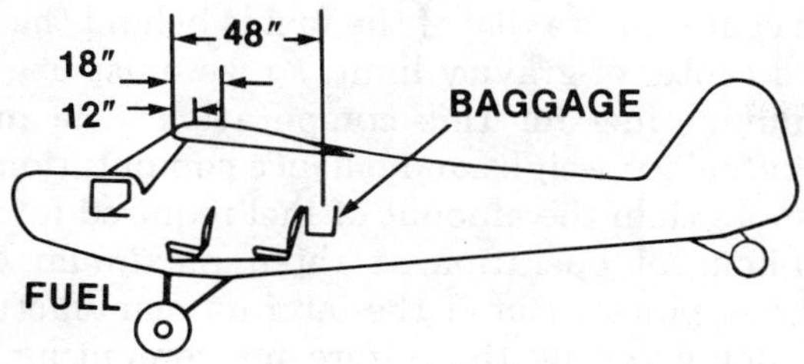

AIRPLANE WITH PILOT & FULL FUEL = 1045 POUNDS
WITH A CENTER OF GRAVITY OF +10
FIND BALLAST NEEDED IN BAGGAGE COMPARTMENT
AT +48 TO MOVE C.G. BACK TO +12

$$\text{BALLAST NEEDED} = \frac{1045 \times 2}{36} = 58 \text{ POUNDS}$$

PROOF

ITEM	WEIGHT	ARM	MOMENT
AIRPLANE	1045	10	10,450
BALLAST	58	48	2,784
	1103	11.99	13,234

Fig. 2-31 Use of temporary ballast to correct for an unusual flight condition

B. Temporary Ballast

For certain flight conditions, it may be necessary to carry temporary ballast to keep the aircraft within the allowable center of gravity limits. Some tandem-seat trainers must be flown solo from the rear seat because, with one occupant in the front seat and a full tank of fuel ahead of the front seat, the center of gravity will be ahead of the forward limit.

If a pilot wants to fly solo from the front seat, he must carry enough ballast in the baggage compartment behind the rear seat to bring the loaded center of gravity into range. The formula used, as we see in Fig. 2-31, is the same as was used to find the amount of permanent ballast required. In this instance, we want to move the center of gravity aft two inches. The loaded weight of the aircraft is 1.045 pounds and the loaded CG is +10 inches. The center of the baggage compartment is 36 inches behind the forward center of gravity limit. When we work this problem, we find that a weight of 58 pounds must be carried in the baggage compartment for solo flight from the front seat. This ballast must be clearly marked, stating that it is to be carried in the aircraft only when the aircraft is being flown solo from the front seat. As a practical consideration, it must be assured that the baggage compartment is structurally strong enough to carry this amount of weight.

C. Shifting Weight

Large aircraft having several rows of seats and more than one baggage compartment may be kept in balance without adding ballast, by shifting some of the weight that is carried. For example, in Fig. 2-23, we have a large aircraft with a baggage compartment at station 26 and one at station 246. We need to find the amount of weight we must shift to bring the center of gravity back 1.5 inches.

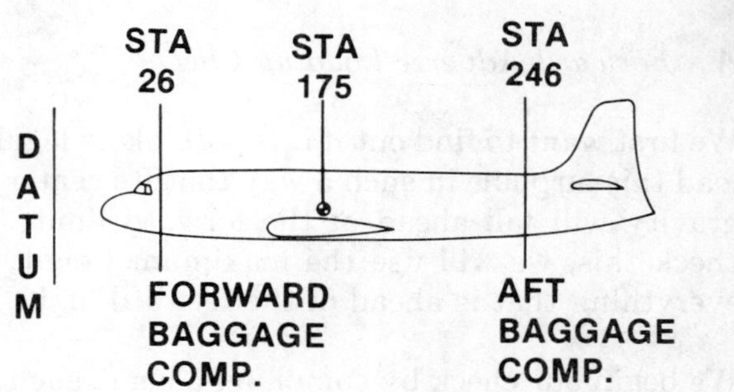

AIRCRAFT LOADED WEIGHT = 8200 POUNDS
AIRCRAFT LOADED C.G. = 175
FIND WEIGHT THAT MUST BE MOVED FROM FORWARD BAGGAGE COMPARTMENT TO AFT COMPARTMENT TO MOVE C.G. TO +176.5

$$\frac{\text{WT. SHIFTED}}{\text{TOTAL WEIGHT}} = \frac{\text{CHANGE IN C.G. REQUIRED}}{\text{DISTANCE WEIGHT IS SHIFTED}}$$

$$\text{WT. SHIFTED} = \frac{\text{TOTAL WEIGHT} \times \text{C.G. CHANGE REQUIRED}}{\text{DISTANCE WT. IS SHIFTED}}$$

$$\text{WT. SHIFTED} = \frac{8200 \times 1.5}{220} = 55.9 \text{ POUNDS}$$

Fig. 2-32 Shifting weight to move the operational center of gravity

The ratio of amount of weight shifted to the total weight of the aircraft is proportional to the ratio of the change in center of gravity required to the distance the weight is shifted. Using the formula in Fig. 2-32, we find that by shifting 55.9 pounds of baggage from the front to the rear baggage compartment, we will shift the center of gravity aft by 1.5 inches.

VIII. Adverse-Loaded Center of Gravity

It is sometimes necessary to determine whether or not it is possible to load an aircraft in such a way that it will fall outside of the allowable center of gravity limits.

In Fig. 2-23 we have a typical four-place airplane powered by an engine of 230 METO horsepower. It carries 12 quarts of oil, 50 gallons of fuel, and four occupants and has a rear baggage compartment that is limited to 120 pounds of baggage. The forward center of gravity limit is at +52 inches, and the aft limit is at +58 inches. The airplane has an empty-weight of 2,024 pounds and an empty-weight center of gravity of +52.5. The maximum allowable gross weight for this airplane is 3,150 pounds.

A. Forward Adverse-Loading Check

We first want to find out if it is possible to legally load this airplane in such a way that its center of gravity will fall ahead of the forward limit. To check this, we will use the maximum weight of everything that is ahead of the forward limit.

We begin our check by compiling a chart such as we have in Fig. 2-34. We enter the airplane with its weight, its arm, and its moment. It is necessary that we carry a full sump of oil, so we enter the full weight of the oil and its arm and moment.

The center of gravity of the fuel is behind the forward center-of-gravity limit, so we must use the minimum fuel for this computation. The minimum fuel for weight and balance computations is no more than the amount of fuel required for one half-hour of operation at rated maximum continuous power. But if the aircraft manufacturer has not given us this figure for computing the weight and balance, it is permissable to use 1/12 gallon of gasoline for each maximum-except-takeoff (METO) horsepower. Aviation gasoline has a nominal weight of six pounds per gallon, so we can find the weight of the minimum fuel by dividing the METO horsepower by two. In our example we have an engine with 230 METO horsepower, so our minimum fuel weighs 115 pounds.

We must have a pilot, and since the center of gravity of the front seat is behind the forward center-of-gravity limit, we want only one occupant in the front seat. The standard weight of a person is 170 pounds.

The rear seats and the baggage compartment are both behind the forward center-of-gravity limit, so we do not want any occupants in the rear seats nor do we want any baggage in the baggage compartment.

When we compute the weight and moment of the airplane, full oil, minimum fuel, and one front-seat occupant, we find a total weight for the maximum forward center-of-gravity condition to be 2,331.5 pounds, and a most forward center of gravity of +52.54 inches. This is behind the forward limit, and so it is not possible to properly load the airplane and exceed the forward center-of-gravity limit.

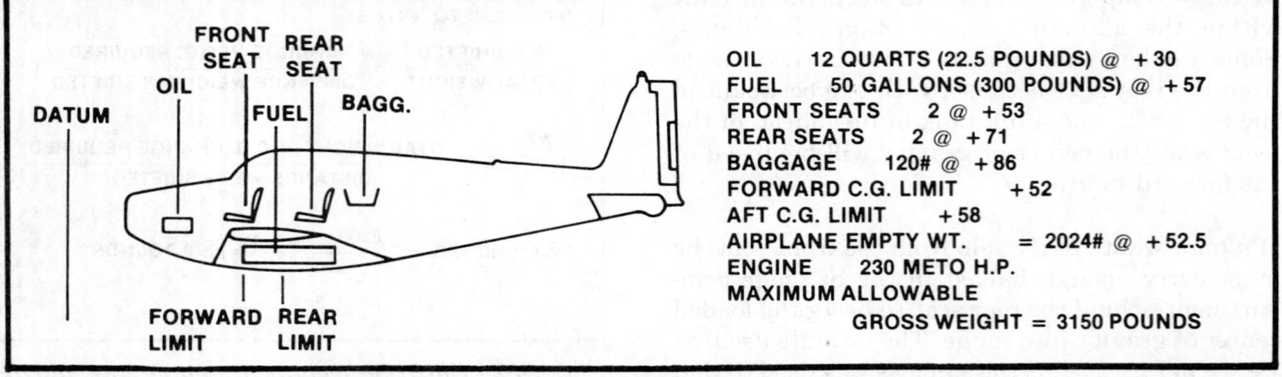

Fig. 2-33 Airplane for use in adverse-loaded center of gravity

ITEM	WEIGHT	ARM	MOMENT
AIRPLANE	2,024	52.5	106,260
OIL	22.5	30	675
FUEL (MIN.)	115	57	6,555
F. SEAT (PILOT)	170	53	9,010
	2,331.5	52.54	122,500

Fig. 2-34 Forward adverse-loaded center of gravity check

B. Rearward Adverse-Loading Check

We can check the rearward loading in the same way. We use the airplane and the full sump of oil. Since the center of gravity of the fuel is ahead of the rear limit, we use minimum fuel. The front seat is ahead of the rear limit, so we must use only one occupant in the front seat; this, of course, is the pilot. Both the rear seat and the baggage compartment are behind the rear center-of-gravity limit, so we use two occupants in the rear seat and the maximum allowable amount of baggage, 120 pounds.

When we find the total moment and divide this by the total weight, we have a weight for maximum rearward center-of-gravity conditions of 2,791.5 pounds and the most rearward center of gravity of +56.23 inches. This is within the allowable limits of both weight and balance.

ITEM	WEIGHT	ARM	MOMENT
AIRPLANE	2,024	52.5	106,260
OIL	22.5	30	675
FUEL (MIN.)	115	57	6,555
F. SEAT (PILOT)	170	53	9,010
R. SEAT (MAX.)	340	71	24,140
BAGG. (MAX.)	120	86	10,320
	2,791.5	56.23	156,960

Fig. 2-35 Aft adverse-loaded center of gravity check

C. Maximum Gross Weight Condition Check

We know now that the airplane cannot be loaded in such a way that it will fall outside of either of the center of gravity limits, so now we want to find that if we can exceed the maximum gross weight.

We add the total weight and moment of the airplane, full oil, full fuel, maximum baggage, and four occupants. The maximum weight is 3,146.5 pounds, which is less than the maximum allowable gross weight. The fully loaded center of gravity is +56.09 inches, which is well within the allowable center of gravity range.

ITEM	WEIGHT	ARM	MOMENT
AIRPLANE	2,024	52.5	106,260
OIL	22.55	30	675
FUEL (FULL)	300	57	17,100
F. SEAT (MAX.)	340	53	18,020
R. SEAT (MAX.)	340	71	24,140
BAGG. (MAX.)	120	86	10,320
	3,146.5	56.09	176,515

Fig. 2-36 Maximum gross weight center of gravity check

IX. Weight and Balance Changes After An Alteration

One of the most important weight and balance problems the A&P mechanic will have to perform is to find the new empty weight and empty-weight center of gravity after he has altered an aircraft.

Normally, if the weights of the items installed or removed from the aircraft are known and the location of the alteration is known, a new empty weight and empty-weight center of gravity can be computed without having to re-weigh the aircraft.

In the example given in Fig. 2-37, we have an alteration that requires the removal of an old radio from the instrument panel, along with its power supply that is mounted in the baggage compartment. In its place there will be a smaller radio installed in the instrument panel. This new radio does not require a separate power supply. At the same time this alteration is made, an anti-collision beacon is to be installed on the tail.

In Fig. 2-37B we have the chart that allows us to compute the change in the empty weight and the empty-weight center of gravity. The empty weight of the airplane is 2,024 pounds and its empty-weight center of gravity is +52.5 inches. The radio that is removed from the instrument panel has a negative weight of 16 pounds at an arm of +49 inches. The power supply weight is also negative, −12, at an arm of +86. The radio that is to be installed has a positive weight of +9 pounds at a positive arm of +49 inches. The anti-collision beacon has a weight of +4 pounds at an arm of +127 inches. The sum of the positive moments is +107,209 pound-inches and the negative moments are −1,816 pound-inches, or a total moment of +105,393 pound-inches. The new total weight is 2,009 pounds. When the total moment is divided by the total weight, we find the new empty-weight center of gravity to be 52.46 inches aft of the datum.

(A)

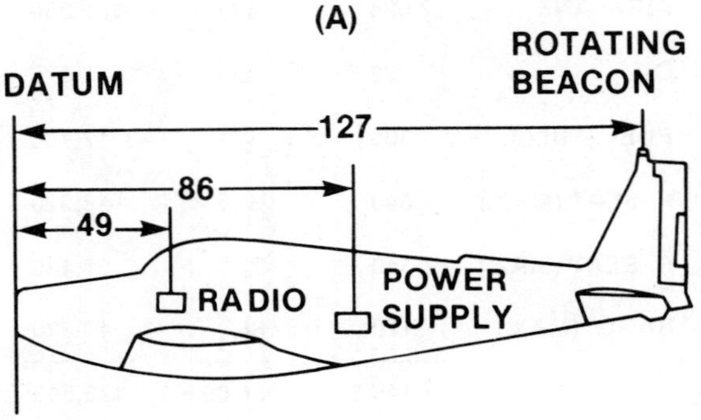

(B)

ITEM	WEIGHT	ARM	MOMENT
AIRPLANE	2,024	+ 52.5	+ 106,260
REMOVE RADIO	(−) 16	+ 49	(−) 784
REMOVE POWER SUPPLY	(−) 12	+ 86	(−) 1,032
ROTATING BEACON	(+) 4	+ 127	+ 508
RADIO	+ 9	+ 49	+ 441
	2,009	52.46	+ 105,393

Fig. 2-37 Weight and balance changes resulting from an alteration

X. Helicopter Weight and Balance

Computing the empty weight and the empty-weight center of gravity of a helicopter is done in the same way as is done for an airplane. The only difference is that the allowable center-of-gravity range for a helicopter is far more limited than that for an airplane.

The helicopter is lifted off the floor with jacks and is leveled according to the leveling means specified in the Type Certificate Data Sheets. The position of the main rotor is quite critical when weighing a helicopter, and for this detail the information furnished by the manufacturer must be followed exactly.

Because the balance of a helicopter is so critical, provisions are made for shifting the center of gravity. Some helicopters have a rack along which the battery may be moved forward or rearward, and others have ballast stations where ballast may be installed to bring the center of gravity within the required range.

XI. Loading and Weight Distribution

A. Small Airplane

It is the responsibility of the aviation maintenance technician to be sure that the weight and balance information on the aircraft is correct and is updated. Any time equipment is added or removed, or any repair or alteration is made that could affect either the empty weight of the aircraft or its empty-weight center of gravity, this change must be recorded in the weight and balance information, and all of the old computations must be plainly marked so the pilot can easily see that they have been superseded.

The flight manual for the airplane normally has a weight and balance loading graph and a moment envelope such as those we have in Figs. 2-39 and

Fig. 2-38 Weight and balance computations of a helicopter are performed in the same way as those for an airplane. But the critical nature of helicopter flight makes accurate placement of the center of gravity more critical on a helicopter than it is on an airplane.

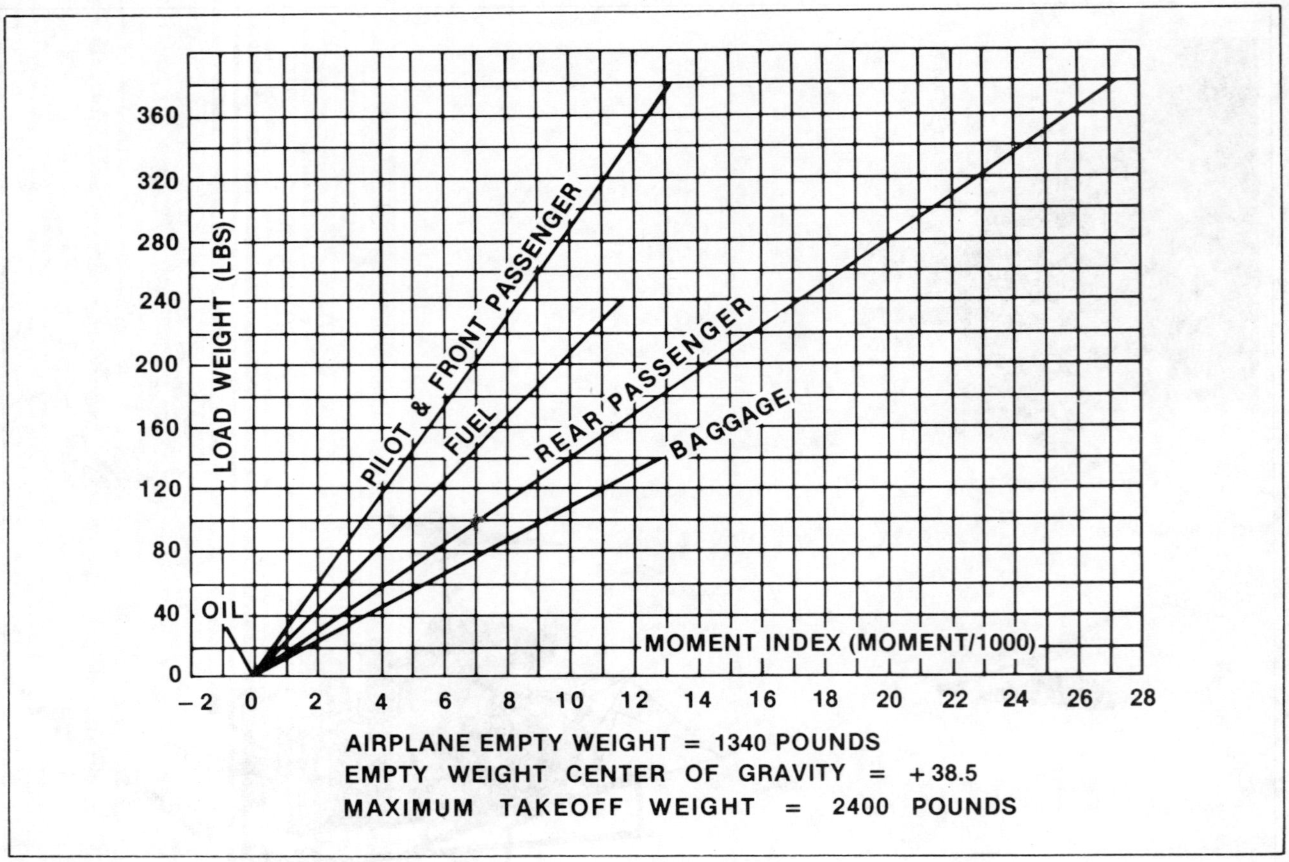

Fig. 2-39 Loading graph for determining the moment index

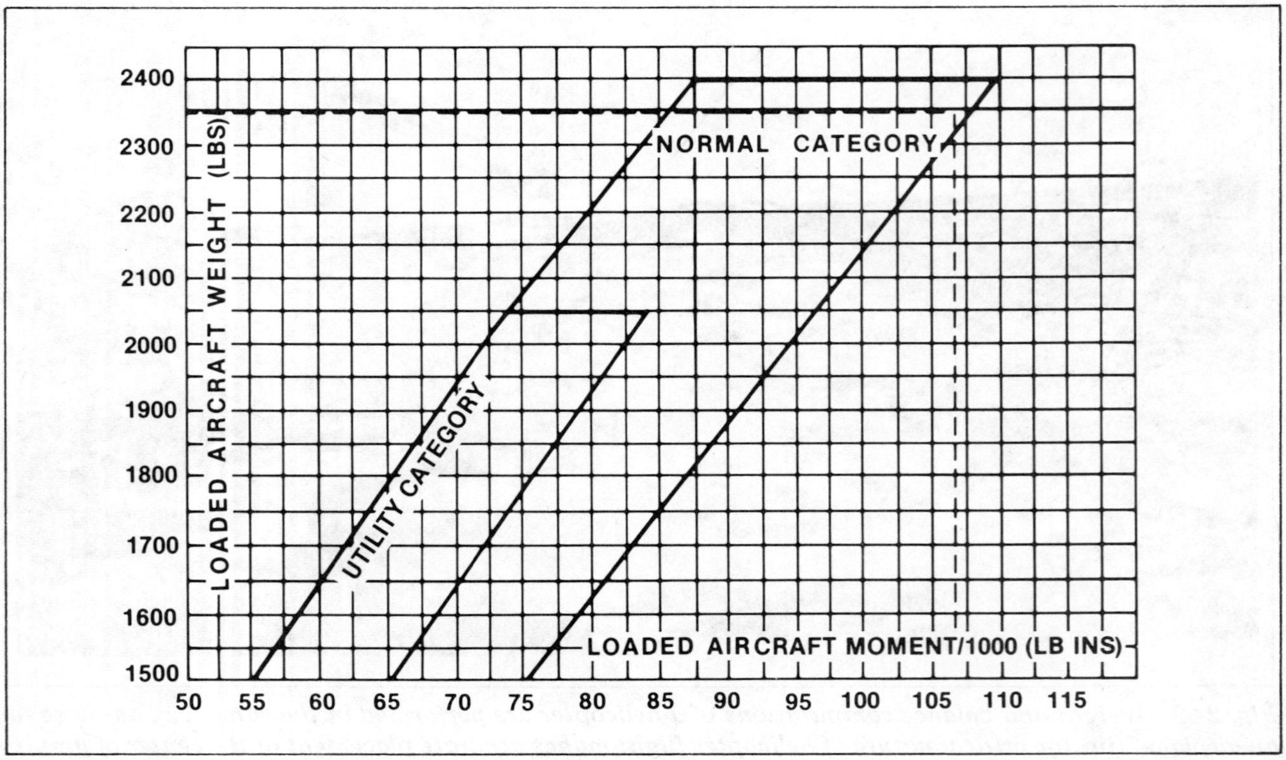

Fig. 2-40 Loaded moment envelope

2-40. Before each flight where the weight and balance may be critical, the pilot should complete a form similar to the one in Fig. 2-41.

A typical loading problem would be to determine whether or not an airplane is within its legally loaded envelope when it carries four standard-weight occupants (170 pounds each), full fuel, full oil, and 80 pounds of baggage.

You will notice that in Fig. 2-39, rather than using moments, we use moment index. This is simply the moment divided by an appropriate reduction factor to get a number that is convenient to use.

We record the empty weight of the airplane and its moment index in the chart of Fig. 2-41. We know that we have eight quarts, or two gallons, of oil on board, and since oil weighs 7.5 pounds per gallon we locate 15 pounds on the vertical scale on Fig. 2-39 and follow it over to the oil curve. We then follow a vertical line from the intersection down to −0.3 on the moment index scale. There are to be two occupants, or a weight of 340 pounds, in the front seats and the moment index for this is +12. The rear seat also carries 340 pounds, and its moment index is +24.5. The 80 pounds of baggage has a moment index of +7.3, and the 40 gallons of fuel weighs 240 pounds and has a moment index of +11.5. The total weight is 2,355 pounds, and the total moment index is +106.6.

To determine whether or not the aircraft is loaded so that its center of gravity falls within the allowable range, we must transfer the information we have just computed to the graph of Fig. 2-40. Draw a horizontal line through the 2,355-pound weight line and a vertical line through the 106.87 pound-inch/1000 moment index line. These two lines must intersect within the envelope. The actual intersection is within the envelope, near the rearward center of gravity limit.

You will notice in Fig. 2-40 that this airplane is approved for both the normal and the utility categories. In the normal category the allowable gross weight is 2,400 pounds and the most rearward center of gravity is 45.83 inches aft of the datum (moment divided by weight). In the utility category, the maximum gross weight is only 2,050 pounds and the most rearward center of gravity is 40.85 inches aft of the datum.

ITEM	WEIGHT	MOMENT INDEX
AIRPLANE	1340	51.6
OIL (8 QTS)	15	−0.3
FRONT SEAT	340	12.0
REAR SEAT	340	24.5
BAGGAGE	80	7.3
FUEL (40 GAL.)	240	11.5
TOTAL	2355	106.6
LIMITS	2400	SEE CHART

Fig. 2-41 Finding the total moment index of a loaded airplane

B. Air Taxi Loading

Airplanes that are operated under Part 135 of the Federal Aviation Regulations, that governs Air Taxi and Commercial Operators, must compile a passenger and cargo manifest for each flight. This allows the pilot to know his exact takeoff weight and whether or not his airplane is loaded within the allowable center of gravity limits.

The operator is given the choice of using either the actual weight of the passengers and crew, or of using standard weights for the occupants and the baggage.

A typical passenger and cargo manifest is shown in Fig. 2-42, in which we are using the actual weight of the passengers and the moment indexes found in Fig. 2-43.

The fuel weight and moment indexes found in Fig. 2-44 are entered in the appropriate blanks as well as the weight and moment indexes of the crew, the oil, and the empty airplane. The total weight and moment index is the sum of these individual values. The maximum allowable weight for this flight is 10,000 pounds.

PASSENGER-CARGO LIST	WEIGHT	SEAT COMPT.	INDEX
	180	1	126.0
	170	2	119.0
Passenger names	140	3	126.0
	150	4	135.0
entered here	200	5	220.0
	160	6	176.0
	210	7	273.0
	130	8	169.0
	120	9	180.0
	140	10	210.0
	130	11	221.0
PASSENGER—CARGO—MAIN CABIN TOTAL			

WEIGHT	FUEL TANK	INDEX
300	R—MAIN	270.0
300	L—MAIN	270.0
120	R—AUX	144.0
120	L—AUX	144.0
	NOSE	
840	TOTAL	828.0

BAGGAGE COMPT	100	NOSE	20.0
BAGGAGE COMPT	200	REAR	380.0
PILOT—COPILOT	330		165.0
OIL	85		7.7
FUEL	840	TOTAL	828.0
EMPTY WEIGHT	6,150		5842.5
TAKEOFF WEIGHT	9,435		9198.2
TAKEOFF LIMITS	10,000		8467.9 9694.5

AIRCRAFT N 123QC
FLIGHT 45
ROUTE OKC - MKC
DATE

I HEREBY CERTIFY THAT THE ABOVE TAKEOFF WEIGHT AND INDEX IS WITHIN LIMITS AS SPECIFIED IN THE FLIGHT OPERATIONS MANUAL.

PILOT IN COMMAND

Fig. 2-42 Passenger and cargo manifest

WEIGHT	PILOT & CO-PILOT	SEATS 1 & 2	SEATS 3 & 4	SEATS 5 & 6	SEATS 7 & 8	SEATS 9 & 10	SEAT 11
100	50.0	70.	90.0	110.0	130.0	150.0	170.0
110	55.0	77.0	99.0	121.0	143.0	165.0	187.0
120	60.0	84.0	108.0	132.0	156.0	180.0	204.0
130	65.0	91.0	117.0	143.0	169.0	195.0	221.0
140	70.0	98.0	126.0	154.0	182.0	210.0	238.0
150	75.0	105.0	135.0	165.0	195.0	225.0	255.0
160	80.0	112.0	144.0	176.0	208.0	240.0	272.0
170	85.0	119.0	153.0	187.0	221.0	255.0	289.0
180	90.0	126.0	162.0	198.0	234.0	270.0	306.0
190	95.0	133.0	171.0	209.0	247.0	285.0	323.0
200	100.0	140.0	180.0	220.0	260.0	300.0	340.0

OCCUPANTS MOMENT/100

Fig. 2-43 Moment indexes for occupants

WEIGHT	NOSE	MAIN	AUX.
30	6.0	27.0	36.0
60	12.0	54.0	72.0
90	18.0	81.0	108.0
120	24.0	108.0	144.0
150	30.0	135.0	180.0
180	36.0	162.0	216.0
210	42.0	189.0	252.0
240	48.0	216.0	288.0
270	54.0	243.0	324.0
300	60.0	270.0	360.0
330	66.0	297.0	396.0
360	72.0	324.0	432.0
390	78.0	351.0	468.0
420	84.0	378.0	504.0
450	90.0	405.0	540.0
480	96.0	432.0	576.0
510	102.0	459.0	612.0
540	108.0	486.0	648.0
570	114.0	513.0	684.0
600	120.0	540.0	720.0

FUEL

MOMENT / 100

Fig. 2-44 Moment indexes for fuel

GALLON	WEIGHT	MOMENT
2	17	1.5
4	34	3.1
6	51	4.6
8	68	6.1
10	85	7.7

Fig. 2-46 Moment indexes for oil

By referring to Fig. 2-47 we may find the minimum and maximum moment indexes that give us the allowable forward and aft center-of-gravity limits for each weight. In our example, the loaded weight is 9,435 pounds, and in order to find the forward and aft limits we must interpolate between the figures for 9,400 and 9,450 pounds. We see in Fig. 2-47 that the difference in the minimum index is 89.8 index numbers for each 100 pounds, and we are 35 pounds, or 35%, above 9,400 pounds; so we must add 35% of 89.8 index numbers to get the minimum index number of 8,467.9. The maximum index interpolation requires 102.8 index numbers for each 100 pounds.

WEIGHT	NOSE BAGGAGE	COMPT. A	COMPT. B	COMPT. C	COMPT. D	COMPT. E	REAR BAG.
20	4.0	16.0	20.0	24.0	28.0	32.0	38.0
40	8.0	32.0	40.0	48.0	56.0	64.0	76.0
60	12.0	48.0	60.0	72.0	84.0	96.0	114.0
80	16.0	64.0	80.0	96.0	112.0	128.0	152.0
100	20.0	80.0	100.0	120.0	140.0	160.0	190.0
120	24.0	96.0	120.0	144.0	168.0	192.0	228.0
140		112.0	140.0	168.0	196.0	224.0	266.0
160		128.0	160.0	192.0	224.0	256.0	304.0
180		144.0	180.0	216.0	252.0	288.0	342.0
200		160.0	200.0	240.0	280.0	320.0	380.0
300		240.0	300.0	360.0	420.0	480.0	
400		320.0	400.0	480.0	560.0	640.0	
500		400.0	500.0	600.0	700.0	800.0	

BAGGAGE MOMENT / 100

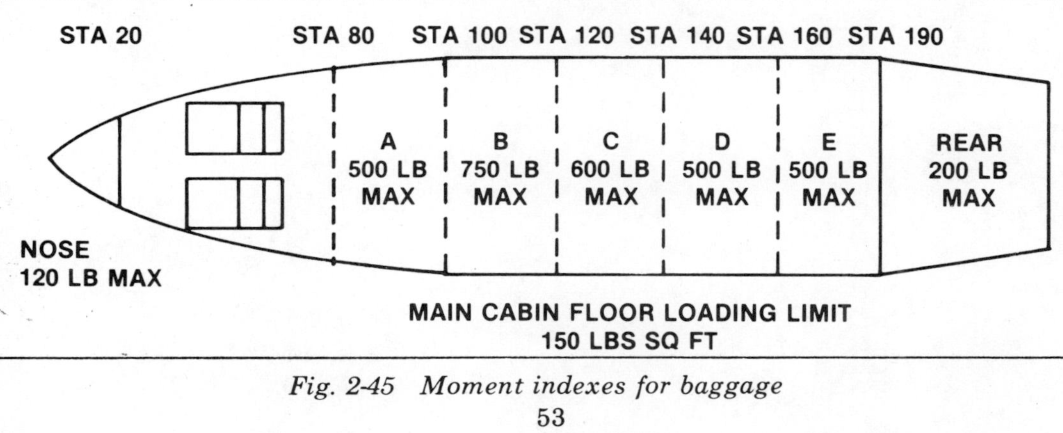

Fig. 2-45 Moment indexes for baggage

Thirty-five percent of this is 36.0, so the maximum index for 9,435 pounds is 9,694.5. Both our loaded weight and loaded moment index are within the allowable limits.

WEIGHT	MIN. INDEX	MAX. INDEX	WEIGHT	MIN. INDEX	MAX. INDEX	WEIGHT	MIN. INDEX	MAX. INDEX
7000	6282.5	7192.5	8000	7180.0	8220.0	9000	8077.5	9247.5
7050	6327.4	7243.9	8050	7224.9	8271.4	9050	8122.4	9298.9
7100	6372.3	7295.3	8100	7296.8	8322.8	9100	8167.3	9350.3
7150	6417.1	7346.6	8150	7314.6	8374.1	9150	8212.1	9401.6
7200	6462.0	7398.0	8200	7359.5	8425.5	9200	8257.0	9453.0
7250	6506.9	7449.4	8250	7404.4	8476.9	9250	8301.9	9504.4
7300	6551.8	7500.8	8300	7449.3	8528.3	9300	8346.8	9555.8
7350	6596.6	752.1	8350	7494.1	8579.6	9350	8391.6	9607.1
7400	6641.5	7603.5	8400	7539.0	8631.0	9400	8436.5	9658.5
7450	6686.4	7654.9	8450	7583.9	8682.4	9450	8481.4	9709.9
7500	6731.3	7706.3	8500	7628.8	8733.8	9500	8526.3	9761.3
7550	6776.1	7757.6	8550	7673.6	8785.1	9550	8571.1	9812.6
7600	6821.0	7809.0	8600	7718.5	8836.5	9600	8616.0	9864.0
7650	6865.9	7860.4	8650	7763.4	8887.9	9650	8660.9	9915.4
7700	6910.8	7911.8	8700	7808.3	8939.3	9700	8705.8	9966.8
7750	6955.6	7963.1	8750	7853.1	8990.6	9750	8750.6	10018.1
7800	7000.5	8014.5	8800	7898.0	9042.0	9800	8795.5	10069.5
7850	7045.4	8065.9	8850	7942.9	9093.4	9850	8840.4	10120.9
7900	7090.3	8117.3	8900	7987.8	9144.8	9900	8885.3	10172.3
7950	7135.1	8168.6	8950	8032.6	9196.1	9950	8930.1	10223.6
						10000	8975.0	10275.0

THE DIFFERENCE IN MIN. INDEX FOR EACH 100 POUNDS IS 89.8 INDEX NUMBERS
THE DIFFERENCE IN MAX. INDEX FOR EACH 100 POUNDS IS 102.8 INDEX NUMBERS

TO FIND LIMITS FOR AN SPECIFIC WEIGHT: FOR EXAMPLE 9435 FIND LIMITS FOR 9400
8436.5 — 9658.5
FOR 9435 POUNDS, ADD 35% OF 89.8 OR 31.4 TO 8436.5 = 8467.9
ADD 35% OF 102.8 OR 36.0 TO 9658.3 = 9694.5

Fig. 2-47 Chart for determining the center of gravity for any specific weight.

SECTION III

Fluid Lines and Fittings

I. Importance of Fluid Lines in Aircraft

As aircraft become more versatile and more effective, they also become more sophisticated. Many of today's airplanes have hydraulic landing gear that require thousands of pounds of force to retract. They have boosted flight controls requiring hydraulic fluid under high pressure to be carried throughout the aircraft, and the fuel systems through which the huge quantities of fuel flow to the engines are built up of complex lines, valves, and fittings.

The plumbing used to carry the vital fluids must be strong, as some of the fluids are under a normal pressure of up to 3,000 psi and have transient pressure surges rising far above this. These lines must be light in weight and must use fittings that produce a minimum drop in the pressure of the fluid flowing through them. Provisions must be made by the use of flexible hose for relative motion between the airframe and such shock-mounted components as the engines and the instrument panels.

II. Rigid Fluid Lines

A. Materials

Many of the fluid lines used in early aircraft were made of copper tubing, and the vibration encountered in flight would harden the copper and cause it to crystallize and eventually break. To prevent this happening, these lines were per-

Fig. 3-1 Modern aircraft depend on both rigid and flexible fluid lines to carry fuel, oil, and hydraulic fluids, some of which are under pressures of thousands of pounds per square inch.

iodically removed from the aircraft, heated red-hot with a torch, and quenched in water. This would soften the copper so that it could be used again.

Aluminum alloy and stainless steel have almost completely replaced copper for use in aircraft. Low-pressure systems such as instrument air or ventilating air are often made of commercially pure aluminum (1100-H14-half-hard), or aluminum having a small percentage of manganese, 3003-H14. Fuel and engine lubricating systems and hydraulic or pneumatic systems that operate in the 1,000 to 1,500 psi range often use lines made of 5052-0 aluminum. This alloy, even in its -0, or annealed, state is about one and three-quarters times as strong as the commercially pure aluminum in its partially hardened condition.

Occasionally 2024-T aluminum alloy is used for fluid lines because of its higher strength, but it is more difficult to blend and flare without cracking, and more care is required when it is used.

Stainless steel is often used for high-pressure hydraulic and pneumatic lines and for lines carrying high-pressure oxygen.

Most of the critical fluid lines in our modern aircraft are fabricated at the factory and have all of the necessary bends made and all of the proper end-fittings installed. These lines are specified and identified by their part number. It is extremely important, when replacing a fluid line, to use only a line having the same part number as that originally installed. If a line must be made up in the shop, the aircraft manufacturer can provide information regarding the material of which the line is made. Aircraft tubing is normally identified with the specification number and the alloy designation stamped on the outside of the tubing.

B. Size Designation

Rigid tubing is identified not only by its material but also by its size.

The outside diameter of the tube is specified in increments of 1/16 inch. All of the fittings that are used, for example, on rigid tubing having a quarter-inch outside diameter are called −4 fittings. Refer to Fig. 3-2.

The wall thickness of the tube determines its strength, and to find the inside diameter of the tube, twice the wall thickness must be subtracted from the outside diameter.

TUBE O.D.	FITTINGS DASH NO.
1/8	− 2
1/4	− 4
3/8	− 6
1/2	− 8
5/8	−10
3/4	−12
1	−16

Fig. 3-2 The outside diameter of rigid tubing is identified by its dash number. The dash number is the numerator of the fractional diameter in sixteenths of an inch.

III. Fabricating Rigid Tubing

A. Cutting

After the correct length of tubing is determined, it should be cut off square with a roller-type tubing cutter, such as the one in Fig. 3-3. This type of cutter supports the tube on hardened steel rollers, while a sharp steel cutting wheel is screwed down against the tube. The cutter is rotated around the tube, and as the blade cuts into the metal, it is turned down further into the tube. This type of cutter produces a cutoff that is smooth and square, but it does leave burrs inside of the tube. These burrs must be completely removed with the reamer blade on the cutter, or with a special reamer such as the one seen in Fig. 3-4. The end of the tube must be polished so there will be no sharp edges that could produce stress concentrations which would cause cracks when the tube is flared.

If no tubing cutter is available, a fine-tooth hacksaw can be used. Be sure to use a blade whose teeth are fine enough that at least two teeth will be on each wall of the tube at all times.

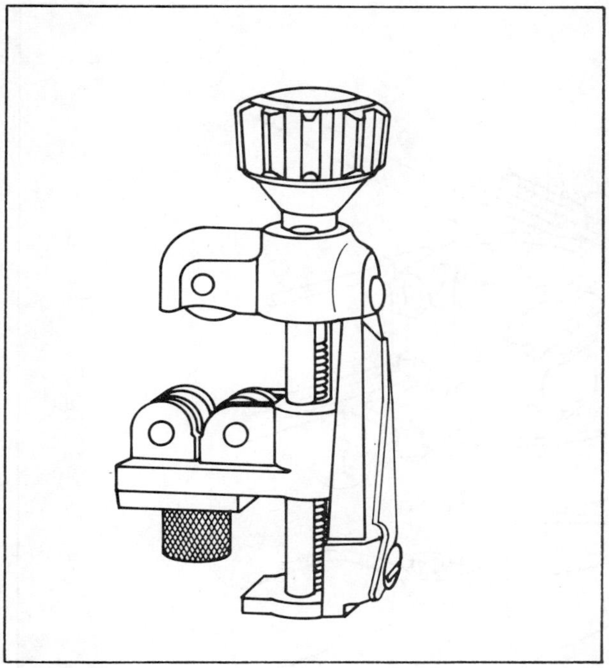

Fig. 3-3 Roller-type tubing cutter used for cutting small diameter rigid tubing. A hardened steel reamer is attached to the back of the cutter to remove the burrs that this type of cutter tends to leave on the inside of the tube.

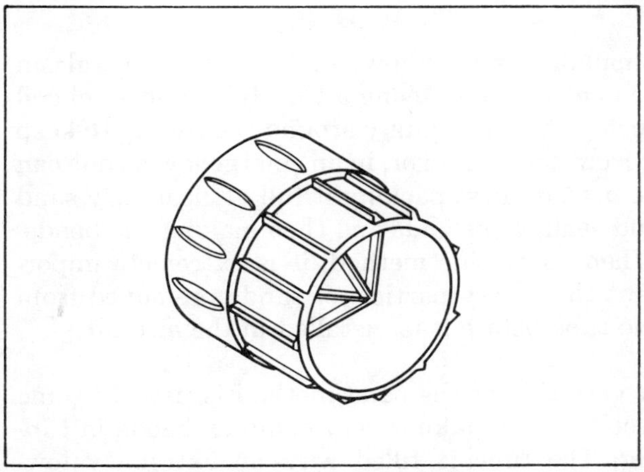

Fig. 3-4 Special type of reamer that can remove burrs from both the inside and the outside of the end of a piece of tubing. One side of the reamer cleans the outside of the cut, and the other side cleans the inside.

It is actually better to cut stainless steel tubing with a hacksaw than with a tubing cutter, because the saw work-hardens the tube to a lesser degree than the cutter. After the tube has been cut and burred, blow it out with compressed air to remove all of the metal chips that could become embedded in the tube when it is flared.

B. Bending

The tubing used for aircraft fluid lines has quite thin walls and will collapse in the bend if adequate precautions are not taken. Fig. 3-5*A* shows a piece of tubing that was bent using no precautions, and the resulting bend is severely kinked, restricting the flow of any fluid through the line. The tube in *B* is wrinkled and, while not restricting the flow so badly as a kink, wrinkles will still cause an opposition and will severely weaken the tube in the bend.

In view *C* we see the tube as it appears after being successfully bent. The bend should be smooth, with no marks from the bending tools, and the minor diameter in the bend should be no less than 75% of the diameter of the tube.

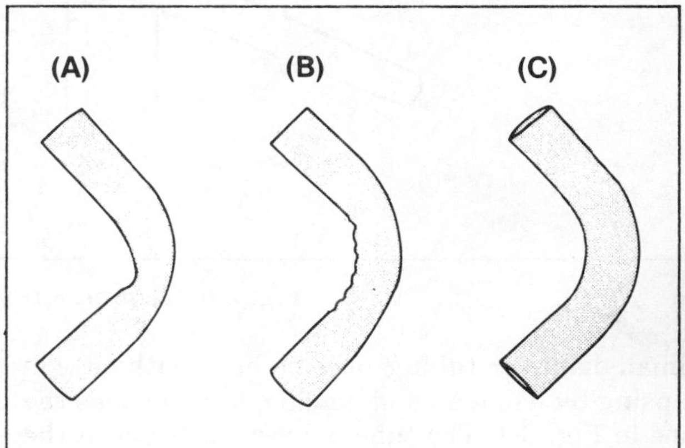

Fig. 3-5 A — Kinked tubing, caused by making the bend without using any bending tools. B — A wrinkled bend, caused by using an improper bending tool. C — A proper bend. The minor diameter in the bend must be no less than 75% of the original diameter of the tube.

Production bending is done by the use of a tube bender similar to the one seen in Fig. 3-6. The tube is clamped between the radius block and a clamp bar, and the radius block is turned by a gear driven by the handle. As the radius block turns, the tube is bent between the radius block and a guide bar. The number of degrees of bend is determined by the amount the radius block is turned. Thin-wall tubing is kept from collapsing by using a mandrel, which is a smooth round-end bar that fits into the tube at the point the bend is being formed. It holds the tube so it cannot collapse.

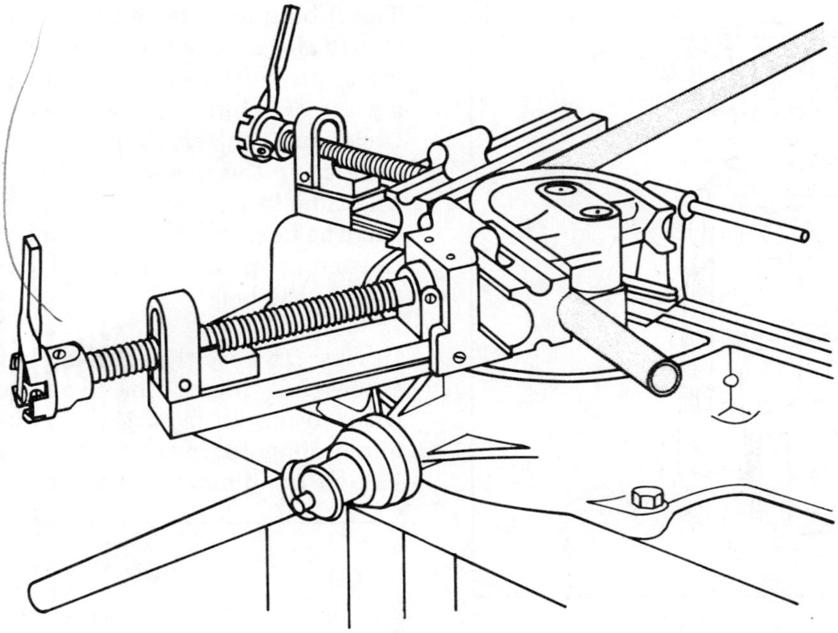

Fig. 3-6 A production-type tube bender

Small-diameter tubing may be bent without collapsing by using a hand bending tool, such as the one in Fig. 3-7. The tube is inserted between the radius block and the slide bar and is held in place by the clip. The slide bar handle can then be moved down to bend the tubing to any angle that is needed. The number of degrees of bend is read on the scale of the radius block, opposite the incidence mark on the slide bar.

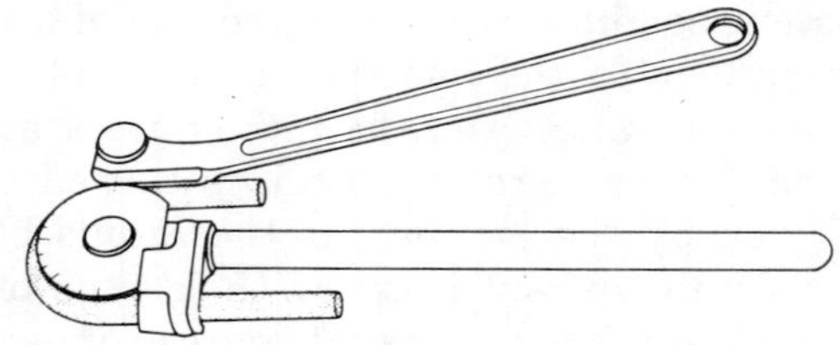

Fig. 3-7 A hand-type tube bender that is used for small-diameter tubing.

Small-diameter, thin-wall tubing of soft metal can be bent by hand, using a tightly wound steel coil spring that fits snugly around the tubing to keep it from collapsing; or, in an emergency, a tube can be bent by first packing it full of clean, dry sand and sealing the ends and then making the bends. When using this method, it is extremely important that every particle of sand is removed from the tube before it is installed in the aircraft.

A variation of the sand method is used in some factories for making very complex bends in tubing. The tube is filled with an extremely low-melting-point metal alloy such as Wood's metal or Cerrobend. These alloys are melted in boiling water, and the tube is filled with the molten alloy and is bent. When the bending operation is completed, the tube is placed in a vat of boiling water, and the alloy melts and runs out of it.

C. *Methods of Joining Rigid Tubing*

1. *Flared fittings*

Much of the rigid tubing used in modern aircraft is joined, or is connected to the components by

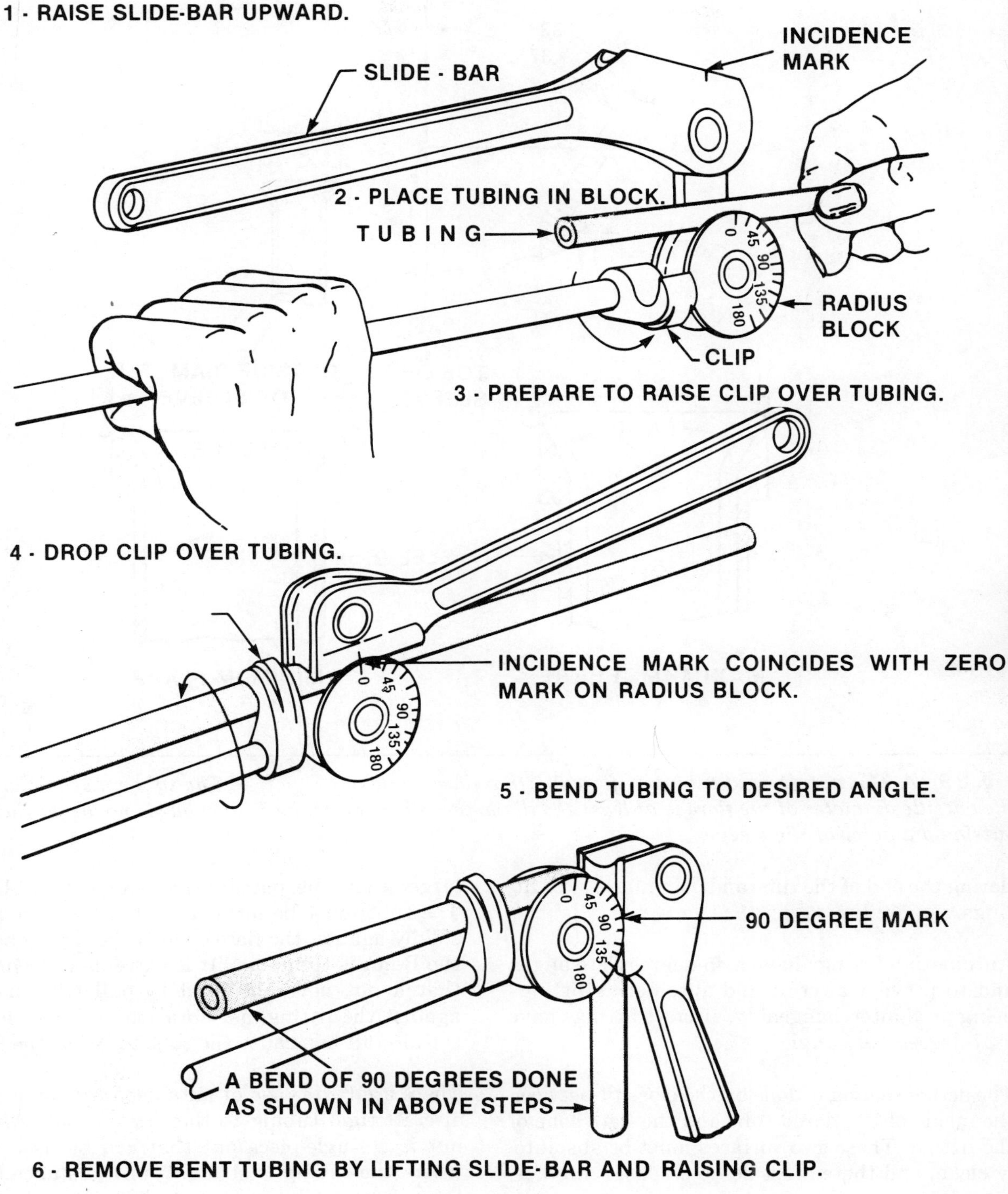

Fig. 3-8 Using a hand-held tubing bender

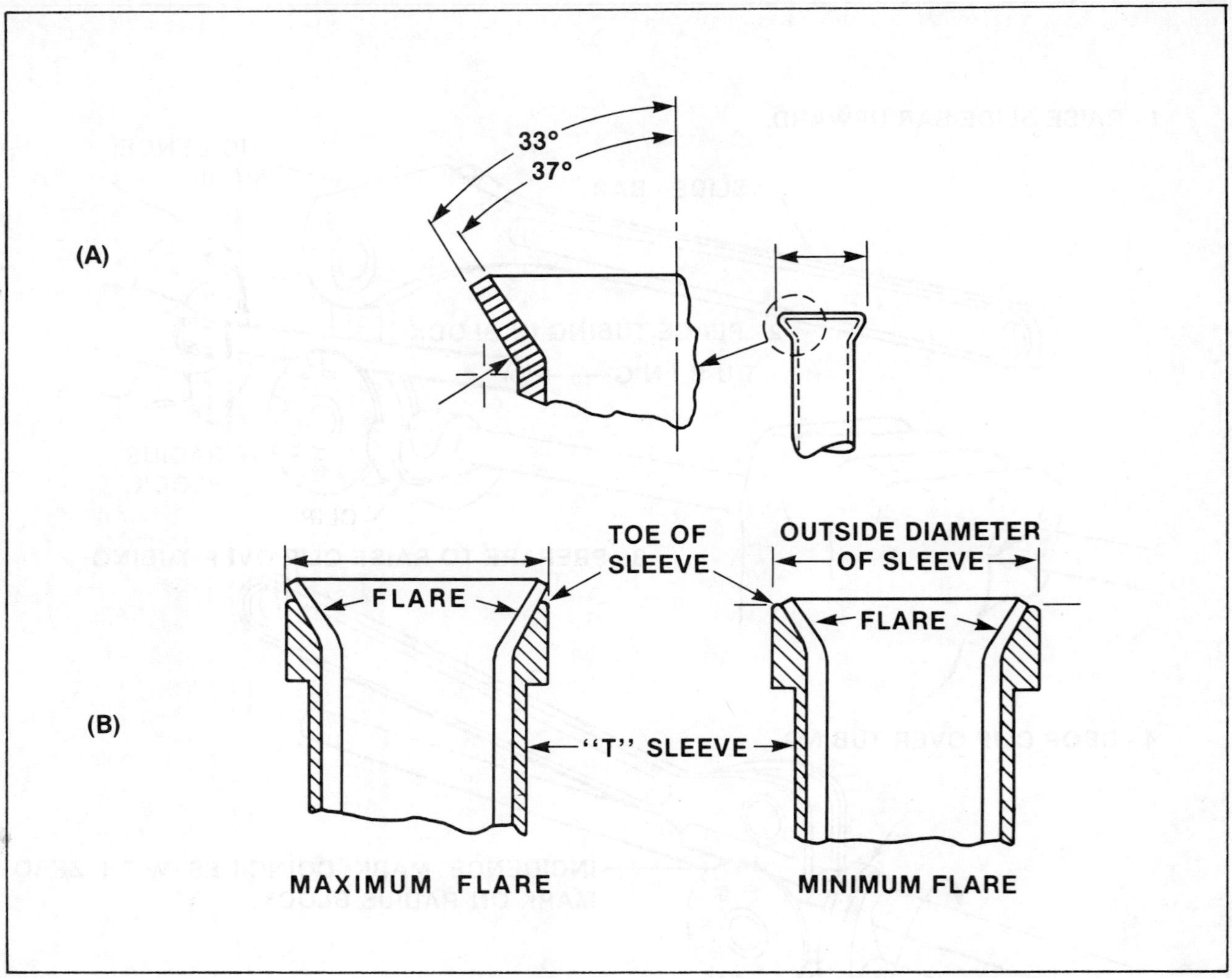

Fig. 3-9 A — Aircraft rigid tubing is flared with a flare angle of 37°. B — The tube must be flared so the outside diameter of the flare is at least the diameter of the toe of the sleeve, but is no larger than the outside diameter of the sleeve.

flaring the end of the tube and using flare-type fittings.

Automotive fittings have a 45-degree flare angle, and to prevent aircraft and automotive fittings being used interchangeably, aircraft fittings have a 37-degree flare angle.

The actual sealing is done by the close fit between the inside of the flared tube and the flare cone of the fitting. These two surfaces must be absolutely clean, and there must be no cracks, scratches, or nicks on either surface.

Before the tube is flared, a B-nut and sleeve are slipped over the end, and a flaring tool is used to form a flare whose outside diameter is at least that of the inside of the toe of the sleeve, and is no larger than the outside diameter of the sleeve. The tube must be formed so that the flare rests solidly against the flare cone of the fitting before the B-nut is tightened. It is extremely important that a nut never be used to pull the tube up against the fitting, as vibration or expansion of the aircraft can cause the tube to fail at the flare.

There are two types of nuts used for connecting aircraft rigid tubing to the fittings. The AN817 nut is a single-piece nut that can be used only where the tubing is straight for a distance back from its end. This nut can be slipped back along the tube while the end is being flared.

Where there is a bend near the end of the tube, the single-piece nut cannot be used and the AN818 nut and the MS20819 sleeve must be used. The hole in this nut is large enough to slip over the tail of the sleeve, and so it can easily slip around the bends in the tubing. The nut and sleeve are slipped over the tubing before it is flared, and when the connection is assembled, the nut forces the sleeve against the tube to form a tight seal between the flare of the tube and the flare cone of the fitting.

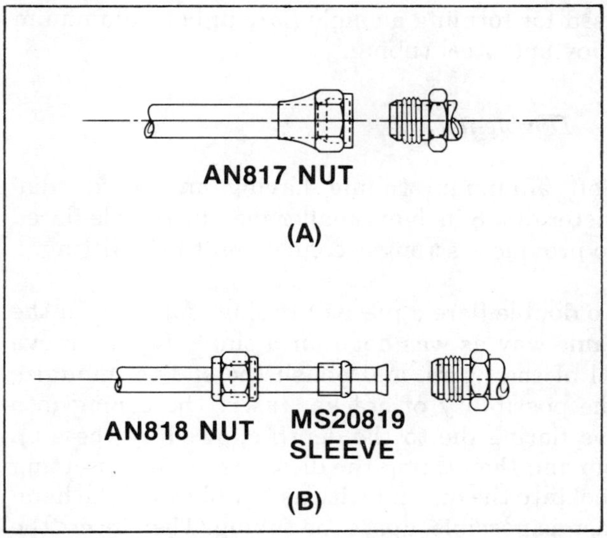

Fig. 3-10 A — The AN817 nut is a single-piece nut and can be used only when there are no bends near the end of the tubing. These nuts are not as satisfactory as the nut and sleeve combination, as they tend to wipe, or iron, the flare as it is being tightened. B — The AN818 nut with the MS20819 sleeve is the most popular type of nut found on rigid tubing. These nuts are commonly called B-nuts.

Tube O.D. (inches)	Wrench torque range for tightening tube nuts (inch pounds)	
	Alum. alloy 1100-H14, 5052-0	Steel
3/16		30-70
1/4	40-65	50-90
5/16	60-80	70-120
3/8	75-125	90-150
1/2	150-250	155-250
5/8	200-350	300-400
3/4	300-500	430-575
1-	500-700	550-750
1-1/4	600-900	
1-1/2	600-900	

Fig. 3-11 Recommended torque values for tightening tube nuts on aircraft rigid tubing

The torque values shown in Fig. 3-11 should be used when tightening the nuts on a flared tube. Overtightening can force the sleeve into the tubing and damage it, causing it to leak.

a. Single flare

A single flare may be formed on a piece of rigid tubing with either an impact-type flaring tool, or one having the rolling action of the flaring cone.

Fig. 3-12 shows an impact-type flaring tool. The tube is cut off square and its ends are polished, and it is clamped between the halves of the vise blocks with only about 1/16 inch of the tube sticking up above the blocks. The blocks are clamped in a vise, and the pin-type flaring tool is driven into the tube with as few sharp blows of a hammer as are needed to flare the tube end out against the bevel cut in the blocks. It is important that as few blows as possible be used, as many light blows will work-harden the tubing.

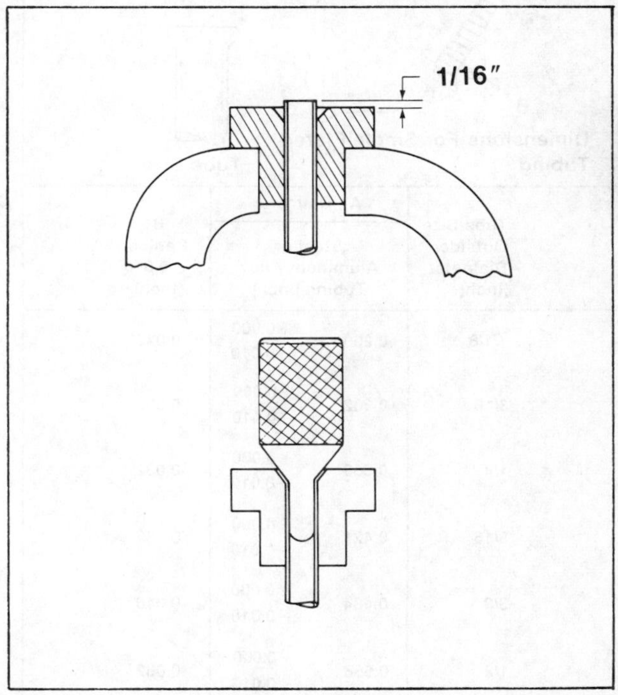

Fig. 3-12 An impact-type flaring tool

Rolling-type flaring tools are quite popular in aviation maintenance shops because they are entirely self-contained and it is easy to make a good flare with this type of tool. The tool pictured in Fig. 3-13 will flare tubing from 1/8 to 3/4 inch outside diameter. To use this tool, the dies are rotated until the two halves of the correct size hole are aligned and the tube is inserted against

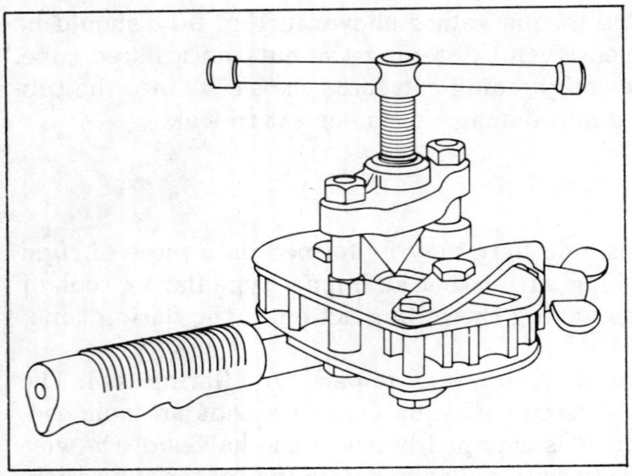

Fig. 3-13 A rolling-type flaring tool

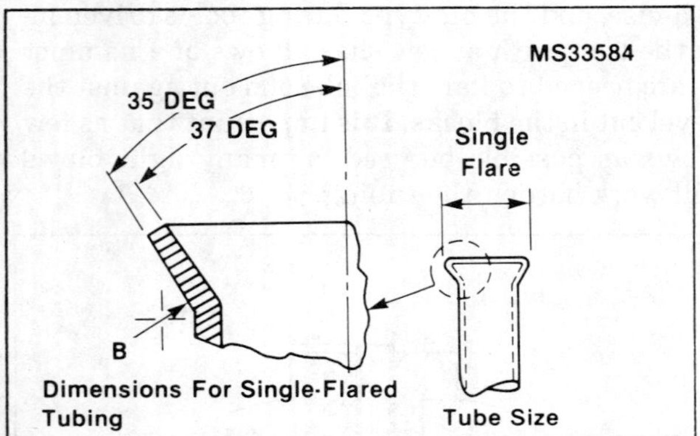

Tube Size Outside Diameter [Inch]	A - Diameter Steel or Aluminum Alloy Tubing [Inch]		B Radius + -0.010 [Inch]
1/8	0.200	+ 0.000 − 0.010	0.032
3/16	0.302	+ 0.000 − 0.010	0.032
1/4	0.359	+ 0.000 − 0.010	0.032
5/16	0.421	+ 0.000 − 0.010	0.032
3/8	0.484	+ 0.000 − 0.010	0.046
1/2	0.656	+ 0.000 − 0.010	0.062
5/8	0.781	+ 0.000 − 0.010	0.062
3/4	0.937	+ 0.000 − 0.010	0.078
1	1.187	+ 0.000 − 0.015	0.093

Fig. 3-14 Dimensional tolerances for a single flare on a piece of aircraft rigid tubing

the stop. The dies are clamped together and the flaring cone is turned into the end of the tube. Rollers in the cone burnish the metal as it is expanded into the die, and a smooth flare results. When the flare is formed sufficiently, the handle is reversed and the dies released, and the tube can be removed from the tool.

Fig. 3-14 is a chart of the dimensional tolerances used for forming a single flare in both aluminum alloy and steel tubing.

b. Double-flare

Soft aluminum tubing having an outside diameter of 3/8 inch or smaller may be double-flared to provide a stronger connection to the fitting.

To double-flare a piece of tubing, cut it off in the same way as was done for a single flare, remove all of the burrs, and polish the end to minimize the possibility of cracks. Insert the tubing into the flaring die to the depth allowed by the stop pin and then clamp the dies. Insert the upsetting tool into the die and with as few blows of the hammer as possible, upset the tubing. Then insert the flaring tool and strike it with the hammer to fold the metal down into the tubing and form the double-flare.

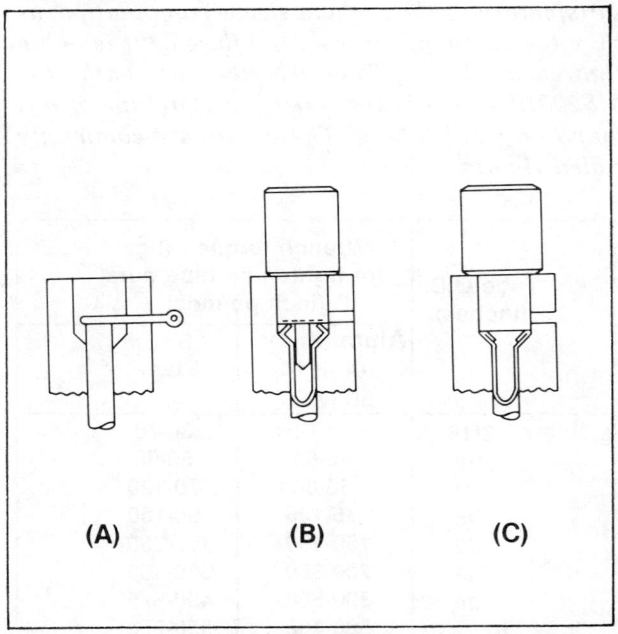

Fig. 3-15 A — Insert the tube to the stop pin and clamp the dies together. B — Insert the upsetting tool and strike it with a hammer to upset the end of the tube. C — Insert the flaring cone and strike it with the hammer to form the flare.

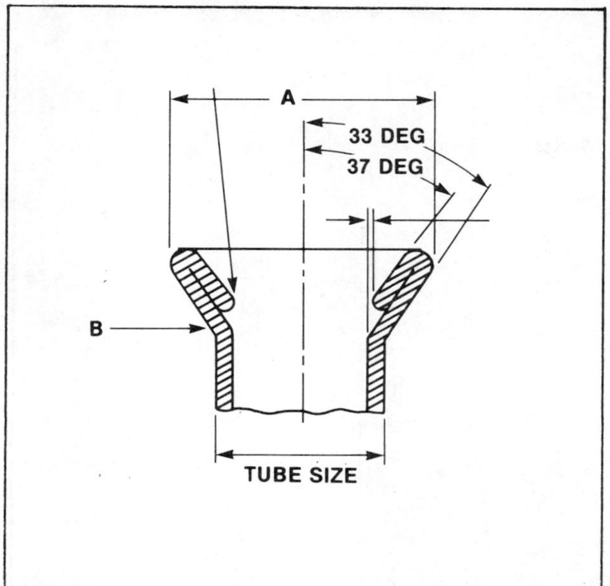

Tube Size Nominal Outside Diameter [in.]	A Diameter +0.010 −0.010 [in.	B Radius +0.010 [in.	Wall Thickness [in.	Minimum Inside Diameter [in.]
1/8	0.224	0.032		
3/16	0.302	0.032	0.028	0.114
			0.035	0.100
1/4	0.359	0.032	0.028	0.178
			0.035	0.159
5/16	0.421	0.032	0.035	0.224
			0.049	0.198
3/8	0.484	0.046	0.028	0.310
			0.035	0.288
			0.049	0.261

Fig. 3-16 Dimensional tolerances for a double flare on a piece of aircraft rigid tubing

2. *Flareless fittings*

It is difficult to flare the heavy wall tubing that is required for some of the higher pressure lines in aircraft hydraulic and pneumatic systems, and the "bite-type" flareless fitting has been designed for these applications.

Cut off the end of the tube square, and remove all of the burrs and polish its end. Then slip a nut and ferrule over the end of the tube and screw the nut onto the presetting tool, as seen in Fig. 3-17. Be sure the tube is square against the bottom of the tool, and then screw the nut down by hand until it tightens the ferrule against the tool. Now, using a wrench, turn the nut from one to one and three-quarters of a turn. The exact amount depends on the size of the tubing and this should be done according to the manufacturer's recommendations.

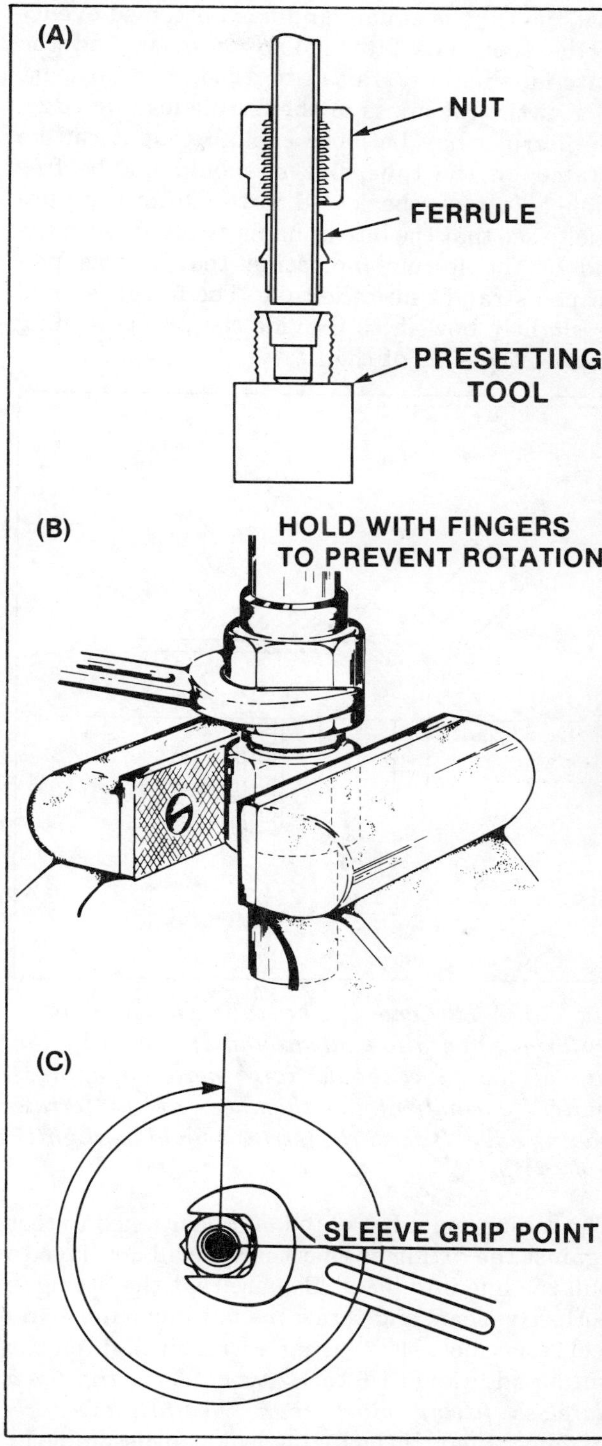

Fig. 3-17A— Install the nut and ferrule on the end of the tubing. B — Seat the end of the tubing firmly against the bottom of the presetting tool. C — Turn the nut down until resistance is felt, and then turn it 1-3/4 turn beyond this point.

Remove the tube from the presetting tool and examine the ferrule. The tube should have a uniform indentation completely around its end, indicating that the tube is square and has bottomed evenly in the tool. The bite should be even, and the material should be raised to a height of at least 50% of the thickness of the ferrule leading edge. The ferrule may be loose enough that it can be rotated on the tube, but it should not be free enough to move back and forth. Rotate the ferrule to see that the bite remains straight, with the end of the ferrule indicating that it was performed straight into the tube. The ferrule should be slightly bowed, to provide the proper sealing against the edge of the fitting.

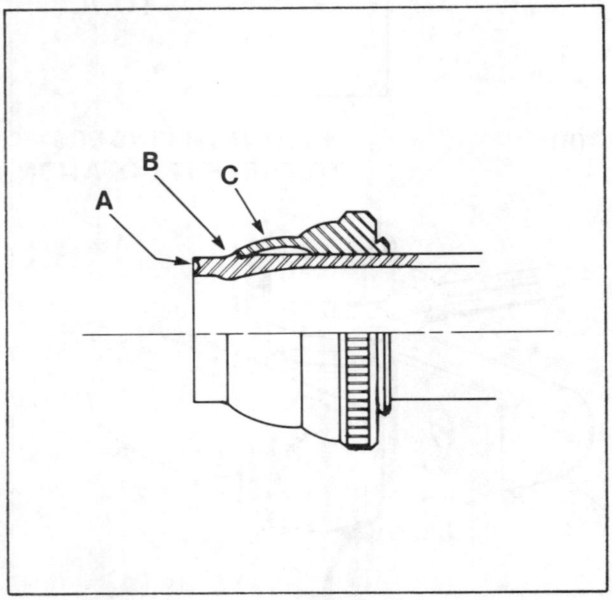

Fig. 3-18 The end of the tube should have a uniform indentation all around its end (A), the bite of the tube should raise material approximately one-half of the thickness of the ferrule leading edge (B), and the ferrule should be slightly bowed (C).

The tubing must be routed and positioned to rest against the fitting so that the nut will not have to pull the line into place. Be sure that the fitting is perfectly clean, and screw the nut down until you feel resistance. Now, using a wrench, tighten the nut an additional 1/6 to 1/3 turn. *Never tighten a flareless fitting more than 1/3 turn after resistance is felt.* If the fitting leaks, disassemble it, and check to be sure that the tube has been properly preformed; then recouple it and properly tighten the nut. If it continues to leak, the end of the tube will have to be cut off and a new ferrule preformed onto the tube.

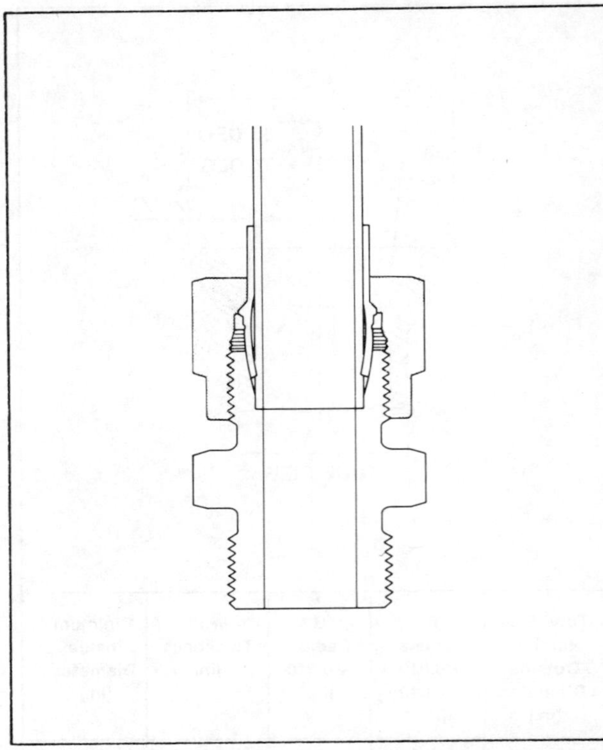

Fig. 3-19 A properly connected flareless tube fitting

3. Beading and hose clamps

Large-diameter lines carrying such low-pressure fluids as engine return oil and cooling air may be joined with a rubber hose slipped over the ends of the tube and held in place with screw-type hose clamps.

Before slipping the hose over the end of the tube, the tube must be beaded by using either a power beader or a hand beading tool, such as the one seen in Fig. 3-20*A*.

After the hose is slipped over the beads, center the hose clamps between the ends of the hose and the beads. Tighten the clamps finger-tight and then one and one-half to two complete turns with a wrench or a pair of pliers. Tightening the hose clamps too tight will cause "cold flow" of the hose. This is a condition in which the rubber is squeezed so tight that it loses it resilience, and its sealing qualities are impaired.

IV. Flexible Fluid Lines

When there is relative motion between the components that are joined by a fluid line, a flexible hose must be used.

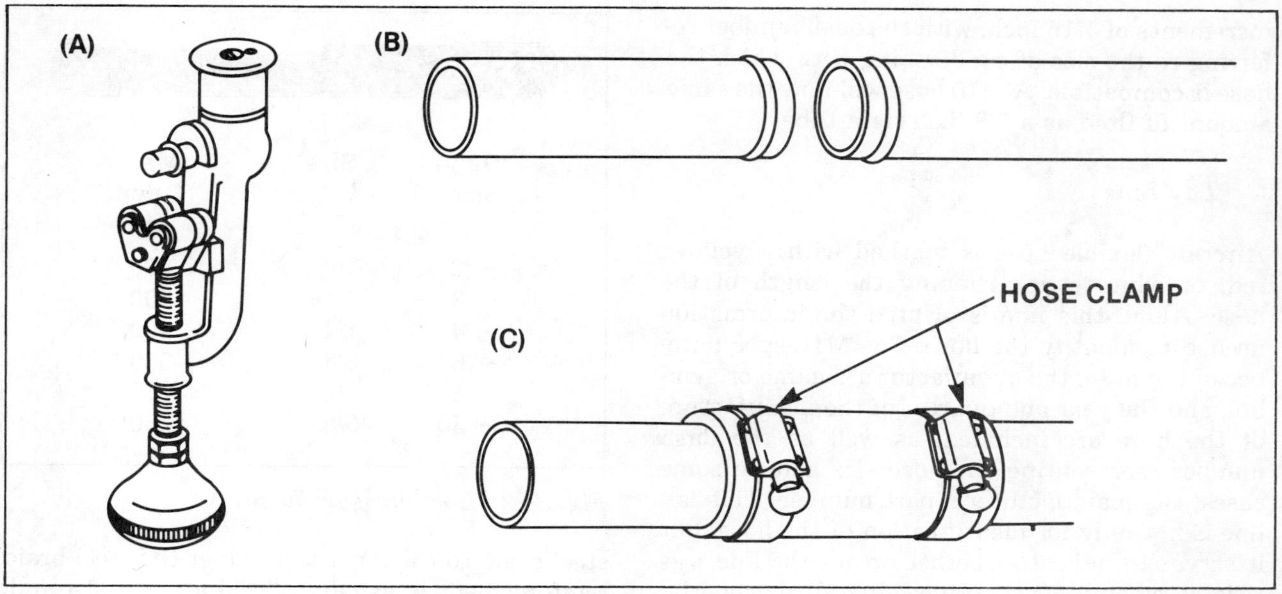

Fig. 3-20 A — Hand-type tube beading tool. B — The ends of the tube are beaded before the hose is slipped over them. C — The hose clamps are centered between the end of the hose and the bead.

A. Flexible Hose Construction

1. Inner liner

The inner tube, or the liner, of a flexible hose carries the fluid, and it must have a minimum porosity and must be chemically compatible with the material being carried. It must be smooth so that it will offer the minimum resistance, or opposition, to the flow, and it must be flexible throughout the entire range of operating temperatures it will encounter.

There are basically four different synthetic compounds used for the inner liner of aircraft fluid lines. Neoprene is a form of synthetic rubber that has a good resistance to abrasion and can be used for many fluid line applications, but its resistance to petroleum-based oils is not as good as some of the other compounds.

Buna-N has a better resistance to petroleum products than neoprene and is used for applications in which petroleum fluids are carried.

Butyl is not suitable for use with petroleum products, but it is excellent as an inner tube for fluid lines carrying phosphate esterbase hydraulic fluids such as Skydrol.

Teflon is a synthetic resin compound by DuPont that can be used as an inner tube for fluid lines. It has an extremely wide usable temperature range, extending from −65° F to as high as +450° F. Teflon's unique wax-like surface has a minimum opposition to fluid flow, and it has a very long service life if it is properly used.

2. Reinforcement

The strength of a hose is determined by the reinforcement which is placed over the inner tube. Reinforcements are made of cotton, rayon, or polyester fabric, or of either carbon-steel wire or stainless steel wire braid.

The design of the reinforcement is quite critical, as the hose has a tendency to increase its diameter and decrease its length when pressure is applied. The proper design of the reinforcement can minimize these dimension changes.

3. Outer cover

A protective outer cover, usually made of fabric, rubber, or stainless steel braid, is put over the reinforcement to protect the hose from such physical damage as abrasion or scuffing.

B. Measurement of Flexible Fluid Lines

The size of a *rigid* fluid line is measured by its *outside* diameter, but the size of a flexible line is determined by its *inside* diameter, measured in increments of 1/16 inch, with the dash number referring to the size of rigid tubing with which the hose is compatible. A −10 hose will flow the same amount of fluid as a 5/8 inch rigid tube.

C. Lay Line

Aircraft flexible hose is marked with a yellow, red, or white stripe running the length of the hose. Along this line is printed the information needed to identify the hose. The MIL-spec number of the hose, the manufacturer's name or symbol, and the year and quarter of the manufacture of the hose are included, as well as the dash number representing the hose size and, in some cases, the manufacturer's part number. This lay line is not only for identification of the hose, but it serves to indicate whether or not the line was twisted when it was installed. Installed properly, this line will run straight and not spiral around the hose.

Fig. 3-21 The lay line printed along a flexible hose provides the needed information about the hose and serves to show whether or not the hose has been twisted when it was installed.

D. Types of Flexible Fluid Lines

1. Low-pressure hose

Low-pressure air or vacuum hoses and the lines to some aircraft instruments are not required to carry high pressures, and they may be of the low-pressure type. These hoses have a seamless inner tube and a reinforcement made of a single layer of cotton braid. An outer cover of ribbed or smooth rubber is used to protect the reinforcement from physical abrasion. The maximum operating pressure of these hoses varies with their size and is shown in Fig. 3-22.

2. Medium-pressure hose

Medium-pressure hose is generally called 1,500 psi tubing, but its maximum operating pressure varies with its diameter. It has a seamless inner tube and a cotton braid impregnated with a rubber-type compound to securely bond the single

Dash Size	Size I.D.	Max. Oper. psi
− 2	1/8	600
− 3	3/16	500
− 4	1/4	400
− 6	3/8	300
− 8	1/2	250
− 10	5/8	250

Fig. 3-22 Low-pressure hose

steel braid to the inner tube. Over the steel braid reinforcement is usually placed a braid of rough oil-resistant cotton.

If the hose is to be used with any of the petroleum products, its inner liner is made of synthetic rubber and its outer braid is gray-black. But, if the hose is to be used with Skydrol or any of the phosphate-ester synthetic hydraulic fluids, the inner liner is made of synthetic Butyl rubber and the outer braid is colored green. The lay line has the word SKYDROL on it, along with the other information.

Dash Size	Size I.D.	Max. Oper. psi
− 3	1/8	3,000
− 4	3/16	3,000
− 5	1/4	3,000
− 6	5/16	2,000
− 8	13/32	2,000
− 10	1/2	1,750
− 12	5/8	1,500
− 16	7/8	800
− 20	1-1/8	600
− 24	1-3/8	500
− 32	1-13/16	350
− 40	2-3/8	250
− 48	3	200

Fig. 3-23 Medium-pressure hose

3. *High-pressure hose*

Two or more steel braids are used for high-pressure hose, which has a maximum pressure rating of 3,000 psi, and the outer cover on high-pressure hose is smooth, to distinguish it from medium-pressure hose. A synthetic rubber inner liner is used with hose made to carry petroleum products, and the outer cover on these hoses is black and has a yellow lay line. A hose designed to carry Skydrol hydraulic fluid has a Butyl rubber inner liner and a green outer cover. The lay line for this hose is white.

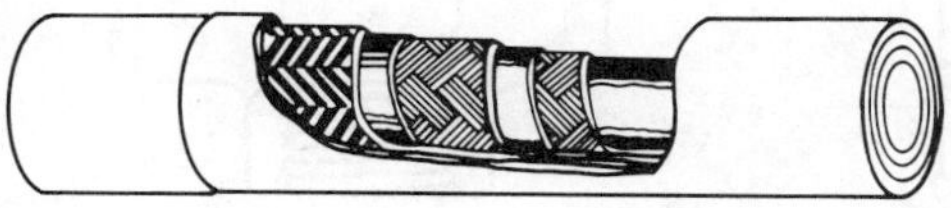

Dash Size	Size I.D.	Max. Oper. psi
−4	7/32	3,000
−6	11/32	3,000
−8	7/16	3,000
−10	9/16	3,000
−12	11/16	3,000
−16	7/8	3,000

Fig. 3-24 High-pressure hose

4. *Hose of Teflon*

The synthetic resin produced by DuPont and known as Teflon has many advantages for use in aircraft fluid lines. Teflon is unaffected by fuels, lubricating oils, coolants, or solvents that are used in aircraft systems, and it also finds extensive use in missiles to carry the fuels and oxidizers.

A seamless extruded inner liner of Teflon resin is covered with a stainless steel braid for medium-pressure applications, and with either two braids or with a double spiral wrap of stainless steel wire for high-pressure lines.

It is important the minimum bend radius restrictions be observed when Teflon hose is installed. After the hose has been in service, it will "take a set," which means that it will become somewhat rigid, and if the hose is removed from the aircraft, it must not be bent against this set that it has taken.

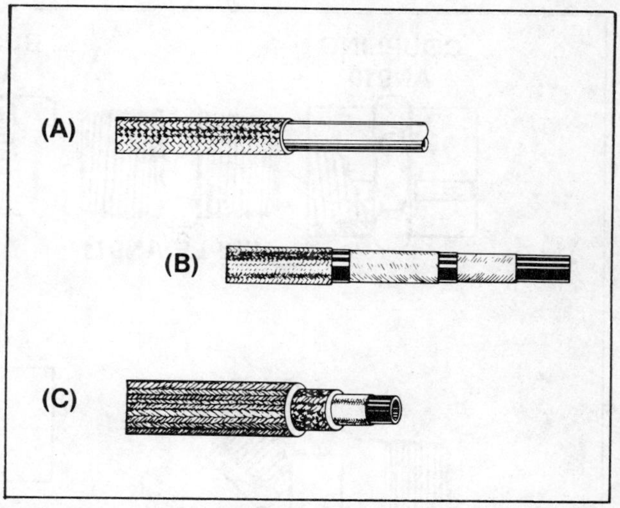

Fig. 3-25 A — Medium-pressure hose of Teflon B — High-pressure hose of Teflon, with two layers of spiral wrap stainless steel wire wound in opposite directions. A stainless steel braid encases the entire tube. C — A medium-pressure hose of Teflon protected from heat and abrasion by a two-layer asbestos loom fireshield.

V. *Fluid Line Fittings*

A. *Attachment to Components*

There are two basic ways in which fluid line fittings may be attached to a component; by tapered pipe threads and by universal bulkhead fitting.

1. *Tapered pipe threads*

Tapered pipe threads seal by the wedging action between the tapered male thread and the tapered

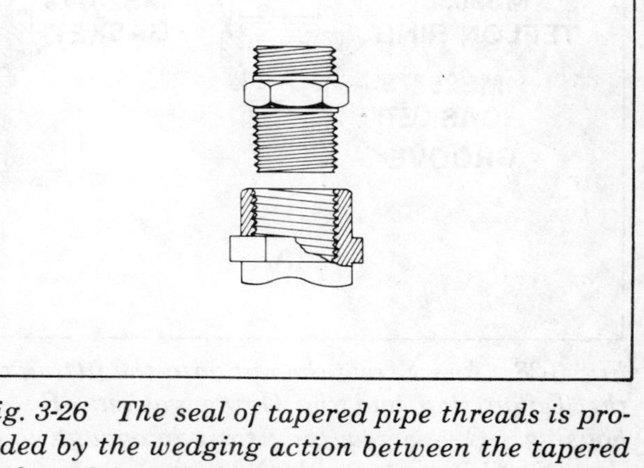

Fig. 3-26 The seal of tapered pipe threads is provided by the wedging action between the tapered male and female threads.

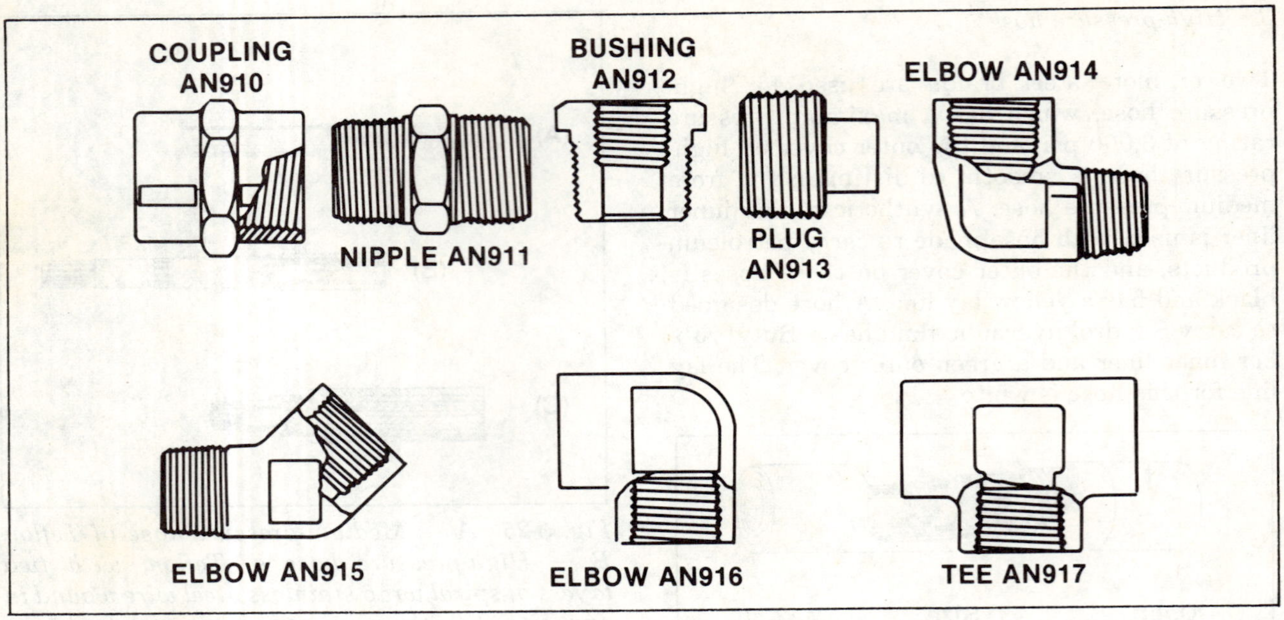

Fig. 3-27 Typical tapered pipe thread fittings. The size of a pipe fitting is measured by the **inside** *diameter of an iron pipe having the same threads used on the fitting.*

female threads. These threads taper 1/16 inch to the inch. Either a small amount of thread lubricant is placed on the male thread, or these threads are wrapped with a thin strip of Teflon tape to assure that the fittings will screw together tight enough to form a complete seal. Care must be taken when screwing tapered fittings into cast aluminum or magnesium housings that the fitting is not screwed tight enough to crack the casting.

2. *Universal bulkhead fitting*

These fittings have the advantage that they can be oriented in exactly the correct direction and then tightened in this position so they will not leak.

Before installing a fitting of this type, an AN6289 nut must be screwed onto the fitting so that it comes up just above the groove cut into

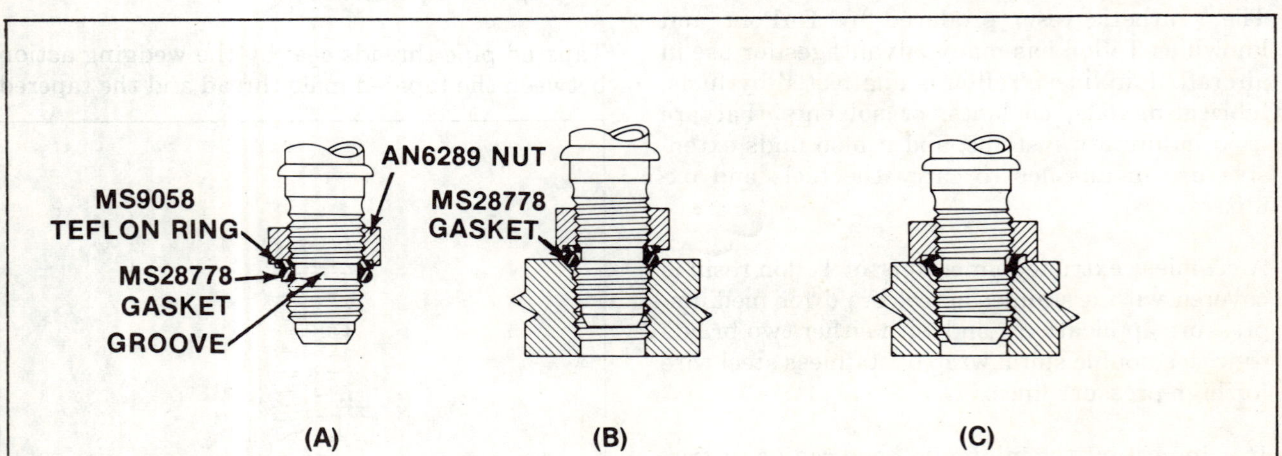

Fig. 3-28 A — Screw the nut onto the fitting until it is above the cutout portion of the threads. Slip on the Teflon ring and the O-ring gasket. B — Screw the fitting down until the O-ring contacts the housing. C — Screw the fitting in enough to align it with the hose. Hold the fitting and turn the nut down until it contacts the housing.

the threads. An MS9058 Teflon ring and an MS28778 O-ring gasket of the correct size are carefully slipped over the lower threads into the groove. Now, rotate the fitting and the nut together into the housing until the O-ring just contacts the face of the housing. Hold the nut with a wrench and turn the fitting in for 1-1/2 turns, to center the O-ring in the groove, and then use up to one more full turn to align the fitting with the hose that is to attach to it. When the fitting is aligned, hold it and tighten the nut with a wrench until the nut contacts the housing. This will force the O-ring down into the chamfered hole and will make a fluid-tight seal.

B. Connection of Fluid Lines

1. Flared fittings

Automotive flare-type fittings have a 45-degree flare cone angle, and it is relatively easy to distinguish these fittings from aircraft fittings which use a 37-degree flare cone. There are, however, two types of aircraft fittings, both of which have essentially the same flare cone angle, but unfortunately these fittings are not interchangeable. The older AC fitting was used extensively through World War II, and some of these fittings may still be found. They look a great deal like the more recent AN fittings, except that the threads go all the way to the flare cone on the AC fitting, while on the AN fitting there is a shoulder between the end of the threads and the cone. Aluminum alloy fittings of the AN design are dyed blue, while AC fittings are dyed either gray or yellow.

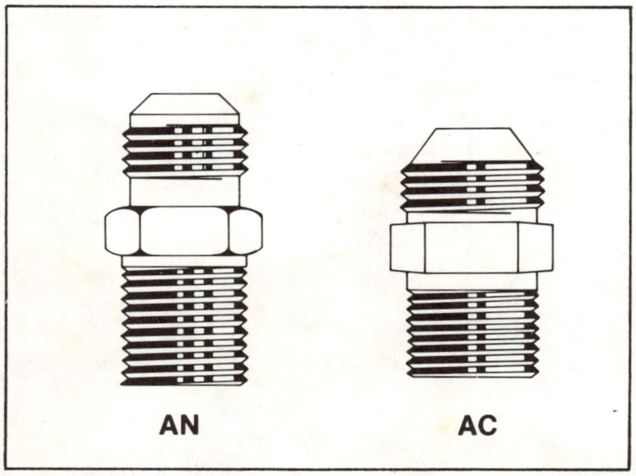

Fig. 3-29 The threads on an AC fitting extend all the way to the flare cone while there is a step between the end of the threads and the flare cone on an AN fitting.

In Fig. 3-30, we see some typical elbows, tees, crosses, and unions used for connecting tubing with flare-type fittings. The female tee and elbows accept the bulkhead-type union or elbow (AN832 and AN833). A fluid-tight seal is accomplished with these fittings by using an MS28778 O-ring gasket forced into the chamfered hole in the female tee or elbow.

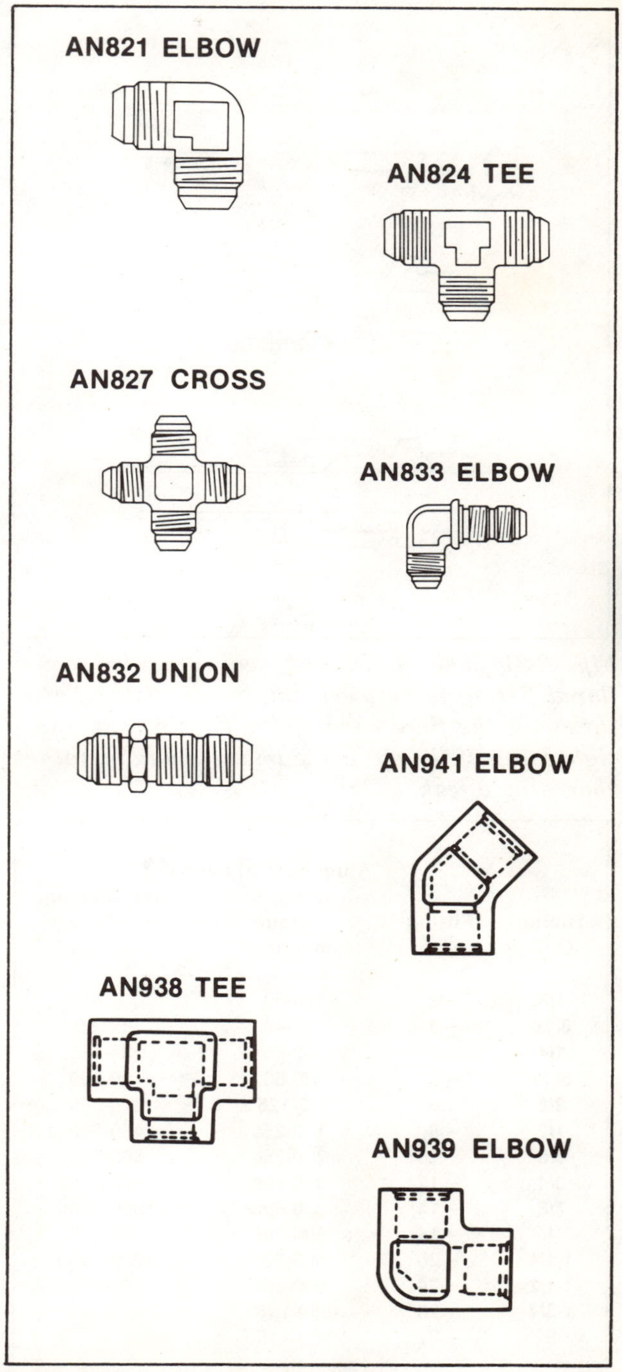

Fig. 3-30 Typical AN fittings for connecting flared tubing

The seal for a flared fitting is provided by the flare cone of the fitting and the inside of the flare of the tube. Excessive tightening of the nut can seriously distort the flare and intensify a leak, rather than stop it. The recommended torque for flare-type fittings is found in Fig. 3-32 and these values should not be exceeded.

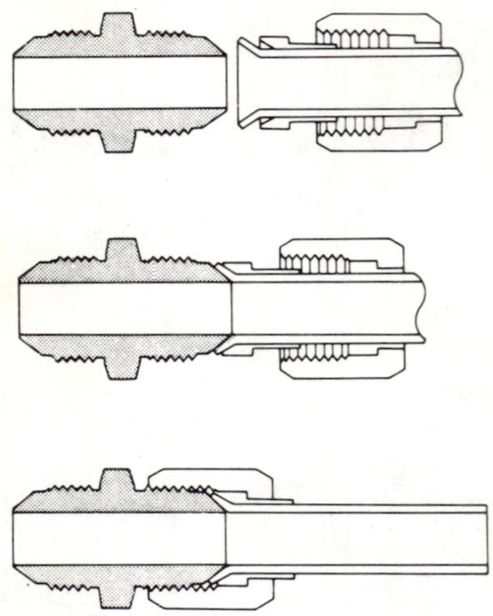

Fig. 3-31 The fluid-tight seal provided by a flared fitting is between the inside of the flared tube and the flare cone of the fitting. Excessive tightening of the nut can damage the flare, rather than stop a leak.

Tubing O.D.	Fitting Size	Aluminum Alloy Tubing, Nut Torque inch-lbs.	Steel Tubing, Nut, Torque inch-lbs.
1/8	−2	20-30	
3/16	−3	30-40	90-100
1/4	−4	40-65	135-150
5/16	−5	60-85	180-200
3/8	−6	75-125	270-300
1/2	−8	150-250	450-500
5/8	−10	200-350	650-700
3/4	−12	300-500	900-1000
7/8	−14	500-600	1000-1100
1	−16	500-700	1200-1400
1-1/4	−20	600-900	1200-1400
1-1/2	−24	600-900	1500-1800
1-3/4	−28	850-1050	
2	−32	950-1150	

Fig. 3-32 Recommended torque for flare-type fittings

2. *Flareless fittings*

Flareless fittings are available in a wide variety of shapes and styles, and in Fig. 3-33 we see some typical fittings.

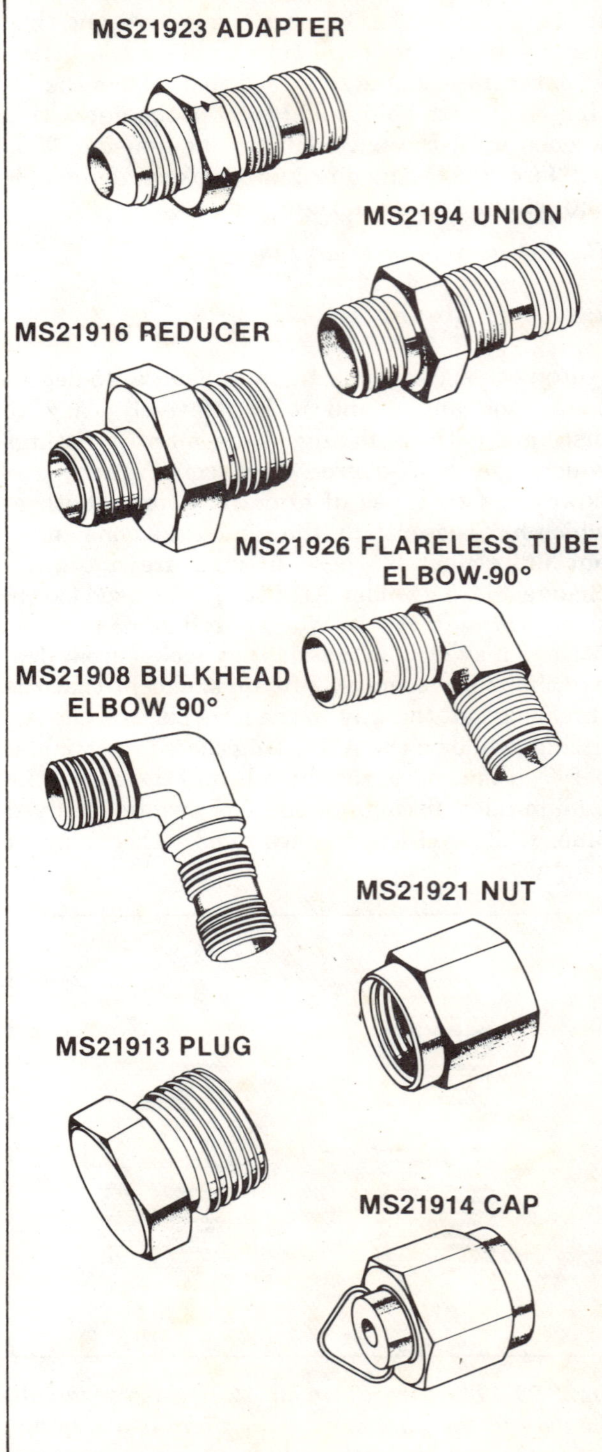

Fig. 3-33 Typical MS-type flareless fittings

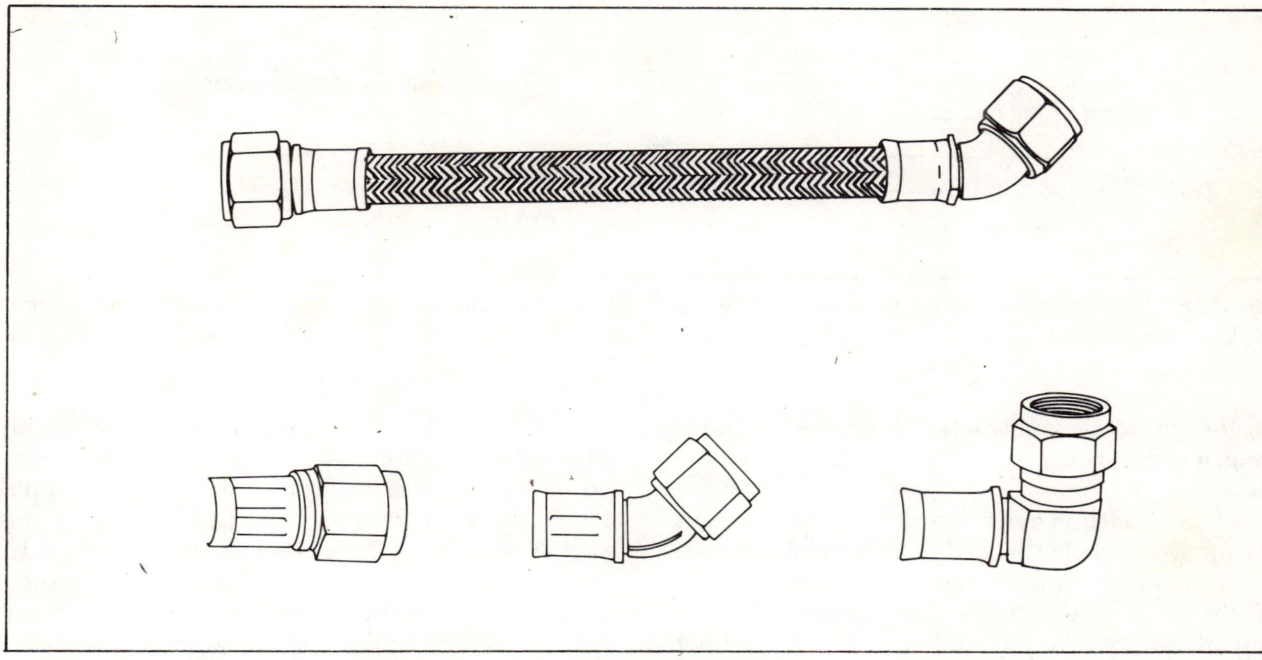

Fig. 3-34 Flexible hose having swaged-on end fittings

3. Fittings for flexible hose

a. Swaged-on fittings

There are two basic types of fittings available for flexible hose. Swaged-on fittings are installed at the facility that manufactures the hose, and these fittings cannot be removed and reused. If the hose is replaced, the entire hose assembly must be scrapped and a new hose, complete with end fittings, must be purchased and installed.

b. Intermediate-pressure hose

Reusable fittings normally have three pieces — a socket, a nipple, and a nut. The socket is screwed into the end of the hose by turning it to the left. This left-hand thread drives the socket tight onto the tube as the nipple is screwed into it.

A nipple, around which the nut is placed, screws into the socket and forces the tube tight between the outside of the nipple and the inside of the socket. This squeezing action provides a good strong physical attachment between the hose and the fitting, and forms a leak-proof seal.

The nut is held onto the hose by the nipple and is free to turn. The inside of the nipple forms the sealing surface which mates with the flare cone of the fitting, and the nut pulls these two sealing surfaces together.

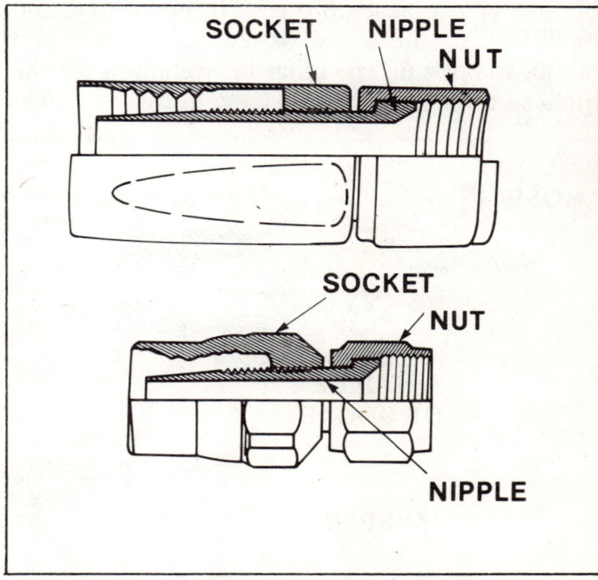

Fig. 3-35 The reusable end fitting for an intermediate pressure hose is made up of a socket, a nipple, and a nut.

The A&P may, himself, fabricate a flexible hose with reusable fittings. To do this, the length of the hose must first be determined. The length between the ends of the fittings must be found and between five and eight percent added to this distance for changes in dimensions of the tube with pressure variations, and for expansion and contraction of the aircraft. Twice the distance between the end of the nipple and the inside of the

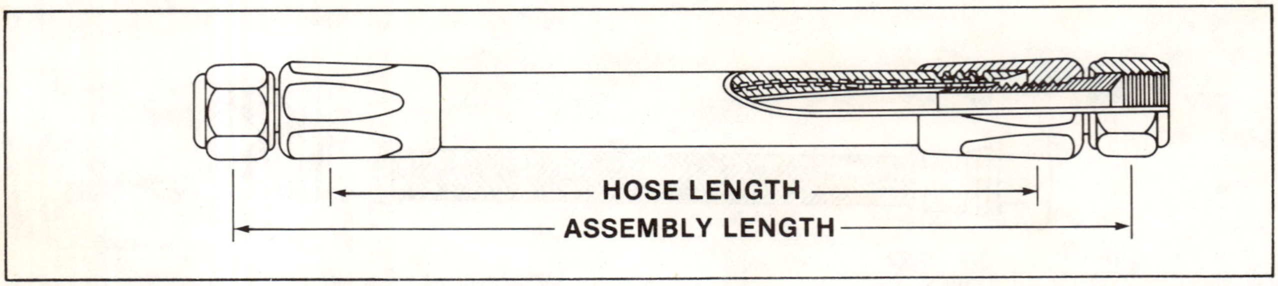

Fig. 3-36 The length required for a hose is the distance between the fittings, plus the required 5% to 8% of this length for slack, minus twice the distance between the end of the nipple and the end of the hole in the socket.

socket must be subtracted from the required length of the hose.

The hose length is marked and both ends are cut off square. A socket is held in a vise and the hose is screwed into it until it bottoms against the end of the hole. The hose is then backed off just enough to prevent the rubber in the inner liner obstructing the hole in the hose.

The end of the assembly tool is lubricated and forced into the hose and worked in and out and around, to open up the inner liner enough for the nipple to force its way in.

The nipple and nut are assembled on the assembly tool and tightened so the nipple can be screwed into the socket by turning the nut. The nipple is then screwed in until the nut bottoms on the socket, and then it is backed off from 1/32 to 1/16 inch to allow the nut to turn freely on the tube assembly. When the assembly tool is screwed out of the hose, the inside of the hose should be blown out with compressed air and the entire hose inspected as to its physical condition. The hose is then hydrostatically tested to a pressure that is normally twice the working pressure of the hose, and this pressure is held for between one and five minutes.

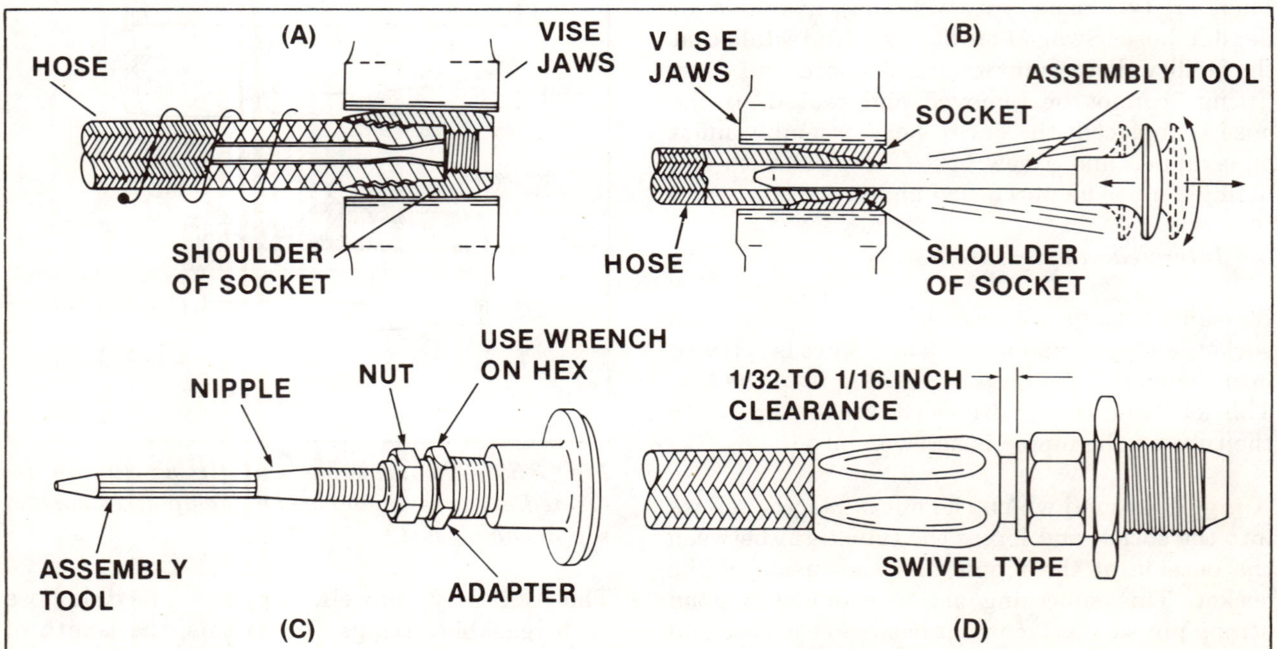

Fig. 3-37 A— Screw the end of the hose into the left-hand threads of the socket. Leave a small amount of clearance between the end of the hose and the end of the hole in the socket. B — Lubricate the assembly tool and work it into the end of the hose to open out the end and force the tube tight into the socket. C — Screw the nipple and nut onto the assembly tool and force the nipple into the socket. Screw the nut down tight against the socket. D — Back the nut off about 1/32 to 1/16 inch and remove the assembly tool from the tube.

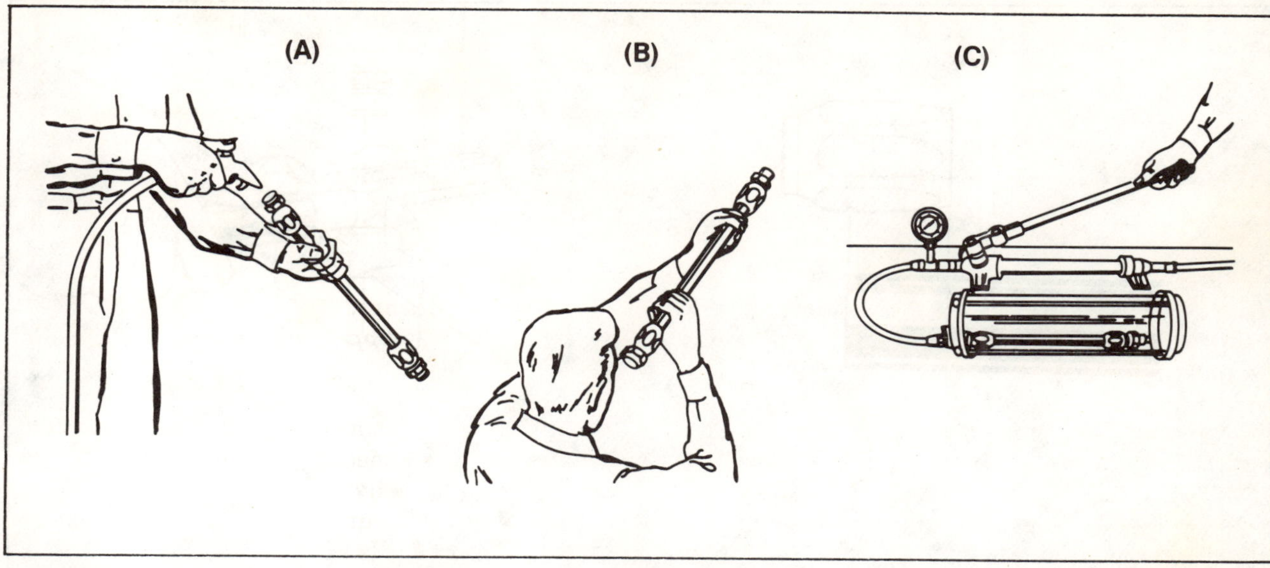

Fig. 3-38 A — Blow out the completed hose assembly with compressed air. B — Examine the hose for any foreign matter inside of the tube. It is possible for the nipple to shave off thin slices of the inner liner. C — Hydrostatically test the hose assembly to twice its normal operating pressure, and hold this pressure for between one and five minutes.

c. *Low-pressure hose*

Reusable end fittings may be assembled onto low-pressure hose in much the same way as is done with high-pressure hose, but the soft rubber inner liner makes the use of an assembly tool unnecessary. A fitting can be screwed into the nut to hold the nipple, and the socket is slipped over the end of the hose rather than being screwed on. When the nut is tightened against the socket, it should be backed off just enough that the nut can turn freely.

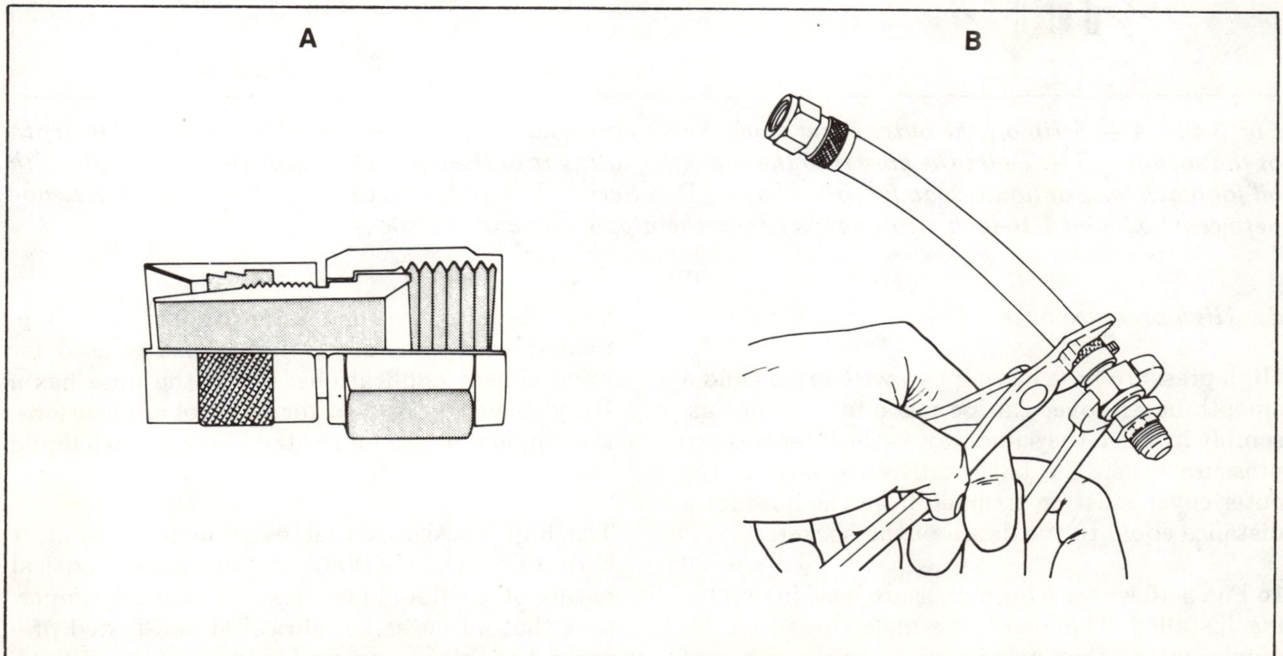

Fig. 3-39 A — Low-pressure hose end fittings slip over the end of the hose, rather than screw on as the intermediate- and high-pressure hoses do. B — A fitting screwed into the nut will hold the nipple so it can be screwed into the socket.

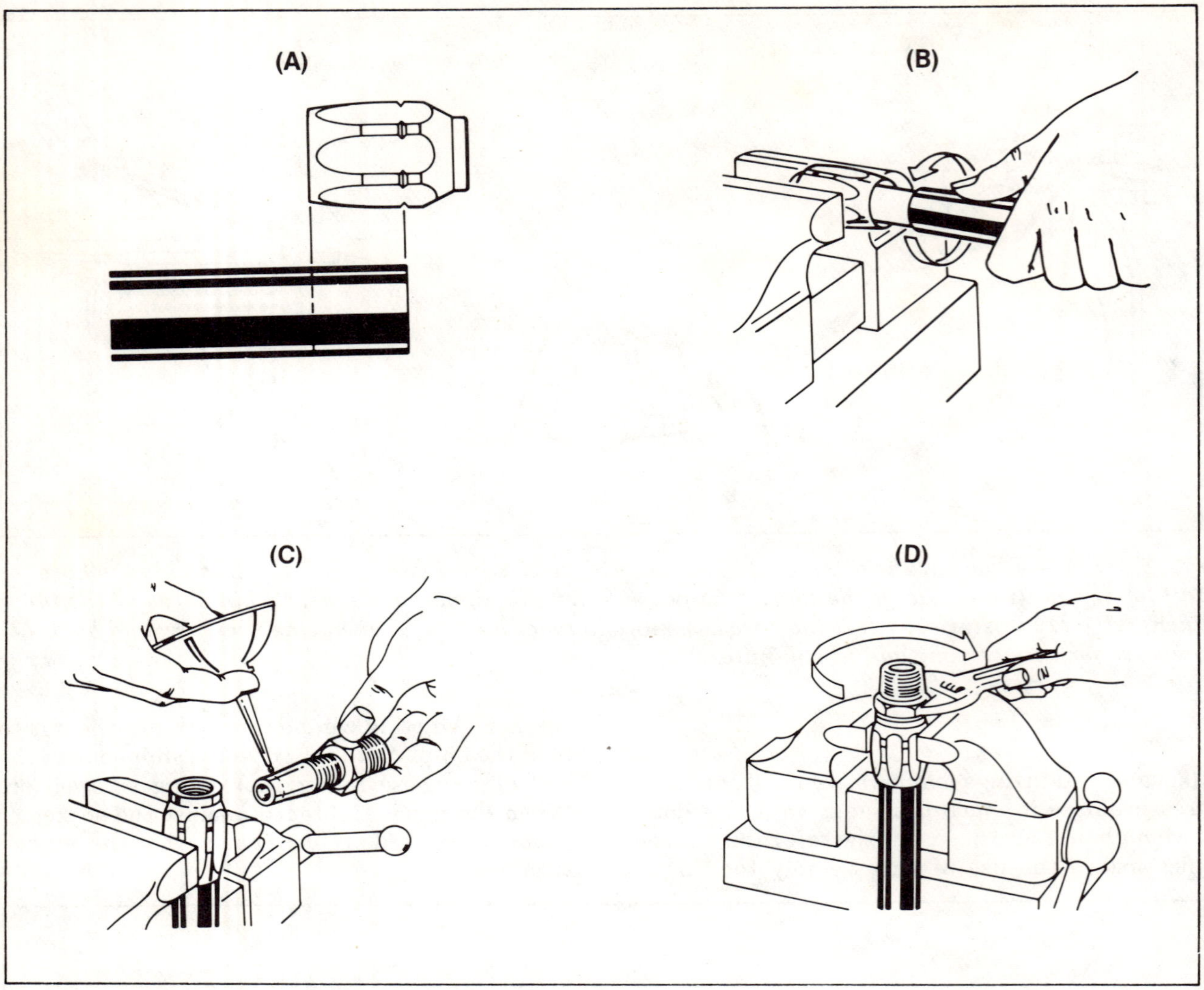

Fig. 3-40 A — Strip off the outer cover from the high-pressure hose a distance equal to the inside depth of the socket. B — Twist the hose into the socket, rotating it to the left. C — Lubricate the nipple with oil for black hose or liquid soap for green hose. D — Screw the nipple into the socket and the hose. Leave between 1/32- and 1/16-inch clearance between the nipple hex and the socket.

d. *High-pressure hose*

High-pressure hose having two wire braids and a smooth outer cover can be made into a hose assembly in much the same way as the intermediate pressure hose. The basic difference is that the outer cover must be stripped from the hose for a distance equal to the depth of the socket.

In Fig. 3-40 we see a high-pressure hose fitting being installed. This one has a male thread on the nipple, rather than provisions to hold a nut, and so an assembly tool is not needed. The nipple is lubricated and screwed directly into the socket and into the hose.

The nipple or the installation tool may be lubricated with mineral oil if the hose is used for mineral-base applications, but if the hose has a Butyl liner and is used for Skydrol applications, the nipple must be lubricated liberally with liquid soap.

The high pressures that exist in most aircraft hydraulic systems and the extremely critical nature of the fluid lines make it extremely important that all hoses be fabricated and tested properly. For this reason, and because of the difficulty in assembling some of the larger hoses, it is common practice for all replacement hoses to be purchased from the factory which produced the aircraft or from approved repair stations that are equipped to fabricate and test these critical lines.

VI. Fluid Line Installation

A. Routing and Securing

All fluid lines should be routed through the aircraft in such a way that they will have the shortest practical length. But for the sake of appearance and ease of attachment, all of the lines should follow the structural members of the aircraft and should be secured with the appropriate clamps to the structure. In Fig. 3-41 we have the approximate spacing of the supports required for rigid fluid lines.

TUBE O.D.	DISTANCE BETWEEN SUPPORTS
1/8 - 3/16	9"
1/4 - 5/16	12"
3/8 - 1/2	16"
5/8 - 3/4	22"
1 - 1-1/4	30"
1-1/2 - 2	40"

Fig. 3-41 Spacing between supports for rigid aircraft tubing

It is important that there be no possibility of the line chafing against any control cable or portion of the aircraft structure, and it must not be allowed to come in contact with electrical wiring bundles or conduit carrying electrical wires. If it is impossible to separate fluid lines from electrical wire bundles, the wire bundle must be routed above the fluid line, and it must be clamped securely to the structure to prevent it contacting the fluid line. Under no circumstances should a wire bundle be supported by a fuel line or any line carrying flammable fluid.

All bends must be located in such a way that they can be supported by clamps, so that the stress from vibration and from expansion and contraction is not placed on the fitting. All runs of tubing must have at least one bend between the fittings to absorb these strains.

All lines carrying fuel must be bonded to the structure by using integrally bonded line support clamps.

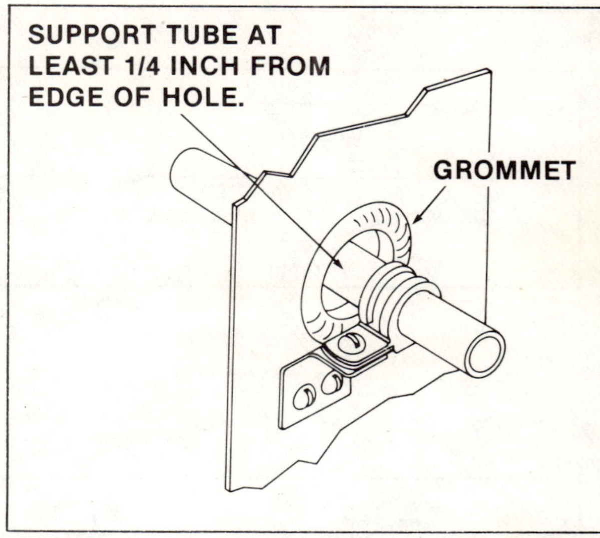

Fig. 3-42 Support all fluid lines where they pass through a hole in the structure.

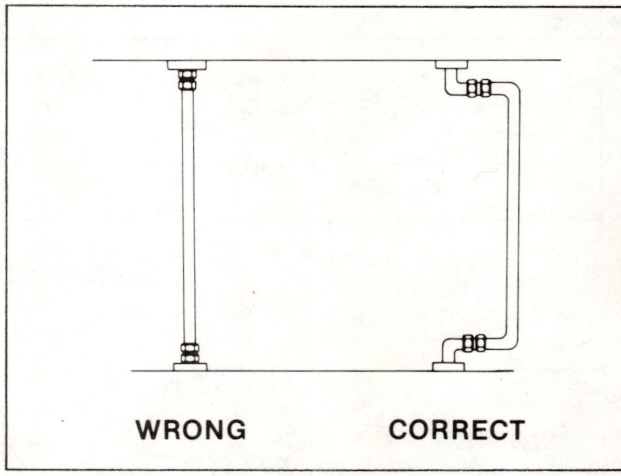

Fig. 3-43 All tubing should have at least one bend to absorb the strains from vibration and from the dimensional changes of the aircraft structure due to temperature changes.

Flexible fluid lines must have between five and eight percent slack, to allow for the change in dimensions of the hose caused by the pressure and for vibration and expansion of the airframe. It is important that the lay line be straight when the hose is installed. Any spiraling of the lay line is an indication that the hose was twisted when it was installed, and it will be under an undue amount of strain when there is pressure in the line.

If a fluid line must pass near a hot exhaust manifold, the line must be protected by a suitable fire shield. For this purpose an asbestos braid may be fastened around the hose.

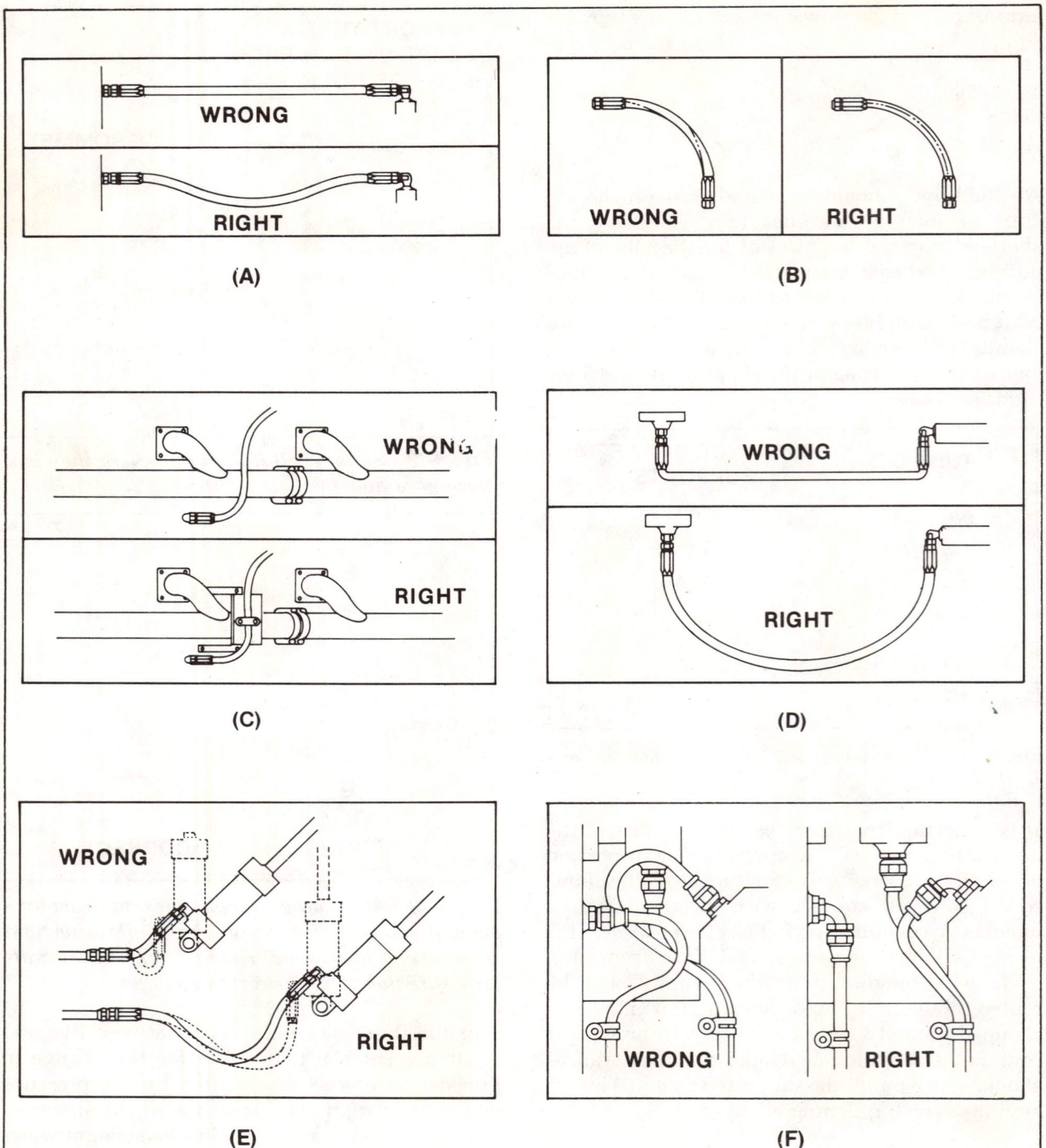

Fig. 3-44 A — A flexible hose must not be stretched tight between two fittings. It must have between five and eight percent slack. B — The lay line must run straight along the hose, indicating that the hose was not twisted when it was installed. C — When a flexible line is routed near a hot exhaust manifold it must be secured to the aircraft structure so that it cannot touch the hot metal. D — Be sure to observe the minimum bend radius when installing hose. It would have been better to use a 90° elbow for one of these fittings. E — Be sure to allow sufficient slack when attaching a hose to a movable cylinder. F — Always use the proper fittings to prevent unwieldy lengths of hose.

It is important that the movement of actuating cylinders be considered when attaching their fluid lines. There must be no kinking of the lines at either end of the travel of the cylinders or in any intermediate position.

The minimum bend radius must be observed when installing flexible line to be sure that the hose is not excessively strained.

If a fluid line must be routed through a passenger or crew compartment or through a baggage compartment, it must be installed in such a way that it cannot be used as a hand hold. Where lines pass through a wheel well, they must be protected from abrasion from rocks and dirt that may be thrown up into the well when the airplane is on the ground.

DASH NUMBER	MINIMUM BEND RADIUS
– 4	3"
– 5	3-3/8"
– 6	5"
– 8	5-3/4"
– 10	6-1/2"
– 12	7-3/4"
– 16	9-5/8"

MINIMUM BEND RADIUS FOR INTERMEDIATE-PRESSURE FLEXIBLE HOSE

Fig. 3-45 Minimum bend radii for intermediate-pressure flexible fluid lines

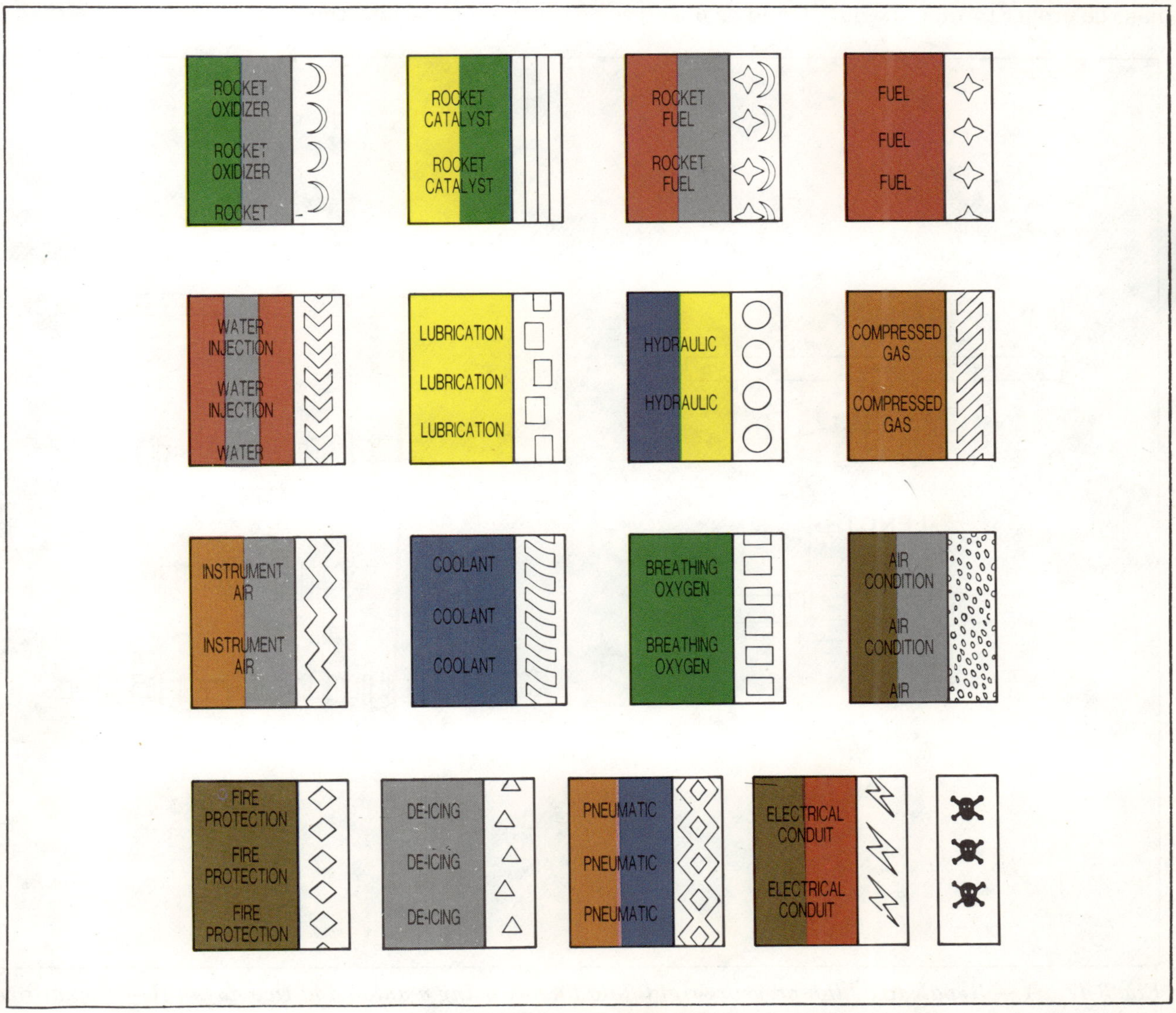

Fig. 3-46 Color codes for fluid lines

All fluid lines should be clearly marked with a color code to identify their contents. In Fig. 3-46 we see the standard color codes for fluid lines. Each line, along with the colored stripes, is marked with a series of symbols that allows a color-blind technician to identify the lines. In addition to these codes, it is proper for some of the lines carrying fuel to be marked with the word FLAM to identify the line as carrying a flammable fluid. Lines carrying materials that are physically dangerous may be marked with the word PHDAN.

B. Servicing

If a fluid line leaks at a fitting, all of the pressure should be removed from the line and the fitting checked for the proper torque. It is never proper to overtorque a fitting in the attempt to stop a leak. If a fitting leaks after it is assured that it has the proper torque, the nut should be unscrewed and the sealing surfaces carefully examined. If there is any damage, the fitting will have to be cut off, and if there is enough line to install a new fitting, this must be done. If the line is not long enough to allow this type of repair, the line will either have to be replaced or the line cut off and a new section of line spliced on, using the appropriate unions and nuts.

If there is minor damage to a piece of high-pressure tubing, the damage may be cut out and a union and the correct nuts installed on the tube.

If there are scratches in a piece of rigid tubing that do not exceed ten percent of the tube wall thickness, and they are not in the heel of the bend, they may be burnished out by using a smooth tool to move the metal raised up beside the scratch back into the depression.

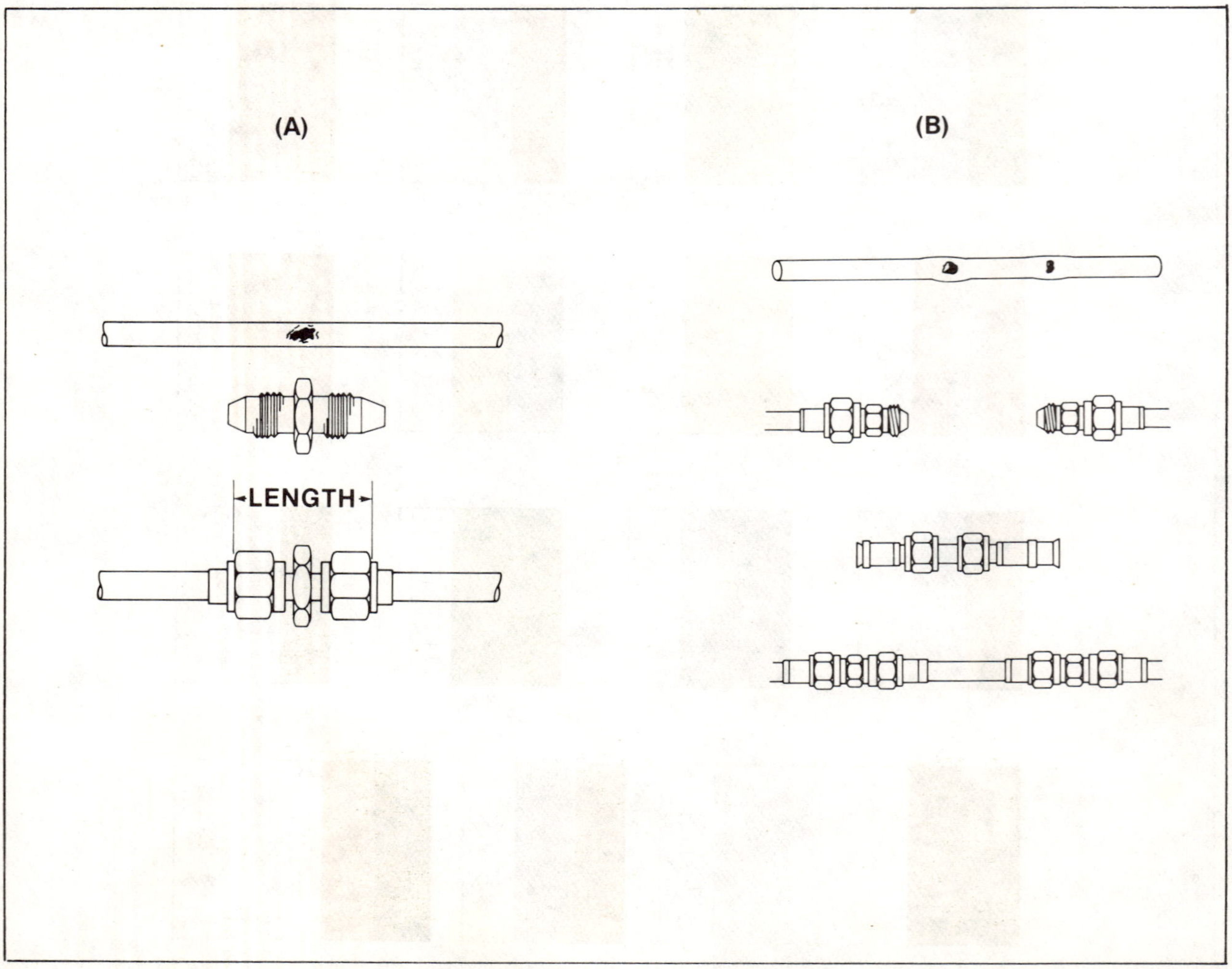

Fig. 3-47 A — Repair to a high-pressure rigid fluid line by using a union and two nuts B — Repair to a high-pressure fluid line by splicing in a new section of line

AVIATION
TECHNICIAN
INTEGRATED
TRAINING
PROGRAM

EA-ITP-G5

Aircraft Hardware
Hand Tools and Measuring Devices

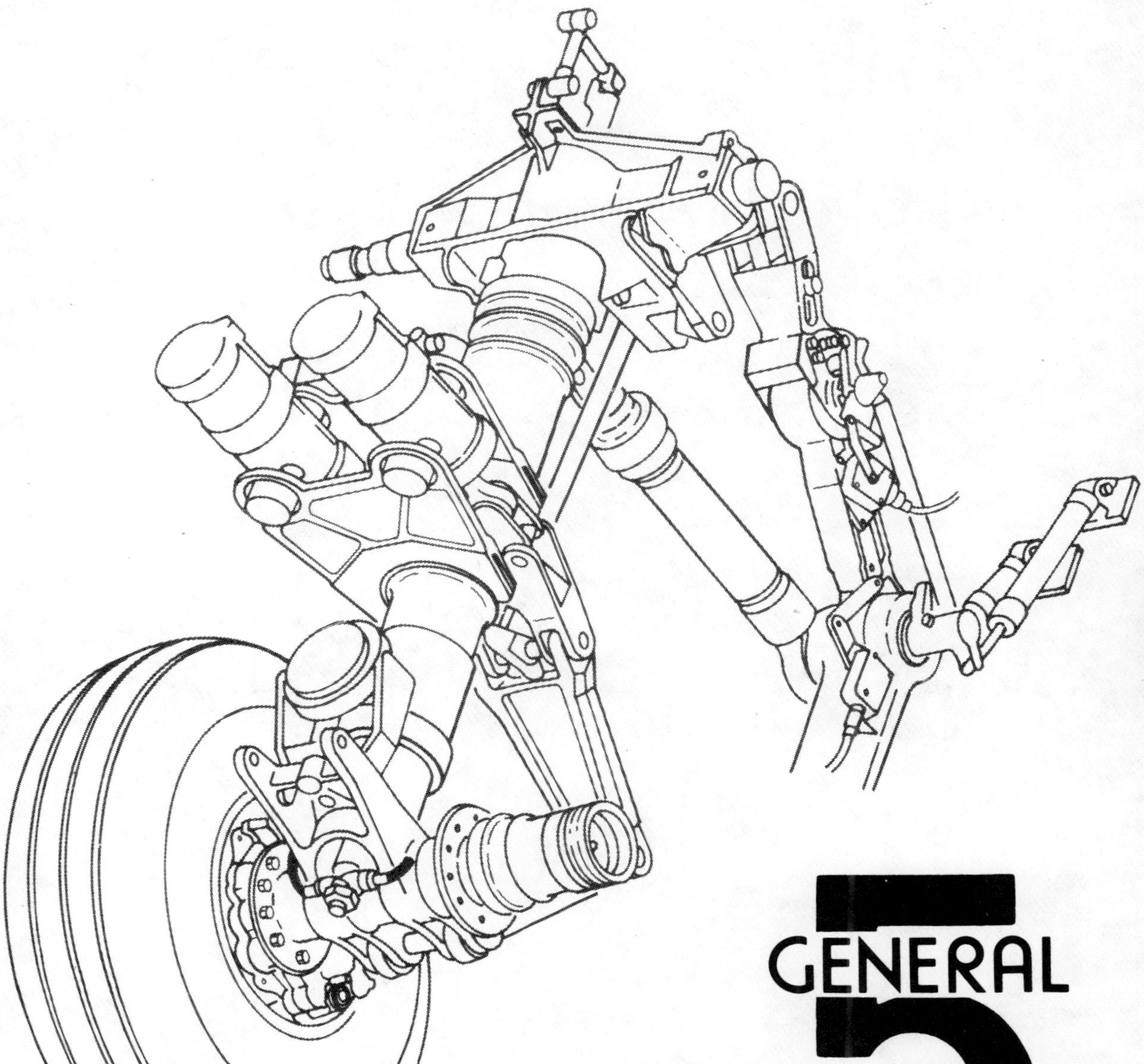

GENERAL
5
SECTION

EA-ITP-G5

TRAINING MANUAL
GENERAL SECTION
BOOK 5

AIRCRAFT HARDWARE
HAND TOOLS AND MEASURING DEVICES

by
Dale Crane

AVIATION MAINTENANCE PUBLISHERS, INC.

INTEGRATED TRAINING PROGRAM
FOR THE
AVIATION MAINTENANCE TECHNICIAN

For Sale by: Aviation Maintenance Publishers, Inc.
P.O. Box 890 Basin, Wyoming 82410
Tel: (307) 568-2413

International Standard Book Number 0-89100-128-X

Aviation Maintenance Publishers, Inc.
211 South Fourth Street, Basin, Wyoming 82410

Printed in the United States of America

TABLE OF CONTENTS

Preface

Part 65 of the Federal Aviation Regulations require that an applicant for a Mechanic Certificate with an Airframe or Powerplant rating have a general knowledge of the construction and maintenance of the aircraft covered by the rating he seeks. And each applicant must prove this knowledge by passing a written, an oral and a practical examination, each given in five sections: General, Airframe Structures, Airframe Systems and Components, Powerplant Theory and Maintenance, and Powerplant Systems and Components.

Certificated Aviation Maintenance Technician Schools all operate under individual Federal Aviation Administration approvals, and all have in their approved curriculum enough information to satisfy the FAA that anyone completing their course will meet the knowledge requirements for certification as a mechanic.

In 1972, the Federal Aviation Administration published a series of three Advisory Circulars providing background information for mechanic certification applicants. These are entitled: *Airframe and Powerplant Mechanics General Handbook, AC 65-9; Airframe and Powerplant Mechanics Powerplant Handbook, AC 65-12;* and *Airframe and Powerplant Mechanics Airframe Handbook, AC 65-15.* These three handbooks were written to cover the area over which a mechanic applicant was to be tested, and they were an attempt by the FAA to coordinate the information available to the applicants with the tests that were given for certification. In 1976, the three handbooks were revised into their present A revision format: AC 65-9A, AC 65-12A, and AC 65-15A.

Individuals attempting to certificate under FAR Part 65, as well as schools operating under FAR Part 147, have found that these handbooks actually do not follow the material used in the FAA examinations, and because of this, plus their lack of study helps, they fail to effectively provide the broad base of knowledge needed for certification and for a foundation on which to build in the aviation maintenance industry.

Realizing the need for updated information, presented in a format that is in keeping with the latest developments in educational technology, the Aviation Maintenance Publishers, Inc., is offering an Integrated Training Program (ITP) that presents the basic information needed for mechanic certification, in a form that will provide a good foundation in aviation maintenance knowledge. Developed in coordination with Crane Aeronautical Activities, Inc., the ITP training manuals, workbooks, study guides, and audio-visuals are designed to follow the FAA written examinations and are written to the level required for the FAR Part 147 schools.

All of the material can be used by an individual preparing for certification on his own, or by schools preparing groups of individuals for certification under FAR Part 65, but it has been prepared primarily for the Aviation Maintenance Technician Schools operating under FAR Part 147. Not only has every attempt been made to meet the requirements of Part 147 in producing this series, but the authors have followed many of the recommendations of the Aviation Mechanic Occupation Study conducted by the University of California, Los Angeles, under a contract from the U.S. Department of Health, Education, and Welfare.

One of the most basic premises of aviation maintenance is that all operations must be carried out according to FAA-approved data, and all information of a specific nature furnished by the manufacturer of an aircraft, powerplant, or component must be followed in detail. These training manuals are, of necessity, general in nature and cover a wide scope of aircraft, powerplants, and components. Because of this, it must be understood that they should *never* be allowed to take precedence over specific information furnished by the manufacturer.

The authors and publishers of this Integrated Training Program wish to express our appreciation to the various manufacturers, Aviation Maintenance Technician Schools, and FAA personnel who have aided our efforts to provide this coordinated training material to meet the challenge of better training for aviation maintenance technicians in the eighties.

SECTION I

Aircraft Hardware

I. *Specifications and Standards*

Before an aircraft is issued a Type Certificate by the Federal Aviation Administration, it must demonstrate that it conforms to all of the requirements specified for airworthiness with regards not only to its performance but to its structural strength and integrity as well. To be sure that every individual aircraft that is produced from a certain design will meet the same standards, all of the material from which it is built as well as every piece of hardware that will be used must be clearly described, so that the aircraft can be duplicated in detail.

To avoid the need for writing elaborate specifications for all of the hardware and fasteners, these parts have already been described in standard specifications. There are quite a number of standards used in the aircraft manufacturing industry, but the three most generally used in aircraft construction are the AN (Air Force-Navy) standards, NAS (National Aerospace Standards), and MS (Military Standards). These are the ones an aviation technician will most often encounter.

AMS	Aeronautical Material Specifications
AN	Air Force-Navy
AND	Air Force-Navy Design
AS	Aeronautical Standard
ASA	American Standards Association
ASTM	American Society for Testing and Materials
MS	Military Standard
NAF	Naval Aircraft Factory
NAS	National Aerospace Standard
SAE	Society of Automotive Engineers

Fig. 1-1 Standards used for aircraft materials.

When an MS 20470-AD4-4 rivet is required, specifications have already been written for it and are available to both the aircraft manufacturer and to the manufacturer of the rivet. These specifications stipulate the type of material that is to be used and its strength and hardness, as well as the detailed shape and dimensions of the rivet. By using these specifications and calling for standard hardware, aircraft manufacturers are able to build reproducible aircraft at the lowest cost.

II. *Types of Hardware*

A. *Solid Rivets*

If we were to count the individual pieces of hardware used in an aircraft, we would find that rivets far outnumber all other fasteners in use. These small components are often taken for granted because there are so many of them, and we usually do not notice them unless we have occasion to remove or install one.

1. *Material*

Aircraft rivets are usually made of an alloy of aluminum, and the type of alloy is identified by a letter in the rivet specification and, physically, by a mark on the head of the rivet.

a. *1100 Aluminum (A)*

This is commercially pure aluminum, and a rivet made of this material is too soft and does not have the strength required for structural applications on aircraft. These rivets are designated by the letter A, and there is no identifying mark on

RIVET MATERIAL	CODE	HEAD MARK	HEAT TREAT	SHEAR STRENGTH P.S.I.
1100 ALUMINUM ALLOY	A	NO MARK	NO	10,000
2117 ALUMINUM ALLOY	AD	DIMPLE	NO	30,000
2017 ALUMINUM ALLOY	D	TEAT	YES	34,000
2024 ALUMINUM ALLOY	DD		YES	41,000
5056 ALUMINUM ALLOY	B		NO	27,000
CORROSION RESIST. STEEL	F	NO MARK	NO	75,000
MONEL	M	DOUBLE DIMPLE	NO	55,000

Fig. 1-2 Aircraft rivet identification.

their head. A-type rivets are used only in non-structural parts of an aircraft such as fairings and furnishings inside the cabin.

b. 2117 Aluminum alloy (AD)

There are more rivets made of 2117 aluminum (formerly known as A-17ST) than any other alloy, and these rivets have sufficient strength for structural use. Since they do not need to be heat-treated, as do some of the other types of structural rivets, they can be driven in the condition they are received from the supplier.

AD rivets are identified by a small dimple in the center of their head.

c. 2017 Aluminum alloy (D)

D-rivets are one of the type called icebox rivets because they must be heat-treated before they can be used. They are used for structural applications where their high strength overcomes the inconvenience of their requiring heat treatment. D-rivets are identified by a raised teat in the center of their head.

These rivets are called icebox rivets because of the unique heat-treatment characteristic of this alloy. When an aluminum alloy is heated to a specific temperature and quenched in water or in brine, it will not get hard immediately as steel does, but will remain soft for several hours and will gradually become hard and gain its full strength. This process is called aging. Rivets made of an alloy such as 2017 can have their aging process retarded by removing them from the quench bath and immediately storing them in a sub-freezing icebox until they are ready to drive. When these rivets warm up to room temperature, they will become hard, but they may be re-heat-treated as many times as is necessary without impairing their strength.

d. 2024 Aluminum alloy (DD)

DD-rivets are the strongest of the aluminum alloy rivets and are identified by two raised dashes on their head. Like D-rivets, DD-rivets are

HEAT TREATING TIME FOR ICEBOX RIVETS FOR AIR FURNACE		
ALLOY	TEMPERATURE	TIME
2024	910° - 930°F (488° - 499°C)	1 hour
2017	925° - 950°F (496° - 510°C)	1 hour

Fig. 1-3 Heat treating time for icebox rivets for air furnace.

icebox rivets, and after they have been heat-treated and quenched, they must be stored in a sub-freezing icebox until they are ready to be driven.

e. 5056 Aluminum alloy (B)

Many aircraft parts are made of magnesium, and to avoid a dissimilar metal contact that could lead to corrosion, magnesium structure is riveted with 5056 rivets which contain about five percent magnesium as an alloying agent. These rivets are identified by a raised cross on their head.

f. Corrosion-resistant steel (F)

Rivets of this material are used for fastening corrosion-resistant steel sheets for such applications as firewalls and exhaust shrouds. They do not have any marking on their head.

g. Monel (M)

Monel rivets are identified with two recessed dimples in their head and are used in some instances in place of corrosion-resistant steel rivets, where their somewhat lower shear strength is not a detriment.

2. Head design

The specification number of a rivet identifies its head design. In the past there have been many different head designs that have required the mechanic to have a large assortment of rivet sets on hand. But in the interests of standardization there are now seldom more than four different types of heads found, and these have been further reduced to two that are by far the most generally used. The two most popular head designs are the 100-degree countersunk AN426, or, as it has been reclassified, MS20426. These rivets are used where the reduced wind resistance provided by flush rivets is more important than the increased cost of dimpling or countersinking the skin necessary to accept these rivets.

The AN470 or, as it is now known, the MS20470, is a universal head rivet that is approved for replacing any of the protruding head rivets. The design of the head, which is crowned and relatively flat on top, is strong enough that it can be used in place of the round- or flat-head rivets, and at the same time, its shape does not create an excess of wind resistance when it is used on the outside of an aircraft skin.

The AN430 rivet is a round-head rivet that is used where the increased thickness of the head gives additional tensile strength to a properly installed rivet. The shape of the head, however, is not efficient for use on the outside of an aircraft skin, as it causes too much wind resistance.

An AN442 rivet has a flat head and is used in the original contruction of the internal structure of an aircraft, where the rivet is upset with a production-type squeeze-riveter, or compression-riveter. They are seldom used in repair work, as it is permissable to replace them with MS20470 rivets.

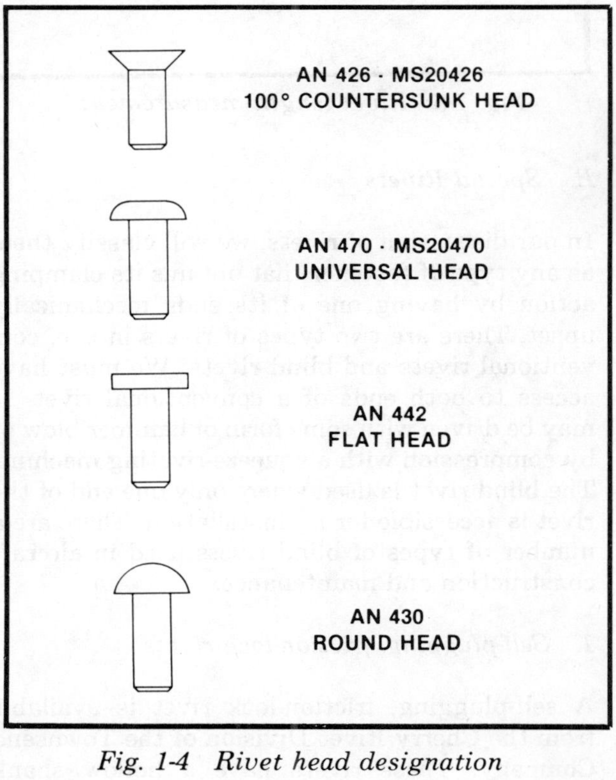

Fig. 1-4 Rivet head designation

3. Size designations

In the code used to describe an aircraft rivet, after the head and the material have been designated, a number is used to indicate the shank diameter in 1/32-inch increments, and the length is noted by a dash number in increments of 1/16-inch.

A rivet designated as an MS20470AD4-4 is a universal head rivet made of 2117 aluminum alloy and is 4/32 (1/8) inch in diameter and 4/16 (1/4) inch long.

The length of a rivet is measured from the part of the head that is flush with the surface of the material being riveted, to the end of the shank. An MS20426 rivet is measured from the *top* of the head to the end of the shank; protruding rivets are measured from the *bottom* of their head to the end of their shanks.

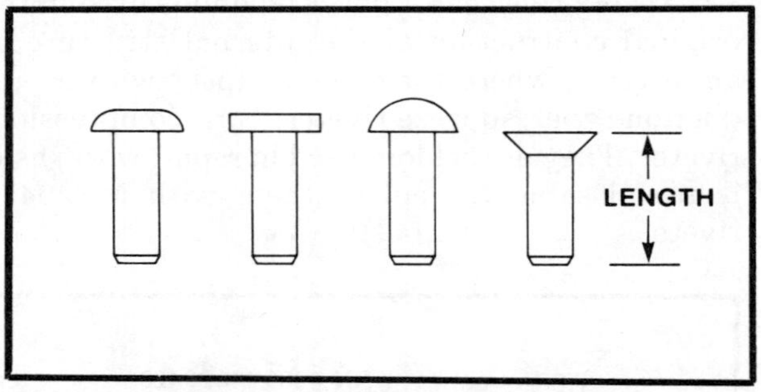

Fig. 1-5 Rivet length measurement

B. Special Rivets

In our discussion of rivets, we will classify them as any type of fastener that obtains its clamping action by having one of its ends mechanically upset. There are two types of rivets in use, conventional rivets and blind rivets. We must have access to both ends of a conventional rivet. It may be driven with some form of hammer blow or by compression with a squeeze-riveting machine. The blind rivet is used where only one end of the rivet is accessible for its installation. There are a number of types of blind rivets used in aircraft construction and maintenance.

1. Self-plugging, friction-lock rivets

A self-plugging, friction-lock rivet is available from the Cherry Rivet Division of the Townsend Company. These rivets have a hollow shank through which a stem with a tapered end protrudes, Fig. 1-6. To install these rivets, drill a hole and remove all of the burrs. Choose a rivet having the correct grip length as determined by a special grip-length gage, Fig. 1-7. Insert the rivet into the hole and, using a special puller, pull the stem

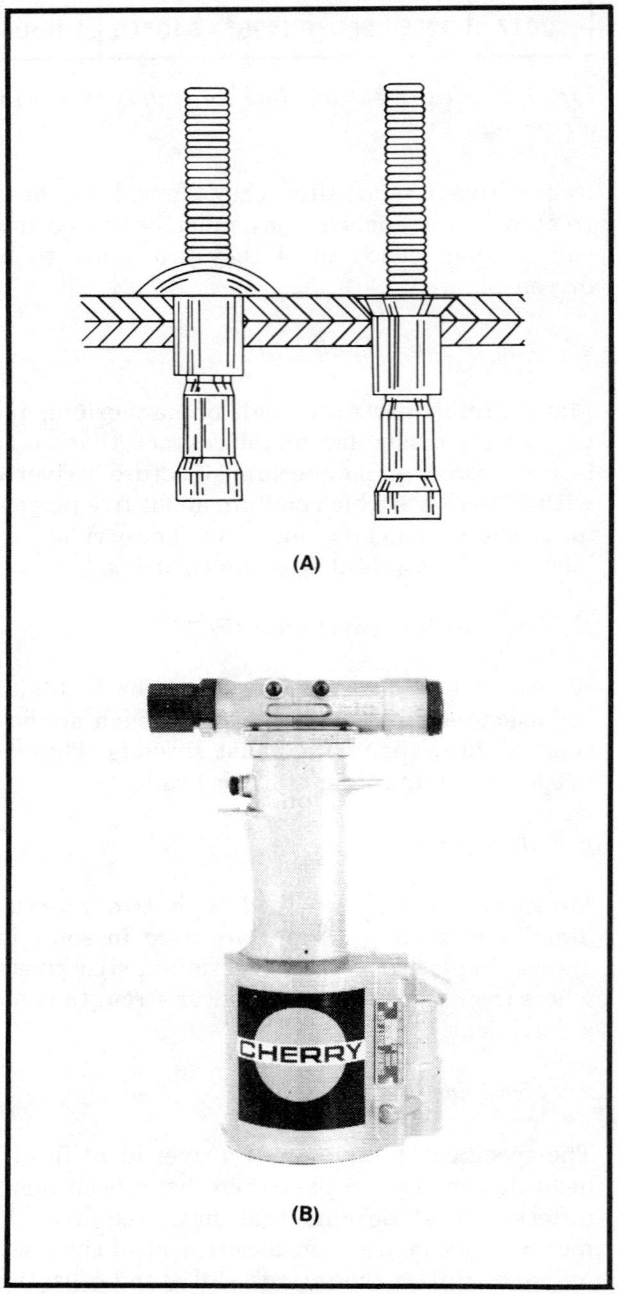

Fig. 1-6 Cherry self-plugging rivets A — Self-plugging rivets may be used to replace solid rivets, but they must generally be one size larger in order to have the required strength. B — Cherry rivets are pulled with either a pneumatic puller, such as this one, or with a hand puller.

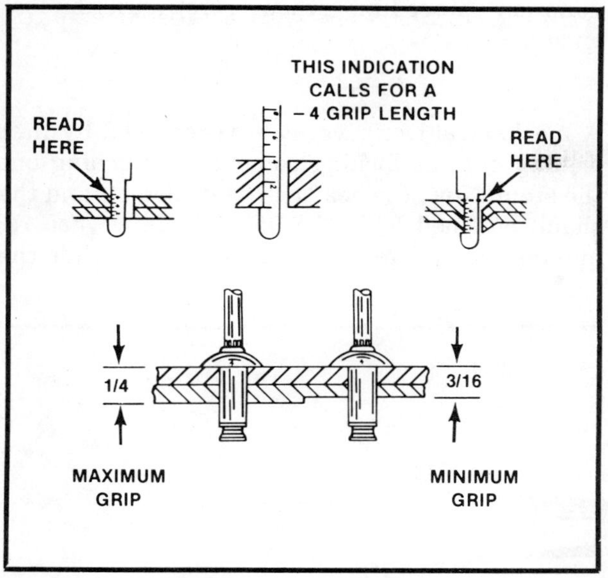

Fig. 1-7 A special gage is used to find the proper grip length for blind rivets. A chart furnished by the manufacturer indicates the maximum and minimum grip length for each dash number of rivet.

through the shank. When the tapered end of the stem upsets the shank against the bottom of the metal, it will break off. Friction holds it tightly wedged inside the shank, and the broken end of the stem can be trimmed smooth with a special trimmer. This trimmer resembles a pair of diagonal cutters except that the cutting edge produces a smooth cut that is flush with the top of the rivet head.

2. *Self-plugging, mechanical-lock rivets*

The bulbed Cherrylock rivet is quite similar to the friction-lock rivets except that a locking collar is swaged into a groove in the stem that locks it in place when the rivet is pulled. In Fig. 1-8 we see the steps in upsetting a bulbed Cherrylock rivet. The hole is prepared as we have previously described, and the correct grip length of rivet is chosen. When the puller is fastened onto the stem and pulled, it first upsets the hollow shank against the bottom of the material and pulls the sheets of metal tightly together. Continued pulling breaks off a shear ring around the end of the

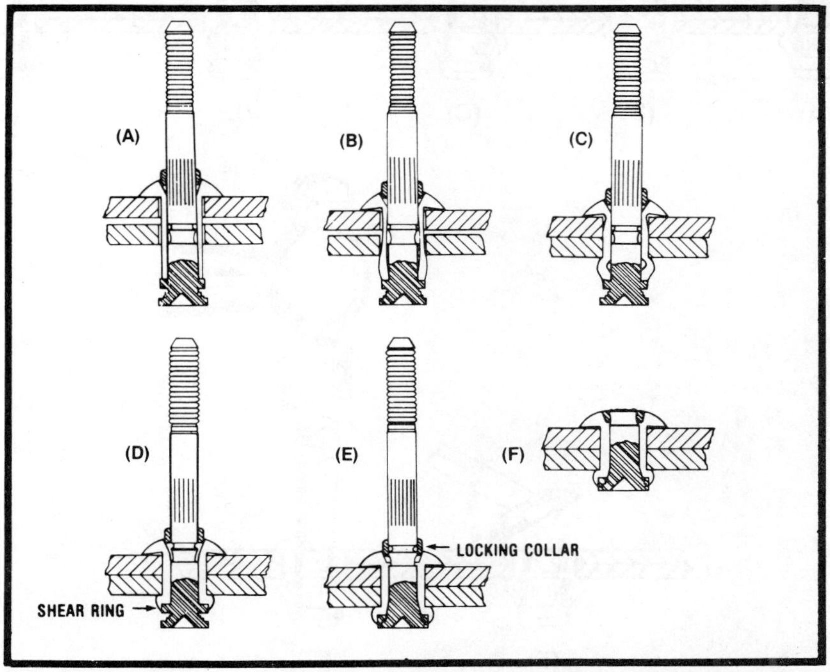

Fig. 1-8 Bulbed Cherrylock rivets are a form of blind rivet which locks the stem in place as it breaks off. A — This type of rivet has a strong clamping action and will pull the sheet gap together. B — As the stem is pulled up through the hollow shank, the clamping action begins. C — As the pulling continues, the upset head is formed and the manufactured head seats firmly against the top sheet. D — When the upset head is fully formed, the shear ring will break away from the stem, allowing the stem to move up into the shank. E — The locking collar can now be forced into its recess by the ram action of the pulling head. F — Further pulling after the locking collar has seated will shear the stem flush with the rivet head.

stem, so the stem can pull up into the shank far enough for a groove around the stem to enter the rivet head, and the locking collar will be swaged into the groove. Further pulling on the stem will break it off just outside the collar. In this type of rivet the stem is firmly locked into the shank and it cannot fall out. Mechanical-lock rivets can normally be used to replace solid-shank rivets without having to use a rivet of a size larger, as is the common procedure when using friction-lock rivets.

A mechanical-lock rivet can be removed by first drilling out the locking collar and punching out the stem. Now, the head can be drilled off and the shank punched form the hole. Be sure when removing this or any other type of rivet that the

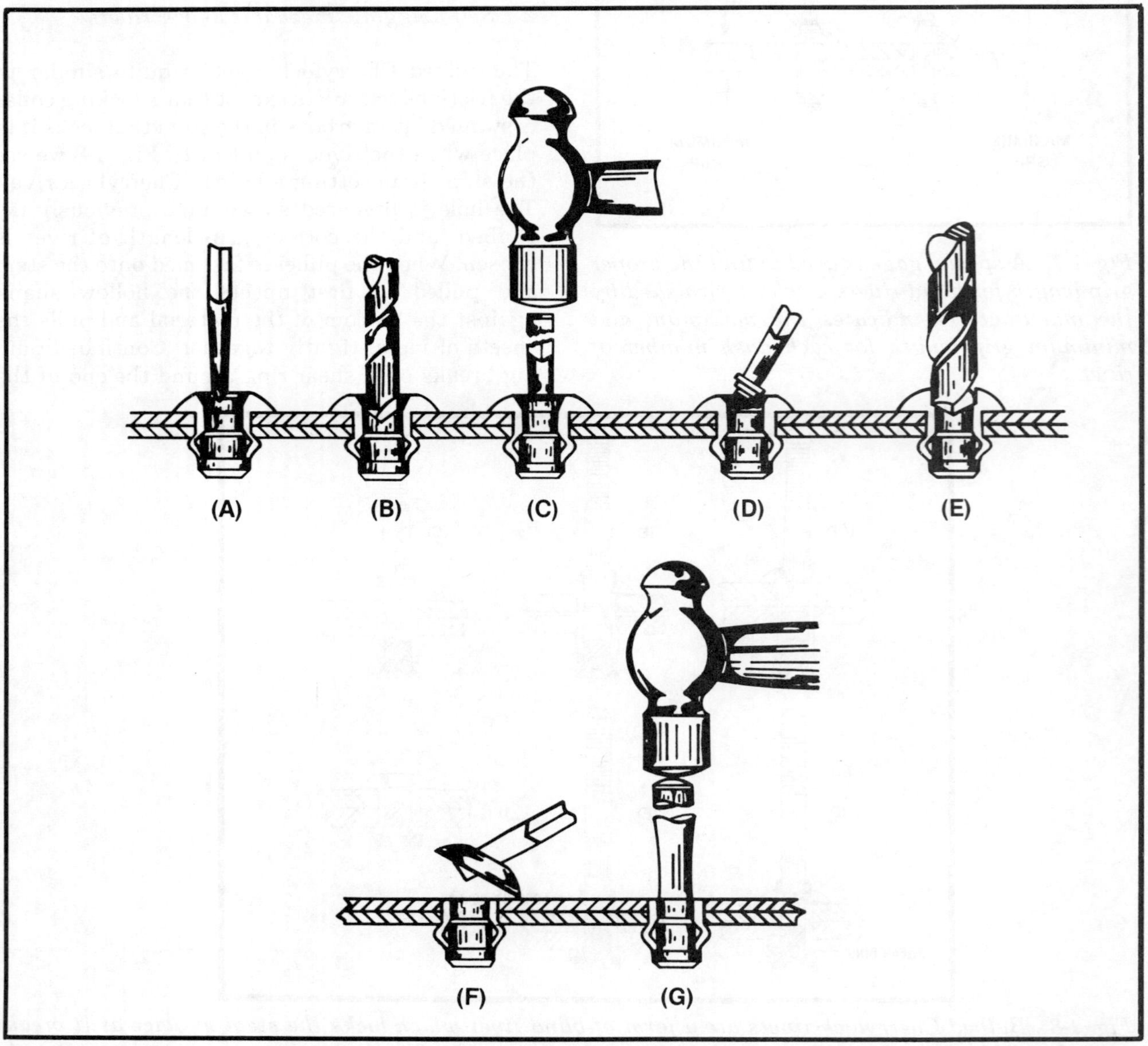

Fig. 1-9 Removing a mechanically locked blind rivet. A — Mark the center of the broken shank with a center punch. B — Using a drill of the same size as the stem, drill through the lock ring. C — Punch the stem out using a small pin punch. D — Lift out the remains of the locking ring. E — Drill almost through the head with a drill slightly smaller than the rivet shank. F — Punch the rivet head off with a cape chisel or lift it off with a pin punch. G — Drive the hollow shank of the rivet from the skin with the proper size pin punch

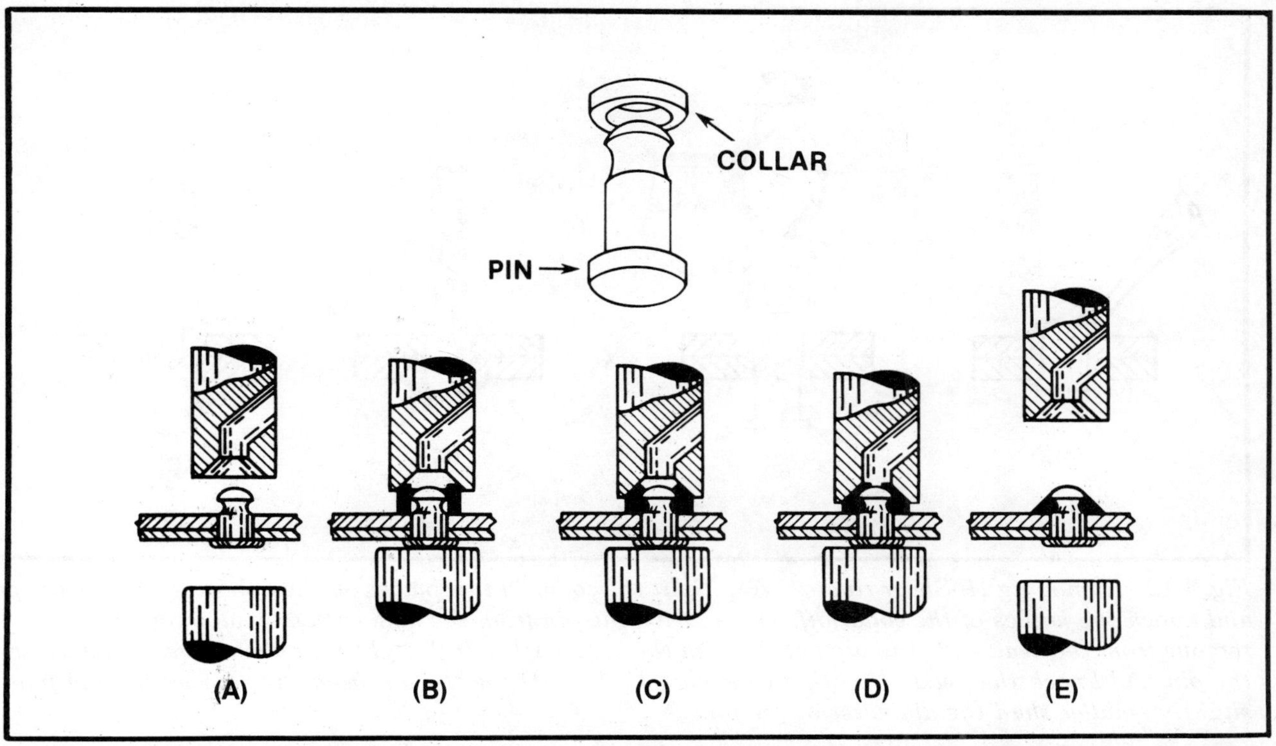

Fig. 1-10 Installation of Hi-Shear rivets. A — Select a pin with the proper grip length and insert it in the hole. Support it with a heavy bucking bar. B — Place the collar over the pin and drive it down with the proper set held in a conventional rivet gun. C — Swage the collar until it is fully seated in the groove in the pin. D — Continue to drive the set until the excess collar shears off. E — When the collar is sheared off there will be only a smooth conical collar gripped into the pin.

skin is solidly supported when the stem is driven out.

3. High-strength pin rivets

There are several types of high-strength pin rivets in common use in aircraft construction. These rivets all have high-strength steel pins that are held in place, not by upsetting the shank, but by having a collar swaged into grooves cut around their shank.

a. Hi-Shear rivets

These patented rivets are used for such applications as landing gear attachment where the high shear strength of a steel bolt is needed, yet where the speed of installation and light weight of a rivet are an advantage.

A steel pin is chosen with the proper grip length for the metal being joined, and it is inserted into the hole in the material that has been drilled and has had all of the burrs removed. An aluminum alloy collar is slipped over the pin and is swaged into the groove.

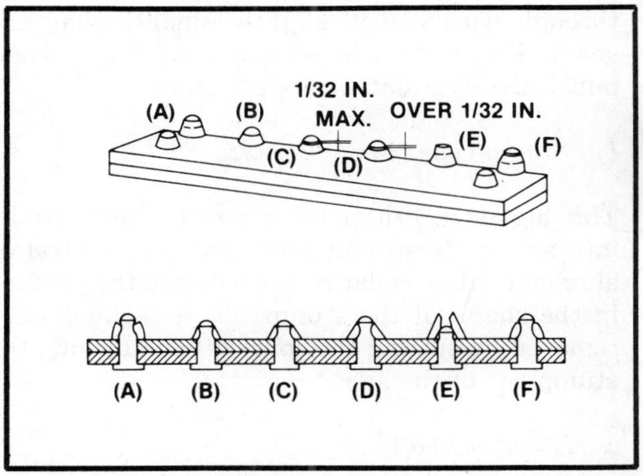

Fig. 1-11 Inspection of Hi-Shear rivet installation. A — Underdriven—continue driving until collar is overdriven 1/32-inch maximum. B — Collar driven to point of cutoff—acceptable. C — Collar overdriven 1/32-inch—correct. D — Collar overdriven more than 1/32-inch—remove and replace. E — Pin too short—remove and replace with next longer size. F — Pin too long—remove and replace with next shorter size.

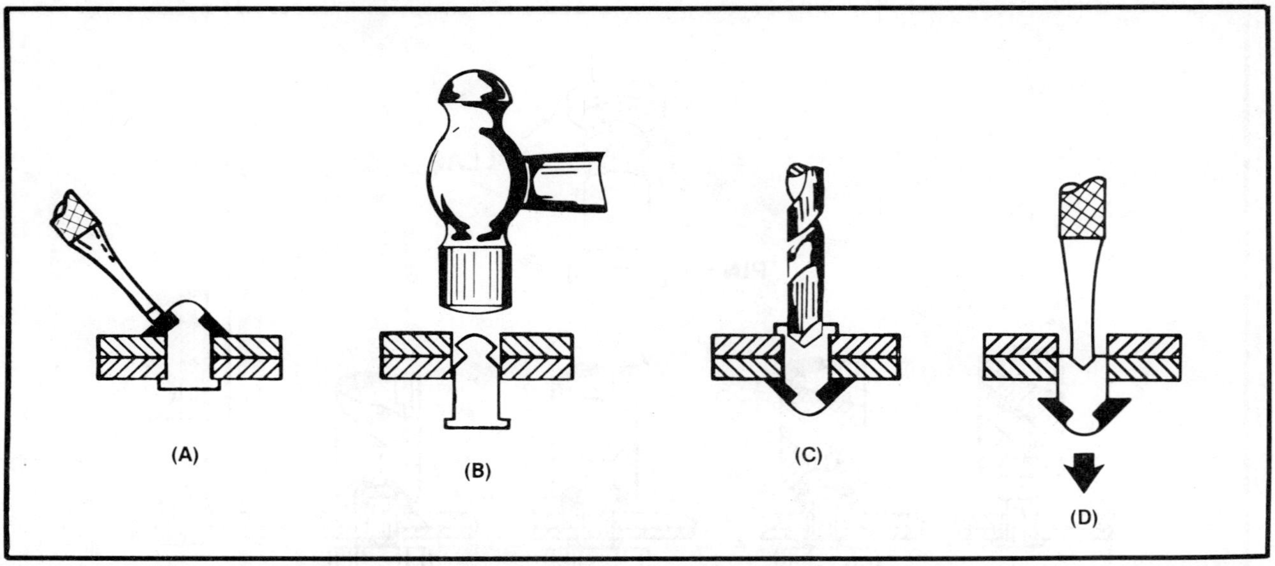

Fig. 1-12 Removing Hi-Shear rivets. A — Split the collar in two places on the opposite sides of the pin and knock the halves of the collar off. B — Drive the pin from the hole with a small hammer. C — To remove from the head end, drill almost through the head with a drill slightly smaller than the diameter of the pin and knock the head off with a cape chisel. D — Drive the pin from the skin with a pin punch slightly smaller than the diameter of the pin.

Hi-Shear rivets may be removed from either side. If the collar is on the side that is accessible, it is cut off with a small, sharp chisel and the pin is driven out of the hole. If the collar side cannot be reached, the manufactured head may be center punched in its exact center and the head drilled through with a drill slightly smaller than the shank. Knock the head off with a small chisel and punch the rivet out with a pin punch.

b. Stump-type Huck-lockbolt

This high-strength fastener uses a stump driven into an interference-fit hole, and a mild steel or aluminum alloy collar is swaged onto the grooves in the shank of the stump. These lockbolts are removed by cutting the collar off and driving the stump out of the hole.

c. Blind lockbolt

A lockbolt may be used to fasten two pieces of material together when there is access to only one side. The hole is prepared in the usual way, and the inner collar of the blind lockbolt is put through the hole and a special puller fastened onto the grooved stem. A femal die forces down on the outer collar as the stem is pulled out. When the stem is pulled, it forces a tapered sleeve into the inner collar, which spreads it out against the

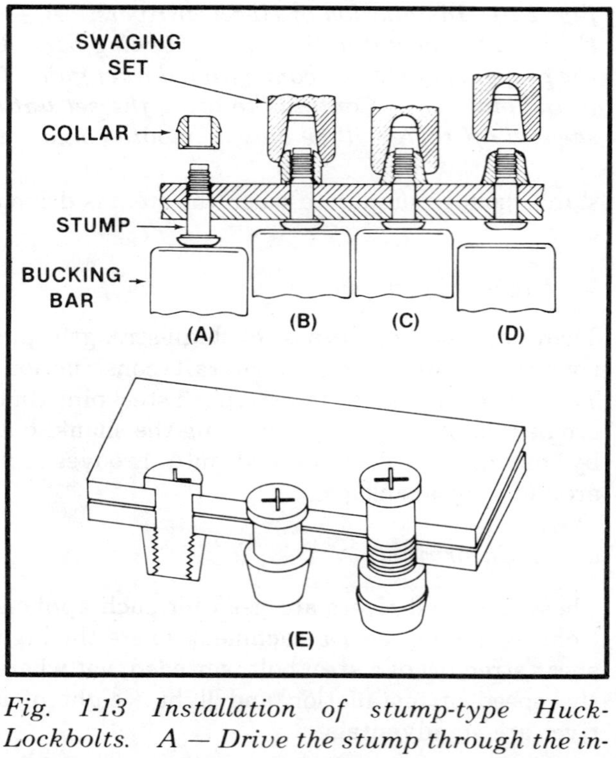

Fig. 1-13 Installation of stump-type Huck-Lockbolts. A — Drive the stump through the interference fit hole. B — Drive the collar down over the grooves in the stump with a swaging set in a rivet gun. C — Continue driving until the set bottoms on the material. E — A properly swaged collar will lock into the grooves of the stump.

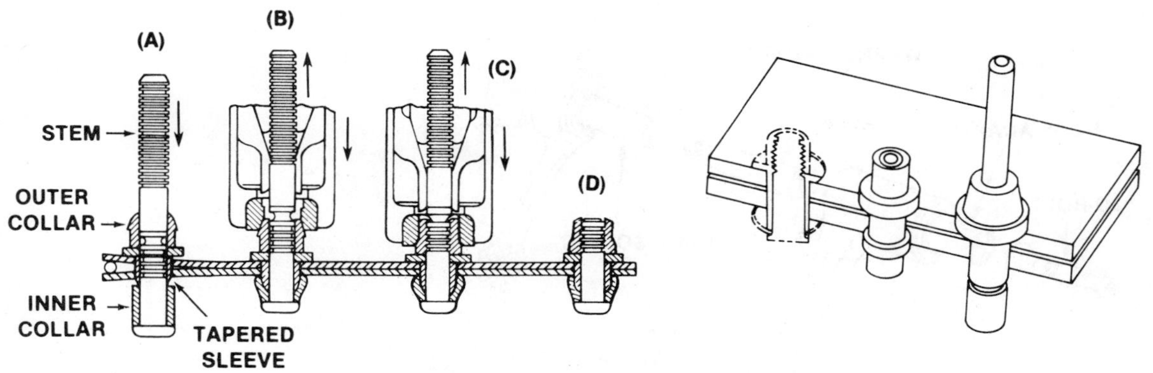

Fig. 1-14 Installation of blind-type Huck-Lockbolts. A — Insert the blind lockbolt through the hole in the metal. B — Attach the puller to the stem and pull the stem up through the collar. This forces the tapered sleeve into the inner collar and swells it out against the inside skin. C — Continued pulling swages the outer collar into the grooves in the stem. D — When the collar is fully swaged into the grooves, the stem will break off at the break-off groove.

lower skin, forming the upset head. As the pulling continues, the die forces the outer collar into the grooves in the stem, and when sufficient force is exerted to completely swage the outer collar, the stem breaks off at a special break-off groove.

d. Jo-Bolts

These are patented high-strength structural fasteners that are used in close-tolerance holes where the strength requirements are high, but where the physical clearance does not allow the use of standard AN, MS, or NAS bolts.

The flush-head Jo-Bolt is seen in Fig. 1-15, and the tools used for driving it are seen in Fig. 1-16. The hole for the Jo-Bolt is drilled, reamed, and countersunk, and the Jo-Bolt is inserted and held tightly in place with the nose adapter of either the hand tool or the power tool. The wrench adapter grips the diving flat of the bolt and screws it up through the nut. As the bolt pulls up, it forces the sleeve up over the tapered outside of the nut and forms the blind head on the inside of the work. When the driving is complete, the driving flat of the bolt will break off.

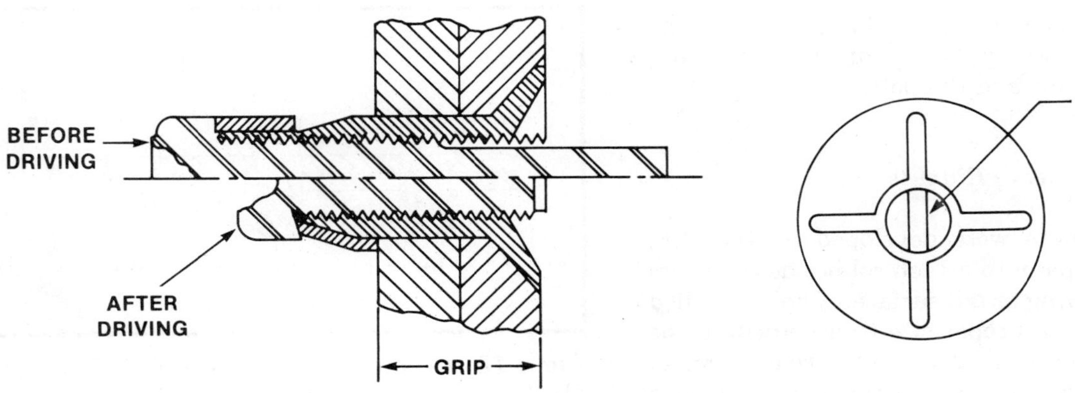

Fig. 1-15 Flush-type Jo-Bolt.

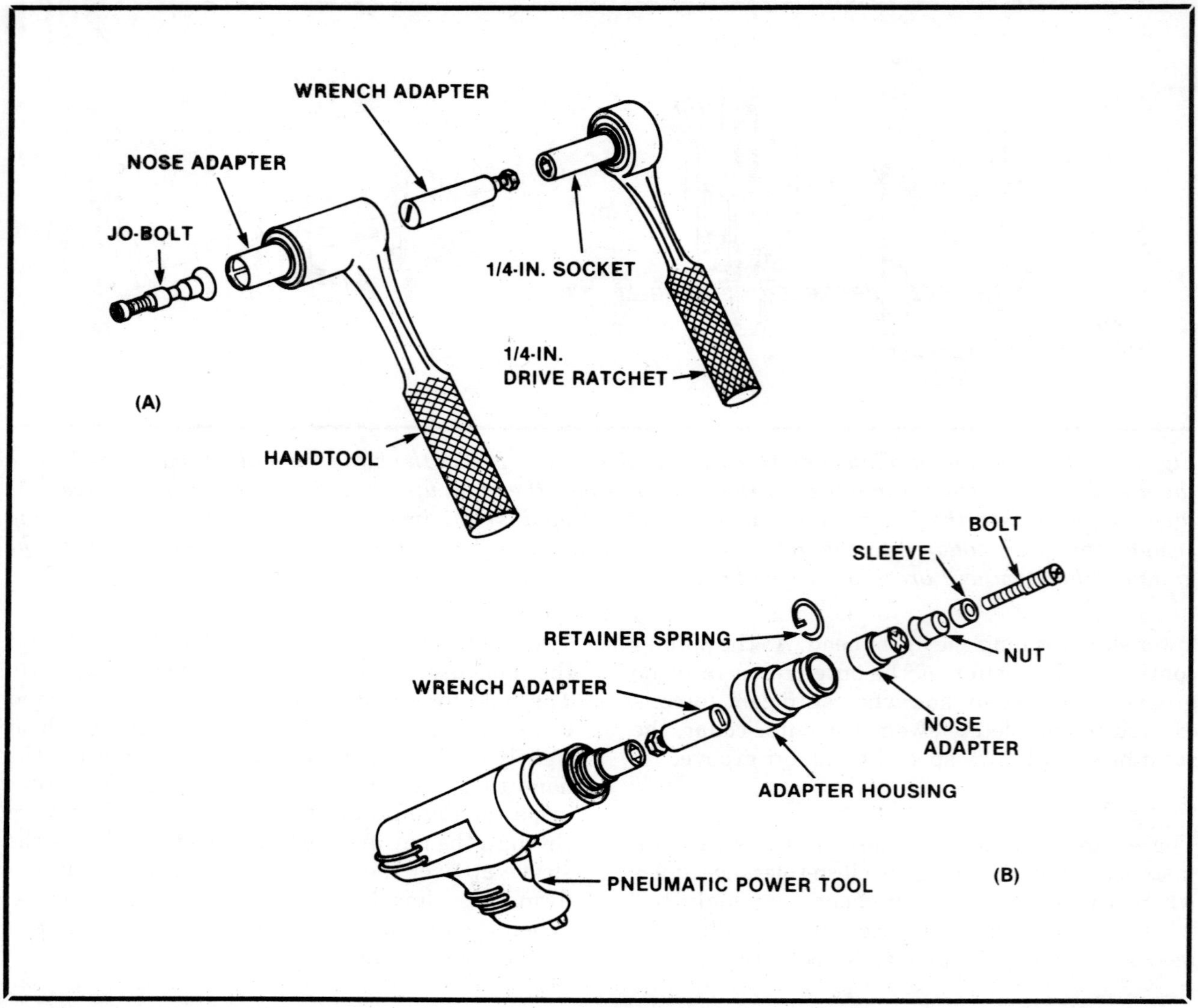

Fig. 1-16 Installation tools for flush-type Jo-Bolts. A — Hand tools B — Pneumatic power tools

Jo-Bolts are removed by drilling through the head of the nut and punching out the remaining portion of the nut and the bolt.

4. Threaded rivets - Rivnuts

Goodrich Rivnuts were developed by the B.F. Goodrich Company to attach rubber deicer boots to an aircraft wing or tail surface. A hole is drilled in the skin of the proper size to accomodate the Rivnut, and a special cutter is used to cut a small notch in the circumference of the hold; this locks the Rivnut into the skin to prevent it turning when it is used as a nut. A Rivnut with the proper grip length is screwed onto the puller and is inserted into the hole with the key on the Rivnut in

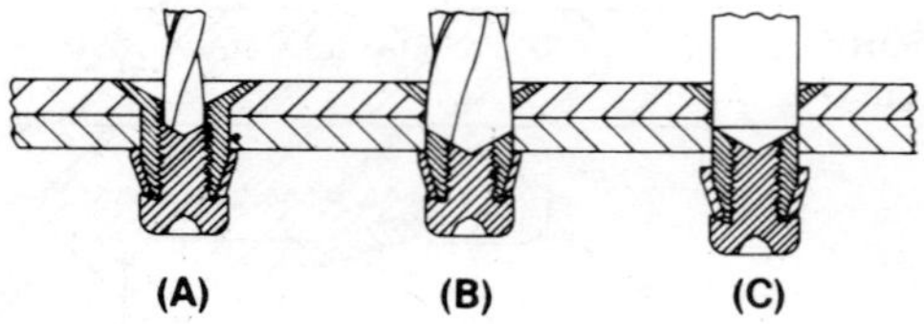

Fig. 1-17 Removal of flush- type Jo-Bolts. A — Drill pilot hole into bolt slightly deeper than the inside of the head of the nut. B — Using a proper size drill, drill to the depth of the pilot hole. C — Drive the shank and blind head from the hole using a pin punch of the proper size.

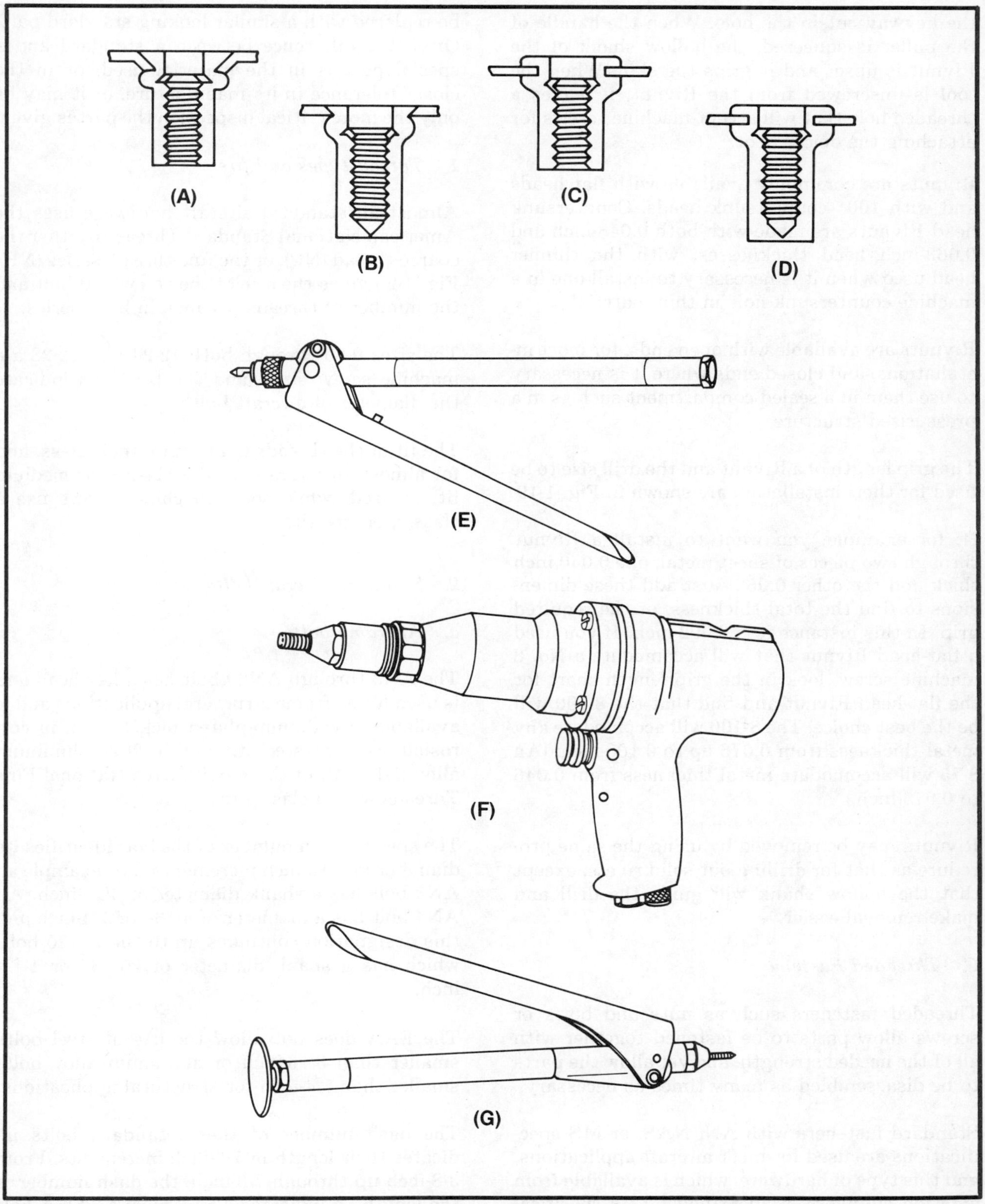

Fig. 1-18 Goodrich Rivnuts and installation tools. A — Countersunk Rivnut with open end. B — Countersunk Rivnut with closed end. C — Flat-head Rivnut with open end. D — Flat-head Rivnut with closed end. E — Key cutting tool F — Pneumatic Rivnut header G — Hand-operated Rivnut header

the keyway cut in the hole. When the handle of the puller is squeezed, the hollow shank of the Rivnut is upset and it grips the skin. When the tool is unscrewed from the Rivnut, it leaves a threaded hole that will accept machine screws for attaching the deicer boot.

Rivnuts are commonly available with flat heads and with 100° countersunk heads. Countersunk head Rivnuts are made with both 0.048-inch and 0.063-inch head thicknesses, with the thinner head used when it is necessary to install one in a machine countersunk hole in thin material.

Rivnuts are available with open ends, for most installations, and closed ends where it is necessary to use them in a sealed compartment such as in a pressurized structure.

The grip length of a Rivnut and the drill size to be used for their installation are shown in Fig. 1-19.

If, for example, you want to install a Rivnut through two pieces of sheet metal, one 0.040-inch thick and the other 0.052, just add these dimensions to find the total thickness, or the required grip. In this instance it is 0.092-inch. If you need a flat-head Rivnut that will accomodate a No. 8 machine screw, look in the grip length chart for the flat-head Rivnut and find that the 8-100 will be the best choice. The 8-100 will accomodate any metal thickness from 0.076 up to 0.100-inch. (An 8-75 will accomodate metal thickness from 0.046 to 0.075-inch.)

Rivnuts may be removed by using the same procedure as that for drilling out solid rivets, except that the hollow shank will guide the drill and make removal easier.

C. Threaded Fasteners

Threaded fasteners such as nuts and bolts or screws allow parts to be fastened together with all of the needed strength, and yet allow the parts to be disassembled as many times as necessary.

Standard fasteners with AN, NAS, or MS specifications are used for most aircraft applications, and this type of hardware, which is available from aviation suppliers, is identified by its specification number. Special fasteners may resemble standard parts, but have a different part number. These special fasteners *must* be replaced with those having the same part number and may not be replaced with a similar looking standard part. Often the difference between a standard and a special part is in the material used, or in the closer tolerance in its manufacture, or it may be only the more critical inspection the part is given.

1. Thread types and fits

Almost all standard aircraft hardware uses the American National Standard Thread in either the coarse-thread (NC), or the fine-thread, series (NF). Fig. 1-20 gives the size of the screw and bolt and the number of threads per inch in both series.

The sizes 0-80 through both 12-24 and 12-28 are machine screw sizes, and the fractions indicate the diameter of aircraft bolts.

The fit of the threads is given by their class, and for almost all aircraft bolts a class-3, or medium fit, is used, while some machine screws use a class-2, or free fit.

2. Standard aircraft bolts

a. Airframe bolts

The AN3 through AN20 bolt has a hex-head and is used for airframe structural applications and is available in cadmium-plated nickel steel, in corrosion-resistant steel (C), and in 2024 aluminum alloy (DD). All of these bolts have National Fine Threads with a class-3 fit.

The specification number of the bolt identifies its diameter in 1/16-inch increments. For example an AN3 bolt has a shank diameter of 3/16-inch. An AN4 bolt has a diameter of 4/16-, or 1/4-inch and this designation continues up to the AN20 bolt, which has a shank diameter of 20/16-, or 1-1/4 inch.

The FAA does not allow the use of steel bolts smaller than 3/16-inch or aluminum alloy bolts smaller than 1/4-inch for structural applications.

The dash number of these standard bolts indicates their length in 1/8-inch increments. From 3/8-inch up through 7/8-inch, the dash number is the length (-4 indicates 4/8, -6 means 6/8). A bolt one inch long, however, is not a -8 bolt, as would be expected (as it is 8 eighths), but is a -10, which means it is one inch plus zero eighths of an inch long. A bolt 1-1/2 inch long would be a -14, or one

		HEAD MARKING					
		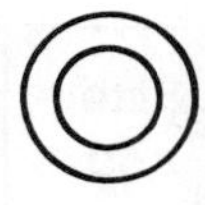					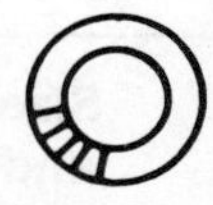
FLAT HEAD	GRIP NO.	45	75	100	120	140	160
	GRIP RANGE	0.010 – 0.045	0.046 – 0.075	0.076 – 0.100	0.101 – 0.120	0.121 0.140	0.141 0.160
100° CSK 0.048 HEAD	GRIP NO.	91	121	146	166	186	206
	GRIP RANGE	0.048 – 0.091	0.092 – 0.121	0.122 – 0.146	0.147 – 0.166	0.167 – 0.186	0.187 – 0.206
100° CSK 0.063 HEAD	GRIP NO.	106	136	161	181	201	221
	GRIP RANGE	0.063 – 0.106	0.107 – 0.136	0.137 – 0.161	0.162 – 0.181	0.182 – 0.201	0.202 – 0.221

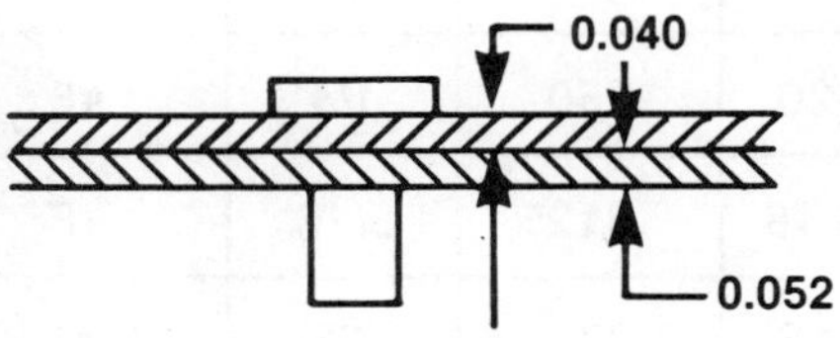

GRIP = 0.040 + 0.052 = 0.092

FOR AN 8-32 SCREW USE
AN 8K100 RIVNUT

8 = 8-32 SCREW
K = KEYED RIVNUT
100 = 0.076 – 0.100 GRIP RANGE

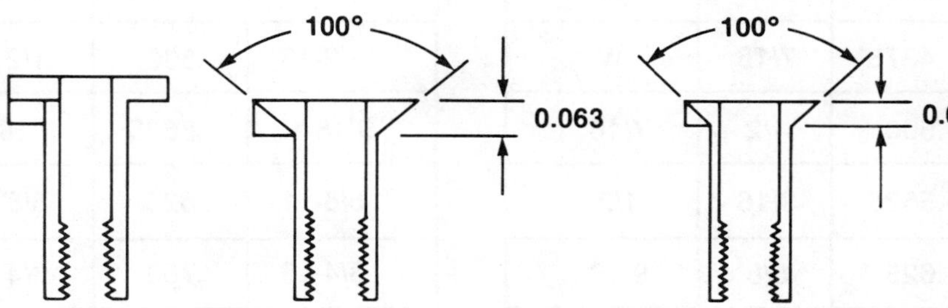

RIVNUT IDENTIFICATION

8 K B 161

- 161 — GRIP RANGE (0.137 - 0.161)
- B — CLOSED-END RIVNUT
- K — KEYED RIVNUT
- 8 — THIS RIVNUT IS FOR AN 8-32 SCREW

SCREW SIZE	OUTSIDE DIAMETER
6-32	3/16″
8-32	7/32″
10-32	1/4″

Fig. 1-19 Rivnut Identification

NATIONAL FINE THREAD SERIES MEDIUM FIT. CLASS 3 (NF)			
Size and Threads	Dia. of body for thread	Body Drill	Tap Drill Size
0-80	.060	52	3/64
1-72	.073	47	#53
2-64	.086	42	#50
3-56	.099	37	#46
4-48	.112	31	#42
5-44	.125	25	#38
6-40	.138	27	#33
8-36	.164	18	#29
10-32	.190	10	#21
12-28	.216	2	#15
1/4-28	.250	F	#3
5/16-24	.3125	5/16	I
3/8-24	.375	3/8	Q
7/16-20	.4375	7/16	W
1/2-20	.500	1/2	7/16
9/16-18	.5625	9/16	1/2
5/8-18	.625	5/8	9/16
3/4-16	.750	3/4	11/16
7/8-14	.875	7/8	51/64
1-14	1.000	1.0	49/64

NATIONAL COARSE THREAD SERIES MEDIUM FIT. CLASS 3 (NC)			
Size and Threads	Dia. of body for thread	Body Drill	Tap Drill Size
1-64	.073	47	#53
2-56	.086	42	#51
3-48	.099	37	5/64
4-40	.112	31	#44
5-40	.125	29	#39
6-32	.138	27	#36
8-32	.164	18	#29
10-24	.190	10	#26
12-24	.216	2	#17
1/4-20	.250	1/4	#8
5/16-18	.3125	5/16	#F
3/8-16	.375	3/8	5/16
7/16-14	.4375	7/16	U
1/2-13	.500	1/2	27/64
9/16-12	.5625	9/16	31/64
5/8-11	.625	5/8	17/32
3/4-10	.750	3/4	41/64
7/8-9	.875	7/8	49/64
1-8	1.000	1.0	7/8

Fig. 1-20 Body and tap drill sizes for American National Screw Threads

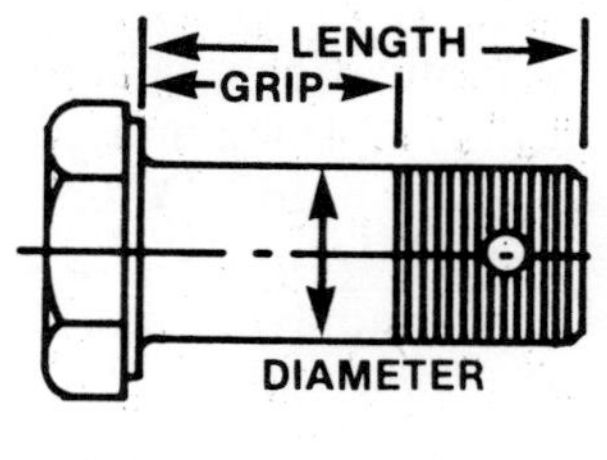

Fig. 1-21 AN3 through AN20 airframe bolts

inch plus four eighths. This type of measurement continues, with the first digit indicating the number of inches and second the number of eighths.

The material of which the bolt is made is identified by marks on its head. If the bolt is made of nickel-steel and is cadmium plated, it will have an asterisk or a cross in the center of its head. If it is made of corrosion-resistant steel, it will have one dash mark on its head; and if it is an aluminum alloy bolt, it will have two dash marks.

Aluminum alloy bolts are coded with the letter DD in place of the dash, and corrosion-resistant steel bolts are coded with the letter C. For example, an AN4DD14 bolt is a 1/4-inch diameter aluminum-alloy bolt, 1-1/2 inch long. An AN5C36 bolt is made of corrosion-resistant steel, has a diameter of 5/16-inch, and is 3-3/4 inches long.

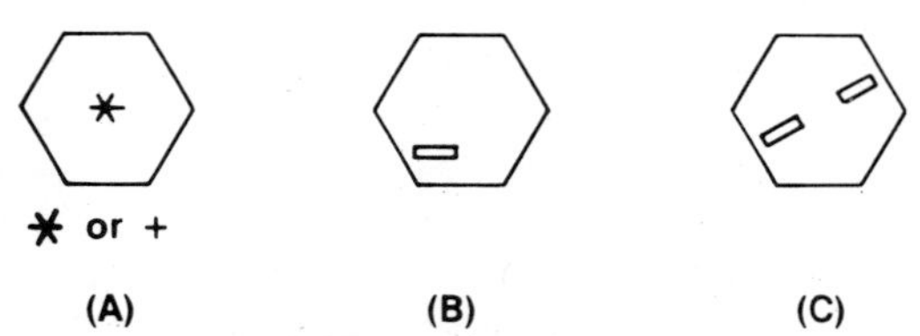

Fig. 1-22 The mark on the head of a bolt identifies the material of which it is made. A — Cadmium-plated nickel steel B — Corrosion-resistant steel C — 2024 aluminum alloy

When hardware was first standardized, almost all nuts were safetied onto the bolts with cotter pins, and all of the bolts had holes drilled near the end of the shank to accomodate the pin. When self-locking nuts became popular, many standard AN bolts were made without the hole in the shank. An un-drilled bolt is identified by the letter A following the dash number. An AN6C12A bolt is made of corrosion-resistant steel, is 3/8-inch in diameter, 1-1/4 inches long, and does not have a hole drilled in its shank.

Some AN bolts, such as those used to fasten a propeller into a flanged shaft, must be safetied by passing safety wire through holes in the bolt head. A bolt drilled for this type of safetying has the letter H preceding the number indicating its length. An AN6H34A would identify the bolt as being made of nickel-steel, being 3/8-inch in diameter, 3-1/2 inches long, and having a hole drilled in its head, but none in the shank.

b. Drilled-head engine bolts

AN73 (MS20073) through AN81 (MS20081) bolts are hex-head, nickel-steel bolts, similar in appearance to the AN3 through AN20 series, except they have a thicker head that is drilled with a small hole across each of the flats and an intersecting axial hole in the center of the head. An AN73 bolt has a diameter of 3/16-inch, and the dia-

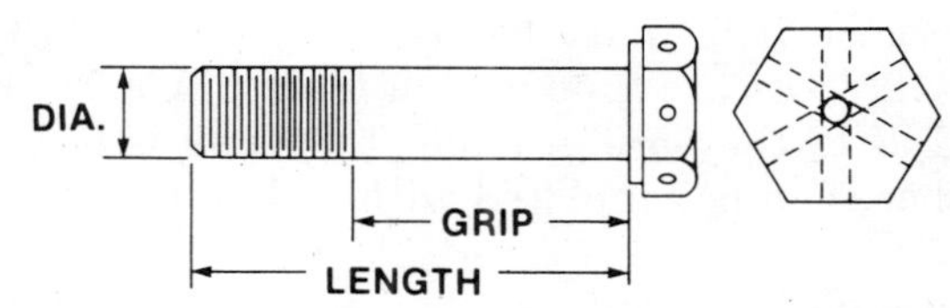

AN73 (MS20073)	3/16 DIAMETER
AN74 (MS20074)	4/16 (1/4) DIAMETER
AN75 (MS20075)	5/16 DIAMETER
AN76 (MS20076)	6/16 (3/8) DIAMETER
AN77 (MS20077)	7/16 DIAMETER
AN78 (MS20078)	8/16 (1/2) DIAMETER
AN79 (MS20079)	9/16 DIAMETER
AN80 (MS20080)	10/16 (5/8) DIAMETER
AN81 (MS20081)	11/16 DIAMETER

Fig. 1-23 AN73 through AN81 drilled-head engine bolts.

meters continue up in 1/16-inch increments to the AN81, which has a diameter of 11/16-inch. The length of these bolts is specified the same way as the lengths of the airframe bolts.

Drilled-head engine bolts are available in both fine-and coarse-thread types. An AN75-16 would identify a fine-thread engine bolt, 5/16-inch diameter and 1-3/4 inches long. But if it were an AN75A16, it would be the same size, but it would have coarse threads.

Since all of these bolts are safetied by passing safety wire through the holes in their head, none of them have cotter pin holes drilled in their shank.

c. *Close-tolerance bolts*

Standard AN3 through AN20 bolts are manufactured with a diameter tolerance of +0.0000 − 0.0025 for AN3 bolts to +0.0000 − 0.0055 for AN16 bolts. This tolerance assures that the bolts will fit in the holes made with a drill of the same nominal diameter. But where a bolt is subjected to a pounding load, or where it is used in a structure that is both riveted and bolted, a close-tolerance bolt must be used. These AN173 to AN186 bots are ground to a fit of ±0.0005-inch, and after they are ground, the shank is greased rather than being plated with cadmium.

Close-tolerance bolts are measured in the same increments of diameter and length as the standard airframe bolts, and they are identified with a triangle around the cross or the asterisk on their head. An AN175-26A is a 5/16-inch diameter close-tolerance bolt, 2-3/4 inches long, with no hole drilled in its shank.

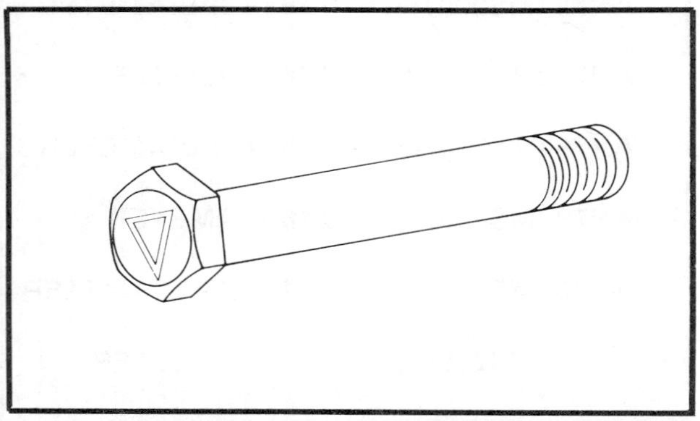

Fig. 1-24 AN173 through AN186 close-tolerance bolts.

d. *Clevis bolts*

All of the bolts we have discussed to this point may be used for either shear or tensile loads. But there are some applications where a bolt may be loaded in shear only. For example, a control cable must be attached to a control horn with a bolt that is loose enough to allow the cable to pivot freely as the surface moves, but it must not be able to fall out. For these applications a clevis

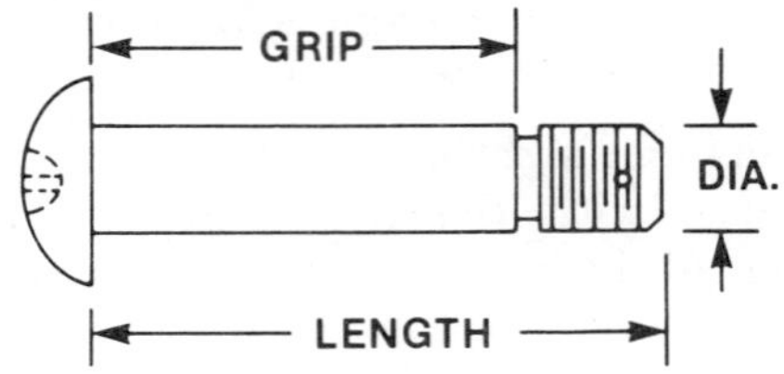

AN21	6-40 MACHINE SCREW
AN22	8-36 MACHINE SCREW
AN23	10-32 MACHINE SCREW
AN24	1/4-28
AN25	5/16-24
AN26	3/8-24
AN27	7/16-20
AN28	1/2-20
AN29	9/16-18
AN30	5/8-18
AN32	3/4-16
AN34	7/8-14
AN36	1-14

AN24 - 14A
- A — SHANK NOT DRILLED FOR COTTER PIN
- 14 — 14/16 (7/8) INCH LONG
- 4 — 1/4″ DIAMETER
- AN2 — CLEVIS BOLT

Fig. 1-25 AN21 through AN36 clevis bolts.

bolt is used. The AN21 through AN36 clevis bolt has a domed head slotted or recessed to accept a screwdriver. It resembles a screw except that there is only a short portion of the shank that is threaded, and there is a small notch between the threads and the shank.

The diameter of a clevis bolt is given in 1/16-inch increments, according to the table of Fig. 1-25. The length of a clevis bolt is more critical than that of the other types of bolts, and it is measured in 1/16-inch increments with the dash number indicating the length in 1/16ths of an inch. An AN29-20 is a 9/16-inch diameter clevis bolt 20/16 (1-1/4) inch long.

The nut on clevis bolts may be of either the self-locking type or it may be safetied onto the bolt with a cotter pin. If the clevis bolt is *not* drilled for a cotter pin, the letter A follows the dash number. An AN22-8A clevis bolt would require a self-locking nut, as it is not drilled for a cotter pin.

e. Internal wrenching bolts

MS20004 through MS20024 bolts are high-strength steel bolts that are used primarily for tensile loads. These bolts have a radius between the head and the shank, and when they are installed in steel parts, the hole must be counterbored to seat the radius. When they are installed in aluminum alloy, an MS200002C washer must be used under the head to provide the needed bearing area.

The strength of these internal wrenching MS bolts is higher than that of standard AN bolts and for this reason AN bolts must never be used to substitute for them.

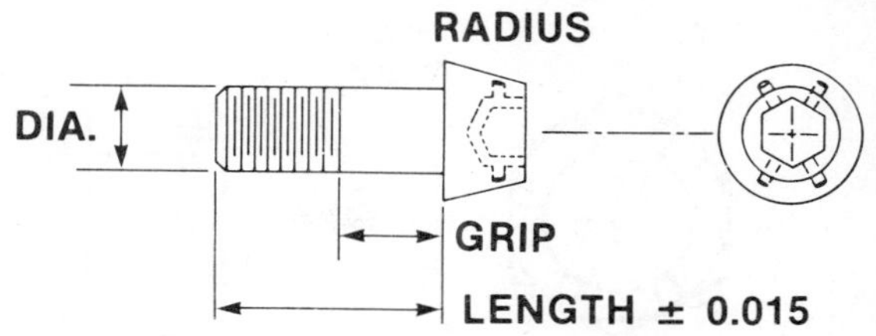

MS20004	1/4-28
MS20005	5/16-24
MS20006	3/8-24
MS20007	7/16-20
MS20008	1/2-20
MS20009	9/16-18
MS20010	5/8-18
MS20012	3/4-16
MS20014	7/8-14
MS20016	1-14
MS20018	1-1/8-12
MS20020	1-1/4-12
MS20022	1-3/8-12
MS20024	1-1/2-12

Fig. 1-26 High-strength internal wrenching bolts, MS20004 through MS20024.

3. Nuts

a. Non-self-locking nuts

All nuts used in aircraft construction must have some way of locking them to prevent their becoming loose and falling off. Many nuts are held on the bolt by passing a cotter pin through a hole in the bolt shank and through slots, or castellations, in the nut. Others have some form of locking insert to grip the threads of the bolt, and a few depend on the tension of a spring-type lock-washer to hold the nut tight enough against the threads to prevent its vibrating loose.

(1) AN310 aircraft castle nut

These fine-thread nuts are designed to fit on the standard air frame bolt (AN3 - AN20) with a class-3 fit, and they may be used when the bolt is subjected to either shear or tensile loads.

The size of the nut is indicated by the dash number which is the size of the bolt it fits. For example an AN310-6 nut fits an AN6 bolt which has a diameter of 3/8-inch.

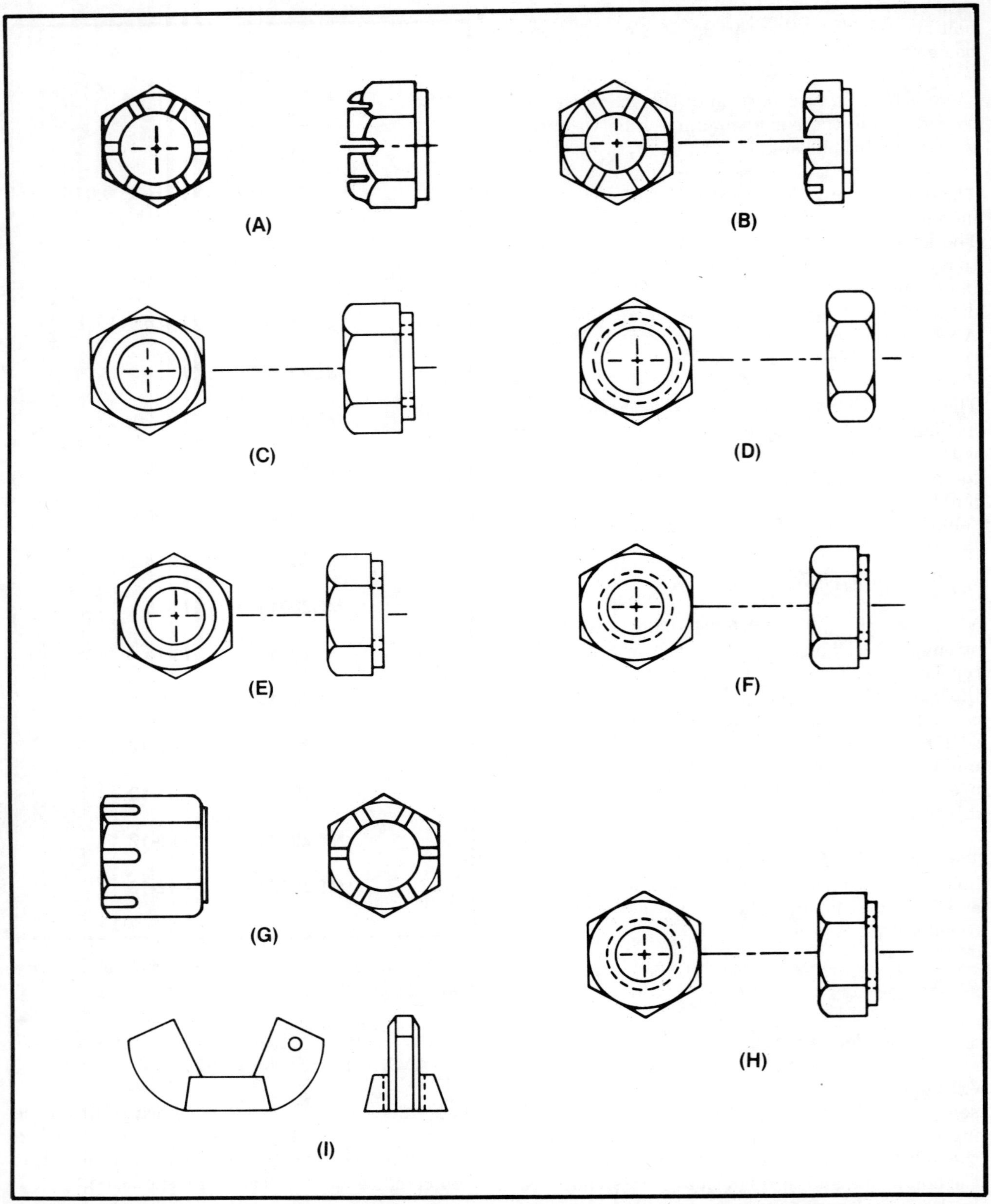

Fig. 1-27 Standard aircraft nuts A — AN310 aircraft castle nut B — AN320 shear castle nut C — AN315 aircraft plain nut D — AN316 aircraft check nut E — AN340 machine screw nut-coarse thread F — AN345 machine screw nut-fine thread G — AN355 slotted engine nut H — AN360 plain engine nut I — AN350 wing nut

AN310 nuts are available in cadmium-plated nickel steel, in corrosion-resistant steel, and in 2024 aluminum alloy. The nickel steel nuts do not use any material designation code, but a corrosion-resistant steel nut is an AN310C6, and one made of aluminum alloy is identified as an AN310D6 nut.

(2) AN320 aircraft shear castle nut

These nuts are made of the same material and have the same thread as the AN310 nut, except that they are much thinner and are used only for shear loads on clevis bolts. An AN320-6 nut is a shear castle nut that would be properly used on an AN26 clevis bolt. An aluminum alloy (2024) nut is identified as an AN320D6.

(3) AN315 aircraft plain nut

These fine-thread nuts do not have any provisions for locking them to the bolt and must depend on either jamming a check nut against them, or by using a spring-type lock washer between the nut and the material in order to apply a spring force to prevent the nut shaking loose. AN315 nuts may be used with either tensile or shear loads and are available in nickel steel, corrosion-resistant steel, and aluminum alloy. They are made with both right- and left-hand threads. An AN315-7R is a nickel steel nut with right-hand threads that fits an AN7 bolt. An AN315C4L is a one-quarter-inch-diameter corrosion-resistant steel plain nut with left-hand threads. The letter D in the designation AN315D-6R indicates that the nut is made of 2024 aluminum alloy.

(4) AN316 check nut

A check nut is used to jam against a plain nut to lock it onto a bolt so it cannot back off. The AN316 nut is made of cadmium-plated steel and is available in both right- and left-hand threads. An AN316-4R is a right-hand check nut that fits a quarter-inch (AN4) bolt, while an AN316-4L has a left-hand thread for a quarter-inch left-handed bolt.

(5) AN340 coarse-thread machine screw nut

These nuts are made in machine screw sizes from number 2 up through quarter inch, and they are available in carbon steel, corrosion-resistant steel (C), brass (B), and 2024 aluminum alloy (DD). A nut identified as an AN340B6 is a brass nut that fits a 6-32 machine screw. An AN340DD416 is an aluminum alloy nut that will fit a 1/4-20 bolt.

(6) AN345 fine-thread machine screw nut

These nuts are similar to the AN340 nuts except they have national-fine series threads. They are available in cadmium-plated carbon steel, corrosion-resistant steel (C), commercial brass (B), and 2024 aluminum alloy (DD). An AN345B6 is a brass nut that will fit a 6-40 machine screw. An AN345DD416 is an aluminum alloy nut that will fit a 1/4-28 bolt.

(7) AN355 slotted engine nut

This nut is designed for use on an aircraft engine and is not approved for airframe use. It is made of heat-treated steel and has a class-3 fit in national fine threads and is available in sizes from AN355-3 (3/16-inch) to AN355-12 (3/4-inch).

(8) AN360 plain engine nut

This nut is similar to the AN355 except that it does not have the slots for a cotter pin and it has a black rustproof finish. It is not approved for airframe use. An AN360-7 is a plain engine nut that fits a 7/16-inch bolt.

(9) AN350 wing nut

Wing nuts are used when it is necessary to remove a unit frequently and without the use of any tools. They are made of either cadmium-plated steel or brass and are available from sizes that fit a number six machine screw up to sizes that fit a 1/2-inch bolt. All of these nuts have national fine threads with a class-2 fit. Nuts for machine screw sizes are designated by just the number of the screw; an AN350-6 nut fits a 6-40 machine screw, while nuts that fit bolts have the bolt size given in 1/16-inch increments followed by the number 16. An AN350-616 nut, for example, fits a 3/8 (6/16) inch bolt.

b. Self-locking nut

(1) Low-temperature self-locking nuts

AN365 self-locking nuts are used on bolts or machine screws and are prevented from vibrating loose by an elastic insert above the threads. This fiber or plastic insert has a hole slightly smaller than the diameter of the threads on which it fits.

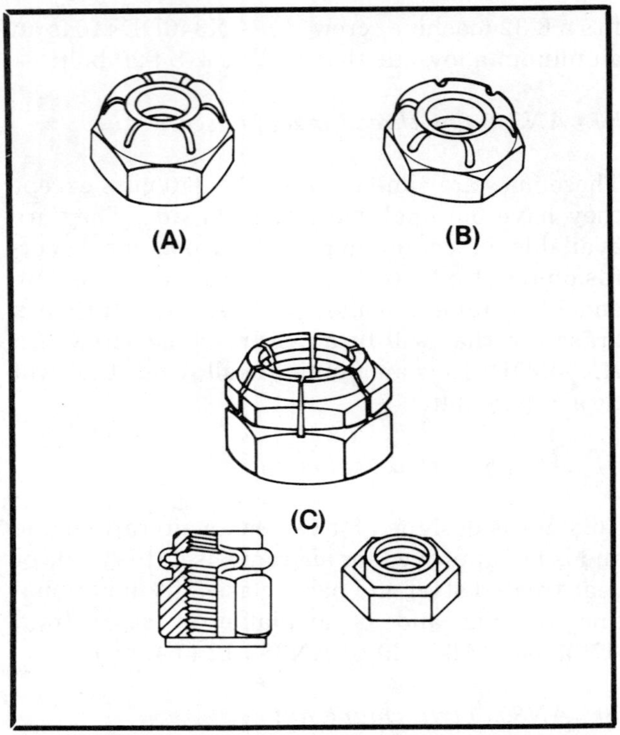

Fig. 1-28 Self-locking nuts A — AN365 (MS20365) low-temperature self-locking nut. B — AN364 (MS20364) low-temperature shear self-locking nut. C — AN363 high-temperature self-locking nuts.

The class-3 fit of the nut allows it to run down on the threads easily until the bolt enters the insert, and then a wrench must be used to turn the nut.

When the bolt threads enter the insert, it tries to force it out of the nut. Since the hole is smaller than the bolt, however, it exerts a strong downward push on the nut that removes all of the axial play between the threads of the bolt and the nut and creates so much friction between all of the threads that are in contact that the nut cannot vibrate loose.

The bolt does not actually cut threads into the insert, but rather forces its way into the resilient material, and because there are no permanent threads made in the insert, these nuts may be reused many times. They are usable as long as there is enough friction between the nut and the bolt that the nut cannot be turned down by hand, requiring a wrench instead. A tap must never be run through a self-locking nut to make it easier to screw onto the bolt, as this would destroy the locking ability of the nut.

A self-locking nut must be screwed down onto the bolt until at least all of the chamfer on the end of the bolt protrudes through the insert. If the bolt is not chamfered, at least one thread and not more than three threads should protrude through the nut.

Some bolts have their shank drilled for a cotter pin, and these bolts may be used with self-locking nuts, provided the bolt is 5/16-inch or larger in diameter and the edges of the cotter pin hole are chamfered so there are no burrs around the hole.

Low-temperature nuts should not be used in any location where the temperature will exceed 250 degrees F. They are approved for use on engines only in those locations specified by the engine manufacturer.

Self-locking nuts should not be used in any location where the nut or the bolt is subject to rotation. This includes such applications as control hinge pins.

(2) High-temperature self-locking nuts

In applications where temperatures exceed 250 degrees F, high-temperature locknuts such as the AN363 can be used. Rather than using a fiber of plastic insert to provide the locking action, there are several ways the nut may be made to grip the bolt. Some of these nuts have a portion of the end slotted and the slots swaged together. This gives the end of the nut a slightly smaller diameter than the body, and the threads will grip those of the bolt. Others have the end of the nut squeezed into a slightly oval shape, and as the bolt screws up through the threads it must make the hole round. This provides the gripping action.

Since both types of self-locking nuts are available in both NF and NC threads, the dash number of the nut specifies both the diameter and the number of threads per inch. We see the dash numbers for both AN365 and AN366 nuts in Fig. 1-29.

(3) Special forms of self-locking nuts

(a) AN364 shear self-locking nuts

These nuts resemble the AN365 self-locking nut, but they are thin and are approved only for shear loads. They are made to be used on clevis bolts that do not have drilled shanks.

DASH NUMBER	SIZE AND THREADS	
– 440	4-40	MACHINE SCREW (NC)
– 448	4-48	MACHINE SCREW (NF)
– 632	6-32	MACHINE SCREW (NC)
– 640	6-40	MACHINE SCREW (NF)
– 832	8-32	MACHINE SCREW (NC)
– 840	8-40	MACHINE SCREW (NF)
– 1024	10-24	MACHINE SCREW (NC)
– 1032	10-32	MACHINE SCREW (NF)
– 420	1/4-20	NC
– 428	1/4-28	NF
– 518	5/16-18	NC
– 524	5/16-24	NF
– 616	3/8-16	NC
– 624	3/8-24	NF
– 714	7/16-14	NC
– 720	7/16-20	NF
– 813	1/2-13	NC
– 820	1/2-20	NF
– 918	9/16-18	NF
– 1018	5/8-18	NF
– 1216	3/4-16	NF
– 1414	7/8-14	NF
– 1614	1-14	NF

Fig. 1-29 Self-locking nut dash numbers and sizes.

(b) Anchor nuts

Inspection plates and access doors that are fastened to an aircraft with screws have anchor nuts secured to the inside of the structure into which these screws fit. Some of the most commonly used anchor nuts are the two-lug anchor nut, the one-lug anchor nut, and the corner anchor nut.

To make the installation of an access door easier where there are a great number of screws, the floating anchor nut is used. These nuts fit loosely into a basket which is riveted to the skin and the nut floats enough to align itself with the screw. To speed the production of aircraft, ganged anchor nuts are installed around inspection plate openings. These are floating-type anchor nuts, and there are a number of them in a channel that is riveted to the structure. Each nut floats in the channel with enough play that the screws can move the nut enough to align it.

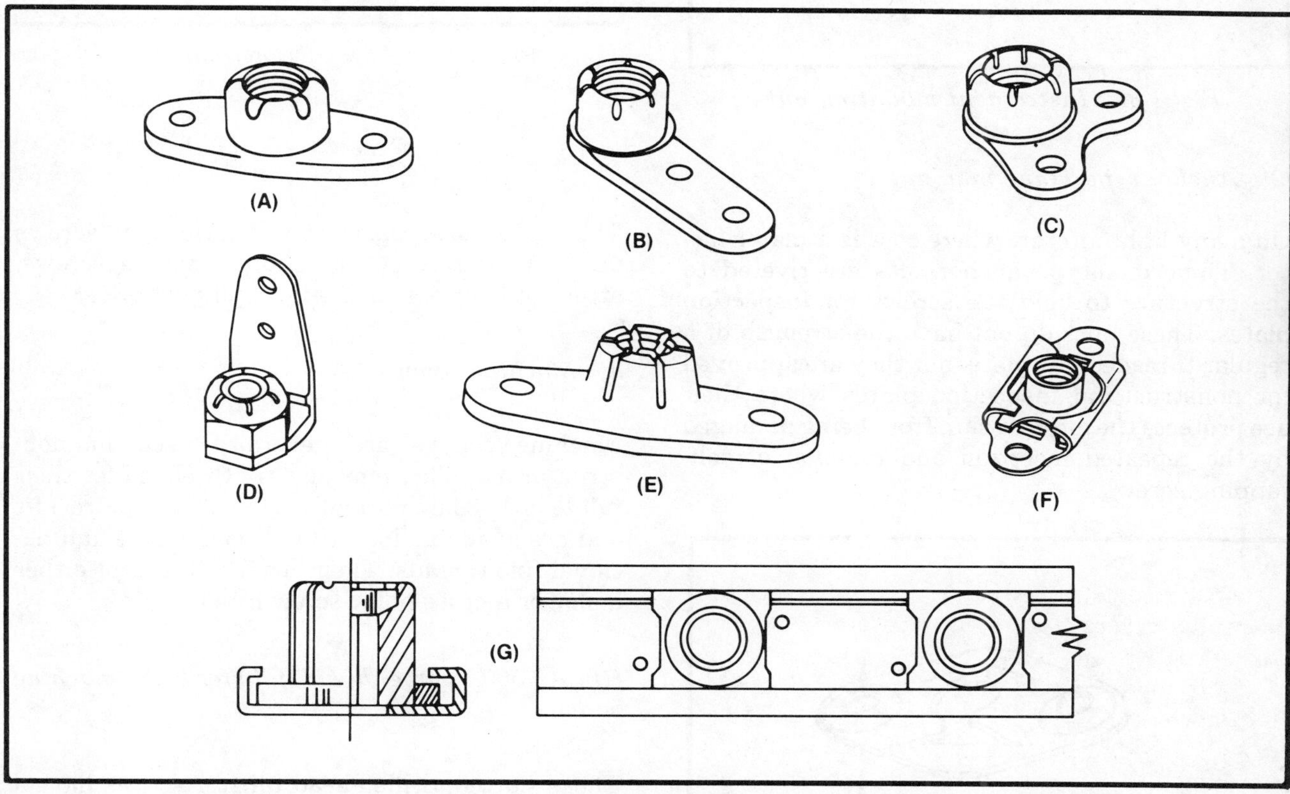

Fig. 1-30 Anchor nuts A — AN366F two-lug anchor nut B — NAS1025 one-lug anchor nut C — NAS1027 corner anchor nut D — AN256F right-angle anchor nut E — AN362F high-temperature two-lug anchor nut F — Two-lug floating anchor nut G — Ganged anchor nuts (Channel nuts)

c. *Tinnerman nuts*

(1) *Instrument mounting nuts*

This special type of nut may be used either in the instrument pannel or in the instrument case itself to mount aircraft instruments in the panel. A non-magnetic phosphor bronze cage holds a brass nut in alignment behind the hole and accepts the screw that holds the instrument in the panel. If the instrument is rear mounted, the legs of the nut are long enough to pass through the instrument case, but if the instrument is front mounted, the nut fastens into the screw hole in the instrument panel.

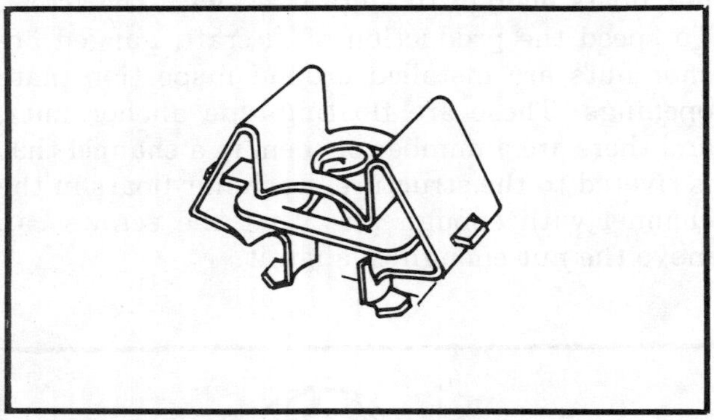

Fig. 1-31 Instrument mounting nut

(2) *Anchor-type Tinnerman nut*

On many light aircraft where cost is a major factor, Tinnerman-type anchor-nuts are riveted to the structure to hold the screws for inspection plates. These nuts do not have the strength of a regular threaded nut plate, but they are approved for nonstructural inspection plates where their use protects the aircraft skin from being damaged by the repeated insertion and removal of self-tapping screws.

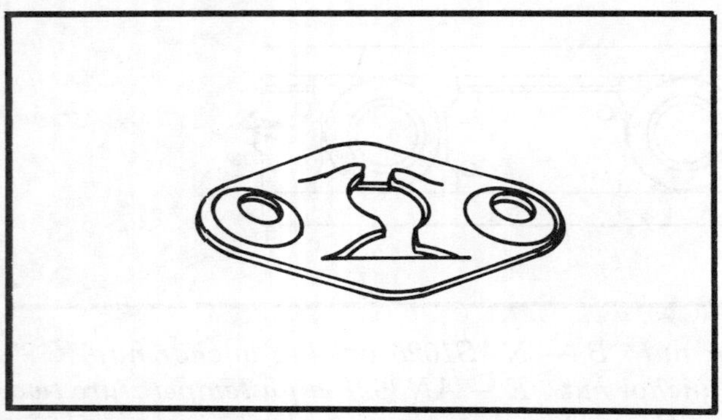

Fig. 1-32 Anchor-type Tinnerman nut

(3) *U-type Tinnerman nuts*

The cowling of many light aircraft is held on with self-tapping sheet metal screws, and to prevent the holes being enlarged from repeated use of these screws, a U-type Tinnerman nut can be slipped over the edge of the inside cowling so that it straddles the screw hole. The screw passes through the hole in the outer cowling and into the nut, and when it is tightened, the spring action of the nut holds the screw tight.

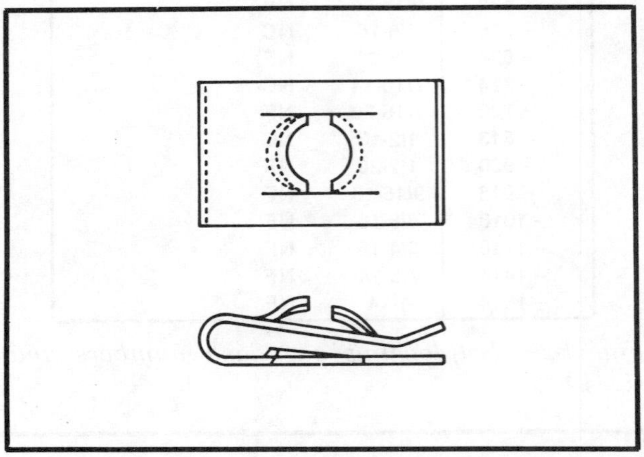

Fig. 1-33 U-type Tinnerman nut

4. *Screws*

There are three basic classifications of screws used in aircraft construction: machine screws, structural screws, and self-tapping screws.

a. *Machine screws*

Machine screws are generally used in nonstructural applications and are threaded for their full length. They normally have a class-2 (free) fit and are available in both national coarse and national fine threads. Their heads will accept either a slot or a cross-point screwdriver.

(1) *AN500 coarse-thread fillister-head machine screw*

These slotted, drilled-head fillister screws have a class-3 coarse thread and have been superseded by MS35265 screws. They are available in sizes from 4-40 up to 1/4-20 and have no clearly defined grip.

(2) AN501 fine-thread fillister-head machine screw

This screw had been replaced by the MS35266 and is like the AN500, except that it has fine threads.

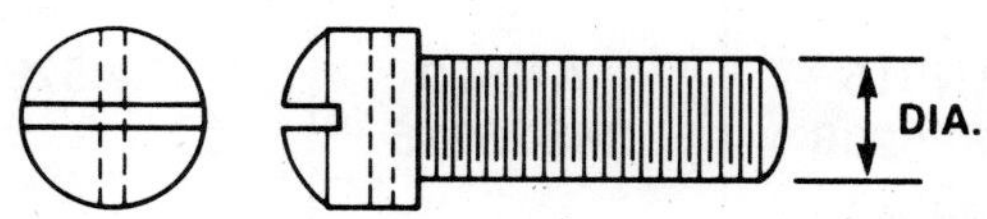

AN500 MS35265	AN501 MS35266	AN502	AN503
2-56	----	----	----
3-48	----	----	----
4-40	----	----	----
5-40	----	----	----
6-32	----	----	6-32
8-32	----	----	8-32
10-24	10-32	10-32	----
1/4-20	1/4-28	1/4-28	1/4-20
5/16-18	5/16-24	5/16-24	5/16-18
3/8-24	3/8-24	----	----

Fig. 1-34 Fillister head machine screws

(3) AN505 coarse-thread 82° flat-head machine screw

These countersunk screws have been superseded by the MS35190 screw. They are made of cadmium-plated carbon steel, and are available with a recessed head for cross-point screwdrivers.

(4) AN510 fine-thread 82° flat-head machine screw

These screws are like the AN505 screw except they have fine threads.

(5) AN515 coarse-thread round-head machine screw

The MS35206 is now used instead of the AN515. It is made of cadmium-plated carbon steel and may have either a slotted or a recessed head. This screw is available in brass as the AN515R-B which has been superseded by the MS35214. This non-magnetic black oxide finished brass screw is used to mount instruments into their panel.

(6) AN520 fine-thread round-head machine screw

These screws are similar to the AN515 except that they have fine threads. The AN520R-B is the recessed-head brass screw that has been replaced by the MS35215.

(7) AN526 truss-head fine-thread machine screw

These large-head machine screws are available in cadmium-plated carbon steel in both slotted and cross-recessed heads.

(8) AN507 100° flat-head machine screw

These screws have been replaced by MS24693 screws. The MS2469S is made of cadmium-plated carbon steel and the MS24693BB is made of black oxide coated brass. These screws are available in sizes from 4-48 through 3/8-16. You will notice that some of these sizes are of the national coarse series and some have national fine threads.

b. Structural screws

(1) AN502 fine-thread fillister-head screw

This fillister-head screw is similar in appearance to the AN500 except for the cross mark on its head indicating that it is made of high-strength steel.

(2) AN503 coarse-thread fillister-head screw

This screw is similar to the AN501 except for the cross mark on its head and the fact that it is made of high-strength steel.

(3) AN509 100° flat-head screw

These structural screws have been replaced by MS24694 screws. They are made of heat-treated carbon steel and are cadmium plated. They are distinguished from the AN507 screw by the X marked on their head.

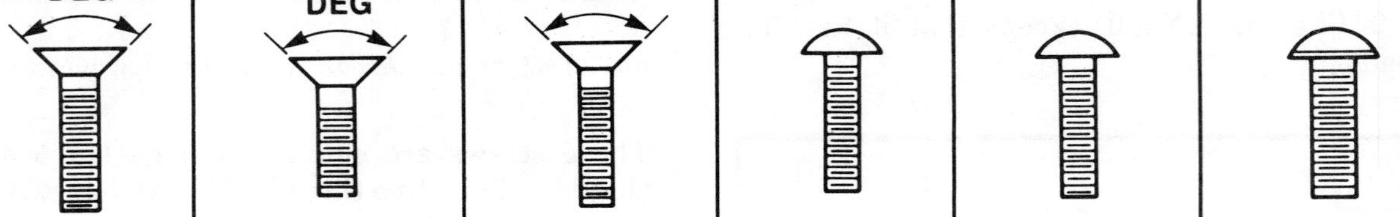

(A) 82 DEG	(B) 100 DEG	(C) 82 DEG	(D)	(E)	(F)
AN505 MS35109	AN507 MS24693	AN510	AN515 MS35206 MS35214	AN520 MS35215	AN526
-----	-----	0-80	-----	0-80	-----
-----	-----	1-72	-----	1-72	-----
-----	-----	2-64	-----	2-64	-----
2-56	-----	-----	2-56	-----	-----
-----	-----	3-56	-----	3-56	-----
3-48	-----	-----	3-48	-----	-----
-----	4-48	4-48	-----	4-48	-----
4-40	4-40	-----	4-40	-----	-----
-----	-----	5-44	-----	5-44	-----
5-40	-----	-----	5-40	-----	-----
-----	6-40	6-40	-----	6-40	6-40
6-32	6-32	-----	6-32	-----	6-32
-----	8-36	8-36	-----	8-36	8-36
8-32	8-32	-----	8-32	-----	8-32
-----	10-32	10-32	-----	10-32	10-32
10-24	10-24	-----	10-24	-----	10-24
-----	1/4-28	1/4-28	-----	1/4-28	1/4-28
1/4-20	1/4-20	-----	1/4-20	-----	1/4-20
-----	5/16-24	-----	-----	-----	-----
5/16-18	5/16-18	-----	5/16-18	-----	-----
-----	3/8-24	-----	-----	-----	-----
3/8-16	3/8-16	-----	3/8-16	-----	-----

Fig. 1-35 Machine screws A — AN505 (MS35109) 82° countersunk head, coarse thread. B — AN507 (MS24693) 100° countersunk head, coarse and fine thread. C — AN510 82° countersunk head, fine thread. D — AN515 (MS35206 and MS35214) round head, coarse thread. E — AN520 (MS35215) round head, fine thread. F — AN526 Truss-head, fine thread.

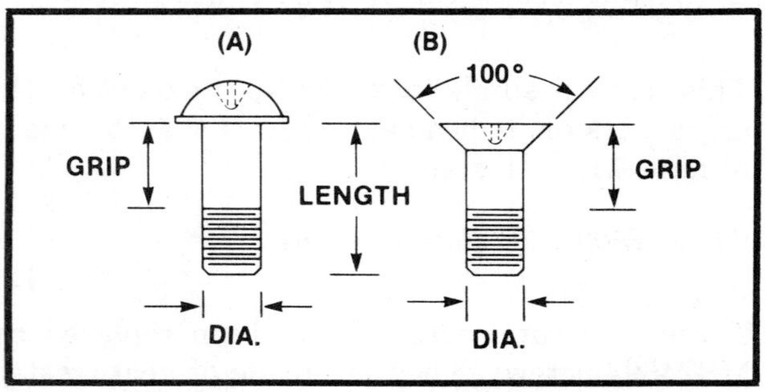

Fig. 1-36 Structural screws A — AN525 washer head B — AN509 (MS24694) 100° countersunk head

(4) AN525 washer-head screw

These structural screws have a large head whose area is increased with a washer formed onto the head. They are made of cadmium-plated high-strength steel.

c. Self-tapping screws

These are coarse-threaded screws that are used to hold thin sheets of metal, plastic, or plywood together. The type-A screw has a gimlet (sharp)

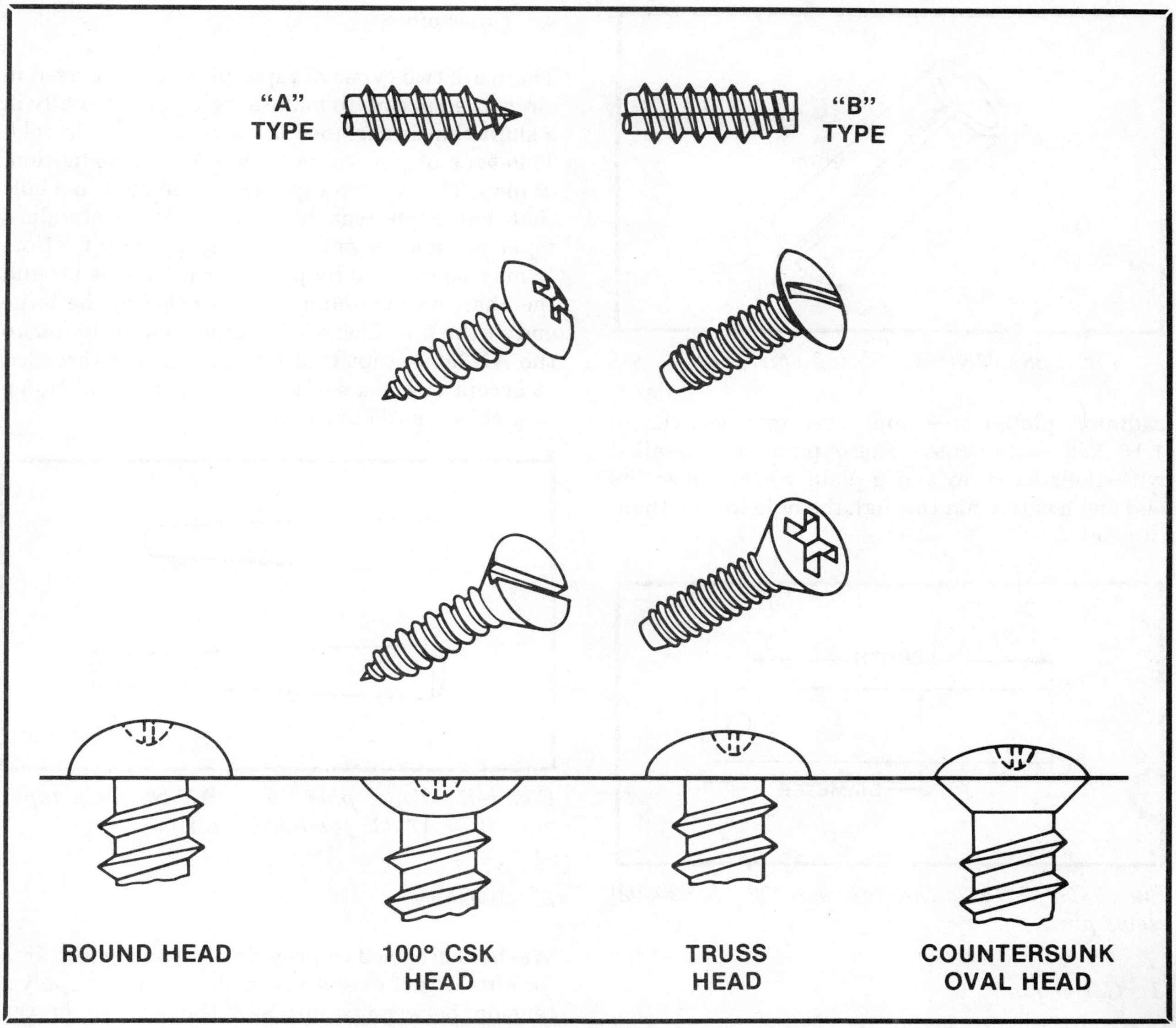

Fig. 1-37 Self-tapping sheet metal screws.

point, and type B has a blunt point with threads that are slightly finer than those of type A.

There are four types of heads available on self-tapping screws: a round head, a truss head which is rounded, but considerably thinner than a round head, a countersunk head, which has a 100° head angle and is flat on top, and the countersunk oval screw which also has a 100° head angle. Instead of a flat top, however, the countersunk oval screw is slightly rounded.

An AN530 screw is a round-head self-tapping screw, and the AN531 screw is a countersunk, flat-head self-tapping screw. Both of these AN screws have type-B ends.

D. Pins

1. Rollpin

These pressed-in pins are used to provide a pivot for a joint where the pin is not likely to have to be removed. The pin is made of spring steel and is rolled into a cylinder. When it is pressed into the hole, it compresses, and the spring action holds it tight against the edge of the hole. It may be removed by driving it from the hole with a proper size pin punch.

2. Clevis pin

Clevis, or flat-head, pins are used for hinge pins in some aircraft control systems. They are made of

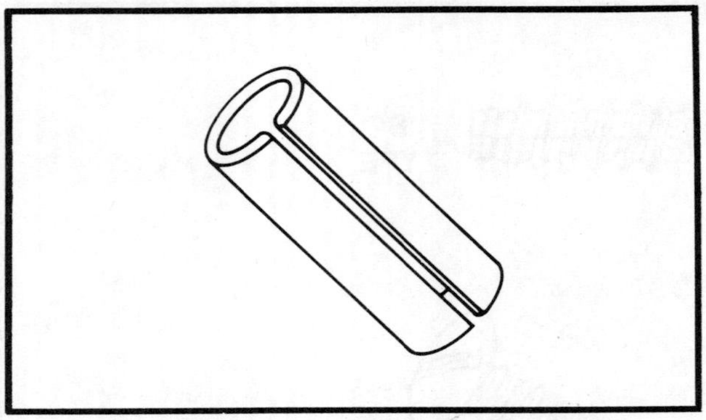

Fig. 1-38 MS16562 Spring steel rollpin

cadmium-plated steel and have grip lengths in 1/16-inch increments. These pins are installed with their head up and a plain washer over the end and a cotter pin through the hole to lock them in place.

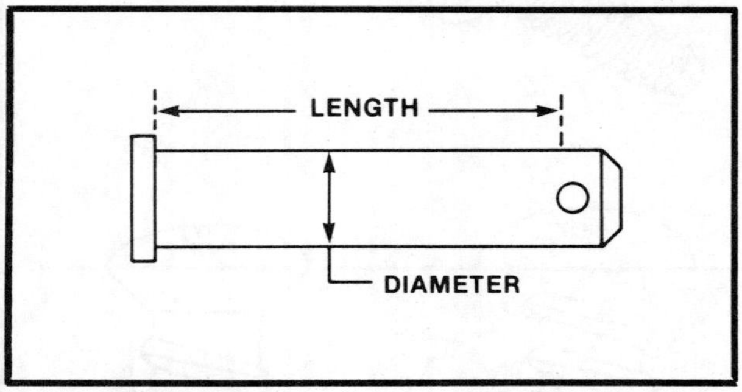

Fig. 1-39 AN392 through AN406 (MS20392) clevis pin

3. Cotter pin

Castellated nuts are locked onto drilled bolts by passing a cotter pin through the hole and the castellations and spreading the ends. They are made of either cadmium-plated carbon steel or of corrosion-resistant steel.

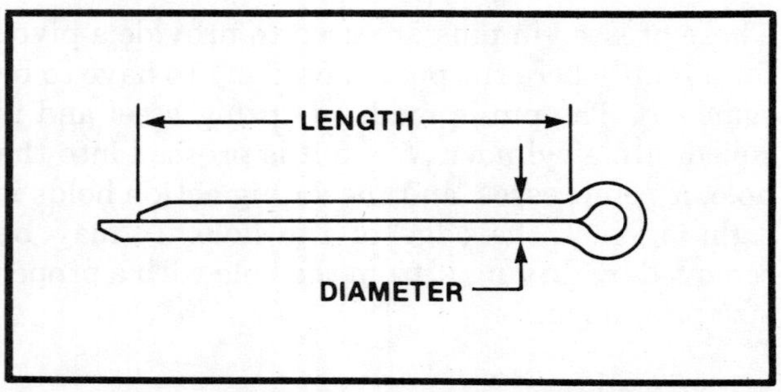

Fig. 1-40 AN380 (MS24665) cadmium-plated carbon-steel cotter pin AN381 (MS24665) corrosion-resistant steel cotter pin

4. Taper pin

There are two types of taper pins that are used in aircraft structure to make a tight joint, usually in a shaft that is designed to carry shear loads only. This type of pin will not allow any loose motion, or play. The AN385 taper pin is forced into a hole that has been reamed with a Morse standard taper pin reamer and is held in place by friction. It may be safetied by passing safety wire around the shaft and through a hole drilled in the large end of the pin. The AN386 taper pin is similar to the AN385 except that its small-end is threaded to accept either a self-locking shear nut (AN364) or a shear castle nut (AN320).

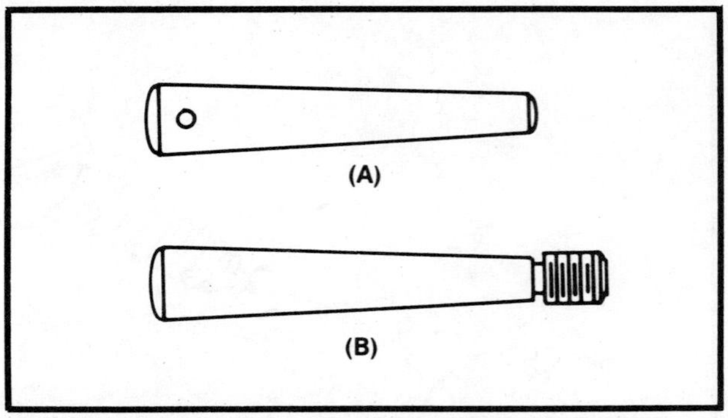

Fig. 1-41 Taper pins A — AN385 plain taper pin B — AN386 threaded taper pin

E. Washers

Washers are used to provide bearing surface area for nuts, to act as spacers or shims, and to apply a tension between a nut and the surface of the material to prevent the nut vibrating loose.

1. Plain washers

The AN960 washer is the one most generally used to provide a smooth surface between a nut and the material being clamped. These washers are made of cadmium-plated steel, of commercial brass (B), of corrosion-resistant steel (C), and of 2024 aluminum alloy (D). They are commonly available in sizes that range from those that fit a number two machine screw to those that fit a one-inch bolt.

A light series of AN960 washers are available that are one-half the thickness of the regular series. If the castellations of an AN310 nut do not line up with the cotter pin hole when the nut is properly torqued, a light series washer may be

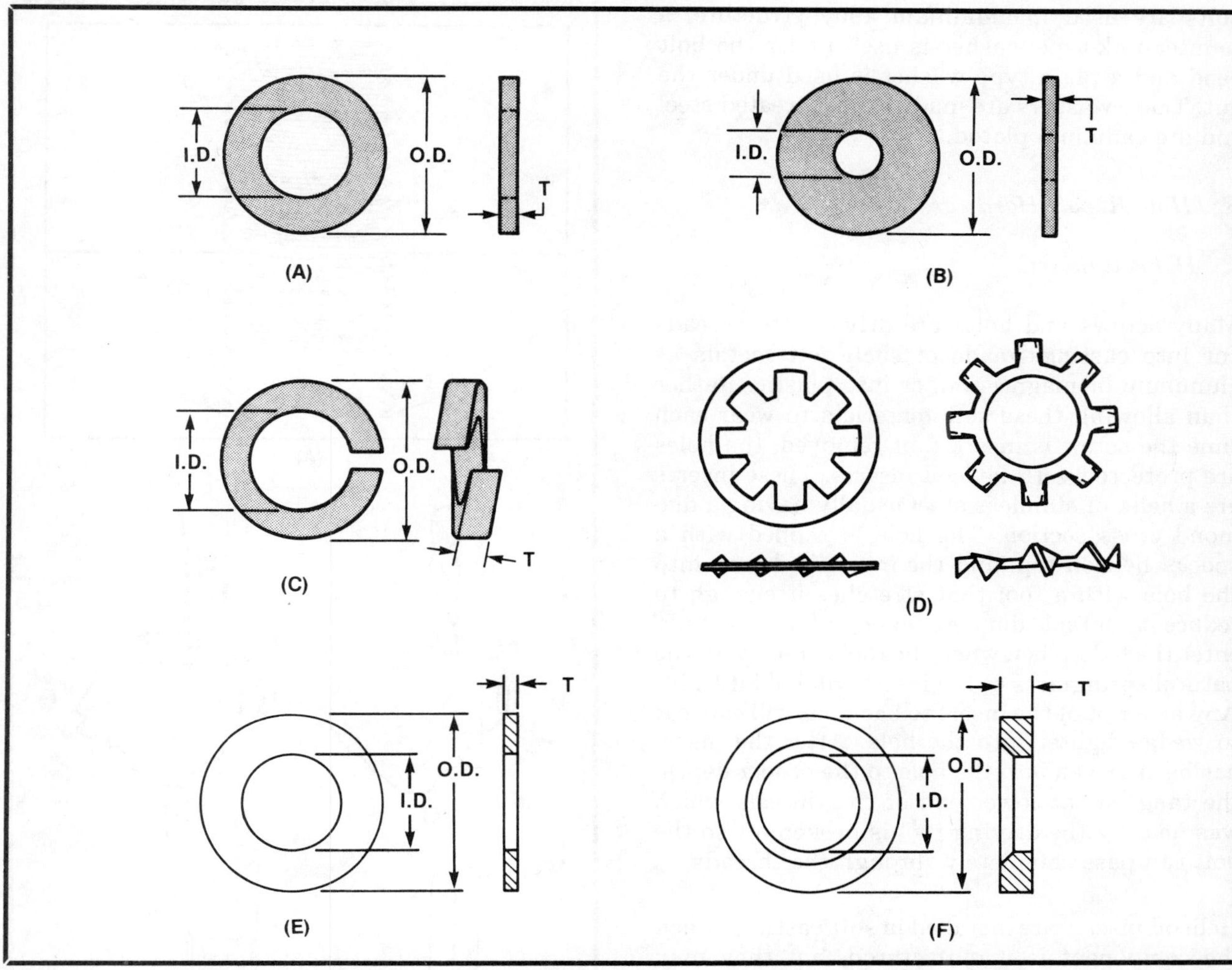

Fig. 1-42 Aircraft washers A — AN960 plain washer B — AN970 large area washer C — AN935 split lock washer D — AN936 shakeproof lock washer E — MS20002 high strength plain washer F — MS20002 high strength countersunk washer.

substituted for the regular washer so that the holes will match.

AN970 washers are large area washers that are used on wooden structures to spread the load of the fastener over a wider area than would be possible with the AN960 washer. These washers are all made of cadmium-plated steel and are available with inside diameters from 3/16- to 1/2-inch.

2. Lock washers

In some instances it is not convenient for a machine screw or a bolt to use a self-locking nut, or to lock the nut with a cotter pin. For these applications, if the joint is not structurally critical, a lock washer may be used between the nut and the surface.

Lock washers are made of steel and are twisted so that when the nut is tightened, the spring action of the washer will create a strong friction force between the threads of the bolt and those in the nut.

There are two types of lock washers used in aircraft contruction. The AN935 split lock washer is available in sizes that will fit from a number four machine screw to a one-half-inch bolt. The shakeproof lock washer, AN936, is thinner and is available with both internal and external teeth.

3. Special washers

MS20002 flat washers are used under the head of high-strength internal wrenching bolts such as the MS20004 through MS20024. When these bolts are used in aluminum alloy structure, a countersunk-type washer is used under the bolt head and a plain type washer is used under the nut. These washers are made of heat-treated steel and are cadmium plated.

F. Hole Repair Hardware

1. Helicoil inserts

Many screws and bolts are driven into threads cut into castings made of such soft metals as aluminum or magnesium, or into plastic. Rather than allowing these soft materials to wear each time the screw is inserted or removed, the holes are protected with helicoil inserts. These inserts are a helix of stainless steel usually having a diamond cross section. The hole is tapped with a special helicoil tap, and the insert is driven into the hole with a tool that stretches it enough to reduce its outside diameter enough for it to easily enter the hole. Then when the tool is removed, the natural springiness of the insert will hold it tight. Any attempt of the insert to back out will cause it to wedge tighter into the hole. After the insert has been driven into the hole to the proper depth, the tang on the lower end of the threads which was held by the driving tool is broken off so the bolt can pass completely through the threads.

Helicoil inserts are installed in soft castings when the component is manufactured, but they may also be installed in the field when the threads in a casting are stripped out. If the threads in a hole are stripped, first determine whether increasing the diameter of the hole for installation of the insert will weaken the casting. If an insert can be used, drill out the stripped threads with the drill furnished in the helicoil installation kit and tap the hole with the special tap. Run the gage through the new threads to be sure that they will accomodate the insert. Then place the insert on the installation tool and screw it into the tapped hole. When the insert is below the surface of the casting to the extent specified in the installation instructions, remove the tool and break off the tang.

When any of this type of insert is used to accomplish a repair, it must be remembered that the repair must be approved by the component manufacturer or by the FAA, and the installation instructions furnished by the insert manufacturer must be followed in detail.

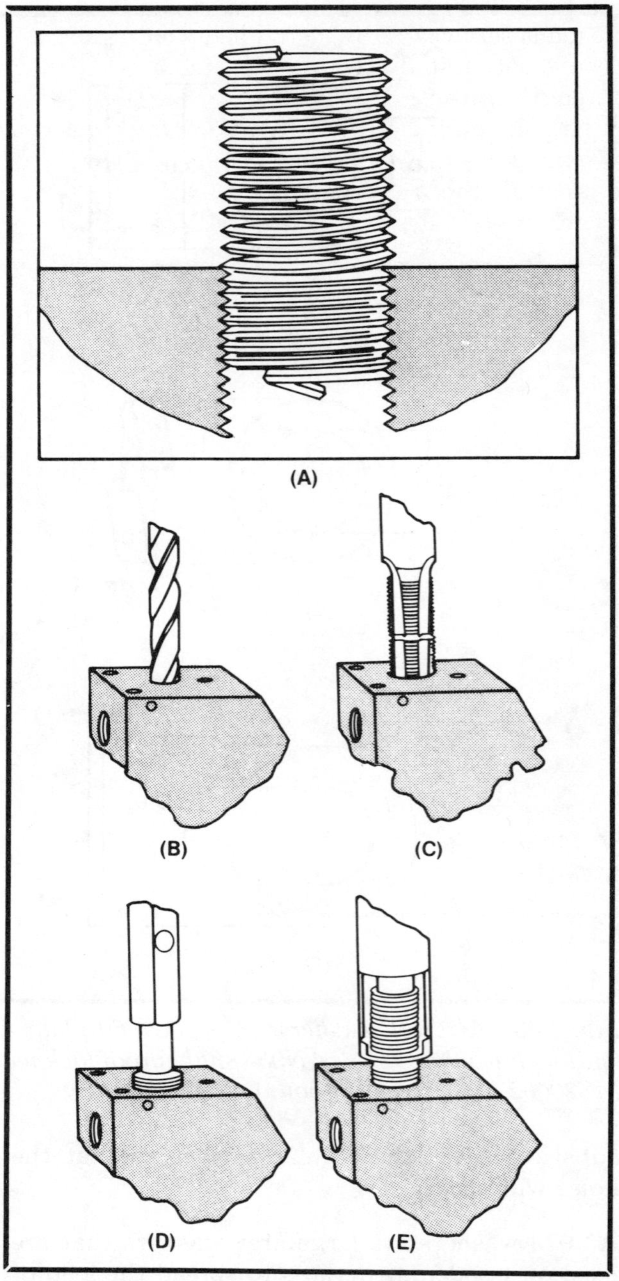

Fig. 1-43 Helicoil insert A — The Helicoil insert is a special stainless steel helical insert that screws into a threaded hole to restore damaged threads to their original size. B — The damaged threads are drilled out. C — New threads are cut with a special tap. D — The threads are gaged with a thread gage. E — The Helicoil insert is screwed into the hole with a special inserting tool.

2. Acres sleeves

If a hole has worn oversize or if there is corrosion in the hole that can be cleaned out, the part may be made serviceable by installing an Acres sleeve. You must first determine that there is enough strength in the material to allow the hole to be enlarged, and if there is, it may be drilled out approximately 1/64-inch larger than the original diameter, and the special aluminum alloy or stainless steel sleeve may be installed in the hole.

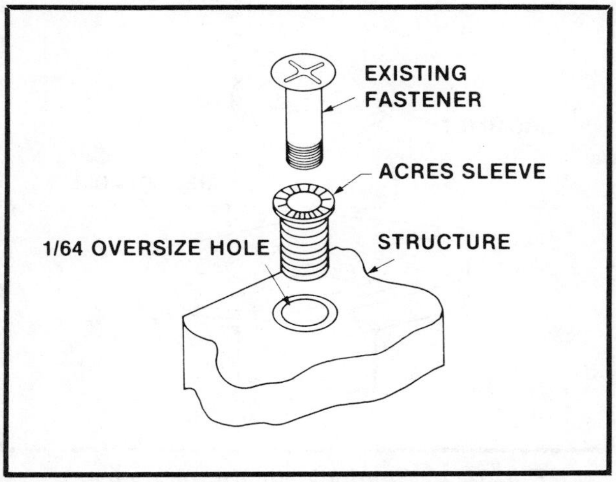

Fig. 1-44 Acres sleeves are used to repair holes that have been worn oversize or corroded. The hole is drilled oversize and the sleeve is pressed or bonded in place and the original fastener is installed.

For certain applications the sleeve may be bonded into the hole with an adhesive, but for others, it is held in the hole by friction and is secured with the regular fastener.

G. Cowling Fasteners

1. Dzus fasteners

Cowling, as well as many inspection access doors that must be opened frequently, may be held with a special type of fastener that requires only a quarter of a turn to lock or unlock. Dzus fasteners are one of the most commonly used types. A hard spring-steel wire is riveted across the opening on the fixed part of the cowling, and a stud in the door is held in place with a metal grommet. When the door is closed, the slot in the stud straddles the spring. Turning the stud a quarter of a turn pulls the spring up into the slanted slot and locks it as the spring passes over the hump in the slot.

When a cowling is fastened with Dzus fasteners, care must be taken that the stud in every fastener straddles the springs rather than passing beside them. In order to be sure that all of the fasteners are properly locked, the slots should all be lined up. When Dzus fastener is being fastened, a

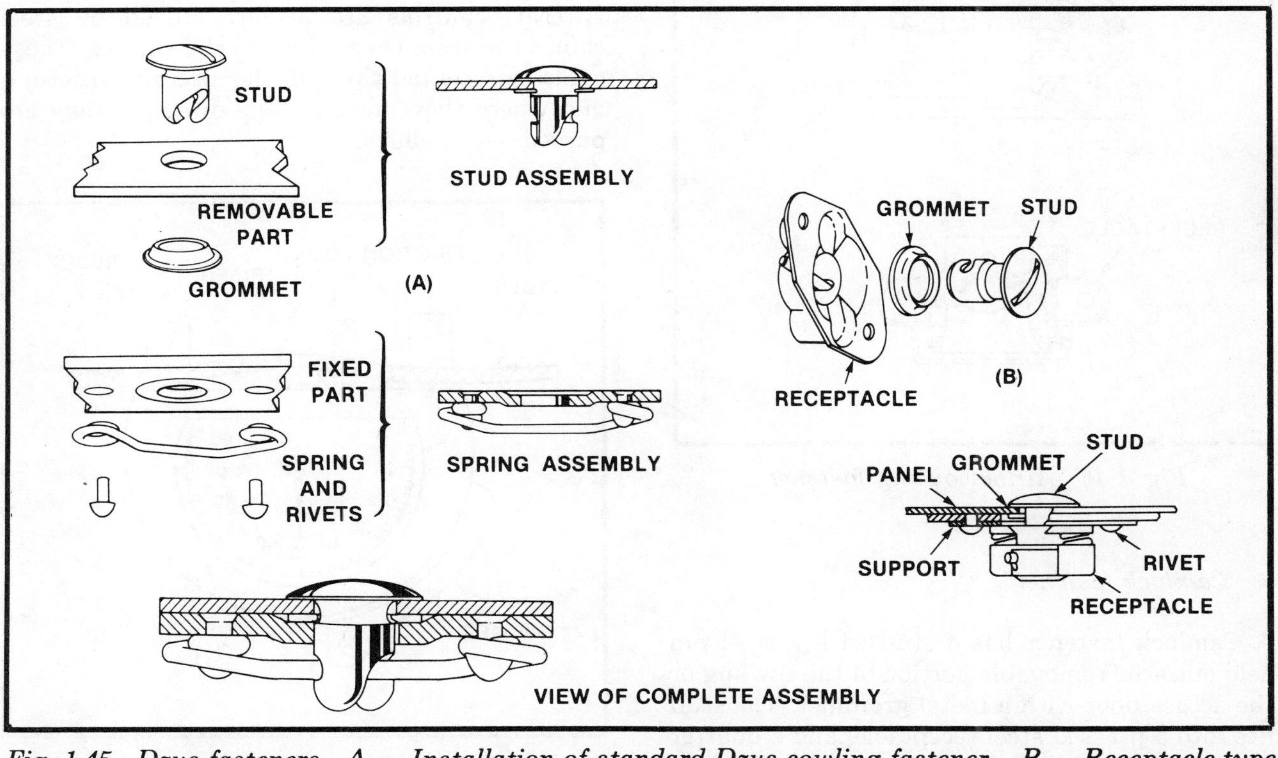

Fig. 1-45 Dzus fasteners A — Installation of standard Dzus cowling fastener. B — Receptacle-type Dzus fastener.

distinct click will be heard when the spring drops over the hump into the locked position. To aid in assuring that no stud misses the spring, special receptacle-type fasteners are available that guide the stud over the spring.

2. Airlock fastener

An Airlock fastener uses a steel stud and cross-pin in the removable cowling or door and uses a sheet-spring-steel receptacle in the stationary member. To lock this type of fastener, the stud is slipped into the receptacle and rotated a quarter of a turn. The pin drops into an indentation in the receptacle spring and holds the fastener locked.

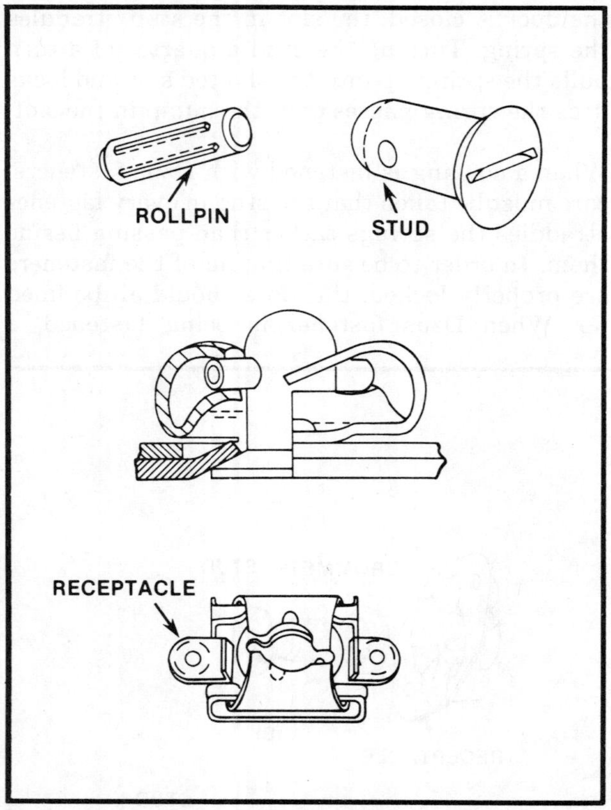

Fig. 1-46 Airloc cowling fastener

3. Camlock fastener

A Camlock fastener has a stud with a steel pin held onto the removable portion of the cowling or the access door with a metal grommet. The stud fits into a pressed steel receptacle, and a quarter of a turn locks the steel pin in a groove in the bottom of the receptacle.

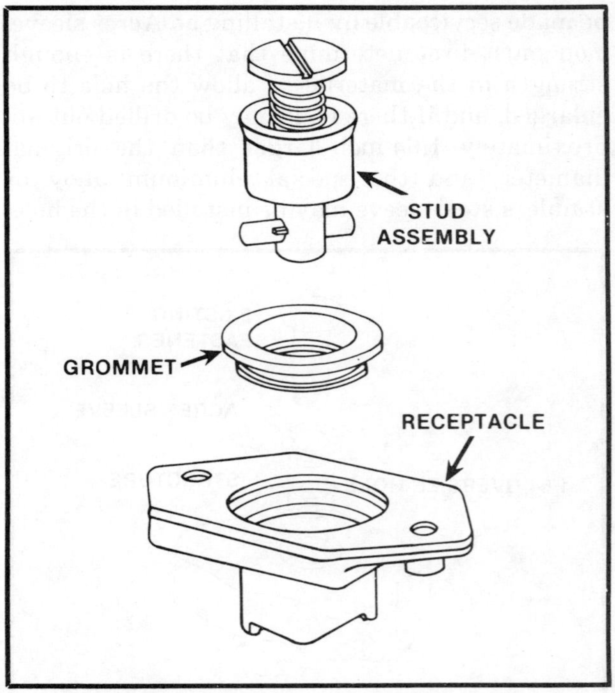

Fig. 1-47 Camlock cowling fastener

H. Control Cables and Terminals

1. Steel control cable

Aircraft controls are usually moved by steel cables run from the cockpit to the surface. These cables are routed through the aircraft structure, and where they must change direction they are passed over pulleys.

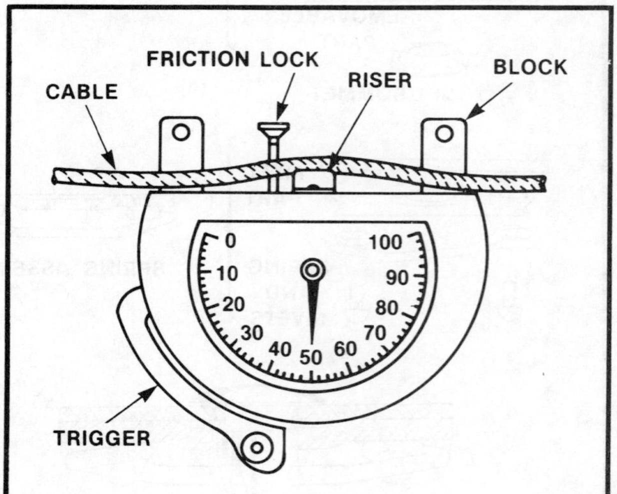

Fig. 1-48 A tensiometer is used to check the tension of aircraft control cables.

One of the disadvantages of a cable system for control movement is the fact that the temperature drops as the airplane goes to altitude, and the vast amount of aluminum in the structure contracts so much more than the small mass of steel in the control cable that the cables lose their tension. Large aircraft have a rather complex automatic tensioning system to keep the control cable tension relatively constant as the aircraft contracts and expands with temperature changes. But small aircraft must have their cable tension adjusted as a compromise, so they will not be too tight when the airplane is hot or too loose when it is cold.

a. Non-flexible cable

For installation where the cable does not pass over any pulleys, non-flexible cable may be used. This is available in either the 1 x 7 or the 1 x 19 configuration. The 1 x 7 cable is made up of seven individual wires, and the 1 x 19, of 19 individual wires. Non-flexible cable is available in both galvanized carbon steel and stainless steel.

b. Flexible cable

Steel cable made up of seven strands of seven wires each is called 7 x 7 or flexible cable, and is available in 1/16- and 3/32-inch sizes in both galvanized carbon steel and in stainless steel. Both types are preformed. This means that in its manufacture each strand is formed into a spiral shape, so that when the cable is wound all of the strands will stay in their position and the cable will not spread out when it is cut. Preforming gives the cable greater flexibility and relieves the bending stresses when the strands are woven into the cable.

c. Extra-flexible cable

Seven by nineteen cable is the most widely used cable in sizes from 1/8-inch up. It is extra flexible and is made up of 133 individual wires wound in seven strands, with each strand having 19 wires. These cables are preformed and are available in both galvanized and stainless steel. The galvanized cable has more resistance to fatigue than the stainless steel, but in applications where corrosion is a factor, stainless steel is used.

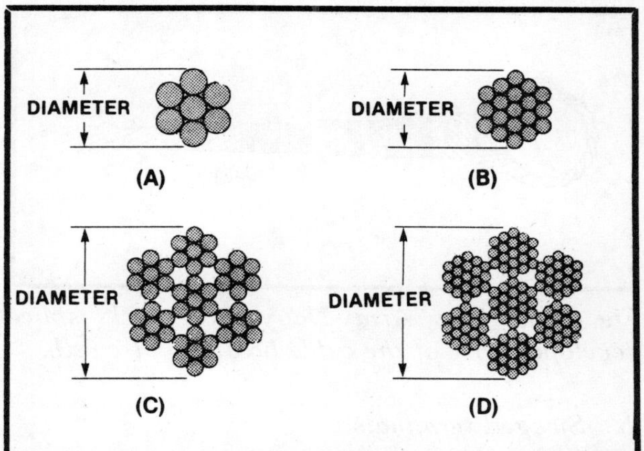

Cable Size	Type	Breaking Strength — Lbs. Galvanized	Stainless
1/16″	1 × 7	500	500
	7 × 7	480	480
3/32″	1 × 19	1200	1200
	7 × 7	920	920
1/8″	1 × 19	2100	2100
	7 × 19	2000	2000
5/32″	1 × 19	3300	3300
	7 × 19	2800	2400
1/4″	1 × 19	8200	8200
	7 × 19	7000	6400

Fig. 1-49 Steel control cable
A - 1 × 7 non-flexible cable
B - 1 × 19 non-flexible cable
C - 7 × 7 flexible cable
D - 7 × 19 extra-flexible cable

2. Cable termination

a. Woven splices

In the early days of aviation most of the control cables were terminated with woven splices, such as either the Army-Navy five-tuck splice or the Roebling roll. Because both types of woven splices require a great deal of hand work, and they develop only 75% of the cable strength, these terminals have been replaced with swaged-type terminals.

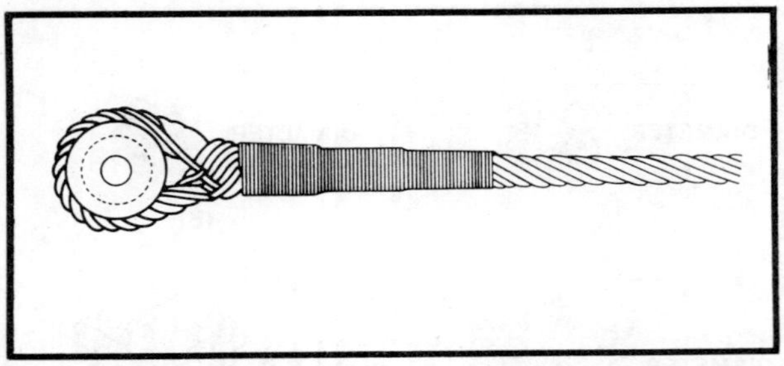

Fig. 1-50 The Army-Navy five tuck splice developed 75% of the cable breaking strength.

b. Swaged terminals

The terminals that are most universally used in large aircraft manufacture are the MS-type terminals we see in Fig. 1-51. To install these terminals, the cable is cut and inserted into the end of the terminals, and either a hand- or a power-swaging tool forces the metal of the terminal down into the cable and forms a joint that is at least as strong as the cable itself.

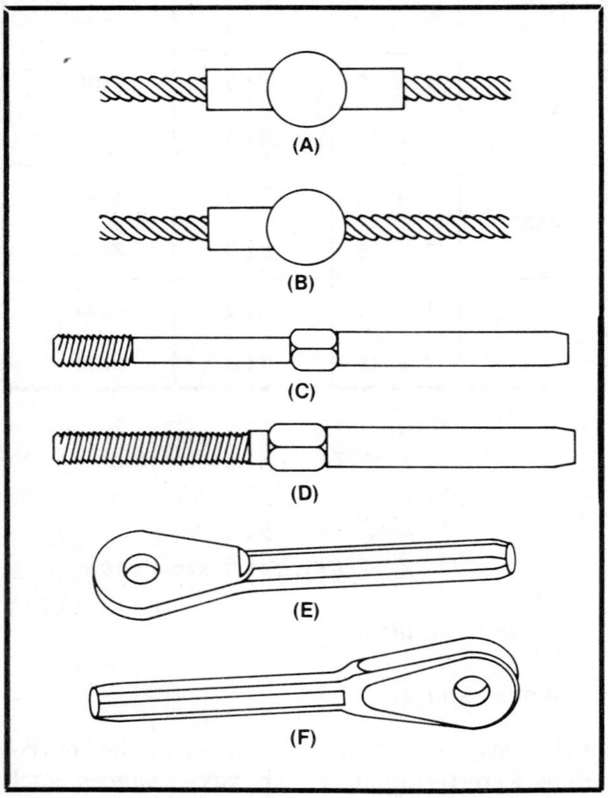

Fig. 1-51 Swaged cable terminals A — MS20663C ball and double shank B — MS20664C ball and single shank C — MS21259 long-threaded stud end D — MS21260 short-threaded stud end E — MS20667 fork end F — MS20668 eye end

To be sure that the terminal is properly swaged, a measurement is made of the terminal before it is swaged with a go no-go gage. The swaging process must decrease its diameter to the extent that the go end of an "after" go no-go gage will pass over it, but the no-go end will not. As a check to be sure that the cable does not pull out of the terminal, a small mark of paint is placed over the end of the terminal and onto the cable. If this paint is ever broken, there is evidence that the cable has slipped inside of the terminal.

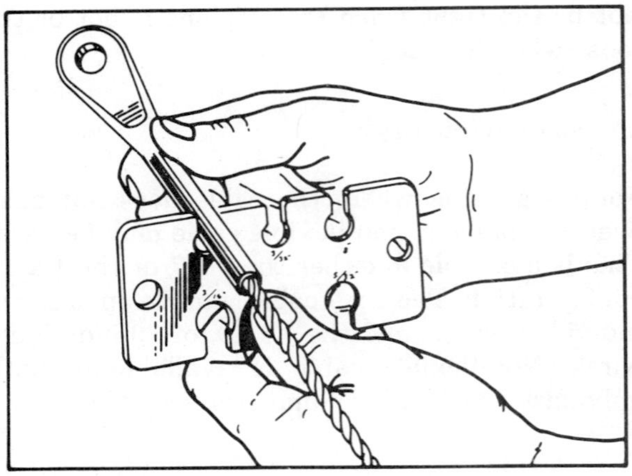

Fig. 1-52 Checking swaged terminal with a go no-go gage for proper swaging.

c. Nicopress oval sleeves

Many of the light aircraft use Nicopress sleeves squeezed onto the control cables to form the end terminals. When the cable is wrapped around an AN100 thimble and properly squeezed with the correct Nicopress squeezer, the terminal will have at least the strength of the cable.

To install this type of terminal, the copper sleeve is slipped over the end of the cable and the end doubled back and slipped into the other side of the sleeve. An AN100 thimble of the proper size is placed inside the loop, and the cable is pulled tight around the thimble with about 1/16-inch of cable sticking from the sleeve. Three squeezes are taken with the squeezing tool — the first squeeze is in the middle of the sleeve, the second is on the end nearest the thimble, and the last squeeze is on the end of the sleeve nearest the free end of the cable. Using this sequence will assure that the thimble is gripped tightly into the cable loop.

After the sleeve is squeezed, it is checked with a gage to be sure that it has been forced far enough down into the cable.

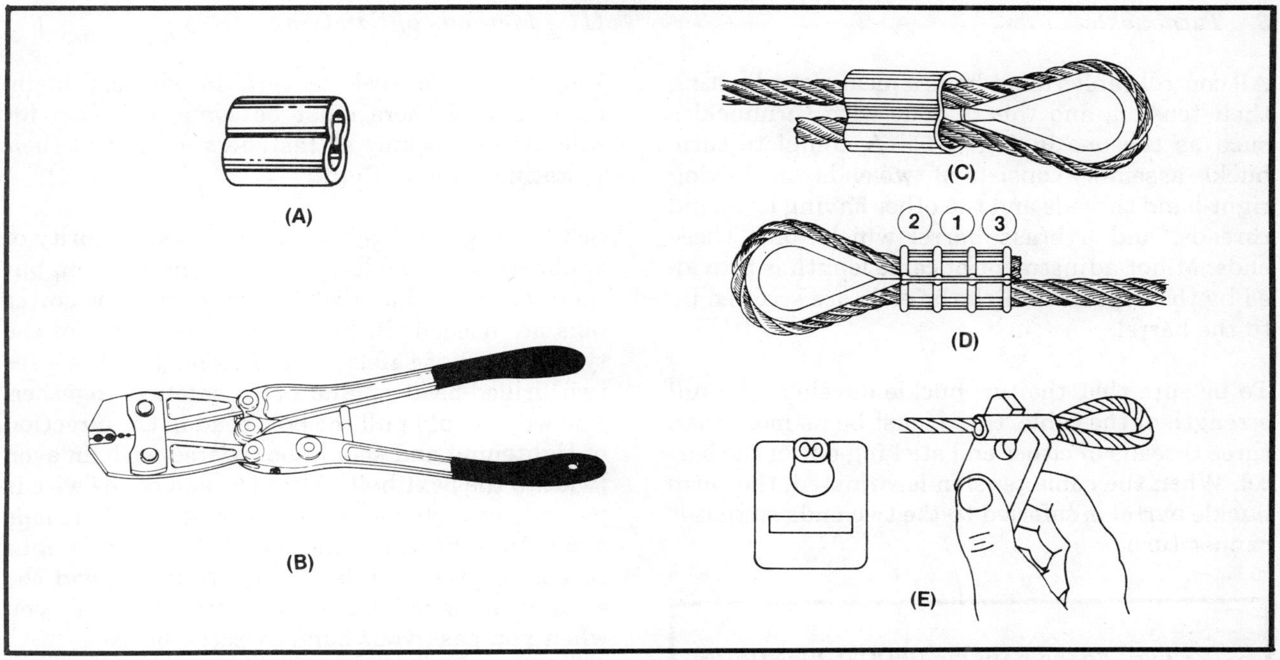

Fig. 1-53 Nicopress oval sleeve cable termination A — Nicopress sleeve before it is placed on the control cable. B — Hand tool for compressing Nicopress sleeves. C — Proper crimping sequence. The sleeve is slipped on the cable and the end bent back over the thimble and stuck through the sleeve. Three squeezes are taken, the center squeeze first, then the squeeze nearest the thimble and finally the squeeze on the end. D — The compressed sleeve is checked with a go no-go gage to determine if it has been properly installed.

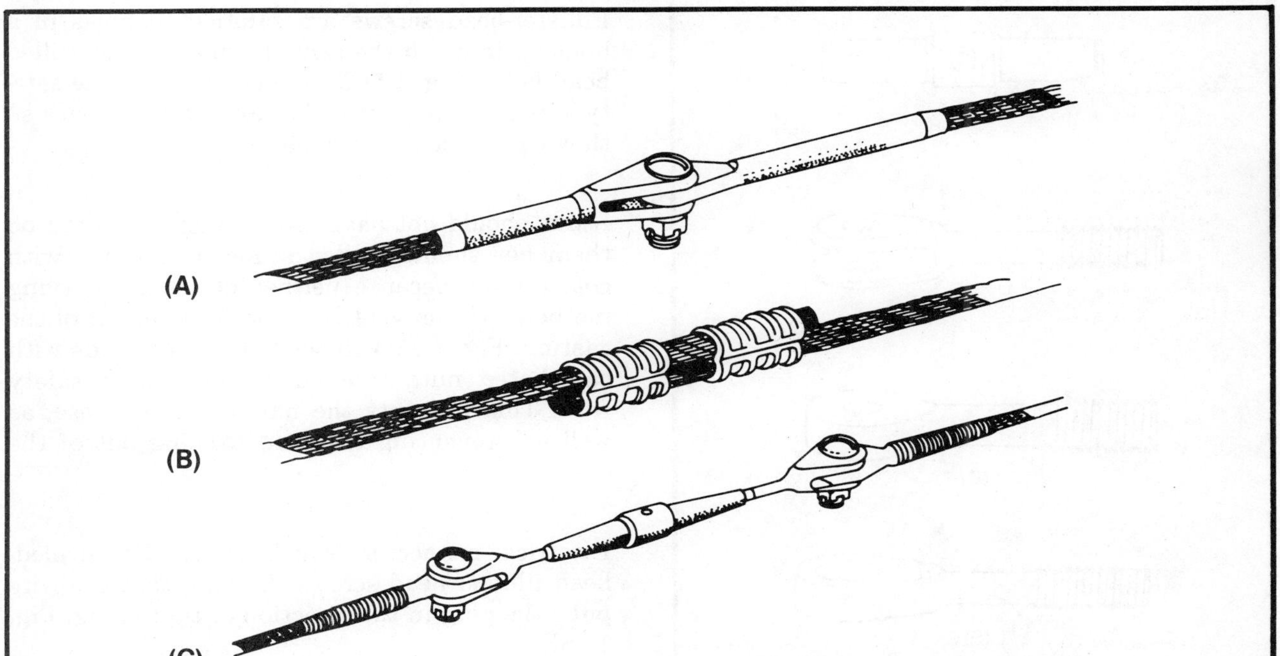

Fig. 1-54 Control cable splices A — Swaged fork and eye. B — Nicopress compression sleeves. C — Woven splices connected with a turnbuckle.

3. *Turnbuckles*

All control cables must have a means of adjusting their tension, and this is done with turnbuckles such as the one in Fig. 1-55. A complete turnbuckle assembly consists of two ends, one having right-hand threads and the other having left-hand threads, and a brass barrel which joins these ends. Minor adjustment of cable length is provided by the depth to which the ends are screwed into the barrel.

To be sure that the turnbuckle develops the full strength of the cable, there must be no more than three threads of either end sticking out of the barrel. When the cable tension is adjusted, the turnbuckle barrel is safetied to the two ends so that it cannot turn.

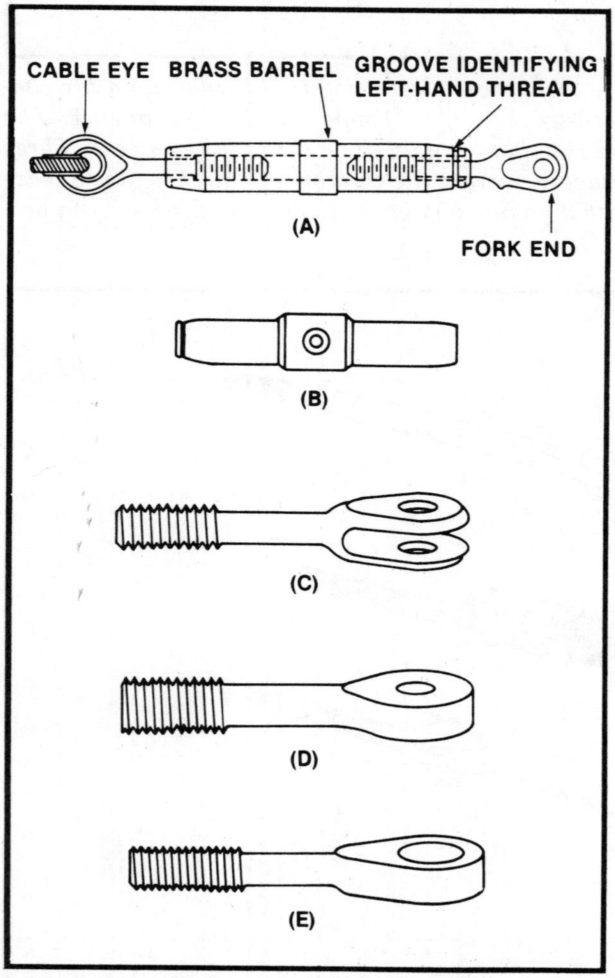

Fig. 1-55 Turnbuckle nomenclature A — Assembled turnbuckle B — Turnbuckle barrel C — Turnbuckle fork D — Turnbuckle pin eye E — Turnbuckle cable eye

III. *Methods of Safetying*

Vibration is a real concern in aircraft maintenance, and there must be some provision for safetying or locking all fasteners to prevent their vibrating loose in flight.

Self-locking nuts are used for the vast majority of applications in modern aircraft construction, but there are still places where safety wire or cotter pins are needed. In Fig. 1-56 we see some of the typical uses of safety wire. In Fig. 1-56A we see two drilled-head engine bolts safetied together. The wire should pull the bolt head in the direction of tightening and should be twisted with an even twist to the next bolt. After the end of the wire is passed through the head of the second bolt it is again twisted, this time for about three or four turns, and the wire is cut off. Always bend the ends of the wire back where they cannot cut you when you pass your hand over the bolts.

If there are a number of bolts, such as used to mount a propeller, you may safety them in groups of three. It is usually not a good idea to tie more than three together with one piece of wire, Fig. 1-56B.

Fillister-head screws are safetied to holes in a housing in much the same manner as the drilled head bolts, Fig. 1-56C. Remember, anytime safety wire is cut off, the ends should be bent back so they cannot cut your hand.

Studs should not have self-locking nuts used on them nor should drilled studs be safetied with cotter pins, because neither of these safetying methods will prevent the stud backing out of the casting. Fig. 1-56D shows two drilled studs with castellated nuts safetied together with safety wire. This prevents the nut becoming loose, as well as preventing the stud backing out of the casting.

Electrical connectors can be safetied to drilled-head fillister-head screws. Be sure that the wire pulls the plug in the direction of tightening, Fig. 1-56E.

Turnbuckles may be safetied in a number of ways. Those used in control systems where the cable is 1/16- or 3/32-inch diameter may be safetied by the single-wrap method, Fig. 1-57A,

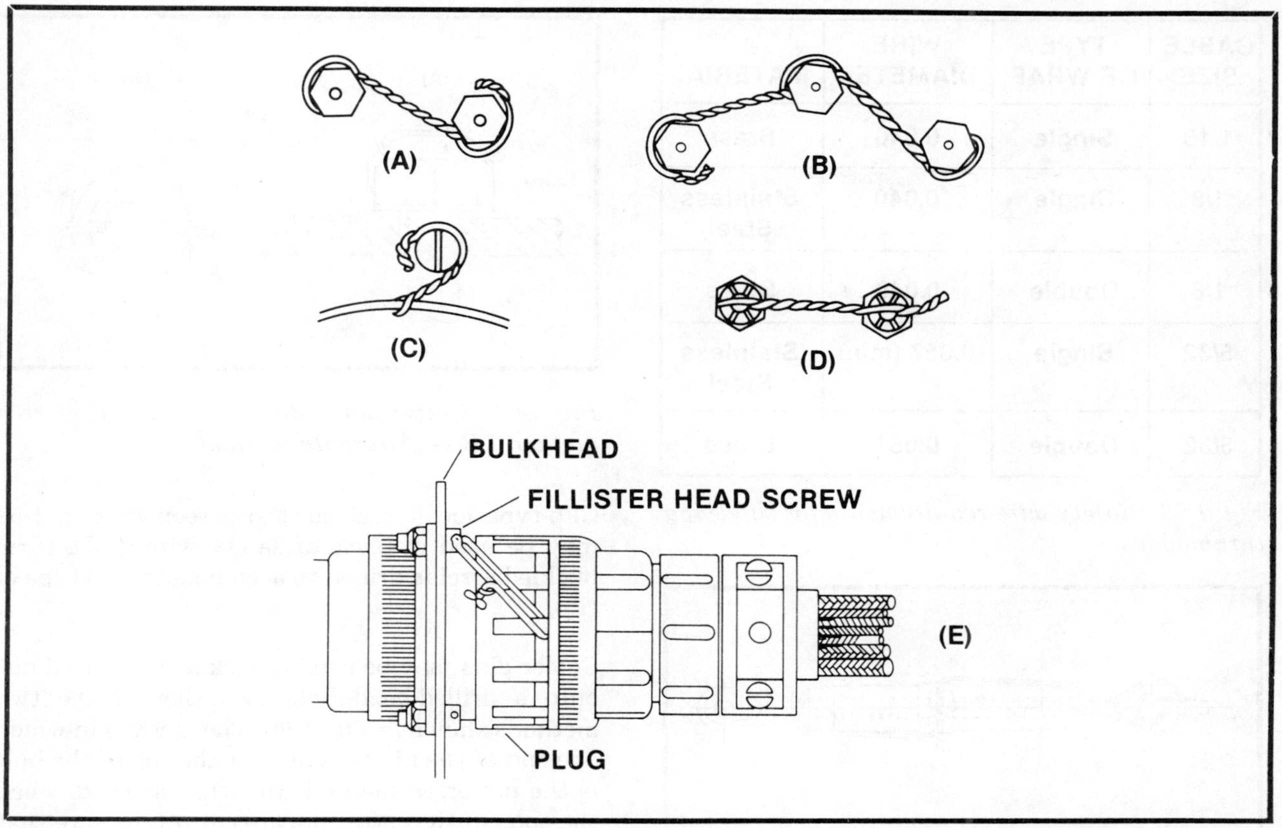

Fig. 1-56 Safety wiring procedures A — Safetying together two drilled head bolts. B — Safetying together three drilled head bolts. C — Safetying a fillister head screw to the structure. D — Safetying castellated nuts on two studs. E — Safetying an electrical plug to a fillister head screw.

using either brass or copper safety wire, 0.040 diameter. Turnbuckles used with 1/8-inch cable may be safetied with a single wrap if 0.040 stainless steel or monel safety wire or 0.057 copper or brass wire is used. Forty-thousandths-inch copper or brass wire may be used if the turnbuckle is double-wrap safetied.

Turnbuckles in 5/32-inch control cable can be double-wrap safetied with 0.040 stainless steel or 0.051 copper or brass, or they may be single wrapped with 0.057 stainless steel wire.

Before safetying a turnbuckle, it must be adjusted to the correct cable tension, and there must not be more than three threads showing on either side of the barrel. Wrap the wire around the turnbuckle in the manner shown in Fig. 1-57 and finish the safety wiring with at least four turns around the shank of the turnbuckle before it is cut off.

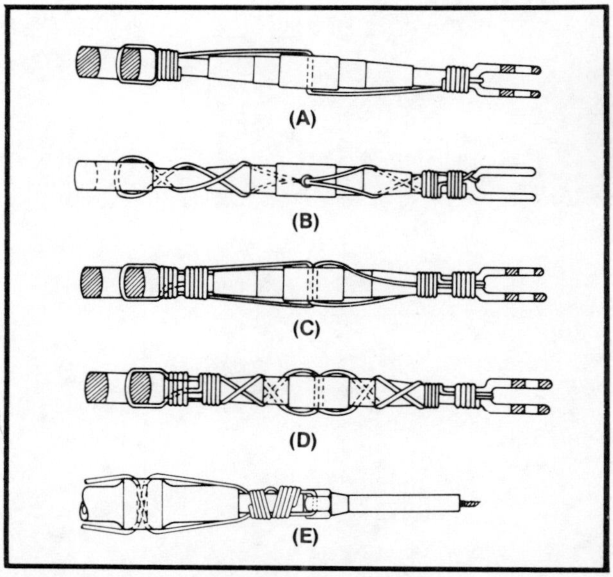

Fig. 1-57 Safetying turnbuckles with safety wire A — Single-wrap B — Single-wrap spiral C — Double-wrap D — Double-wrap spiral E — Double-wrap tied to a swaged cable stud

CABLE SIZE	TYPE OF WRAP	WIRE DIAMETER	MATERIAL
1/16	Single	0.040	Brass
1/8	Single	0.040	Stainless Steel
1/8	Double	0.040	Brass
5/32	Single	0.057 (min)	Stainless Steel
5/32	Double	0.051	Brass

Fig. 1-58 Safety wire requirements for safetying turnbuckles

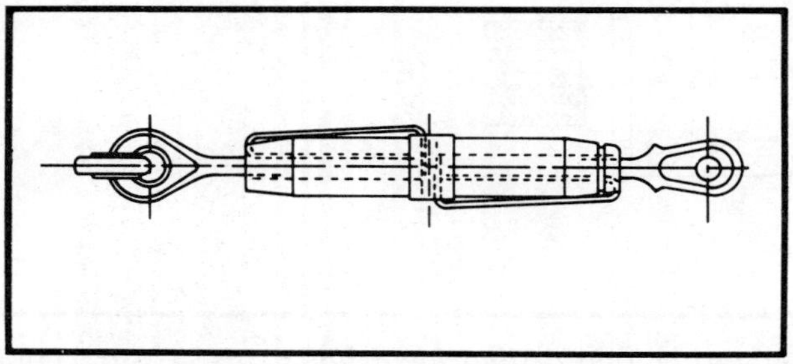

Fig. 1-59 Turnbuckle safetied with clips.

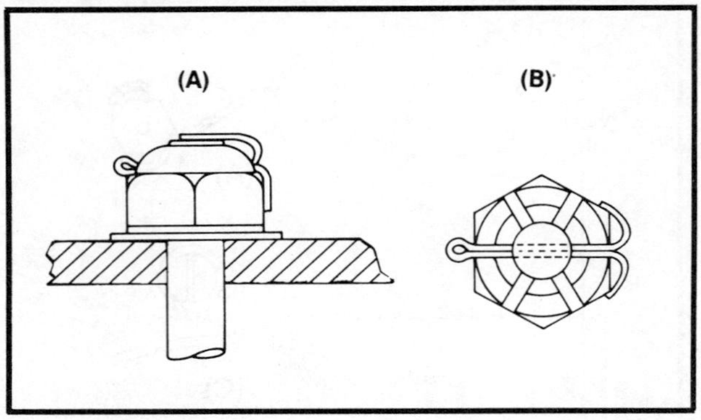

Fig. 1-60 Cotter pin installation A — Preferred method B — Alternate method

Clip-type locks such as those seen in Fig. 1-59 may be used in place of safety wire if the turnbuckle barrel is drilled to accomodate this type of clip.

Cotter pins may be used to lock a castellated nut onto a drilled-head bolt by either of the two methods shown in Fig. 1-60. The method in which one end of the pin is bent over the top of the bolt is the preferred method. In either method, when the ends of the cotter pin are cut off, be sure that they are bent back so they cannot cut you if you rub your hand over the end of the bolt.

SECTION II

Hand Tools And Measuring Devices

Fig. 2-1 An A&P mechanic has a large investment in his tool chest. Having the correct tool for the job makes work more productive.

Technical progress has been made in our society only as man has had tools with which to change his environment. At first he had rocks or stones, to pound with, and then by chipping a sharp edge on his stone he was able to dig and cut. A handle was added to the stone to give increased leverage, and his crude tool became more effective. As metal was discovered and made workable, it replaced wood and stone for many applications, especially when ways were found to harden the metal. It has, throughout history, been quite evident that no culture has progressed beyond its ability to produce and use tools.

Aviation maintenance is totally dependent on the use of tools, and all practicing technicians have made a sizable investment in their tools. In professional aviation maintenance, time is such a critical factor that good economics demands that the efficient technician have a large variety of high quality tools so he can have the right tool for the job.

I. Types of Hand Tools

A. Pounding Tools

1. Ball peen hammers

One of the most familiar tools to a mechanic is the ball peen hammer. These hammers range in weight from as small as one ounce to as large as two to three pounds. One face is flat, and the other is in the form of a ball whose purpose is to peen over rivets in commercial sheet metal work. Of course, this method of securing rivets is not used for aircraft sheet metal work.

The face of a ball peen hammer is quite flat rather than slightly crowned as the face of a claw hammer, and because of this it is not practical to use a ball peen hammer to drive nails. The flat face does not allow control in keeping the nail from bending.

The ball peen hammer is designed to pound on hardened steel, and for this reason the hardness of its face is controlled so it will not chip.

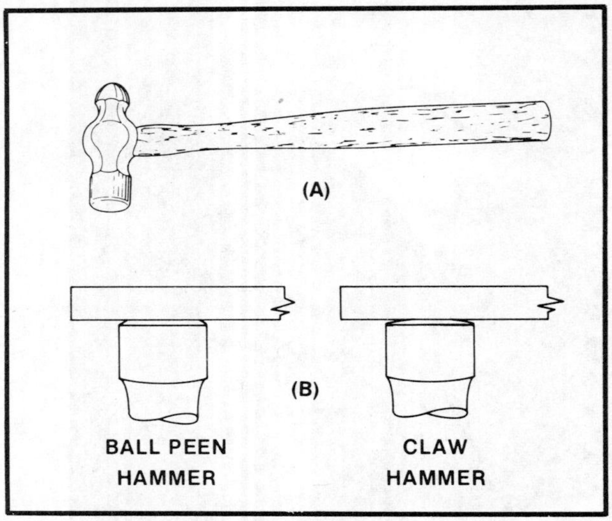

Fig. 2-2 A — The ball peen hammer is the most commonly used driving tool in the mechanic's tool chest. B — The face of a ball peen hammer is flat whereas the face of a claw hammer is slightly crowned to give control when driving nails.

2. Cross peen and straight peen hammers

Commercial sheet metal and automotive body work often requires that the metal be bent by hammering. This may be done by using either cross peen or straight peen hammers. Instead of a ball end, the heads of these hammers have a wedge-type end that can be used to either crease the metal to start a bend, or can be used to straighten out a rolled edge of a piece of metal.

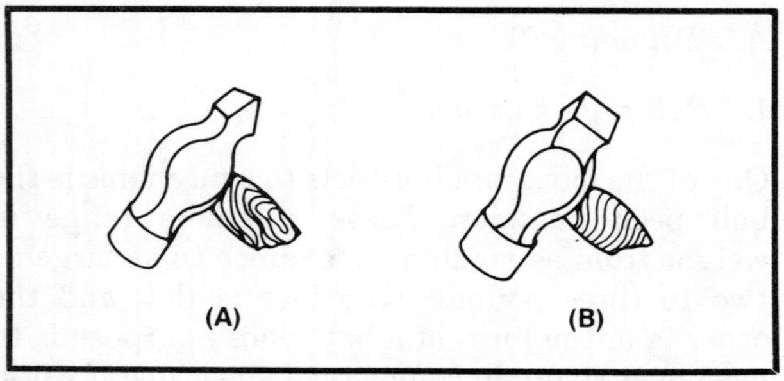

Fig. 2-3 A — Cross peen hammer B — Straight peen hammer

The wedge of the cross peen hammer is at right angles to the handle, while the wedge of a straight peen hammer is parallel to the handle.

3. Claw hammers

While we cannot consider this to be a special aviation tool, the carpenter's claw hammer is a most useful tool around the shop. The head of this hammer is slightly crowned so the nail can be controlled as it is driven, and the face of this hammer is usually hardened until it is more brittle than the face of a ball peen hammer. For this reason it should not be used to hammer on hardened steel parts, as there is a possibility of the hammer chipping.

There are two types of claws used on these hammers. Some claws curve so nails can be pulled out easily, and some are straight so they can be used to split wood as well as pull nails.

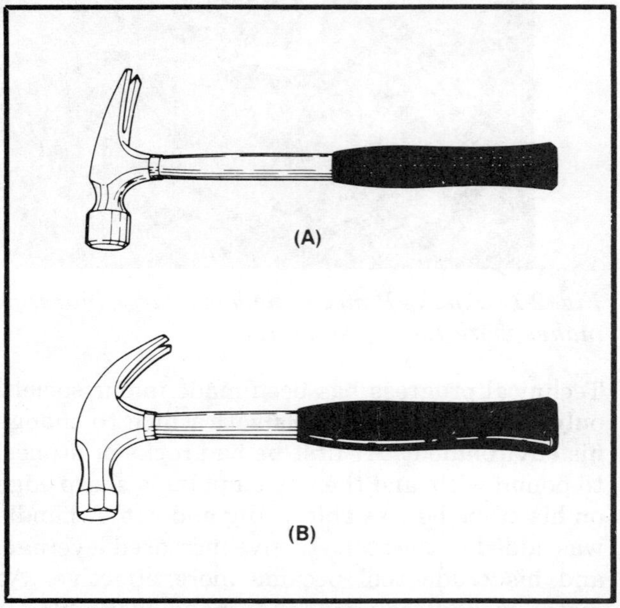

Fig. 2-4 A — The straight claw of this hammer is used to split wood. B — The curved claw of this hammer makes it easy to pull nails.

4. Sledge hammers

Sledge hammers have two flat faces and are used for heavy pounding operations such as driving stakes into the ground.

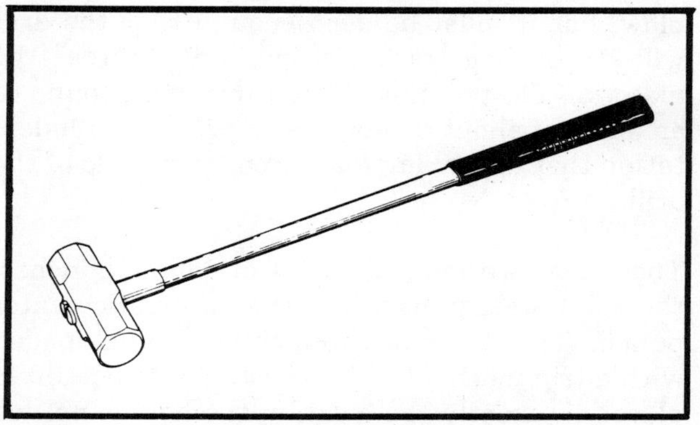

Fig. 2-5 Sledge hammer

5. Body hammers

Body hammers, sometimes called planishing hammers, have large smooth faces and are used to smooth out sheet metal when forming it. These hammers do not have a great deal of weight and are used specifically for removing small dents and for smoothing or stretching the metal.

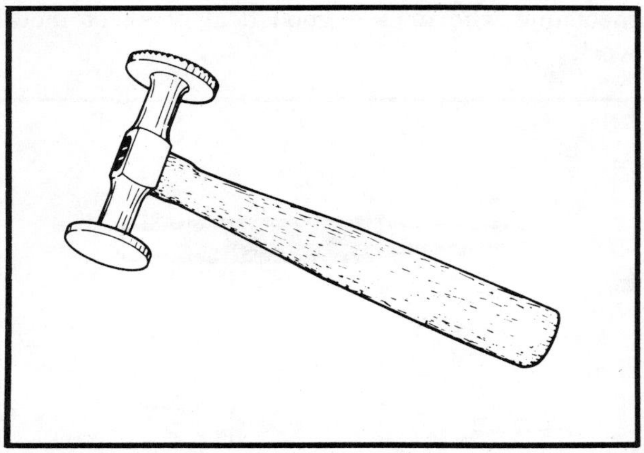

Fig. 2-6 Body hammer

6. Mallets

Much forming and shaping of soft aluminum alloy is done with soft-faced mallets. The original soft mallets used in aviation consisted of a roll of rawhide held in a clamp that provided the necessary weight and rigidity to increase the effectiveness of the mallet.

As the state of the art of synthetic materials has developed, plastic mallets have just about replaced those made of rawhide. Some of these mallets have a hard plastic tip and some a more resilient rubber-like tip. Many of these mallets have replaceable tips so that when a tip is worn out, the hammer need not be discarded; only the tip must be replaced.

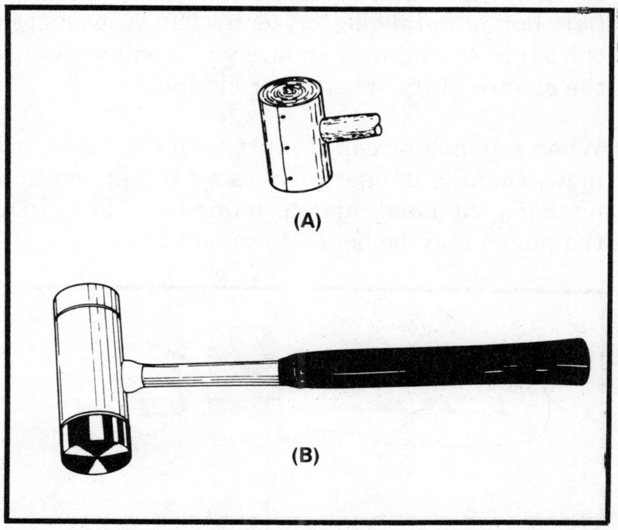

Fig. 2-7 A — Rawhide mallet B — Plastic mallet with replaceable tips

7. Punches

While not actually a pounding tool in themselves, punches direct the blow of the hammer to place the force exactly where it is needed. By concentrating the force of the blow to the small area covered by the end of the punch, the pressure is greatly increased.

It is extremely important when using punches that you always wear safety glasses or goggles to protect your eyes and face in the event that either the hammer or the punch should chip.

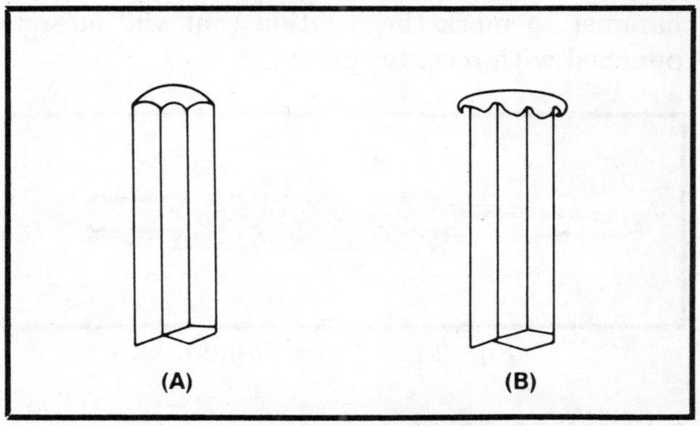

Fig. 2-8 A — The properly ground end of a punch or a chisel is slightly crowned to concentrate the hammer blows. B — When the end of a punch or chisel is mushroomed from hammer blows it must be ground to restore its original shape.

Continued hammering on a punch shank will upset the end and form a mushroom. This must be removed and the end of the punch ground to the shape it originally had. The end is not ground flat, but is parabolic, or somewhat crowned, so the hammer will hit it squarely and will minimize the chance of its splitting or chipping.

When a punch or chisel must be hit a very hard blow, there is danger of missing the punch and hitting your hand, and to minimize this chance the punch may be held in a punch holder.

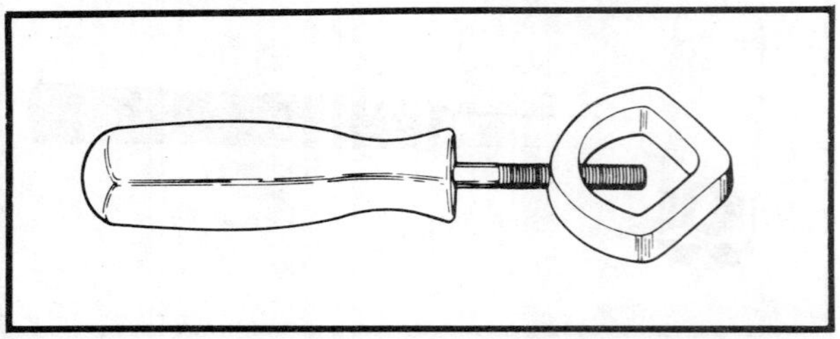

Fig. 2-9 A punch or chisel may be clamped in a punch holder when it is being used.

There are several types of punches the mechanic must be familiar with; some are used to mark locations, some to indent the material so a drill will start in exactly the correct location, and some to punch bolts, rivets, or pins from their holes. Each punch is designed for a specific purpose, and each should be used only for the purpose for which it is designed.

a. Prick punches

These small, sharp-pointed punches are used to transfer dimensions and locations onto the metal that is to be drilled. They are tapped with a small hammer to mark the location that will later be punched with a center punch.

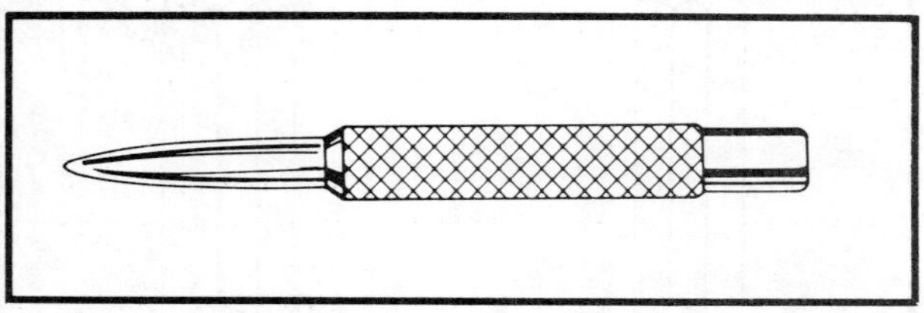

Fig. 2-10 Prick punch

b. Center punches

A relatively sharp-pointed punch is used to make an indentation in the metal where a hole is to be drilled. If the hole is made too deep, the metal may be distorted by an unnecessary hard hammer blow, but it must be deep enough that the drill will start to cut rather than wander across the material. The point of a center punch is ground to an angle of about 60 degrees to provide an indentation that approximates the cutting angle of the drill.

There are two general types of center punches; the solid steel punch, and the automatic center punch. The automatic punch is spring loaded with a trip mechanism in its handle. The point is placed in exactly the location the hole is needed and the handle is pushed in. When the proper force is reached, the trip mechanism will release and the point will be struck with a sharp blow from within the handle. Rotating the end of the handle adjusts the compression on the spring and controls the amount of impact applied to the point. This type of punch allows punch marks to be made that are accurately located and consistently deep, and its relatively low cost and the speed with which holes can be marked with an automatic center punch make this type of punch a vital addition to the tool collection of any mechanic who does a good deal of sheet metal work.

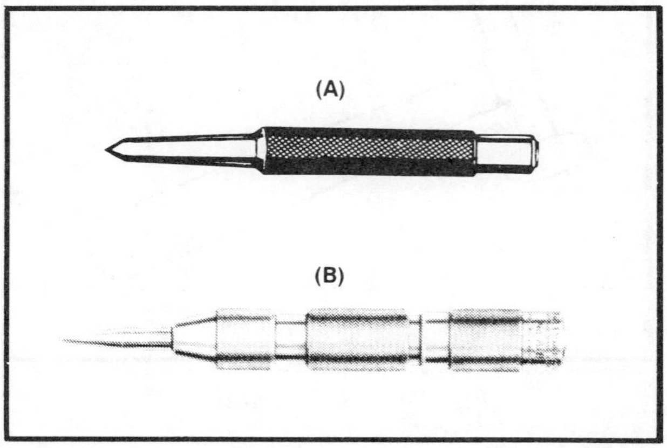

Fig. 2-11 A — Center punch B — Automatic center punch

c. Starting punches

Prick punches and center punches both have sharp points and should never be used to drive a pin or rivet from a hole, as they will spread the end and make it more difficult to drive out. A special punch having a tapered shank is used to start the fastener from its hole, and it is driven with this punch until the tapered shank just about fills the hole; then the job is finished with a pin punch.

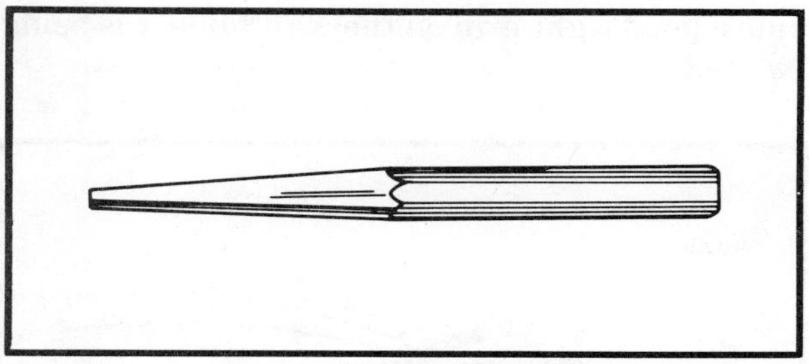

Fig. 2-12 *Starting punch*

d. Pin punches

One of the most useful punches for a sheet metal worker is the pin punch. When a rivet is removed from an aircraft structure, the head is drilled through and is knocked off of the rivet shank with a small chisel; then the shank is punched out of the hole with the proper size pin punch. These punches are available in the sizes of all of the rivets, and it is important that the correct size punch be used. The thin sheet metal must be properly backed up to prevent bending it when the rivet shank is driven out.

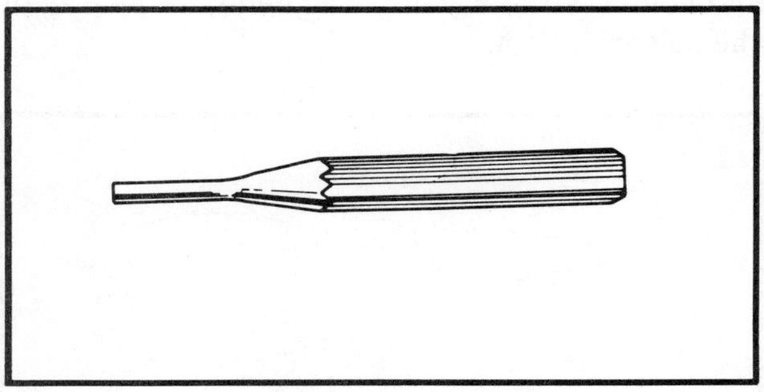

Fig. 2-13 *Pin punch*

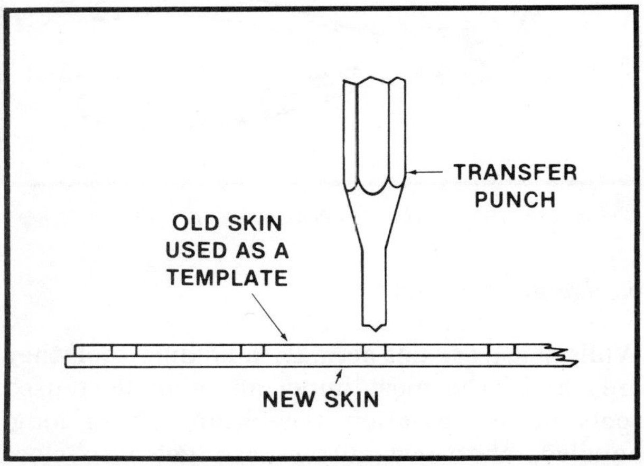

Fig. 2-14 A transfer punch has a shank the size of the rivet hole being transferred and a sharp point in its exact center.

e. Transfer punches

A special form of marking punch is the transfer punch, used to mark rivet holes when using an old skin as a template for laying out a new skin. The diameter of the shank is the same as that of the rivet hole, and the end is ground so that the mark on the new skin will be in the actual center of the hole.

Transfer punches are made as both straight and automatic punches.

B. Holding Tools

1. Slip-joint pliers

Slip-joint pliers are the most commonly used holding tool found in an aviation mechanic's tool chest. These pliers come in lengths from around four inches to more than nine inches, with the six-inch size the most commonly used. They are called slip-joint because of the double hole for their pivot, which allows the jaws to work in either of two positions. This vastly increases the range of material that may be gripped.

While these are exceptionally versatile tools, they should be used only for the purpose for which they are intended. They should never be used as a wrench to turn a nut, as they invariably round off the corners of the nut so the proper size wrench cannot later be used. This is especially true of tubing nuts which are made of soft metal such as aluminum or brass.

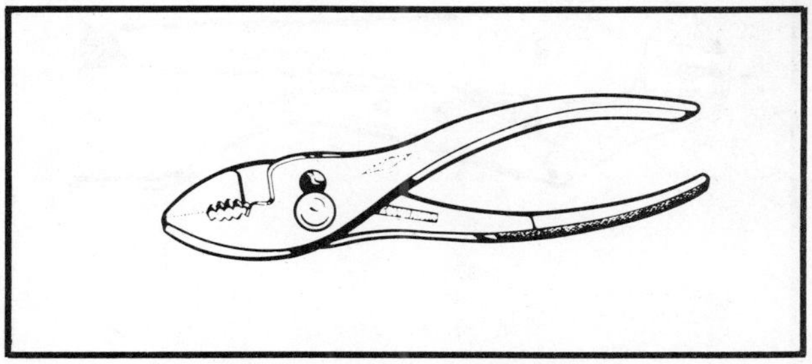

Fig. 2-15 *Slip-joint pliers*

2. Interlocking-joint pliers

These are commonly called water pump pliers because of their ability to open up to such a wide jaw opening that they can be used to tighten the packing gland nut around a water pump shaft. These pliers are available in lengths from around eight inches up to about 20 inches.

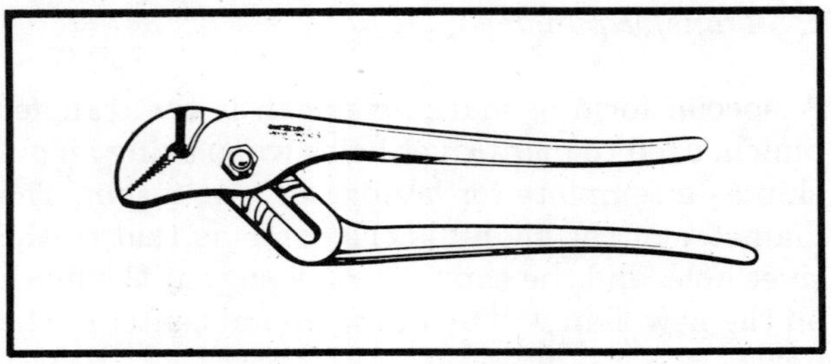

Fig. 2-16 Waterpump pliers

3. Vise-grip pliers

This is the registered trade name of the Petersen Mfg. Co. for a special compound-action type of pliers. The opening of the jaws may be adjusted by a knurled screw in the end of the handles, and when the handles are squeezed together, compound leverage multiplies the effort and applies a tremendous force to the jaws. A toggle action clamps the jaws together so they will not open when the handles are released. The jaws are released by a small lever in one of the handles.

These versatile tools come in a wide variety of lengths and jaw styles. Some are designed to hold pipes, some to cut wires, and others to pinch off hoses. Special forms of Vise-grip pliers are made with sheet metal-bending jaws, and some even with jaws that serve as welding clamps and C-clamps.

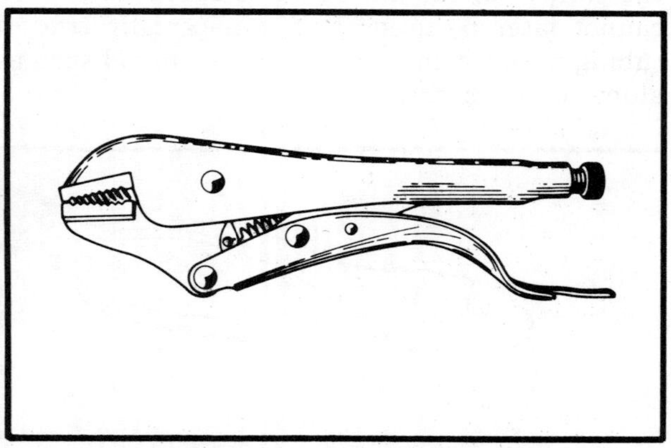

Fig. 2-17 Vise-grip pliers

4. Duckbill pliers

One type of pliers that finds special use in aviation maintenance is the duckbill pliers. These long-handled flat-nose pliers are used to twist safety wire. The jaws have serrations that grip the wire, and the handles are long enough to provide a good tight grip on the wire while it is being twisted.

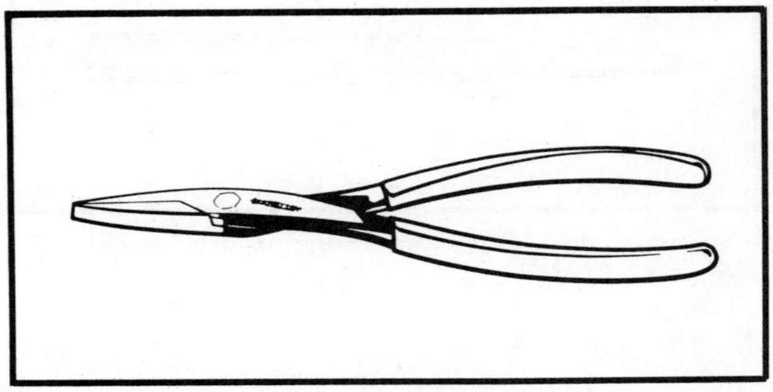

Fig. 2-18 Duckbill pliers

5. Needle nose pliers

Several different designs of long nose pliers are used in aviation maintenance, especially in electrical and electronic work to grip and hold small components and wires while they are being installed or removed. Some of these pliers have long thin jaws that are bent almost at right angles to the handles. to allow them to grip the component and hold it without your hand being in the way of the soldering iron.

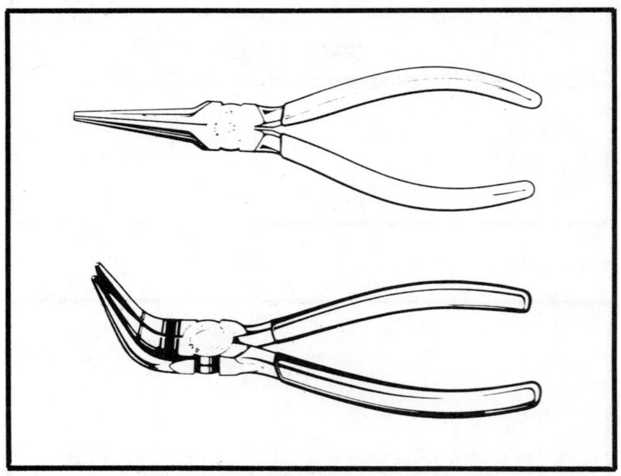

Fig. 2-19 Needle nose pliers

6. Diagonal cutters

While these are not actually a holding tool, they are one of the most important and most used tools of the aviation technician. These long-handled, short-jaw cutters can get into close quarters to clip safety wire after it has been twisted with duck bill pliers. Diagonals, or dikes, as they are commonly called, are available from

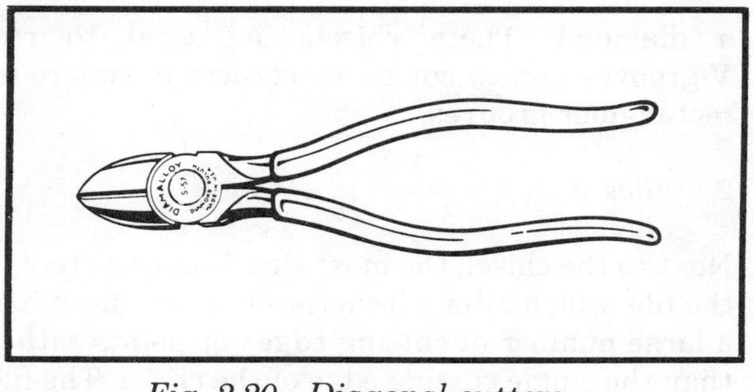

Fig. 2-20 *Diagonal cutters*

the extremely small size used in delicate electronic and instrument work up to about nine inches in length.

One special form of cutter that is quite similar to diagonal cutters is the flush cutter. These cutters leave a perfectly smooth edge on the part they cut and are used for cutting the stem off of certain types of blind rivet whose head is upset by pulling the stem through the hollow shank until it breaks off.

7. *Wire twister/cutters*

A special tool for use by the aviation technician is the wire twisting and cutting tool. This special tool combines the features of diagonals with those of duck bills and has a special twister built into the handle. The wire is cut to length and inserted into the bolt head or into the hole in the nut that is to be safetied. The wire is crossed and gripped with the flat serrated jaws, and the handles are locked together. When the knob in the handle is pulled, a spiral rod in the handle rotates the pliers and twists the wire. When the twisting is completed and the wire is secured to the structure or to the next fastener, the ends of the wire can be cut off with the cutters.

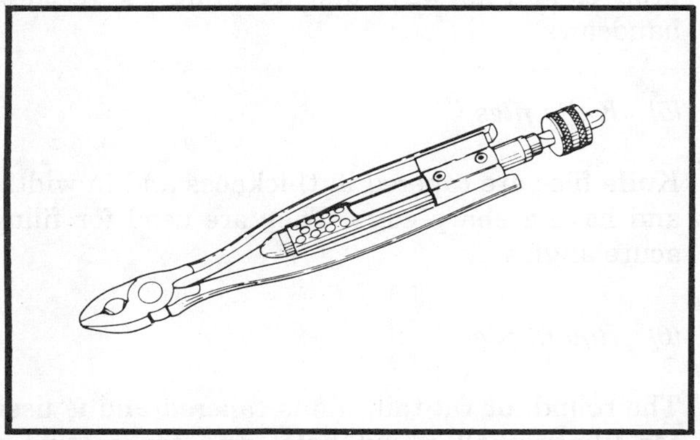

Fig. 2-21 *Safety wire cutter/twisters*

C. *Cutting Tools*

1. *Chisels*

The chisel is the simplest form of cutting tool and is made of high grade tool steel that has been carefully heat-treated and tempered. It is struck with a hammer to chip, or cut, the metal.

a. *Flat, or cold chisels*

These are the most common type of chisel used by the aviation mechanic. They are usually made from square or octagonal stock, ranging from 5/16-inch to 11/16-inch across, and are forged so that the cutting edge is slightly wider than the shank. The cutting edge is ground to an angle of about 70 degrees, and to cut or chip metal the edge is held flat against the material, and the depth of the cut is controlled by varying the angle between the chisel and the work. The cutting edge of the cold chisel is ground very slightly convex, so that the greatest stress from the hammer blows is directed into the center of the chisel, while the lesser stresses are transmitted to the sides of the cutting edge.

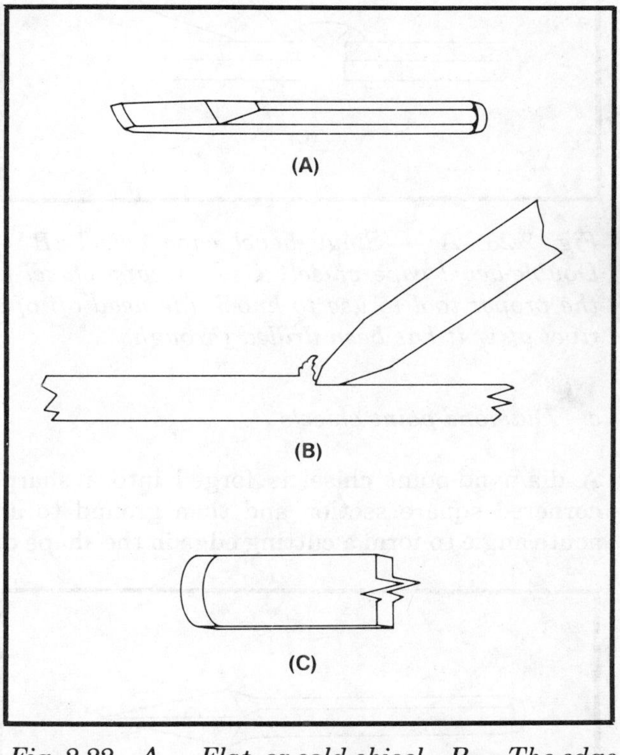

Fig. 2-22 A — Flat- or cold-chisel B — The edge of a cold-chisel is ground to approximately 70°. C — The cutting edge of a cold-chisel is ground convex.

b. Cape chisels

The cape chisel is forged from the same type of tool steel as the cold chisel, but its cutting edge is much narrower and may have either a single or a double bevel. These chisels are used to cut keyways and channels and are most familiar to the aviation mechanic as the proper chisel to use to knock the head off of rivets after the head has been drilled through. Their narrow cutting edge is less likely to damage the skin than the wider cold chisel.

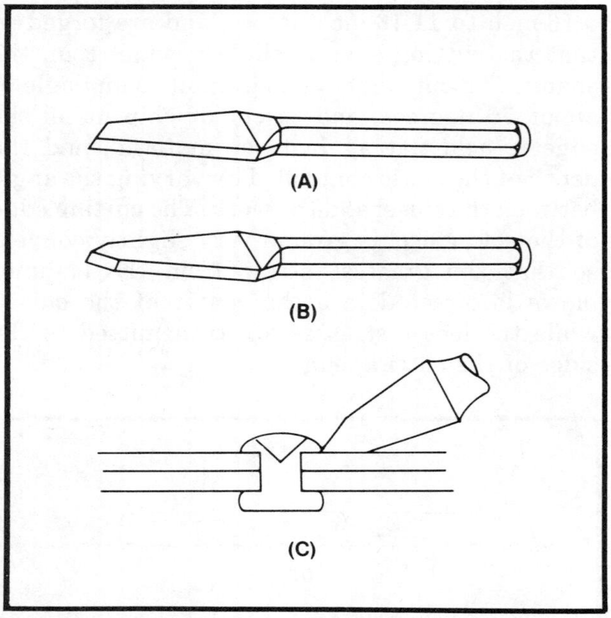

Fig. 2-23 A — Single-bevel cape chisel B — Double-bevel cape chisel C — A cape chisel is the proper tool to use to knock the head off of a rivet after it has been drilled through.

c. Diamond-point chisels

A diamond-point chisel is forged into a sharp-cornered square section and then ground to an acute angle to form a cutting edge in the shape of a diamond. These chisels are used to cut V-grooves and to cut sharp corners in square or rectangular grooves.

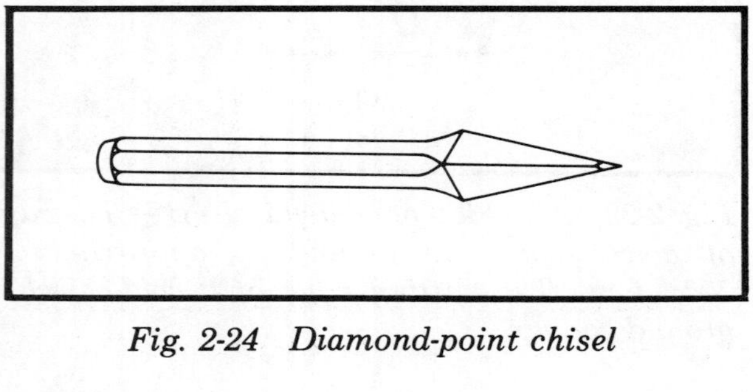

Fig. 2-24 Diamond-point chisel

2. Files

Next to the chisel, the most simple cutting tool is the file, which differs from the chisel in that it has a large number of cutting edges or points rather than the single cutting edge of the chisel. The file is pushed across the material by hand, and as it moves, the teeth act as small chisels, each removing small chips.

Files are classified with regard to their kind, their length, and their cut.

a. Kinds of files

(1) Mill files

This file is uniformly thick, but tapered in its width.

(2) Hand files

Hand files are uniform in width but taper in thickness and some hand files have at least one smooth, or safe, edge. Hand files are used for smoothing flat surfaces.

(3) Half-round files

Having one flat side and one curved, half-round files are tapered in both thickness and width.

(4) Three-square (triangular) files

These files have three cutting sides and are tapered. They are used for filing sharp corners in grooves and keyways and for filing the teeth of handsaws.

(5) Knife files

Knife files are tapered in thickness and in width, and have a sharp edge. They are used for filing acute angles.

(6) Round file

The round, or rat-tail, file is tapered and is used for filing small round holes and for cutting a radius in a groove or a keyway.

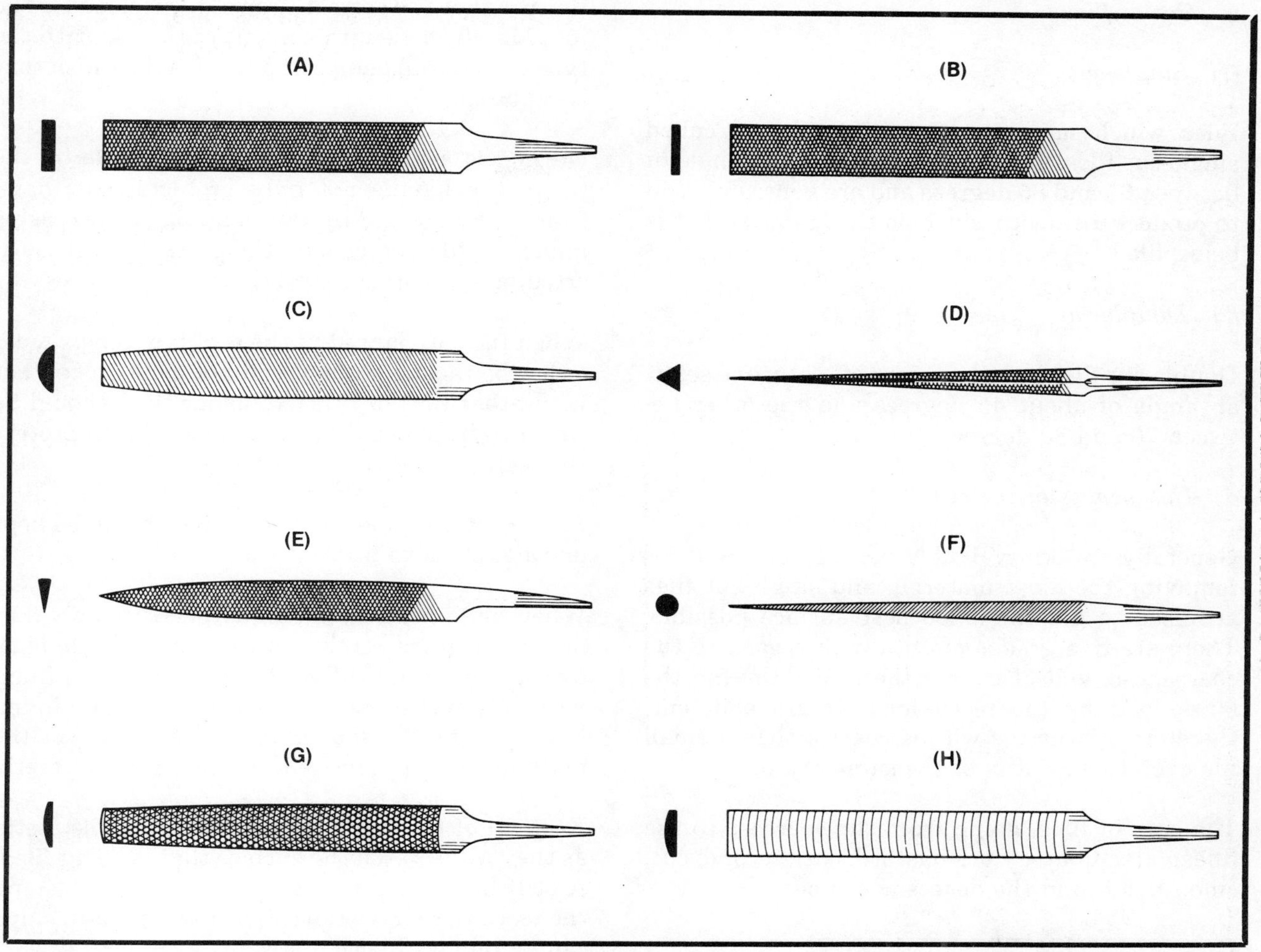

Fig. 2-25 Commonly used files for aviation maintenance A — Mill file B — Hand file C — Half-round file D — Three-square file E — Knife file F — Round file G — Wood rasp H — Vixen file

(7) Wood rasps

Rasps have individual teeth cut into their surfaces rather than rows of teeth. They are usually half-round and tapered. Wood rasps are used to remove wood where it is not practical to use a saw or a plane. The surface they leave is quite rough and must be smoothed with a file or with sandpaper.

(8) Vixen files

These files have curved teeth and are used to produce a smooth finish by slicing off very small amounts of material from the surface over which they are moved. Vixen files do not taper in either thickness or width and are often used in a special file holder which slightly arches the file for more concentrated cutting pressure.

b. Length of files

Most files used in aviation maintenance vary in length from about six to 16 inches. This is the length of the body of the file and does not include the tang, that part of the file that fits into the handle.

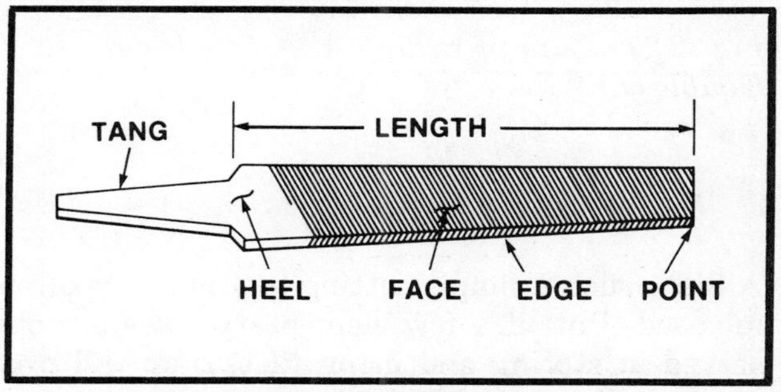

Fig. 2-26 Dimensions and nomenclature of a file.

c. Cut of files

(1) Single-cut

Files which have one row of teeth are called single-cut files. These teeth are cut at an angle of between 65 and 85 degrees and are generally used to produce a smooth finish on the material that is being filed.

(2) Double-cut

Double-cut files have two rows of teeth, one set at an angle of about 45 degrees and the other between 70 and 80 degrees.

d. Coarseness of the cut

Generally speaking, double-cut files are used for removing the most material, and single-cut files are used to get the smoothest surface possible. There are five grades of cuts with regard to the coarseness, and of course the larger the file the larger will be the teeth for any grade of cut; therefore a large file will be coarser than a small file even though it is of the same cut.

The cuts of files, going from the coarsest to the finest are: Coarse cut, bastard cut, second cut, smooth cut, and the dead smooth cut.

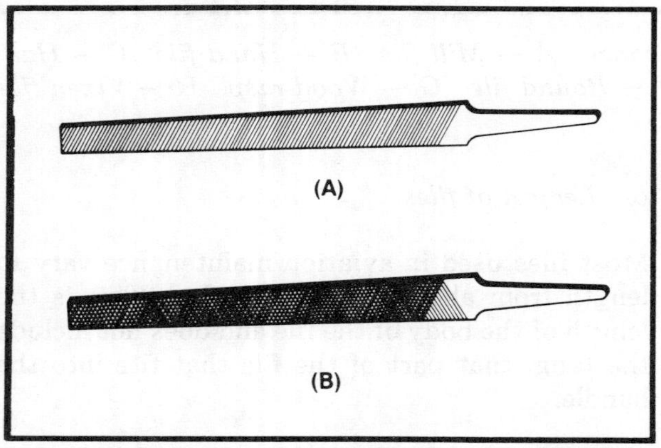

Fig. 2-27 Cuts of a file A — Single-cut B — Double-cut

e. Use of files

A file is such a simple cutting tool that it is often misused. But, if a few elementary rules are observed in storing and using files, they will produce excellent results and will give long service.

Be sure that the correct type of file is used for the job. Match the length and cut of the file with the type of material being cut and with the finish that is required.

Never use a file unless there is a handle on the tang. The handle not only protects your hand from being gouged by the tang, but it also helps guide the file and control the pressure that is exerted on the cutting stroke.

When files are stored in the tool box they should either be racked so they will not come in contact with other files or other tools, or they should be wrapped in paper or plastic envelopes to protect the teeth.

Files should not be oiled, but they should be kept dry and clean to prevent their rusting.

When filing hard metal, apply pressure only during the forward stroke, as drawing the file back over the material will dull the teeth. When filing very soft metal such as lead or soft aluminum, however, the file should be drawn back over the material to help remove the chips from the teeth.

Keep all of the chips cleaned out from the teeth, as they will scratch the surface that is being filed. A stiff brush or a file card can be brushed along the teeth to remove the chips, and stubborn chips can be removed with a steel pick.

Always use a slow, even stroke when filing, and use the entire length of the file.

Hold the material that is being filed firmly in a vise so that the file will not chatter as it moves over the work.

Enough pressure should be held on the file as it is moved over the work to prevent its slipping. If the file slips it will be dulled.

Rotate a round file, and file with a slicing motion when using a half-round file to get a smoother cut and to avoid dips and hollows in the material.

When using a file for cross filing, the handle should be held in the right hand with the thumb on the top of the handle and the fingers below the handle. The left hand should hold the end of the file with the thumb on top of the file and the fingers below it. With the file held in this manner it is moved smoothly across the material with pressure being applied only on the forward stroke.

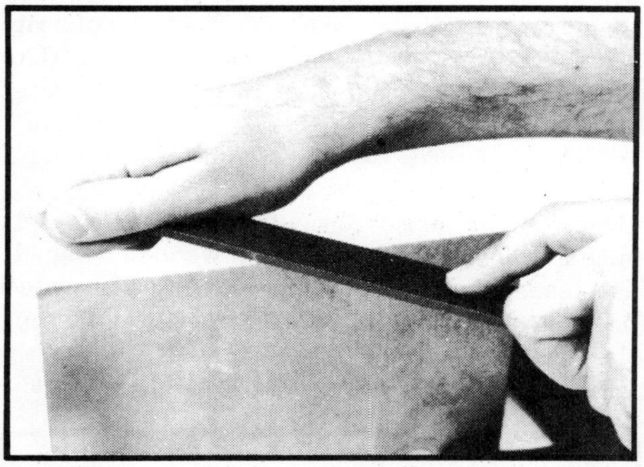

Fig. 2-28 In cross filing the file is moved lengthwise across the work.

An extremely smooth surface may be produced by using a smooth or dead-smooth single-cut file and draw filing. In this process, the file is grasped by both of its ends and is moved sideways across the work.

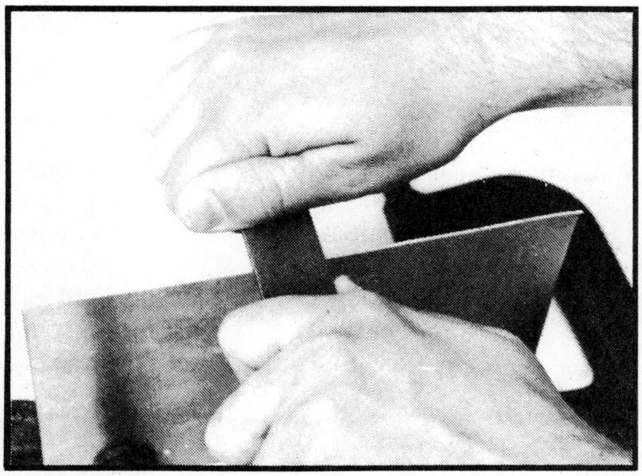

Fig. 2-29 In draw filing, the file is moved sidewise across the work.

3. Shears

Long straight cuts across a piece of sheet metal are made on squaring shears which may be either foot or power operated, but the fabrication of small parts requires a great deal of hand cutting and fitting.

The tinner's snips, or, as they are more commonly called, tin snips, are found in almost every sheet metal worker's tool chest. These snips range in length from about seven inches up to 12 inches, and while they are basically used for making straight cuts, they can also be used to cut curves to either the left or the right.

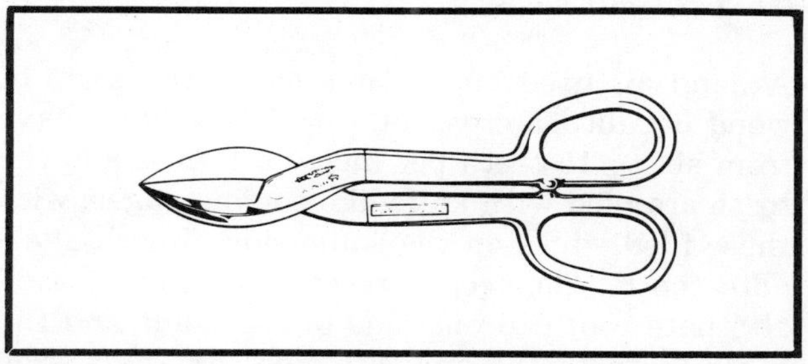

Fig. 2-30 Tinner's, or tin snips.

Compound action shears are known throughout the industry as aviation snips or "Dutchman" snips. These shears have serrated cutting edges and a leverage multiplication that allows relatively thick sheet metal to be cut without requiring an excessive force to be exerted on the handles.

Aviation snips come in a set of three, one that cuts straight, one that cuts to the right, and one that cuts to the left. The straight snips are color coded with yellow handles, the right-hand snips have green handles, and those that cut to the left have red handles.

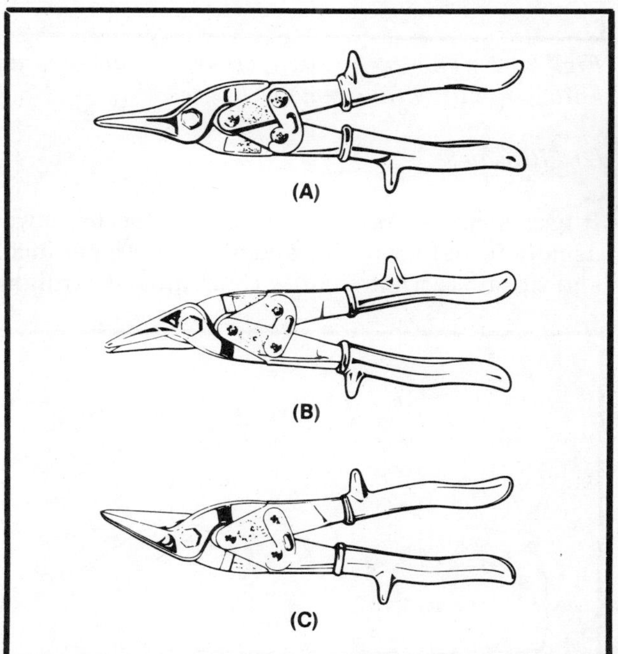

Fig. 2-31 A — Straight-cut aviation snips B — Aviation snips that cut to the right C — Aviation snips that cut to the left.

4. Saws

a. Wood saws

(1) Crosscut saws

A handsaw used for cutting across the grain of wood is called a crosscut saw. These saws have from six to 12 teeth per inch, or points, and the teeth are filed with knife-like cutting edges, with these filed edges on the same side of each alternate tooth. The teeth are set by bending each alternate tooth to one side of the blade and the other teeth to the opposite side. As the saw is pushed through the wood, the teeth will cut the fibers, and the set causes the width of the cut, or the kerf, to be wider than the blade so the blade will not bind in the cut.

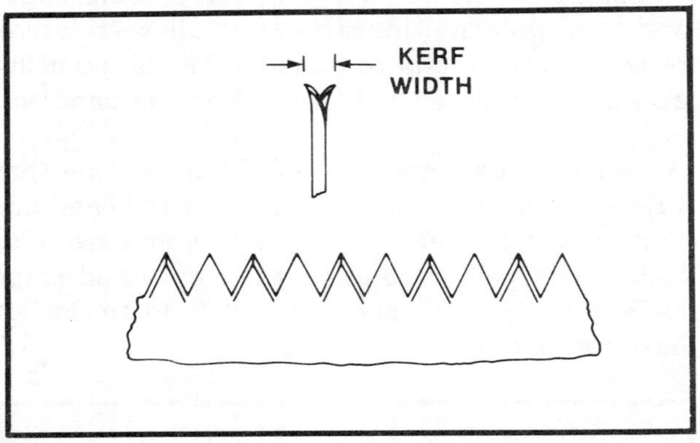

Fig. 2-32 The teeth of a crosscut saw act as knives to cut through wood fibers.

(2) Ripsaws

Ripsaws have fewer teeth than crosscut saws, usually from four to five and a half teeth per inch, and ripsaw teeth are cut with an almost straight leading edge while the back of the blade tapers. The teeth are filed flat across the front so they will act as chisels rather than as knives, the way the teeth of a crosscut saw act. The teeth of a ripsaw have a much smaller amount of set than those of a crosscut saw.

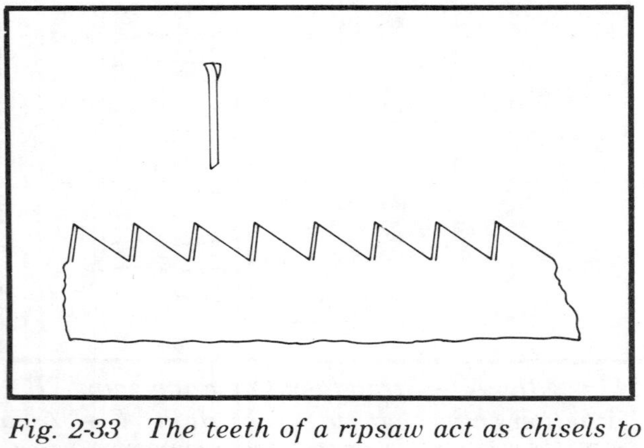

Fig. 2-33 The teeth of a ripsaw act as chisels to dig the wood fibers.

(3) Backsaws

Backsaws are used for cutting precision parts such as those used in making a wooden wing rib. The blades of these saws are stiffened with a steel back so they will cut absolutely straight, and they have from 18 to 32 teeth per inch with very little set, so that they will make a smooth cut.

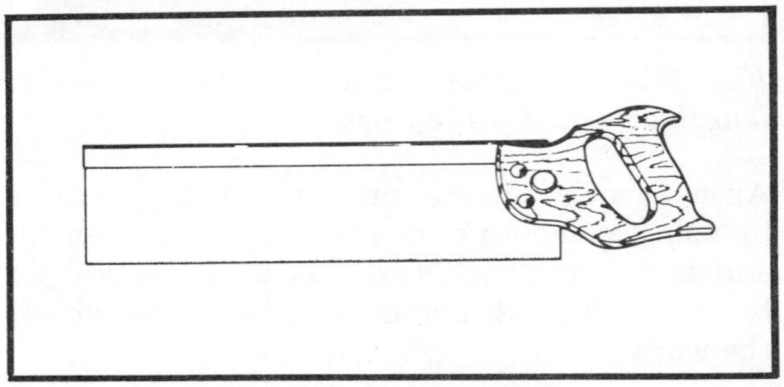

Fig. 2-34 Backsaw

(4) Keyhole saws

When it is necessary to cut a large hole with only a small hole to start with, the keyhole saw is used. This thin tapered blade is inserted into the drilled hole and the cut is made as is needed. Most keyhole saws bought today have replaceable blades that range from 24 teeth per inch for cutting plastics to eight teeth per inch for cutting wood.

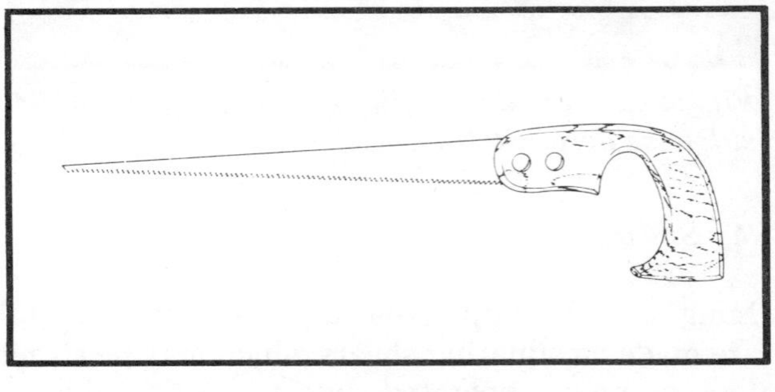

Fig. 2-35 Keyhole saw

b. Metal saws — hacksaws

The hacksaw is the most widely used metal-cutting hand saw. It consists of an adjustable frame that holds the small flexible steel blade under tension.

Hacksaw blades are made of either carbon steel or high-speed molybdenum steel. The back of a molybdenum blade is annealed to make it less brittle, while the teeth are hardened so they will cut steel. These high-speed blades are considerably more costly than the carbon steel blade, but their much longer life makes them the most economical choice for use by the aviation mechanic.

Hacksaw blades are available in both 10- and 12-inch lengths and have from 18 to 32 teeth per inch. In Fig. 2-36 we see the proper way to choose the blade that is suitable for the job. If the work is wide and is not too hard, use as few teeth as possible, as this will give a fast cut with plenty of chip clearance. For harder materials, a blade with more teeth is needed for a good cut, and for thin materials, or odd-shaped materials having thin sections, a blade should be chosen that will allow at least two teeth to be on the metal at all times. A great deal of aircraft structure is made of thin wall tubing, much of which is chrome molybdenum steel. This extremely tough steel can be cut only with high-speed hacksaw blades, and it is

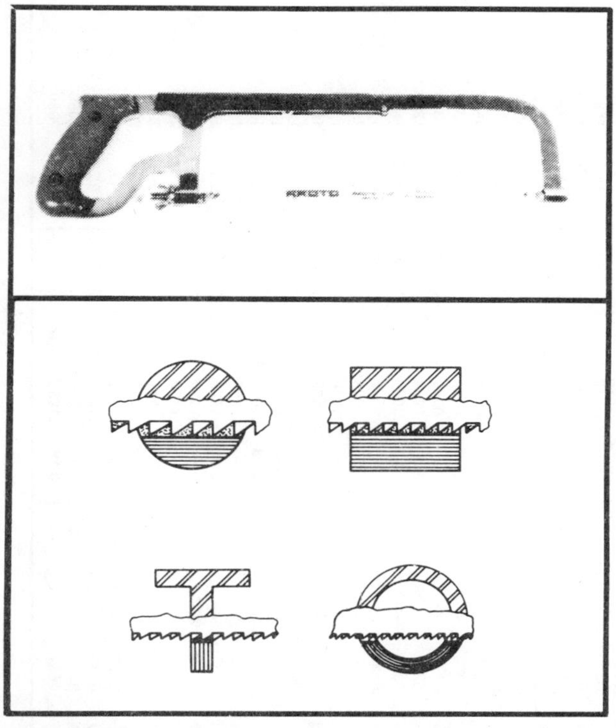

Fig. 2-36 A hacksaw should be selected that allows at least two teeth to rest on the work at all times.

important to have at least two teeth on each side of the tube as it is cut.

When using a hacksaw, use as much of the blade as possible, and make long, slow, and steady strokes. As with almost all cutting tools, the cutting is done only on the forward stroke of the blade, and the blade should be lifted for the return stroke. Short, fast strokes with equal pressure on both the forward and backward stroke will rapidly dull the blade and will cause it to break.

5. Drills

a. Wood bits

(1) Auger bits

These spiral-type auger bits are turned by a bow-type brace to cut a hole in wood. These bits are available in 1/16-inch increments, from 1/4-inch to one inch. They range in length from about seven to ten inches. The feed screw in the end of the bit screws into the wood and pulls the cutting edges into the wood. The chips that are cut loose are carried out of the hole by the spiral flutes.

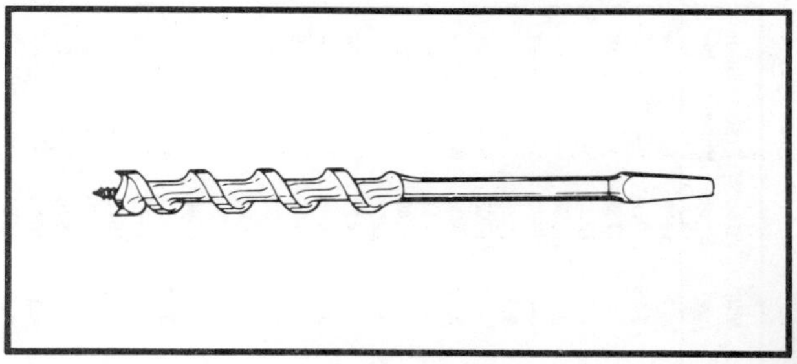

Fig. 2-37 Wood auger bit

(2) Flat wood-boring bits

Large holes in wood can be bored with an electric drill by using the flat wood-boring bits. A tapered pilot in the center of the cutting edge keeps the bit centered as the hole is being cut.

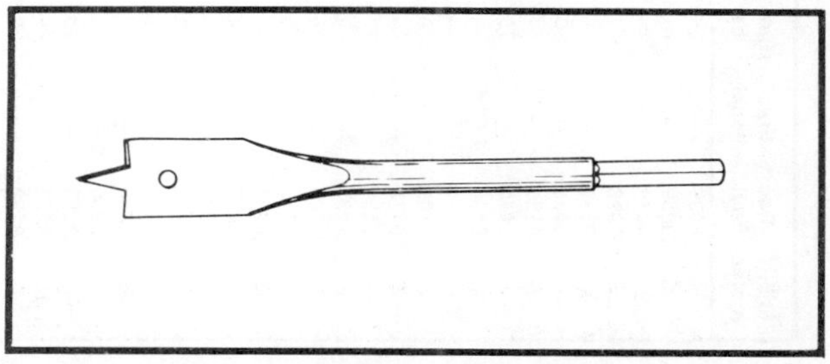

Fig. 2-38 Flat wood boring bit

Milli-meter	Dec. Equiv.	Frac-tional	Num-ber
.1	.0039	----	----
.15	.0059	----	----
.2	.0079	----	----
.25	.0098	----	----
.3	.0118	----	----
----	.0135	----	80
.35	.0138	----	----
----	.0145	----	79
.39	.0156	1/64	----
.4	.0157	----	----
----	.0160	----	78
.45	.0177	----	----
----	.0180	----	77
.5	.0197	----	----
----	0.200	----	76
----	.0210	----	75
.55	.0217	----	----
----	.0225	----	74
.6	.0236	----	----
----	.0240	----	73
----	.0250	----	72
.65	.0256	----	----
----	.0260	----	71
----	.0280	----	70
.7	.0276	----	----
----	.0292	----	69
.75	.0295	----	----
----	.0310	----	68
.79	.0312	1/32	----
.8	.0315	----	----
----	.0320	----	67
----	.0330	----	66
.85	.0335	----	----
----	.0350	----	65
.9	.0354	----	----
----	.0360	----	64
----	.0370	----	63
.95	.0374	----	----
----	.0380	----	62
----	.0390	----	61
1.0	.0394	----	----
----	.0400	----	60
----	.0410	----	59
1.05	.0413	----	----
----	.0420	----	58
----	.0430	----	57
1.1	.0433	----	----
1.15	.0452	----	----
----	.0465	----	56
1.19	.0469	3/64	----
1.2	.0472	----	----
1.25	.0492	----	----
1.3	.0512	----	----
----	.0520	----	55
1.35	.0531	----	----
----	.0550	----	54
1.4	.0551	----	----

Milli-meter	Dec. Equiv.	Frac-tional	Num-ber
1.45	.0570	----	----
1.5	.0591	----	----
----	.0595	----	53
1.55	.0610	----	----
1.59	.0625	1/16	----
1.6	.0629	----	----
----	.0635	----	52
1.65	.0649	----	----
1.7	.0669	----	----
----	.0670	----	51
1.75	.0689	----	----
----	.0700	----	50
1.8	.0709	----	----
1.85	.0728	----	----
----	.0730	----	49
1.9	.0748	----	----
----	.0760	----	48
1.95	.0767	----	----
1.98	.0781	5/64	----
----	.0785	----	47
2.0	.0787	----	----
2.05	.0807	----	----
----	.0810	----	46
----	.0820	----	45
2.1	.0827	----	----
2.15	.0846	----	----
----	.0860	----	44
2.2	.0866	----	----
2.25	.0885	----	----
----	.0890	----	43
2.3	.0905	----	----
2.35	.0925	----	----
----	.0935	----	42
2.38	.0937	3/32	----
2.4	.0945	----	----
----	.0960	----	41
2.45	.0964	----	----
----	.0980	----	40
2.5	.0984	----	----
----	.0995	----	39
----	.1015	----	38
2.6	.1024	----	----
----	.1040	----	37
2.7	.1063	----	----
----	.1065	----	36
2.75	.1082	----	----
2.78	.1094	7/64	----
----	.1100	----	35
2.8	.1102	----	----
----	.1110	----	34
----	.1130	----	33
2.9	.1141	----	----
----	.1160	----	32
3.0	.1181	----	----
----	.1200	----	31
3.1	.1220	----	----
3.18	.1250	1/8	----

Milli-meter	Dec. Equiv.	Frac-tional	Num-ber
3.2	.1260	----	----
3.25	.1279	----	----
----	.1285	----	30
3.3	.1299	----	----
3.4	.1338	----	----
----	.1360	----	29
3.5	.1378	----	----
----	.1405	----	28
3.57	.1406	9/64	----
3.6	.1417	----	----
----	.1440	----	27
3.7	.1457	----	----
----	.1470	----	26
3.75	.1476	----	----
----	.1495	----	25
3.8	.1496	----	----
----	.1520	----	24
3.9	.1535	----	----
----	.1540	----	23
3.97	.1562	5/32	----
----	.1570	----	22
4.0	.1575	----	----
----	.1590	----	21
----	.1610	----	20
4.1	.1614	----	----
4.2	.1654	----	----
----	.1660	----	19
4.25	.1673	----	----
4.3	.1693	----	----
----	.1695	----	18
4.37	.1719	11/64	----
----	.1730	----	17
4.4	.1732	----	----
----	.1770	----	16
4.5	.1771	----	----
----	.1800	----	15
4.6	.1811	----	----
----	.1820	----	14
4.7	.1850	----	13
4.75	.1870	----	----
4.76	.1875	3/16	----
4.8	.1890	----	12
----	.1910	----	11
4.9	.1929	----	----
----	.1935	----	10
----	.1960	----	9
5.0	.1968	----	----
----	.1990	----	8
5.1	.2008	----	----
----	.2010	----	7
5.16	.2031	13/64	----
----	.2040	----	6
5.2	.2047	----	----
----	.2055	----	5
5.25	.2067	----	----
5.3	.2086	----	----
----	.2090	----	4

Milli-meter	Dec. Equiv.	Frac-tional	Num-ber
5.4	.2126	----	----
----	.2130	----	3
5.5	.2165	----	----
5.56	.2187	1/32	----
5.6	.2205	----	----
----	.2210	----	2
5.7	.2244	----	----
5.75	.2263	----	----
----	.2280	----	1
5.8	.2283	----	----
5.9	.2323	----	----
----	.2340	----	A
5.95	.2344	15/64	----
6.0	.2362	----	----
----	.2380	----	B
6.1	.2401	----	----
----	.2420	----	C
6.2	.2441	----	----
6.25	.2460	----	D
6.3	.2480	----	----
6.35	.2500	1/4	E
6.4	.2520	----	----
6.5	.2559	----	----
----	.2570	----	F
6.6	.2598	----	----
----	.2610	----	G
6.7	.2638	----	----
6.75	.2657	17/64	----
6.75	.2657	----	----
----	.2660	----	H
6.8	.2677	----	----
6.9	.2716	----	----
----	.2720	----	I
7.0	.2756	----	----
----	.2770	----	J
7.1	.2795	----	----
----	.2811	----	K
7.14	.2812	9/32	----
7.2	.2835	----	----
7.25	.2854	----	----
7.3	.2874	----	----
----	.2900	----	L
7.4	.2913	----	----
----	.2950	----	M
7.5	.2953	----	----
7.54	.2968	19/64	----
7.6	.2992	----	----
----	.3020	----	N
7.7	.3031	----	----
7.75	.3051	----	----
7.8	.3071	----	----
7.9	.3110	----	----
7.94	.3125	5/16	----
8.0	.3150	----	----
----	.3160	----	O
8.1	.3189	----	----
8.2	.3228	----	----

Milli-meter	Dec. Equiv.	Frac-tional	Num-ber
----	.3230	----	P
8.25	.3248	----	----
8.3	.3268	----	----
8.33	.3281	21/64	----
8.4	.3307	----	----
----	.3320	----	Q
8.5	.3346	----	----
8.6	.3386	----	----
----	.3390	----	R
8.7	.3425	----	----
8.73	.3437	11/32	----
8.75	.3445	----	----
8.8	.3465	----	----
----	.3480	----	S
8.9	.3504	----	----
9.0	.3543	----	----
----	.3580	----	T
9.1	.3583	----	----
9.13	.3594	23/64	----
9.2	.3622	----	----
9.25	.3641	----	----
9.3	.3661	----	----
----	.3680	----	U
9.4	.3701	----	----
9.5	.3740	----	----
9.53	.3750	3/8	----
----	.3770	----	V
9.6	.3780	----	----
9.7	.3819	----	----
9.75	.3838	----	----
9.8	.3858	----	----
----	.3860	----	W
9.9	.3898	----	----
9.92	.3906	25/64	----
10.0	.3937	----	----
----	.3970	----	X
----	.4040	----	Y
10.32	.4062	13/32	----
----	.4130	----	Z
10.5	.4134	----	----
10.72	.4219	27/64	----
11.0	.4330	----	----
11.11	.4375	7/16	----
11.5	.4528	----	----
11.51	.4531	29/64	----
11.91	.4687	15/32	----
12.0	.4724	----	----
12.30	.4843	31/64	----
12.5	.4921	----	----
12.7	.5000	1/2	----
13.0	.5118	----	----
13.10	.5156	33/64	----
13.49	.5312	17/32	----
13.5	.5315	----	----
13.89	.5469	35/64	----
14.0	.5512	----	----
14.29	.5625	9/16	----

Milli-meter	Dec. Equiv.	Frac-tional	Num-ber
14.5	.5709	----	----
14.68	.5781	37/64	----
15.0	.5906	----	----
15.08	.5937	19/32	----
15.48	.6094	39/64	----
15.5	.6102	----	----
15.88	.6250	5/8	----
16.0	.6299	----	----
16.27	.6406	41/64	----
16.5	.6496	----	----
16.67	.6562	21/32	----
17.0	.6693	----	----
17.06	.6719	43/64	----
17.46	.6875	11/16	----
17.5	.6890	----	----
17.86	.7031	45/64	----
18.0	.7087	----	----
18.26	.7187	23/32	----
18.5	.7283	----	----
18.65	.7344	47/64	----
19.0	.7480	----	----
19.05	.7500	3/4	----
19.45	.7656	49/64	----
19.5	.7677	----	----
19.84	.7812	25/32	----
20.0	.7874	----	----
20.24	.7969	51/64	----
20.5	.8071	----	----
20.64	.8125	13/16	----
21.0	.8268	----	----
21.03	.8281	53/64	----
21.43	.8437	27/32	----
21.5	.8465	----	----
21.83	.8594	55/64	----
22.0	.8661	----	----
22.23	.8750	7/8	----
22.5	.8858	----	----
22.62	.8906	57/64	----
23.0	.9055	----	----
23.02	.9062	29/32	----
23.42	.9219	59/64	----
23.5	.9252	----	----
23.81	.9375	15/16	----
24.0	.9449	----	----
24.21	.9531	61/64	----
24.5	.9646	----	----
24.61	.9687	31/32	----
25.0	.9843	----	----
25.03	.9844	63/64	----
25.4	1.0000	1	----

Fig. 2-39 Twist drill sizes

b. Metal drills

There are three sets of twist drills that find common use in aviation maintenance. Number drills range in size from 0.0135 for the number 80 drill to 0.2280 for the number one drill. Fractional drills are available from 1/64-inch (0.0156) to 1/2-inch (0.500). Letter drill sizes are all larger than number drills and range from the *A* drill (0.2340) to the *Z* drill (0.4130). The only size drill that is available in two sets is the letter *E* drill and the 1/4-inch drill, both of which have a diameter of 0.2500 inch.

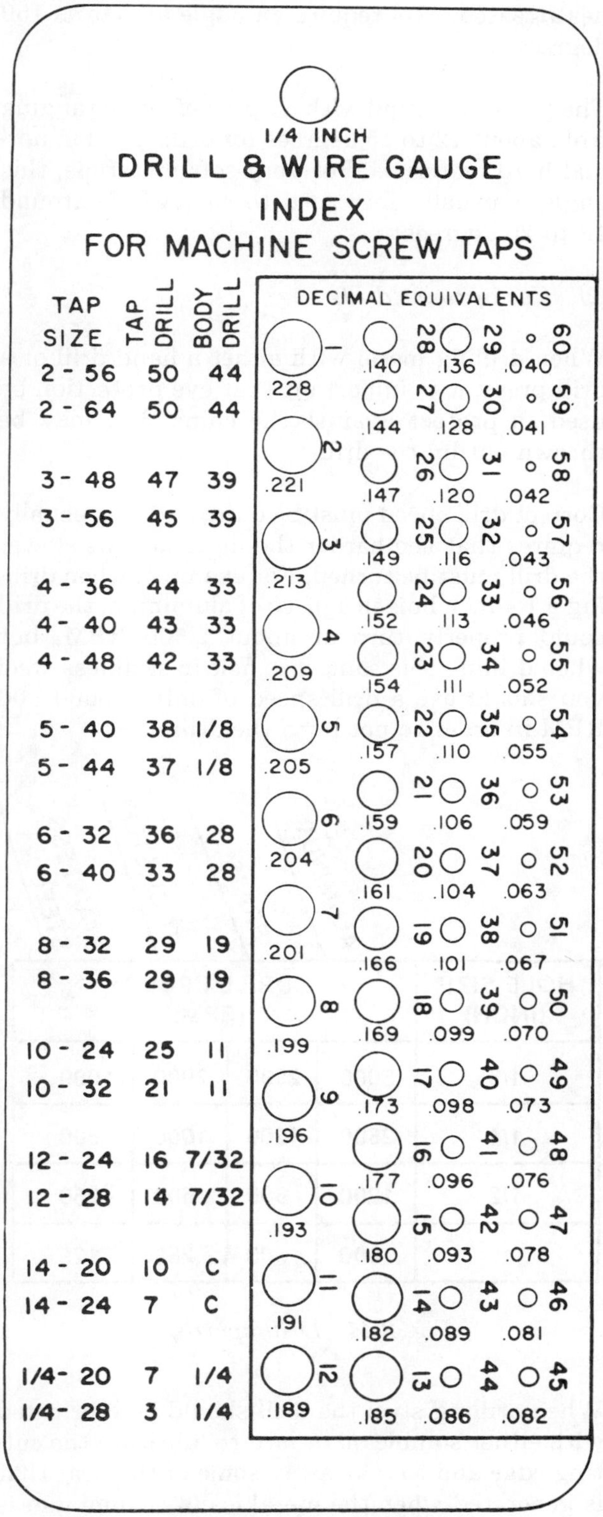

Fig. 2-40 Drill and wire index

(1) Drill nomenclature

Twist drills are made of both carbon steel and high-speed steel. The carbon steel drills cost less, but they will overheat and lose their hardness when they are used to drill very hard or tough materials. For this reason high-speed drills are by far the most economical for use in aviation maintenance.

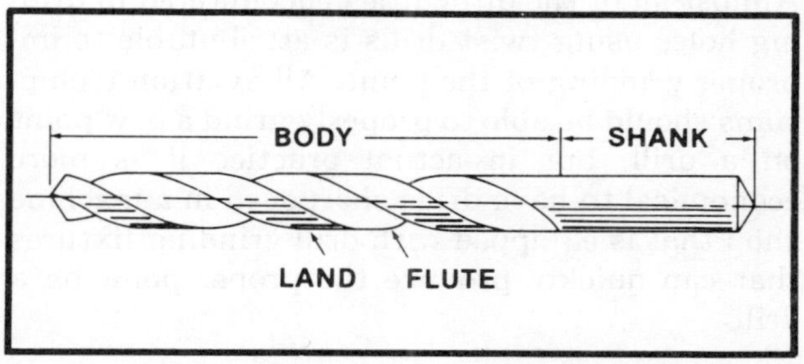

Fig. 2-41 Twist drill nomenclature

(a) The shank

A twist drill is made up of three parts; the shank, the body, and the point. The shank is that portion of the drill that is held in the chuck. Drills used in aircraft maintenance all have straight shanks, while drills turned by the large drill presses in machine shops usually have tapered shanks.

(b) The body

The body of a twist drill has spiral flutes milled from the point to the shank. The chips that are cut by the cutting edge are carried out of the hole through these flutes and lubricants can reach the point through them. The material between the flutes is called the land and the edge of the land nearest the cutting edge of the point is the full size of the drill and is called the margin. Immediately behind the margin, the land is ground away slightly to provide body clearance.

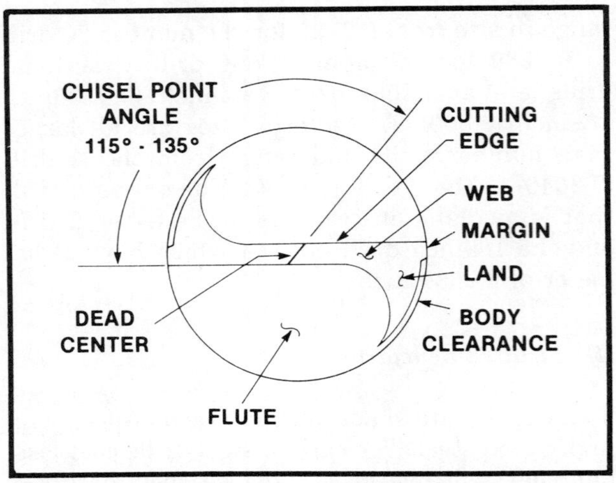

Fig. 2-42 Drill point nomenclature

(c) The point

Almost all of the difficulties encountered in drilling holes using twist drills is attributable to improper grinding of the point. All aviation technicians should be able to properly grind a new point on a drill, but in actual practice it is more economical to have drills sharpened in a machine shop that is equipped with drill grinding fixtures that can quickly produce the proper point on a drill.

The drill must be ground so that its two cutting edges, or lips, are exactly the same length, and meet in the exact center of the drill. For most materials the cutting edges are ground at an angle of 59 degrees from the central axis of the drill, which gives an included angle of 118 degrees. For drilling soft material, this angle can be made a good deal steeper, up to as much as 90 degrees for drilling lead, wood, or very soft aluminum. Hard materials such as stainless steel, or heat-treated steel require an angle as flat as 150 degrees.

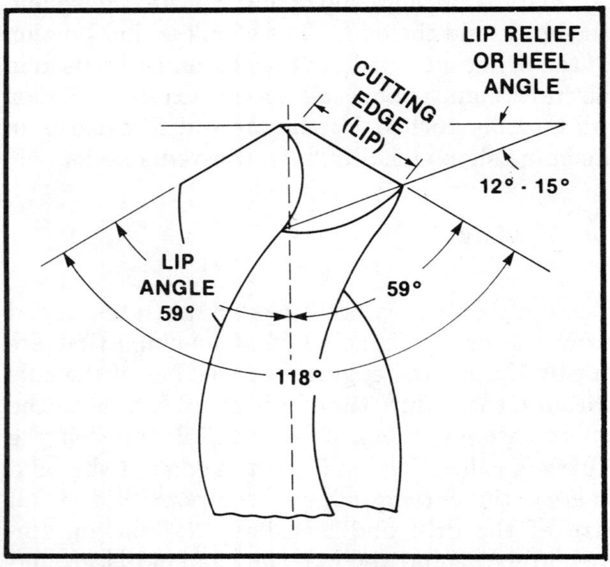

Fig. 2-43 Drill point angles

The point is ground with a lip relief angle ranging from about 12 to 15 degrees for drills used in normal hard materials. For very soft materials, this angle is usually increased to somewhere around 18 to 20 degrees.

(2) Use of twist drills

When drilling metal with either a hand drill or a drill press, it is important that eye protection be used to protect against the chips that may be thrown out by the drill.

Correct drill speed must be used, and this usually requires that the harder the material, the slower the drill must be turned. For example, when drilling a 1/4-inch hole in a piece of aluminum, the drill could properly turn at about 2,500 RPM, but when drilling this same size hole in stainless steel you should use a drill speed of only around 500 RPM in order to not burn the drill.

HOLE SIZE (INCH)	DRILL SPEED (RPM)			
	ALUMINUM	BRASS	TOOL STEEL	STAINLESS STEEL
1/8	5000	2500	2000	1000
1/4	2500	1200	1000	500
1/2	1200	600	500	250
1	600	300	250	125

Fig. 2-44 Drill speeds

When drilling steel the drill should be lubricated with either soluble oil or lard to lubricate the cutting edge and to take away some of the heat that is generated when the metal is cut. Aluminum is normally drilled without the use of any lubricant.

6. *Hole cutters*

a. *Hole saws*

For cutting holes in sheet metal that are larger than can be drilled, circular saws may be used. These saws are available in diameters from 9/16-inch up to 4-1/8 inch. The arbor for these saws have a round shank that fits into a 1/4-inch electric or air drill motor have a special pilot drill is centered in the arbor. These pilot drills have a very short body and a long shank so that the smooth shank is in the metal when the saw is cutting. This keeps the pilot hole from becoming enlarged and holds the saw centered.

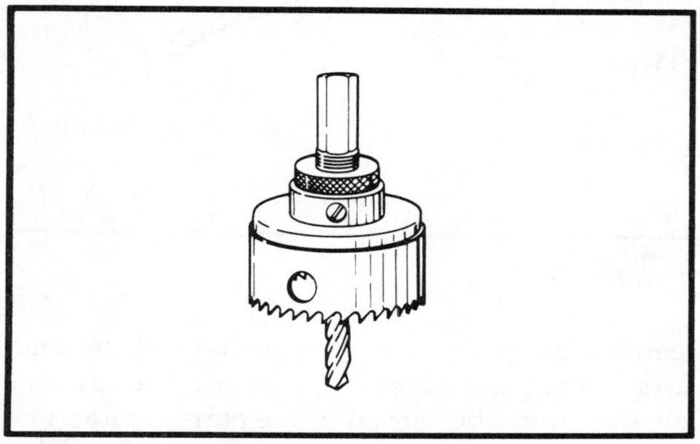

Fig. 2-45 Hole saw

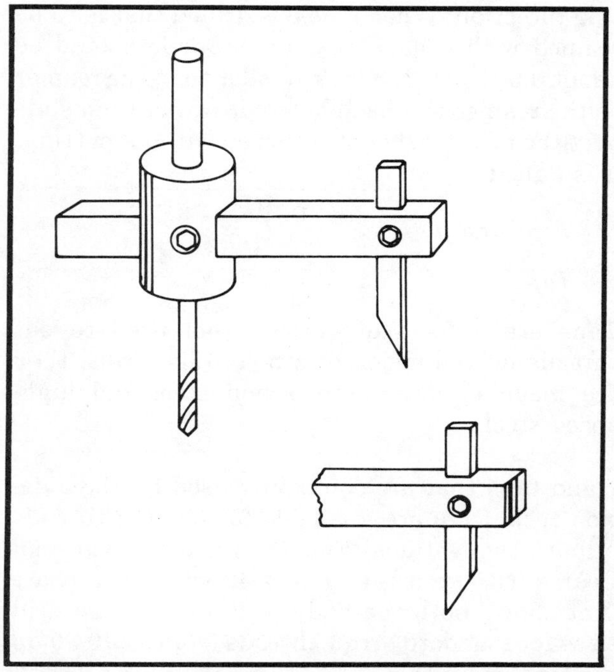

Fig. 2-46 A — Fly cutter B — The cutting tool of a fly cutter may be reversed to place the bevel on the correct piece of material.

b. *Fly cutters*

The fly cutter is an adjustable hole cutter that is properly turned by a drill press. The cutting tool is set so that its radius from the center of the pilot drill is exactly that required for the radius of the hole. The pilot drill cuts the center hole, and the cutting tool is fed very slowly into the metal. The cutting tool may be reversed in the holder so that the edge of the hole being cut can have either straight or beveled edges.

When using a fly cutter, it is extremely important that the metal being cut be backed with scrap wood, and it must be firmly clamped to the table of the drill press. If the cutter is fed into the metal too fast, there is danger of the cutter grabbing the metal rather than cutting it.

c. *Countersinks*

Almost all flush rivets used in aircraft construction have a head angle of 100 degrees. If the metal into which the rivet is driven is thin enough, it is dimpled, but thick metal must be countersunk so the head of the rivet will be flush with the surface of the skin.

The most generally used countersink is the adjustable stop countersink seen in Fig. 2-47B. A cutter of the correct size for the rivet is placed on the cutter shaft and the proper size pilot is inserted into the center of the shaft. The body and the stop are screwed onto the shaft, and the entire assembly is put into the chuck of a 1/4-inch air or electric drill motor. Trial holes are countersunk in scrap metal and the stop is adjusted until the diameter of the countersunk hole is exactly correct for the rivet. A fiber collar covers the end of the stop, so that it will not scratch the skin when the countersink is being used.

7. *Reamers*

Drills cannot make holes of sufficient accuracy for certain applications. For example, if a structure is to be held together with both rivets and bolts, there must be no relative motion between the bolt and its hole, and for this reason close tolerance bolts must be used. The holes for these bolts are drilled undersize and reamed to a size that will require a light press fit to get the bolt into the hole.

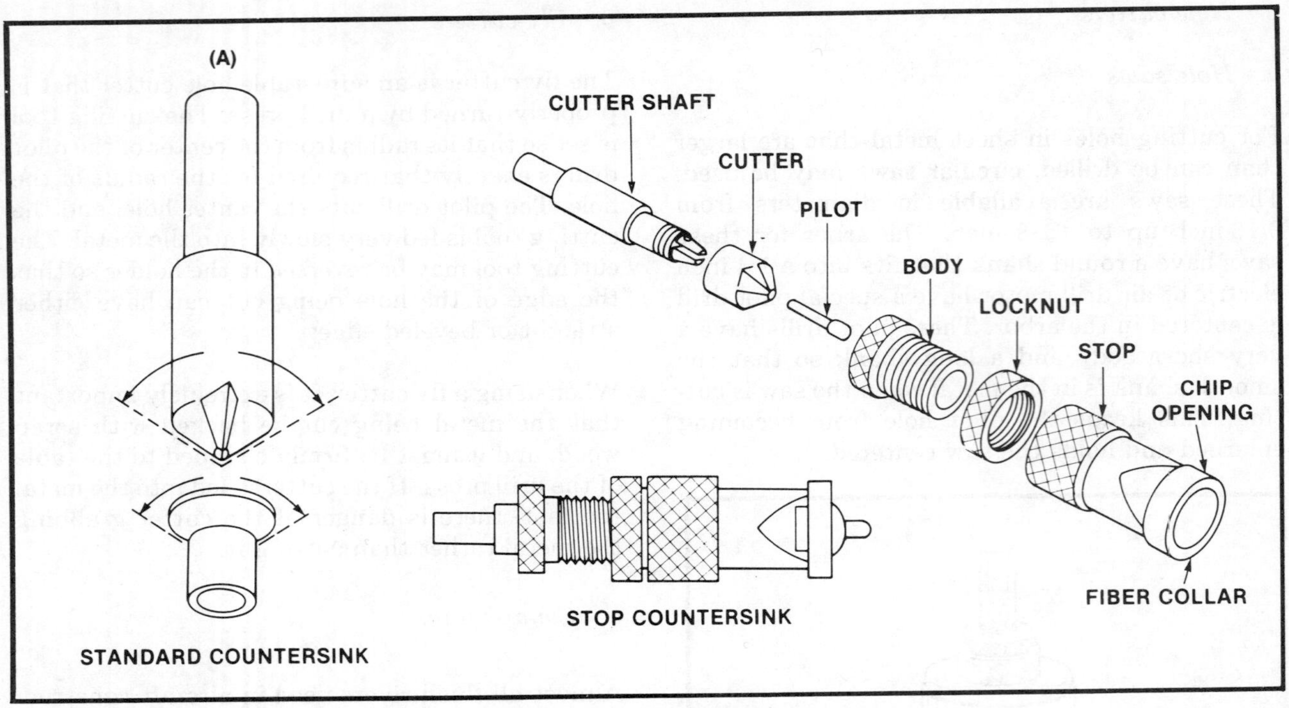

Fig. 2-47 Countersinks

There are two basic types of hand reamers used by the aviation technician, the solid reamer and the adjustable or expansion reamer.

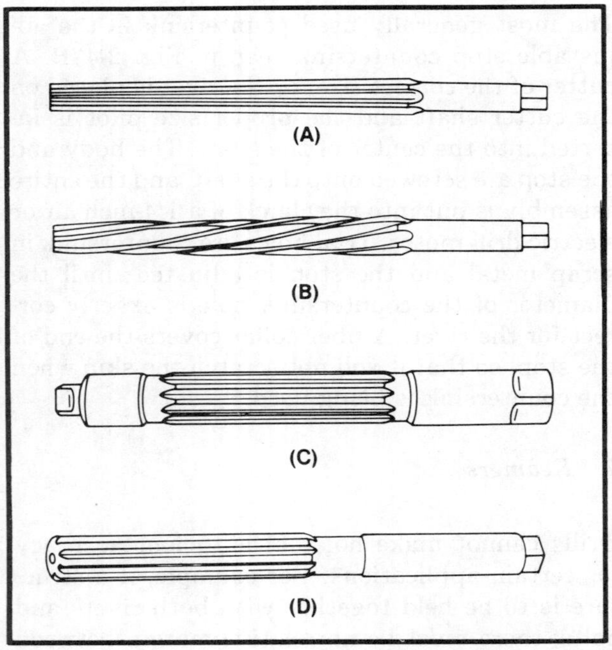

Fig. 2-48 Reamers A — Tapered reamer with straight flutes. B — Tapered reamer with spiral flutes. C — Expansion reamer with straight flutes. D — Bottoming reamer with straight flutes.

Reamers are delicate tools and have sharp and hardened cutting edges. As they are fed into the hole they must be turned in the correct direction for cutting, and when they are removed from the hole, the turning should be continued in the cutting direction. When a hole is drilled that is to be reamed with a hand reamer, the hole should be about 0.001 to 0.003 inch smaller than the reamer so the reamer will be able to cut rather than slide over the metal. When it is turned without cutting, it is dulled.

8. Taps and dies

a. Taps

Taps are a form of cutting tool used to cut threads on the inside of a hole. Like drills, they are made in both high-carbon steel and high-speed steel.

Hand taps that are commonly used by the aviation technician are available for cutting threads of both the National Fine (NF) and the National Coarse (NC) series. In Fig. 2-49 we have a chart that shows both the body drill and the tap drill sizes for standard screw threads from size 0-80 up through one inch.

The fit between the screw and nut for an American National Thread is listed by its class. A

NATIONAL FINE THREAD SERIES MEDIUM FIT. CLASS 3 (NF)			
Size and Threads	Dia. of body for thread	Body Drill	Tap Drill Size
0-80	.060	52	3/64
1-72	.073	47	#53
2-64	.086	42	#50
3-56	.099	37	#46
4-48	.112	31	#42
5-44	.125	25	#38
6-40	.138	27	#33
8-36	.164	18	#29
10-32	.190	10	#21
12-28	.216	2	#15
1/4-28	.250	F	#3
5/16-24	.3125	5/16	I
3/8-24	.375	3/8	Q
7/16-20	.4375	7/16	W
1/2-20	.500	1/2	7/16
9/16-18	.5625	9/16	1/2
5/8-18	.625	5/8	9/16
3/4-16	.750	3/4	11/16
7/8-14	.875	7/8	51/64
1-14	1.000	1.0	49/64

NATIONAL COARSE THREAD SERIES MEDIUM FIT. CLASS 3 (NC)			
Size and Threads	Dia. of body for thread	Body Drill	Tap Drill Size
1-64	.073	47	#53
2-56	.086	42	#51
3-48	.099	37	5/64
4-40	.112	31	#44
5-40	.125	29	#39
6-32	.138	27	#36
8-32	.164	18	#29
10-24	.190	10	#26
12-24	.216	2	#17
1/4-20	.250	1/4	#8
5/16-18	.3125	5/16	#F
3/8-16	.375	3/8	5/16
7/16-14	.4375	7/16	U
1/2-13	.500	1/2	27/64
9/16-12	.5625	9/16	31/64
5/8-11	.625	5/8	17/32
3/4-10	.750	3/4	41/64
7/8-9	.875	7/8	49/64
1-8	1.000	1.0	7/8

Fig. 2-49 Body and tap drill sizes for American National Screw Threads

class 1 fit is a loose fit and is used where ease of assembly is the major concern. A class 2 fit is called a free fit, and it is standard for non-critical screw products. Almost all standard threaded aircraft hardware has a class 3 fit which is called a medium fit. This fit is a fine balance between a precision fit and one that is practical and economical in production. A close fit is the class 4 fit which produces a snug fit that may require selective assembly to get parts that will screw together without using a wrench. A wrench-tight, or interference fit, is the class 5 fit. The maximum allowable nut size is smaller than the minimum allowable bolt size, and this assures an interference fit with all combinations.

Hand taps come in sets of three; a taper tap that is used to start the threads in the hole, a plug tap to cut threads nearly to the bottom of blind holes, and a bottoming tap that cuts them closer to the bottom than a plug tap. The taper tap has at least six of its threads tapered back before the full diameter of the tap is reached. The plug tap has between three and five threads tapered, and the bottoming tap has only the bottom 1-1/2 threads chamfered.

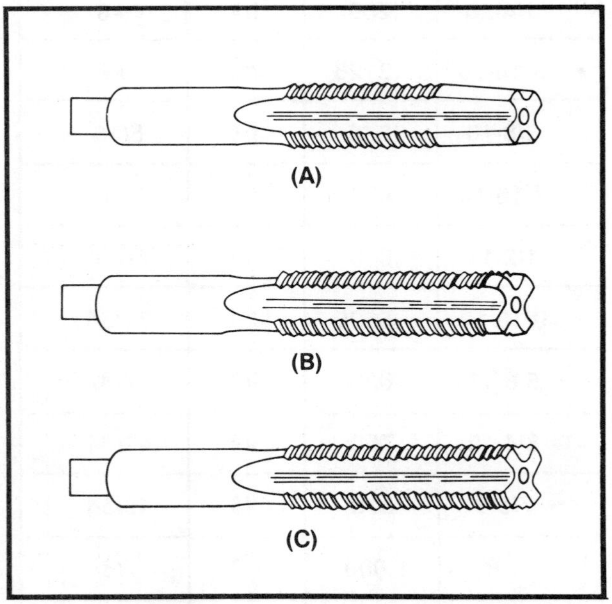

Fig. 2-50 Types of taps A — Taper tap B — Plug tap C — Bottoming tap

Pipe threads that are used in some fluid line fittings are tapered, and their seal is obtained by the wedging action between the male and female threads. The size of a pipe thread is not measured by the outside diameter as it is with the standard screw thread, but corresponds to the inside diameter of an iron pipe. In Fig. 2-51 we see the relationship between the nominal size of an NPT pipe thread and the outside diameter of the pipe or the male fitting.

NOMINAL SIZE (INCHES)	THREADS PER INCH	PIPE O.D. (INCHES)	TAP DRILL SIZE
1/8	27	0.405	R
1/4	18	0.540	7/16
3/8	18	0.675	37/64
1/2	14	0.840	23/32
3/4	14	1.050	59/64
1	11.5	1.315	1 - 5/32

Fig. 2-51 National Taper Pipe Thread sizes

Small taps are held with a T-handle tap wrench, while the larger taps are held with a solid handle tap wrench, which must be turned with two hands.

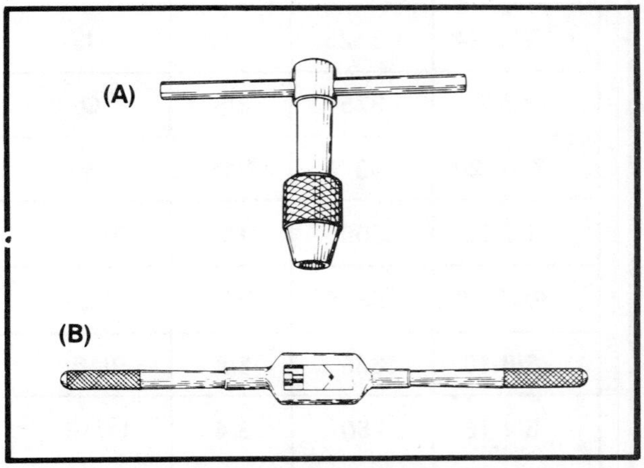

Fig. 2-52 Handles for taps A — T-handle B — Solid handle

b. Dies

Threading dies are used to cut threads on the outside of a rod or a pipe. The die resembles a nut with clearance holes for the chips to fall out and to provide sharp edges of the female threads to cut male threads.

Aircraft technicians find use for two basic types of dies. The most simple is the solid hexagonal rethreading die, which is most commonly called a thread restorer or thread chaser. These dies are screwed onto a male thread that has been damaged and turned down with an appropriate size wrench. The die will straighten out any flattened threads and remove the minimum amount of material.

Split dies are round and must be held in a die stock. They have an adjusting screw that allows the split to be forced apart to change the fit of thread cut by the die. When the spring action of the die forces the sides together, the male threads that are cut will have a loose fit, but when the die is spread apart with the adjusting screw, the threads will have a tight fit.

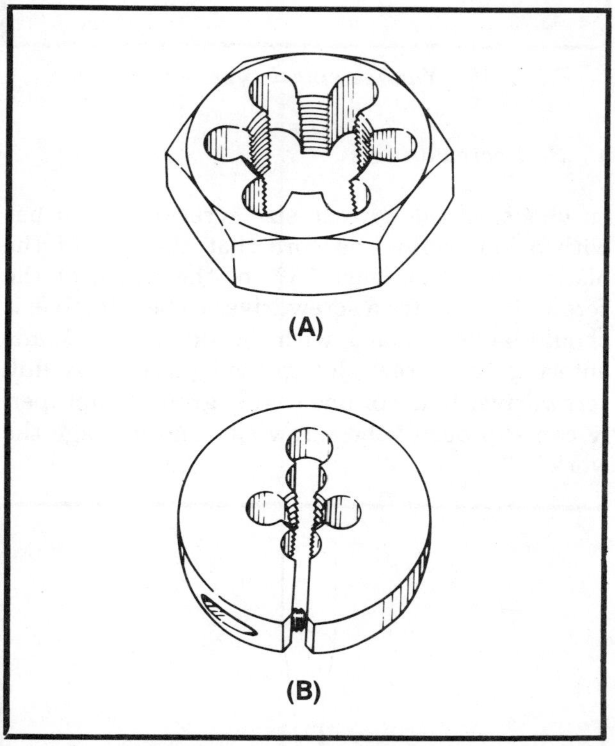

Fig. 2-53 Thread cutting dies A — Thread chasing or restoring die B — Round split die

D. Turning Tools

1. Screwdrivers

There are no tools in the mechanic's tool collection with which he is more familiar, yet which are more consistently misused, than the screwdriver. Their handy shape and wide assortment of sizes make them usable as punches, chisels, pry bars, and drifts. But screwdrivers should be used only for the purpose for which they are made — to drive screws.

Screwdrivers are available with both wood and high-impact plastic handles, and they have blades in a wide variety of lengths and shank and blade sizes. Standard screwdrivers are usually available from the "stubby," which has a blade length of about 1-1/2 inches, up to those with blades of 10 to 12 inches. Special application screwdrivers are available with blades up to 20 inches long.

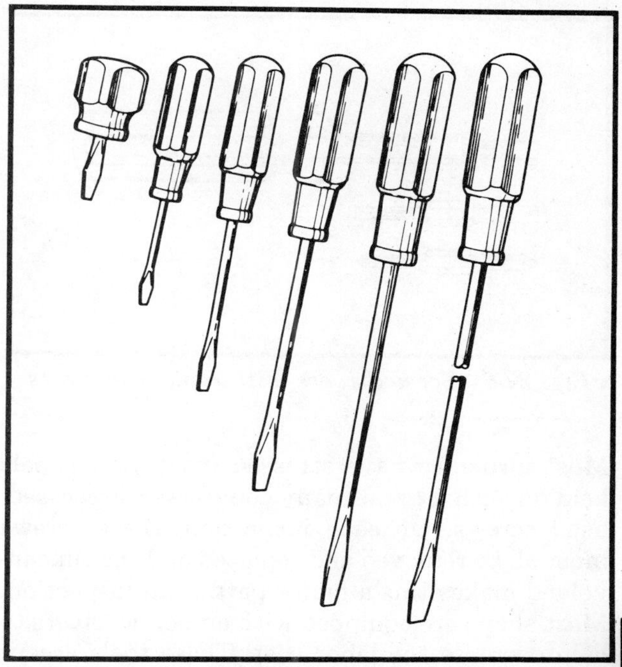

Fig. 2-54 Assorted slot screwdrivers

Offset screwdrivers are available for turning screws where there is no clearance for a straight screwdriver. These are made in two types, one with two blades at right angles to each other, and one with four blades oriented in 45 degree increments.

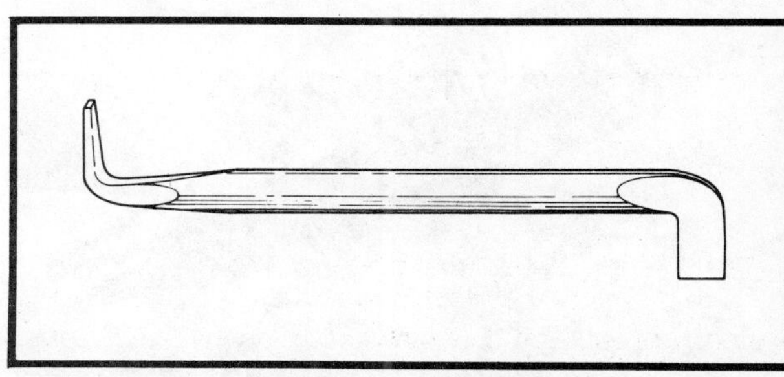

Fig. 2-55 Offset slot screwdriver

In order to have the correct tool available without carrying such a large assortment of handles, there are two special types of screwdrivers that are popular with line mechanics working away from their main tool chest. The reversable-blade screwdriver has a blade for a slot screw on one end and a tip on the other end that will fit a recessed-head screw. Another handy type of screwdriver has a hollow shaft that is magnetized to hold 1/4-inch hex bits. With one handle and an assortment of bits, the mechanic is set for whatever type and size of screw he may encounter.

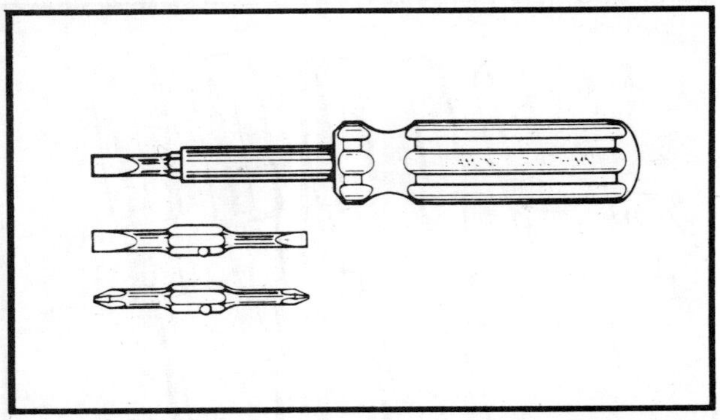

Fig. 2-56 Screwdrivers with replaceable bits

Most airplanes have stressed inspection panels held on with a great many countersunk recessed-head screws. On each inspection, these screws must all be removed and replaced and the time involved makes this a major part of an inspection. Most shops are equipped with air screwdrivers to help decrease the labor cost. These tools accept standard 1/4-inch screwdriver bits, and when the bit is inserted into the screw and one of the buttons on the handle is pressed, the bit is spun to remove the screw. To replace the screw, the screw is started and the other button is pressed. The

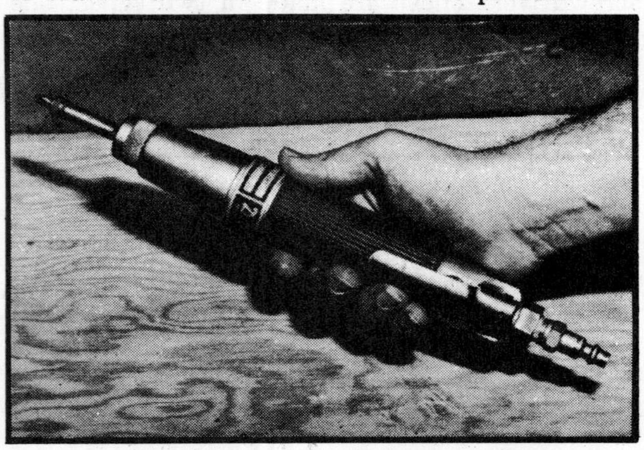

Fig. 2-57 Air driven power screwdriver

screw is driven into the structure, and at a pre-set torque, the chuck slips, preventing overtorquing the screw.

For shops that do not have air screwdrivers, the Yankee ratchet-type screwdriver serves as a fair substitute. This screwdriver will accept replaceable tips, or bits, and the bit is spun by working the handle in and out over its spiral shaft.

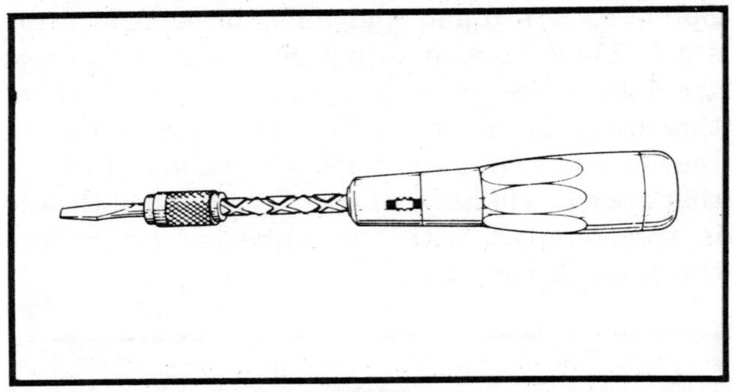

Fig. 2-58 Yankee spiral type screwdriver.

a. Slot screwdrivers

In choosing the correct size screwdriver to use with a slot screw, be sure that the size of the blade will fill at least 75% of the width of the screw slot; and for a screwdriver to be effective, it should be kept sharp, with the sides of the blade entering the screw slot ground parallel. A dull screwdriver blade or one that is ground improperly can slip out of the screw slot and damage the work.

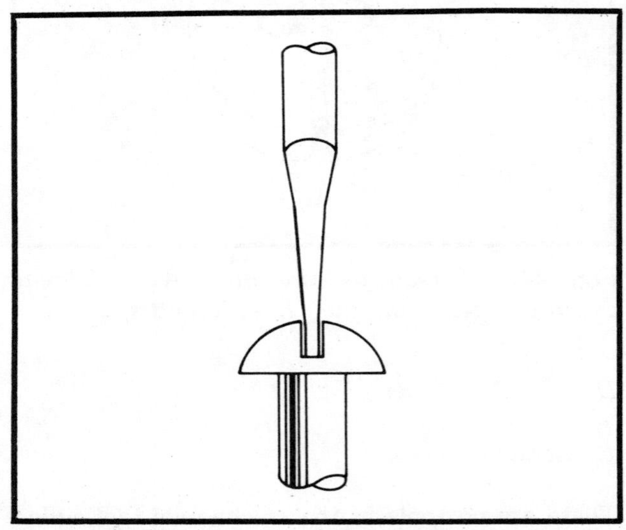

Fig. 2-59 The end of the blade of a slot screwdriver should be ground parallel and should fill approximately 75% of the width of the slot.

The shanks of some screwdrivers are square, so a wrench can be used to provide additional leverage for screws that require excessive force. Other screwdrivers have a hex-shaped bolster where the blade enters the handle, on which to use a wrench for increased leverage.

b. Cross-point screwdrivers

(1) Phillips

There are two types of cross-point screwdrivers used in aviation maintenance that look similar, but care must be taken that only the correct screwdriver be used with them. The cross in the Phillips screw head is cut with a double taper and the sides of the cut ae not exactly parallel. The Phillips screwdriver has a relatively blunt point to fit into the flat-bottomed hole.

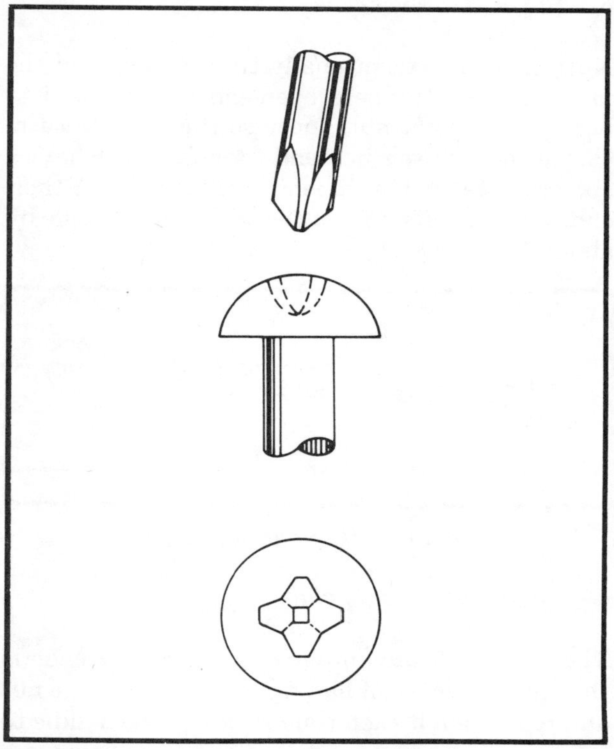

Fig. 2-60 Phillips screwdriver

(2) Reed and Prince

The Reed and Prince screw slot has straight sides and forms a more perfect cross than the Phillips, and the width of the slot is less than that of a comparable size Phillips. There is only one taper to the point of a Reed and Prince screwdriver.

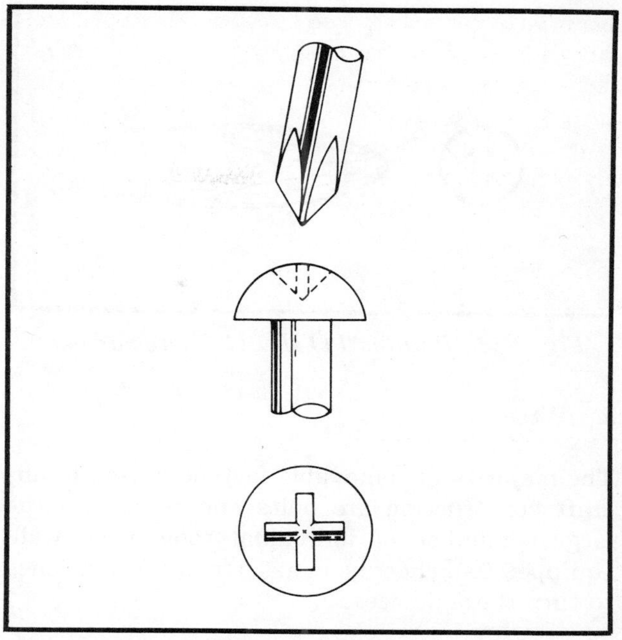

Fig. 2-61 Reed and Prince screwdriver

(3) TRI-WING

Screws with a special type of screw head are produced by the Phillips Screw Company and are used by some of the airlines. These screws have three slots instead of four, as found on the regular cross-point screw. It is identified by its registered trade name TRI-WING.

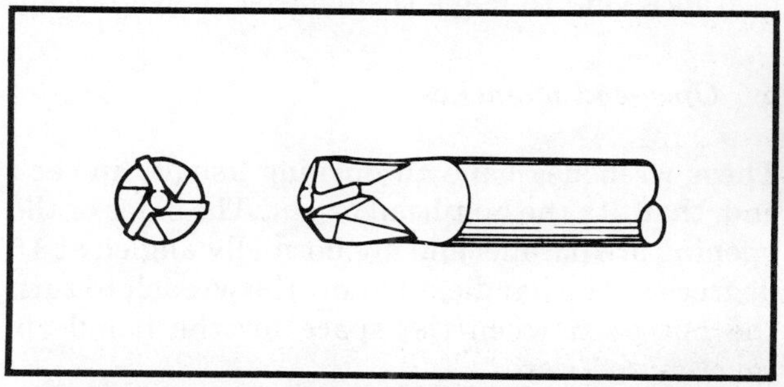

Fig. 2-62 Phillips TRI-WING screwdriver

(4) POSIDRIV

This is a cross-point screw that resembles the Phillips screw except for the bottom of the recess. It has been designed and produced by the Phillips Screw Company to provide a tighter and more positive connection between the screwdriver tip and the recess in the screw, providing less chance for the bit of a power screwdriver to slip out of the screw head.

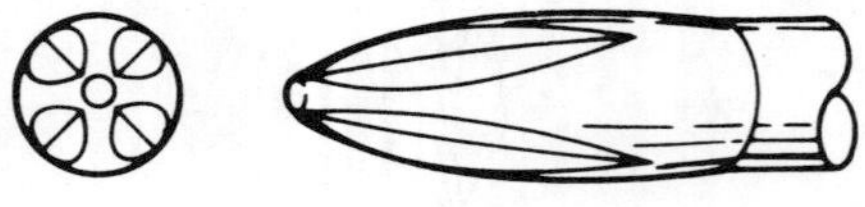

Fig. 2-63 Phillips POSIDRIV screwdriver

2. Wrenches

The majority of removable fasteners used in aircraft construction are bolts and nuts, and the largest number of individual tools in a well-equipped tool chest is an assortment of wrenches to turn these devices.

American aircraft have used nuts and bolts with English sizes, but now that the metric system is becoming standard in the United States, metric size wrenches are also needed for aviation maintenance.

Most wrenches are forged of chrome-vanadium steel which is an extremely tough alloy. After they are forged and all of the burrs have been removed, wrenches are plated with hard chrome to protect them from rust, to improve their appearance, and to make them easier to keep clean.

a. Open-end wrenches

These wrenches have an opening usually in each end, that fits the bolt head or nut. The sides of the opening are parallel and are normally angled at 15 degrees to the handle, to allow the wrench to turn the nut even when the space for the handle is severely restricted.

The standard combination of sizes used for both English and metric open-end wrenches is given in Fig. 2-64. Of course there are many other combinations available.

While the 15-degree head angle is standard, there are many other angles available. One very popular set of wrenches has the same size opening on both ends; one opening is angled at 30 degrees, while the opening on the other end is angled at 60 degrees.

ENGLISH (INCHES)	METRIC (mm)
1/4 — 5/16	6 — 7
3/8 — 7/16	8 — 9
1/2 — 9/16	10 — 11
5/8 — 11/16	12 — 13
3/4 — 13/16	15 — 18
7/8 — 15/16	21 — 24
1 — 1-1/8	

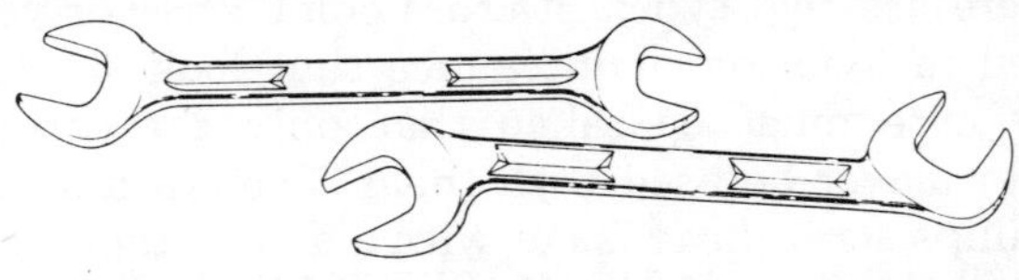

Fig. 2-64 Open-end wrenches

b. Box-end wrenches

Nuts that are exceptionally tight can spread the jaws on even the best open-end wrench, and to break these tight nuts loose so they can be turned, we should use a box-end wrench. These have a six- or a twelve-point opening broached into their end, which is offset from the axis of the handle by about 15 degrees.

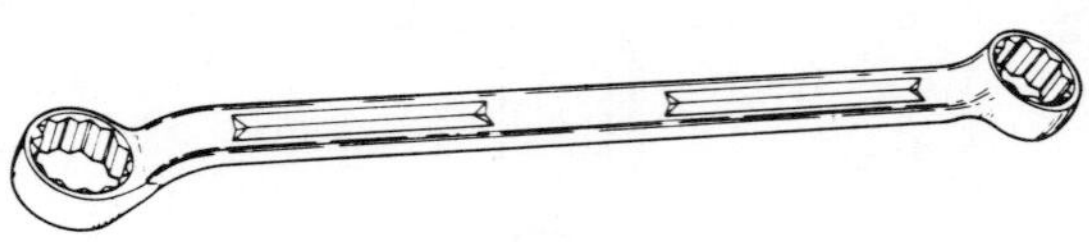

Fig. 2-65 Box-end wrenches

c. Combination wrenches

The biggest disadvantage of a box-end wrench is the inconvenience of having to lift it off of the nut and reposition it each time we move the handle to the extent the surrounding structure allows. An open-end wrench is much easier to slip off and on the nut. To have the advantage of both a box-end and an open-end wrench, the combination wrench is available. This popular configuration of wrench has a box end broached on one end, and an open end of the same size on the other end. The hard nuts are broken loose with the box end and then are turned to remove them with the open end. The box end is usually angled 15 degrees to the handle to allow clearance for your hands between the wrench and the work, and the open end is offset by 15 degrees to the axis of the handle to allow a new grip on the nut each 15 degrees of handle movement.

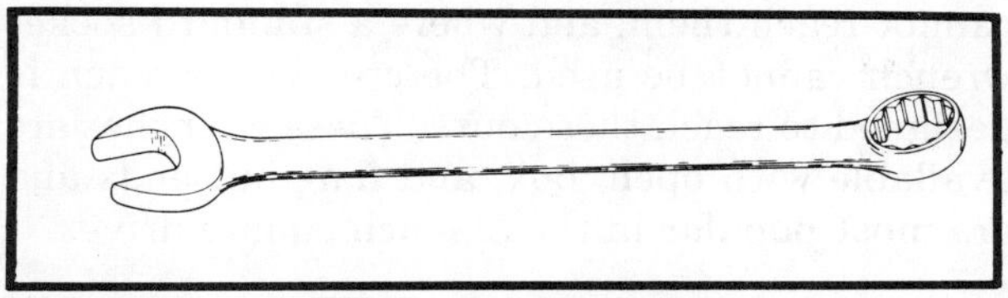

Fig. 2-66 Combination wrenches

d. Flare nut wrenches

Aircraft fluid lines are connected to the components with flare nuts, and while these nuts should never be so tight that an open-end wrench could be sprung, they are often situated in locations where the swing of the wrench handle is severely restricted. Because of this, a special type of box-end wrench is used. A slot is cut into the box end to allow the wrench to slip over the tube, and then the hex of the nut can be engaged each 15 degrees of handle movement.

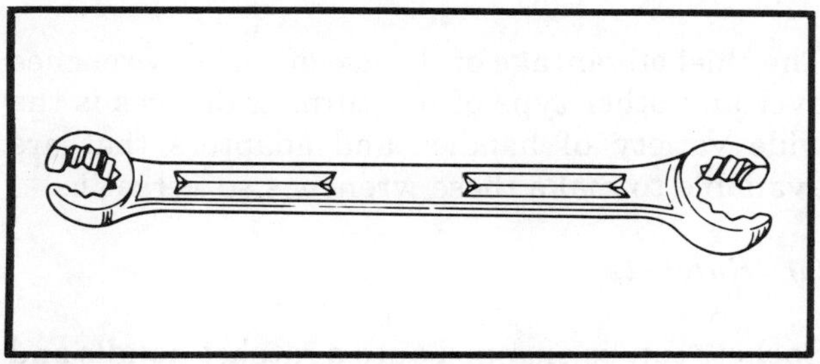

Fig. 2-67 Flare nut wrenches

e. Ratcheting open-end wrenches

A special design of open-end wrench allows nuts to be turned down without removing the wrench. One of the jaws of an open end is ground away with just enough left to engage one edge of the nut.

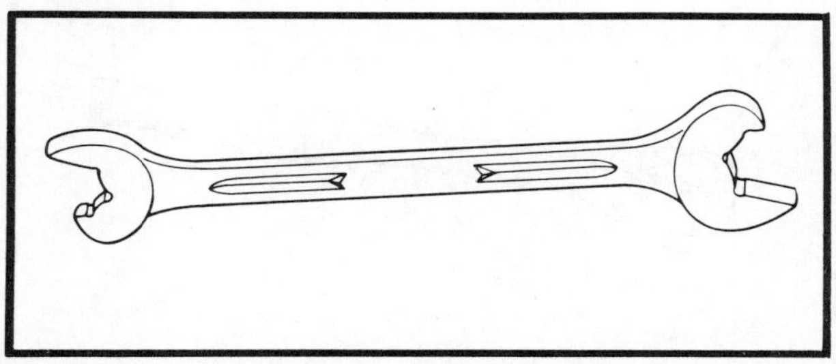

Fig. 2-68 Ratcheting open end wrenches

f. Ratcheting box-end wrenches

One of the handiest wrenches to be devised in recent years is the ratcheting box-end wrench. A box end is set into a handle and is allowed to turn in one direction, but locked in the opposite direction. The wrench is slipped over the nut and worked back and forth to remove the nut. The wrench is just turned over to reverse the direction of the driving action.

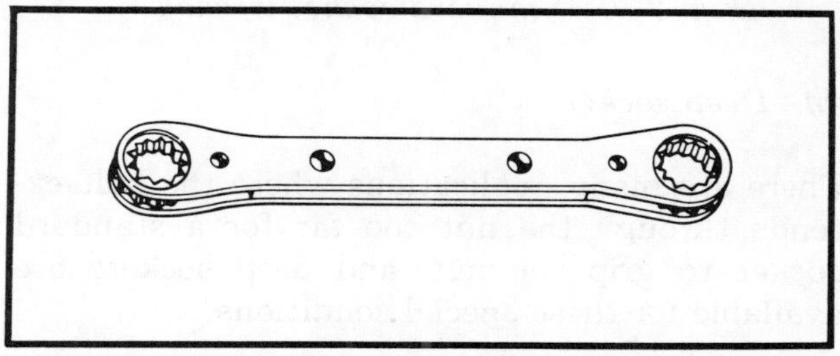

Fig. 2-69 Ratcheting box end wrench

g. Socket wrenches

The invention of the removable-handle socket wrench has been a major innovation in mechanics' hand tools. A small socket is broached in one end with a six- or twelve-point opening to fit the nut, and the other end has a square hole to fit a drive handle. A large number of these sockets can fit into a relatively small tool chest, and the wide variety of handles and extensions that are available makes these the most adaptable tool available.

Socket sets are available in a wide variety of drive sizes, but in aviation maintenance the 1/4-inch square drive and the 3/8-inch square drive are the most popular. One-half inch, 3/4-inch, and one inch square drive are available, and for extremely large work, there are square drives as large as 2-1/2 inches.

(1) Sockets

(a) Standard sockets

Standard sockets are available in all of the popular drive sizes and are available with either six- or twelve-point openings. They are deep enough to fit over a bolt head or a nut if too much of the shank does not protrude.

Four- and eight-point sockets are available to turn square head pipe plugs.

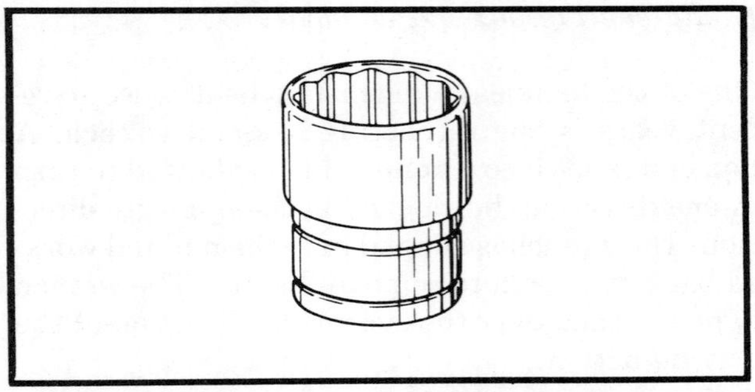

Fig. 2-70 Standard socket wrench

(b) Deep sockets

There are many applications where the bolt extends through the nut too far for a standard socket to grip the nut, and deep sockets are available for these special conditions.

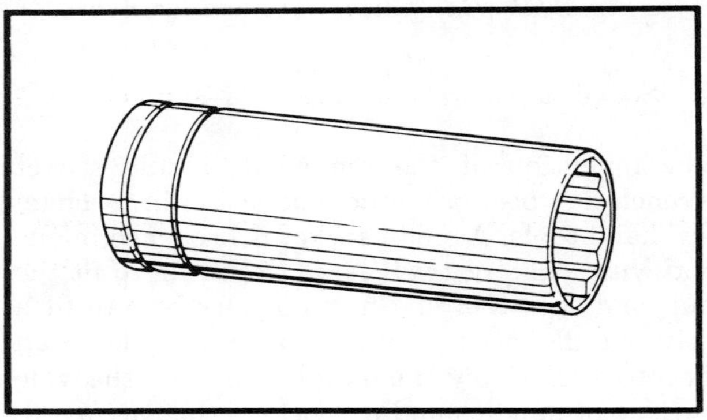

Fig. 2-71 Deep socket

(c) Flex-sockets

Universal joints are available to fit between the handle and the socket, but this additional length would be in the way in many applications. For these occasions, flex-sockets, or universal sockets, are available. These sockets are made with both six- and twelve-point openings and in the drive sizes that are most used in aviation maintenance.

Fig. 2-72 Flex socket

(d) Crowfoot sockets

Nuts are sometimes placed on aircraft in such locations that a box-end or an open-end wrench cannot reach them, and where a standard socket wrench cannot be used. The crowfoot wrench is designed to reach these nuts. These wrenches are available with open, box, and flare-nut ends and are most popular in the 3/8-inch square drive.

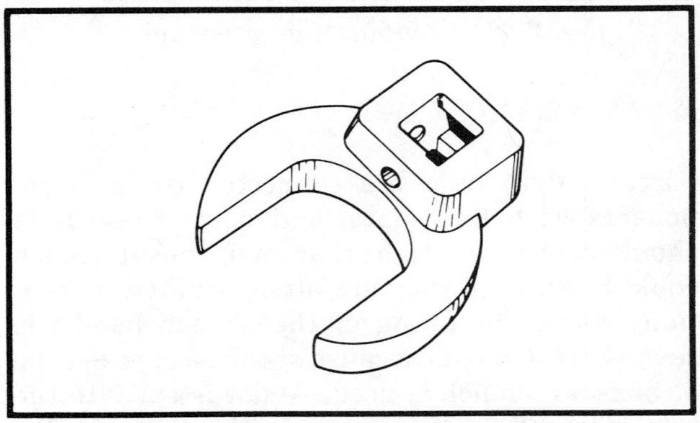

Fig. 2-73 Crowfoot wrench

(2) Handles and adapters

The chief advantage of the use of socket wrenches over any other type of nut-turning devices is the wide variety of handles and adapters that are available to make these wrenches so versatile.

(a) Ratchets

Sockets can be snapped onto a ratchet handle and used to turn nuts with a minimum of handle movement. Rachets are available with long or short handles, and with either solid or flexible heads. When selecting a ratchet, be sure to choose one that has a small amount of handle movement between the positions at which the pawl grips the drive gear.

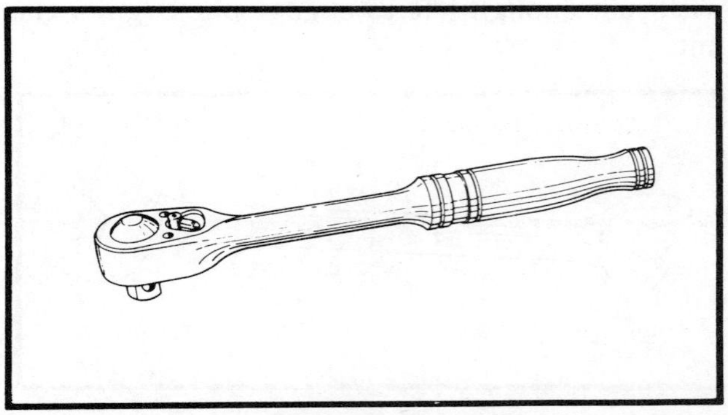

Fig. 2-74 Ratchet handle

A ratchet adapter is available that fits onto an extension or onto a non-ratcheting handle to convert it into a reversible ratcheting wrench.

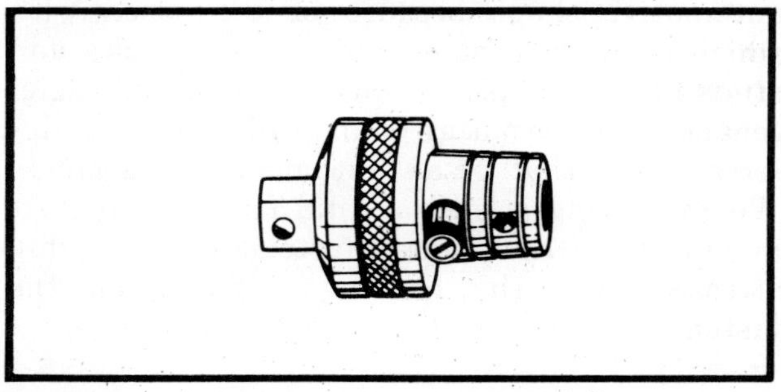

Fig. 2-75 Ratchet adapter

(b) Breaker bar

When a nut is extremely hard to break loose, and more force is required than you would want to apply to a ratchet handle, the socket may be placed on a breaker bar and the required amount of force applied. While it is often done in practice, it is not a recommended procedure to slip a piece of pipe over the handle of a breaker bar to increase the leverage. When more force is required than can be applied with a breaker bar, use a socket and handle of the next larger drive size.

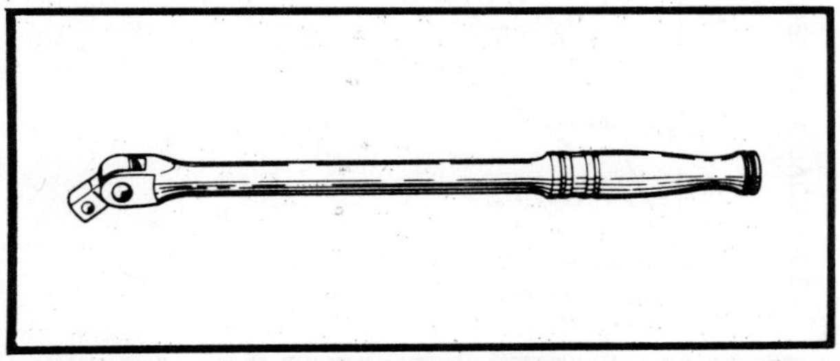

Fig. 2-76 Breaker bar

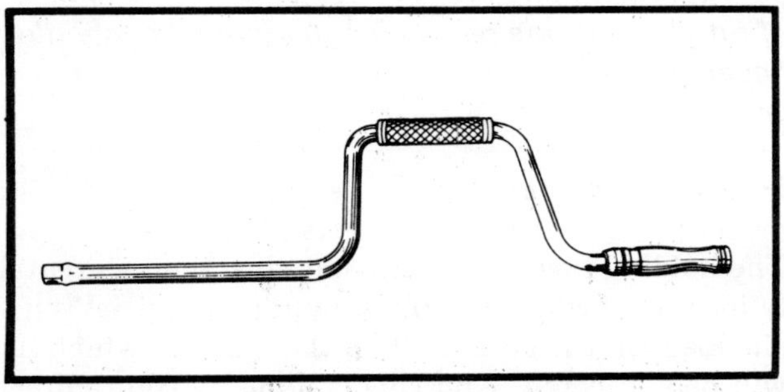

Fig. 2-77 Speed handle

(c) Speed handle

In aviation maintenance, time is an expensive commodity and any tool that will decrease the time required for an inspection or repair must be used. A speed handle, which resembles a bow-type brace, can have a socket or a screwdriver bit snapped onto its end, and screws or nuts can be turned much faster than they could be with a conventional screwdriver or with a ratchet wrench.

(d) Extensions

Straight bar-type extensions may be used to put a socket quite a distance away from the wrench handle. These extensions are available in lengths from less than two inches up to some that are two to three feet long. Some extensions are made of flexible stranded steel that are encased in two spirals of wire wound in opposite directions.

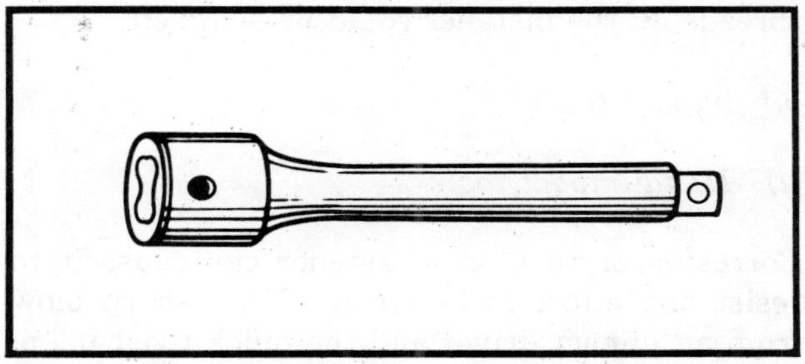

Fig. 2-78 Extension for use with socket wrenches

(e) Universal joints

Small universal joints with a male square drive on one end and a female socket on the other are available to fit between the handle and the socket, or between the handle or socket and an extension.

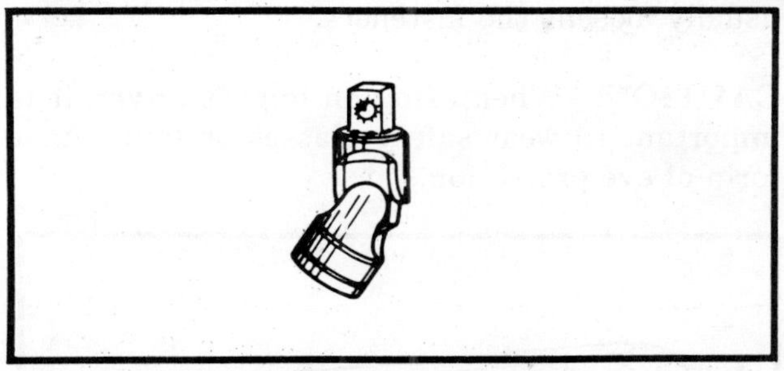

Fig. 2-79 Universal joint

(f) Adapters

Adapters are available to adapt between two sizes of sockets and handles. Adapters can fit a 1/4-inch drive socket onto a 3/8-inch drive handle or vice-versa, and there are adapters between the

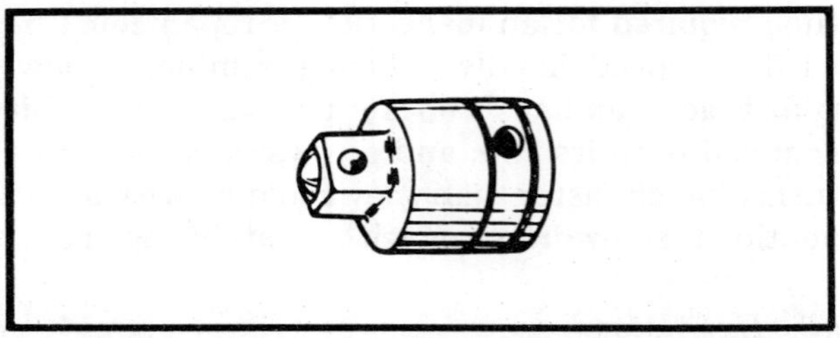

Fig. 2-80 Adapter to adapt a 1/4" drive socket to a 3/8" drive handle.

3/8-inch and 1/2-inch drive components. You must use good judgment when adapting these tools because, for example, the added length of a 3/8-inch drive breaker bar could exert enough force on a 1/4-inch drive socket that the socket could be broken or the adapter could be broken, or the threads on the fastener could be stripped.

h. Impact tools

(1) Hand impact tools

Corrosion or rust on a fastener can cause it to resist any effort to loosen it. But a sharp blow from an impact driver and the quick twist it imparts will usually break the most stubborn fastener loose. A handy impact driver set for aviation maintenance consists of a driver and an assortment of special six-point impact sockets and bits for the sizes and types of screws that are most often found on airplanes.

To use the impact driver, put the proper bit or socket on the driver and place it on the fastener. Then strike the driver with a sharp hammer blow. The combination of sharp blow and sudden twist usually loosens the fasteners.

CAUTION: When using an impact driver, it is important to wear safety glasses or some other form of eye protection.

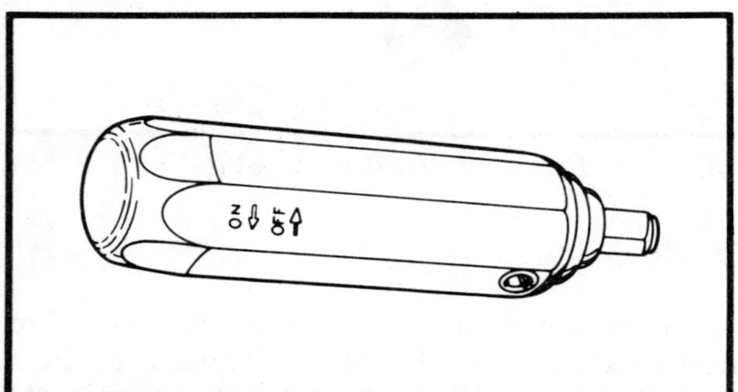

Fig. 2-81 Hand impact driver

(2) Power impact tools

Air-driven impact tools are commonly used in automotive shops because of the speed with which they drive nuts onto or off of bolts and studs. Time is also essential in aviation maintenance, but the nature of an airplane is quite different from that of an automobile or a truck. Weight saving is such a vital factor in aircraft manufacture that most fasteners are highly stressed, and the housings into which the fasteners are attached are usually quite thin.

An impact wrench does not apply a smooth and even torque to a fastener, but, rather, it applies its force in a series of jerks or impacts. Because of this, an impact wrench set to limit its torque at a specific level will actually apply a much higher peak torque than that for which the wrench is set. Because of these torque spikes, or peaks, an impact wrench must never be used on any fastener whose torque is critical.

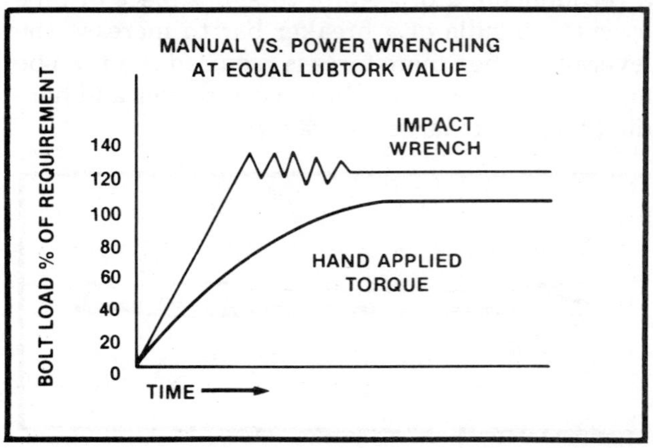

Fig. 2-82 The torque produced by a power impact tool is applied in a series of blows or spikes of torque. For this reason impact wrenches should never be used on fasteners whose torque is critical.

i. Torque wrenches

The holding power of a threaded fastener is greatly increased when it is placed under an initial tensile load that is greater than the loads to which it will be subjected. And this is done by tightening the bolt or the nut to a pre-determined torque. This pre-load is applied to the bolt with a torque wrench.

A torque wrench is a precision measuring tool that measures the amount of force required to

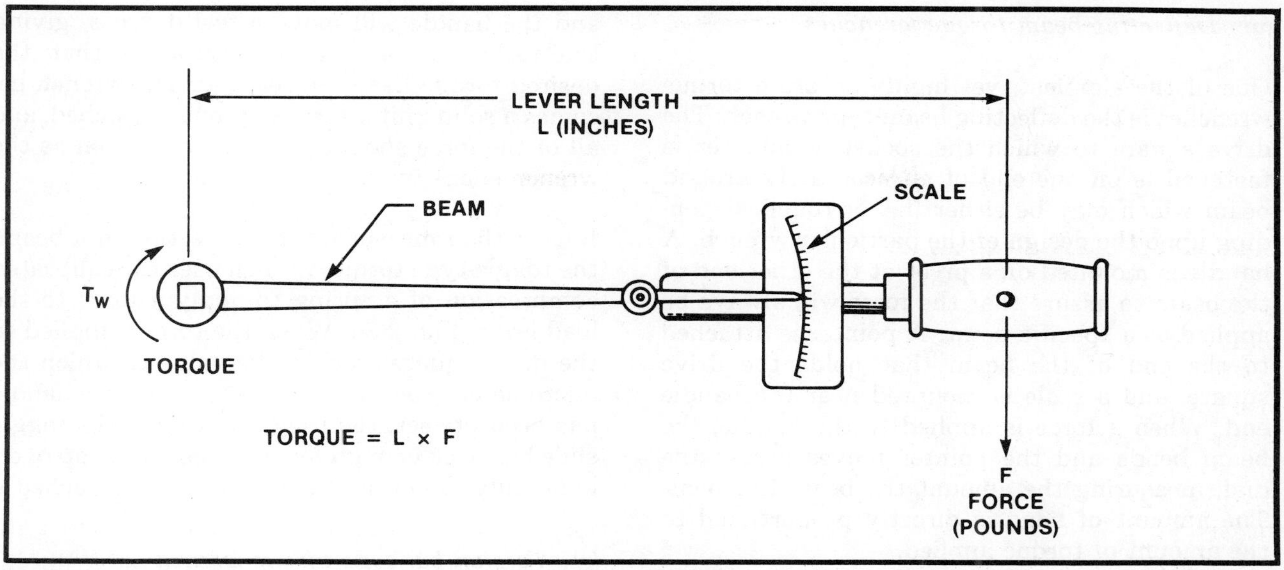

Fig. 2-83 The torque measured by a torque wrench is the product of the length between the drive square of the wrench and the point of application of the force on the handle.

turn a fastener. Under controlled conditions, the amount of force required to turn the fastener is directly related to the tensile stress within the fastener.

The amount of torque, measured in inch-pounds (or in foot-pounds or meter- or centimeter-kilograms) is the product of the force required to turn the fastener multiplied by the distance between the center of the fastener and the point at which the force is applied.

A torque wrench, such as the one we see in Fig. 2-83, has a length that is permanently established between the center of the drive square on which the socket is placed and a pivot in the handle. The force applied is measured by the amount the beam deflects, and we know from Hooke's law, which is a very basic law of physics, that the amount a beam deflects is directly related to the force applied. If the lever, for example, is exactly 12 inches long and a force of 30 pounds is applied to the handle, there will be a torque of 360 inch-pounds on the fastener.

(1) Types of torque wrenches

There are three basic types of torque wrenches that are most generally found in aviation maintenance shops; the deflecting beam-type, the toggle-type, and the torsion bar-type.

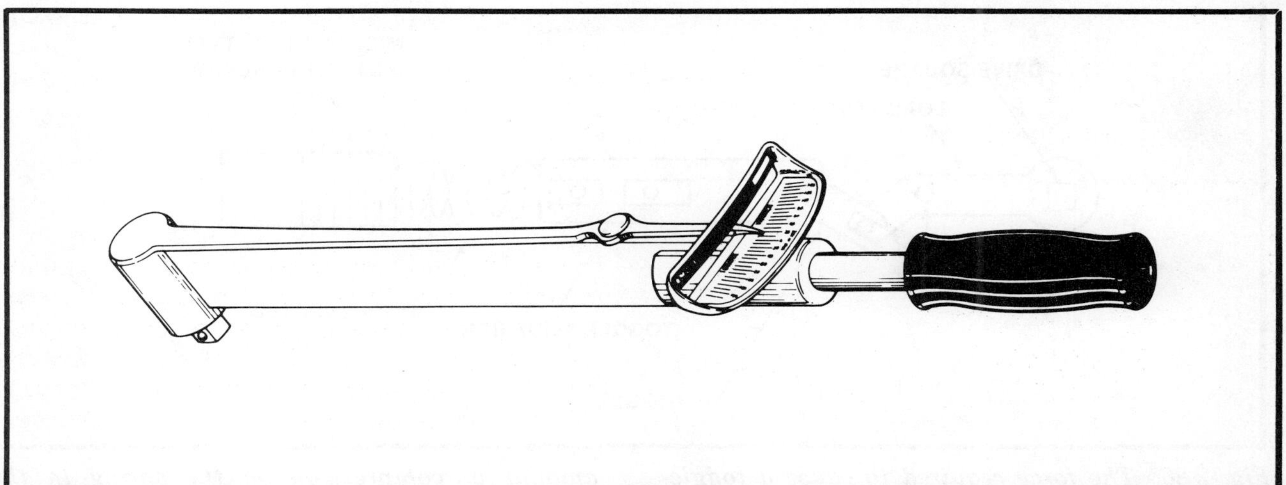

Fig. 2-84 Deflecting beam torque wrench

(a) Deflecting-beam torque wrenches

One of the simplest, yet highly accurate torque wrenches is the deflecting beam-type wrench. The drive square to which the socket or adapter is fastened is on one end of an accurately ground beam which may be either flat or round, depending upon the design of the particular wrench. A handle is mounted on a pivot at the other end of the beam to assure that the force will always be applied at a specific point. A pointer is attached to the end of the beam that holds the drive square, and a scale is mounted near the handle end. When a force is applied to the handle, the beam bends and the pointer moves across the dial, measuring the amount the beam has bent. The amount of bend is directly proportional to the amount of torque applied.

(b) Toggle-type torque wrenches

The toggle-type torque wrench can be pre-set to the desired torque before it is put on the fastener, and when this torque is reached a sound is heard and the handle will move a few degrees, giving the technician a positive indication that the desired torque has been reached. The wrench becomes a solid unit after the torque is reached, and all of the force should be released as soon as the wrench snaps over.

Rather than measuring the deflection of a beam, the toggle-type torque wrench uses the calibrated compression of a spring to apply a force to the load lever, Fig. 2-86. When the torque applied to the drive square reaches the value to which the micrometer-type adjusting screw in the handle has been pre-set, the toggle will force the toggle slide bar back enough for the toggle to snap over, indicating the desired torque has been reached.

(c) Torsion bar-type torque wrench

A bar will accurately deflect in torsion as well as bending when a force is applied, and this principle is made use of in the torsion bar-type torque wrench. The drive square of this type of wrench is an accurately ground torsion bar which has a rack

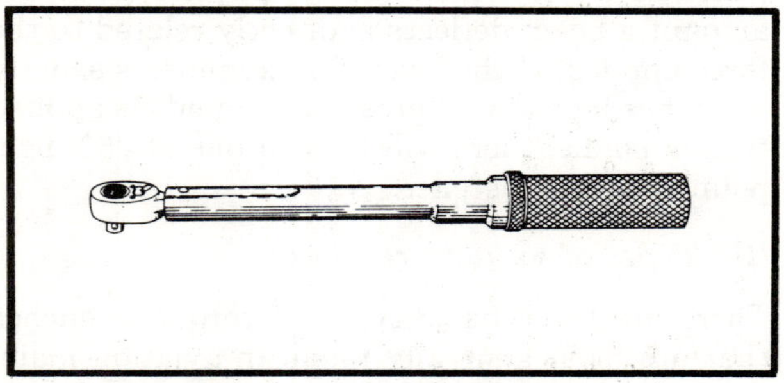

Fig. 2-85 Toggle-type torque wrench

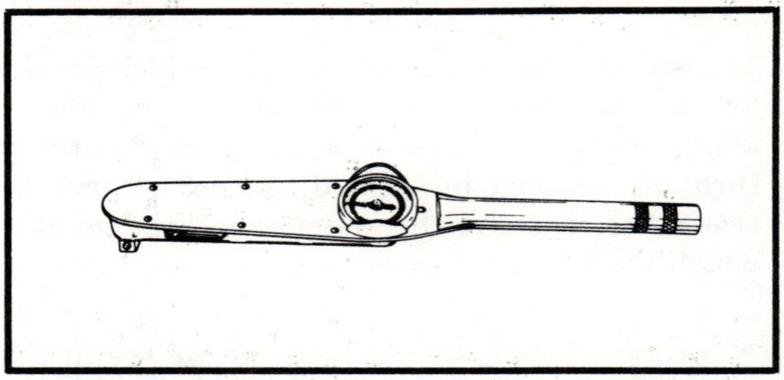

Fig. 2-87 Torsion-bar type torque wrench

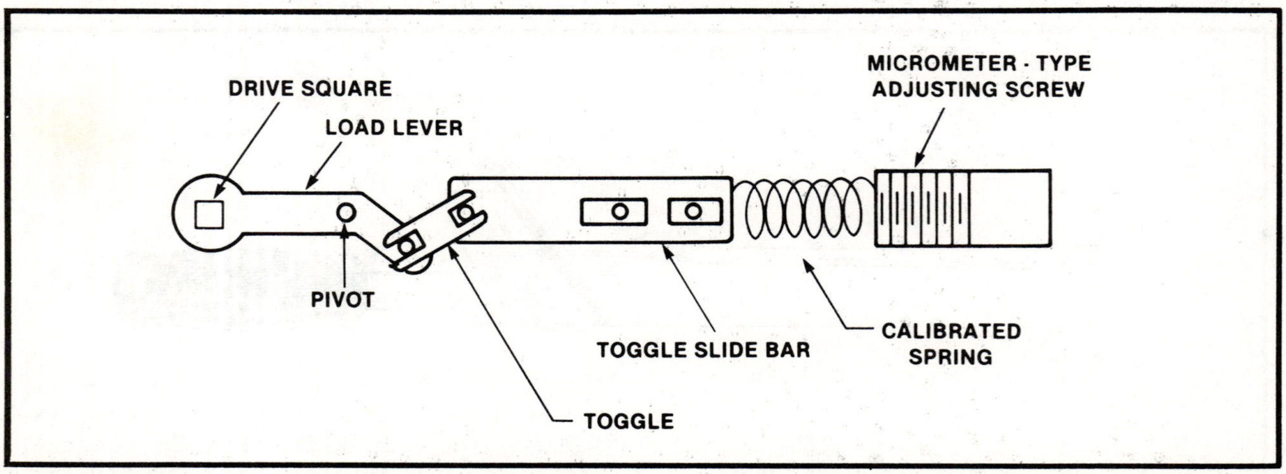

Fig. 2-86 The force required to cause a toggle-type torque wrench to snap is determined by the amount of compression on the spring in the wrench handle.

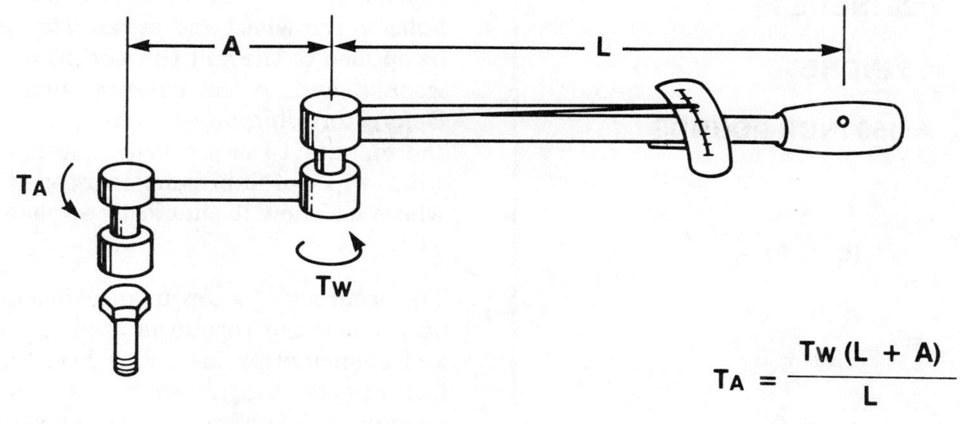

Fig. 2-88 Formula for finding the amount of torque applied to a fastener when an adapter is used.

gear on one end. When the bar is twisted, the rack is moved across a pinion gear in a dial indicator which indicates the amount the bar has deflected. This deflection is calibrated in such units of torque as inch-pounds, foot-pounds, or in meter-or centimeter-kilograms.

(2) Extending the range of a torque wrench

It is often necessary to use an adapter on a torque wrench to get to a particular fastener, and when these adapters are used, the indication of the wrench must be modified to find the torque that is actually being applied.

In Fig. 2-88 we have a torque wrench whose length between the drive square and the handle pivot is 20 inches, and we have a five-inch extension on the wrench. To find the torque applied to the fastener when we have an indication of 120 inch-pounds on the wrench, we use the following formula:

$$T_A = \frac{T_W \times (L + A)}{L}$$

When we work this problem we see that when the wrench reads 120 inch-pounds, we actually have 150 inch-pounds of torque on the fastener.

We can find the amount of torque the wrench should indicate to have a given torque on the fastener by using this formula:

L = 20 INCHES

A = 5 INCHES

T_W = 120 INCH-POUNDS

$$T_A = \frac{T_W (L + A)}{L}$$

$$= \frac{120\ (20 + 5)}{20}$$

$$= \frac{120 \times 25}{20}$$

= 150 INCH-POUNDS

Fig. 2-89 Determining the amount of torque applied to a fastener when an adapter is used.

$$T_W = \frac{T_A \times L}{(L + A)}$$

Using the same figures as before, we find that in order to get 150 inch-pounds of torque on the fastener when we have a five-inch extension on the wrench, we must pull on the handle until we get an indication of 120 inch-pounds on the wrench.

$$L = 20 \text{ INCHES}$$

$$A = 5 \text{ INCHES}$$

$$T_A = 150 \text{ INCH-POUNDS}$$

$$T_W = \frac{T_A \times L}{(L + A)} = \frac{150 \times 20}{20 + 5} = \frac{3000}{25} = 120 \text{ INCH-POUNDS}$$

Fig. 2-90 Determining the torque reading on the wrench needed to produce a given amount of torque when an adapter is used.

(3) Use of the torque wrench

If a fastener is undertorqued, there is danger that the joint will be subjected to unnecessary loads being transmitted into the structure which can cause a premature failure. If the fastener is overtorqued, the threads are overstressed and can fail.

Unless the table of torques used for a specific fastener specifies otherwise, the torque values given are for clean dry threads.

When a self-locking nut is torqued, the nut should be run down on the threads until it nearly contacts the washer, and the amount of torque required to run the nut down should be measured and this value added to the amount of torque needed for the fastener. The torque needed to turn the nut down is called the friction drag torque.

Some engine overhaul manuals require that all fasteners be lubricated with oil as a lubricant, and the table of torque values specifies the torque needed for lubricated threads.

There is no part of an aircraft that is subjected to more pounding loads than the wheels and brakes, and one brake manufacturer recommends that a special type of thread lubricant be used on all bolts in the wheel and brake. This lubricant is to be applied to the bolt threads, to both sides of the washer, and to the bearing surface of the nut. Before torquing any fastener, be sure to consult the manufacturer's service manual to determine what type of lubricant is needed, if any, and where and how it should be applied.

The accuracy of a torque measurement is assured only when the torque is applied with a smooth and even motion, and it is because of this that impact-type torque wrenches should never be used on any fastener whose torque is critical.

If a nut is accidently overtorqued, it should be loosened and then retorqued to the proper value. Never back off a nut or a bolt and leave it in that condition.

If a cotter hold does not line up when torquing a castellated nut, another nut should be substituted to see if a slot will line up in it.

Torque wrenches are precision measuring tools and should be periodically checked for accuracy. Torque analyzers are available in most cities so that your torque wrenches can be calibrated. An accurate wrench will let you know that you are actually applying the torque your wrench indicates.

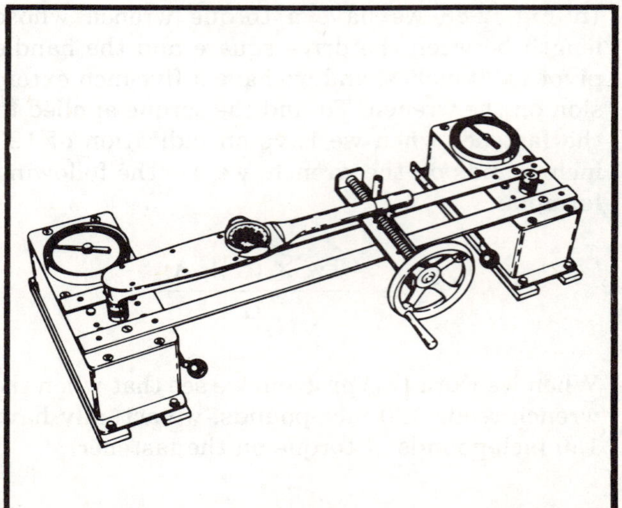

Fig. 2-91 Torque wrench calibrator

E. *Measuring and Layout Tools*

1. *Scales or rules*

Steel scales or steel rules are found in almost all tool chests in both 6-inch and 12-inch lengths. They are used for sheet metal layout and for making measurements where the accuracy of a vernier or a micrometer caliper is not needed.

These scales are made of either tempered carbon steel or of satin-finished stainless steel. They may be graduated in either the fraction or the decimal system of English measurement or in metric measurement, with some scales having graduations in both systems. Fractionally graduated scales usually have one scale divided in increments of 1/32-inch and the other in 1/64-inch increments. Decimal scales have one scale in tenths or fiftieths of an inch and the other scale in increments of 1/100-inch. Metric graduations are in centimeters and millimeters.

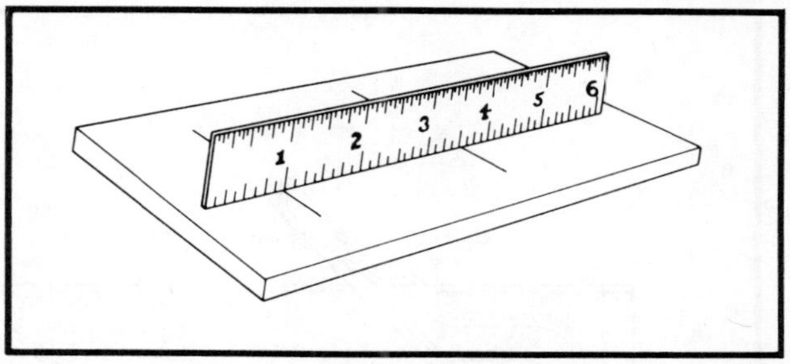

Fig. 2-92 The proper use of a steel scale requires that the end of the scale not be used for the measurement.

Scales are available in both the flexible form, about 0.015 inch thick, and the rigid form with a thickness of about 0.040 inch.

The hook rule is a special form of scale that is useful to measure the dimensions of a part having a radius on its edge. The inside of the hook is exactly in line with the end of the rule.

DECIMAL AND METRIC EQUIVALENT OF INCHES					DECIMAL AND METRIC EQUIVALENT OF INCHES				
Inches			Decimal Equivalent	Millimeter Equivalent	Inches			Decimal Equivalent	Millimeter Equivalent
1/64			0.0156	0.397	33/64			0.5156	13.097
	1/32		0.0313	0.794		17/32		0.5313	13.494
3/64			0.0469	1.191	35/64			0.5469	13.891
		1/16	0.0625	1.588			9/16	0.5625	14.287
5/64			0.0781	1.985	37/64			0.5781	14.684
	3/32		0.0938	2.381		19/32		0.5938	15.081
7/64			0.1094	2.778	39/64			0.6094	15.478
		1/8	0.1250	3.175			5/8	0.6250	15.875
9/64			0.1406	3.572	41/64			0.6406	16.272
	5/32		0.1563	3.969		21/32		0.6563	16.669
11/64			0.1719	4.366	43/64			0.6719	17.067
		3/16	0.1875	4.762			11/16	0.6875	17.463
13/64			0.2031	5.159	45/64			0.7031	17.860
	7/32		0.2188	5.556		23/32		0.7188	18.238
15/64			0.2344	5.953	47/64			0.7344	18.635
		1/4	0.2500	6.350			3/4	0.7500	19.049
17/64			0.2656	6.747	49/64			0.7656	19.446
	9/32		0.2813	7.144		25/32		0.7813	19.842
19/64			0.2969	7.541	51/64			0.7969	20.239
		5/16	0.3125	7.937			13/16	0.8125	20.636
21/64			0.3281	8.334	53/64			0.8281	21.033
	11/32		0.3438	8.731		27/32		0.8438	21.430
23/64			0.3594	9.128	55/64			0.8594	21.827
		3/8	0.3750	9.525			7/8	0.8750	22.224
25/64			0.3906	9.922	57/64			0.8906	22.621
	13/32		0.4063	10.319		29/32		0.9063	23.018
27/64			0.4219	10.716	59/64			0.9219	23.415
		7/16	0.4375	11.112			15/16	0.9375	23.812
29/64			0.4531	11.509	61/64			0.9531	24.209
	15/32		0.4688	11.905		31/32		0.9688	24.606
31/64			0.4844	12.303	63/64			0.9844	25.004
		1/2	0.5000	12.700			1.0	1.0000	25.400

Fig. 2-93 Decimal and metric equivalents

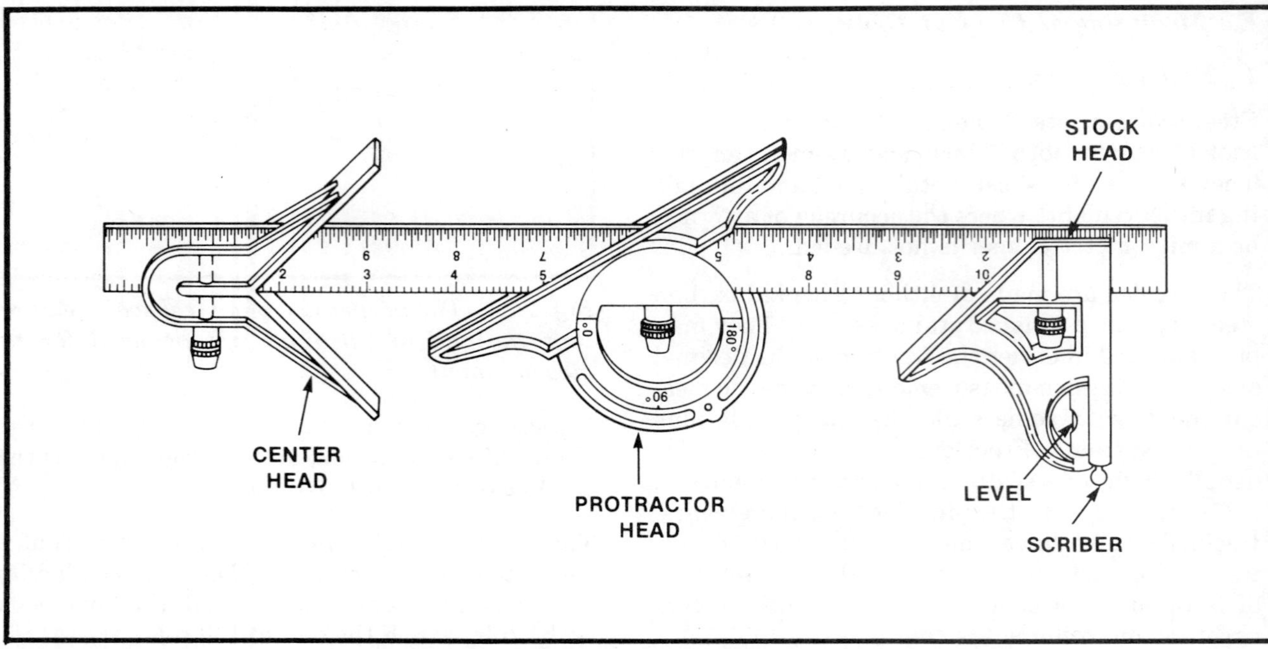

Fig. 2-94 Combination set

2. *Combination set*

A handy elaboration of the steel rule is the combination set, which has a 12-inch steel rule as its foundation and three heads that are movable up and down the rule and may also be removed from the rule.

a. *Stock head*

This is the head on a combination set that is used as a square. One side of the head is exactly perpendicular to the blade (the steel rule), and the other side of this head is at 45 degrees to the blade. A bubble level and a scriber are made into the stock head.

b. *Protractor head*

A bubble level and a 360-degree protractor are made into the protractor head. The head has a lock which may be released so that the protractor can rotate within the frame. This makes the head particularly useful for measuring control surface travel. The surface is locked in its streamline position, and the protractor head is placed on the surface and the protractor turned until the bubble is centered. Now the angular difference between level can be read, as indicated by the bubble and the surface. Unlock the control and move it to its extreme of travel and again center the bubble. The difference between the two readings is the amount of control surface movement.

c. *Center head*

This head is used for finding the center of a circular object such as a piece of round bar stock. Put the blade into the head and place the head over the end of the material. The blade exactly bisects the angle between the two faces of the head, and a line scribed along the edge of the blade passes through the center of the material. Move the head around the material about 90 degrees and make another mark along the edge of the blade. These two marks will cross in the center of the material.

3. *Dividers and calipers*

a. *Dividers*

Dividers are a layout tool having two legs with needle-sharp points joined in the center with a pivot. To accurately lay out a dimension on a piece of metal, first cover the metal with a layout dye and mark the base line in the dye with a very sharp scriber. Then make a very light prick punch mark on the base line from which all measurements will be made. Set the dividers with a steel scale to the distance desired, and then, with one point in the prick punch mark, scratch through the transfer dye on the base line with the other leg. This is a far more accurate method of laying out a distance than measuring directly from the steel scale.

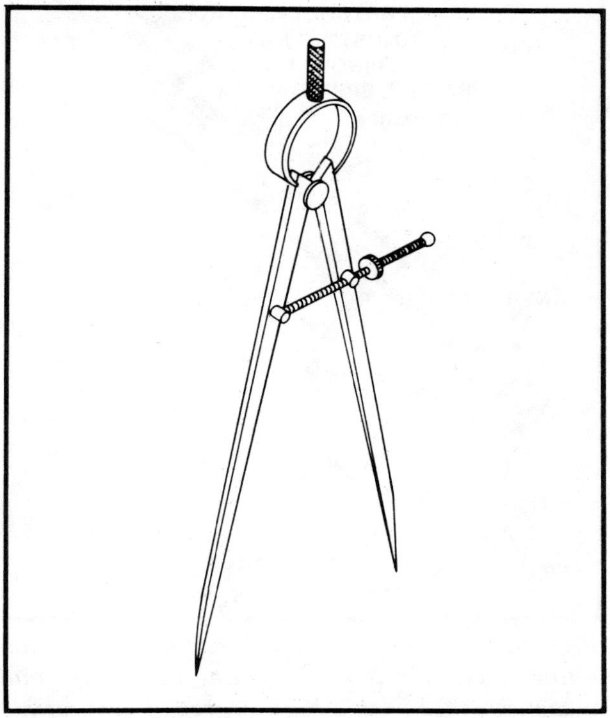

Fig. 2-95 Dividers

b. Calipers

(1) Inside caliper

When the accuracy of a micrometer caliper is not needed, a pair of inside calipers can be used to measure the diameter of a hole. The calipers are adjusted until they fit exactly across the widest part of the hole, and then the distance between the points is measured with the steel scale.

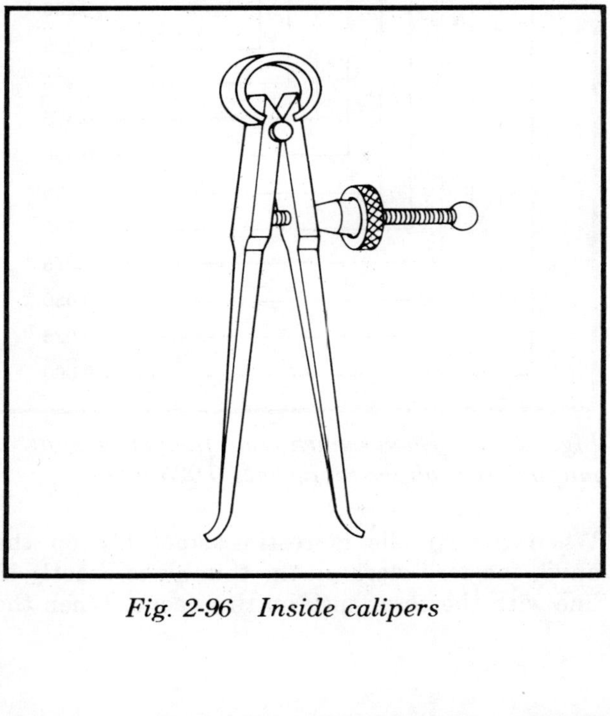

Fig. 2-96 Inside calipers

(2) Outside caliper

The outside dimension of a piece of material may be measured with an outside caliper. The legs are adjusted until they fit exactly across the dimension to be measured, and then they are taken from the material and the distance between the points is measured with a steel scale.

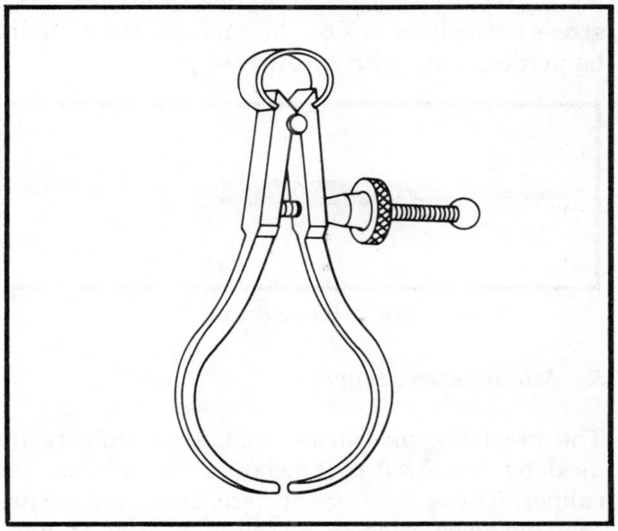

Fig. 2-97 Outside calipers

(3) Hermaphrodite caliper

These calipers have one sharp-pointed leg, similar to that used on dividers, and one leg of an outside caliper. These calipers are used to scribe a mark along a piece of material that is a specific distance from the edge. The material is covered with layout dye, and the distance between the two points is adjusted to the distance the line should be from the edge of the part. The caliper point is moved along the edge as the sharp point is drawn across

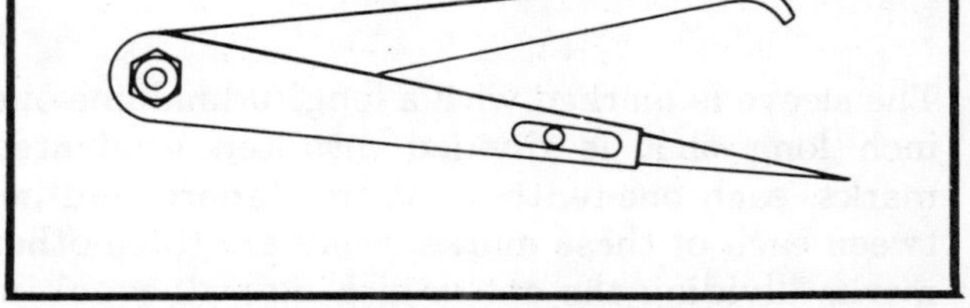

Fig. 2-98 Hermaphrodite calipers

the surface. The dye will be marked with a very fine line the correct distance from the edge.

4. Scribers

Layout of metal parts is done by covering the part with a layout dye and scribing through it with a very sharp-pointed scriber, or the points of dividers. The scribers used most often have needle-sharp points and may be made of hard steel or may be carbide tipped.

Bend lines on sheet metal should never be marked with a scriber as even these fine scratches cause stress concentrations on the surface. They should be marked only with a soft lead pencil.

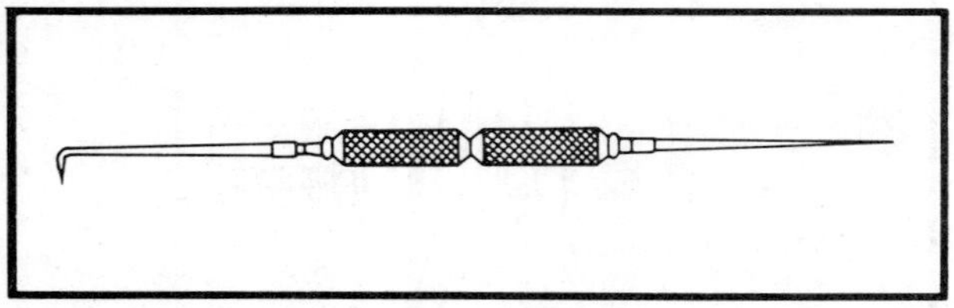

Fig. 2-99 Scriber

5. Micrometer calipers

The precision measuring tool most universally used by the A&P technician is the micrometer caliper. These instruments are used to measure the thickness of sheet metal, as well as to measure the fits of almost all of the parts in an engine.

A micrometer caliper has as its foundation a frame which resembles a C-clamp. On one side of the frame is a hardened and polished anvil and on the opposite side is a graduated sleeve with precision ground internal threads. Into these threads there is screwed a spindle which is turned with a thimble.

The threads in a micrometer caliper are ground to extreme accuracy and have a pitch of 40 threads to the inch. One revolution of the thimble advances the spindle twenty-five thousandths of an inch.

The sleeve is marked with a longitudinal line one inch long that is divided into ten graduated marks, each one-tenth of an inch apart, and between each of these marks, there are three other marks dividing the entire inch into 40 equal increments.

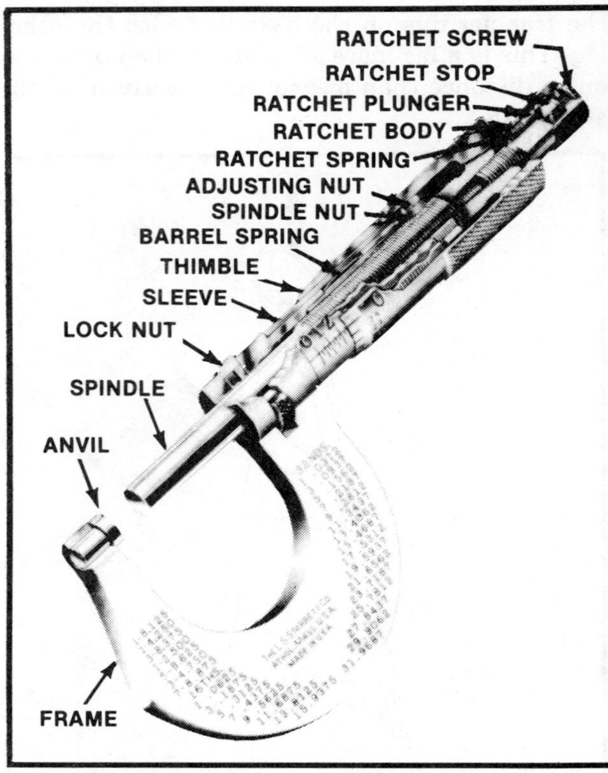

Fig. 2-100 The micrometer caliper is the most universally used tool for making relatively small measurements with a high degree of accuracy.

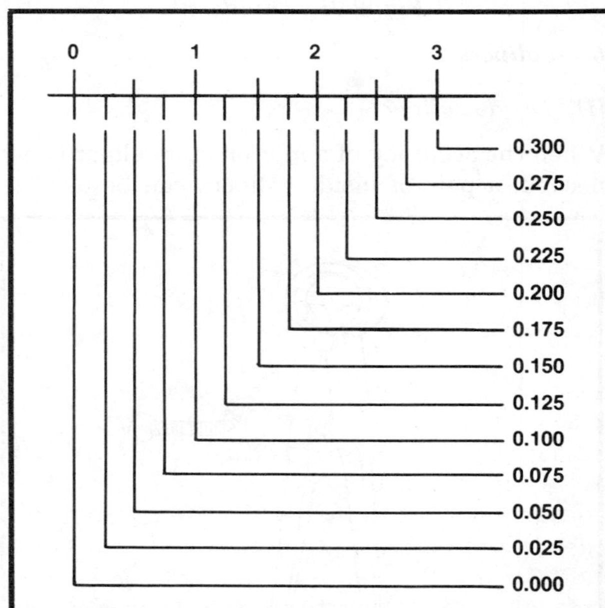

Fig. 2-101 Each mark on the sleeve of a micrometer caliper represents 0.025 inch.

When the spindle is resting smoothly on the anvil, the zero mark on the thimble is exactly in line with the zero mark on the sleeve. When the

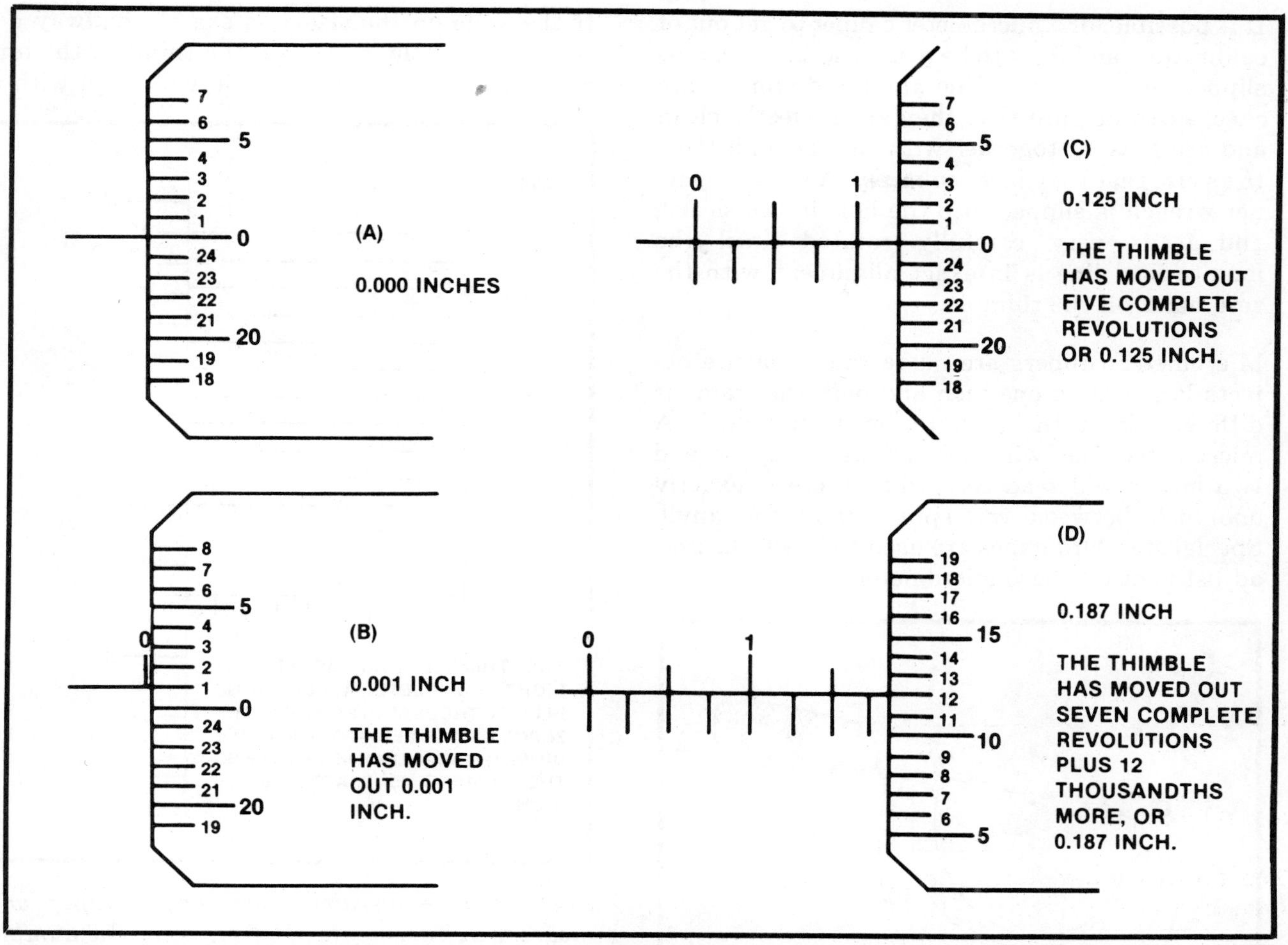

Fig. 2-102 Micrometer caliper reading A — 0.000 inch B — 0.001 inch C — 0.125 inch D — 0.187 inch

zero mark on the thimble is aligned with the longitudinal mark on the sleeve, but is one mark out from zero, there is a space of twenty-five thousandths of an inch (0.025) between the spindle and the anvil. Two marks out gives us 0.050 space; three marks, 0.075; and so on.

There are 25 graduated marks around the circumference of the thimble, which allows us to find the distance between the marks on the sleeve. In Fig. 2-102*A* the spindle is against the anvil and there is a reading of zero inches. In *B* of the same figure, the thimble has rotated out one-thousandths of an inch, until the first mark on its circumference is aligned with the longitudinal mark on the sleeve and there is 0.001-inch between the spindle and the anvil. In *C*, the thimble has been screwed out exactly five turns, and there is 5 × 0.025-inch, or 0.125-inch, between the spindle and the anvil. In *D*, the thimble has moved out seven full rounds, or 0.175-inch, and 12-thousandths of an inch beyond. The total space between the spindle and anvil is 0.175-inch plus 0.012-inch, or 0.187-inch.

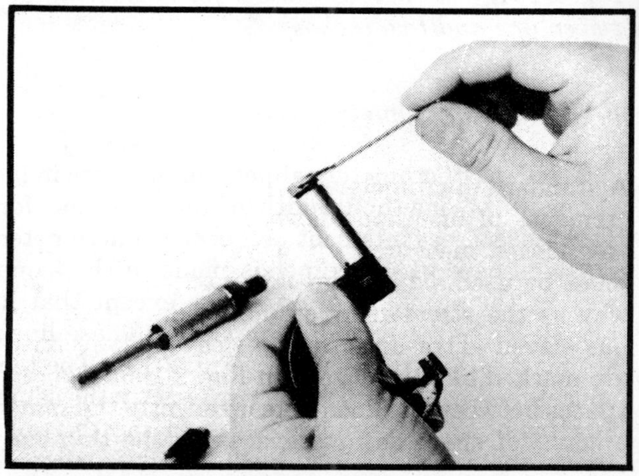

Fig. 2-103 A micrometer caliper may be calibrated by carefully rotating the sleeve until its longitudinal line exactly aligns with the zero mark on the thimble.

It is possible for a micrometer caliper to get out of calibration, and for it to be reset, the sleeve can be slipped on the frame. The anvil and spindle are checked to be sure that they are perfectly clean and are pressed together with just enough force to assure that they meet squarely. A special spanner wrench is slipped into the hole in the sleeve, and it is very carefully rotated until the longitudinal line is in exact alignment with the zero mark on the thimble.

Micrometer calipers are made that measure objects larger than one inch and only the frame is different from the zero to one-inch model. A micrometer that will measure between one and two inches will read zero when there is exactly one inch between the spindle and the anvil. Special standard gages are used to check the zero adjustment on these micrometers.

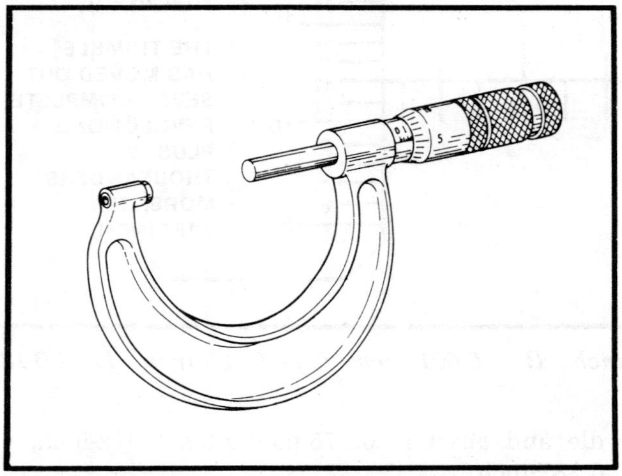

Fig. 2-104 Micrometer caliper for measuring between one and two inches.

6. *Vernier micrometer calipers*

A standard micrometer caliper will measure in increments of one-thousandth of an inch, and for any closer measurement a vernier micrometer must be used. This caliper is made in the same way as the standard micrometer, except that it has eleven extra lines on the sleeve. These lines are marked like those seen in Fig. 2-105. The ten spaces between the lines take up exactly the same amount of space as nine spaces on the thimble. When the zero mark on the thimble is aligned with the longitudinal scale on the sleeve, both of the zero marks on the vernier scale are aligned with marks on the thimble, and no other mark on the vernier scale lines up.

If the scale on the sleeve is exactly halfway between two of the marks on the thimble, the line marked 5 on the vernier scale will align with a

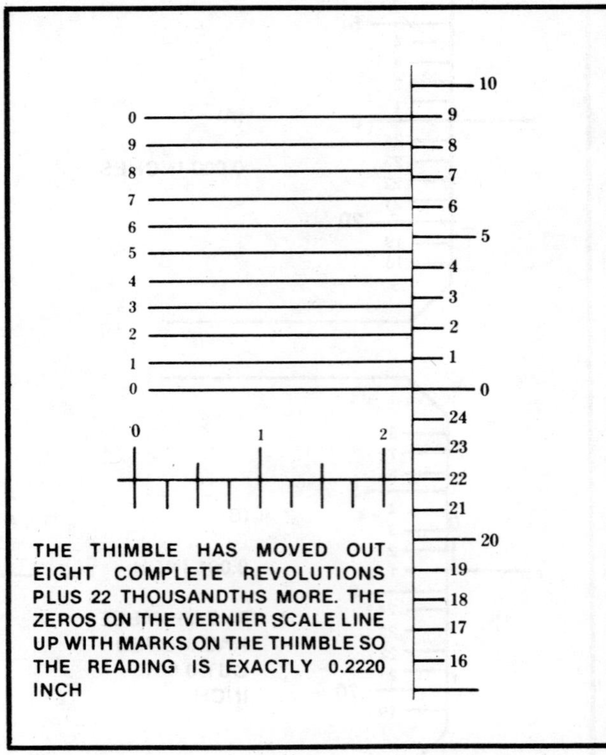

Fig. 2-105 A vernier micrometer caliper will measure in increments of 0.0001 inch. The number on the vernier scale that is lined up with a line on the thimble indicates the number of ten thousandths of an inch to add to the reading on the thimble.

mark on the thimble; no other mark will line up. The micrometer is read in the usual way, and 0.0005 is added to the reading.

7. *Inside micrometers*

A standard micrometer caliper head can be used not only as a part of an outside micrometer caliper, but as an inside micrometer as well. The inside micrometers shown in Fig. 2-106 use a single micrometer head and various extensions to cover the range of measurement that is needed.

Inside micrometers are read in exactly the same way as outside micrometers. Adjust the caliper until it is exactly the same size as the opening you are measuring, and then remove it. Read the measurement indicated by the micrometer head and add the length of the extension.

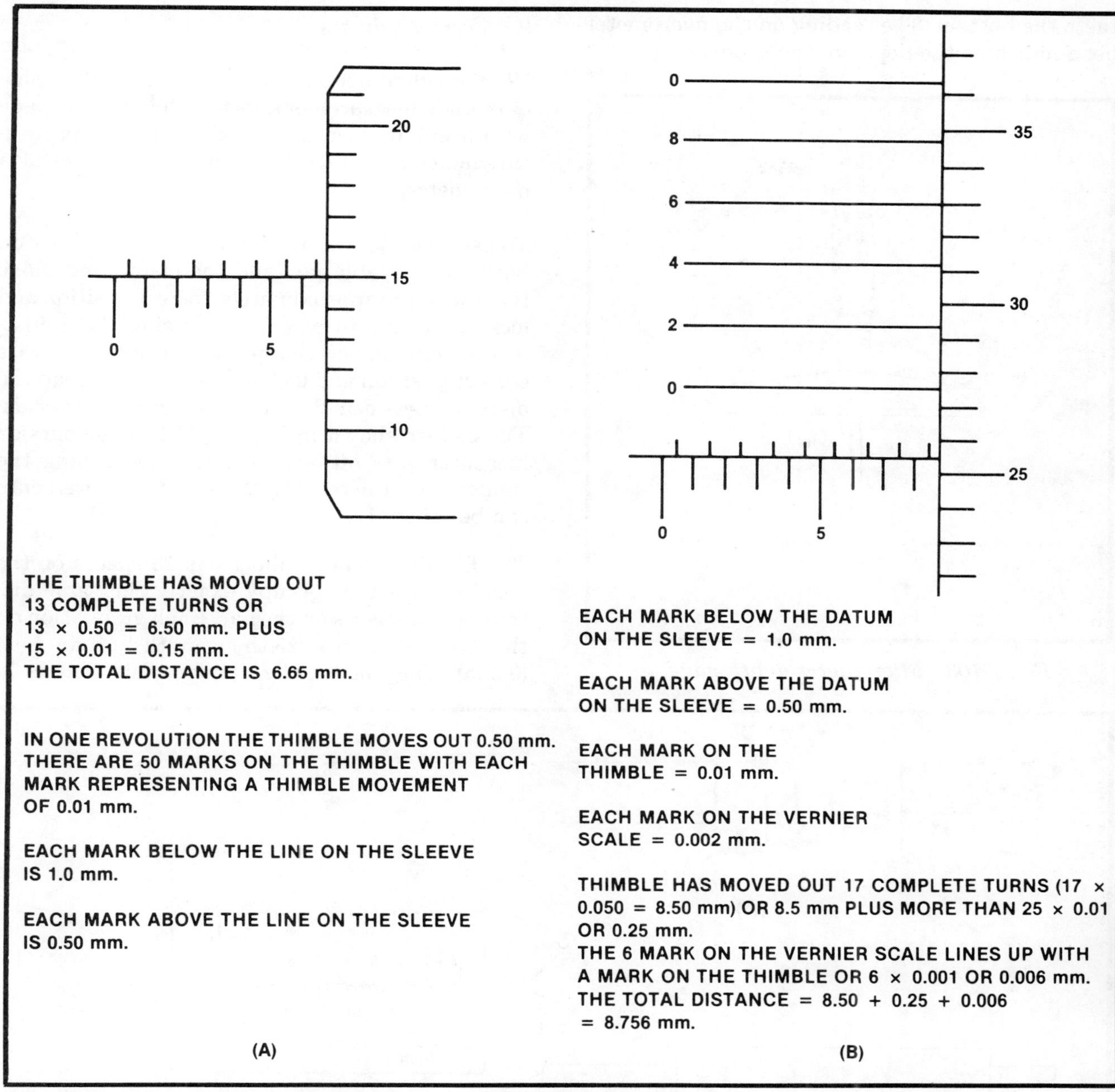

Fig. 2-106 Metric micrometer caliper A — One revolution of the thimble moves it out 0.50 mm B — In the metric vernier micrometer caliper, each mark on the vernier scale represents 0.002 mm. These instruments can be read to an accuracy of one thousandth of a millimeter.

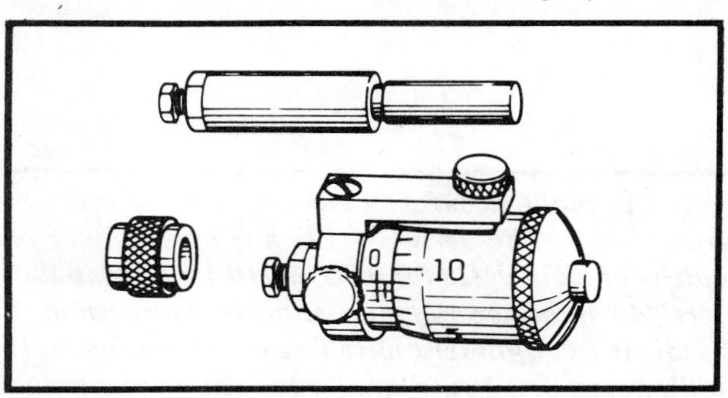

Fig. 2-107 Inside micrometer caliper.

8. *Micrometer depth gage*

The depth of grooves or recesses in a part can be accurately measured with a micrometer depth gage. These devices have a standard micrometer head mounted in a precision ground bar. When the spindle is exactly flush with the surface of the bar, it reads zero. To measure the depth of a groove or a recess, the bar is placed across the groove and the spindle screwed down until it contacts the bottom. The reading on the micrometer head indicates the depth of the groove.

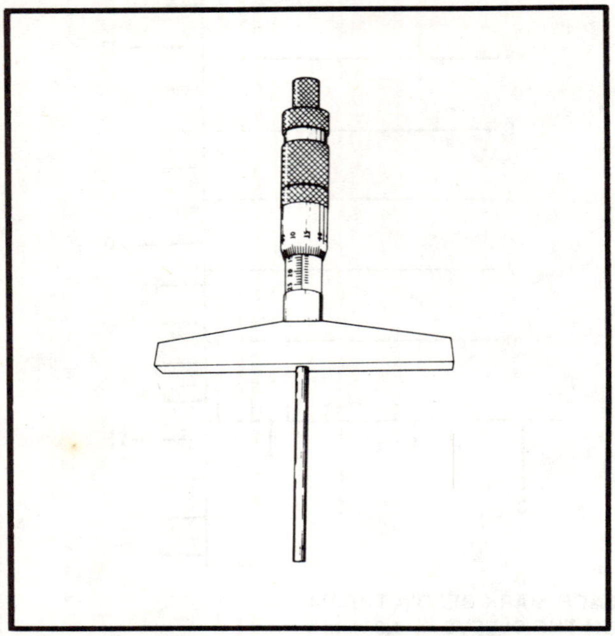

Fig. 2-108 Micrometer depth gage

9. Vernier calipers

Vernier calipers are used to make accurate inside or outside measurements faster than can be made with a micrometer caliper, and for making measurements that exceed the practical range of a micrometer.

To use a vernier caliper, loosen the lock screws on both the movable jaw and the clamp and move the jaw to the approximately correct position and lock the clamp in place. Now, using the adjustment screw on the clamp, move the jaw into its correct position and lock it. You can now read the distance between the jaws on the vernier scale. The caliper shown in Fig. 2-109 has the outside measurements on one side, and by turning the caliper over, the reading for inside measurements can be taken.

The English vernier caliper has 25 spaces on the vernier scale that occupy exactly the same distance as 24 spaces on the bar. Each graduation on the bar represents 25-thousandths of an inch (0.025). The line on the vernier scale that is in

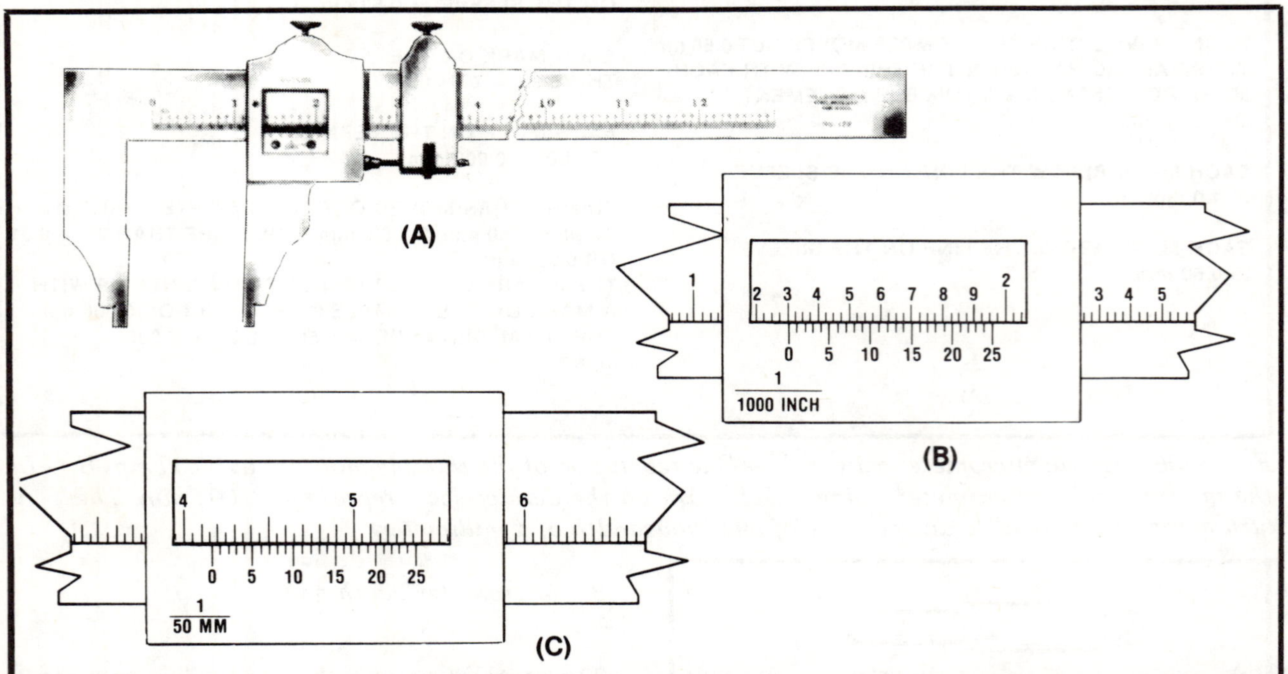

Fig. 2-109 Vernier caliper A — A vernier caliper is used to make accurate inside or outside measurements faster than can be made with a micrometer caliper. B — The vernier scale has moved over one inch, plus three-tenths of an inch, plus fourteen thousandths of an inch (1.00+0.300+0.014=1.314 inches). C — This metric vernier has moved four centimeters (40 millimeters) plus slightly more than 1.5 millimeters more. The line marked nine on the vernier scale is in alignment with one of the marks on the bar which indiates that the jaws have moved 9/50 of a millimeter, or 0.18 millimeter beyond the last mark. The total reading is then 41.68 millimeters or 4.168 centimeters.

alignment with one of the marks on the bar indicates the number of thousandths of an inch we must add to the measurement indicated on the bar before the zero on the vernier scale.

The caliper shown in Fig. 2-109B shows that the measurement is more than 1.3 inches, and to find out exactly how much more, we must find which of the marks line up. In this example, only the 14 mark on the vernier scale lines up with one of the marks on the bar, so the reading would be 1.000 + 0.300 + 0.014, or 1.314 inches.

A metric vernier caliper is read in exactly the same way. Each number on the bar represents one centimeter, or 10 millimeters, and each graduation represents one-half millimeter. The graduations on the vernier scale represent divisions of 1/50 mm. The vernier caliper in Fig. 2-109C shows that the measurement is more than 41.5 millimeters, and since the nine mark on the vernier scale is the only mark that lines up with a mark on the bar, the scale has moved 9/50 of a millimeter, or 0.18 mm, beyond the last mark. The total reading is therefore 40 mm + 1.5 mm + 0.18 mm, or a total of 41.68 mm. This is the same as 4.168 centimeters.

NOTES

NOTES

AVIATION
TECHNICIAN
INTEGRATED
TRAINING
PROGRAM

EA-ITP-G6

Aircraft Structural Materials
Corrosion and Its Control
Nondestructive Inspection

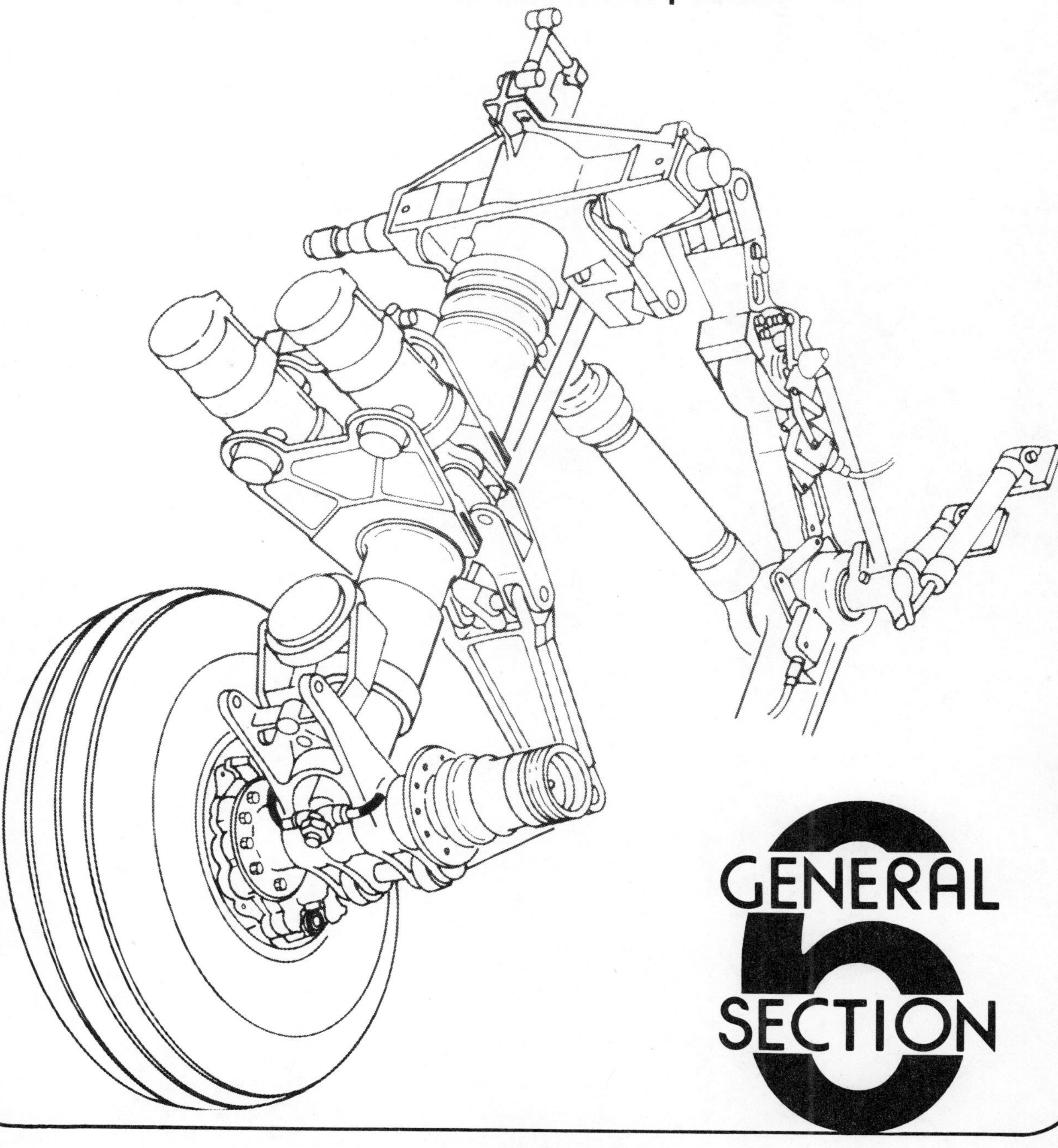

GENERAL
6
SECTION

EA-ITP-G6

TRAINING MANUAL GENERAL SECTION BOOK 6

AIRCRAFT STRUCTURAL MATERIALS CORROSION AND ITS CONTROL NONDESTRUCTIVE INSPECTION

by
Dale Crane

AVIATION MAINTENANCE PUBLISHERS, INC.

INTEGRATED TRAINING PROGRAM
FOR THE
AVIATION MAINTENANCE TECHNICIAN

International Standard Book Number 0-89100-130-1

Aviation Maintenance Publishers, Inc.
211 South Fourth Street, Basin, Wyoming 82410

Printed in the United States of America

Table of Contents

Preface

Part 65 of the Federal Aviation Regulations require that an applicant for a Mechanic Certificate with an Airframe or Powerplant rating have a general knowledge of the construction and maintenance of the aircraft covered by the rating he seeks. And each applicant must prove this knowledge by passing a written, an oral and a practical examination, each given in five sections: General, Airframe Structures, Airframe Systems and Components, Powerplant Theory and Maintenance, and Powerplant systems and Components.

Certificated Aviation Maintenance Technician Schools all operate under individual Federal Aviation Administration approvals, and all have in their approved curriculum enough information to satisfy the FAA that anyone completing their course will meet the knowledge requirements for certification as a mechanic.

In 1972, the Federal Aviation Administration published a series of three Advisory Circulars providing background information for mechanic certification applicants. These are entitled: *Airframe and Powerplant Mechanics General Handbook, AC 65-9; Airframe and Powerplant Mechanics Powerplant Handbook, AC 65-12;* and *Airframe and Powerplant Mechanics Airframe Handbook, AC 65-15.* These three handbooks were written to cover the area over which a mechanic applicant was to be tested, and they were an attempt by the FAA to coordinate the information available to the applicants with the tests that were given for certification. In 1976, the three handbooks were revised into their present A revision format: AC 65-9A, AC 65-12A, and AC 65-15A.

Individuals attempting to certificate under FAR Part 65, as well as schools operating under FAR Part 147, have found that these handbooks actually do not follow the material used in the FAA examinations, and because of this, plus their lack of study helps, they fail to effectively provide the broad base of knowledge needed for certification and for a foundation on which to build in the aviation maintenance industry.

Realizing the need for updated information, presented in a format that is in keeping with the latest developments in educational technology, the Aviation Maintenance Publishers, Inc., is offering an Integrated Training Program (ITP) that presents the basic information needed for mechanic certification, in a form that will provide a good foundation in aviation maintenance knowledge. Developed in coordination with Crane Aeronautical Activities, Inc., the ITP training manuals, workbooks, study guides, and audiovisuals are designed to follow the FAA written examinations and are written to the level required for the FAR Part 147 schools.

All of the material can be used by an individual preparing for certification on his own, or by schools preparing groups of individuals for certification under FAR Part 65, but it has been prepared primarily for the Aviation Maintenance Technician Schools operating under FAR Part 147. Not only has every attempt been made to meet the requirements of Part 147 in producing this series, but the authors have followed many of the recommendations of the Aviation Mechanic Occupation Study conducted by the University of California, Los Angeles, under a contract from the U.S. Department of Health, Education, and Welfare.

One of the most basic premises of aviation maintenance is that all operations must be carried out according to FAA-approved data, and all information of a specific nature furnished by the manufacturer of an aircraft, powerplant, or component must be followed in detail. These training manuals are, of necessity, general in nature and cover a wide scope of aircraft, powerplants, and components. Because of this, it must be understood that they should *never* be allowed to take precedence over specific information furnished by the manufacturer.

The authors and publishers of this Integrated Training Program wish to express our appreciation to the various manufacturers, Aviation Maintenance Technician Schools, and FAA personnel who have aided our efforts to provide this coordinated training material to meet the challenge of better training for aviation maintenance technicians in the eighties.

SECTION I

Aircraft Structural Materials

The evolution of the airplane has followed the development of structural materials.

The earliest flying machines were made of strips of wood or of bamboo braced with hard steel wires, with the lift-producing surfaces covered with linen or cotton fabric. As technology developed following the first world war, the fuselage as well as the tail surfaces and the landing gear structure was made of welded steel tubing. And finally the stressed-skin form of sheet metal construction gave us a structure that was both sufficiently strong and suitably streamlined.

As exotic as some of our present structural materials are, they still are not the final answer to the required strength with light weight, because the progress we make in our space age technology continues to give us better materials for more efficient flying machines.

In structural materials we have gone through the evolution from the first airplanes being made primarily of non-metal materials, through the era of the all-metal airplane, to the present generation of high-strength plastics. We will discuss each of these materials and see the way they apply to aircraft construction and maintenance.

A. Metals

1. Terminology

a. Ferrous metal

Ferrous metal is one that contains iron as its chief ingredient. It is used in aircraft structure where high strength is the basic requirement. Pure iron has very little use in aircraft structure, and it is converted into steel by the removal of most of the carbon. The characteristics of the steel are determined and controlled by the type and amount of alloying elements that are added to the iron in the process of converting it into steel.

b. Strength

The strength of a metal is one of its most important characteristics. The term "strength" sums up several of its attributes and is generally its ability to withstand stresses without failing. There are a number of classifications of strength that we must understand in our study of aircraft structure.

(1) Tensile strength

This is the ability of a material to resist the forces that tend to lengthen or stretch it. The ultimate tensile strength is the maximum stress the material will withstand. Tensile strength is measured in pounds per square inch and, for most metals, in thousands of pounds per square inch, abbreviated KPSI.

(2) Yield strength

The amount of stress needed to cause a metal to deform is called yield strength. When a tensile load is applied to a material, it will resist any deformation until its yield point is reached. At this time it will actually stretch without any increase in the stress that is applied. When it stretches, its molecular structure changes enough to increase its strength, and it will resist any more deformation until the ultimate load is reached, at which time the material will break.

(3) Shear strength

This is the strength of a material that resists the forces that tend to slide one portion of the body past the other.

(4) Bearing strength

Bearing strength is the strength of a material that resists the forces that tend to tear it at the point of the applied load. When selecting the proper size rivet for a structural application, we must try to match the shear strength of the rivet with the bearing strength of the sheet metal. If the shear strength of the rivet is too high for the sheet, the sheet will fail, or will tear, before the rivet shears off. This will result in a much more extensive repair than would be required if the rivets were to shear without the sheet tearing.

c. Hardness

The hardness of a metal refers to its ability to resist cutting, penetration, or abrasion. The tensile strength of steel relates directly to its hardness, but for most of the non-ferrous metals, this relationship is not absolute. A metal may be hardened by heat-treating or by work-hardening, and it may be softened by a process called annealing.

d. Malleability

The ability of a material to be bent, formed, or shaped without cracking or breaking is called malleability. Generally speaking, the hardness of a metal and its malleability are opposite characteristics. When a part must be formed into a complex shape and at the same time must have high strength, it is annealed, or softened, and then formed. After forming is complete, it is heat-treated to increase its strength. The metal may be fully annealed when the forming is started, but the hammering and shaping may harden it to such an extent that it must be re-annealed before its forming can be completed.

e. Ductility

This characteristic is similar to malleability, except that it refers primarily to the ability of metal to be shaped by drawing or pulling on it. The characteristic of ductility is especially important in the manufacture of wire.

f. Brittleness

Brittleness is the opposite of ductility and malleability and is the characteristic which causes a material to shatter when it is bent or otherwise deformed. Cast iron and fully hard steel, such as we use in the manufacture of files, are quite brittle and will break rather than bend.

g. Conductivity

(1) Thermal conductivity

This is the characteristic of a material to carry heat with a minimum loss. Most metals are good conductors of heat, while most of the non-metals we use in aircraft construction are "insulators," or poor conductors of heat.

(2) Electrical conductivity

Almost all metals are good conductors of electricity. This means that they will allow a flow of electrons with a minimum of opposition or resistance. Some metals have a relatively high resistance and become quite hot when electricity flows through them. These metals are used for electrical heater elements.

h. Thermal expansion

Almost all forms of matter change their physical size as their temperature changes, and a knowledge of these changes is important for the aircraft technician. For example, most aircraft structure is made of aluminum alloy which has a rather large change in its dimensions with a change in temperature. But steel, which is used for control cables, has a much smaller change in its dimensions, and if the control cable tension is properly set when the aircraft is hot, the cables will become dangerously loose when the airplane flies at a high altitude for a long time and contracts from the extreme cold.

i. Elasticity

A material that returns to its original shape when the deforming force is removed is said to be an elastic material. Elasticity is an extremely valuable characteristic for material used in aircraft construction, because it allows parts to be distorted without being permanently deformed.

All metals have a point beyond which they will not return to their original shape after the deforming force is removed. This is known as the elastic limit. Soft materials such as lead, copper, and aluminum have very low elastic limits, while the elastic limit of hard spring steel is very high.

j. Toughness

Toughness is the desirable characteristic of a material to resist tearing or breaking when it is bent or stretched. Hammer faces and wrenches must be tough as well as hard to make them useful as tools.

k. Fusibility

We are concerned with the fusibility of a material when we consider its weldability. Steel melts, or becomes a liquid, in the neighborhood of 2,500° F so that it can be fused or welded.

l. Density

Density is actually the mass-per-unit volume of a material, and throughout the discussion of metals we use the density of a material to compare weight. One cubic centimeter of pure water weighs one gram, and is said to have a density of one. Aluminum has a density of 2.7, so a cubic centimeter of aluminum weighs 2.7 grams. In the English system, a cubic inch of pure water weighs 0.03611 pound, and aluminum with its density of 2.7 will weigh 2.7 times this amount, or 0.0975 pound. The density of some of the more commonly used metals is shown in Fig. 1-1.

Metal	Density	Weight of One Cubic Inch (Pounds)
Aluminum	2.7	0.0975
Titanium	4.4	0.159
Steel	7.9	0.285
Copper	8.9	0.321
Lead	11.4	0.412
Gold	19.3	0.697

Fig. 1-1 Densities of typical aircraft metal materials

2. Non-ferrous metals

a. Aluminum and its alloys

If it were not for the developments that have been made in aluminum and its alloys, it is highly doubtful that we would have made as much progress as we have with our modern transportation systems. Steel has adequate strength for use as an aircraft structural material, but its weight restricts its use to the basic skeleton for truss structure and to highly stressed fittings. Pure aluminum has the necessary light weight and is highly corrosion-resistant, but it has too little strength for use as a structural material.

(1) Alloying agents

One of the first successful aluminum alloys for aircraft use was developed in World War I by the Germans and was known as Duralumin. It was pure aluminum alloyed with copper, magnesium, and manganese, and it had the strength and light weight needed for their fleet of huge lighter-than-air Zeppelins. It had the serious drawback, however, of being highly susceptible to corrosion.

Modern aluminum alloys are no less susceptible to corrosion, but we have gained enough knowledge in controlling corrosion that it has become a limitation that we can live with.

In Fig. 1-2, we have a list of the more common alloying agents used with aluminum, with the alloy number assigned to the various combinations. We see from this list that the alloys may be classified as heat-treatable or non-heat-treatable.

Non-heat-treatable alloys may be softened, or annealed, by heating them to a specified temperature and allowing them to cool slowly, but they can be hardened only by cold-working them by such processes as rolling or drawing. Heat-treatable alloys can be both softened and hardened with heat.

(2) Heat-treatable alloys

When aluminum is alloyed with copper or zinc, its characteristics change to the extent that it can be hardened as well as softened by the use of heat. These alloys can develop strength as high as 76,000 psi, but they have the disadvantage that they are susceptible to corrosion.

The most popular of the heat-treatable aluminum alloys is alloy 2024 which contains copper, manganese, and magnesium. It is used for most of the structural applications in modern aircraft.

When greater strength is needed than is possessed by 2024, alloy 7075, in which zinc is the predominant alloying element, may be used. It is stronger than 2024, but is more difficult to work.

NON-HEAT TREATABLE ALLOY	PERCENT OF ALLOYING ELEMENTS					
	COPPER	SILICON	MANGANESE	MAGNESIUM	ZINC	CHROMIUM
1100						
3003			1.2			
5052				2.5		0.25
HEAT-TREATABLE ALLOY						
2017	4.0		0.5	0.5		
2117	2.5			0.3		
2024	4.5		0.6	1.5		
6061	0.25	0.6		1.0		0.25
7075	1.6			2.5	5.6	0.3

Fig. 1-2 Alloying elements used with aircraft structural aluminum alloys

(3) Non-heat-treatable alloys

Commercially pure aluminum is identified as alloy 1100, and an almost identical metal, except for the addition of a small amount of manganese, is known as 3003. These metals are lightweight and are not too susceptible to corrosion, but they do not have a great amount of strength and are not heat-treatable. Thus, their use in aircraft is limited to non-structural components such as fairings and streamlined enclosures that carry little or no loads.

Alloy 5052 is perhaps the most important of the non-heat-treatable aluminum alloys. It contains about 2.5% magnesium and a small amount of chromium, and it is used for welded parts such as gasoline or oil tanks and for rigid fluid lines. Its strength may be increased to as high as 41,000 psi by cold working.

(4) Thermal treatment

(a) Annealing

Both heat-treatable and non-heat-treatable aluminum alloys may be softened by this procedure: The temperature of the alloy is raised to the annealing temperature shown in Fig. 1-3 and held there until the alloy has reached this temperature throughout. The metal is then cooled to 500° F at a rate of about 50° F per hour. Below 500° F, the rate of cooling is not important.

Annealing removes the effect of previous heat treatment and of strain hardening. When aluminum alloy is being formed, it will become strain hardened, and often before the forming can be completed, the metal must be annealed, sometimes more than once.

(b) Solution heat treatment

Solution heat treatment is a hardening process for aluminum alloy in which the metal is heated to a temperature at which the alloying agents enter into a uniform solid solution with the matrix of the base metal. The material is held at this elevated temperature until the solution is uniform throughout, and then it is removed from the furnace and rapidly quenched before the alloying elements have an opportunity to precipitate out.

THERMAL TREATMENT				
ALLOY	ANNEALING	SOLUTION HEAT TREAT.	PRECIPITATION HEAT TREAT.	
	TEMP ~ °F	TEMP ~ °F	TEMP ~ °F	TIME ~ HOURS
1100	650°			
3003	775°			
5052	650°			
2017	775°	940°		
2117	775°	940°		
2024	775°	920°	375°	7 - 9
6061	775°	970°	350°	6 - 10
7075	775°	870°	250°	24 - 28

Fig. 1-3 Thermal treatment of aluminum alloys.

Fig. 1-3 includes the temperatures to which the various aluminum alloys must be heated for the alloying agents to enter into the required solid solution. The metal must be held in the furnace until this temperature is reached throughout, and then it is removed and immediately quenched.

It is extremely important that no more than about ten seconds elapse between the time the metal is removed from the furnace and the quench. Any delay will allow the metal to cool enough that some of the alloying agents will precipitate out. This precipitation occurs along the grain boundaries and drastically lowers the resistance of the material to intergranular corrosion.

The metal may be heated in either a vat of molten sodium or potassium nitrate or, more commonly, in a hot-air furnace. In whatever method of heating used, the temperature must be controlled to within about plus or minus ten degrees Fahrenheit of the specified temperature. When sheet metal is removed from the furnace, it may be quenched by either immersing it in cold water or by spraying it with cold water. Large forgings may be quenched in hot water to minimize the thermal shock that could cause cracking. It is normal for thin sheet metal to warp and distort when it is quenched, and it must be straightened immediately after it is removed from the quench.

Heat-treatable aluminum alloys are comparatively soft when they are first removed from the quench, but with time some of the alloying agents will precipitate out of the solution and the metal will become hard and will gain strength. This aging, as it is called, occurs at room temperature in alloys containing copper, and these alloys will gain about 90% of their strength in the first half-hour after removal from the quench. In about four or five days they will become fully hard. This aging process can be slowed by storing the metal at a sub-freezing temperature immediately after it is removed from the quench.

(c) Precipitation heat treatment

Most aluminum alloys will age naturally after they are removed from the quench. This means that they will gain strength as a result of a certain portion of the alloying agents precipitating out of the solution with time at room temperature. But the strength of the alloy can be further increased if it is reheated to a temperature considerably lower than that required for solution heat treatment and held at this temperature for a given period of time. The time and temperature are shown in Fig. 1-3. This process is called artificial aging, or precipitation heat treatment.

With this treatment, the grain structure of the metal is changed as the alloying agents precipitate from their solid solution, and the metal gains in both strength and hardness. When the metal has been held at this elevated temperature for the proper time, it should be allowed to cool in still air to room temperature.

(d) Strain hardening

Both heat-treatable and non-heat-treatable aluminum alloys may have their strength and hardness increased by cold working, or strain hardening. Non-heat-treatable alloys are hardened in the manufacturing process when they are rolled to the desired dimensions, and if they are hardened too much, they can be partially annealed to the degree of hardness that is required.

Heat-treatable alloys may have their strength increased by rolling after they have been solution-heat-treated.

(5) Temper designations

The letter -0 indicates that the metal is in its annealed condition. This applies to both heat-treatable and non-heat-treatable alloys.

The designation -F indicates that the metal is in its as-fabricated condition and there has been no control over its temper. In cast alloy, -F indicates that it is in the as-cast condition.

Heat-treatable alloys have their condition of hardness designated by the letter -T followed by one or more numbers:

- -T3 Solution-heat-treated, followed by strain hardening. Different amounts of strain hardening of the heat-treated alloy are indicated by a second digit. -T36 has been solution-heat-treated and has had its thickness reduced by 6% by cold rolling.
- -T4 Solution-heat-treated, followed by natural aging at room temperature to a stable condition.
- -T5 Artificially aged after an elevated-temperature, rapid-cool-fabrication process such as extrusion or casting.
- -T6 Solution-heat-treated, followed by artificial aging (precipitation-heat-treated).
- -T7 Solution-heat-treated, followed by stabilizing to control its growth and distortion.
- -T8 Solution-heat-treated, strain-hardened, and then artificially aged.
- -T9 Solution-heat-treated, artificially aged, and then strain-hardened.

The temper designation of non-heat-treatable alloys is indicated by the letter -H followed by one or more numbers:

- -H1 Strain-hardened temper produced by cold working the metal to the desired dimensions. The second digit indicates the degree of hardness:
  - -H12 Quarter-hard
  - -H14 Half-hard
  - -H16 Three-quarter hard
  - -H18 Full-hard
  - -H19 Extra-hard
- -H2 Strain-hardened and partially annealed, to remove some of the hardness. The second digit indicates the degree of hardness and is the same as used in the -H1 classifications.
- -H3 Strain-hardened and stabilized

b. Magnesium and its alloys

Magnesium is one of the lightest metals to have sufficient strength and suitable working characteristics for use in aircraft structure. It has a density of 1.74, compared with 2.69 for aluminum. This means that it weighs only about 2/3 as much as aluminum.

Magnesium is obtained primarily by electrolysis of sea water or of brine from deep wells, and does not have sufficient strength in its pure state for use as a structural metal. But when it is alloyed with zinc, aluminum, thorium, zirconium, or manganese, it develops strength characteristics that make it quite useful.

The American Society for Testing Materials (ASTM) has developed a classification system for magnesium alloys that consists of a series of letters and numbers to indicate the alloying agents and the temper condition. In Fig. 1-4 we have the ASTM identification of AZ31A-T4 magnesium alloy with the meaning of each of the characters.

Magnesium is used as an aircraft structural metal primarily because of its light weight. It has two rather serious drawbacks that have had to be overcome to a great extent before it was able to be used successfully. Magnesium is highly susceptible to corrosion, and it has quite a tendency to crack, which contributes to its difficulty in forming and limits its use for thin sheet metal parts.

Magnesium is used for castings and is available in its wrought form as sheet, bar, tubing, and extrusions. It has excellent machining characteristics, and its difficulty in forming is overcome to a great extent by forming the parts while the metal is hot.

Magnesium will burn, and caution must be taken when grinding it and when taking very fine cuts while machining. If a fire should occur, it should be extinguished by smothering it with dry sand or some other dry material that will exclude air from the metal and will cool the surface. No form of water should be used, as it will only intensify the fire.

ALLOYING ELEMENTS	TEMPER CONDITION	EXAMPLE AZ31A-T4	
A — ALUMINUM	F — AS FABRICATED	A	CHIEF ALLOYING AGENT–ALUMINUM
E — RARE EARTH	O — ANNEALED	Z	SECONDARY ALLOYING AGENT–ZINC
H — THORIUM	H24 — STRAIN HARDENED AND PARTIALLY ANNEALED	3	3% ALUMINUM
K — ZIRCONIUM	T4 — SOLUTION-HEAT-TREATED	1	1% ZINC
M — MANGANESE	T5 — ARTIFICIALLY AGED ONLY	A	THIS IS THE FIRST ALLOY STANDARDIZED WITH THESE PARTICULAR PERCENTAGES
Z — ZINC	T6 — SOLUTION HEAT-TREATED & ARTIFICIALLY AGED	T4	THIS ALLOY HAS BEEN SOLUTION-HEAT-TREATED

Fig. 1-4 Magnesium alloy designations

The corrosion problem is minimized by treating the surface of the metal with chemicals to form an oxide film that will prevent oxygen reaching the metal. When oxygen is excluded from the surface, no salts of corrosion can form.

c. Titanium and its alloys

Titanium is one of the metals that have been developed in our search for a material that will retain its strength and corrosion resistance at elevated temperatures.

Titanium ores are found in abundance in the earth's crust, and while the element was discovered in 1791, it was not until after World War II that a practical process was developed in which titanium metal could be refined from the ores.

Titanium has a density of 4.5, which is between that of aluminum (2.69) and stainless steel (7.9). It may be alloyed with manganese, molybdenum, aluminum, or chromium, and when heat-treated, these alloys can develop tensile strength comparable to high-strength alloy steel.

Titanium is non-magnetic and is highly corrosion-resistant and maintains its strength to temperatures as high as 800° F. Above about 1,000° F, however, titanium will combine with oxygen, so for welding or for other thermal treatment where these temperatures are encountered, it must be blanketed with an inert gas such as helium or argon.

The ability of titanium to maintain its strength and corrosion resistance at elevated temperatures has made it useful as a skin for aircraft in locations near the engine where the high temperatures would damage aluminum alloys. It also finds use as the skin for supersonic aircraft which becomes quite hot from the friction caused by the supersonic airflow.

Turbine engine compressor blades and vanes are often made of titanium alloy because of its good resistance to creep, at the temperatures and stresses encountered by these components.

Machining and fabrication processes used with titanium are different from those used with either stainless steel or aluminum alloy, but procedures have been established that have overcome their fabrication problems, and we are now able to exploit the tremendous advantages of this metal.

d. Nickel and its alloys

(1) Monel

Monel is a high-nickel-content alloy that contains about 68% nickel and 29% copper, along with small amounts of iron and manganese. It can be welded and it has very good machining characteristics. Certain types of monel, especially those which contain small percentages of aluminum, can be heat-treated to a tensile strength equivalent to that of steel.

Monel is effectively used for gears and for parts that require high strength and toughness, as well as for parts in exhaust systems that require high strength and corrosion resistance at elevated temperatures.

(2) Inconel

The International Nickel Company, Inc., has a series of high-strength, high-temperature alloys that contain approximately 80% nickel and 14% chromium, along with small amounts of iron and other elements. Inconel is similar in appearance to stainless steel and is used for hot-section parts of turbine engines because of its ability to maintain its strength and corrosion resistance under the extremely high-temperature conditions found there.

e. Copper and its alloys

Neither copper nor its alloys find much use as structural metal in aircraft construction because of weight, but copper is the primary metal used for electrical wiring because of its excellent electrical and thermal conductivity.

When copper is alloyed with zinc, aluminum, and tin, it develops a high strength and is often used for bushings and for bearing surfaces. An alloy of copper and zinc is normally called brass, while an alloy of copper and tin is called bronze.

Copper tubing was at one time used extensively for fluid lines in aircraft, but because of its weight and the extremely unfortunate characteristic copper has of becoming brittle when subjected to vibration, it has been replaced by aluminum alloy for this purpose. Brass fittings for fluid lines have also been replaced with aluminum alloy for most applications and with steel where high strength is needed.

3. Ferrous metals

Any metal which contains iron as its chief constituent is called a ferrous metal. In aircraft structure we are chiefly concerned with steel, which is iron into which carbon of up to about 1.7% has been mixed.

a. Iron

Iron is a chemical element which, in its pure form, is a fairly soft malleable and ductile metal that is easy to form or shape. It is silvery white in color and is quite heavy, having a density of 7.9. Iron readily combines with oxygen to form iron oxide, which is more commonly known as rust.

In practical use we seldom see pure iron, but rather see it mixed with various forms of carbon and other alloying agents or impurities. When iron is poured from the furnace into molds, it is known as cast iron, and it normally contains more than two percent carbon and usually has silicon mixed in it.

Cast iron is seldom used in aircraft construction because of its low strength-to-weight ratio, but there are some applications in engine manufacturing, such as valve guides, where its porosity allows it to hold a lubricant film. Iron also has good wear characteristics which are important in this application. It is also familiar to us in engine maintenance as the brittle material of which piston rings are made.

Iron is obtained from ore, a form of iron oxide, by mixing the ore with coke and limestone and passing extremely hot air through the mixture. The coke burns and forms superheated carbon monoxide which absorbs much of the oxygen from the ore, and the molten iron drops to the bottom of the furnace. The limestone reacts with the impurities in the iron and the coke to form a slag which floats on top of the molten iron and is poured off from the iron.

When the iron is removed from the furnace it is in a form known as pig iron. It may be remelted and cast into cast-iron components, or it may be taken to some of the steel-making furnaces or converters to convert the iron into steel.

b. Steel

Pig iron from the blast furnace is melted in a special furnace or a converter where oxygen is forced through the molten metal so that it can unite with the carbon and burn it out. When the carbon is burned out, a controlled amount of carbon is put back in, along with other elements that are needed to get the particular characteristics of the steel we want. The molten steel is poured from the furnace into molds where it solidifies to form ingots. These steel ingots are placed in a soaking pit where they are heated with burning gas or oil until their temperature is uniform at about 2,200° F throughout. They are then taken from the soaking pit and passed through steel rollers in the rolling mill to form plate or sheet steel.

Much of the steel we use in aircraft construction is made in electric furnaces, which allow much better control of the alloying agents than the gas-fired furnaces. These electric furnaces are charged with scrap steel, limestone, and other materials that are used as flux to mix with the impurities in the steel. Carbon electrodes are lowered into the steel, and electric arcs form between the steel and the carbon. The intense heat from the arc melts the steel. The impurities from the steel mix with the flux and are drawn off the molten metal. When the impurities have been removed, controlled quantities of whatever elements are needed to alloy with the steel are added, and the furnace is tipped to pour the liquid metal into the molds.

(1) Alloying agents in steel

(a) Carbon

Carbon is the most important element found in steel, mixing with the iron to form compounds of iron carbides called cementite. It is the carbon in the steel that allows it to be heat-treated to obtain varying degrees of hardness, strength, and toughness. The greater the carbon content, the better the heat-treatability and therefore the higher the tensile strength and the hardness. However, increasing the carbon content decreases the malleability and the weldability of the steel.

A high carbon content, up to about 0.95%, may be used where extreme hardness is needed and malleability is not of primary concern. High-carbon steel is used in the manufacture of cutting tools and springs. Low-carbon steel which contains about 0.20% carbon is used where a great deal of forming or deep drawing is needed, and where strength is of minor concern. Medium-carbon steel, one having between 0.25% and 0.50% carbon, is used for applications where both strength and ductility or formability are required.

(b) Sulfur

Sulfur is one of the more undesirable constituents in steel, and as much of it as possible must be removed in the refining process. Sulfur causes steel to be brittle when it is being rolled or forged, as it mixes with the iron to form iron sulfides which are in their liquid state at the temperatures required for this type of fabrication.

The effect of any sulfur that is not removed from the steel may be counteracted by using manganese. This converts it into manganese sulfide, which is not detrimental to the steel.

(c) Manganese

Manganese is alloyed with steel to eliminate some of the oxides and the sulfur from the steel, making it a clean, tough, and uniform metal. Manganese also improves the forging characteristics of the steel by making it less brittle at the rolling and forging temperatures.

(d) Silicon

When this non-metallic element is alloyed with steel, it acts as a hardener. When it is used in small quantities, it also improves the ductility of the steel.

(e) Phosphorous

This element raises the yield strength of steel and improves the resistance of low-carbon steel to atmosheric corrosion. No more than 0.05% phosphorous is normally used in steel, as more of it will cause the metal to become brittle when it is cold.

(f) Nickel

Nickel adds strength and hardness to steel and increases its yield strength. And nickel, used as an alloy, slows the critical rate of hardening when the steel is heat-treated. This increases the depth of hardening and produces a finer grain structure. Nickel also reduces the tendency of steel to warp and scale when it is heat-treated.

Nickel is one of the chief ingredients in corrosion-resistant, or stainless, steel.

(g) Chromium

This hard, high-melting-point element is alloyed with steel to increase its strength and wear resistance, as well as its resistance to corrosion. It also gives steel a good measure of hardness. Chromium is usually used in conjunction with other elements such as molybdenum, nickel, and vanadium to give steel the special qualities that are needed in aircraft structure.

Chrome-vanadium alloy is used when an extremely tough steel is needed for applications such as wrenches and ball bearings. Chrome-molybdenum steel has the strength, toughness, and light weight needed to make it the primary steel for aircraft structure and for high-strength engine components such as cylinders. Nickel-chrome steel is the primary alloy used in the manufacture of aircraft hardware, and because of its strength and hardness at elevated temperatures, tungsten-chrome steel is used for high-speed cutting tools.

In addition to its important function as an alloying element in the manufacture of steel, chromium may be electrolytically deposited on cylinder walls and bearing journals to provide a hard, wear-resistant surface.

(h) Molybdenum

One of the most widely used alloying elements for aircraft structural steel is molybdenum. It reduces the grain size of the steel and increases both its impact strength and its elastic limit. Molybdenum steels have some of the best wear resistance and fatigue strength of any of the alloys. This accounts for their popularity for high-strength structural members and for engine cylinder barrels.

Chrome-molybdenum steel responds very well to heat treatment, machines easily, and is readily welded.

(i) Vanadium

When extremely high strength is required of a steel along with toughness and good ductility, chrome-vanadium steel is used. Vanadium, used in amounts up to about 0.20%, improves the grain structure and increases both the ultimate tensile strength and the toughness.

Most wrenches are made of chrome-vanadium steel.

(j) Titanium

Titanium metal has found an important place in aircraft structure where high strength at elevated temperatures is needed. When a small amount of titanium is alloyed with stainless steel, it reduces the brittleness of the steel caused by high temperatures and makes a good material for exhaust stacks.

(k) Tungsten

Tungsten, which has an extremely high melting point, brings some of this characteristic into the steels with which it is alloyed. Tungsten alloys are used for breaker contacts in magnetos and for cutting tools. Tungsten steels retain their hardness even when they operate red hot.

(2) SAE classification of steels

The Society of Automotive Engineers (SAE) has classified steel alloys with a four-digit numerical index system. The first digit identifies the principal alloying element in the steel, the second digit, the percent of this alloying element, and the last two digits, the percentage in hundredths of a percent, of the carbon in the steel.

TYPE OF STEELS	NUMERALS AND DIGITS
Carbon Steels	1xxx
Plain Carbon Steels	10xx
Free Cutting Steels	11xx
Manganese Steels	
[Manganese 1.60 to 1.90%]	13xx
Nickel Steels	2xxx
3.50% nickel	23xx
5.00% nickel	25xx
Nickel Chromium Steels	3xxx
9.70% nickel, 0.07% chromium	30xx
1.25% nickel, 0.60% chromium	31xx
1.75% nickel, 1.00% chromium	32xx
3.50% nickel, 1.50% chromium	33xx
Corrosion and heat resisting	30xxx
Molybdenum Steels	40xx
Chromium Molybdenum Steels	41xx
Nickel Chromium Molybdenum Steels	43xx
Nickel Molybdenum Steels	
1.75% nickel, 0.25% molybdenum	46xx
3.50% nickel, 1.50% chromium	48xx
Chromium Steels	5xxx
Low chromium	51xx
Medium chromium	52xxx
Corrosion and heat resisting	51xxx
Chromium Vanadium Steels	6xxx
1.00% chromium	61xxx
National Emergency Steels	8xxx
Silicon Manganese Steels	9xxx
2.00% silicon	92xx

Fig. 1-5 SAE numerical identification of steel

(3) Carbon steel

Pig iron contains more than two percent carbon, and when it is refined into steel, the carbon is burned out and then a controlled amount of carbon is put back into the molten iron. Low-carbon steels contain between 0.10 and 0.30 percent carbon and are classified as SAE 1010 to SAE 1030 steel. This low-carbon steel does not have sufficient strength for aircraft structural applications, and its use is limited to secondary structure where the loads are not high. Low-carbon steel is easily welded and machines readily, but it does not accept heat treatment satisfactorily.

Medium-carbon steels contain between 0.30 and 0.50 percent carbon. The increased carbon makes them accept heat treatment.

High-carbon steels which are used for springs, files, and some cutting tools contain between 0.50 and 1.05% carbon. These steels can be heat-treated until they are extremely hard.

(4) Alloy steels

(a) Chrome-molybdenum steel

The most familiar steel to the aviation maintenance technician is SAE 4130 steel, commonly known as chrome-moly steel. It has, as we can see from the SAE classification, approximately one percent molybdenum and 0.30% carbon. It machines readily, is easily welded by either gas or electric arc, and responds to heat treatment. Heat-treated SAE 4130 steel has an ultimate tensile strength of about four times that of SAE 1025 steel, which makes it an ideal choice for landing gear structures and engine mounts.

The toughness and wear resistance of chrome-molybdenum steel makes it a good material for engine cylinders and other highly stressed engine parts.

(b) Nickel steel

SAE 2330 steel which contains three percent nickel and 0.30% carbon is used for aircraft hardware such as bolts, nuts, rod ends, and pins.

(c) Nickel-chromium steel

Nickel toughens steel, and chromium hardens it. Both elements give steel desirable characteristics for use in high-strength structural applications. Nickel-chrome steels such as SAE 4130, 3250, and 3435 are used for forged and machine parts where high strength, ductility, shock resistance, and toughness are required.

(d) Stainless steel

This is a classification of corrosion-resistant steels that have large amounts of chromium and nickel alloyed with the steel. These steels do not normally use the SAE classification but are identified by a three-digit system.

The 200 and 300 series of stainless steels are known as austenitic steels and contain a large percentage of chromium and nickel and, in the case of the 200 series, some manganese. These steels are heated to a temperature above their critical range and held until a structure known as austenite forms. Austenite is a solid solution of peralite, an alloy of iron and carbon, and gamma iron, which is a non-magnetic form of iron. Neither the 200 nor the 300 series of stainless steel can be hardened by heat treatment and both are non-magnetic.

The 400 series of stainless steel is a martensitic steel and is a straight chromium alloy. It is magnetic and it may be hardened by rapidly quenching it from an elevated temperature.

(5) Heat treatment of steel

Iron can exist in three solid structural forms, depending upon its temperature. At temperatures up to 1,625° F (911° C), iron is in its *alpha* form which has a certain arrangement of its molecules. When the iron is heated to between 1,625° F and 2,538° F (911° C and 1,392° C), the molecular arrangement changes and the iron is said to be in its *gamma* form. The molecular arrangement again changes at temperatures from 2,538° F to 2,800° F (1,392° C to 1,538° C), and it becomes *delta* iron. Pure iron changes to its liquid form, or melts, at 2,800° F (1,538° C).

Steel exists in two basic forms that are of interest in our consideration of heat treatment. *Ferrite* steel is *alpha* iron into which some carbon has dissolved, and it exists at temperatures below its critical temperature. At its critical temperature, which varies with the alloy, the steel turns into *austenite* steel which consists of *gamma* iron into which carbon can dissolve. Below its critical temperature, the carbon which exists in the steel in the form of iron carbides is scattered throughout the iron matrix, or its molecular structure, as a physical mixture. When the steel is heated to its critical temperature, this carbon dissolves into the matrix and becomes a solid solution rather than a physical mixture. The steel has now become austenite, and the iron is in its gamma form.

When the temperature of the steel drops below its critical value, the carbide particles precipitate out of the solution. If the steel is cooled slowly, the particles will be quite large and the steel will be soft, but if it is cooled very rapidly by quenching, the carbon particles will be extremely fine and will effectively bind the molecular structure of the steel together, making it hard and strong.

The critical temperature and the ultimate strength the steel will develop vary with the alloying agents, but the most important factor is the amount of carbon. Low-carbon steel will not heat-treat satisfactorily because of the small amount of carbon, but as the amount of carbon is increased, the steel gains the ability to be hardened and strengthened by heat treatment. This occurs up to about 0.80% carbon. Beyond this, the hardness does not increase, but its wear resistance improves with an increase in the amount of carbon.

(a) Annealing

If steel is heated above its critical temperature and is held there until the carbon goes into solid solution with the matrix, and is then allowed to cool very slowly, the carbon that precipitates out will be in large particles that do not bind the matrix tightly, and the steel will be soft and ductile and may be easily formed, or worked. The temperatures and cooling rates vary with the alloy, but for annealing they are normally above 1,600° F. The steel may be cooled by leaving it in the furnace but shutting off the heat so it will cool slowly, or by packing the steel in hot sand or ashes to insulate it so the heat will be conducted away slowly.

(b) Normalizing

Steel that has been forged, welded, or machined usually has stresses left within the structure that could cause failure, so these stresses must be relieved by a process known as normalizing. The steel is heated to a temperature above its critical temperature and held until this temperature is uniform throughout. Then it is removed from the furnace and allowed to cool in still air. The particles of carbon that precipitate out are not as large as those formed when the steel is annealed, but the molecular stresses are relieved.

(c) Hardening

If the steel is heated above its critical temperature so the carbon can enter into a uniform solution with the iron matrix, and is then cooled rapidly by quenching it in water, oil, or brine, the carbon will precipitate out in extremely small particles. These tiny particles will effectively bind the matrix tightly together and increase both the strength and hardness of the steel.

The speed of the quench is determined by the quenching medium. Oil provides the slowest quench, and brine, the most rapid.

(d) Tempering

When steel is hardened by quenching, it is usually too hard and brittle for the use for which it is intended, and the rapid cooling usually causes stresses within the steel that must be relieved. The hardened steel may be tempered, or have some of the hardness drawn out, by heating it to a temperature considerably below its critical temperature and holding it there until it has reached this temperature throughout. It is then allowed to cool to room temperature in still air.

/1/ Determining the temperature of steel

If we need to heat-treat steel and do not have a pyrometer to measure its temperature, we can make a reasonable estimate of its temperature by observing its color. In Fig. 1-6, we see the rela-

Color of Steel	Temperature of steel	
	[°F]	[°C]
Faint red	900	482
Blood red	1050	566
Dark cherry	1075	579
Medium cherry	1250	677
Cherry or full red	1375	746
Bright red	1550	843
Salmon	1650	899
Orange	1725	941
Lemon	1825	996
Light yellow	1975	1079
White	2200	1204
Dazzling white	2350	1288

Fig. 1-6 Color chart of steel at various temperatures.

tionship between the color of steel and its temperature. For example, if we want to anneal a piece of chrome-molybdenum steel (SAE 4130), we look up its annealing temperatures in the chart of Fig. 1-8 and see that it is between 1,525° and 1,575° F. In Fig. 1-6 we find that steel has a temperature of 1,550° F when it is heated to a bright red. To harden this same steel, we would heat it until it is not quite so bright as it was for annealing and then quench it in oil.

Color of Oxide	Carbon steel temp.	
	[°F]	[°C]
Pale yellow	428	220
Straw	446	230
Golden yellow	469	243
Brown	491	255
Brown dappled w/purple	509	265
Purple	531	277
Dark blue	550	288
Bright blue	567	297
Pale blue	610	321

Fig. 1-7 Color chart for oxides on carbon steel at various temperatures.

When steel is hardened it becomes extremely brittle, and we can temper, or draw, it to relieve some of this hardness and brittleness. We first polish a part of the steel so we can see its bright, shiny metal surface. We then heat the part and watch this area to see the way the oxides that form change color. The first oxide to form is a pale yellow, then the oxides progress through brown into the blues. When the oxide on SAE 4130 steel is between a bright and a pale blue, the heat should be removed and the steel allowed to cool in still air. This will temper it to a tensile strength of around 200,000 psi.

/2/ Tempering small tools

We find a practical example of tempering steel when we make screwdriver blades, chisels, or punches from tool-steel rod. This steel has a high carbon content and is usually obtainable in its normalized condition. After we have shaped the tool, we heat it to a cherry red color and quench it in oil. Then we polish a portion and re-heat it until the proper oxide color appears. When we quench the steel again, it should have the proper hardness.

TOOL	OXIDE COLOR
Scribe or hammer face	Pale yellow
Center punch	Golden yellow
Cold chisel or drift	Brown
Screwdriver	Purple

(e) Case hardening

Wearing surfaces in aircraft engines and landing gear systems are often hardened, while the core of the material is allowed to remain soft and tough. There are two ways this is done. The first is a process known as carburizing, in which controlled amounts of carbon are added to the surface of low-carbon steel. Nitriding is the other means of case hardening, and the hard surface is produced in this process by converting the surface of the metal into an extremely hard aluminum nitride.

/1/ Carburizing

There are three methods that can be used to increase the carbon content of the surface of a low-carbon steel. Pack carburizing is done by enclosing the part in a fire-clay container and packing it with a carbon-rich material such as charcoal. The container is sealed and placed in a furnace where it is heated to 1,700° F (927° C). Carbon monoxide gas forms from the heated charcoal, and since it cannot escape, its carbon combines with the gamma iron in the surface of the steel. The depth to which this carbon penetrates depends upon the soaking time.

Gas carburizing is similar to pack carburizing, except that the carbon monoxide is produced by a gas rather than a solid material. In this process, as with pack carburizing, the carbon from carbon monoxide combines with the gamma iron and forms a high-carbon surface.

Liquid carburizing produces the high-carbon surface when the part is heated in a molten bath of sodium cyanide or barium cyanide which supplies the needed carbon.

Any portion of the part that should not be case-hardened may be copper-plated. The carbon will not infuse into the copper-plated surface.

Steel No.	Temperatures						Quenching Medium [N]	Tempering [Drawing] Temperatures For Tensile Strength									
	Normalizing		Annealing		Hardening			100,000		125,000		150,000		180,000		2000,000	
	°F	°C	°F	°C	°F	°C		°F	°C	°F	°C	°F	°C	°F	°C	°F	°C
1020	1650-1750	899-954	1600-1700	871-927	1575-1675	856-912	Water	....	....	....	....	....	....	....	....	....	....
1022 [X1020]	1650-1750	899-954	1600-1700	871-927	1575-1675	856-912	Water	....	....	....	....	....	....	....	....	....	....
1025	1600-1700	871-927	1575-1650	856-899	1575-1675	856-912	Water	[A]	[A]	....	....	....	....	....	....	....	....
1035	1575-1650	856-899	1575-1625	856-885	1525-1600	829-871	Water	875	468	....	....	....	....	....	....	....	....
1045	1550-1600	843-871	1500-1600	843-871	1475-1550	800-843	Oil or water	1150	621	....	....	....	....	[N]	[N]	....	....
1095	1475-1550	800-843	1450-1500	788-816	1425-1500	773-816	Oil	[B]	[B]	....	....	1100	593	850	454	750	399
2330	1475-1525	800-829	1425-1475	773-800	1450-1500	788-816	Oil or water	1100	593	950	510	800	427	....	....	....	....
3135	1600-1650	871-899	1500-1550	816-843	1475-1525	800-829	Oil	1250	677	1050	566	900	482	750	399	650	343
3140	1600-1650	871-899	1500-1550	816-843	1475-1525	800-829	Oil	1325	718	1075	579	925	496	775	413	700	371
4037	1600	871	1525-1575	829-856	1525-1575	829-856	Oil or water	1225	662	1100	593	975	524	....	....	....	....
4130 [X4130]	1600-1700	871-927	1525-1575	829-856	1575-1625	856-885	Oil [C]	[D]	[D]	1050	566	900	482	700	371	575	301
4140	1600-1650	871-899	1525-1575	829-856	1525-1575	829-856	Oil	1350	732	1100	593	1025	551	825	440	675	357
4150	1550-1600	843-871	1475-1525	800-829	1500-1550	816-843	Oil	....	....	1275	690	1175	635	1050	566	950	510
4340 [X4340]	1550-1625	843-885	1525-1575	829-856	1475-1550	800-843	Oil	....	....	1200	649	1050	566	950	510	850	454
4640	1675-1700	912-927	1525-1575	829-856	1500-1550	816-843	Oil	....	....	1200	649	1050	566	750	399	625	329
6135	1600-1700	871-927	1550-1600	843-871	1575-1625	856-885	Oil	1300	704	1075	579	950	510	800	427	750	399
6150	1600-1650	871-899	1525-1575	829-856	1550-1625	843-885	Oil	[D][E]	[D][E]	1200	649	1000	538	900	482	800	427
6195	1600-1650	871-899	1525-1575	829-856	1500-1550	816-843	Oil	[F]	[F]	....	....	....	....	....	....	....	....
NE8620					1525-1575	829-856	Oil	....	....	1000	538	....	....	....	....	....	357
NE8630	1650	899	1525-1575	829-856	1525-1575	829-856	Oil	....	....	1125	607	975	524	775	413	675	413
NE8735	1650	899	1525-1575	829-856	1525-1575	829-856	Oil	....	....	1175	635	1025	551	875	468	775	454
NE8740	1625	885	1500-1550	816-843	1500-1550	816-843	Oil	....	....	1200	649	1075	579	925	496	850	....
30905	....	....	[G][H]	[G][H]	[I]	[I]	....	....	....	....	....	....	....	....	....	....	....
51210	1525-1575	829-856	1525-1575	829-856	1775-1825 [J]	967-995	Oil	1200	649	1100	593	[K]	[K]	750	399	....	....
51335	....	....	1525-1575	829-856	1775-1850	967-1010	Oil	...	...	....	....	....	....	....	....	....	....
52100	1625-1700	885-927	1400-1450	760-788	1525-1550	829-843	Oil	[F]	[F]	....	....	....	....	....	....	....	....
Corrosion resisting										....	....	....	....	....	....	....	....
[16-2][1]	....	....	....	....	....	....	....	[M]	[M]	....	....	....	....	....	....	....	....
Silicon chromium [for springs]	....	....	....	....	1700-1725	927-941	Oil										

NOTES:

[A] Draw at 1150° [621°C] for tensile strength of 70,000 psi.

[B] For spring temper draw at 800° to 900°F [427° to 484°C] Rockwell hardness C-40-45.

[C] Bars or forgings may be quenched in water from 1500° to 1600°F [816° to 871°C].

[D] Air-cooling from the normalizing temperature will produce a tensile strength of approximately 90,000 psi.

[E] For spring temper draw at 850° to 950°F [454° to 510°C] Rockwell hardness C-40-45.

[F] Draw at 350° to 450°F [177° to 232°C] to remove quenching stains. Rockwell hardness C-60-65.

[G] Anneal at 1600° to 1700°F [871° to 927°C] to remove residual stresses due to welding or cold work. May be applied only to steel containing titanium or columbium.

[H] Anneal at 1900° to 2100°F [1038° to 1149°C] to produce maximum softness and corrosion resistance. Cool in air or quench in water.

[I] Hardened by cold work only.

[J] Lower side of range for sheet 0.06 inch and under. Middle of range for sheet and wire 0.125 inch. Upper side of range for forgings.

[K] Not recommended for intermediate tensile strengths because of low impact.

[L] For steel, military specification MIL-S18732. It is recommended that, prior to tempering, corrosion-resisting [16 Cr - 2 Ni] steel be quenched in oil from a temperature of 1875° to 1900°F [1024° to 1038°C], after a soaking period of 1/2-hour at this temperature. To obtain a tensile strength at 115,000 psi, the tempering temperature should be approximately 525°F [274°C]. A holding time that these temperatures of about 2 hours is recommended. Tempering temperatures between 700° and 1100°F [371° to 593°C] will not be approved.

[M] Draw at approximately 800°F [427°C] and cool in air for Rockwell hardness of C-50.

[N] Water used for quenching shall not exceed 65°F [18°C]. Oil used for quenching shall be within the temperature range of 80° to 150°F [27° to 66°C].

Fig. 1-8 Heat treatment procedures for steel

/2/ Nitriding

Nitriding differs from carburizing in that the part to be hardened is first heat-treated to give it the characteristics we want — that is, it is hardened, tempered, and then ground to the proper dimensions. It is then placed in a special retort and heated to a temperature of approximately 1,000° F (538° C), and is surrounded by ammonia gas (NH_3). The high temperature breaks the ammonia down into nitrogen and hydrogen. The nitrogen reacts with the iron and with any of the nitride formers in the steel such as aluminum, chromium, molybdenum, or vanadium to form the appropriate nitrides. The nitrided surface is extremely thin, with its thickness depending on the length of time it is exposed to the ammonia gas. The hardness of the surface gradually decreases with depth until it reaches that of the core.

Aircraft engine crankshafts and cylinder walls are commonly nitrided for increased wear resistance. But these surfaces have one serious problem: they are highly susceptible to pitting corrosion and must be protected from the air with a coating of oil.

(6) Relationship between hardness and strength

As we have seen in the section on the heat treatment of steel, the smaller the carbon particles that are in solution with the matrix of the steel, the harder and stronger the steel will be. It is because of this fact that there is a definite relationship between strength and hardness of steel. This relationship does not necessarily exist, however, with metals other than steel.

There are two systems of measuring the hardness of steel that we find commonly used in shops where steel is heat-treated, the Brinell and the Rockwell systems.

(a) The Brinell hardness system

The Brinell hardness tester uses a hydraulic force to impress a spherical penetrator into the surface of the metal being tested. An extremely hard 10-mm-diameter ball is forced down onto the surface of the metal by a 3,000-kilogram force for steel or a 500-kilogram force for non-ferrous metals. This force is hydraulically applied by a hand pump on the tester, and the amount of the force is read on the pressure gage.

Rockwell C-scale hardness number	Brinell hardness number 10-mm ball, 3000-kg load	Tensile Strength (approximate) in 1000 psi
52	500	262
51	487	253
50	475	245
49	464	239
48	451	232
47	442	225
46	432	219
45	421	212
44	409	206
43	400	201
42	390	196
41	381	191
40	371	186
39	362	181
38	353	176
37	344	172
36	336	168
35	327	163
34	319	159
33	311	154
32	301	150
31	294	146
30	286	142
29	279	138
28	271	134
27	264	131
26	258	127
25	253	124
24	247	121
23	243	118
22	237	115
21	231	113
20	226	110
[18]	219	106
[16]	212	102
[14]	203	98
[12]	194	94
[10]	187	90
[8]	179	87
[6]	171	84
[4]	165	80
[2]	158	77
[0]	152	75

Values in [ ] are beyond normal range.
It is possible that steels of various compositions and processing histories will deviate in hardness-tensile strength relationship from data presented in this table. Above the level of Rockwell C43, deviation increases with increasing hardness and the table shall not be used above RC48 except in the absence of other data specifically approved by the procuring agency.

Fig. 1-9 Hardness and strength relationship for steel.

When the metal is removed from the tester, the diameter of the impression is measured with a special calibrated microscope. This diameter is converted into a Brinell number by using the chart furnished with the tester.

In Fig. 1-9, we have a chart showing the relationship between Brinell numbers and the tensile strength of steel. If the steel has a Brinell number of 400, for example, it has a tensile strength of 201,000 psi.

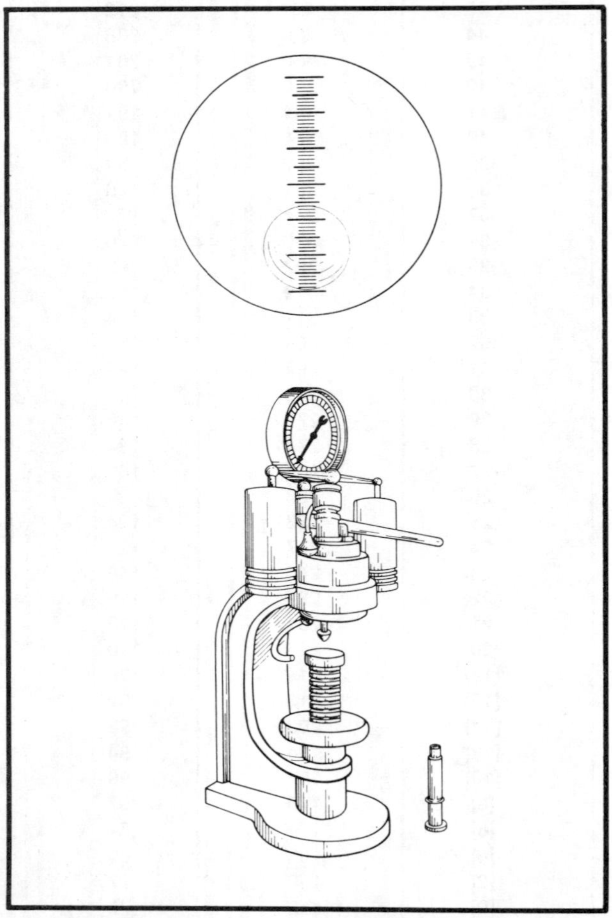

Fig. 1-10 Brinnel hardness tester

(b) The Rockwell hardness system

The Rockwell hardness tester gives us the same information the Brinell tester gives, except that it measures the *depth* to which the penetrator sinks into the material rather than the diameter of the impression.

The material to be tested is placed on the anvil of the tester and is raised up against the penetrator

Alloy Designation	Hardness Temper	Brinell Number 500 kg load 10 mm ball
1100	0	23
	H18	44
3003	0	28
	H16	47
2017	0	45
	T6	105
2024	0	47
	T4	120
5052	0	47
	H36	73
6061	0	30
	T4	65
	T6	95
7075	T6	135

Fig. 1-11 Hardness values for aluminum alloys

by a hand-wheel-operated screw. A 10-kilogram load, called the minor load, is applied and the machine is zeroed. A major load is then applied and the dial on the tester indicates the depth the penetrator sinks into the metal. Instead of indicating the depth of penetration in thousandths of an inch, it indicates in Rockwell numbers on either a red or a black scale. There are three types of penetrators: a conical diamond, a 1/16″ ball, and a 1/8″ ball.

There are also three major loads: 60 kilograms, 100 kilograms, and 150 kilograms. The two most

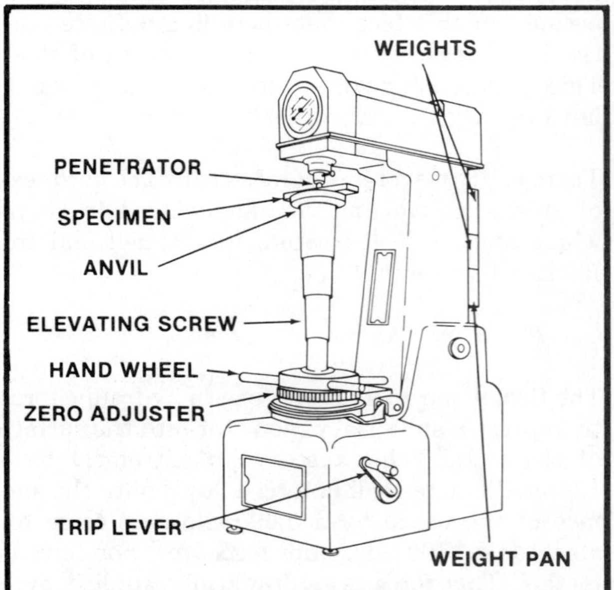

Fig. 1-12 Rockwell hardness tester

commonly used Rockwell scales are the B-scale for soft metals, which uses a 1/16" ball penetrator and a 100-kg major load, and the C-scale for hard metals, which uses the conical diamond penetrator and a 150-kg major load.

Fig. 1-9 gives the relationship between the C-scale Rockwell hardness and the tensile strength of steel. A steel that tests Rockwell C40 has a tensile strength of 186,000 psi.

B. Nonmetallic Structural Materials

1. Wood

There is far less aircraft structure that is made of wood than there has been in the past, simply because of the greater economy of building aircraft of sheet metal. Wooden structures require quite a bit of handwork, and in this day of automation handwork is extremely expensive.

It is possible, however, that an aviation maintenance technician will be involved in the restoration of a classic aircraft or one of the wooden homebuilt aircraft which are built as labors of love, rather than as an economic venture. For these eventualities we will discuss the types of wood that are available for aircraft structural use.

a. Types of wood

There are three basic types of wood used for aircraft structure. Solid wood is used for some aircraft wing spars and, as the name implies, is made of one solid piece cut from the log.

Most of the solid wood we use is cut from the tree by a method known as quarter sawing, Fig. 1-14. When planks are cut from the log in this fashion, the annual rings cross the plank at an angle of more than 45°, and there is less chance of warpage than when they are cut straight across, or plain sawed.

Wing spars require long pieces of extremely high-quality wood that is free from imperfections.

Fig. 1-13 This aircraft uses a great amount of non-metallic material in its construction.

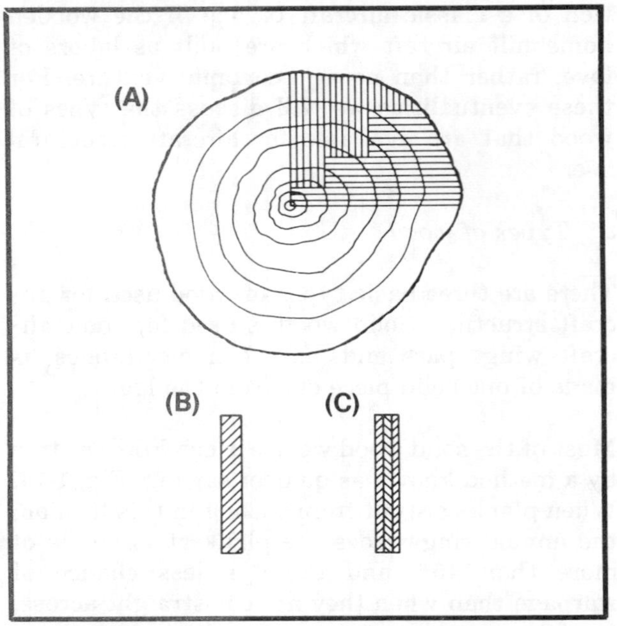

Fig. 1-14 A — Method of sawing a log to produce quarter-sawed planks B — Solid wood spar that has been produced by quarter sawing C — Laminated wood spar made by gluing together strips of quarter-sawed wood with the grain of all pieces running in the same direction

Wood of this size, length, and quality is quite expensive, so many manufacturers use laminated wood spars. These are made up of two or three pieces of thin wood glued together with the grain of all pieces running in the same direction. Each of the laminations can be spliced, so the spars can be made up of pieces of wood that are not large enough for a solid spar. The FAA allows solid and laminated spars to be used interchangeably, as long as they are both of the same high quality and neither has a grain deviation that exceeds the allowable limits.

The third type of wood used for aircraft construction is plywood. This is made up of three or more layers of thin veneer glued together so the grain of each successive layer crosses the others at an angle of 45° or 90°.

b. Species of wood

There are two basic classifications of wood species used for aircraft construction: hardwood and softwood. Hardwoods come from trees having broad leaves, while softwoods come from cone-bearing trees having needle-like or scale-like leaves. These terms are quite deceptive, as some of the so-called hardwoods are softer than some of the woods that are classified as softwoods. The most obvious example of this is balsa, which is extremely light and soft, but, because of its type of leaf, is classified as a hardwood.

(1) Spruce

Sitka spruce is the most important wood for aircraft structure, and because it is available in large sizes and free from defects and because of its high strength-to-weight ratio, it has been chosen by the FAA as the reference wood for aircraft construction.

(2) Douglas fir

The strength properties of Douglas Fir exceed those of spruce, but it is much heavier. It is more difficult to work than spruce, and it has a tendency to split.

(3) Noble fir

This type of fir is slightly lighter than spruce, and it is equal to or superior to spruce in all properties except hardness and shock resistance. It is especially good for such structural parts as spars, spar flanges, and cap strips which must carry heavy loads in bending and compression.

(4) Northern white pine

This wood has excellent working and gluing characteristics, but its strength is slightly lower than that of spruce.

(5) Balsa

An extremely light wood grown in South America, balsa has no structural strength; but it is often sliced across its grain for use as the core for sandwich-type panels where light weight and rigidity are needed.

(6) Mahogany

This hardwood is heavier than spruce but is stronger. Its primary use in aircraft construction is for the face sheets of plywood used for aircraft skin.

(7) Birch

Birch is a heavy hardwood with very good shock resistance characteristics, and it is recommended for the face plies of plywood used for reinforcement plates on wing spars and in the construction of wooden propellers.

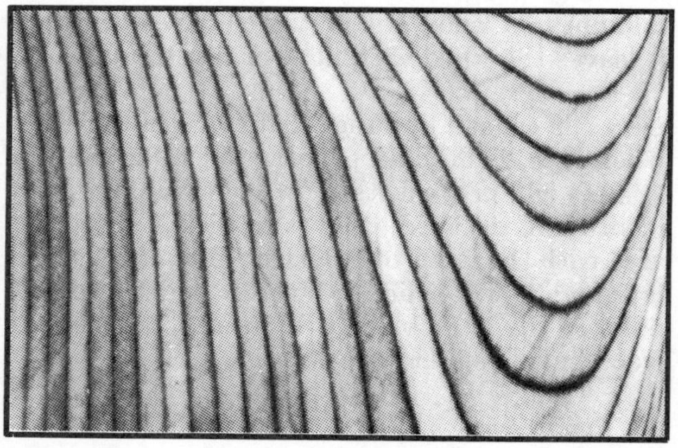

Fig. 1-15 This laminated wood propeller is made of thin strips of birch wood glued together, with the grain of all strips running in the same direction.

c. Quality of wood

Any wood approved for aircraft construction or for the repair of aircraft structure must meet specific requirements with regard to its quality and the defects that the wood is permitted to have.

(1) Grain deviation

Regardless of the species of wood used for aircraft structure, it must have straight grain; that is, all of the wood fibers must be oriented parallel to the longitudinal axis of the material. A maximum deviation of 1:15 from this is allowed. This means that the grain must not slope more than one inch in 15 inches. In some woods it is difficult to determine exactly the way the grain runs, but you can see the grain direction by putting a bit of free-flowing ink on the wood. The ink will be absorbed into the fibers along the grain.

(2) Knots

A knot in a piece of wood is that portion of a branch that becomes a part of the trunk of the tree. It may be round, oval, or spiked in shape, depending on the way the lumber has been sawn.

Sound, hard knots up to 3/8-inch at their maximum dimension are acceptable in aircraft wood with the following restrictions:

1. These knots must not be in the projecting portion of I-beams or along the edges of rectangular or beveled unrouted beams, or along the edges of the flanges of box beams, except in areas of low stress.

2. They must not cause a grain divergence at the edge of the board or in the flanges of a beam more than 1:15.

3. They must be the center third of the beam and no closer than 20 inches to another knot or other defect.

Small pin-knot clusters are acceptable if they do not cause more grain deviation than is allowed. And, even though we see that knots of up to 3/8-inch are allowed, with reservations, any wood having knots larger than 1/4-inch in any direction should be used only with extreme caution. Spike knots — that is, knots which have been sawn parallel to the branch — will weaken the board, and wood having this type of knot is not acceptable.

(3) Pitch pockets

These are small openings within the annual rings that contain resin. They are acceptable in the center portion of a beam, provided they are at least 14 inches apart and measure no more than 1-1/2 inch by 1/8-inch and are no more than 1/8-inch deep.

(4) Compression wood

If a tree was leaning in its growth, its annual rings will not be concentric, and the summer wood, the heavier and darker rings, will be much larger on the lower side of the tree. Wood having extra wide summer-growth rings has been subjected to compressive stresses during its growth, and it is denser and weaker than wood of the same type that grew normally. This "compression wood" cannot be used in aircraft construction or repair.

(5) Compression failure

This must not be confused with compression wood, but is a defect caused when the tree is felled. If the tree falls across a log, some of the wood fibers on the lower side of the tree may be buckled and the wood will be severely weakened.

Compression-failed wood is identified by irregular, thread-like lines *across* the grain, and any wood showing this type of failure has been weakened to the extent that it is unacceptable for use in aircraft construction or repair.

(6) Checks, shakes, and splits

A check is a crack that runs *across* the annual rings of a board and occurs during the seasoning process. A shake is a crack, or separation, that occurs when two annual rings separate and is along the annual rings. A split is a lengthwise separation of the wood caused by the wood fibers tearing apart. Wood having any of these defects is not allowed in aircraft structure.

(7) Stains and decay

Not all stains are evidence of decay, but stain generally accompanies all of the stages of decay. Stains caused by decay usually appear as streaks that may extend several feet along the grain from the decayed area. As a rule, stains that uniformly discolor the annual rings are evidence of decay. Rotted wood characteristically ranges from red to white in color. The stain of red rot in its early stages, for instance, is reddish-purple, and in its more developed stage the heart-wood is honeycombed with small, soft, white pockets. Even in the early stages, decay seriously reduces the toughness of wood, and in its well-developed state it leaves the wood with little or no strength.

Decay is caused by fungi that grow in the wood when it is damp, and proper seasoning and dry storage will prevent its growth. A simple way of identifying decayed wood is to pick a suspected area with the point of a pocket knife; if the wood is sound, it will splinter, but if it is decayed, the point of the knife blade will bring up a chunk of wood but no splinter.

2. Plastics

Plastics has become a household word, since molded resins have been developed to replace so many of the metal and wood items that make our modern lives functional. This type of material first became familiar to us around 1870, when celluloid was first developed. And then a stronger plastic was developed in 1909, when phenol and formaldehyde were mixed to start the family of phenolics that are so useful to us today.

Fig. 1-16 Modern aircraft construction makes extensive use of plastic materials for engine cowling and wheel fairings and for the windshield and side windows.

Plastics, or resins, fall into two natural classifications: thermoplastic and thermosetting. Thermoplastic materials in their normal state are hard, but become soft and pliable when they are heated. When soft, they can be molded or shaped, and will retain their shape when they cool. Unless their heat limit is exceeded, this process can be repeated many times without damaging the material.

Thermosetting materials usually have little strength in themselves, but are reinforced by using them to impregnate linen, paper, or glass cloth. When formed and cured, they develop a hard consistency that will retain its shape unless it is damaged. Reinforced thermosetting resins are strong and lightweight and resist weathering, aging, and chemical attack.

a. Thermoplastic resins

(1) Cellulose acetate

There are two types of transparent thermoplastic materials that are used for aircraft windshields and side windows. Cellulose acetate plastic was used in many of the earlier aircraft because of its transparency and light weight, but it had such a tendency to shrink and to turn yellow that it has been replaced by the more stable acrylic resin plastics.

Cellulose acetate can be distinguished from acrylic by its slightly yellow tint, and if a scrap of it is burned, it will produce a sputtering flame and dark smoke. When the flame is extinguished, the odor left will be unpleasant.

Acetone will soften but not discolor cellulose acetate, but it will not soften acrylic. It will instead leave a white residue on its surface.

(2) Acrylic

Acrylic plastics are identified by such trade names as Lucite or Plexiglas, or in Britain by Perspex. It is considerably stiffer than cellulose acetate, is more transparent, and for all practical purposes is colorless. It burns with a clear flame and produces a fairly pleasant odor.

b. Thermosetting resins

(1) Polyester resin

Both fiberglass cloth and mat have a great deal of strength for their weight, but they have no rigidity. To convert fiberglass into a useful structural material, we can impregnate it with polyester resin and mold it into almost any form we might need.

Polyesters differ from materials which cure by evaporation of an oil or a solvent, since they cure by chemical action rather than by evaporation. They have a complex molecular structure, and in its pure form, polyester resin is thick and unmanageable. But by adding a styrene monomer, we can make polyester workable. Styrene, while acting as a thinner, is actually much more, as it combines with the polyester to alter its characteristics.

If left by itself, a mixture of polyester and styrene will harden into a solid mass, and to prevent this we must add inhibitors. Now, for it to cure, or solidify, when we want it to, we must add a catalyst to suppress the inhibitors. Polyester resin, when mixed with appropriate amounts of catalyst, will harden into a solid mass. The time required for this solidification can be appreciably shortened by the addition of a measured amount of an accelerator. The amount of accelerator needed depends on the ambient temperature and the thickness of the layer of resin. The actual cure, or solidification, of the resin, is caused by heat, and since the heat is generated within the resin by a chemical action between the catalyst and the accelerator, the less surface area there is from which the heat can escape, the more heat will remain in the resin and the faster it will cure. A thick layer of resin will therefore cure more rapidly than a thin layer if the ambient temperature is the same.

(2) Epoxy resin

There are other resins which may take the place of polyester in laminated structure. One of these is epoxy, which is available in many forms and colors, some in the form of a liquid and others as a thick syrup. The form most frequently encountered is that which consists of two thick, syrup-like materials, the resin and the hardener.

The properties which make epoxy useful for aircraft laminated structure are its low percentage of shrinkage, its high strength for its weight, and its ability to adhere to a wide variety of materials. Epoxy can form a tight bond between even such materials as glass and metal.

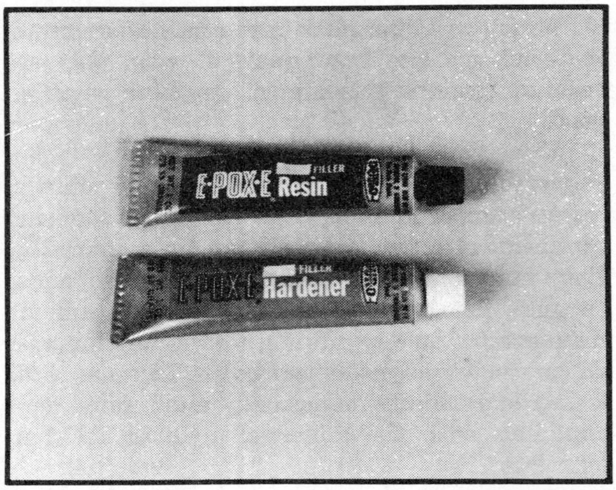

Fig. 1-17 Epoxy resin and hardener, when mixed in approximately equal portions, form a high-strength adhesive and provide a good bond between many types of materials.

Epoxy does not actually use a catalyst as do polyester resins; rather, it uses a hardener, or curing agent, with the resin. The difference between the two materials is most apparent when we look at the amounts of the materials that are used. For polyesters, about one ounce of catalyst is used with 64 ounces of resin. The final weight of the catalyzed material will be about 64 ounces. Epoxies, on the other hand, use a ratio of about one quart of curing agent with four quarts of resin to make five quarts of material.

c. Thixotropic agents

Some plastic resins are extremely sensitive to temperature changes. At temperatures of around 60° F they may be as thick as molasses, while at temperatures of 90° F. they may be as thin as water. Since heat is used to effect the cure, these resins are likely to run off of any vertical or near vertical surface before they have a chance to cure. In order to eliminate this problem we can use a thixotropic agent.

Among the more familiar thixotropic agents used in aviation maintenance are microballoons. These are hollow glass or phenolic balls ranging in diameter from about 10 to 300 microns, so tiny they resemble fine sand. If we were to examine these microballoons under a microscope, we would see that each one is a perfect sphere. The epoxy or polyester resin is mixed as recommended by the manufacturer, and microballoons are carefully folded in to make up a paste having enough body that it will not run off when it is troweled onto the surface. After the resin cures, it can be sanded to the required contour.

d. Reinforcing materials

(1) Cloth

For many years aircraft control cable pulleys have been made from thermosetting resin reinforced with layers of linen cloth. The resin-impregnated layers of cloth are cured under high temperature and pressure in a mold to form pulleys that have high strength, but will not wear the control cable.

Fig. 1-18 Phenol-formaldehyde resin is used to impregnate linen cloth to make control cable pulleys. These pulleys have high strength but do not wear the control cable.

(2) Paper

When layers of paper are impregnated with a thermosetting resin such as phenol-formaldehyde or urea-formaldehyde, they can be molded into either flat sheets or into complex shapes. This material is stiff and hard and is an exceptionally good electrical insulator. It is, however, quite heavy. Electrical terminal strips have traditionally been made of paper-reinforced resin.

(3) Glass fiber

Modern reinforced plastic materials are more often than not made of glass fibers used to reinforce a thermosetting resin. For high strength requirements, the glass fibers may be woven into a cloth. For applications where strength is not as important as cost, the fibers are gathered into a loose mat which is saturated with resin and molded into the required shape.

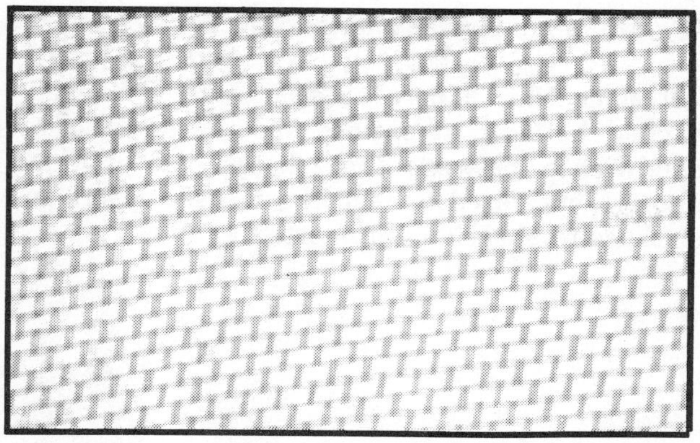

Fig. 1-19 Thin filaments, or fibers, of glass may be woven into a cloth to reinforce thermosetting resins to form various types of aircraft structure.

Fig. 1-20 Thin filaments of glass fibers may be matted together to reinforce resins when the strength of woven cloth is not required.

e. Laminated structural materials

(1) Fiberglass-faced honeycomb

(a) Paper or glass core

The cellular core for laminated honeycomb material may be made of resin-impregnated kraft paper or glass cloth. When this material is cut to the proper shape and bonded between two face sheets of fiberglass cloth that has been impregnated with resin and cured, it makes an electrically transparent material that has a high strength for its weight and is extremely rigid.

(b) Metal core

A cellular material may be made up of extremely thin aluminum and bonded between face sheets of fiberglass-reinforced resin.

Fig. 1-21 A high-strength structural material used to provide an electrically transparent streamlined housing for radar antenna is made of fiberglass-reinforced honeycomb.

(c) Wood core

Balsa wood can be cut across its grain and the end-grain slabs bonded between two face sheets of fiberglass or metal. This provides an exceptionally strong and lightweight structural material that may be used for floors, wall panels, and skins of aircraft.

(2) Metal-faced honeycomb

All of the types of core material that can be sandwiched between fiberglass sheets can also be sandwiched between thin sheets of aluminum alloy. Metal-faced honeycomb is used where resistance to abrasion is important.

SECTION II

Corrosion And Its Control

A. Corrosion — An Electro-chemical Action

In order to understand corrosion, we must understand the basic facts of electro-chemistry. For this is what corrosion actually is — an electro-chemical action.

All matter is made up of atoms and molecules. An atom is the basic unit of a chemical element, while a molecule is a cluster of atoms which make up the smallest identifiable unit of a chemical compound. For example, sodium (Na) is a metallic element and consists simply of atoms. Combined with chlorine (Cl), however, it forms a molecule of sodium chloride (NaCl) which is known to us as common table salt.

An atom consists of a nucleus made of protons which have positive charges and neutrons with no charge. Surrounding the nucleus are electrons, negatively charged particles of electrical energy. If the atom has exactly the same number of electrons as protons, it is considered to be balanced. But, if there are more or fewer electrons than protons, the atom is said to be charged and is called an ion. If there are more electrons than protons, it is a negative ion, but if there are more protons than electrons, it is a positive ion. An ion is unstable, always seeking to lose or gain the extra electron so it can change back into a balanced, or neutral, atom.

Electrons are the particles of electrical energy that flow, or move, in an electrical circuit, and it is the movement of these sub-microscopic charges with which we are concerned.

Metals may be arranged in the order shown in the Electro-Chemical Series Table in Fig. 2-1. This table shows the relative ease with which these metals ionize. The earlier the metal appears in the series, the more active it is; or, as we will see later, the more anodic it will act in the formation of corrosion.

Many metals will become ionized due to galvanic action when brought into contact with even dilute acids, salts, or alkalis such as are found in industrially contaminated air. If an aluminum structure is in contact with moisture having a trace of hydrochloric acid, there will be a chemical reaction between the acid and the aluminum and aluminum chloride and hydrogen will form.

$$2AL + 6HCL \longrightarrow 2ALCL_3 + 3H_2$$

The hydrogen is released as a gas, and aluminum chloride, which is a salt, forms as a white powder on the surface of the metal. This is the visible evidence of corrosion.

Corrosion is an electro-chemical action in which one metal is changed into a chemical salt or, as we sometimes think of it, is eaten away. When two metals are in contact with each other in the

ELECTRO-CHEMICAL SERIES FOR METALS

Arranged in order of Electrode Potential (Nobility) Most Anodic — will give up electrons most easily.

Magnesium
Zinc
Clad 7075 Aluminum Alloy
Commercially Pure Aluminum (1100)
Clad 2024 Aluminum Alloy
Cadmium
7075-T6 Aluminum Alloy
2024-T3 Aluminum Alloy
Mild Steel
Lead
Tin
Copper
Stainless Steel
Silver
Nickel
Chromium
Gold
Most Cathodic — Least Corrosive

Any metal appearing before another in this series is anodic to any metal which follows it, and will be the one corroded when they are subject to galvanic action.

Fig. 2-1 The electro-chemical series for metals

presence of some electrolyte such as hydrochloric acid, the less active metal, will act as the cathode and attract electrons from the anode. It is the anode which is corroded.

Perhaps the easiest way to visualize what is actually taking place is to consider the action of a very simple battery, such as the one we have in Fig. 2-2. If we immerse two metals in an acid, a saline, or an alkaline solution which will act as an electrolyte, a battery is formed. This battery will produce a flow of electrons between the two metals as long as the metals exist; as long as the solution remains acidic, saline, or alkaline; and as long as a conductive path connects the two metals.

In our example, we have a piece of copper and a piece of aluminum in a weak solution of hydrochloric acid and water. In the electro-chemical series, aluminum is considerably more active than copper. When electrons flow from the aluminum through the conductor to the copper, positive aluminum ions are left. Two of these ions attract six negative chlorine atoms from the acid and form two molecules of aluminum chloride, $ALCL_3$, on the surface of the aluminum. This eats away some of the base metal. The six positive hydrogen ions remaining in the acid are attracted to the copper by the electrons which came from the aluminum. These electrons meutralize the hydrogen ions and the six atoms form three hydrogen molecules ($3H_2$) and leave the surface as free hydrogen gas.

One of the basic characteristics of metals is their electrode potential. When two dissimilar metals are placed in an electrolyte, there will exist between these two metals a voltage, or an electrical potential. This potential will force electrons in the more negative material, the anode, to flow to the less negative material, the cathode, if a conductive path is provided.

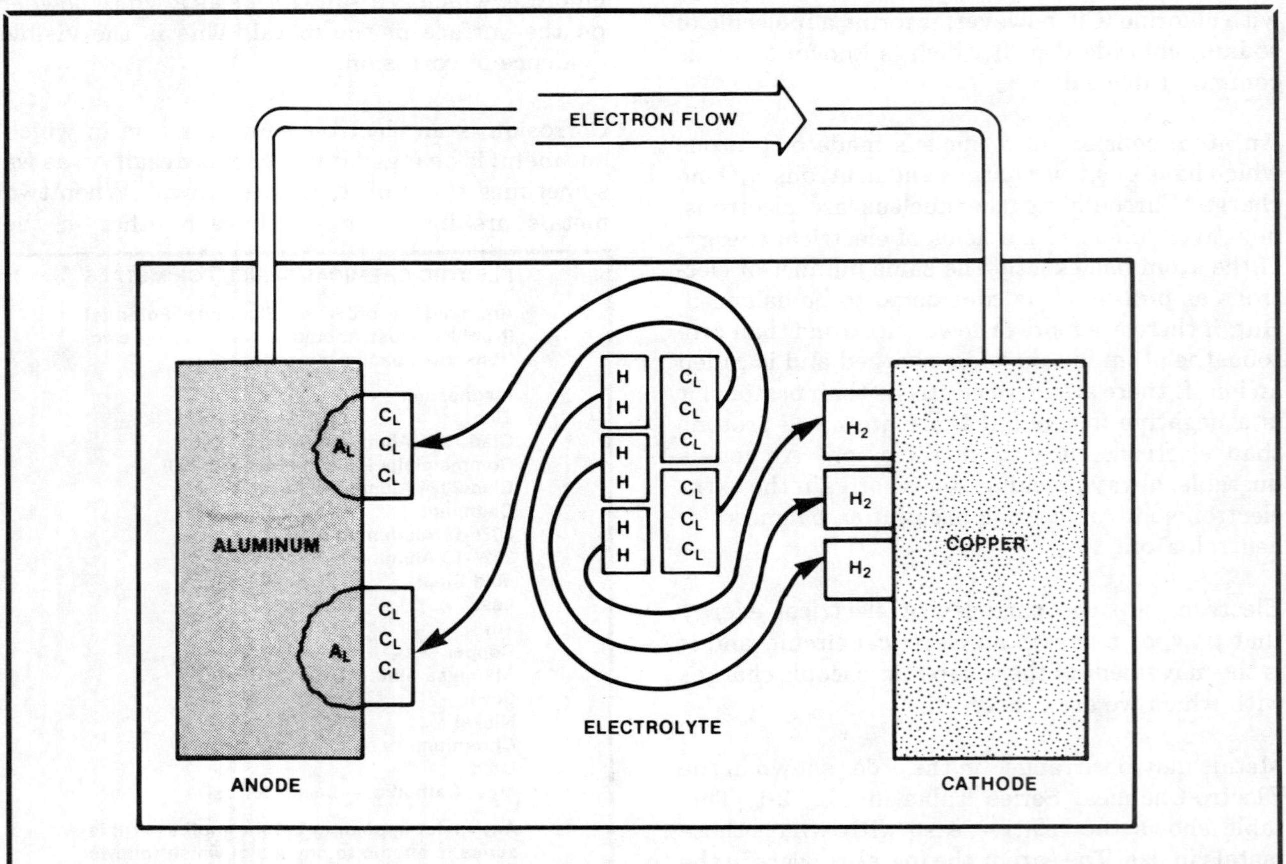

Fig. 2-2 A simple battery explains the formation of corrosion, as electrons leaving the anode attract chlorine ions from the electrolyte to form aluminum chloride, which is the visible evidence of corrosion.

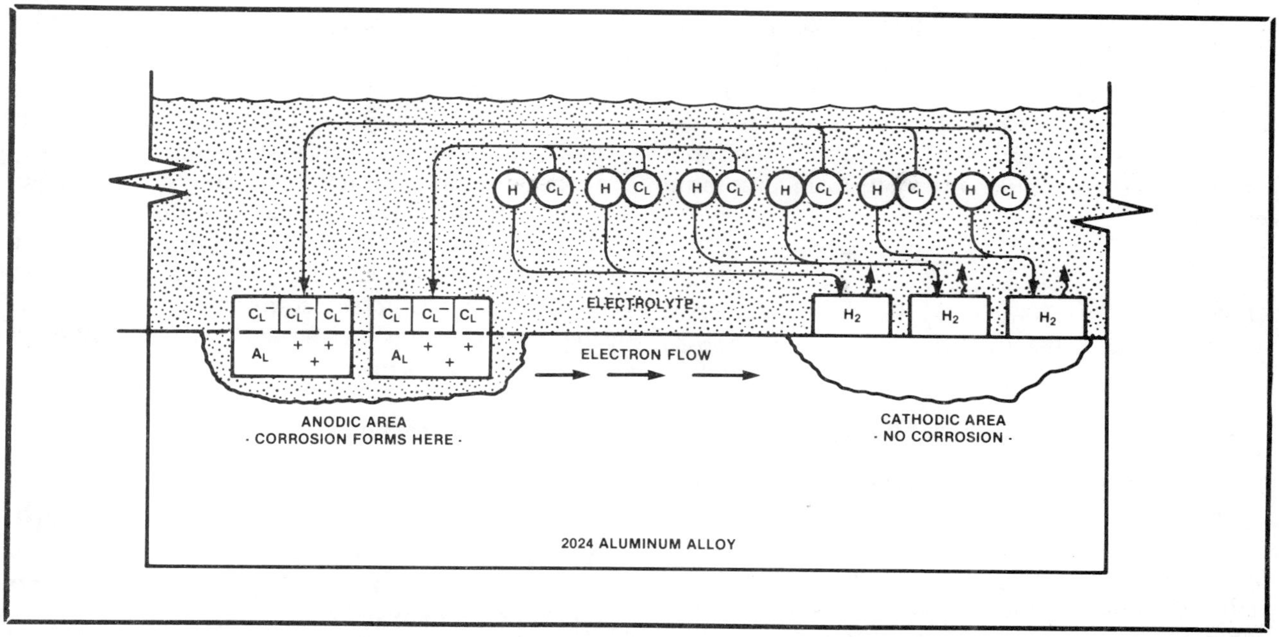

Fig. 2-3 Aluminum alloys have both anodic and cathodic areas. Corrosion will appear at the anodic area when chlorine atoms from the electrolyte join the aluminum to form aluminum chloride.

In Fig. 2-3, we have a piece of aluminum alloy such as alloy 2024, of which most aircraft structure is made. In this material, copper is alloyed with aluminum, and the microscopic grains of the copper and the aluminum serve as the cathode and the anode for our explanation. Aluminum is more negative than copper and will act as the anode in the galvanic action that takes place. Within the metal itself, the forces are such that there will be no flow of electrons between the two alloying agents until an external path is provided to form a complete circuit. This path is provided by the electrolyte which may be a surface film of moisture which contains such pollutants as acids, salts, or other industrial contaminants. In our illustration we will stick with a little hydrochloric acid in the atmosphere.

The electrode potential difference between the aluminum and the copper grains causes positive ions to exist within the aluminum. When the electrolyte film covers the surface, the aluminum ions attract chlorine ions from the hydrochloric acid and form aluminum chloride, the salt of corrosion. Hydrogen ions will be attracted to the copper by the electrons from the aluminum. These hydrogen ions will become neutralized and form molecules (H_2) which leave the surface as a free gas. Corrosion has formed on the anodic aluminum but no corrosion products are evident on the copper, the cathode.

This type of electro-chemical attack produces pits filled with the salts of corrosion, and is usually rather localized. But, if the entire surface is covered with a strong electrolyte, there will be so much corrosion that it will produce a more or less uniformly damaged area. This type of corrosion is called a direct chemical attack.

From this basic introduction, we see the three requirements for the formation of corrosion:

1. There must be areas of electrode potential difference within the metal.

2. There must be a conductive path through the metal between these two areas.

3. On the surface between these two areas, there must be a conductive chemical path containing materials to unite with the metal to form the chemical salts.

Corrosion control, therefore, consists of preventing the chemical action by eliminating one or more of these basic requirements.

B. Types of Corrosion

1. Oxidation

One of the simpler forms of corrosion, and perhaps the one with which we are most familiar, is "dry" corrosion or, as it is most generally known, oxidation. When a metal such as aluminum is exposed to a gas containing oxygen, a chemical reaction takes place on the surface between the metal and the gas. Two aluminum atoms join three oxygen atoms to form aluminum oxide:

$$2AL + 3O \longrightarrow AL_2O_3$$

If the metal is iron or steel, ferrous metals, two atoms of iron will join three atoms of oxygen to form iron oxide, or rust:

$$2Fe + 3O \longrightarrow Fe_2O_3$$

There is one big difference between iron oxide and aluminum oxide. The film of aluminum oxide is unbroken, and once it has formed, further reaction with the oxygen continues at a greatly reduced rate; it almost stops. Iron oxide, on the other hand, forms a porous, or an interrupted, film, and the metal will continue to react with the oxygen in the air until the metal is completely eaten away.

In order to protect iron from dry corrosion, or rusting, the best procedure is to prevent oxygen coming into contact with its surface. This may be done temporarily by covering the surface with oil or grease or, for more permanent protection, with a coat of paint.

Aluminum alloy may be protected from oxidation by the formation of an oxide film on its surface. This film insulates the aluminum from any electrolyte (gas or liquid), and will not, itself, further react with the oxygen.

Fig. 2-4 Iron oxide, more commonly known as rust, forms a porous scale on the surface of the material. It will continue to convert the iron oxide as long as it remains on the surface.

2. *Uniform surface corrosion*

Where an area of unprotected metal is exposed to atmosphere containing battery fumes, exhaust gases, or industrial contaminants, there will be a rather uniform attack over the entire surface. This dulling of the surface is caused by microscopic amounts of the metal being converted into corrosion salts. If these deposits are not removed and the surface protected against further action, there will develop such a rough surface that corrosion pits will form.

3. *Pitting corrosion*

Pitting is a logical result if uniform surface corrosion is left untreated. Pits form as localized anodic areas, and corrosive action may continue until an appreciable percentage of the metal thickness is converted into salts and may, in extreme cases, eat completely through the metal. Pitting corrosion is usually detected by the appearance of clumps of white powder on the surface.

Fig. 2-5 Pitting corrosion forms clumps of powdery salts of corrosion on the surface of the metal as it is corroded.

Fig. 2-6 A photomicrograph of the surface of an aluminum sheet that is pitted by corrosion

4. Intergranular corrosion

Micro-photographs of aluminum alloys show them to be made up of extremely tiny grains held together by chemical bonds; that is, an interaction between the atoms of the various elements. In the process of heat treating, the metal is heated to such a temperature that these alloying agents go into solid solution with each other. When the required temperature is reached uniformly throughout the metal, it is removed from the furnace and must be *immediately* quenched in water to solidify these elements into extremely tiny grains. If quenching is delayed for even a few seconds, the grains will grow, and when finally quenched, may have reached such a size that will produce areas of dissimilar metals large enough to form effective cathodes and anodes so that corrosion can form.

If then, through a pit, corrosive action reaches the boundary between the enlarged grains, it can continue within the metal. The electrolyte needed for this action is supplied from the surface through the corrosion deposits, and the corrosion continues along the grain boundaries.

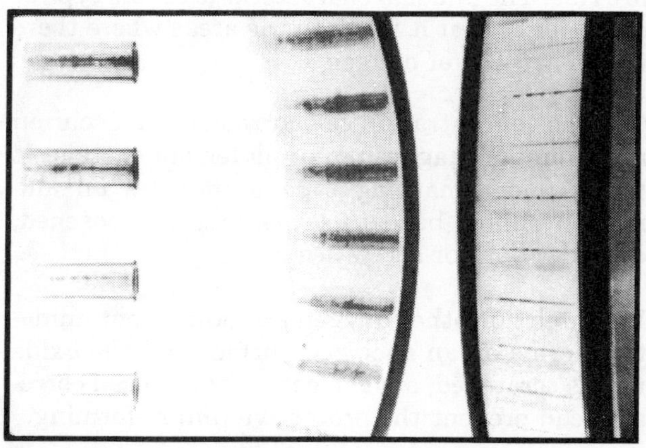

Fig. 2-7 The water marks on this hardened steel bearing race are indications of intergranular corrosion in the hardened metal.

Spot or seam welding can also cause grain enlargement that will leave the metal susceptible to intergranular corrosion.

Beneath the surface, removed from the point of initial penetration, there may occur small blisters. The surface metal over these blisters is quite thin, and when pricked with a knife point, shows a cavity filled with the salts of corrosion.

Intergranular corrosion, since it is within the metal itself, rather than on the surface, is quite difficult to detect without ultrasonic or X-ray equipment, and just about the only practical remedy for intergranular corrosion is replacement of the part.

5. Exfoliation corrosion

This type of corrosion, despite its high sounding name, is simply an extreme case of intergranular corrosion. It occurs chiefly in extruded materials such as channels or angles where the grain structure is more laminar (layer-like) than in rolled sheets or castings. This type of corrosion occurs along the grain boundaries and causes the material to seperate, or delaminate. As with other types of intergranular corrosion, by the time it is evident on the surface, the metal has been damaged beyond salvage.

Fig. 2-8 Exfoliation corrosion is an example of extreme intergranular corrosion in an extruded material.

6. Galvanic corrosion

This common type of corrosion occurs any time two dissimilar metals are connected in such a way that provides a path for electron flow, and their common surfaces are covered with some material that can serve as an electrolyte.

Galvanic corrosion can take place where dissimilar metal skins are riveted together, or where aluminum inspection plates are attached to the structure with steel screws.

When metals of the same galvanic grouping, Fig. 2-9, are joined together, they show little tendency

GALVANIC GROUPING OF METALS

Group I	Magnesium and its alloys, aluminum alloys 5052, 5056, 5356, 6061, and 6063
Group II	Cadmium, zinc, and aluminum and its alloys except those alloys listed in Group I
Group III	Iron, lead, and tin and their alloys except stainless steel
Group IV	Copper, chromium, nickel, silver, gold, platinum, titanium, cobalt, and rhodium and their alloys, and stainless steel

Fig. 2-9 Galvanic grouping of metals

for galvanic corrosion. But metals of one group will corrode when they are held in contact with those in another group. The further apart the groups, the more active the corrosion will be. The material of the lower numbered group will be the anode of the galvanic pair and will be the one that is corroded.

In the example of a steel screw used to hold a 2024 aluminum alloy inspection plate in place, if contaminated moisture gets between the two metals, the aluminum, which is in group II, will be corroded rather than the steel screw which is in group III. If 2024 aluminum alloy and magnesium are riveted together and are covered by an electrolyte, the magnesium from group I will corrode instead of the aluminum which is in group II.

7. Concentration cell corrosion

a. Oxygen concentration cell corrosion

When water (H_2O) covers the surface of an aluminum airplane skin and seeps into the cracks between the lap joints, oxygen concentration cell corrosion may form. Since water in an open area will readily absorb oxygen from the air, it will attract electrons from the metal to form negative hydroxide ions:

Two molecules of water plus one molecule of oxygen plus four electrons will form four negative hydroxide ions.

$$2H_2O + O_2 + 4 \text{ Electrons} \rightarrow 4(OH)^-$$

The electrons required to form these negative ions come from the metal itself. The area between the skins does not give up electrons to the water on *its* surface because there is not enough oxygen there to form hydroxide ions. Its electrons flow instead to the cathodic surface — the open area. The area between the skins, having lost electrons, now contains positive aluminum ions, and this area becomes the anode.

As we have have seen, electrons flow within the metal from the anode to the cathode, leaving positive metal ions in the area between the sheets. These positive aluminum ions attract negative hydroxide ions from the open water and the aluminum corrodes, forming aluminum hydroxide. The unusual characteristic of this type of corrosion is that it forms in the areas where there is a *deficiency* of oxygen.

Oxygen concentration cell corrosion can occur on aluminum or magnesium or on ferrous metals. It forms under marking tape of ferrules on aluminum tubing, beneath sealer that has loosened, or under bolt- or screw-heads.

When dirt or other oxygen-excluding contamination forms on an anodized surface and the oxide film is scratched, oxygen concentration cell corrosion can prevent the protective film re-forming.

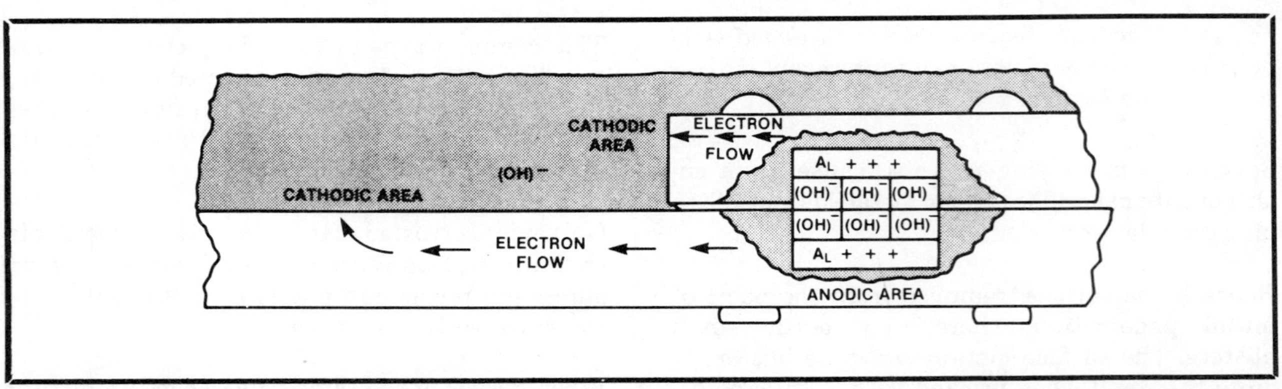

Fig. 2-10 Oxygen concentration cell corrosion forms in the areas where there is a deficiency of oxygen.

Fig. 2-11 Oxygen concentration cell corrosion has formed under the ferrule on this tube.

b. Metallic ion concentration cell corrosion

The electrode potential within a metal is dependent on the different metals that make up the alloy, but a potential difference will occur if an electrolyte having a non-uniform concentration of metal ions covers the surface. In Fig. 2-12, we have the representation of the surface of an aircraft which is covered with a film of water. As was true in oxygen concentration cell corrosion, when water absorbs oxygen, it attracts electrons from the metal and forms negative hydroxide ions. When these electrons left the aluminum, positive aluminum ions were formed. The water can move freely on the surface of the metal, and this movement continually carries the aluminum ions away. But the water between the faying surfaces of the skins does not move, and it therefore has a higher concentration of metal ions than there is in the open. The skin area between the faying surfaces, where the concentration of positive metal ions is the highest, becomes the cathode and attracts electrons from the skin in the open area, which is the anode. As electrons flow from the anode to the cathode, they leave positive aluminum ions on the surface near the supply of negative hydroxide ions. These hydroxide ions join the aluminum ions to form aluminum hydroxide, corrosion.

Notice the difference between the two types of concentration cell corrosion: the metallic ion concentration cell corrosion forms on the *open* surface, while oxygen concentration cell corrosion forms in the *closed* areas between the faying surfaces. Corrosion on airplanes is usually complex and is generally composed of more than one type.

8. Stress corrosion

Another special type of intergranular corrosion is stress corrosion. This occurs when the metal is subjected to a tensile stress in the presence of a corrosive environment. The stresses in the metal may come from improper quenching after heat treatment or from an interference (press) fit of a fastener. Cracks caused by stress corrosion grow rapidly as the corrosive attack concentrates at the end of the crack rather than along its sides.

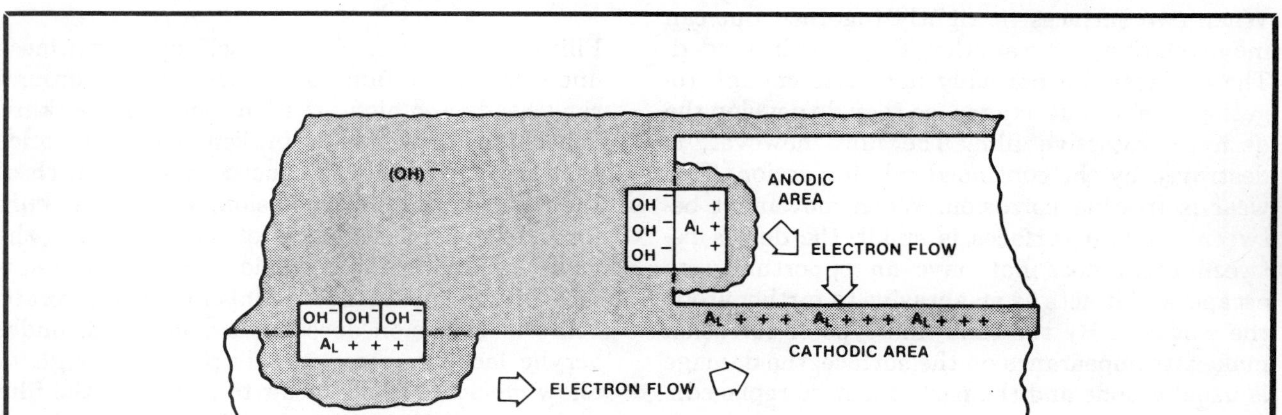

Fig. 2-12 Metallic ion concentration cell corrosion forms in the open at the edge of the lap joint.

Since stress corrosion can occur only in the presence of tensile stresses, one expedient for preventing this type of corrosion in some heat-treated aluminum alloy parts is to shot-peen the surface to provide a uniform compressive stress on the surface. These stresses must be overcome by tensile forces before stress corrosion can form.

Common locations for stress corrosion to form are between rivets in a stressed skin, around pressed-in bushings, or tapered pipe fittings. Careful visual inspection may show up this type of corrosion, but dye penetrant inspection is required to find the actual extent of the crack.

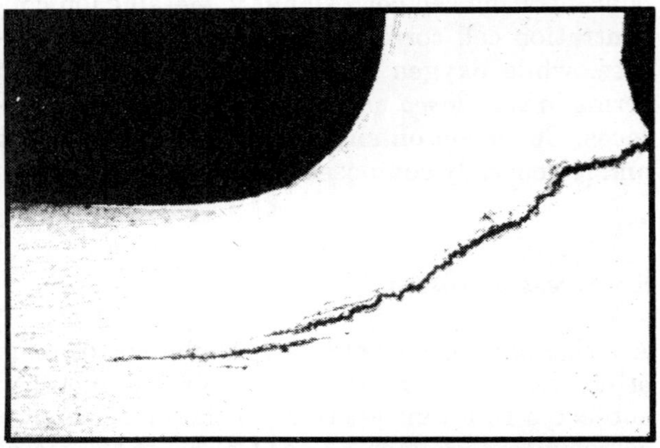

Fig. 2-13 Stress corrosion forms at the end of a crack and rapidly enlarges the crack until the material fails.

9. Fretting corrosion

When two surfaces fit tightly together, but can move relative to one another, they may be eroded. The surfaces are normally not close enough together to shut out oxygen, so they do develop the desired protective film. The film, however, is destroyed by the continued rubbing action. This wear is fretting corrosion. When movement between the two surfaces is small, the debris between them does not have an opportunity to escape, and it acts as an abrasive to further erode the surfaces. By the time this type of corrosion makes its appearance on the surface, the damage is usually done and the parts must be replaced.

Fretting corrosion may occur around rivets in a skin and will be indicated by dark deposits around the rivet heads streaming out behind, giving the appearance of the rivet smoking. Rivets showing this sign of fretting must be drilled out and replaced.

Fig. 2-14 Fretting corrosion causes black residue to trail out behind the rivet, causing it to appear to smoke.

10. Filiform corrosion

Filiform corrosion forms in fine thread- or filament-like lines under a polyurethane enamel finish if the wash primer used on the metal has not been properly cured. A wash primer is a two-part metal preparation material, in which phosphoric acid converts the surface of the metal into a phosphate film that protects the metal from corrosion and provides an excellent bond for the topcoats of the finish. If there is not enough moisture in the air to convert all of the acid, some acid will remain on the metal; then, if a dense polyurethane film is applied, the acid will be trapped and will react with the aluminum alloy to form corrosion.

Filiform corrosion shows itself as a puffiness under the paint film and is first noticed around rivet heads and along the lap joints of the skins. When the paint film is broken, you will notice that the puffiness was caused by the growth of the powdery salts of corrosion. There is no cure for filiform corrosion short of stripping all of the paint, removing the corrosion, treating the surface of the metal, and refinishing the aircraft. Filiform corrosion does not usually form under acrylic lacquer, because it is porous enough to allow moisture from the air to penetrate the film and complete the conversion of any acid that was left on the metal.

It is because of the possiblility of the formation of filiform corrosion that epoxy primers are used under polyurethane finishes when the speed of finishing is not of prime concern.

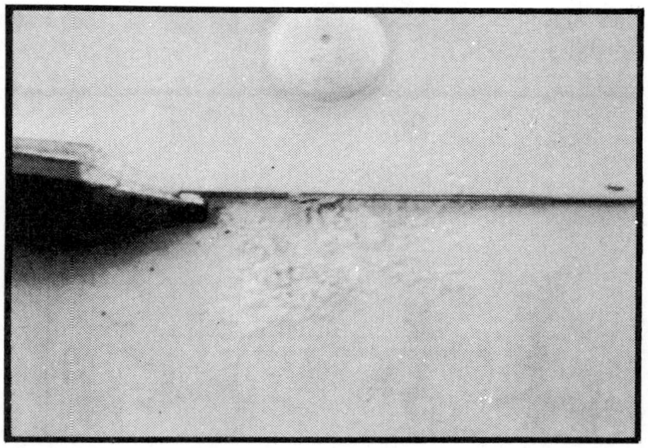

Fig. 2-15 Filiform corrosion forms under the dense film of a polyurethane finish, where it first appears as a puffiness under the finish.

C. Corrosive Agents

1. Acids and alkalis

For corrosion to form on metal, there must be an electrode potential difference within the metal and an electrolyte covering the surface. Almost all acids and alkalis form effective electrolytes as they react with metals to form metallic salts, but some electrolytes are more active than others. Sulfuric acid such as is found in aircraft batteries is especially active in corroding aluminum, while a weak solution of chromic or phosphoric acid may even be used as a surface treatment to prepare metal for painting.

Ferrous metals are subject to damage from both acids and alkalis, but aluminum is more vulnerable to strong alkaline solutions than it is to acids. Aluminum structure, for instance, can be severely corroded if it is allowed to remain in contact with a concrete floor. Water will draw out enough lime from the cement to form an alkaline solution that will corrode the aluminum.

2. Salts

Marine atmosphere and air above some of the industrial areas hold large concentrations of salts. These chemicals will precipitate out of the air and collect on the surface of an aircraft where they will attract water and form an effective electrolyte on the metal. Magnesium is particularly vulnerable to corrosive attack from an electrolyte formed by salt solutions.

3. Mercury

Although it is not too commonly found in any quantity around aircraft, there is a definite possibility that mercury could be spilled in an airplane. Mercury attacks aluminum by a chemical reaction known as amalgamation. In this process, the mercury attacks along the grain boundaries within the alloy, and in an exceedingly short time will completely destroy it. Extreme care must be exercised when removing spilled mercury, as it is "slippery" and will flow through even a tiny crack to get to the lowest part of the structure where it can cause extensive damage. Not only is mercury damaging to an aircraft structure, but mercury and mercury vapors are also dangerous to people. If mercury is spilled, remove every particle with a vacuum cleaner having a mercury trap in the suction line, or with a rubber suction bulb or a medicine dropper. Never attempt to remove mercury by blowing it with compressed air. This will only scatter it and spread the damage.

Brass control cable turnbuckle barrels are especially susceptible to mercury damage. Any sign of mercury discoloration requires the replacement of the barrel.

4. Water

Pure water will react with metals to form corrosion or oxidation, but water holding a concentration of salts or other contaminants will cause much more rapid corrosion. Seaplanes are continually in a battle with the elements, and every precaution must be taken to stay ahead of the formation of corrosion. Seaplanes operating in salt water are especially vulnerable to attack, and when one is taken out of salt water it should be hosed down with large volumes of fresh water to get every trace of salt off of the structure. Seaplane ramps are often in areas where there is a concentration of industrial wastes; this makes the water even more corrosive.

Float bottoms are subject to the abrasive effect of high-velocity water on takeoff and landings. This abrasion tends to damage the natural protective oxide film, and seaplanes must be carefully inspected to detect any damage which would allow water to get to the base metal of the structure.

5. Air

Naturally, it is impossible to isolate a structure from the air in which it exists; but the very presence of air is a factor in the deterioration of the metal. Airborne salts and other chemical compounds settle onto the surface of an airplane and attract moisture from the air. Corrosion then has one of its key requisites — an electrolyte.

6. Organic growths

New developments bring new problems. For years, water which condensed in the fuel tanks produced relatively minor corrosion problems. Small perforated metal containers of potassium dichromate crystals protected the fuel tanks by changing any water into a mild chromic acid solution which inhibited corrosion.

Jet aircraft, however, use fuel with a higher viscosity than gasoline, fuel which holds more water in suspension. They also fly higher than reciprocating engine aircraft, and in high altitude, low-temperature flight conditions, water entrained in the fuel will condense out and collect in the bottom of the tanks. To further complicate matters, this water contains microbes, which are simply microscopic sized plant and animal life. These organic bodies live in the water and feed on the hydrocarbon fuel. The dark insides of the fuel tank promote their growth, and in very short periods of time, these tiny creatures multiply and form a scum inside the tank.

The scum, composed of microbes and the water in which they live, covers the bottom of the tank and holds water in contact with the tank structure where corrosion of the concentration cell type will inevitably form. If the scum forms along the edge of the sealant in an integral fuel tank, the sealant may pull away from the structure, causing a leak and an expensive resealing operation.

It is virtually impossible to prevent the formation of this scum as long as microbes are allowed to live in the fuel, and the most successful solution of the problem has been to use an additive in the fuel which kills these organic growths and prevents the formation of the corrosion-forming scum. In addition to preventing corrosion by its biocidal action, this same fuel additive has the additionally useful function of serving as an antifreeze to prevent the freezing of the water which has been freed from the fuel plugging the fuel screens.

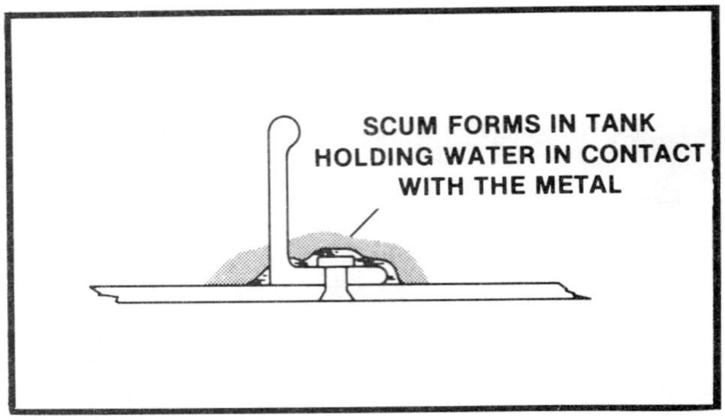

Fig. 2-16 Microbial growth within a turbine aircraft fuel tank forms a scum that holds water against the metal and causes corrosion to form.

D. Detection of Corrosion

1. Visual inspection

Exotic inspection equipment is often needed for certain parts of an aircraft to make the work of the A&P more efficient, but a well-trained eye is still the most effective tool for inspection. Corrosion can often be detected by careful visual inspection of the airplane structure. Corrosion of aluminum or magnesium appears as a white or gray powder along the edges of the skins and around rivet heads. Small blisters appearing under the finish on painted surfaces is an indication of corrosion. Examine the lap joints in the skins for bulges which indicate the presence of corrosion between the faying surfaces. Corrosion salts have more volume than sound aluminum and will push out against the skin.

The complex structure of modern aircraft makes the use of magnifying glasses, mirrors, borescopes, fiber optics, and other optical inspection tools imperative for a good visual inspection.

2. Dye penetrant inspection

Stress corrosion cracks are sometimes difficult, if not impossible, to detect by visual inspection alone, but they may be found by the use of a dye penetrant inspection. This inspection method is effective on both ferrous and non-ferrous metals and on non-porous plastics. The procedure used in dye penetrant inspection consists of spraying a penetrating liquid on the surface to be inspected. The liquid has a very low surface tension and will seep deep into any crack that extends to the surface. The penetrant is left on the surface long enough to penetrate all of the cracks and then the surface is wiped clean. A developer, which is a white, chalky powder, is sprayed over the surface where it acts as a blotter and draws the penetrant out of any cracks into which it has seeped.

The penetrant is usually dyed bright red, and so the cracks appear as red lines on the white surface. Another type of penetrant inspection uses a fluorescent penetrant which is viewed under an ultraviolet, or "black," light. This special light will cause any crack to appear as a green line on the surface. The main limitation of dye penetrant inspection is the fact that it can fail to find cracks that are so full of corrosion products that the dye cannot penetrate. Also, if the crack is filled with oil or grease, the penetrant cannot get in, and there will be no indication of a flaw. Porous or rough surfaces are almost impossible to clean of all of the penetrant, so materials with rough surfaces do not lend themselves to this type of inspection.

3. Ultrasonic inspection

One inspection method that is effectively used for corrosion inspection is that using ultrasonic energy. In this method of inspection, high-frequency pulses of energy, similar to sound waves, only at frequencies far above the audible range, are introduced into the airplane structure. There are two types of ultrasonic indications which may be used for corrosion detection: the pulse-echo and the resonance method.

In the pulse-echo method, a pulse of ultrasonic energy is directed into the structure by a device known as a transducer. This energy travels through the material to its opposite side and then bounces back. When the return pulse is received by the transducer, it is displayed on the screen of a cathode-ray oscilloscope as a spike which establishes a time base, representing the thickness of the material. If there is any change in thickness, such as may be caused by corrosion, the return will occupy a shorter space and will indicate the extent of damage. If there should be a crack or other flaw within the material, such as may be caused by intergranular corrosion, a second spike will appear on the oscilloscope screen which indicates the approximate position of the flaw within the material.

The second method of inspection using ultrasonic energy is the resonance method. This method operates on the principle that for any given thickness of material, there is a specific frequency of ultrasonic energy that will resonate, or produce the greatest amount of return. Variable frequency ultrasonic energy is fed into the transducer, and the output is monitored visually with a meter or audibly with a set of head phones. When the resonant frequency is reached, the meter will read the highest value or the tone will be the loudest in the phones. If the metal has been eaten away by corrosion, its resonant frequency will be different from that of sound metal, and the meter reading or tone volume will be lower.

The resonance method can be used to determine the actual thickness of the material by calibrating the probe with a test speciman of the same type material being tested. Ultrasonic inspection must be conducted by a person highly qualified and equipped for this procedure. False returns can easily disguise a fault, and many special transducers are needed for the different locations to be inspected.

Ultrasonic inspection is primarily a method of inspection by comparison. Good skin is measured, and any change in thickness of the skin being inspected may have been caused by corrosion. Further inspection must be made to determine the actual extent of the corrosion indicated by the ultrasonic inspection.

4. Radiological inspection

Like ultrasonic inspection, radiological inspection such as X-ray is used to determine, from the outside of the structure, if there is any damage on the inside. X-ray is a photographic form of inspection in which extremely high-frequency pulses of electromagnetic radiation are passed through the structure being inspected. Energy at this frequency has the ability to expose photographic film, and as it passes through the structure, areas of high density pass less of the radiation energy and therefore expose the film less. After the exposure, the film is developed like any other photographic negative. In areas where the density of the material is greatest, less energy will penetrate, and this area will appear light on the negative. Areas of lower density pass more energy, expose the film more, and appear darker.

X-ray inspection requires extensive training and experience for proper interpretation of the results. The use of X-ray involves some danger, because exposure to the radiation energy used in this process may cause burns, damage to the blood, and possibly death. Persons around X-ray equipment should wear a radiation monitor film badge which is developed at the end of an exposure time to determine the amount of radiation the wearer has absorbed. A blood count should be made periodically for persons involved in X-ray inspections.

The depth of penetration of X-rays is determined by the amount of power used. This power is normally measured in terms of kilovolts applied to the X-ray tube and may range from about eight kilovolts to as high as 200 kilovolts. The lower power applications, called "soft" X-rays, are used to inspect for corrosion.

E. Corrosion-prone Areas

Modern airplanes are made of thin, reactive metals which can tolerate very little loss of strength. One of the most important functions of the A&P technician on an annual or 100-hour inspection is to check the entire aircraft for indications of corrosion which could degrade the strength of the structure. Almost all parts of an airplane are subject to this type of damage, but certain areas are more prone than others.

1. Engine exhaust area

Both reciprocating and turbine engines generate their power by converting chemical energy from a hydrocarbon fuel into heat energy. Because of the inefficiency of the engine, much of this heat, along with energy-rich gases, passes out of the engine through the exhaust. The gases contain all of the constituents for a potent electrolyte, and because of their elevated temperature, corrosion can form all the more rapidly. Exhaust areas must be carefully inspected and all exhaust residue removed before corrosion has a chance to start. Cracks and seams in the exhaust track are prime areas for corrosion. Fairings on the nacelles, hinges, and inspection door fasteners all contain crevices which invite the formation of corrosion.

2. Battery compartments and vents

Almost all airplanes nowadays have an electrical system and a battery. These batteries store electrical energy by converting it into chemical energy and are therefore active chemical plants, complete with environmental polluting exhausts. Airplanes having lead-acid batteries must have their boxes protected by a material that resists corrosion from the sulfuric acid fumes, and airplanes with nickel-cadmium batteries must have their battery areas protected with an alkaline-resistant finish. These finishes may have a bitumastic (tar) base or a rubber base or, even better, may be polyurethane finishes.

During an inspection, these areas must be carefully checked, especially *under* the battery, and all traces of corrosion must be removed and the area refinished. Both types of batteries should have a sump vent jar containing an absorbent pad moist with a neutralizing agent; bicarbonate of soda for a lead-acid battery or boric acid for a nickel-cadmium battery. These sump vent jars should be checked to see that the pads are moist and that there is no leakage. All of the vent openings should be clear and the intake and exhaust lines free and open. If battery electrolyte is spilled during servicing, it must be cleaned up immediately and the area neutralized. Flush with water and neutralize with bicarbonate of soda for lead-acid batteries, or with boric acid or vinegar for nickel-cadmium installations. The thoroughness of the neutralization can be checked with litmus paper. Blot the water on the surface with a piece of this indicator paper. If the area is acidic, the paper will turn pink; if it is alkaline, it will

Fig. 2-17 The strong acid or alkaline fumes around a battery box can cause corrosion on the adjacent structure.

turn blue. The entire area should be neutral and it should not change the color of the paper.

3. Lavatories and food service areas

Organic materials such as food and human waste are highly corrosive to aluminum surfaces, and where this type of material can be spilled the areas must be inspected with extreme care. Food service areas may be troublesome if there is a possibility of some of the debris from food getting into cracks under or behind the galley where it cannot be removed. While this material, in itself, may not be corrosive, it can hold water which will cause the structure to corrode.

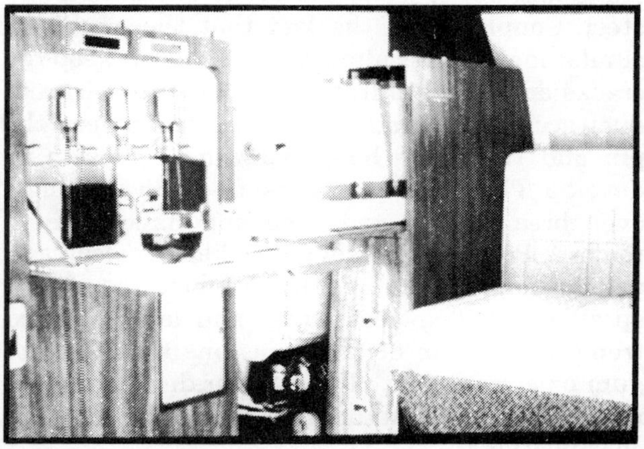

Fig. 2-18 The food service area in an aircraft is subject to corrosion damage if spilled foodstuff traps water against the structure where it cannot be removed.

The lavatory area is an especially important area to check for corrosion. Human wastes are usually acidic and promote corrosion in a hurry if allowed to remain on the skin of the airplane or to get into cracks or seams in the structure. Disinfectants may be used that will cause further damage to the aircraft. Be sure to check the disinfectant carried in the airplane to be sure that it is not of a type that is harmful to aluminum.

The area around and behind the discharge of relief tubes must be inspected carefully for indications of corrosion. These areas should be painted with an acid-proof paint where the discharge contacts the airplane skin.

4. Wheel wells and landing gear

Probably no one area of an airplane is subjected to as much hard service as the wheel well area. On takeoff and landing, debris from the runway surface may be thrown up into this area. This can be especially troublesome in the winter when chemicals are used on runways for ice control. Abrasion may remove the protective lubricants and coatings, and water and mud can freeze and cause damage. Corrosion can take place in any of the electrical components such as antiskid sensors, squat switches, and limit switches. Bolt heads and nuts on magnesium wheels are susceptible to galvanic corrosion, and concentration cell corrosion can form under the marking tape or ferrules on aluminum tubing. Special care must be taken to search out any area where water can be entrapped.

Fig. 2-19 Wheel wells are corrosion-prone areas because abrasion from water and dirt removes the protective finish, and dirt and grease can trap and hold moisture against the aluminum.

5. External skin areas

a. Seams and lap joints

One of the first places corrosion appears on the surface of an aircraft is along the seams and lap joints. It is here that concentration cell corrosion frequently appears and, in clad skins, it is here that the sheared edges expose the alloy without the protection of the pure aluminum. There is also a danger of water or cleaning solvents becoming entrapped in the lap joints and providing an effective electrolyte.

Corrosion may start in a spot welded seam if the spot welding process has left an enlarged grain structure in the metal. The area is then susceptible to corrosion when moisture seeps between the skins. Check carefully for bulging along the spot welds. Hold a straightedge along the row of spots, and if there is corrosion in the seam, the surface will bulge between the spots and show up as a wavy skin. Corrosion in a seam can progress to such a degree that the spots will actually pull apart.

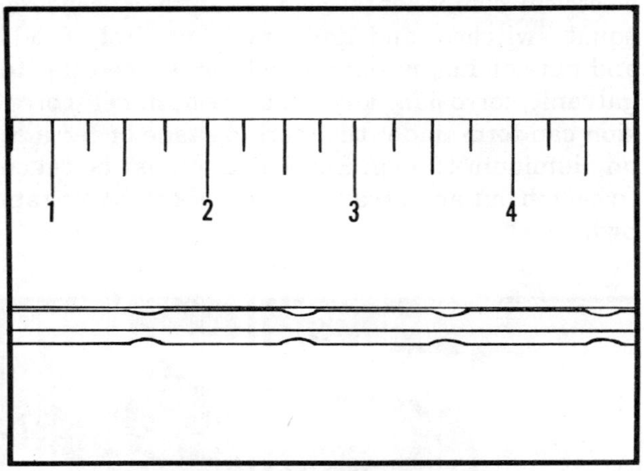

Fig. 2-20 Corrosion in a spot-welded seam may be detected by laying a straightedge over the seam. Corrosion will cause the seam to have a wavy surface.

b. Engine inlet areas

One of the most vital parts of an airplane is the area directly in front of the engine where the air is taken in. In jet aircraft, this area is usually quite large, and air rushes into the engine at a high velocity. Abrasion by this high-velocity air and by contaminants carried in the air will tend to remove any protective coating. Abrasion strips along the leading edge of intake ducts help protect these areas, but careful inspection and maintenance of the finish in these inlet areas is imperative.

6. Inaccessible areas

a. Fuel tanks

There is probably no single place in an aircraft that is less accessible for inspection and repair than the inside of an integral fuel tank; yet, especially in jet aircraft, the fuel tanks are highly susceptible to corrosion formation. As we have discussed in the section on organic growth as a cause of corrosion, we have seen that turbine fuel can hold microscopic organisms in suspension until they have an opportunity to join together and grow into the water-holding scum which attaches to the aluminum alloy skin. Sealants used to convert the structure of a wing into a fuel tank are impervious to fuel, but only *resistant* to water, and it is possible for water to seep through the sealer and cause oxygen concentration cell corrosion. Corrosion under the sealant is extremely difficult to detect and must usually be found with X-ray or ultrasonic inspection from the outside of the wing.

b. Piano hinges

Piano hinges on control surfaces and access doors are ideal locations for the start and development of corrosion. The hinge body is usually made of aluminum alloy, while the pin is of hard carbon steel. Coupled with the fact that these are dissimilar metals, it is almost impossible to keep the cracks and crevices free from moisture. Dirt and dust accumulate and hold moisture between the pin and the hinge body, allowing corrosion to develop. The pin may rust and freeze in the hinge, even breaking off and becoming impossible to remove. Piano hinges should be kept as clean and dry as possible and should be lubricated with a spray which displaces water and leaves an extremely thin film of lubricant; one that will not gum or attract dust. These water-displacing lubricants are manufactured to meet MIL-C-16713 specifications.

Fig. 2-21 A piano hinge must be kept dry and free from dirt which could cause galvanic corrosion because of the different metals used in the hinge.

c. Control surface recesses

Any place on an airplane which is difficult to inspect is an area where corrosion has an opportunity to grow. Some airplanes have areas in the wing or empennage where the movable surfaces recess back into the fixed surface. Hinges are buried back in these cavities and are difficult to lubricate. Special attention must be paid when inspecting these areas to remove every trace of corrosion and provide drains for any water that might collect. A thin film, of the same water-displacing lubricant as is used on piano hinges, may be used to protect the skin lap joints in these areas.

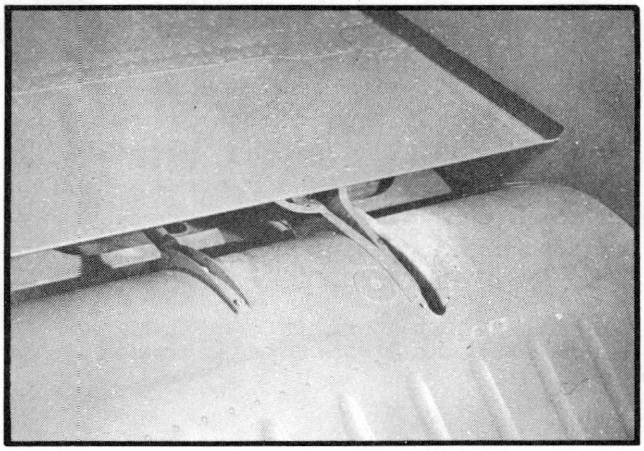

Fig. 2-22 Control surface recesses should be kept clean and dry to prevent the formation of corrosion where it is difficult to inspect.

d. Bilge areas

The bottom of the fuselage, below the floor, is an area where water and all forms of liquid and solid debris can accumulate and cause corrosion to form. These areas are not only ideal for the formation of corrosion because they provide the dissimilar metals and the constant exposure to an electrolyte, but because of their inaccessibility, corrosion can often go undetected until it has caused major damage. Airplanes having areas prone to accumulate water have drain holes provided, but since dirt and other debris also collect here, these holes may become clogged and not do their job. On every inspection, be sure all drain holes are clear, and inspect carefully any area where water might possibly stand. Air-powered vacuum cleaners may be used to remove dirt or water collected in these areas, and a water-displacing liquid spray will form a thin film on the surface of the metal to prevent further contact with the electrolyte.

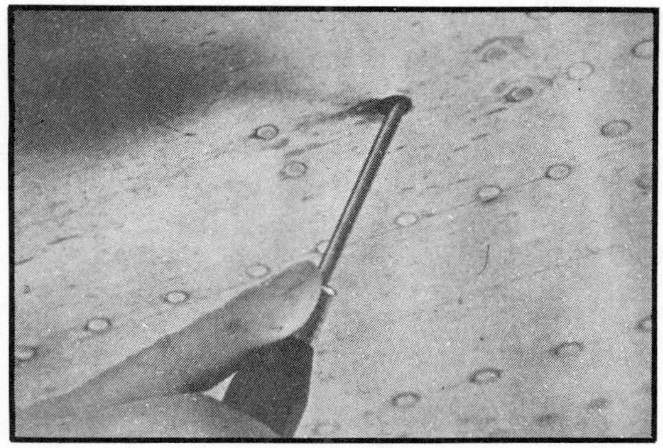

Fig. 2-23 The drain holes in the bottom of a structure must be kept open to prevent moisture collecting and causing corrosion.

e. Landing gear boxes

There are few areas in a modern fixed landing gear airplane that are as highly stressed, yet as difficult to inspect, as the landing gear box structure. The landing gear attaches into the fuselage by a strong, heavy-gage aluminum alloy box structure which is under the floor and is accessible for inspection only through a relatively small access hole. Water can collect in this area if the drain holes should become plugged. Inspection of these boxes should be thorough, all drains opened, and the entire enclosed area generously sprayed with a water-displacing lubricant film.

Fig. 2-24 The landing gear box is one of the most highly stressed areas in an aircraft structure and is one of the least accessible for adequate inspection.

7. Engine mount structure

The heavy current from the starter must return to the battery through the engine mount. This current flowing through joints in an aluminum alloy mount will create the potential difference required for corrosion and cause it to form in these areas.

Welded steel tube mounts are protected from corrosion by filling each section of the tubing with hot linseed oil or other type of tubing oil. As much of the oil as possible is drained out, and the drain hole is plugged with a drive screw or a self-tapping metal screw.

8. Control cables

The cables used in an aircraft control system may be made of either carbon steel or corrosion-resistant steel. If carbon steel cable is left unprotected and water is allowed to enter, it will soon corrode, and the corrosion that forms on the inside of the cable is difficult to detect. If corrosion is suspected, release the tension from the cable and open the strands by twisting them against the lay. This allows you to see between the strands. Cable with any indication of corrosion should be replaced. Corrosion may be prevented by spraying the cable with a water-displacing type of lubricant, or if it is used in a seaplane or exposed to agricultural chemicals, it should be coated with a waxy grease such as Paral-ketone.

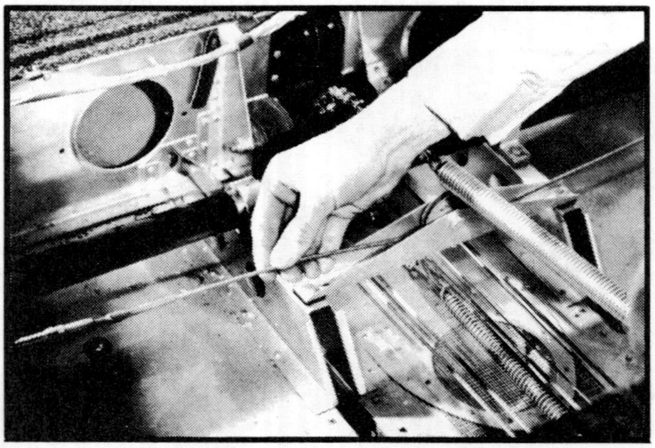

Fig. 2-25 A steel control cable must be carefully inspected to be sure there is no corrosion in the strands. If there is any corrosion, the cable must be replaced.

9. Welded areas

Aluminum welding requires the use of a flux to exclude oxygen from the weld. This flux may contain lithium chloride, potassium chloride, potassium bisulphide, or potassium fluoride. These are extremely corrosive to aluminum, and *all traces* of the flux must be removed after the welding is completed. Welding flux is soluble in water and may be removed with hot water and a non-metallic bristle brush.

10. Electronic equipment

The use of copper, lead, tin, and other metals in electronic wiring and printed circuit boards makes them a target for corrosion. The fact that electrons are continually flowing in these conductors and the extremely low tolerance for corrosion make corrosion control in these areas vital. This control is usually accomplished by sealing the wiring and circuit boards with a transparent film which excludes all oxygen and moisture from the surface. Detection and repair of corrosion in these areas is a highly specialized field and is not usually the job of an A&P technician.

F. Removal and Treatment of Corrosion

1. Surface preparation

a. Cleaning

The first and most important step in corrosion control is a thorough cleaning of the airplane. An emulsion-type cleaner that meets MIL-C-25769 specifications does an excellent job of removing grime, dirt, exhaust residue, and dried oil and grease deposits. The airplane should be parked on a wash rack or in an area where it can be hosed down, and preferably in an area where the sun will not dry the surface before the cleaner has time to penetrate the film. For the main part of the exterior of the airplane, use a 1:5 or a 1:3 dilution of the cleaner with water. Brush or spray the mixture onto the surface and allow it to stand for a few minutes; then rinse it off with a high-pressure stream of warm water.

The engine cowling and wheel well area usually have grease or oil deposits that this treatment will not remove, so they must be soaked with a 1:2 dilution of the emulsion cleaner. After allowing the cleaner to remain on the surface for a few minutes, scrub it with a soft bristle brush to completely loosen the film, and rinse it with a high-pressure stream of warm water.

Stubborn exhaust stains may require a 1:2 mixture of the cleaner with varsol or kerosene. Mix these ingredients into a creamy emulsion and apply it to the surface. Let it stand for a few minutes and then work all of the loosened residue with a bristle brush and hose it off with a high-pressure stream of warm water. This treatment may be repeated if the first time does not remove all of the stain.

Fig. 2-26 An aircraft must be kept as clean as possible, and an emulsifying cleaner is the most effective way of loosening stubborn dirt. After the dirt has been loosened, it should be washed off with high-pressure warm water.

b. Paint removal

After the surface is completely clean and it is determined that corrosion does exist, the damage must be carefully assessed, and a decision must be made as to what action should be taken. All corrosion products must be removed as soon as they are discovered, because corrosion will continue as long as the deposits remain on the surface.

Corrosion under a paint film cannot be thoroughly inspected without first removing all of the paint. But, before using any paint remover with which you are not familiar, first test it on a piece of metal similar to that with which you will be working. If no adverse affects are apparent, it is safe to use. All areas not to be stripped should be masked with heavy aluminum foil to prevent the stripper accidently coming into contact with these areas. Water-rinsable paint remover having a syrupy consistency is usually best for aircraft surfaces. This type of remover is applied with a brush by daubing it on the surface rather than brushing it on. Cover the surface with a heavy coating of remover, and allow it to stand until the paint swells and wrinkles up. This breaks the bond between the finish and the metal.

It may be necessary to reapply the remover; if so, scrape the old paint away with a plastic or aluminum scraper, and apply the second coat of remover. This allows the active chemicals to get to the lower layers of the finish. After all of the finish has swelled up and broken away from the surface, it should be rinsed off with hot water or with live steam. A stiff bristle brush will most probably have to be used around rivet heads and along the seams to get all of the stubborn paint that adheres to these places.

Caution must be observed whenever you use any paint remover. Many solvents used in paint removers attack rubber and synthetic rubber products. Tires, hoses, and seals must be protected from contact with paint removers. These solvents are also highly toxic, and special care must be taken that none of them get onto your skin or in your eyes. If paint remover is spilled or splashed on your skin, flush the area with water *immediately*. If any gets into your eyes, flood them with water, and get to a doctor as soon as possible.

2. Treatment of aluminum alloys

a. Mechanical corrosion removal

After all of the paint has been removed, all traces of the corrosion products, the gray or white powder, must be removed from the surface.

Very mild corrosion may be removed by using a household abrasive cleaner such as Bon-Ami, but be sure that the abrasive does not contain chlorine. Nylon scrubbers such as "Scotch-Brite" pads can also be used to remove this mild corrosion.

More severe but not heavy corrosion can be removed by brushing with aluminum wool or with an aluminum wire brush. **Caution:** Do not use a steel wire brush or steel wool, as traces of the steel will embed in the aluminum and cause severe corrosion. Blasting the surface with glass beads smaller than 500 mesh can be used to remove corrosion from the pits. After using abrasives or brushing, examine the metal with a five- to ten-power magnifying glass to make sure that all traces of the corrosion have been removed.

Severely corroded aluminum alloys must be given more drastic treatment to remove all of the corrosion — rotary files or power grinders using rubber wheels impregnated with aluminum oxide may be used to grind out every trace of corrosion damage. Watch carefully to be sure that the minimum amount of material is removed but that all of the damage has been cleaned out. After an examination with a five- or ten-power magnifying glass shows no trace of corrosion remaining, remove about two thousandths of an inch more material to be sure that the ends of all of the intergranular cracks have been reached. Finish by sanding the area smooth with 280-grit, then 400-grit, abrasive paper. Clean the area with solvent or an emulsion cleaner, and neutralize the surface with an inhibitor.

b. Chemical neutralization

After removing all of the corrosion products possible, treat the surface with a 5% chromic acid solution to neutralize any remaining corrosion salts. After the acid has been on the surface for at least five minutes, it should be washed off with water and allowed to dry. Alodine treatment conforming to MIL-C-5541 will neutralize the corrosion, as well as forming a protective film on the surface of the metal. Its application is discussed under the heading of Surface Oxide Film Chemical Treatment.

c. Protective coating

(1) Cladding

Pure aluminum is considered to be non-corrosive. This is not altogether true, however, because aluminum readily acts with the oxygen in the air to form an oxide film. This film is so dense that it does not allow air to reach the surface of the metal, and there will be no more action once the film has formed. Pure aluminum, though, is not strong enough for aircraft structure and it must be alloyed with other metals to get the desired strength.

Any time we have an alloy, there is the possibility of getting dissimilar metal corrosion, and the surface of the metal must be protected. The aluminum alloy skin of an airplane may be protected from corrosion and at the same time made attractive in appearance by a coating of pure aluminum. This is called "cladding." In the manufacture of clad aluminum, pure aluminum is rolled onto the surface of the alloy. It actually penetrates the alloy and becomes a part of it, accounting for 1.5% to 5% of the total sheet thickness. This cladding is anodic to the core material, and any corrosion that takes place will attack the cladding rather than the core. The oxide film that forms on the cladding is extremely thin, very tight and transparent, and will prevent further corrosion damage.

(2) Surface oxide film

The characteristic of aluminum to form an oxide film on its surface, a film that resists further oxidation, is of real value in protecting the aluminum from corrosion. Metallurgists have found ways of forming films on metal surfaces that are hard, decorative, waterproof, and airtight. There are two ways this film may be formed. It is most generally done in the factories by an electrolytic process known as anodizing. In the field, the film may be formed by a chemical process known as Alodining.

(a) Electrolytic treatment

In the anodizing process, the part to be treated is cleaned in a hot water bath with a special non-caustic soap solution and then becomes the anode in an electrolytic process. Chromic acid and water are used as the electrolyte. After the oxide film has formed, the part is washed in hot water and air-dried. Aluminum treated by this process has not been appreciably affected with regard to its tensile strength, its weight, or its dimensions. The anodic film on aluminum alloy is normally a light gray color, varying to a darker gray for some of the alloys. Some aluminum alloy parts, such as fluid line fittings, are dyed for identification, with the dye applied in the anodizing process where it colors the oxide.

When the oxide film first forms, it is an aluminum hydroxide which is porous and quite soft. In this condition it readily absorbs dyes, but as it converts into aluminum oxide, it becomes hard and non-absorbent and forms the protective film needed to shield the metal from moisture and air to prevent corrosion.

The anodic film on aluminum alloys is an electrical insulator and must be removed before any electrical connection can be made. Bonding straps are often connected directly to an aluminum alloy part, and for this attachment, the anodized film must be removed by sanding or scraping.

(b) Chemical treatment

When small parts are fabricated in the field or when the protective anodizing film has been damaged or removed, the part may have a protective film applied by a chemical rather than an electrolytic method. This process uses a chemical that meets specification MIL-C-5541 and is available under several proprietary names, such as Alodine 1201, or, if an invisible film is needed, Alodine 1001. These materials should be applied to the surface after all traces of the corrosion have been mechanically removed. The surface should be cleaned with a metal cleaner until it will support an unbroken water film. Any break in the film of rinse water indicates the presence of wax, grease, or oil on the surface, and further cleaning must be done. While the surface is still wet with rinse water, brush or spray on a liberal coating of the chemical, allow it to stand for from two to five minutes, and then rinse it off. Keep the surface wet while the chemical is working, because if it is allowed to dry, streaks will appear, and the film will not be adequately protective. Only an area of such size that can be kept wet should be worked at any one time. After the chemical has had its full working time, it should be flushed from the surface with a spray of fresh water. If a swab or a brush is used, care must be taken not to damage the film, which is quite soft while it is wet. Allow the surface to dry, and after it is dry, it is ready to paint. A satisfactory job produces a uniform yellowish-brown iridescent film or an invisible film, depending on the chemical used.

If a powder appears on the surface after the material is dried, it is an indication of poor rinsing or failure to keep the surface wet during the time the chemical was working. If the powder shows up, the part must be re-treated.

WARNING: Rags or sponges used in the application of conversion coating chemicals must be kept wet or thoroughly washed out before they are discarded. Rags drying with the chemical in them can constitute a fire hazard.

(3) Organic film

One of the most universally used corrosion control devices for metal surfaces is a good coat of paint. On porous surfaces, paint adherence is no large problem; but on some metal surfaces, it may be difficult to get the paint to stick. The surface of aluminum must be prepared in order for the paint to have a rough surface on which it can adhere. The surface may be roughened with a mild chromic acid etch, or by the formation of the oxide film by either the electrolytic or the chemical method. The surface may also be mechanically roughened by carefully sanding it with 400-grit sandpaper. When sandpaper is used, it is absolutely imperative that every bit of the sanding dust be removed with a rag damp with thinner before the primer is applied.

For lacquer or enamel, the primer which has been used for years is zinc chromate. This is an inhibiting primer, meaning that the film is slightly porous and water can enter it causing chromate ions to be released and held on the surface of the metal. This ionized surface prevents electrolytic action and inhibits the formation of corrosion. Zinc chromate conforms to specification MIL-P-8585A and may have either a yellow-green or a dark green color. It may be thinned with toluol or some of the proprietary reducers made especially for zinc chromate. The surface to be painted is cleaned of all fingerprints and any traces of oil; then a thin, wet coat of zinc chromate is applied with a spray gun.

The synthetic resin base of the zinc chromate primer provides a good bond between the finish and the metal. It also has the property of being dope-proof; that is, aircraft dope will not cause it to lift. When a repair is made or a part is fabricated, it can be given a good coat of zinc chromate and corrosion will be checked before it ever gets a chance to start.

A wash primer is used in aircraft factories for priming new aircraft before they are painted. This two-part primer consists of a resin and an alcohol-phosphoric acid catalyst. The material is mixed, allowed to stand for a short time to begin its cure, and sprayed onto the surface with a very light tack coat and then a full-bodied wet coat. It cures quickly enough that the topcoats may be sprayed on within less than an hour after the surface is primed.

The new finishes that are available for aircraft use give us beauty and protection far greater than the older, more familiar materials, but the critical nature of their mixing and application makes it imperative that the manufacturer's instructions be followed in detail.

Epoxy primers are one of the most popular primers for use under polyurethane finishes and under any finish where maximum corrosion protection is needed. This is a two-component material that produces a tough, dope-proof sandwich coat between the finish and the surface. Epoxy primers can be used on aluminum, magnesium, or steel, and for maximum corrosion protection, they can be applied over a wash primer.

3. Treatment of ferrous metals

a. Mechanical removal of corrosion

Unlike that which forms on aluminum, the oxide film that forms on ferrous metals is porous and will attract moisture and continue to convert the metal into corrosion, or rust, and so all traces of this rust must be removed to stop it. The most effective method of removing rust is by mechanical means. Abrasive paper and wire brushes, both hand- and power-driven, may be used, but the most effective way of getting all of the corrosion products from unplated steel parts is by abrasive blasting, which gets to the bottom of the corrosion pits. Sand, aluminum oxide, or glass beads may be used effectively. If a part has been plated, either with cadmium or with chromium, care must be taken to protect the plating, as it is usually impossible to restore it in the field.

Highly stressed steel parts such as those found in landing gears and engines must be cleaned with extreme care. If corrosion is found on these parts, it should be removed immediately by removing the absolute minimum amount of material. A fine stone, fine abrasive paper, or even pumice may be used. Wire brushes should not be used, as they cause minute scratches which produce stress concentrations and potentially weaken the part. If abrasive blasting is used, it must be done with caution, using only very fine-grit abrasive or glass beads.

After all of the corrosion has been removed, the rough edges of the pits or damage must be faired with a fine stone or with 400-grit abrasive paper, and the surface primed as soon as possible. A dry, clean surface is an ideal setting for corrosion, and if not protected immediately will allow new damage to begin. Zinc chromate primer may be used to protect most freshly cleaned steel surfaces.

b. Surface treatment

(1) Plating

(a) Nickel or chrome plating

This electrolytically deposited coating of nickel or chromium forms an airtight coating over the surface and excludes air and moisture from the base metal. There are two types of chrome plating used in aircraft construction: decorative and hard chrome. Decorative chrome is used primarily for its appearance and surface protection, while hard chrome is used to form a wear-resistant surface on piston rods, cylinder walls, and other parts which are subject to abrasion. Parts to be plated with hard chrome are normally ground undersize and plated back to the proper dimensions. Engine cylinder walls are often plated with a porous chrome whose surface has thousands of tiny cracks which hold the oil and thus aid in lubrication.

(b) Cadmium plating

While chrome plating protects steel by excluding moisture from its surface, cadmium plating protects it in an entirely different way, by sacrificial corrosion. In the electro-chemical series shown in Fig. 2-1, we see that cadmium is below steel. This means that if these two metals are involved in an electrolytic action, the anodic material, the cadmium, will be corroded, while the cathodic material, the steel, will not be damaged.

Almost all steel aircraft hardware is cadmium-plated. This soft, silvery-gray metal is electroplated onto the steel to a minimum thickness of 0.005 inch. It provides an attractive finish as well as protection against corrosion. While the cadmium plating on a part is scratched through to the steel, galvanic action takes place and the cadmium corrodes. If the steel should corrode, action would continue until the part is ruined, but the oxides which form are on the surface of the cadmium and are similar to those which form on aluminum. They are dense, airtight, and watertight. No further action can take place once the initial film has formed. This type of protection is known as sacrificial corrosion.

(c) Galvanizing

Treating steel parts such as firewalls with a coating of zinc is called galvanizing. The protection afforded by this process is similar to that provided by cadmium plating — it is sacrificial corrosion. When scratched through, zinc will corrode and form an airtight oxide film. Steel is galvanized by passing it through vats of molten zinc and then rolling it smooth through a series of rollers.

(d) Metal spraying

Aircraft engine cylinders are sometimes protected from corrosion by spraying molten aluminum on their surface. The steel cylinder barrel is sand-blasted absolutely clean, then aluminum wire is fed into an acetylene flame, and high-pressure compressed air blows the molten aluminum onto the surface. Corrosion protection afforded by this treatment is sacrificial corrosion, such as is provided by cadmium and zinc coating.

(2) Organic coating

Since paint is an effective and widely used corrosion protection system, its film must not be allowed to break down, and so the surface to be painted must be properly prepared. Dry abrasive blasting will remove all of the surface oxides and roughen the surface enough to provide a good bond for the paint. Parts which have been cadmium-plated must normally have their surface etched with a five-percent solution of chromic acid before the primer will adhere. After a clean, dry surface has been prepared, a thin, wet coat of zinc chromate primer should be sprayed on and allowed to dry. The final finish can usually be applied after about an hour.

4. Treatment of magnesium alloys

a. Mechanical removal of corrosion

Magnesium is one of the most active metals commonly used in aircraft construction, but because of its excellent weight-to-strength ratio, designers will accept its corrosiveness. Magnesium alloys do not naturally form a protective film on their surfaces the way aluminum does, so special care must be taken that the chemically or electrolytically deposited film is not destroyed.

Magnesium corrosion occupies more space than the metal so it will raise the paint, or if it forms between skin lap joints, it will swell the joints. When corrosion is found on a magnesium structure, all traces must be removed and the surface treated to inhibit further corrosion. Since magnesium is anodic to almost all of the commonly used aircraft structural metals, corrosion should not be removed with metallic tools. They will leave contaminants embedded in the metal and will cause further damage. Stiff non-metallic bristle brushes or nylon scrubbers may be used to remove corrosion from the surface or from the pits. Deep pits must be cut out with sharp steel-or carbide-tipped cutting tools or scrapers. Carborundum wheels or paper must not be used because of their tendency to contaminate the surface and cause galvanic corrosion.

If abrasive blasting is done, use only glass beads which have been used for nothing but magnesium. Scale or residue left in the abrasive from other metals can embed in the magnesium and cause further damage.

b. Surface treatment

After all of the corrosion possible has been removed from the surface, a chromic acid pickling solution which conforms to MIL-M-3171A Type 1 (Dow No. 1) should be applied. A satisfactory substitute for this solution may be made by adding about fifty drops of battery acid (sulfuric acid) to a gallon of ten percent chromic acid solution. Apply this to the surface with rags and allow it to stand for about ten or fifteen minutes; then rinse it thoroughly with hot water.

A treatment which forms a more protective film is a dichromate conversion treatment, Dow No. 7, which conforms to MIL-M-3171A Type IV. This solution is applied to the metal and allowed to stand until a golden brown oxide film forms uniformly on the surface. Rinse the surface with cold water and dry it with an air blast. The oxide film is extremely soft when it is wet, and it must be protected from excessive wiping or touching until it dries and hardens. This film is continuous and protects the magnesium surface from corrosion by excluding all electrolyte from its surface.

Like aluminum alloys, magnesium can also have a film deposited on its surface by electrolytic methods. Anodizing magnesium by the Dow No. 17 process produces a hard, surface oxide film which serves as a good base for further protection that can be provided by a good coat of paint.

Magnesium is such an active metal and magnesium skins are usually so thin that it is absolutely essential that only the proper solutions and proper procedures be used for corrosion treatment. Because of this, rather than mixing your own pickling and conversion coating solutions, you should use prepared chemicals meeting the appropriate MIL specifications, and you must follow the manufacturer's recommendations in detail.

G. Corrosion Prevention

The best way of actually preventing the formation of corrosion is by eliminating one or more of its basic requirements. We can prevent the electrode potential difference within the metal, or if a potential difference does exist, we can insulate the pathway between the areas and prevent the accumulation of any electrolyte that would form a conductive path on the surface of the metal.

1. Cleanliness

A clean, polished surface does not provide a good place for corrosion to start. But dirt, grease, or a partially broken paint film will hold electrolyte in contact with the surface of the metal and will promote corrosion.

Airplanes should be kept clean and waxed. A good emulsion-type cleaner will remove the dirt film and industrial contaminants which attract moisture to the surface. It is especially important that the inside of the airplane be kept clean, with particular attention paid to the areas where moisture is most likely to collect. All drain openings must be kept open, and deposits which have formed from the engine exhausts must be removed before they cake on the surface.

2. Corrosion-inhibiting film

Ferrous metals are susceptible to corrosion damage because they form porous oxides which allow moisture and oxygen to enter and remain in contact with the base metal so corrosion can continue. To prevent this, a coating of metal which forms a continuous non-porous oxide film is plated or sprayed on the surface of the steel. Cadmium or zinc are both less noble (more corrosive) than steel and will be the one to corrode when the two are covered by an electrolyte. The oxides of both of these metals are non-porous and will prevent oxygen reaching the steel, thus protecting it.

Aluminum alloys corrode easily because of the electrode potential difference between the alloying agents and the aluminum. But pure aluminum combines with oxygen in the air to form a non-porous oxide film which excludes oxygen and moisture from the base metal and prevents further action. In order to have the strength of an alloy with the corrosion resistance of pure aluminum, a thin coating of pure aluminum may be rolled onto the surface of the alloy. If the slight loss of strength caused by cladding is not desirable, unclad aluminum alloy may be protected by an electrolytically deposited or a chemically formed non-porous oxide film. This film, in addition to protecting the metal, forms a good base for paint.

Magnesium alloys are protected in the same way as aluminum by the use of a non-porous film produced either electrolytically or chemically. Corrosion forms where the film has been disrupted at such places as rivet holes, skin edges, or scratches in the surface. And a good protective coating of paint is vital for the exclusion of oxygen or moisture from the surface.

3. Dissimilar metal insulation

It is often required that different metals be held in contact with each other. When this is the case, dissimilar metal, or galvanic, corrosion can take place. In order to minimize this danger, some form of insulation must be provided.

a. Fasteners

When steel fasteners are used in an aluminum structure and are subjected to a corrosive environment, the holes should be drilled and countersunk, treated with a conversion coating material such as Alodine, and then primed with zinc chromate. The bolts or screws should be coated with the primer and installed while they are wet. This treatment does not insure a completely insulated joint, but it does exclude moisture.

b. Skin lap joints

In any joint between dissimilar metal skins, special care must be taken to see that there is no good conductive path between the metals. After the skins have been cut to size, fitted, and drilled, coat each faying surface with a least one coat of zinc chromate primer. Allow the primer to dry completely between each coat, and attach the skin with rivets wet with primer. If one of the skins is made of magnesium, at least two coats of zinc chromate primer should be applied to each surface and a strip of three-mil (0.003-inch thick) pressure-sensitive vinyl tape applied between the faying surfaces. If, for reasons of high temperature or relative motion between the skins, this tape is impractical, a couple of additional coats of zinc chromate primer should be used. Dip the rivets in zinc chromate and drive them while they are wet.

c. Sealers and sealants

The higher speeds reached with modern aircraft have brought out requirements for aerodynamic sealers used to fair the edges of skins on the wing surfaces. Airplanes flying at high altitudes require that the cabin be pressurized, so every place where plumbing or electrical wiring passes through a bulkhead, these openings must be sealed with some form of plastic sealer. Integral fuel tanks are actually a part of the wing structure, and every rivet joint must be sealed to prevent the fuel leaking.

Electrical connectors are potted, or sealed with a special sealer, to prevent water entering the plug and causing the terminals to corrode, which would cause intermittant electrical contact.

All of these sealers have specific requirements and are used for unique operating conditions. When replacing any sealer, be very sure to use exactly the same material as was originally used, or a substitute *specifically* approved by the airframe manufacturer. These materials are usually of the two-component type, having one container of resin and one of accelerator or catalyst. The mixing sequence, proportions, time, and temperature must be rigidly adhered to. Otherwise, there is likely to be a poor bond between the sealer and the metal.

When a repair is made, every trace of the old sealer must be removed and the surface prepared for the new sealer. Clean the surface with the recommended cleaner and remove every trace of fuel, oil, and fingerprints. Apply the new sealer and feather the edges down exactly as the instructions specify. This type of material is critical and has no tolerance for careless work or non-recommended methods of application.

Fig. 2-27 Fuel tanks are sealed with a special resilient sealer in all of the seams.

SECTION III

Nondestructive Inspection

As aircraft have grown in complexity and in importance to our transportation system, the cost of down time has increased to the point that new methods of inspection have had to be developed to allow the aircraft and engines to be inspected without disassembly. This branch of inspection technique is known as nondestructive inspection and incorporates methods as unsophistocated as mere visual inspection to those as complex as radiographic inspection.

In this text, we will discuss the various systems in the order of their complexity; visual, liquid penetrant, magnetic particle, eddy current, ultrasonic, and finally radiographic inspections.

A. Visual Inspections

The most fundamental method of inspection and one that will never be replaced but only augmented, is the visual inspection. It is limited, of course, in that nothing can be inspected by this method unless it is uncovered and made visible.

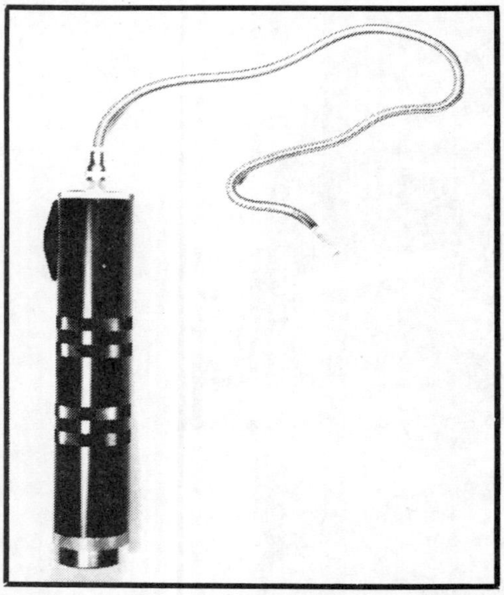

Fig. 3-1 A special flashlight with an extension for the bulb makes it possible to inspect in hard-to-see locations.

The basic tools required for visual inspections are a good light and usually some form of magnifying glass. Flashlights are normally used to give spot-type illumination to the area being inspected, and to make flashlights more useful for inspecting in difficult locations, special inspection lights are available with long flexible extensions that place the bulb inside areas that would otherwise be dark.

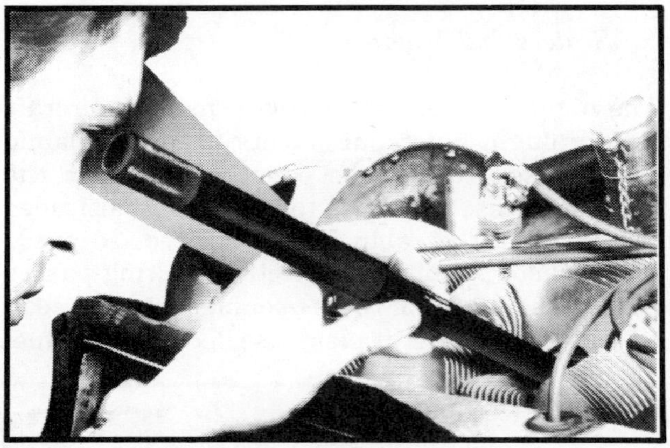

Fig. 3-2 A borescope is a special optical tool used to see inside of engine cylinders.

Borescopes are used to inspect the insides of engine cylinders through a spark plug hole. A borescope is actually an optical device similar in principle to a telescope but with a built-in light source that illuminates the area being inspected.

One of the more useful types of borescopes for inspecting extremely difficult locations is the fiber optic borescope. This uses a bundle of optical glass fibers to transmit light from a special cold light source to the end of the scope which may be worked well up inside the engine or the structure being inspected. A special attachment for this instrument allows a single-lens reflex camera to be used to photographically record what is seen through the scope. Fiber optic borescopes are available in lengths up to almost four feet.

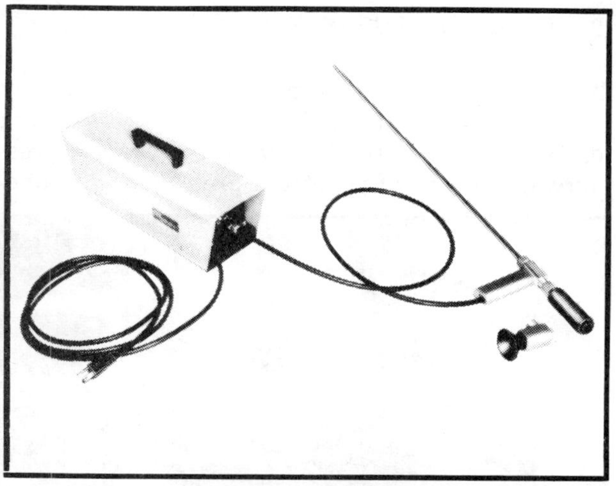

Fig. 3-3 A fiber optic borescope uses the principle of the light transmission through flexible fibers of glass to see inside inaccessible areas.

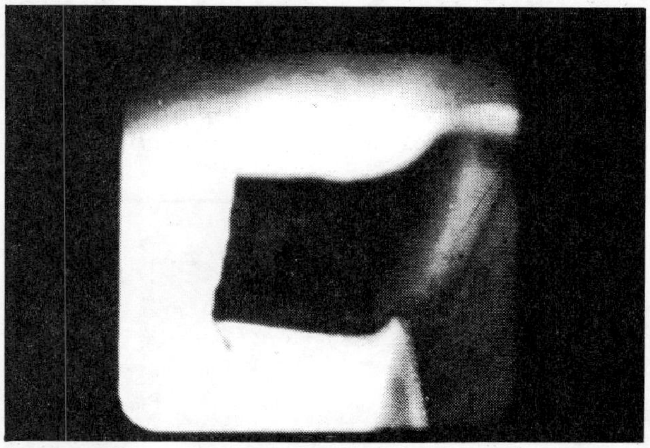

Fig. 3-4 A borescope may be fitted with a special single-lens reflex camera to take pictures inside the structure. This photograph of a cracked first stage high-pressure turbine blade was taken inside the engine.

B. *Liquid Penetrant Inspection*

Liquid penetrant inspection is a method of nondestructive inspection that is suitable for locating cracks, porosity, or any other type of discontinuity that extends to the surface. It is usable on both ferrous and non-ferrous metals as well as non-porous plastic material.

This inspection method is based on the principle of capillary attraction. The area to be inspected is covered with a penetrating liquid; that is, one that has a very low viscosity and low surface tension. After the penetrant has been allowed to remain on the surface long enough to completely penetrate any fault that extends to the surface all of the excess is washed off, and the surface is covered with a developer. The developer is a material that will, by the process of reverse capillary action, blot the penetrant out of cracks or other faults where it will form a visible line in the developer.

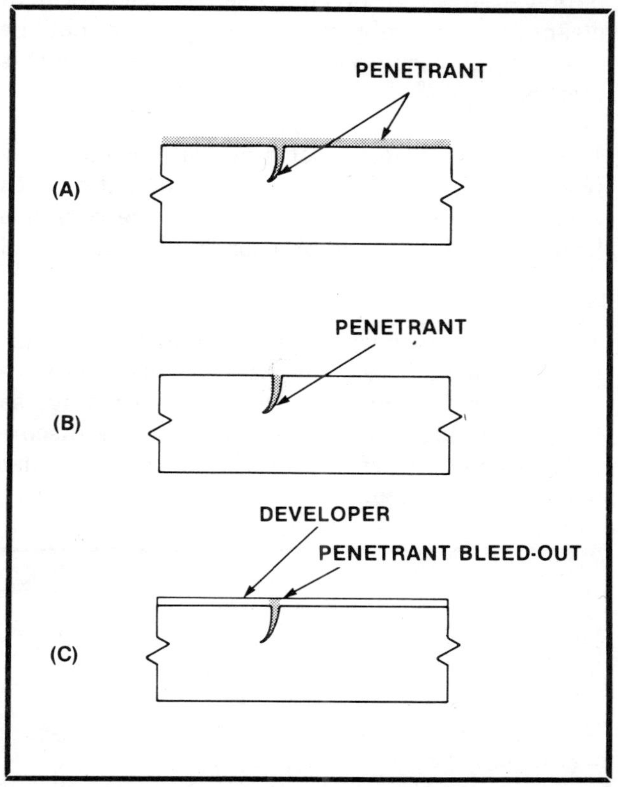

Fig. 3-5 A — A penetrating liquid is spread over the surface of the material being examined. B — After the penetrant is allowed to soak into any cracks open to the surface, it is washed off of the surface. C — An absorbent developer is sprayed over the surface where it blots out any penetrant from the crack. The crack then shows up as a bright line on the white surface of the developer.

There are two types of indications used by the liquid penetrant inspection method: the fluorescent and the colored dye. An ultraviolet, or black, light is used with the fluorescent penetrant, and any flaw will show up as a green line. With the colored dye method, faults show up as red lines on the white surface of the developer.

1. *Preparation*

One of the most important considerations when using liquid penetrant inspection is the cleanliness of the surface being inspected. Only when the surface is perfectly clean can the penetrant be assured of getting into any cracks or faults. The best method of cleaning the surface is with a vapor degreaser which is particularly effective in removing all traces of oil and grease. Special care must be taken, however, as some metals may be damaged by the solvents used in this form of cleaning, and only materials that cannot be damaged by the vapors should be cleaned by this method.

If vapor degreasing is not practical, the part may be cleaned by scrubbing it with a solvent or with a strong detergent solution. After the part has been cleaned, it must be thoroughly rinsed and dried.

Parts to be inspected by the liquid penetrant method should not be cleaned by abrasive blasting or by scraping or heavy brushing, as these methods tend to close any of the discontinuities on the surface and will hide a defect that should be detected.

2. Penetrant application

The penetrant may be applied to the surface by immersing the part in the liquid or by swabbing or brushing it onto the surface. However, pressure spray cans of penetrant are the most conve-

Fig. 3-6 Spray cans of cleaner, penetrant, and developer make dye-penetrant inspection an effective and handy method of field inspection.

Fig. 3-7 Large parts are inspected by the liquid penetrant inspection method in shops where vats of penetrating liquid and emulsification agent are used. A dry developer is used in this shop.

nient method of applying the penetrant to small areas for localized inspection. Whichever system is used, the area to be inspected must be completely covered with the penetrating liquid and the penetrant allowed to remain on the surface for the recommended length of time. This time is called the dwell time of the penetrant and is determined by the size and shape of the discontinuities we are looking for. Thin cracks require a longer dwell time than large and more open cracks. The dwell time may be decreased if the part is heated, but care must be taken that it does not get hot enough to evaporate the penetrate from the faults.

3. Removal of surface penetrant

There are three types of penetrating liquid that may be used, and the type used determines the way the penetrant on the surface is to be removed.

a. Water-soluble penetrants

The easiest penetrant to remove is one which is water soluble. If this type has been used, the penetrant on the surface of the part is flushed away with water at a pressure of between 30 and 40 psi with an adjustable spray nozzle held about 45 degrees to the surface. When removing this type of penetrant, be sure that the penetrant is not washed out of the faults.

b. Post-emulsifying penetrants

These penetrants are not water soluble, but may be treated with an emulsifying agent that will allow the penetrant to be washed from the surface by rinsing it with water. The main advantage of this type of penetrant over the straight water-soluble type is the fact that the depth of emulsification can be controlled by varying the emulsifier dwell time. By using the proper dwell time, the penetrant on the surface can be emulsified and rinsed off, while that in the faults will be untouched by the rinse water.

c. Solvent-removable penetrants

Some penetrants are neither water soluble nor emulsifiable, and these must be removed from the surface of the part with a solvent specifically recommended for this material. The excess penetrant should be wiped from the surface with an absorbent towel and then the surface cleaned with clean towels damp with the solvent. The surface should never be sprayed with nor immersed in the solvent, as it will wash the penetrant out of the fault or dilute it enough that it will not make a proper indication in the developer.

4. Application of the developer

Just as there were three types of penetrants, there are three kinds of developers that may be used to draw penetrants from the faults. All three types do the same job, but their applications differ.

a. Dry developer

The penetrant will begin to bleed out of any fault as soon as the surface penetrant has been removed, and to pinpoint the location of the fault, we must cover the surface to be inspected with a developer as soon as possible. Dry developer is a loose talcum powder-like material that adheres to the penetrating liquid and acts as a blotter to draw the penetrant out of any faults that reach the surface.

The part to be inspected must be perfectly dry. It may be immersed in a bin of the loose powder, or the dry powder may be brushed over the surface with a soft brush or blown over the surface with a powder gun. After the powder remains on the surface for the recommended time, the excess is removed with a low-pressure flow of air.

Fig. 3-8 After this part was removed from the dry developer and examined under black light, the fault that showed up was marked with a felt-tip marking pen so it would be visible during the repair procedure.

The penetrant used with a dry developer may have been treated with a fluorescent dye or with a colored dye. Parts on which fluorescent penetrant has been used are examined under black light, in which faults will appear as a green indication as the light causes the dye to fluoresce, or glow. Colored dye penetrants are usually red, and any faults will appear as red marks, clearly visible on the surface.

b. Wet developer

A wet developer is similar in its action to the dry developer. It is a white powder mixed with water, and it may be either flowed over the surface or have the part immersed in it. Wet developer may be applied as soon as the surface penetrant has been rinsed off and the part is then air-dried and inspected in the same way as a part on which dry developer has been used. Wet developers may be used with penetrants treated with either fluorescent or colored dyes.

c. Non-aqueous developer

For field maintenance, the most generally used developer is the non-aqueous type, which is a white chalk-like powder suspended in a solvent vehicle. This developer is normally applied from a

Fig. 3-9 Colored dye is used in this penetrating liquid, so the part may be examined in white light.

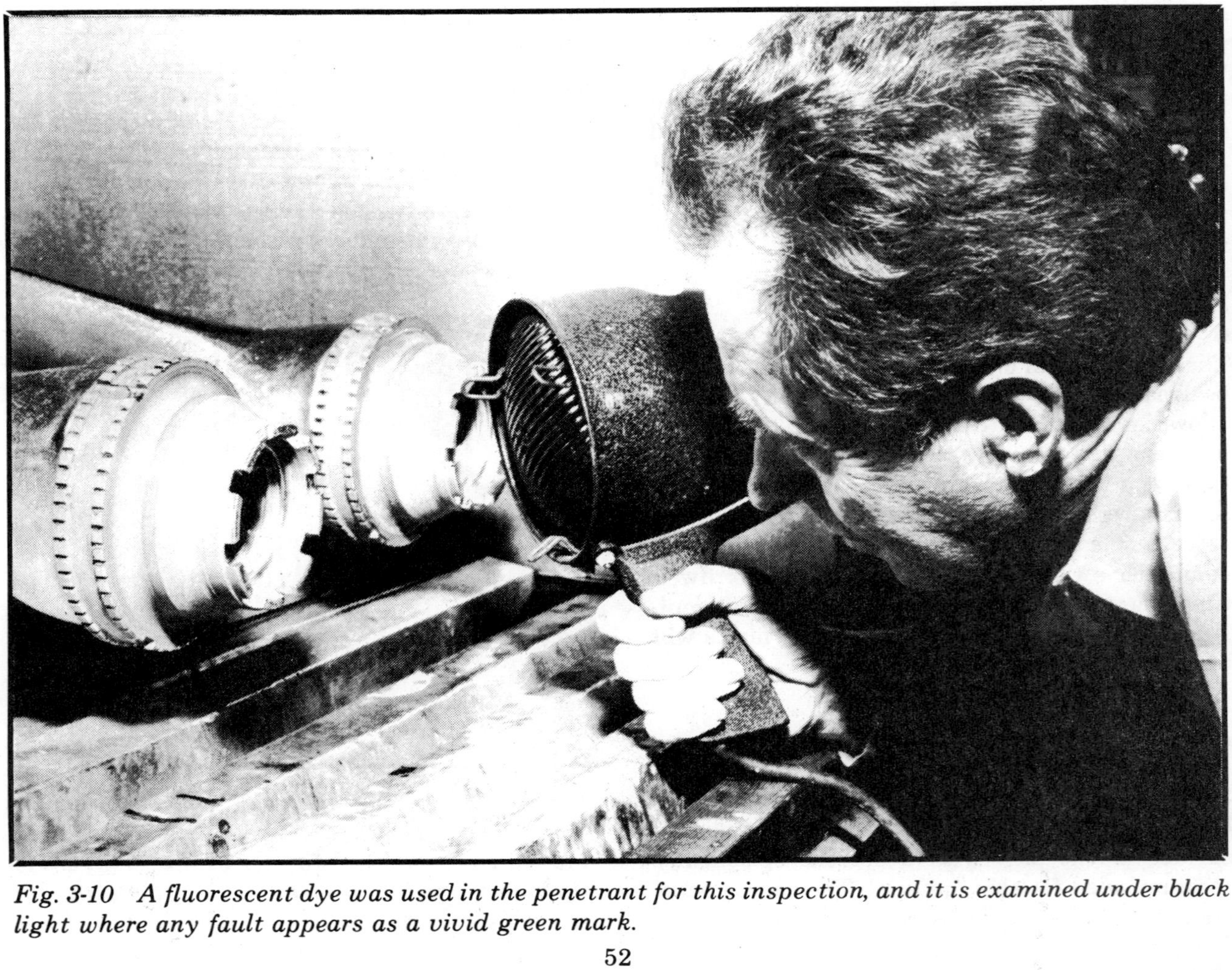

Fig. 3-10 A fluorescent dye was used in the penetrant for this inspection, and it is examined under black light where any fault appears as a vivid green mark.

pressure spray can or it may be sprayed onto the surface with a paint gun.

The part to be inspected must be thoroughly dry and the developer sprayed on in a thin most coat, which dries rapidly and pulls out any penetrant that has seeped into a fault. The penetrant stains the developer and is easily seen with the black light if a fluorescent penetrant is used, or with a white light if the dye was colored.

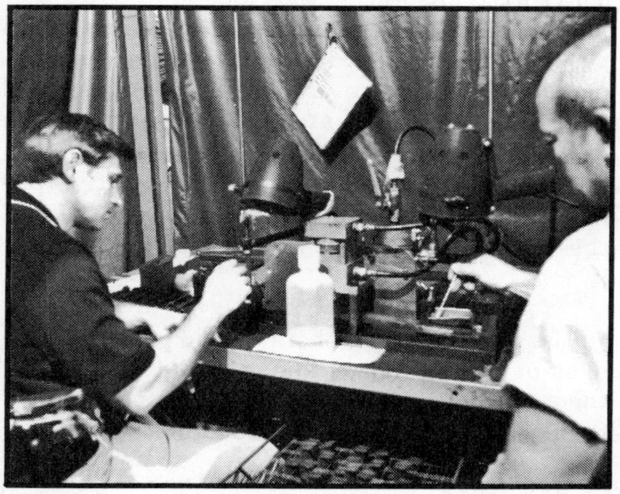

Fig. 3-11 Wink Zyglo is a method of inspecting for cracks that open up only under stress. The part is flexed, and any faults will appear as intermittent flashes of the fluorescent dye.

C. Magnetic Particle Inspection

The nondestructive inspection method that is most generally used for the inspection of ferrous metal parts is magnetic particle inspection. In this method of inspection, the part is magnetized and a material which contains magnetic particles is poured or sprayed over the surface. If there are any discontinuities in the material, either on the surface or subsurface, north and south magnetic poles will form across the fault and the magnetic particles will collect between the poles.

1. Principle of magnetism

a. Permanent magnets

Iron, or material containing a large amount of iron, can be magnetized. Within the material, there are many weak magnetic fields, and when the material is not magnetized these fields are oriented in such a random fashion that they cancel each other, and as a result there is no overall magnetic field of the material. Now, if the material is subjected to a strong external magnetic field, the domains, as these internal magnetic fields are called, will align themselves with the external field and the part will become magnetized.

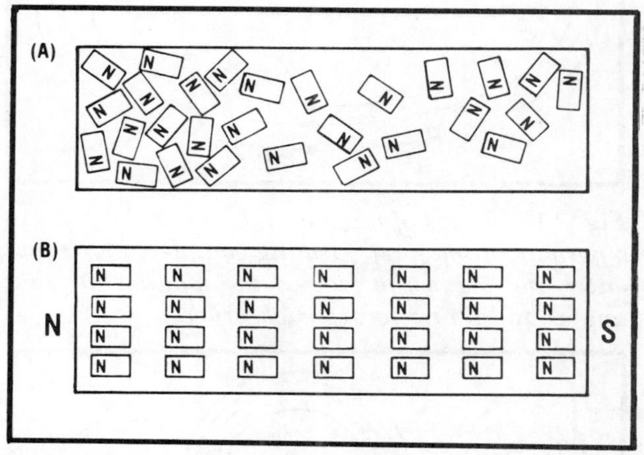

Fig. 3-12 A — In an unmagnetized piece of steel, the magnetic domains are oriented in such a random fashion that they cancel each other and there is no resultant magnetic field. B — When the material is magnetized, the domains are all aligned and the material has north and south magnetic poles.

If the domains remain in alignment when the magnetizing force is removed, the part becomes a permanent magnet. Lines of magnetic flux are invisible lines of force that are considered to leave the north pole of the magnet at right angles to its surface and travel to the south pole. These lines repel each other so that they spread out as they leave the north pole and become most widely separated midway between the poles; then they come together again as they enter the magnet at its south pole, again at right angles to its surface. The lines then travel through the magnet, parallel to each other. All of the lines of force form complete loops and never cross one another.

If the bar magnet we have in Fig. 3-14A is bent into a circle with a small space between its ends, there will be lines of flux within the magnet and its ends will form a north and a south pole. However, if instead of there being a space between its ends, the magnet were a complete unbroken circle, there would be lines of flux within the material, but there would be no poles.

Now, if there is a surface flaw or a break in the ring, magnetic poles will appear, one on either side of the break, and there will be a leakage field across the break that will attract magnetic material. This is the principle that is used in magnetic particle inspection.

b. Electromagnets

Any time a conductor carries a flow of electrons (current), there will be the same kind of lines of magnetic flux surrounding the wire that we found passing from one pole to the other in a permanent magnet. If we grasp a conductor with our left hand in such a way that the thumb points in the direction of electron flow (from the negative terminal of the power source to the positive terminal), the fingers will encircle the conductor in the same direction that the lines of flux encircle it.

In the same way that there was no polarity when lines of flux flowed within the continuous ring, there will be no polarity caused by the lines of flux encircling a straight conductor.

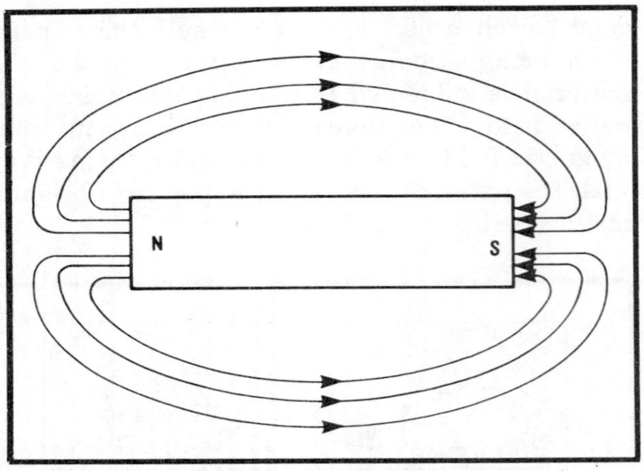

Fig. 3-13 Lines of magnetic flux link the poles of a permanent magnet, forming complete loops that enter and leave the face of the magnet at right angles and do not cross each other.

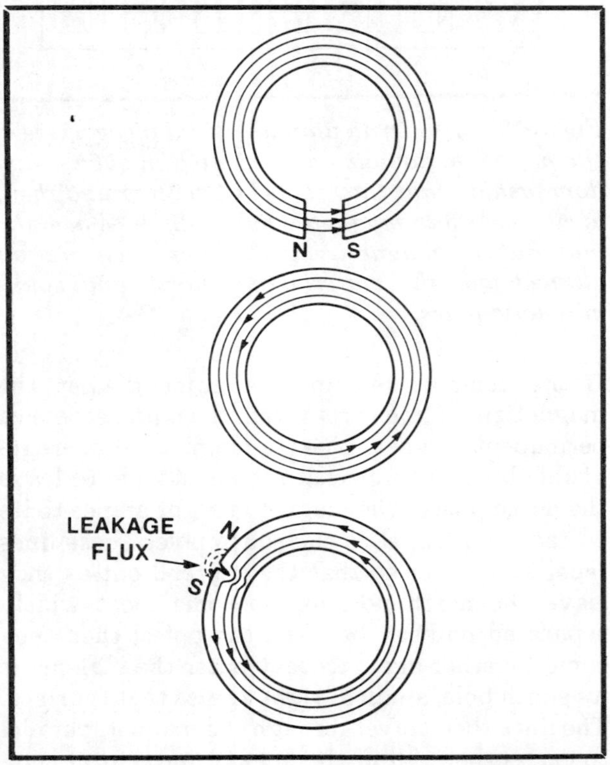

Fig. 3-14 A — A bar magnet bent into the form of a circle with a space between its ends will have a concentration of magnetic force between its ends. It will have north and south magnetic poles. B — If the magnet were joined into a continuous circle, there would still be lines of flux, but no poles. C — If there is a break, or fault, either on the surface or subsurface, the flux will be interrupted and there will be leakage flux on the outside of the ring that will attract the magnetic oxide.

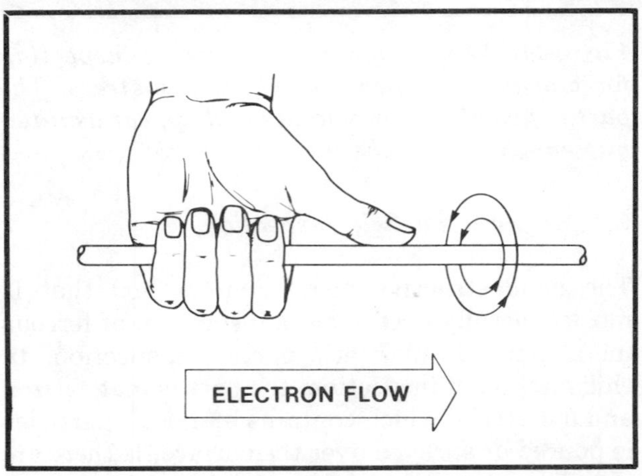

Fig. 3-15 Lines of magnetic flux encircle a conductor carrying electrical current. The direction of lines of flux is determined by the left-hand rule.

But, if we form the conductor into a coil, the lines of flux will be concentrated within the coil and will spread out outside of the coil. They will behave in the same way as the lines of flux around and through the permanent magnet. The strength of this magnetic field is determined by the amount of current flow in amperes and the number of turns in the coil. It is measured in ampere-turns.

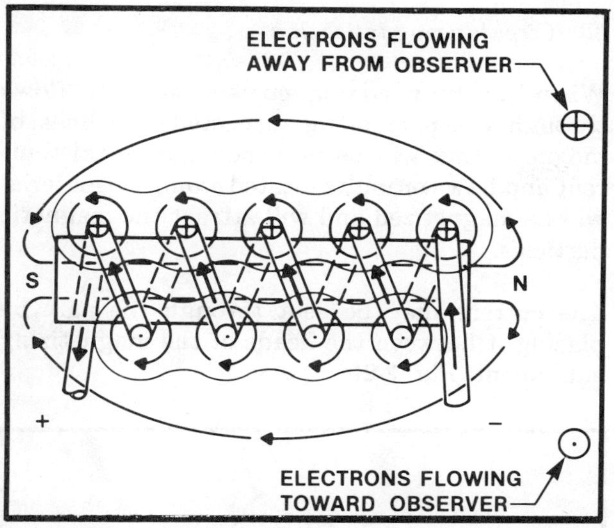

Fig. 3-16 The lines of magnetic flux encircling a conductor may be increased by winding the conductor into the form of a coil. The coil will act as a magnet with north and south poles.

c. Magnetic terms

(1) Magnetic flux

These are invisible lines of magnetic force that exist between the poles of a magnet and they are assumed to flow externally from the north pole of the magnet to its south pole. One line of magnetic flux is called one maxwell and is symbolized by the Greek letter phi (Φ).

(2) Flux density

Flux density is the number of lines of magnetic flux per unit area, and its unit is the gauss. One gauss is equal to one maxwell per square centimeter.

(3) Residual magnetism

This is the magnetism that remains in a material after the magnetizing force is removed.

(4) Permeability

The ease with which a material will conduct lines of magnetic flux is called its permeability.

(5) Retentivity

Retentivity is the ability of a material to retain some residual magnetism after the magnetizing force is removed.

(6) Coersive force

This is the amount of reverse magnetizing force required to remove the residual magnetism from a material.

(7) Hysteresis loop, or B-H curve

If we plot the flux density in a material against the magnetizing force, we get a curve similar to the one in Fig. 3-17. We begin with no flux and no magnetizing force at point *A*, and then as we increase the magnetizing force the flux density increases, rapidly at first. It then levels off until at point *B* the material is saturated, and there can be no more flux density regardless of the amount of magnetizing force applied. Now, when we decrease the magnetizing force, the flux will follow the curve *B-C* rather than dropping back to *A*. The material will retain a flux density of *A-C* when the magnetizing force is reduced to zero. This is the residual magnetism. In order to demagnetize the part, we must apply a coersive force of *A-D* to reduce the flux density to zero.

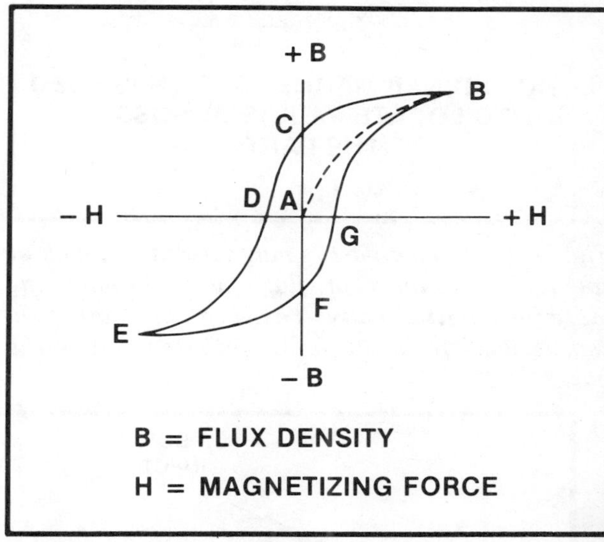

Fig. 3-17 The hysteresis loop of a magnetic material shows the relationship between the magnetizing force and the flux density.

d. Magnetic orientation

In order to detect a crack by the magnetic particle inspection method, the part must be magnetized in such a way that the lines of flux will cross the fault as nearly perpendicular as possible. In order to do this, there are two methods of magnetization we may use: longitudinal and circular.

(1) Longitudinal magnetization

In longitudinal magnetization, the magnetizing current flows either through a coil in which the part is placed, Fig. 3-18, or the current may flow through a coil around a soft iron yoke. In either method, the magnetic field is oriented along the material and magnetic fields will form on either side of faults that are located *across* the material.

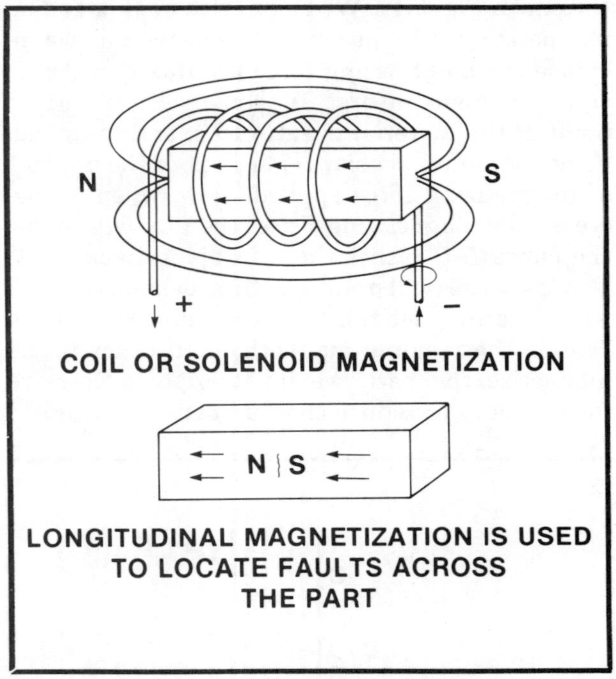

Fig. 3-18 When a part is magnetized in a coil, or solenoid, the lines of flux pass through the material longitudinally. This type of magnetization is used to locate faults oriented across the part.

(2) Circular magnetization

When the magnetizing current actually flows through the part being inspected, the lines of magnetic flux will be perpendicular to this current and flaws or faults located *along* the material will be magnetized and will attract the magnetic particles.

The current may be sent through the part by placing it between the heads of the magnetizing equipment, Fig. 3-20

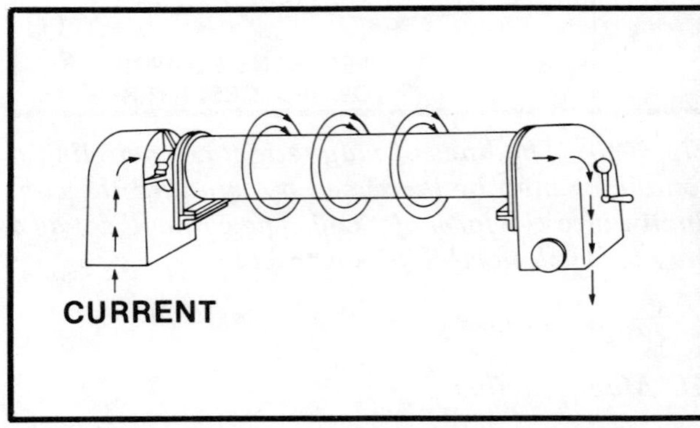

Fig. 3-20 When current passes through a part, the lines of flux encircle the part, and it is said to be circularly magnetized. Faults extending lengthwise along the part may be detected by this method.

If the part is tubular, a conductive rod may be placed between the heads of the magnetizing machine to carry the current, and the tubular object is slipped over this rod. Nuts and similar small parts may be magnetized in this way.

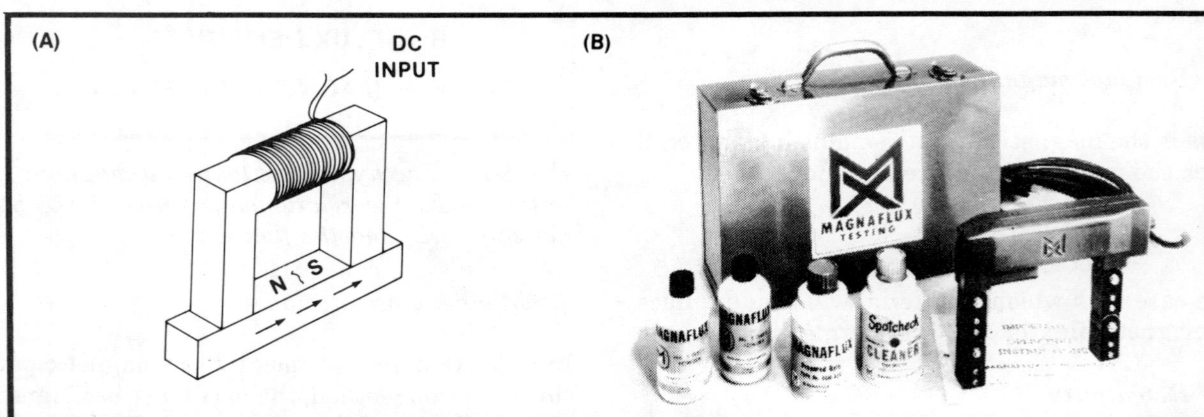

Fig. 3-19 A — When a yoke is used for magnetization, the flux passes through the part longitudinally and inspects for faults **across** *the part. B — A portable setup for magnetic particle inspection using yoke magnetization.*

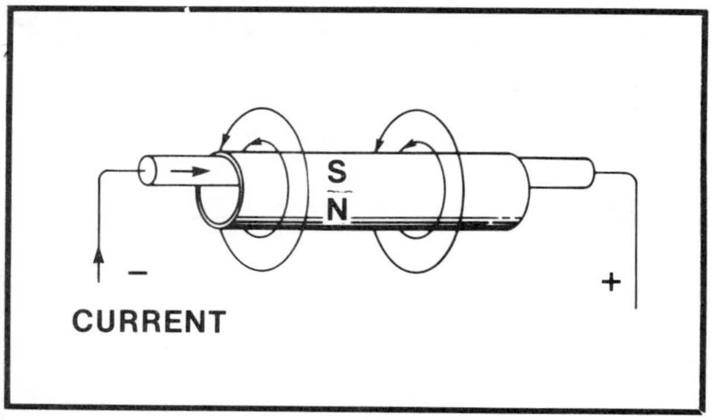

Fig. 3-21 When a circular or tubular part is placed over a current-carrying conductor, the part will be circularly magnetized.

Large flat objects may be circularly magnetized by using test prods which are held firmly against the surface and current passed through them. The magnetic field will be oriented as shown in Fig. 3-22, and will be, as we see, perpendicular to current that flows between the prods.

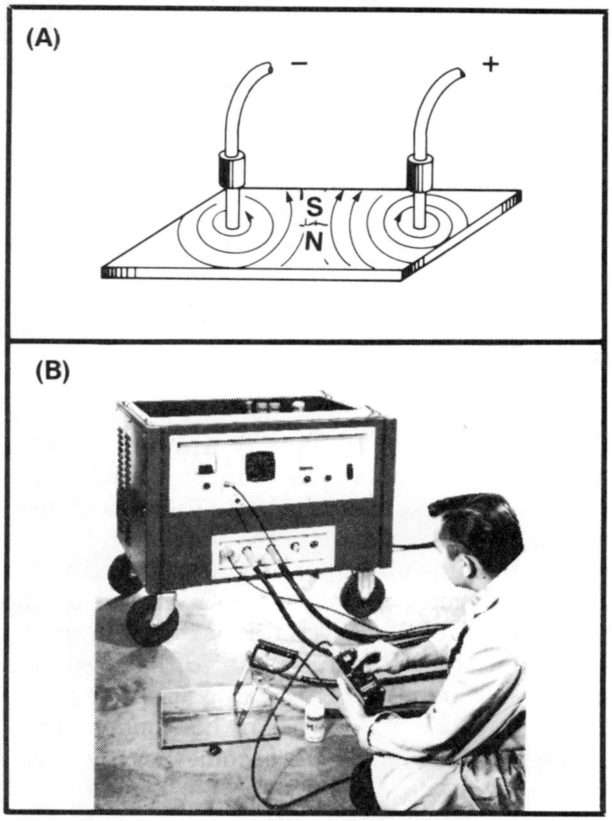

Fig. 3-22 A — Current flowing through the part from high-current probes magnetizes the part circularly and will detect cracks or faults that are in line with the probes. B — Magnetic particle inspection equipment using probes to magnetize a flat plate to test for cracks across a weld.

e. Magnetic field strength

(1) Solid non-magnetic conductor

When we pass current through a solid non-magnetic conductor, the magnetic field will build up from zero in the center of the conductor to maximum at its outside surface, and then it will decrease as the distance from the conductor increases. At a distance of twice the radius, it will be one-half its maximum strength; at three times the radius, it will be 1/3 the maximum strength, and so on.

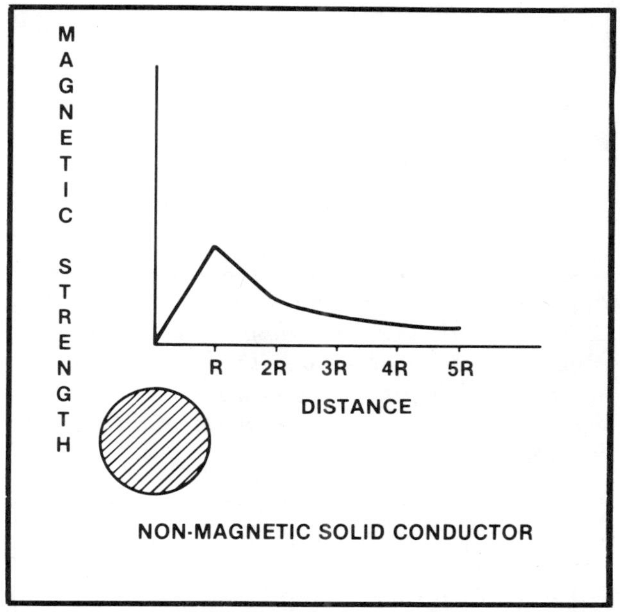

Fig. 3-23 The magnetic field strength caused by direct current flowing through a solid non-magnetic conductor is greatest at the surface of the part.

(2) Solid magnetic conductor

If we flow current through a magnetic conductor, its field strength will behave similarly to that of a non-magnetic conductor, except that at its surface the magnetic strength will be much greater, with its final strength related to the permeability of the material.

As we move away from the surface of the conductor, the field strength drops off from maximum in the same way it did for a non-magnetic conductor. At a distance of twice the radius from the center of the conductor, the field strength will be equal to one-half of the field strength at the surface of a non-magnetic conductor of the same size and shape as the magnetic conductor being magnetized.

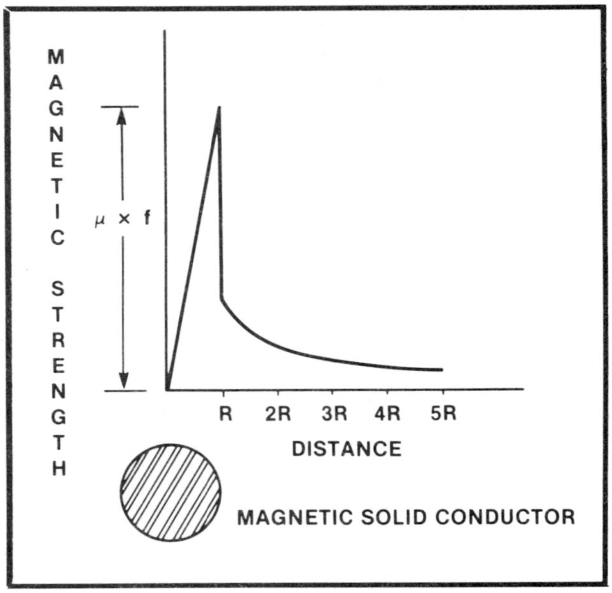

Fig. 3-24 The magnetic field strength caused by direct current flowing through a solid magnetic conductor is greatest on its surface, and the magnetic strength is determined by both the magnetizing force and the permeability of the material.

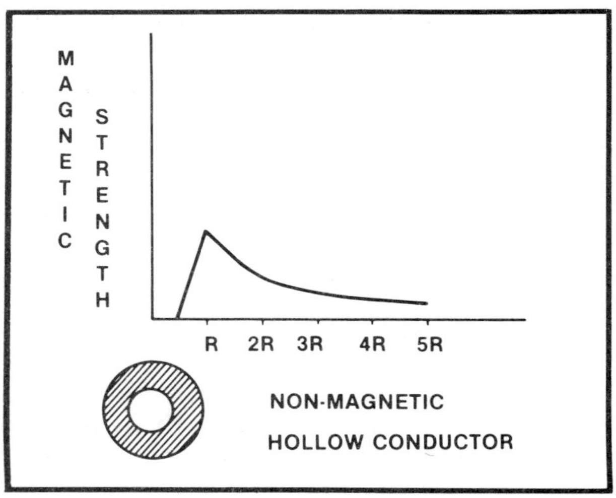

Fig. 3-25 The magnetic field strength caused by direct current flowing through a hollow non-magnetic conductor is zero on its inner surface and greatest on its outer surface.

(3) Hollow non-magnetic conductor

The field strength around a hollow non-magnetic conductor acts in much the same way as that surrounding a solid non-magnetic conductor, except that within the hollow center there is no magnetic field at all. The field builds up from the inside surface and reaches its maximum at the outside surface.

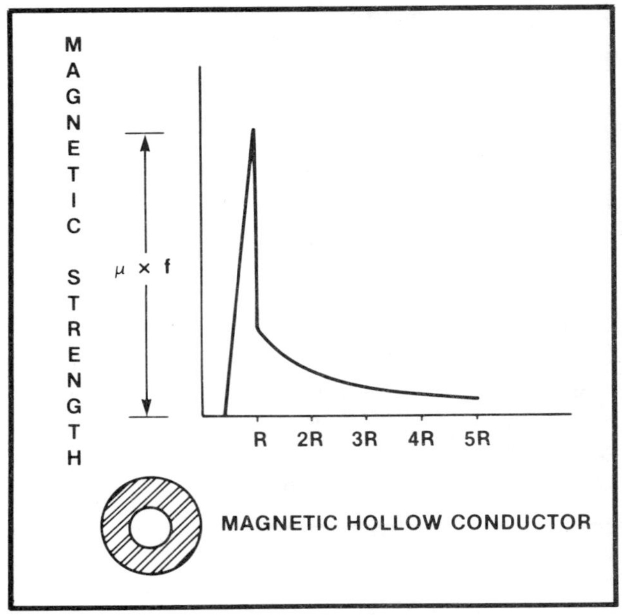

Fig. 3-26 The magnetic field strength caused by direct current flowing through a hollow magnetic conductor is zero on its inner surface and greatest on its outer surface. Its magnetic strength is determined by both the magnetizing force and the permeability of the material.

(4) Hollow magnetic conductor

The field strength surrounding a hollow magnetic conductor behaves in the way we would predict from studying hollow non-magnetic conductors and solid magnetic conductors. There is no field within the hollow center of the conductor, but the strength builds up from the inner surface until it reaches a maximum which is determined by the permeability of the material, and then it drops off at the surface to a value equal to that of a non-magnetic conductor of the same size and shape. Then as the distance from the conductor increases, the field strength becomes less.

(5) Hollow magnetic conductor magnetized by current flowing through a central conductor

With this arrangement, such as is used for circular magnetization, the field strength follows the curve shown in Fig. 3-27.

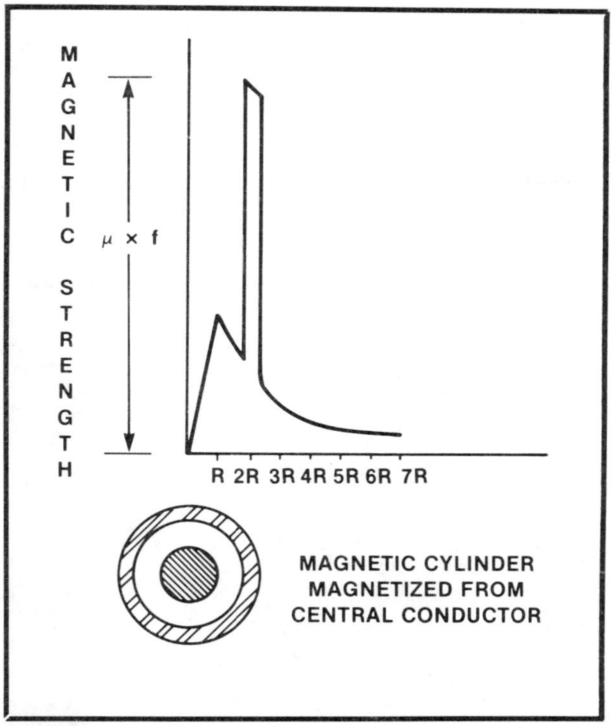

Fig. 3-27 The magnetic field strength on a cylinder surrounding a conductor carrying direct current varies from the center of the current-carrying conductor, as is seen here.

f. Methods of magnetization

(1) Direct current

Straight direct current at voltages from 110 to 440 volts has excellent penetrating qualities and is suitable for magnetizing parts in coils and with yokes, but it has the disadvantage that it is difficult to change its value as is required for inspecting objects of different sizes.

(2) Half-wave rectified direct current

A rectifier in the low-voltage, high-current side of a transformer will produce a pulsating direct-current output. Commercial frequency alternating current is readily available in the shop, and by controlling the input AC, the output DC may be adjusted to any value the test calls for.

Half-wave DC has as good penetrating qualities as straight DC, and its pulsating nature helps distribute the magnetic particles so they will arrange themselves over any fault.

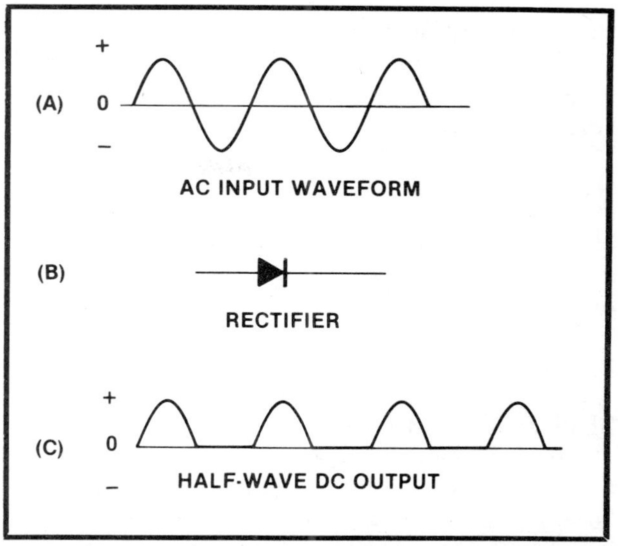

Fig. 3-28 A — AC input waveform to the magnetizing equipment B — Single-diode-type rectifier C — Pulsating half-wave direct-current waveform

(3) Alternating current

Remember from our discussion of the principle of magnetization that an object is magnetized when the magnetic domains within the material align with the external magnetizing force. If the magnetizing force is produced by alternating current, it will reverse each cycle, and the magnetic polarity will change.

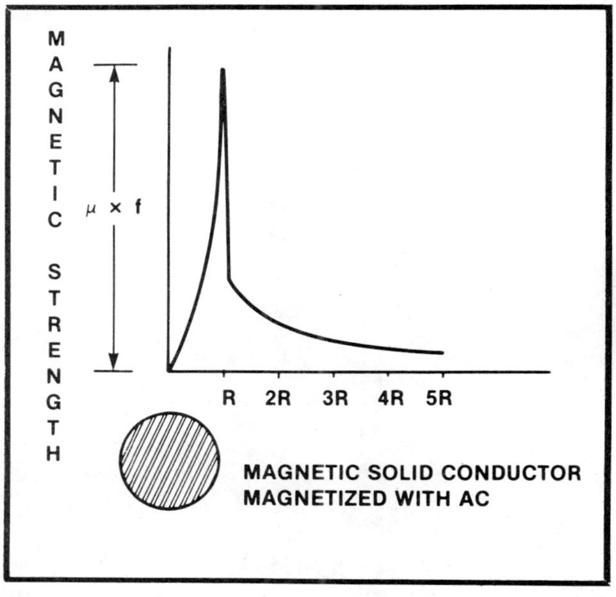

Fig. 3-29 The magnetic field strength around a magnetic conductor carrying alternate current is concentrated at its surface.

The magnetic field produced by AC differs from that produced by DC in that, as we see in Fig. 3-29, the field strength is almost totally concentrated on the surface of the material.

If a part is subjected to an AC field, and the flow of current is suddenly interrupted, the part will become magnetized, but the penetration will be very shallow. Therefore AC magnetization is primarily used for locating surface faults.

2. Testing medium

The medium that is used to indicate the presence of a fault by magnetic particle inspection must be ferromagnetic, it must be finely divided, must have a high permeability and a low retentivity, and for operator safety, it must be non-toxic.

There is no one medium that is best for all applications, but in general, these materials are extremely fine iron oxides that are dyed gray, black, or red, or are treated with a dye that causes them to fluoresce when they are illuminated with an ultraviolet light.

The iron oxides may be used dry and dusted onto the surface with a bulb or shaker can, or they may be mixed with kerosine or some other light oil and flowed over the surface.

3. Testing methods

a. Residual magnetism

If a part is magnetized and the magnetizing force is removed before the testing medium is applied, the part is said to be tested by the residual method.

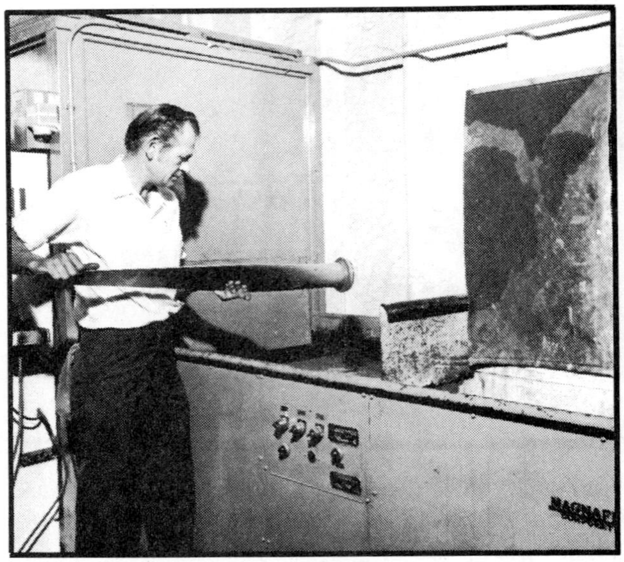

Fig. 3-30 This propeller blade has been magnetized and is being inspected by the residual method.

Fig. 3-31 This crankshaft is magnetized by flowing current through it from the two heads of the magnetizing machine. The oxide is suspended in a light oil and is flowed over the magnetized part.

b. Continuous magnetization

If the part is still subjected to the magnetizing force when the testing medium is applied, the inspection is being done by the continuous process.

4. Inspection

The type of dye that is used with the magnetic particles determines the type of inspection that must be used. If gray, black, or red dye is used, the inspection is made in white light, but if the fluorescent dye is used, the part must be inspected using a black light in a dark booth.

Fig. 3-32 These small engine parts are being inspected under black light after they have been magnetized and covered with the oxide.

5. Demagnetization

After the part has been inspected, and it can be returned to service, it must be thoroughly demagnetized. This may be done by either AC or DC demagnetization.

a. AC demagnetization

In order to demagnetize a part, we must subject it to a magnetizing force that is opposite to the force used to magnetize it. This will swing the magnetic domains around. If the magnetizing force is AC, the domains will alternate their polarity, and if the part is slowly removed from the field while the current is still flowing, the reversing action will progressively become weaker and the domains will be left with random orientation; or in other words, the part will be demagnetized.

AC, as we will remember, does not penetrate the surface very deeply, and so for complete demagnetization of some parts, we must resort to some other method, such as DC demagnetization.

b. DC demagnetization

For deeper penetration than is possible with AC demagnetization, we can use DC. Place the part in a coil through which we can pass more current than was initially used to magnetize the part. Flow a shot of current through the coil and then reverse the direction of flow and decrease the amount. Give it another shot of current. Continue to reverse the direction of flow and decrease the amount until the lowest value of current you can flow has been reached.

Check for the presence of any residual magnetism with a magnet strength indicator. The parts must be demagnetized to within the limits specified in the appropriate overhaul manual.

D. Eddy Current Inspection

Eddy current inspection works on the principle of determining the ease with which a material will accept induced current. An alternating current is induced into the material being tested. This AC is measured to determine the characteristics of the material.

The ease with which a material will accept the induced eddy currents is determined by four properties: its conductivity, its permeability, its mass, and by the presence of any voids or faults.

The conductivity of a metal varies with its alloy type, grain size, degree of heat treatment, and its tensile strength, and an eddy current test may be used to sort metals with regard to any of these characteristics. A comparison probe is held on the material we are using as a standard and the test probe on the metal we are testing. The meter is balanced to a null indication when the test probe is on the reference material, and when the probe is placed on the test material, the meter will indicate off of null if the conductivity of the test material is different from that of the reference.

The permeability of a material is a measure of its ability to accept lines of magnetic flux. If a bridge-type eddy current tester is zeroed with the test probe on a reference material, it will indicate

Fig. 3-33 The eddy current probe is being zeroed by placing it inside of a piece of material that has the same characteristics as the propeller blade being inspected.

an off-zero reading when the test probe is placed on a material whose permeability is different from that of the reference. The meter in Fig. 3-34 indirectly measures the current flowing in the probe, and the probe current is proportional to the current that is induced into the test specimen.

The mass of the material being tested determines the ease with which eddy currents can flow. The probe is placed on the surface of the test material and is zeroed, and the meter will move off zero when the probe is moved over any part of the material whose mass is different. If the probe passes over an area whose opposite side has been eaten away with corrosion, the meter will indicate this decrease in mass by a deflection of the meter needle.

The presence of any discontinuities, or faults, within the material will cause the meter to deflect. The meter is zeroed with the probe over

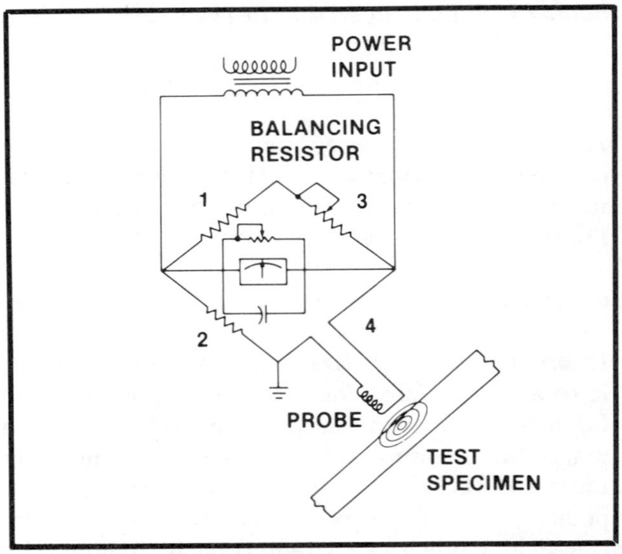

Fig. 3-34 The absolute method of eddy current inspection measures the amount of current required to induce a given amount of eddy current into the test specimen.

sound material, and as the probe is moved over a surface free from faults, it will remain steady, but as soon as it passes over a crack or other fault, the needle will move, indicating its presence.

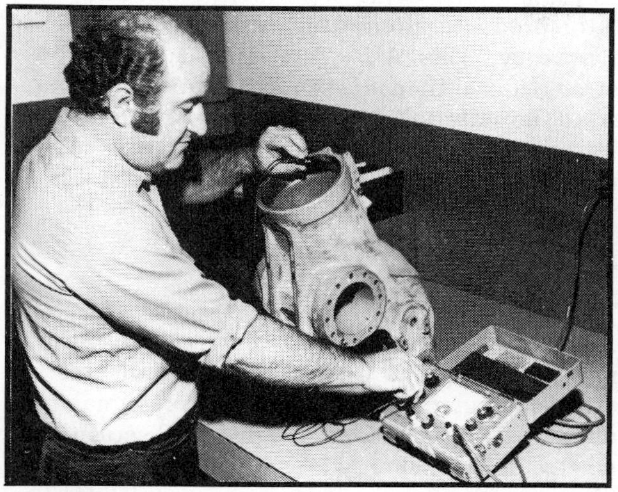

Fig. 3-35 When the eddy current probe is passed over any fault in the material being tested, the meter will deflect.

1. Absolute method of inspection

The bridge-type eddy current equipment is used to measure the characteristics of a material by measuring the amount of probe current that flows when current is being induced into the test specimen. The meter is zeroed by balancing the bridge when the probe is held in contact with a sound piece of test material that is the same as the material being tested. When the probe is placed on the test specimen, it will cause an off-zero meter reading if the characteristics of the test specimen differ from those of the reference material.

2. Comparison method of inspection

This method uses a double-coiled probe, and instead of being zeroed with a standard piece of material, it indicates any difference in characteristics between the material under the reference probe and that under the test probe.

E. Ultrasonic Inspection

Sound waves are produced by a vibrating body. When the particles of the medium in which the body vibrates oscillate longitudinally about a fixed mean position, the waves produced are called longitudinal waves. If the vibrations are at right angles to the direction of the sound, they are called transverse, or shear, waves. Shear waves that travel along the surface of a material but do not appreciably extend into the material are known as surface, or Rayleigh, waves.

Ultrasonic waves that are used for nondestructive inspection vary in frequency from 200 kilohertz to 25 megahertz, and they may be reflected,

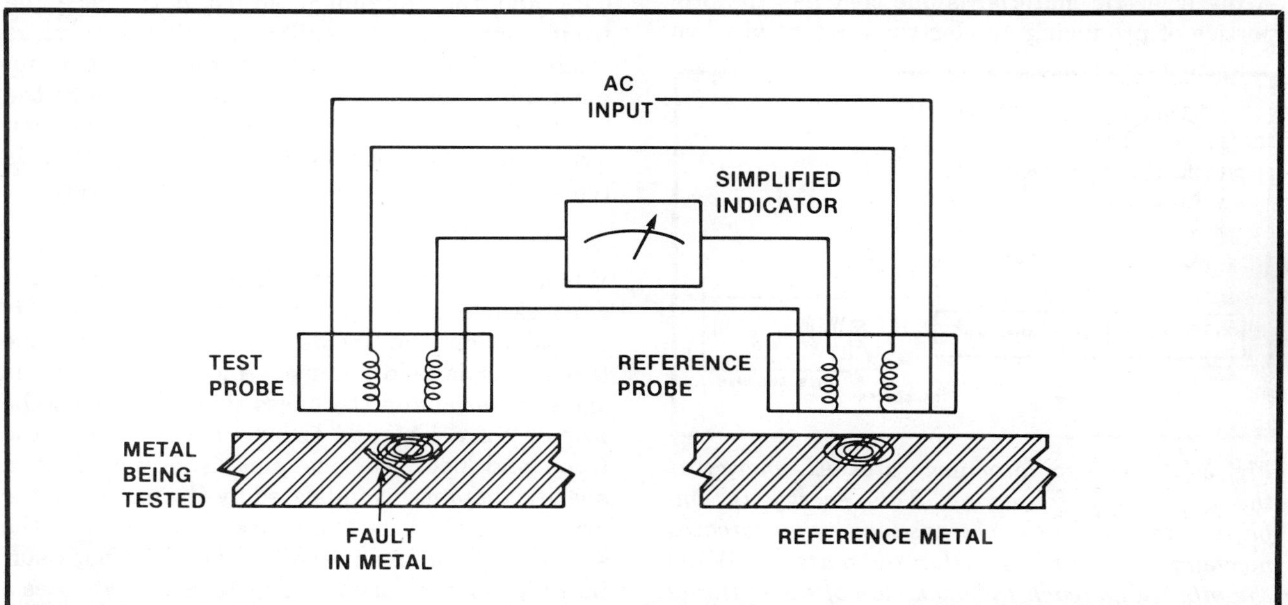

Fig. 3-36 The comparison method of eddy current inspection compares the eddy current induced into the reference material with the eddy current that flows in the current being tested.

focused, or refracted. This sound energy will propagate through a solid or a liquid material in much the same manner as a focussed beam of light travels in air, and it is not absorbed to any great extent. This property makes sound energy usable, not only in nondestructive inspection, but for sonar operation and for ultrasonic cleaning.

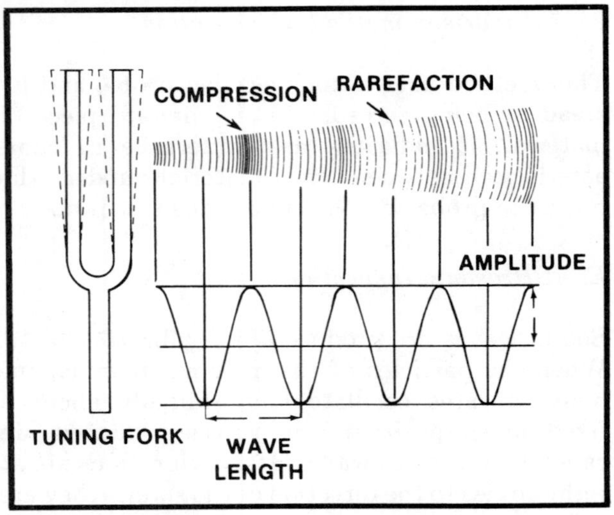

Fig. 3-37 Energy used for ultrasonic inspection consists of longitudinal waves that cause a compression and a rarefaction of the medium through which it passes.

1. Piezoelectricity

A piezoelectric material is one that has the properties of producing an electrical potential when it is pressed, bent, or otherwise distorted, and when it is subjected to an electrical potential it is physically distorted. This makes it useful as a transducer for introducing physical vibrations into the material being tested. An electronic oscillator produces alternating current of the proper frequency. This AC is amplified to the proper strength, and then it is sent to a transducer coupled to the material being tested. The energy will be introduced into the material and it will be vibrated at the frequency the oscillator produces.

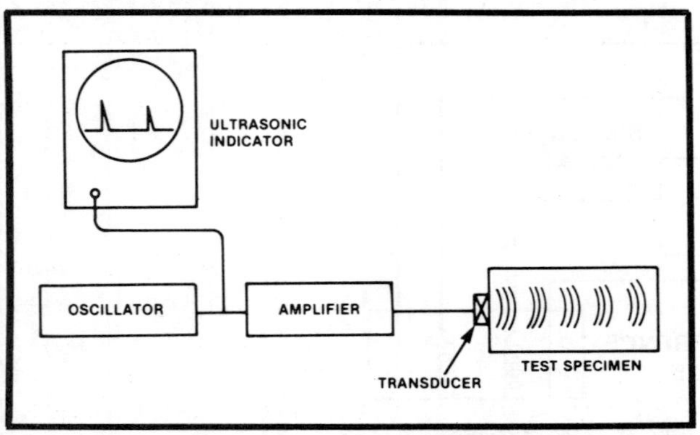

Fig. 3-38 An oscillator generates AC energy at the appropiate frequency. This energy is amplified and sent to a transducer that creates mechanical vibrations within the material. When the vibrations reach the back side of the material and bounce back, they are received by the transducer where they create an electrical inpulse that is seen on the cathode ray surface.

2. Fault indication

A cathode ray oscilloscope, which uses a cathode ray tube (CRT), is used as a fault indicator for ultrasonic inspection. A time base signal produces a straight line across the CRT screen, and when a pulse of energy is sent into the material, a pip, or peak, occurs on this horizontal line. The energy travels through the material until it reaches its back side where it is reflected back to the transducer. The energy received by the transducer will cause it to produce a second pip on the time base line. There is a definite relationship between the physical distance between the front and back surface of the test piece and the distance between the pips on the time base line.

A calibrated grid may be placed in front of the CRT and the equipment adjusted to place the front surface pip on a calibration line to serve as reference. A second adjustment on the test equipment allows us to vary the distance between the front and the rear surface pips so that we can establish a distance between the pips that is representative of the thickness of the material.

When an ultrasonic test is being made on an aircraft skin to determine whether or not there is corrosion on the inside of the structure, the transducer is held against a piece of skin of the same material and thickness that is known to be free of corrosion, and the distance between the front and back surface pips is adjusted to a specific distance as indicated by the marks on the grid. Now, when the transducer is moved over the skin being tested, any movement of the back surface pip closer to the front surface pip indicates a decrease in skin thickness that could possibly be caused by corrosion. Further investigation must be made.

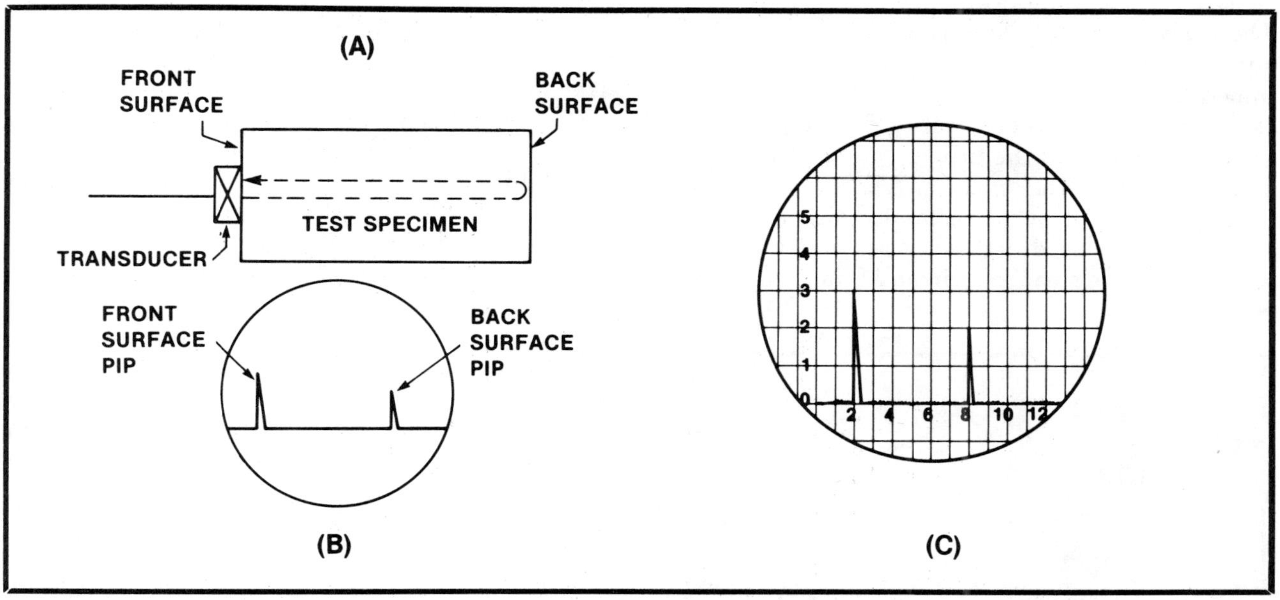

Fig. 3-39 A — Vibrations pass through the material and bounce off of its back surface. B — A pip is recorded on the time-base line to represent the front surface of the material, and another pip represents the back surface. C — The distance between the front and back surface pips and the height of the pips may be adjusted to coincide with the marks on the transparent grid that is placed in front of the cathode ray tube.

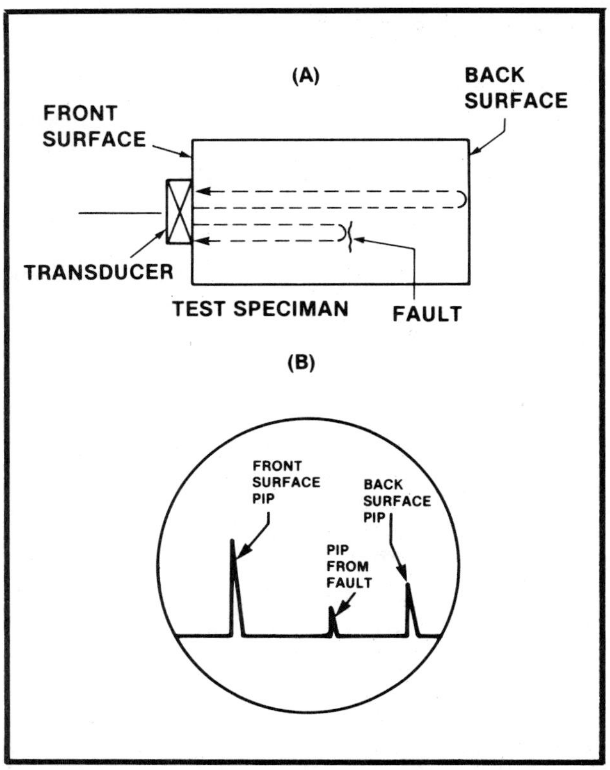

Fig. 3-40 A — The ultrasonic energy will bounce off of a fault. B — The reflection from the fault will create a pip on the time-base line between the pipe representing the front and the back surfaces.

3. Introduction of energy into the material being tested

Ultrasonic inspection is a highly specialized field that requires an intimate knowledge of the methods used, as well as the behavior of sound waves in materials of varying density and elasticity. In a text such as this, it is impossible to cover more than the fundamentals, and a person who wants to specialize in nondestructive inspection can get specialized training in the intracies of ultrasonic inspection, as well as the other types of inspection, at schools conducted by the manufacturers of the test equipment.

There are three basic ways in which ultrasonic energy can be introduced into the test specimen. The first is by direct contact on only one side of the material. The energy is transmitted from this point and the return echo is also received here. A second method uses a transducer on both sides of the material; one to introduce the pulse into the material, and the other to receive the signal and send it to the CRT. The third way of getting the energy into the material is an immersion method in which the test specimen is immersed in water and the transducer beams its energy through the water to the surface of the part.

The type of wave — longitudinal, transverse, or surface — that is directed to the material is determined by the orientation of the transducer with the surface, and this orientation is brought about by using wedges of the proper shape between the transducer and the surface. These wedges are usually made of acrylic plastic, and a film of water or oil is used between the transducer, the wedge, and the surface to insure a more intimate contact.

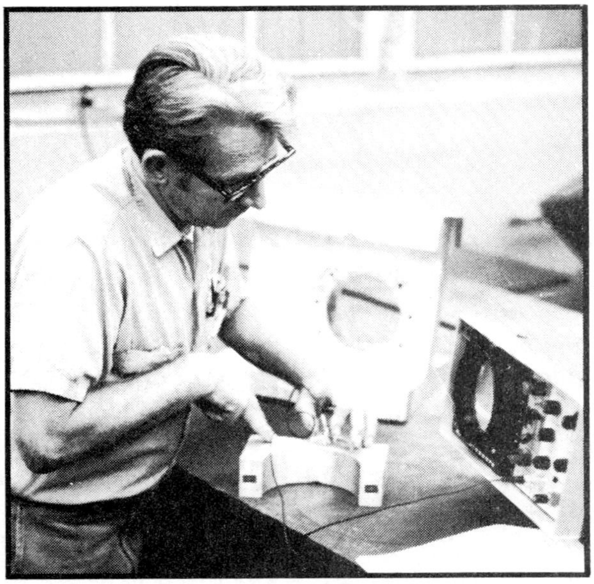

Fig. 3-41 The ultrasonic transducer mounted on an acrylic adapter beams the waves of ultrasonic energy into the material being tested. Here the transducer is being adjusted with a reference material.

F. Radiographic Inspection

One of the most important methods of nondestructive inspection we have available to us is radiographic inspection. It allows us to photographically see inside a structure that is opaque to light.

X-rays and gamma rays are forms of high energy, short wavelength electromagnetic waves. They are electromagnetic in nature, with the amount of energy they contain directly related to their wavelength; the shorter the wavelength, the greater the energy. They have no electrical charge nor mass, and they travel in straight lines at the speed of light. They can penetrate matter, and their depth of penetration is dependent upon the energy in the rays.

The characteristics that make them especially useful in nondestructive inspection is that they are absorbed by the matter through which they pass, and the amount of absorption is proportional to the density of the material. They can ionize certain materials, which makes it possible for them to expose photographic film and cause certain materials to fluoresce, or glow.

1. Generation of X-rays

The tube used to produce X-rays consists of a heavy insulating envelope in which is mounted a coil of wire, or cathode, which is heated with electrical current so it will emit electrons. At the other end of the tube is the anode on which the

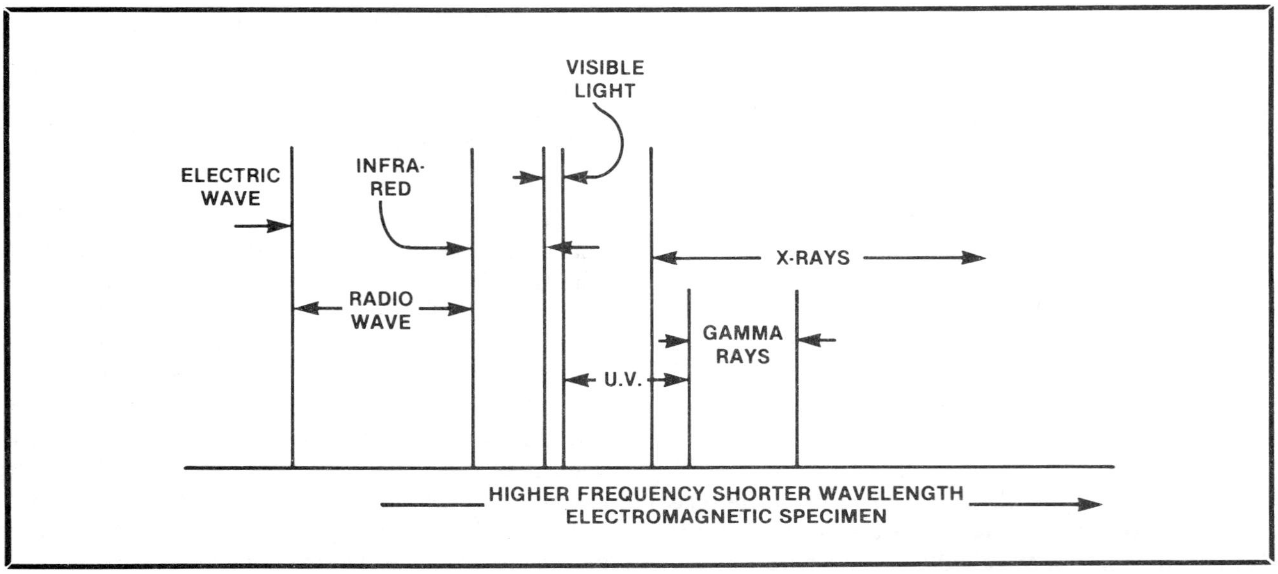

Fig. 3-42 The electromagnetic spectrum

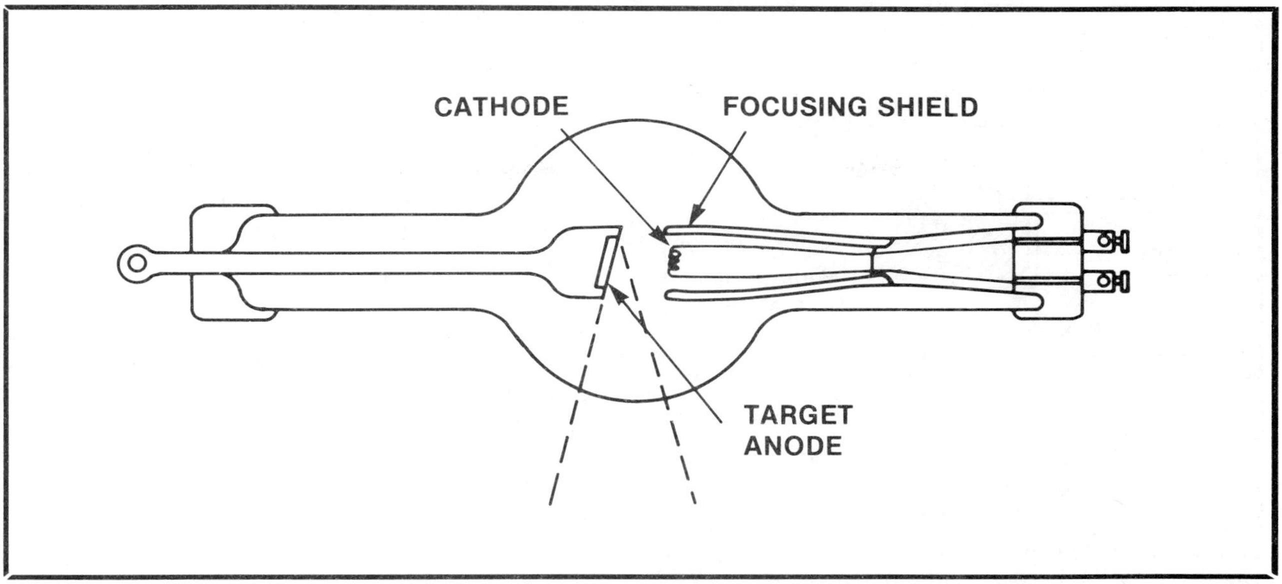

Fig. 3-43 X-rays are produced when electrons from the cathode bombard the target anode and dislodge electrons.

target is mounted. The target is made of a material having a high atomic number, therefore a high density and lots of electrons. A high positive voltage is connected to the anode so it will draw the electrons from the cathode.

X-rays are generated when an electron from the cathode which has been acclerated to a high velocity by the high positive voltage on the anode strikes the target and dislodges an electron. The intensity of the X-ray beam is determined by the number of electrons available, and this is controlled by the amount of current used to heat the cathode. This current is measured in milliamperes.

The energy of the X-ray is determined by its wavelength, which is in turn determined by the velocity of the electrons striking the anode. This velocity is controlled by the voltage applied to the anode which is measured in kilovolts.

2. Classification of X-rays

The intensity of the X-ray falling upon the film or the fluorescent screen is related to the distance between the film and the radiation source, and it obeys the inverse square law. If the distance between the source and the film is doubled, the amount of radiation falling on each unit of area is decreased to one-fourth. If the distance is tripled, it is decreased to one-ninth. In order to get the required amount of radiated energy at the surface of the film, we must control the amount of current supplied to the cathode. Low current (MA), low intensity, and high current, high intensity.

The voltage (KV) supplied to the anode determines the amount of energy in the X-ray, and the higher the voltage, the harder the wave, or the more energy it contains. Low-energy X-rays are called soft X-rays and those with high energy are called hard X-rays.

3. Gamma rays

Gamma rays are composed of photons of radiation energy that are produced by the disintegration of certain isotopes. An isotope is a variety of a chemical element that has the same atomic number and position in the atomic table as the original element. It has the same number of protons in its nucleus and the same number of electrons, but a different number of neutrons in the nucleus. This gives it a different atomic mass and different physical characteristics.

Unlike X-rays which are produced only when electrons are flowing within the tube, gamma rays cannot be shut off, controlled, nor directed, so the equipment in which the isotopes are kept must provide safe radiation-proof storage. The container which holds the isotope is called a camera, and in the one shown in Fig. 3-44 the source is enclosed in a heavy shielding of lead until it is needed to expose the film. The equipment is set

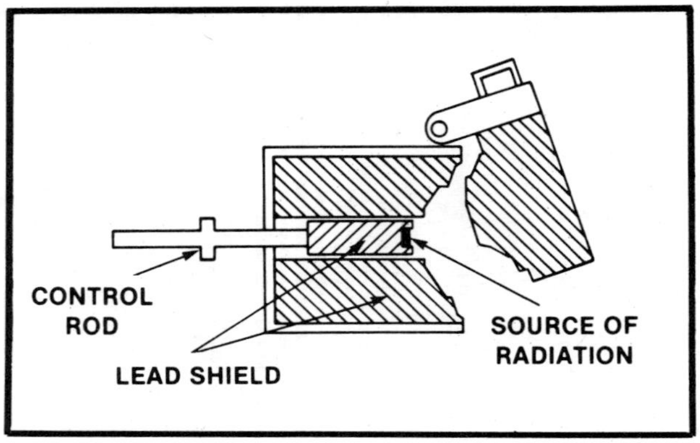

Fig. 3-44 Gamma rays are produced by the disintegration of certain isotopes of material. There is no way of shutting off the radiation from an isotope, so when no radiation energy is required, the isotope must be totally enclosed in a lead isotope camera.

up and the cover over the isotope is moved away to expose the source. With this type of camera, the control rod may extend the source to provide a wider angle of coverage.

4. Radiographic indication

a. Film

For a permanent record of a radiographic inspection, a sheet of photographic film is placed on one side of the object being inspected and the radiation source on the other, and the proper exposure is made.

The film is placed as close to the specimen as possible and the source is oriented in such a way that the radiation will penetrate it to pass an amount of radiation that is proportional to the density of the specimen. The denser the specimen, the less radiation will pass through, and the less the film will be exposed.

Photographic film is composed of a flexible transparent plastic sheet on which a thin coating of gelatin is placed. This gelatin contains an emulsion of extremely fine grains of silver bromide. When this bromide is exposed to photons of energy, either visible light or radiation energy, a latent image is formed on the film.

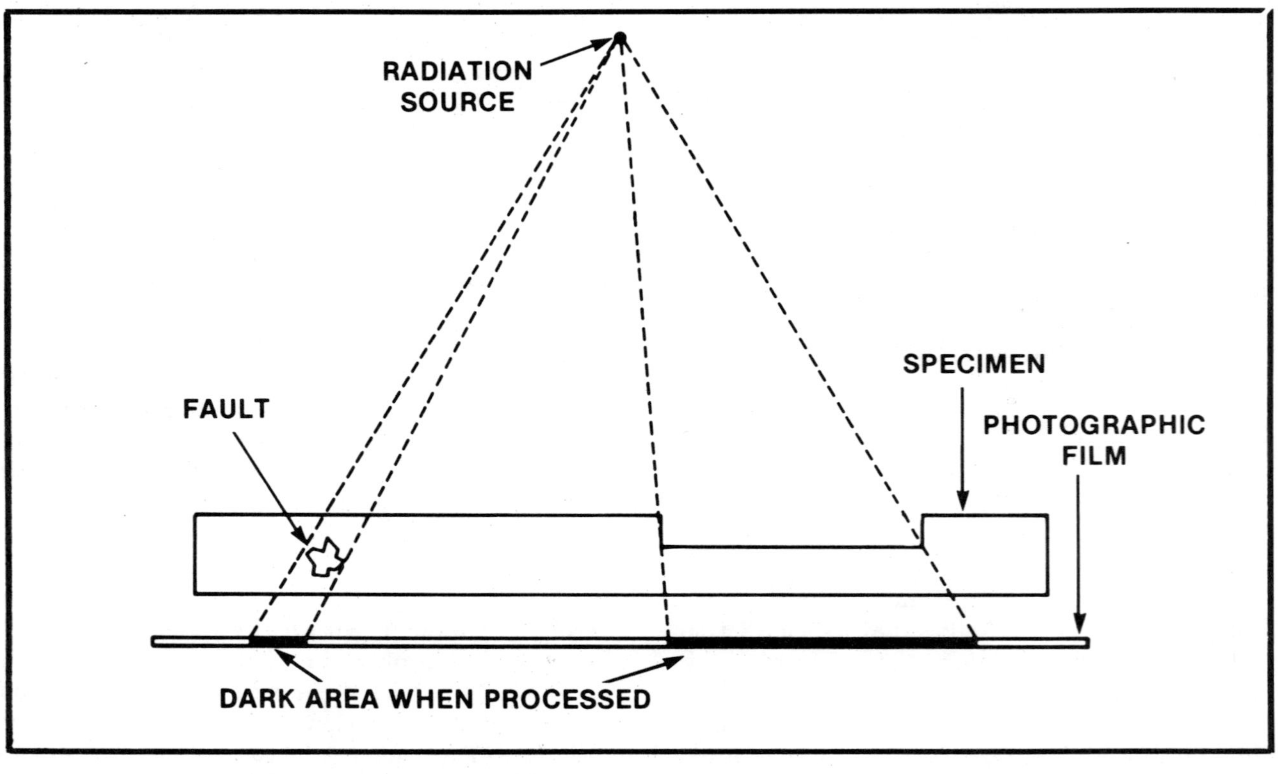

Fig. 3-45 Radiation from X-rays or gamma rays penetrate a specimen and expose a photographic film. The denser the material, the less the exposure will be.

The exposed film is treated with a developer which reduces the silver bromide grains that were touched by the radiation into clumps of black metallic silver. After all of the affected emulsion has been converted, the developing action is stopped with an acid stop bath, and the film is treated with a chemical fixative which removes all of the unexposed silver bromides in the emulsion and makes the film insensitive to light. All of the chemicals are then washed out of the emulsion, and it is dried and is ready for examination.

The film is a negative, and the places which got the most radiation — that is, the less dense areas — will be dark, and those where the density is the greatest will get the least radiation and will be the most clear.

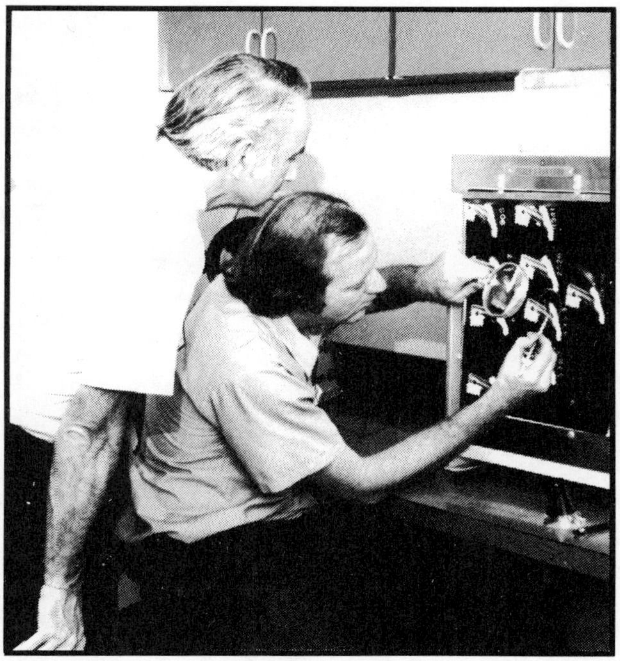

Fig. 3-46 The developed X-ray film is in the form of a photographic negative, with the denser area being lighter and the less dense areas being darker.

b. Fluoroscopy

For high-speed radiographic inspection where no permanent record is required of the results, a fluoroscope may be used. This system is very similar to the photographic process except that the entire equipment is enclosed in a lead box and instead of using photographic film to record the results, the radiation passes through the specimen and causes a fluorescent screen to glow. The indication on the screen is such that the portion of the screen that receives the most radiation, the least dense portion of the specimen, will glow the brightest. The image is viewed through a lead glass viewing window.

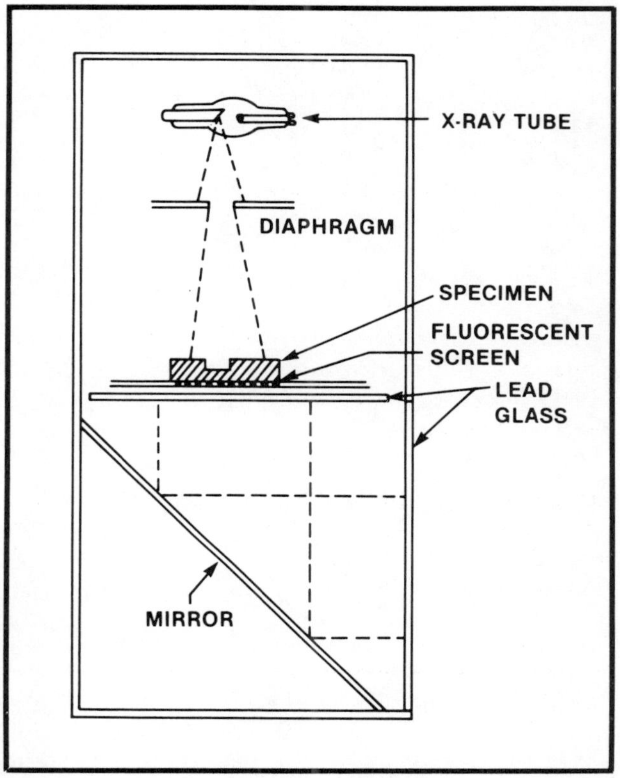

Fig. 3-47 When a permanent record is not required, the part may be viewed on a fluoroscope machine. Here the material is viewed on a fluorescent screen that gives an indiction in much the same way as we would see it on X-ray film.

5. Radiographic safety procedures

Any form of radiation must be carefully monitored, as this form of energy is harmful to human life. The United States Atomic Energy Commission sets up safety procedures regarding the handling of radioactive materials, and their recommendations must be adhered to.

The main reason for the use of radiographic inspection is its ability to penetrate almost anything we must inspect, and it is for this very reason that it is difficult to protect ourselves from its effects. The best shielding we can use is a layer of lead, and we must be very careful that there are no holes in the shielding that will allow the radiant energy to escape. Persons working around X-ray or gamma ray equipment should wear radiation monitor film badges. These are special containers of film that are worn while in the vicinity of the equipment, and at specified intervals, these pieces of film are developed to see if they have been affected by the radiation.

Areas where X-ray or gamma ray equipment is used should be controlled to prevent anyone being exposed to radiation energy. Special signs are required on containers carrying radioactive material.

NOTES

NOTES

AVIATION
TECHNICIAN
INTEGRATED
TRAINING
PROGRAM

EA-ITP-G7

Ground Handling and Servicing
Maintenance Publications
Federal Aviation Regulations
Maintenance Forms and Records

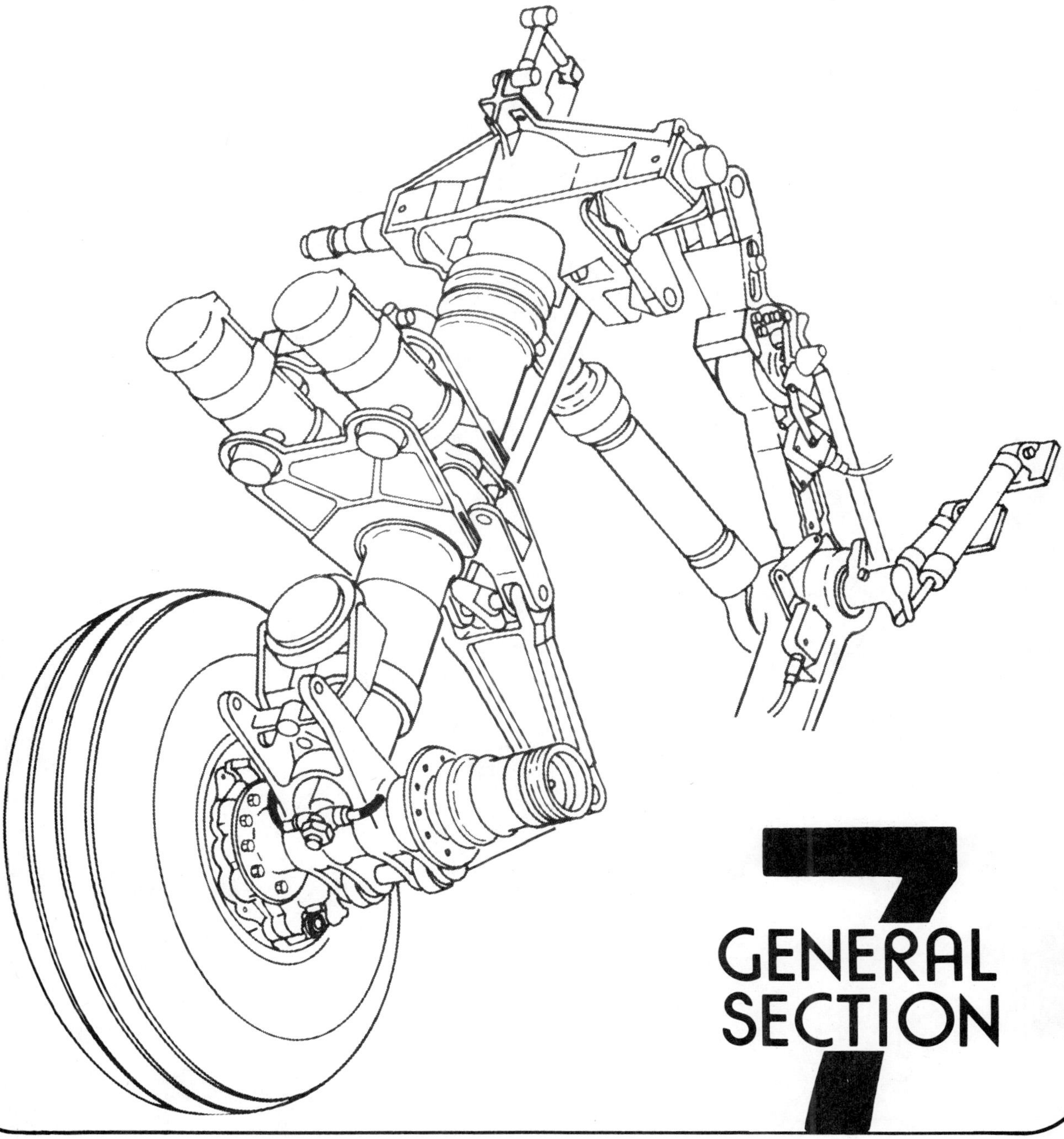

7 GENERAL SECTION

EA-ITP-G7

TRAINING MANUAL
GENERAL SECTION
BOOK 7

GROUND HANDLING AND SERVICING MAINTENANCE PUBLICATIONS FEDERAL AVIATION REGULATIONS MAINTENANCE FORMS AND RECORDS

by Dale Crane

AVIATION MAINTENANCE PUBLISHERS, INC.

INTEGRATED TRAINING PROGRAM
FOR THE
AVIATION MAINTENANCE TECHNICIAN

International Standard Book Number 0-89100-132-8

TABLE OF CONTENTS

Preface

Part 65 of the Federal Aviation Regulations require that an applicant for a Mechanic Certificate with an Airframe or Powerplant rating have a general knowledge of the construction and maintenance of the aircraft covered by the rating he seeks. And each applicant must prove this knowledge by passing a written, an oral and a practical examination, each given in five sections: General, Airframe Structures, Airframe Systems and Components, Powerplant Theory and Maintenance, and Powerplant systems and Components.

Certificated Aviation Maintenance Technician Schools all operate under individual Federal Aviation Administration approvals, and all have in their approved curriculum enough information to satisfy the FAA that anyone completing their course will meet the knowledge requirements for certification as a mechanic.

In 1972, the Federal Aviation Administration published a series of three Advisory Circulars providing background information for mechanic certification applicants. These are entitled: *Airframe and Powerplant Mechanics General Handbook, AC 65-9; Airframe and Powerplant Mechanics Powerplant Handbook, AC 65-12;* and *Airframe and Powerplant Mechanics Airframe Handbook, AC 65-15.* These three handbooks were written to cover the area over which a mechanic applicant was to be tested, and they were an attempt by the FAA to coordinate the information available to the applicants with the tests that were given for certification. In 1976, the three handbooks were revised into their present A revision format: AC 65-9A, AC 65-12A, and AC 65-15A.

Individuals attempting to certificate under FAR Part 65, as well as schools operating under FAR Part 147, have found that these handbooks actually do not follow the material used in the FAA examinations, and because of this, plus their lack of study helps, they fail to effectively provide the broad base of knowledge needed for certification and for a foundation on which to build in the aviation maintenance industry.

Realizing the need for updated information, presented in a format that is in keeping with the latest developments in educational technology, the Aviation Maintenance Publishers, Inc., is offering an Integrated Training Program (ITP) that presents the basic information needed for mechanic certification, in a form that will provide a good foundation in aviation maintenance knowledge. Developed in coordination with Crane Aeronautical Activities, Inc., the ITP training manuals, workbooks, study guides, and audio-visuals are designed to follow the FAA written examinations and are written to the level required for the FAR Part 147 schools.

All of the material can be used by an individual preparing for certification on his own, or by schools preparing groups of individuals for certification under FAR Part 65, but it has been prepared primarily for the Aviation Maintenance Technician Schools operating under FAR Part 147. Not only has every attempt been made to meet the requirements of Part 147 in producing this series, but the authors have followed many of the recommendations of the Aviation Mechanic Occupation Study conducted by the University of California, Los Angeles, under a contract from the U.S. Department of Health, Education, and Welfare.

One of the most basic premises of aviation maintenance is that all operations must be carried out according to FAA-approved data, and all information of a specific nature furnished by the manufacturer of an aircraft, powerplant, or component must be followed in detail. These training manuals are, of necessity, general in nature and cover a wide scope of aircraft, powerplants, and components. Because of this, it must be understood that they should *never* be allowed to take precedence over specific information furnished by the manufacturer.

The authors and publishers of this Integrated Training Program wish to express our appreciation to the various manufacturers, Aviation Maintenance Technician Schools, and FAA personnel who have aided our efforts to provide this coordinated training material to meet the challenge of better training for aviation maintenance technicians in the eighties.

PART I

Ground Handling and Servicing

I. *Shop Safety*

A. *Electrical Safety*

Every shop in which aircraft maintenance is performed depends upon electrical power for its operation, and it is the responsibility of everyone that uses electrical power to be aware of the safety procedures that are required for its use.

1. *Physiological safety*

The human body is a conductor of electricity and when electrical current passes through it, it will cause burns at the points where the current enters and leaves the body and will distrupt the nervous system.

The common 110/120-volt "house current" is particularly dangerous because it affects our nerves in such a way that we are often unable to turn loose of the current-carrying wire.

Most shop equipment operating on 110/120 volt single-phase alternating current is connected with a three-conductor cord. The black insulated wire carries the power; the white wire is the ground and is connected to the earth ground where the power enters the building. The green wire is the equipment ground and connects the housing of the equipment to the earth ground through the long, round pin in the male plug.

An electric drill has an AC motor inside a metal housing. One wire is connected to the power terminal of the motor, and the other terminal connects to the ground through the white wire. Now, if there are only two wires in the cord and the power lead becomes shorted to the housing, the return current will flow to ground through your body when you operate the drill. But, if the drill motor is wired with a three-conductor cord, as it should be, the return current will flow instead through the green wire to ground.

Be sure that all electrical equipment is connected with three-wire extension cords of adequate capacity. Do not use cords that are frayed, or that have any of the wires exposed, and be sure to replace any plugs that are cracked.

Water is a conductor of electricity, so we must be sure that we never handle any electrical equipment while we have a good path to ground through a wet surface or through wet shoes.

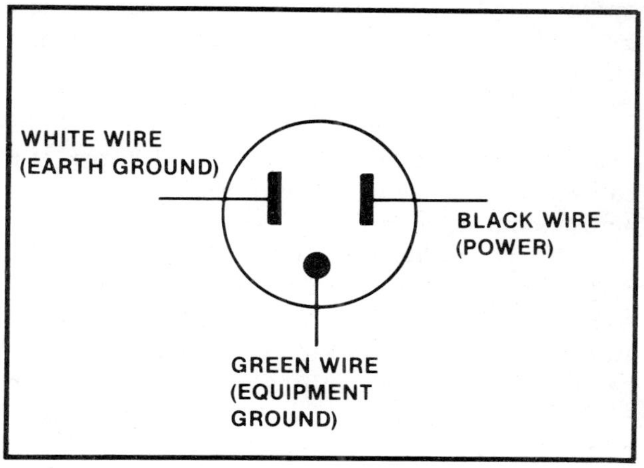

Fig. 1-1 The black wire in a three-conductor extension cord carries the electrical power, the white wire serves as the earth ground, and the green wire connected to the round pin serves as the equipment ground.

2. *Fire safety*

Electrical current flowing through a conductor produces heat, and the higher the resistance of the conductor, the greater the heat. It is possible for a plug on an extension cord to make such a poor connection that its resistance will cause enough heat to start a fire. Do not overload extension cords or wall outlets, as their contacts are not designed to carry more than the current for which they are rated.

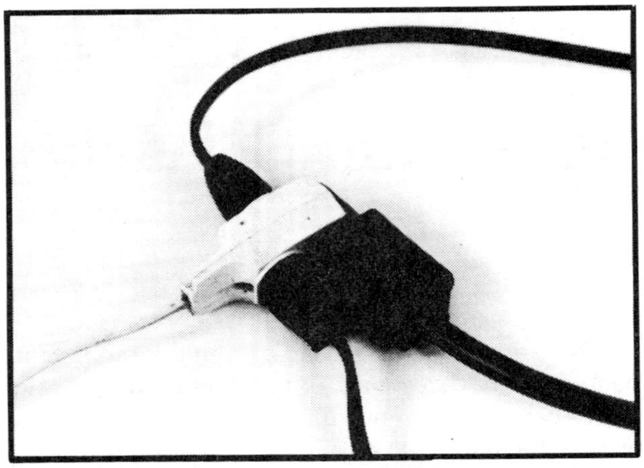

Fig. 1-2 An overloaded extension cord is a fire hazard.

B. Safety Around Compressed Gases

Compressed gasses are found in all aircraft maintenance shops. We use compressed air to operate air drill motors, rivet guns, paint spray guns, cleaning guns, and to inflate tires and shock struts.

Most shop compressed air is held in the tanks and lines under a pressure of about 100 psi. If a concentrated stream of compressed air is blown across a cut in your skin, it is possible for the air to enter your bloodstream and cause severe injury. For this reason, air dusting guns are usually equipped with a restrictor that reduces the pressure at their discharge to 30 psi or less. Be very careful when using compressed air that you do not blow dirt or chips into the face of anyone standing near by.

Fig. 1-3 Be sure the protective cap is screwed on a cylinder containing high-pressure gas, to prevent damage to the valve when the cylinder is not connected into a system.

High-pressure compressed gases are especially dangerous if they are mishandled. Oxygen and nitrogen are often found in aviation maintenance shops stored in steel cylinders under a pressure of around 2,000 psi. These cylinders have brass valves screwed into them, and if a cylinder should be knocked over and the valve knocked off, the escaping high-pressure gas will propel the tank like a rocket. Be sure that all gas cylinders are properly supported, and that the cap is screwed securely on any tank that is not connected into a system to protect the valve from damage.

It is extremely important that oxygen cylinders be treated with special care. Not only do they have all of the problems inherent with other high-pressure gases, but oxygen must never be allowed to come in contact with petroleum products such as oil or grease as oxygen will cause the oil to ignite spontaneously and burn. Never use an oily rag or tools that are oily or greasy to install a fitting or a regulator on an oxygen cylinder.

C. Safety Around Machine Tools

Spinning and cutting tools are commonly found in aviation maintenance shops and are dangerous when they are misused, but pose no particular problem to us when we use them in the way in which they are intended and observe reasonable safety precautions.

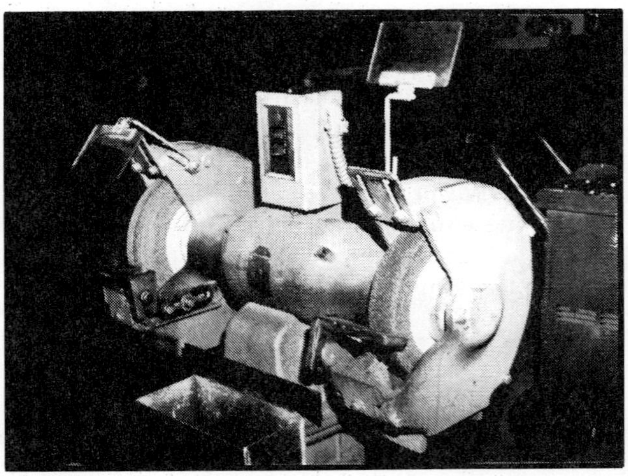

Fig. 1-4 Be sure the eye-protection shields are in place when using a bench grinder.

Do not use any machine tools with which you are not familiar or any tool whose safety features you are unfamiliar with. Many tools are equipped with guards and safety covers that have been put there for a reason — to protect you, the operator. Some of these guards may appear to interfere with the operation of the equipment, but do not disable the guards or remove them. The slight inconvenience they may cause is more than compensated for by the added safety they provide.

Be sure that all cutting and drilling tools are properly sharpened. A dull or improperly sharpened tool can require excessive forces to be used, and as a result the work may be grabbed or thrown out of the machine.

Be sure that the metal is securely clamped to the drill press table when cutting large holes in thin sheet metal. It is possible for the drill or the cutter to grab the metal and spin it around, effectively slicing anything in its way.

Always use eye protection when drilling, grinding, or sawing, as particles of the material or of the grinding wheel may be thrown into your eye.

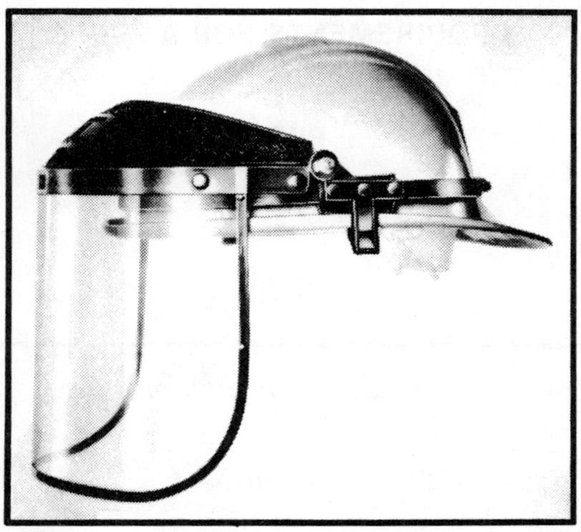

Fig. 1-5 Face shields or goggles should be used when drilling, grinding, or sawing.

A power tool is one of the greatest timesavers we have around a maintenance shop, but we must not allow its convenience to cause us to misuse it. Never be in a hurry around a power tool, never use the tool for a purpose for which it is not intended and, most important of all, *THINK* while using it.

D. *Fire safety*

In aviation maintenance shops we continually have all of the makings for fires, and so it is our responsibility to prevent fires occurring. All combustibles must be stored in the proper type containers and in areas where spontaneous combustion cannot occur. Containers of dope and paint solvents should be stored in a cool area outside of the shop, where there is adequate ventilation.

Neither spilled gasoline nor dried paint overspray should be swept with a dry broom as static electricity can cause a spark and ignite the gasoline or overspray. Always flush these combustible products with water before sweeping them.

Always be aware of the possibility of a fire and provide for exits when putting aircraft in the hangar. Be sure that the fire extinguishers are properly serviced, clearly marked, and never obstructed.

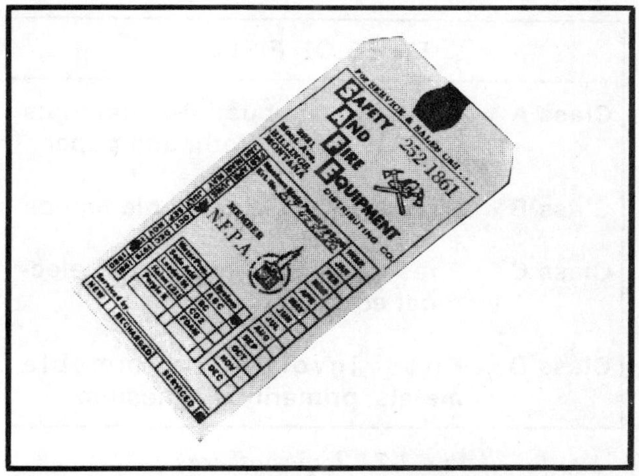

Fig. 1-6 Be sure that all fire extinguishers are properly serviced, are clearly marked to indicate the type of fire for which they are suitable, and are not obstructed.

II. *Fire Protection*

A. *Classification of Fires*

The very nature of aircraft makes them highly susceptible to fire. They carry large amounts of highly flammable fuel as well as oxygen under high pressure and, to complicate matters, smoking is allowed in many aircraft.

In order to understand fires and fire protection, we should look at the classification of fires as defined by the National Fire Protection Association.

Class A fires are those in which solid combustible materials such as wood, paper, or cloth burn. Aircraft cabin fires are usually of this class.

Class B fires involve combustible liquids such as gasoline, oil, turbine fuel, and many of the paint thinners and solvents.

Class C fires are those in which energized electrical equipment is involved. Special care must be exercised in the selection of a fire extinguisher for class C fires.

A Class D fire is one in which some metal such as magnesium is burning, and the use of the improper type of extinguisher can intensify these fires rather than extinguish them.

TYPES OF FIRES	
Class A	Ordinary combustible materials such as wood, cloth, and paper
Class B	Fires involving flammable liquids
Class C	Fires involving energized electrical equipment
Class D	Fires involving flammable metals, primarily magnesium

Fig. 1-7 Types of fires

B. Requirements for Fire

Chemically, a fire is a reaction between a material and oxygen in which the material is reduced to its elements with the release of a great deal of heat. For example, paper is an organic material composed basically of hydrogen and carbon. When it is heated to its kindling temperature, it will unite with the oxygen in the air and will be changed into carbon dioxide and water. Other elements in the paper and incomplete combustion will result in a residue we know as ash or as black carbon.

From this explanation, we can see that there are three requirements for fire to take place. We must have a fuel, which is any material that will combine with oxygen; we must have the oxygen, with which the fuel can combine; and we must have some means of increasing the temperature of the fuel to its kindling point, the point at which combustion can occur.

Different types of fuel have different kindling temperatures. For example, gasoline when raised to a relatively low temperature, will combine with oxygen, while some materials such as wood, for example, must reach a temperature considerably higher than that of gasoline before they will ignite.

The concentration of the oxygen also determines the combustibility of the material. A petroleum product such as oil or grease will ignite at room temperature if it is blanketed with pure oxygen. Even steel, which, for all practical purposes, is not combustible, will actually burn when it is heated red-hot and a stream of pure oxygen is fed into it. This is the principle of oxyacetylene cutting.

REQUIREMENTS FOR A FIRE
FUEL
OXYGEN
HEAT

Fig. 1-8 Requirements for a fire

C. Types of Fire Extinguishers

In order to extinguish a fire, we must either lower the temperature of the material below its kindling temperature, or deprive it of oxygen. All of the fire extinguishers we have in common use work on one or the other of these principles.

1. Water extinguishers

Water can be used only for Class A fires, such as aircraft cabin fires, where there is no electricity involved. Most modern water-type extinguishers consist of a container of water in which some anti-freeze material has been mixed. The water is propelled from the extinguisher by a charge of carbon dioxide. Once the extinguisher is activated, all of the propellant will be discharged and a new cartridge must be installed when the extinguisher is serviced.

In the past, soda-acid extinguishers were used in which the container was inverted to mix sulfuric acid with the water that contained bicarbonate of soda. The soda and acid generated enough carbon dioxide to expell the water from the extinguisher in a stream. This type of extinguisher has been replaced by carbon dioxide extinguishers in most aircraft shops today.

2. Carbon dioxide extinguishers

Carbon dioxide is an inert gas that is carried in a steel cylinder under pressure. When it is released, it expands to many times its compressed volume, and when it does, its temperature drops and it blankets the fire in the form of a white snow. The carbon dioxide gas excludes the air containing oxygen from the fire and the fire dies out. CO-2 fire extinguishers are available from the small two-pound unit that may be mounted in brackets in the cockpits of even small aircraft to the large units that are mounted on wheels and rolled around the shop as they are needed.

For this type of extinguisher to be used, the valve is opened and the CO-2 is directed from the horn to the base of the fire. Carbon dioxide is heavier than air and is electrically non-conductive. It is effective for both Class B and Class C fires.

3. Halogenated hydrocarbon extinguishers

A halogen element is one of the group that contains chlorine, fluorine, bromine, or iodine. When some of the hydrocarbons are combined with halogens, very effective fire extinguishing agents are produced.

Some of the early aircraft fire extinguishers used carbon tetrachloride, which is a form of halogenated hydrocarbon. It is a liquid that may be pumped onto the fire where it becomes a gas and blankets the fire, extinguishing it very effectively.

Carbon tetrachloride is no longer used as a fire extinguisher because of the harmful effect it has on the human body. The fumes from the carbon tetrachloride are harmful to the liver, and its damage is cumulative; that is, continued exposure allows the harmful effects to build up. In addition to this, when carbon tetrachloride is used on an open flame, it converts into a poisonous gas, phosgene.

The most popular fire extinguishing agent for cabin fires in modern aircraft is Halon 1301, also known as Freon 13, which is most useful as a fire extinguishant, and is not harmful to humans in concentrations as high as 20% by volume. In addition to its use for cabin fires, Halon 1301 is extremely effective for extinguishing fires in engine compartments of both piston- and turbine-powered aircraft. In engine compartment installations, the Halon 1301 container is pressurized by compressed nitrogen and is discharged through spray nozzles in the engine compartment.

Fig. 1-9 An installed halogenated hydrocarbon fire extinguisher in an airplane engine compartment.

4. Dry-powder extinguishers

Bicarbonate of soda, ammonium phosphate, or potassium bicarbonate are used as dry-powder extinguishants. Their main effectiveness is with Class B, C, and D fires; that is, fires from liquids and fires in which electrical circuits are energized. They are also the most effective type of an extinguishant for metal fires, such as fires involving burning magnesium.

Dry powder is expelled from the container by compressed nitrogen, and it blankets the fire, excluding oxygen from the fuel, and preventing a reflash that would reignite the fuel after it has been extinguished.

Fig. 1-10 The dry-powder fire extinguisher blankets a fire to exclude oxygen from the fuel.

D. *Checking Fire Extinguishers*

The three most popular types of fire extinguishers for aircraft installation or for use in aircraft maintenance shops are the carbon dioxide and the nitrogen pressurized dry powder or halogenated hydrocarbon type. All of these types have seals over the operating handle that tell us at a glance whether or not the extinguisher has been discharged. The content of a CO-2 extinguisher is determined by its weight, with the weight of the empty container stamped on the filler valve. The nitrogen pressurized extinguishers have pressure gages on them with red and green arcs on their dial. As long as the gage is indicating within the green arc, the gas pressure is sufficient.

Fig. 1-11 The content of a CO-2 fire extinguisher is determined by its weight. The content of the propellant of a dry-powder fire extinguisher is indicated by the pressure gage.

E. *Identifying Fire Extinguishers*

Fire extinguishers should be identified by color-coded marks, to indicate to anyone that may have a need to use them the types of fires for which they are suitable. A green triangle with the letter A identifies an extinguisher that is suitable for Class A (ordinary combustibles) fires; a red square with the letter B identifies an extinguisher that is suitable for Class B fires (flammable liquids); A blue circle with the letter C is for Class C fires (electrical equipment); and a yellow star with the letter D tells us that the extinguisher identified in this way is suitable for Class D fires (combustible metals).

Fig. 1-12 The symbol on the fire extinguisher indicated the type of fire for which it is suited.

III. *Safety on the Flight Line*

A. *Hearing Protection*

The extreme amount of energy released from the fuel by turbine engines has made the flight line of a modern airport a high-noise area. Continued exposure to this noise will permanently damage our ears and impair our hearing.

Most operators require that all personnel on the flight line and in any shop where the noise level is high wear some type of hearing protection device. These may be either external protectors similar to a pair of large earphones, or internal protectors that fit into the auditory canal of the ear. Either type prevents the sound energy damaging the inner mechanism of the ear.

Be sure that the protection you use is adequate for the noise level in which you work. We can build a tolerance that will allow us to become *accustomed* to the noise, but continued exposure to high noise levels, even though they are below the threshold of pain, is still damaging.

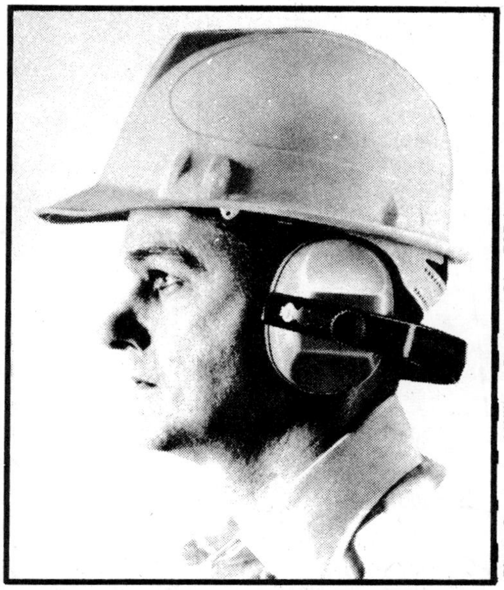

Photo Courtesy of Willson Products Div., ESB Inc.

Fig. 1-13 Hearing protection must be worn in all areas of high noise.

B. Foreign Object Damage

It is extremely important that the flight line of a modern airport be kept clean of any objects that can be sucked into the intake of a jet engine, or that can be blown by the jet blast or the propeller into hangars or into other aircraft.

All maintenance technicians should develop the habit of picking up all loose hardware and rags they find on the ramp and depositing them in suitable containers where they cannot cause any damage.

Even more important than not allowing foreign objects to get drawn into an aircraft is to be sure that we do not become a fatality ourselves, from contact with an aircraft propeller or rotor, or by being burned by the jet exhaust. If it is necessary to approach an aircraft whose engine is running, be sure that you have the attention of the person in the cockpit so he will know where you are and what movements he can expect. The danger of a propeller is so obvious that we hardly need mention it, but sadly enough, many people have been killed by walking into a spinning propeller.

Turbojet aircraft are dangerous from both the front and the rear. In Fig. 1-14 we have a diagram that illustrates the danger areas around a large jet aircraft. When an engine is running, it moves such a huge volume of air that a low-pressure area in front of the intake is produced that is capable of drawing a person into the engine. It will certainly ingest such items as hats, clipboards, and loose items of clothing that are allowed in this area.

Behind the aircraft in the exhaust area, the high-velocity jet of hot air can cause severe damage to people and equipment alike.

C. Safety Around Helicopters

Helicopters are a unique type of aircraft often operated in terrain in which an airplane could not possibly be used, and we must often approach and depart from them when the engine is operating and the rotor turning.

In Fig. 1-15 we see some of the more commonly accepted safety practices to use around operating helicopters. We must remember as aviation maintenance technicians operating with helicopters that many of the passengers we carry into remote areas are not familiar with aircraft and it is our responsibility to watch that they do not endanger either themselves or the helicopter.

IV. Tiedown Procedures

The lightweight construction of an aircraft, along with its airfoil-shaped wings and tail surfaces or rotors, makes it highly susceptible to damage from wind.

The most violent windstorms and the ones that do the most severe damage to aircraft in the United States are tornados that occur during the early summer months, generally in the central states, and hurricanes which occur along the Gulf and East coastal regions, usually during the months of August through October. Localized thunderstorms can occur throughout the year and may appear in almost any part of the country.

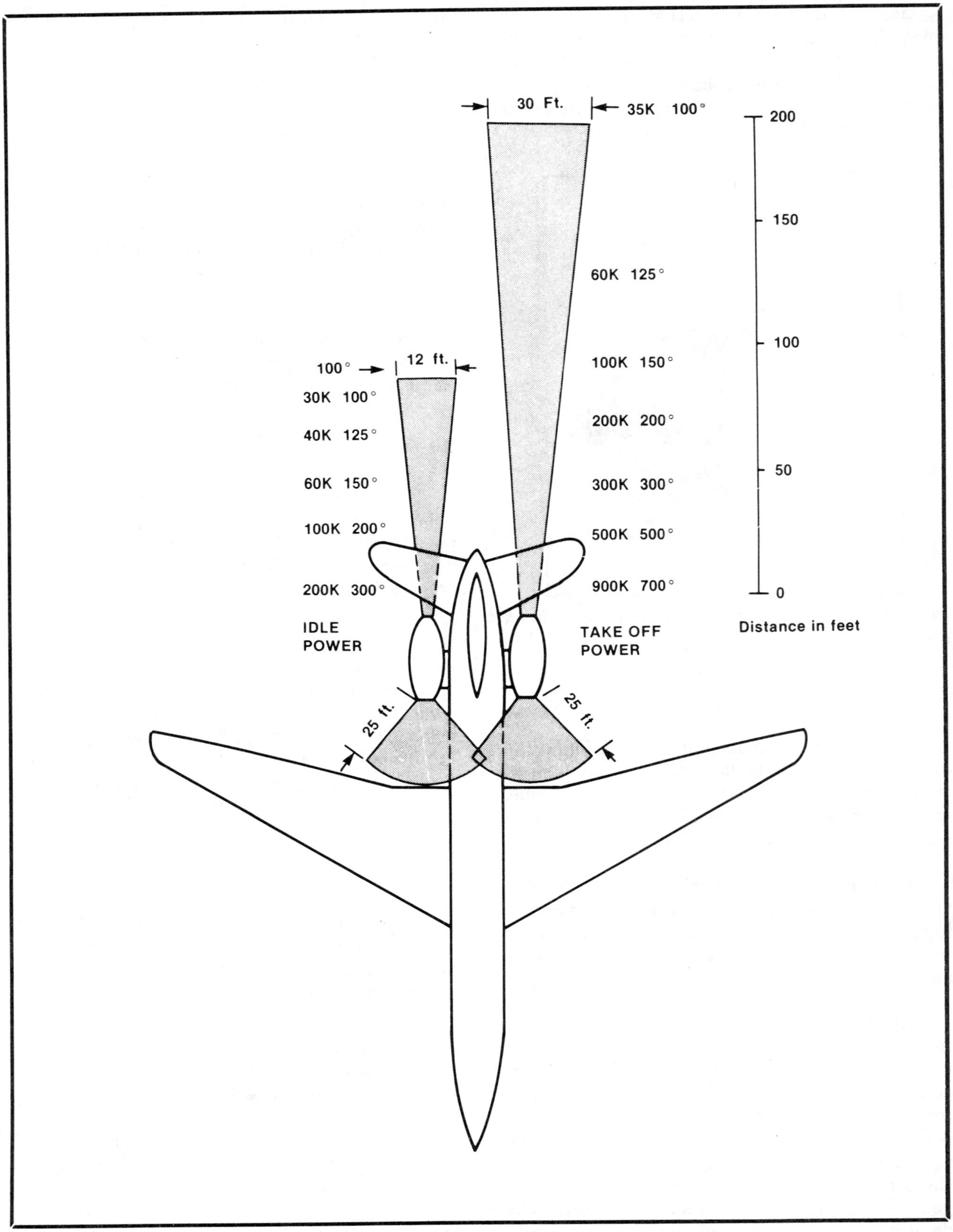

Fig. 1-14 Danger areas around a jet aircraft.

SAFETY AROUND HELICOPTERS

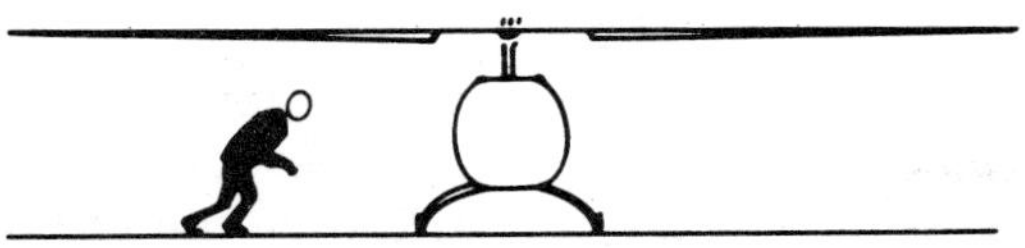

1. Approach or leave machine in a crouching manner (for extra clearance from main rotor).

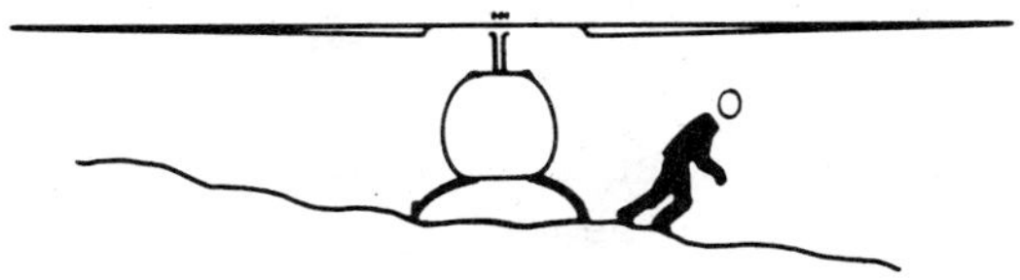

2. Approach or leave on the down slope side (to avoid main rotor).

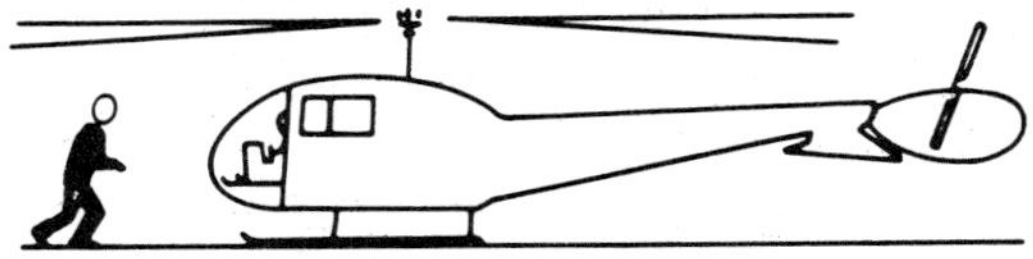

3. Approach or leave pilot's field of vision (to avoid tail rotor).

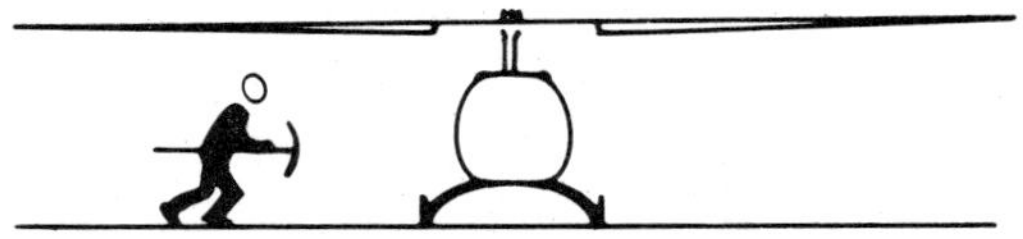

4. Carry tools horizontally, below waist level (never upright or over shoulder).

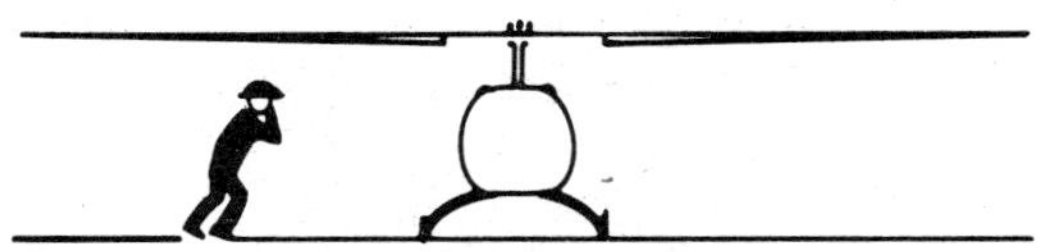

5. Hold onto hard had when approaching or leaving machine, unless chin straps are used.

6. Fasten seat belt on entering helicopter and leave it buckled until pilot signals you to get out.

7. If leaving machine at the hover, get out and off in one smooth, unhurried motion.

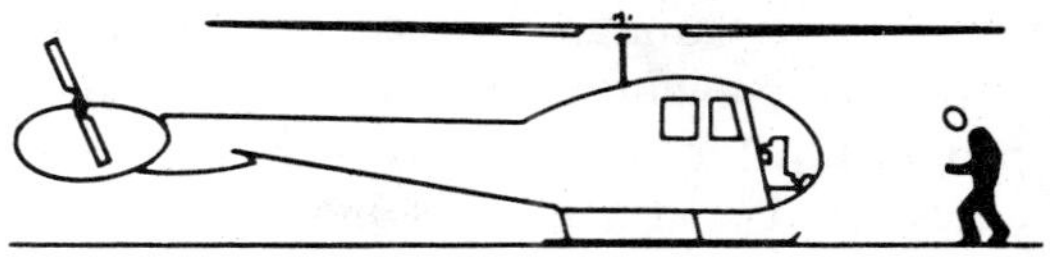

8. Do not touch bubble or any of the moving parts (tail rotor linkage, etc.).

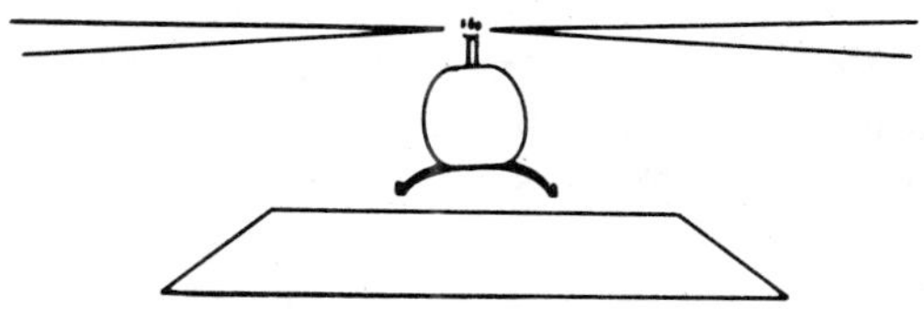

9. Keep landing pad clear of loose articles.

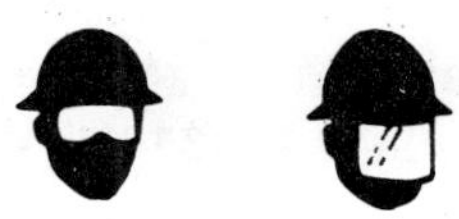

10. Loading assistants should always use plastic eye shields.

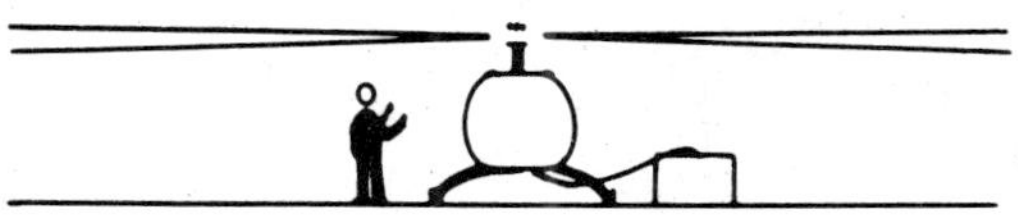

11. After hooking up cargo sling, move forward and to side to signal pilot.

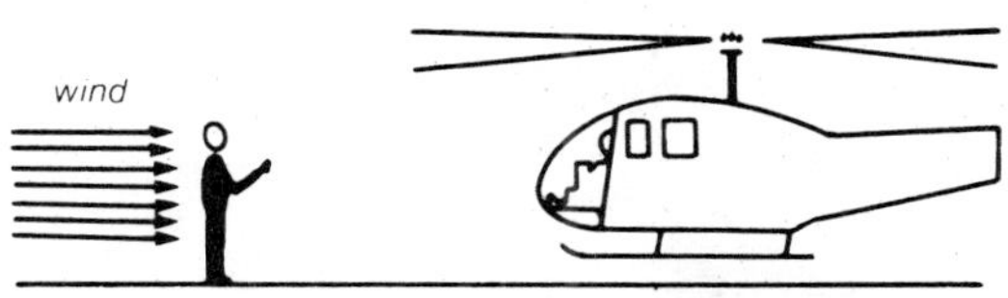

12. When directing machine for landing, stand with back to wind with arms outstretched toward landing pad.

13. When directing pilot by radio, give no landing instructions that require acknowledgement as pilot will have both hands busy.

Fig. 1-15 Safety around helicopter

Regardless of the type of storm, the damage can be severe for aircraft that are not protected, but most of the damage can be minimized if proper protective measures are taken. Naturally the best protection for the aircraft is to fly it out of the area where the storm is predicted to hit, but if this is not possible, the aircraft should be protected by storing it in a storm-proof hangar or by tying it down securely.

A. Preparation of the Aircraft

When a windstorm is anticipated, the aircraft should have all of the controls securely locked to prevent the surfaces banging against the stops and causing structural damage. Most aircraft are equipped with internal control locks that hold the surfaces in a streamlined position, but since these locks secure the cockpit control, there is a possibility that severe forces put on the surface by the wind can damage the control actuating system. To prevent this, control surface battens can be used that hold the surface in a streamline position by clamping it against the fixed surface.

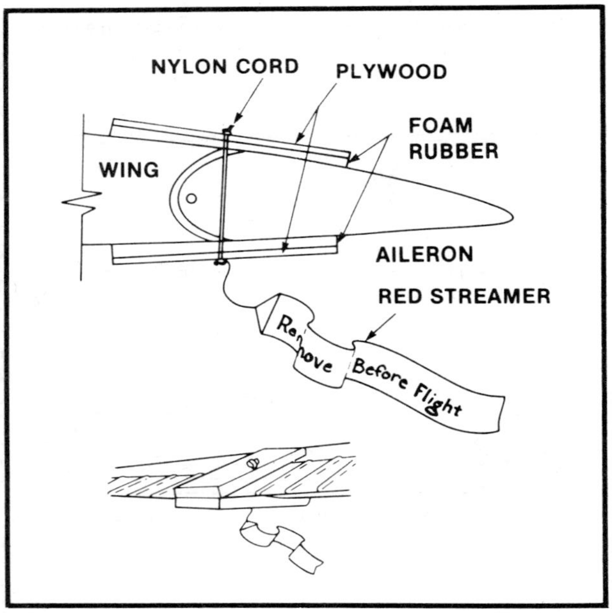

Fig. 1-16 Control battens are used to lock the control surfaces to prevent wind damage. They should have red streamers attached to help prevent their being inadvertently left on the aircraft when it is prepared for flight.

These battens should be lined with one-inch foam rubber, should be painted red, and should have a long red streamer attached to them so that they will be easily seen and not be inadvertently left on the controls when the aircraft is ready to fly.

If a tail wheel aircraft is tied down facing into the wind, the elevators should be locked in a fully up position so the wind will force the tail down; but if it is tied down facing away from the wind, the elevators should be locked fully down, so the wind blowing over them from behind will force the tail down.

If a severe wind is expected, spoiler boards may be lashed to the top surface of the wing to destroy the lift. These spoilers can be made of 2 × 2 boards on which a one-inch strip of foam rubber is attached with contact cement. Holes are drilled through the boards through which nylon rope is threaded. The nylon rope is tied around the wing so it will hold the spoiler parallel with the wing span, about one-fourth of the way back from the leading edge. Special care must be taken when securing the spoilers that the ropes are not pulled tight enough to damage the leading or the trailing edge of the wing. The place where the rope contacts these places may be protected with foam rubber or with scraps of carpet.

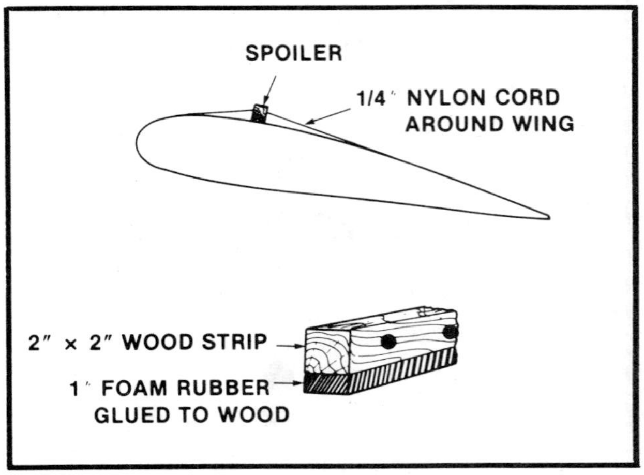

Fig. 1-17 If an airplane must be left outside in a windstorm, spoilers may be lashed to the top of the wing to prevent lift being generated.

The doors and windows should be secured so they cannot be blown open, and all of the engine openings should be covered so no blowing dirt can be forced into either the engine compartment or into the engine itself. The pitot head should also be covered so water or dirt cannot be blown into it.

B. Tiedown Procedures

1. Landplanes

Most airports have a tiedown area with anchors permanently embedded into the hard-surfaced ramp. These may be in the form of individual anchors or tiedown cables. With either method, the aircraft should be secured so that it is headed as nearly into the wind as is practical and should have as much separation between adjacent aircraft as possible.

The nose- or tail-wheel should be locked in the straight-ahead position so the aircraft cannot slew around in the wind, and the tiedowns should be secured to each wing and to the tail.

Aircraft may be tied down with either rope or chain, with ropes normally providing the strongest means of securing the aircraft. Manila rope, however, has a double disadvantage in that it has a tendency to shrink when it gets wet, and it will mildew and rot from exposure to weather and so must be periodically replaced. Nylon is the strongest material for rope, with Dacron and yellow polypropylene also providing sufficient strength for the purpose.

The rope should be attached to the aircraft tiedown rings with a bowline knot. It should not be pulled tight enough to put a strain on the wing, but must be tight enough to prevent the aircraft jerking against the rope. The proper tension allows about one inch of movement. If manila rope is used, a little extra slack must be allowed in the event the rope shrinks.

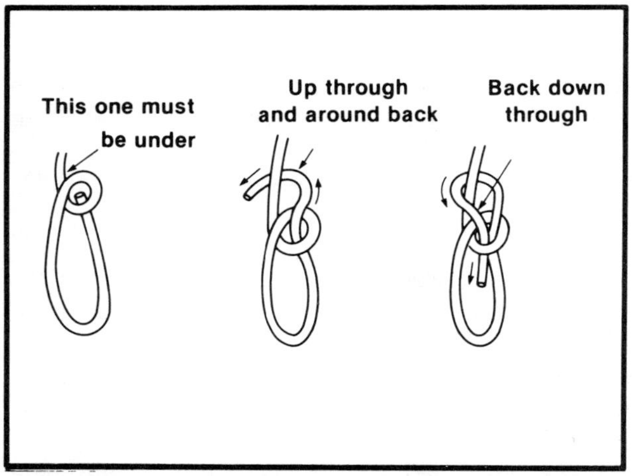

Fig. 1-18 The bowline is the most generally used knot for tying an aircraft with ropes.

Chains are used at many airports. They have a much longer life than rope and are easier to use, but they are not as strong as the proper size rope. If chains are used, they must be secured to the aircraft by passing the chain through the tiedown ring and then sticking one link through a link in the standing chain and fastening it in place with the snap to prevent the link slipping out. Do not allow the snap to take any of the strain, as it is not made for this purpose.

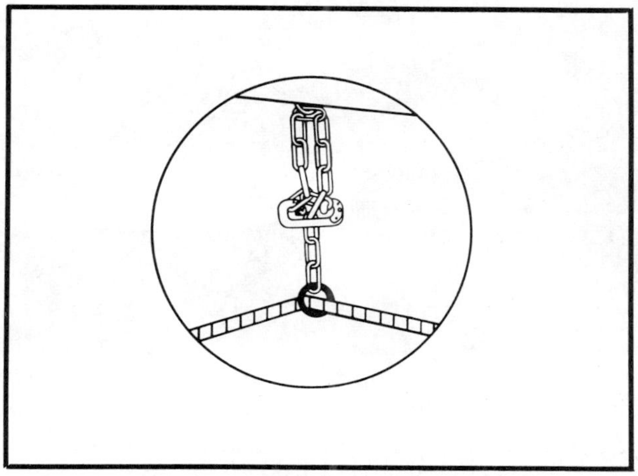

Fig. 1-19 When securing an aircraft with a chain, do not depend upon the snap to carry any strain. The snap should be used only to secure a free link through a standing link.

Airplanes with nose wheels should be tied down to prevent damage from severe wind by using, in addition to the wing ropes, one rope through the nose gear tiedown ring and two ropes through the tail tiedown ring. The ropes from the tail should pull away at about a 45-degree angle to each side of the tail.

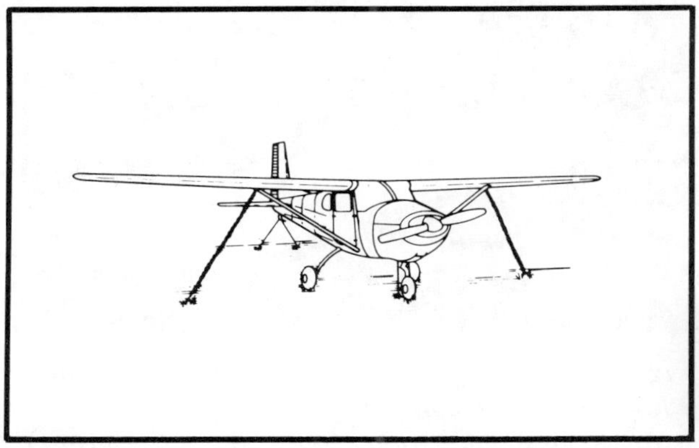

Fig. 1-20 The proper method of securing an airplane with ropes

Fig. 1-21 When a seaplane is secured in a high wind, some of the float compartments may be filled with water to add weight to help hold the airplane down.

When tying an aircraft down, be sure that the wheels are blocked with proper fitting chocks in front of and behind the wheels.

2. Seaplanes

If a seaplane is to be secured against high wind, it may be either towed into shallow water or onto the beach and some of the float compartments should be flooded with water to add weight. This will assist the tiedown ropes in holding it down.

Fig. 1-22 If a ski plane is to be secured outside in a high wind, snow may be packed over the ski and water poured over it to freeze it to the ground. This will assist the tiedown ropes holding the aircraft.

3. Skiplanes

Many times a ski-equipped airplane will be caught out on a trip where it must wait out a high wind condition without the possibility of protecting it from being blown away. When this happens, it is possible to pack loose snow over and around the skis and pour water over them so they will freeze in. The tiedown ropes may also be frozen into ice to secure them.

4. Helicopters

The critical nature of a helicopter rotor system makes it extremely important that the manufacturer's operational manual be followed in detail when tying down a helicopter. A few general considerations, however, may be noted:

The helicopter should be headed into the anticipated wind, clear of any other helicopters, airplanes, or buildings by at least a rotor-span. The wheel brakes should be set and chocks placed in front of and behind the wheels. If the helicopter is mounted on skids, these should be securely tied to the ground tiedown facilities by attaching the ropes to the points specified by the helicopter manufacturer. Position the main and tail rotor blades as specified by the manufacturer and secure them to the helicopter structure by the method described in the operations manual.

Fig. 1-23 Helicopter rotor blades must be secured by the method specified in the helicopter operations manual.

V. Jacking and Hoisting

A. Jacking

1. For single wheel, brake, or tire service

It is often necessary to lift only one wheel from the floor to change a tire or to service a wheel or brake. For this type of jacking, some manufacturers have made provisions on the strut for the placement of a short hydraulic jack. Be very sure when lifting the aircraft by this method that you do not place the jack under the brake housing or in any location that is not specifically approved by the manufacturer. It is usually recommended that both wheels *NOT* be lifted off of the floor at the same time from the landing gear struts.

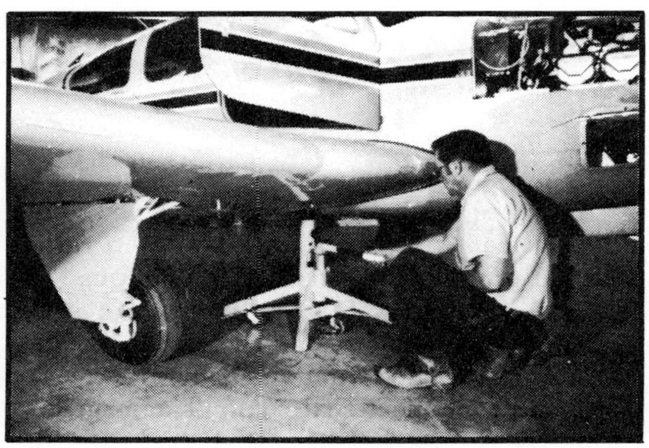

Fig. 1-24 When lifting a single wheel off of the ground, be sure the jack is positioned as recommended by the airframe manufacturer.

Some aircraft have a tendency, when jacked from the struts, for the wheels to move sideways and tilt the jack as the weight is removed from the tire. If this should occur, lower the jack and straighten it, and then raise the wheel again. To prevent the aircraft moving while it is on the jack, the wheels that are not jacked should be securely chocked.

2. For retraction tests and landing gear servicing

The most general way for an aircraft to be jacked is from jack pads located on the main wing spar. Many nose-wheel-type aircraft have a third jack pad located on the fuselage to raise the nose. Other aircraft require the tail to be pulled down to raise the nose.

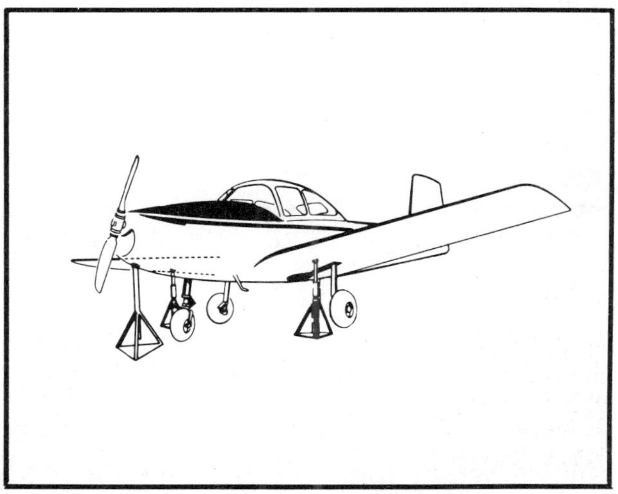

Fig. 1-25 When lifting the aircraft off of the ground for a retraction test, be sure that all of the jacks are raised and lowered evenly to prevent the aircraft tipping and falling off of the jacks.

The most important consideration when jacking an aircraft is to follow the manufacturer's instructions in detail.

Be sure to use the proper jacks and raise them evenly, so that the aircraft will be level, and there will be no tendency for it to slip off of the jacks. Most of the higher-capacity jacks have screw-type safety collars to prevent the jack inadvertently retracting. Be sure that these collars are screwed down as the airplane is raised. Jacks that do not have the screw-type safety will usually have holes drilled in the strut so lock pins can be inserted to guard against the jack retracting. Be sure to use whatever method of safety is provided.

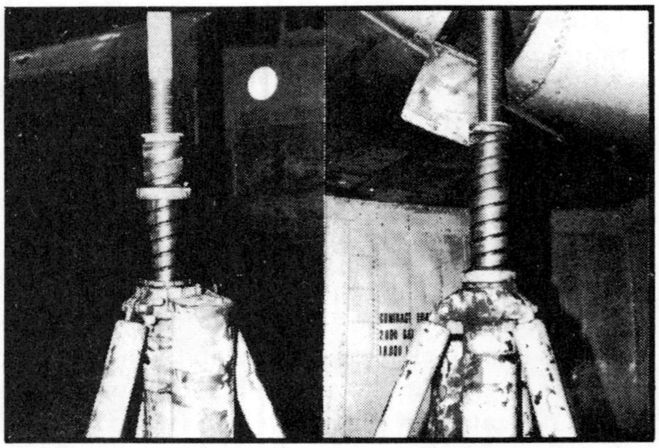

Fig. 1-26 The safety collar on the jack should be used to prevent fluid leakage causing the jack to inadvertently retract.

Many of the lighter aircraft can be jacked from only the main spar position by securing a weighted stand to the tail tiedown ring. Be sure that there is the amount of weight specified by the manufacturer in the stand, or else tie the tail to a tiedown ring embedded in the hangar floor.

Some aircraft can have their tail held down by weights placed on the main spar of the horizontal stabilizer. Before using this procedure, be sure that it is approved by the aircraft manufacturer, and that only the recommended type of weight be used.

Guard against any movement within the aircraft when it is on jacks, as shifting the weight behind the jack could cause the aircraft to tilt enough to fall off of the jack.

Before lowering the aircraft, be sure to remove all of the work stands and equipment that fit comfortably when the aircraft is raised, but which will cause damage if the aircraft is lowered onto it. Be sure that the landing gear is down and locked and then lower the jacks evenly. It is possible for some landing gear to produce a side load on the jacks as the weight is taken by the tires, and this must be watched to prevent this load causing the jack to tip. Be sure that the oleo struts do not bind and hold the aircraft. If they do bind enough to allow the jack to be lowered away from the wing, and the strut should suddenly collapse, it can drop the airplane back onto the jack and cause serious damage.

B. Hoisting

Many times an aircraft must be hoisted rather than jacked to lift it off of the hangar floor. When hoisting, be sure to follow the manufacturer's recommendations in detail. Use a hoist of sufficient capacity, and where necessary, use spreader bars between the cables to prevent side loads on the attachment points.

Fig. 1-27 Be sure to follow the aircraft manufacturer's recommendations when hoisting an aircraft.

VI. Ground Movement of Aircraft

A. Taxiing

Airplanes and helicopters are designed to fly, and movement on the ground is often a rather awkward procedure. Be sure that only qualified persons authorized to taxi the aircraft are at the controls, and before starting the engine, be sure that the area, both in front and behind, is clear of people and equipment that could be damaged, or that could cause damage to the aircraft.

It is usually difficult to assure that there is sufficient clearance between the wings or rotor

Fig. 1-28A Standard hand signals for ground movement of an airplane.

Fig. 1-28B Standard hand signals for ground movement of a helicopter.

and any buildings or other aircraft, so before moving, it is a good policy to station signalmen where they can watch the wings or rotor and any obstructions. In Fig. 1-28 we have the standard signals that may be used between a ground signalman and the pilot of both airplanes and helicopters. Whatever signals are used, be sure that both persons understand exactly what they mean so there can be no misunderstanding at a crucial time. The signalman has the responsibility of remaining in a position so he will be visible from the cockpit, and to assure that he can be seen at all times, he should be certain that he can see the pilot's eyes while he is directing him.

Busy airports usually require radio contact between the aircraft and the control tower when an airplane is taxiied onto any of the taxiways or runways. In the event of lack of radio contact, or radio failure, the control towers are equipped with highly directional light guns so that they can direct a high intensity beam of light to the aircraft. A light code is used that should be familiar to one operating an aircraft on a controlled airport.

Flashing green light—OK to taxi.

Steady red light—Stop.

Flashing red light—Taxi clear of the runway.

Flashing white light—Return to the place you started from.

Alternating red and green flashes—Use extreme caution.

B. Towing

It is often necessary to move an aircraft without using its engines. Large aircraft are towed with a tractor or special towing vehicle and are connected with a special tow bar. Extreme care must be used to not move the aircraft too fast and to be sure that there is always sufficient clearance between the wings and any obstructions. A qualified person should be in the cockpit to operate the aircraft brakes when needed, as the brakes on the towing vehicle are usually insufficient to stop the momentum of a large aircraft that is moving very fast.

The nose gear of most aircraft has a very definite limit to the amount it can be turned, and it is easy when towing to exceed these limits if care is not taken. Some aircraft have a method of disconnecting the torque links so the nose wheel can be swiveled to facilitate maneuvering.

Fig. 1-29 When towing an aircraft, be sure that the limits of nose wheel movement are observed, as the towing stops can be damaged if these limits are exceeded.

Small aircraft can be moved by hand, and damage can be inflicted on these aircraft by improper procedures.

Do not move an airplane by pulling on the propeller. The propeller is designed to move the aircraft through the air, but the thrust it is designed to produce is uniform. Moving the airplane by pulling on one blade will put an assymetrical load on both the propeller and the engine that can be damaging.

Be sure to use a tow bar. These bars normally attach to the nose wheel and are used to apply the force for both moving and steering the airplane.

When pushing an aircraft, be sure that you push only at points that are specified by the aircraft manufacturer as satisfactory for pushing. Do not push on the control surfaces nor in the center of a strut.

VII. Ground Servicing Equipment

A. Electrical Auxilliary Power Units

Almost all large aircraft require more power for starting than their batteries can supply, and

Fig. 1-30 Auxiliary power cart for supplying electrical current when the engine is not running.

for this reason auxilliary power units, or APU's, are available for supplying power for operating electrical equipment when the engine is not running and for supplying the high current demands for starting. Some of the larger APU's also have a supply of 115-volt 400-hertz AC, as well as the DC.

The batteries used in most of the smaller aircraft have a very limited current capacity, and for cold weather starting, the starter can demand more current than the battery is capable of supplying. For this reason, most airports are equipped with battery carts and cables that can be used to assist the aircraft battery for starting. It is extremely important when installing the battery on the chart that its polarity be correctly attached to the cable so it will be the same as that of the battery in the airplane. Reversed polarity can damage the alternator as well as the battery.

B. Hydraulic Power Units

Many aircraft have hydraulically retracted landing gear that must be periodically tested for operation. To do this, the aircraft is lifted off of the hangar floor with jacks and the landing gear is cycled through its reactions and extension cycle until you can be assured that it operates properly without any indication of binding or of excessive play or looseness.

These hydraulic power supplies, or "mules" as they are colloquially called, connect into the aircraft hydraulic system at the engine-driven pumps by the use of line-disconnect fittings. When the fitting is unscrewed from the pump, it seals off both ends and when it is screwed onto the hydraulic power supply, it automatically opens so it can supply pressure for the aircraft.

Be sure that all of the lines are clean so no dirt or contamination can get into the aircraft system. These power supplies do not furnish fluid for the aircraft; rather, they use the fluid in the aircraft. Some of the fluid, however, remains in the pump and lines of the unit, and it is extremely important that the unit only be used with one type of fluid.

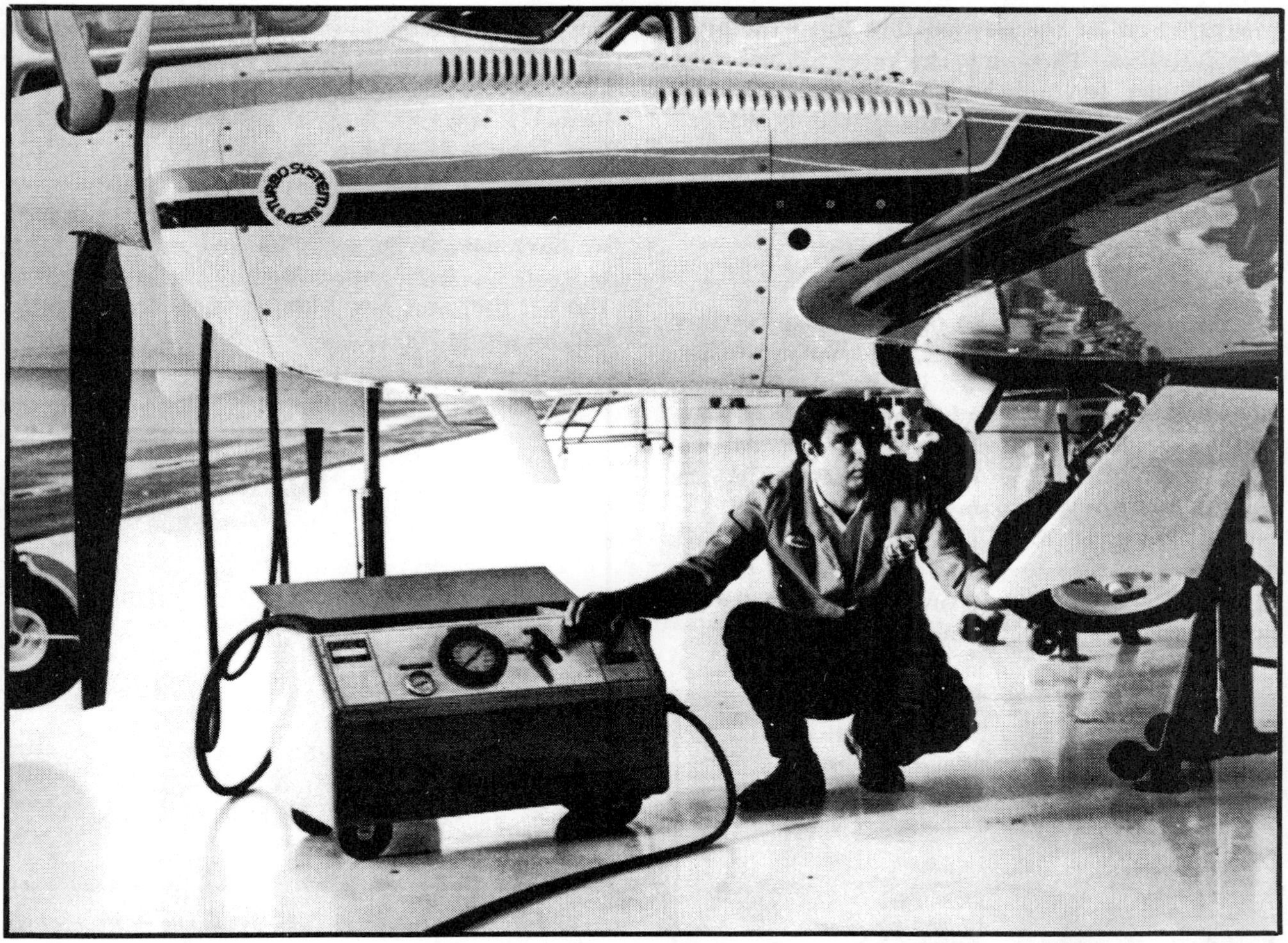

Fig. 1-31 A hydraulic power unit replaces the engine-driven pump for producing hydraulic pressure for retraction tests.

C. Oxygen Servicing Equipment

Modern aircraft fly at altitudes where life support systems are needed, and even though most of these aircraft are pressurized, emergency oxygen must be carried in the event of failure of the pressurization equipment.

Military aircraft almost universally use liquid oxygen, but most civilian aircraft carry high-pressure gaseous oxygen in steel cylinders, or bottles, under a pressure of around 1,800 to 2,000 psi.

These systems are serviced from oxygen carts on which there are usually six high-pressure bottles of oxygen and one of nitrogen. The nitrogen is used for filling hydraulic accumulators and oleo shock struts, and so to prevent inadvertently mistaking it for oxygen it is usually laid on the cart facing the opposite way from the oxygen cylinders.

Fig. 1-32 A service cart for supplying high-pressure oxygen to charge the aircraft oxygen system

The oxygen cylinders are all manifolded together and connected to the aircraft service port. When servicing an oxygen system, first open the valve on the cylinder having the lowest pressure and let the oxygen flow until the pressure stabilizes. Then shut the valve off and open the cylinder having the next lowest pressure. Continue this process until the system is charged from the cylinder having the highest pressure.

VIII. Aircraft Fueling

Aircraft fueling is an operational procedure that is conducted more frequently than any other. It must be done under a wide variety of conditions and is often done under the pressure of time, and so the knowledge of and use of correct procedures are of extreme importance for safety as well as efficiency. In commercial operations, the person doing the fueling is the personal contact between the operator and the customer, and so appearance, attitude, and professionalism are important as well as technical knowledge.

Fig. 1-33 An aircraft being fueled at the fuel tank filler neck. This is called overwing fueling.

A. Types of Fuel

1. Gasoline

Reciprocating-engine aircraft use high-octane aviation gasoline as their fuel. This is a special blend of gasoline with such additives as tetraethyl lead and ethylene dibromine and is dyed to indicate its rating.

It is extremely important when an aircraft is fueled that the grade of fuel required by the engine be used. The Federal Aviation Regulations require that all fuel tanks be clearly marked with the grade of fuel required. In civilian aviation, we have three grades of fuel in common use: Grade 80, formerly called 80/87, is dyed red; grade 100, formerly 100/130, is dyed green; and grade 100 Low-Lead is dyed blue. The military uses 115/145 gasoline for some of their large reciprocating engines and this fuel is dyed purple. In the past we have used 91/96 gasoline that was dyed blue, but this has been superseded by the grade 100 or 100-LL fuel and any blue gasoline found today will be grade 100-LL.

Fig. 1-34 An aircraft fuel tank must be clearly marked with the proper grade of fuel required and the capacity of the tank.

2. Turbine fuel

There are basically two types of turbine fuel in use today, but within these two classifications there are differences because of the operating temperatures to which the aircraft is subjected.

Jet A is a kerosene type of fuel and is used by almost all of the civilian jet aircraft. It has a very low vapor pressure and is suitable for use at the high altitudes flown by jet-powered aircraft.

Jet B is a blended fuel that contains both gasoline and kerosene fractions. The military forces use it because of the tremendous amounts of this fuel that are used daily. Its vapor pressure is higher than that of Jet A, so it must be handled in the same manner, and with the same precautions as aviation gasoline.

Fig. 1-35 Jet aircraft are often fueled from underground hydrants, using a fuel servicing cart or truck.

B. Fire Hazard

1. Volatility

The volatility of a liquid is a measure of its ability to change from a liquid into a vapor, and it is usually expressed in terms of Reid vapor pressure. This is the pressure above the liquid required to prevent vapors escaping from the liquid at a given temperature. Aviation gasoline must have a vapor pressure of seven pounds per square inch, or less, at 100° F. Jet A has a vapor presure of less than 0.1 psi and Jet B has a vapor pressure of between two and three pounds per square inch, both at 100° F.

When an aircraft is fueled, vapors rise from the tank and these vapors can be ignited by a spark. The more volatile the fuel (the higher its vapor pressure) and the higher the outside temperature, the more vapors will be released, and the more caution is required when fueling. Because of the flammable nature of fuel vapors, no fueling or defueling should be done in a hangar or an enclosed area, and spilled fuel must be wiped up or washed away with water as soon as possible. It is extremely important that spilled fuel never be swept up with a dry broom, as the static electricity generated by the broom will ignite the vapors above the liquid fuel.

Fuel should be stored only in approved containers, and these containers must be kept closed and stored in a cool and isolated area that has been approved for fuel storage. Improper storage and handling of fuel can adversely affect insurance coverage.

2. Static electricity

When fuel flows in a hose, static electricity is generated, and an electrical potential difference is built up between the fuel hose and the aircraft fuel tank. If proper precautions are not taken, it is possible for a spark to jump to neutralize this charge. The spark will occur within the fuel vapors that are rising from the tank filler opening, and a serious fire can result.

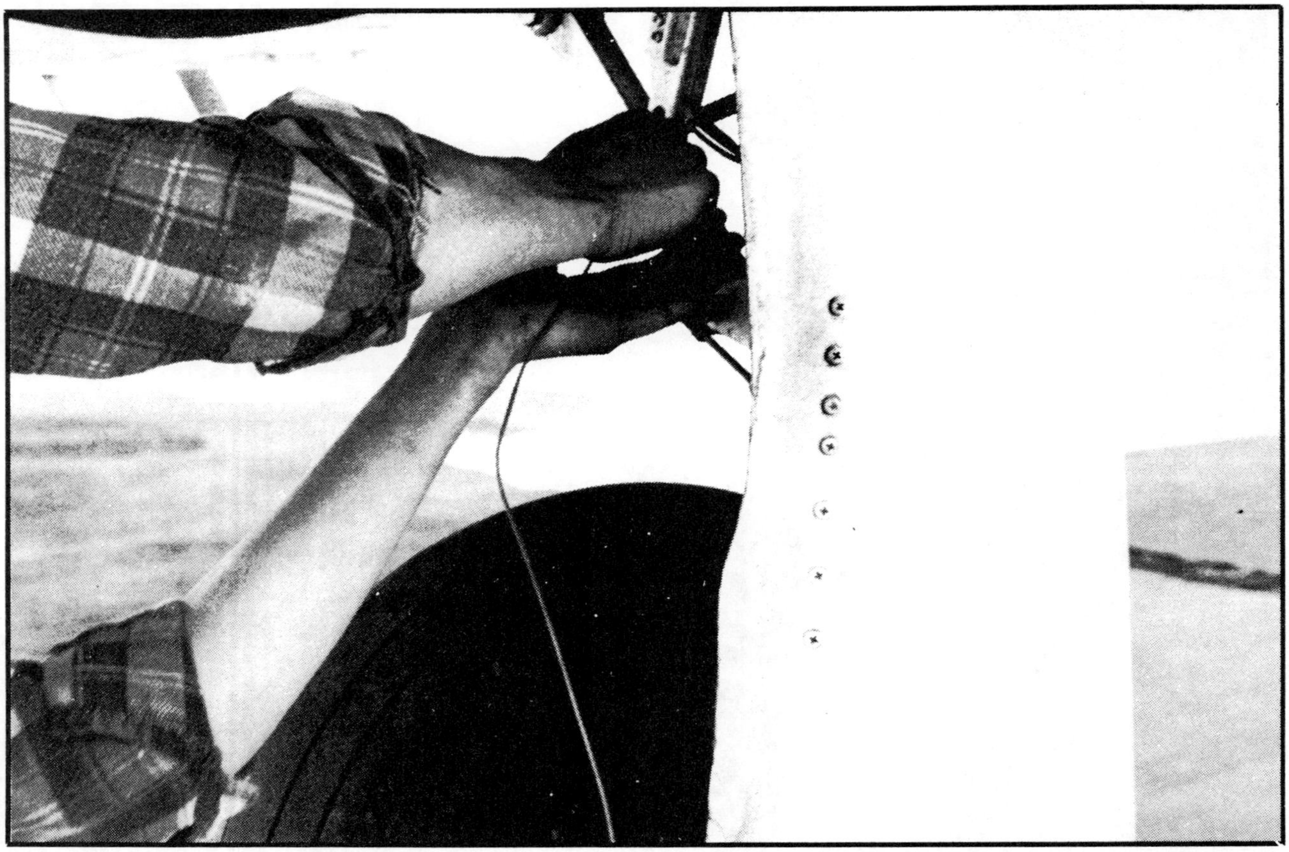

Fig. 1-36 Before opening the fuel tanks or connecting the fuel hose, be sure that the aircraft and the fuel truck are connected together by an electrical bonding cable.

Before beginning the fueling operation, the fuel truck or pit must be connected to the aircraft with a static bonding wire, and before removing the fuel tank cover, the nozzle must be connected to the structure with the bonding wire that is provided for this purpose. When the proper bonding connections are made, the danger of fire from static electricity is minimized.

C. Contamination Control

1. Water

Water has always been one of the major contamination problems with aviation fuel, because it will condense out of the air in the storage tank, the fuel truck, and even in the aircraft fuel tanks themselves.

When the tank truck is filled from bulk storage facilities, elaborate precautions are taken to trap any water before it enters the truck tank. Water is heavier than the fuel, and it will settle to the bottom of the tank where it collects in sumps and is periodically drained out. A floating suction line draws the fuel from near the top, and as it leaves the storage tank it passes through a filter that removes any solid contaminants and any water. These filters are normally of type that will shut off the flow of fuel if water is detected.

The tank trucks are equipped with filters and sight gages that allow the operator to be sure that the fuel being delivered to the aircraft is clear and bright and has no visible trace of water or other contaminants.

If there is any suspicion of water in the fuel, a test may be made by dissolving a capsule of crystals in a small sample of the fuel. If the level of water in the fuel is higher than is allowed, the normally gray crystals will change their color to pink. If this change is noted, the fuel must be circulated through a water entrapment filter before it can be into the aircraft.

2. *Solid contaminants*

It is extremely important that all of the fuel nozzles be protected with properly fitting dust caps when they are not in use. These caps will prevent dirt or dust collecting inside of the nozzle and being dropped into the aircraft fuel tank.

3. *Microbial growths*

Turbine fuel is more viscous than aviation gasoline, and it is possible for it to contain water in a suspended, or entrained, form that will not settle out so it can be trapped in a sump or by a filter. When the aircraft flies in the low-temperature air of high altitude, this water will precipitate out and can collect on the strainers and freeze, actually shutting off the flow of fuel to the engine.

Another serious problem is the fact that microbes (bacteria, fungus, and yeasts) can live in this water and feed on the hydrocarbon fuel. This microbial growth occurs at the interface between the water and the fuel and forms a scum which can hold the water against the structure of the fuel tank and cause corrosion.

The problem of microbial growths in turbine aircraft fuel tanks has been overcome to a great extent by the use of an additive in the fuel that serves two purposes. It lowers the freezing temperature of any water that may precipitate from the fuel, and it acts as a biocidal agent that kills any of the microbial growths that try to live within the water. This additive may be put into the fuel at the refinery, the bulk storage facility, or in some cases may actually be put into the fuel as it is pumped into the aircraft tanks.

Fig. 1-37 A biocidal agent is added to turbine fuel to prevent microbial growths that cause corrosion in the fuel tanks.

D. *Fueling Procedures*

1. *From a fuel truck*

Aircraft may have the fuel pumped directly into the tank from over the wing or from a single point source under the wing. For simplicity of this explanation we will consider the fueling procedures from a fuel truck to be done over the wing, and when we discuss the fueling from a pit we will discuss the under-wing, or single-point method.

Before driving the fuel truck to the aircraft, be sure that the sumps have been drained and that the sight gages show that the fuel is bright and clear. The fire extinguishers must be in place, and they must be fully charged.

Approach the aircraft with the fuel truck parallel to the wings, and stop the truck in front of the aircraft. Set the parking brake on the truck and connect the static bonding wire between the truck and the aircraft.

Put the mat over the wing so the fuel hose will not scratch the finish, connect the static bonding wire between the nozzle and the aircraft, and remove the fuel tank cover. Remove the dust cap from the nozzle, and when inserting the nozzle into the tank be sure that the end of the nozzle does not contact the bottom of the tank, as it will dent the thin metal.

2. *From an underground storage system*

Most of the large airports that service transport aircraft with great amounts of fuel have underground storage tanks and buried fuel lines that allow the aircraft to be fueled without having to carry the fuel to the aircraft in tank trucks. Since most of the aircraft that are fueled from this type of system use under-wing fueling, we will discuss this method here.

A service truck having filters, water separators, and a pump is driven to the aircraft and its inlet hose is connected to the underground hydrant valve. The discharge hose or hoses from the servicer are attached to the fueling ports on the aircraft, and with a properly qualified maintenance person in the aircraft monitoring the fuel controls, the valves are opened and the pumps started. The person monitoring the fuel controls can determine the sequence in which the tanks are filled and can shut off the fuel when a sufficient load has been taken on board.

Fig. 1-38 High rates of fuel flow can be put into an aircraft system from the underwing fueling ports.

E. Defueling

It is sometimes necessary to remove fuel from an aircraft for either maintenance reasons or because of a change in flight plans after the aircraft was serviced. Defueling is carried on in much the same manner as fueling, with the same safety precautions being taken.

Never defuel an aircraft inside a hangar or in any area where there is not adequate ventilation. Be sure that all of the proper safeguards are taken with regard to neutralizing any static electricity that builds up when the fuel flows through the lines.

It is extremely important when defueling an aircraft that the fuel taken from the aircraft be carefully checked for contamination and for the proper identification, before it is pumped into a storage tank where it can be delivered to another aircraft.

If the aircraft is defueled into drums, be sure that the drums are clean and that the bungs are replaced and tightened immediately after the drums are filled.

F. Review of Safety Procedures

Let us review the safety procedures that we must observe when fueling or defueling an aircraft:

Be sure that only the correct grade of fuel is put into the aircraft. Remember that aviation gasoline comes in various grades and the wrong grade can cause severe damage to the engine. Turbine fuel in a reciprocating engine can cause severe detonation and engine failure, and the improper use of aviation gasoline in a turbine engine can also be harmful.

Be sure that the fuel truck or servicer is properly bonded to the aircraft and the fuel nozzle is bonded to the structure before the cover is taken from the fuel tank.

Wipe up spilled fuel or flood it with water. Do not sweep spilled fuel with a dry broom.

Be sure that there are no open fires in the vicinity of the fueling or defueling operation.

Be sure that the fire extinguishers that are available are of a type that are suitable for a Class B fire (either CO-2 or dry powder are most generally used).

Protect the aircraft structure from damage from the fuel hose and from the nozzle.

Be sure that the radio or radar are not used during fueling or defueling, and that no electrical equipment is turned on or off except for the equipment needed for the fueling operation.

When defueling, be sure that the fuel is not contaminated if it is to be used again.

Be sure that the filters in the tank truck or servicer remove all traces of water and contamination and that the fuel pumped into the aircraft is bright and clear.

If a biocidal additive is required, be sure that it is mixed with the fuel in the proper concentration.

If the aircraft is being fueled in the rain, be sure that the tank opening is covered to prevent water getting into the tank.

Be sure that dust covers and caps are placed over the end of the fuel nozzles and any open fuel lines when they are not in use.

Drive the tank truck parallel to the wing of the aircraft, and be sure that the parking brake is set so the truck cannot roll into an aircraft.

When using pressure fueling under the wing, be sure that the pressure used and the delivery rate are those specified by the manufacturer of the aircraft.

If any fuel is spilled onto your body, wash it off with soap and water as soon as possible. Do not wear any clothing on which fuel has been spilled.

IX. Engine Starting Procedures

A. Reciprocating Engines

Starting an aircraft engine is such a specialized procedure and it varies so much within the individual engine and aircraft that no general procedure can be used. Improper starting procedure can cause fire or can damage the engine, so before starting any aircraft engine, be sure that you are adequately checked out by a person familiar with the installation.

There are some general considerations we should take when starting any engine. Before starting any engine, be sure that the area in front of and behind the aircraft is clear. The aircraft should not be parked in an area of loose gravel, since small stones will be sucked up into the propeller and will damage the blades. Be sure that the blast from the propeller does not blow dirt into any hangar or building and does not blow onto another aircraft.

There is always a possibility of fire when starting an engine, and to minimize the possibility of the fire spreading, there should be a fire extinguisher of adequate capacity available. For starting large aircraft where it is not possible to see the engine when it is being started, a fire guard should be stationed near the engine in case it should catch fire. Induction system fires are the type which occur most frequently. If the engine is over-primed and then fires back through the carburetor, the gasoline in the induction system will ignite. When this occurs, the best procedure is to continue cranking and start the engine if possible. The engine will suck the fire into the cylinders so no further damage can occur.

Large radial engines have a problem of oil seeping past the piston rings and getting into the lower cylinders. This oil forms a hydraulic lock and will seriously damage the engine if some of the cylinders should fire and drive the piston onto this oil. To be sure that none of the cylinders have a hydraulic lock, the engine should be turned through by hand or with the starter with the ignition off until all of the cylinders have passed through their compression strokes.

Almost all reciprocating engines are equipped with either a carburetor-heat or an alternate-air position on the carburetor air inlet system. For starting and for ground operation, these controls should be in the cold position, because heated air is not filtered.

Hand propping of aircraft engines is a procedure with which all mechanics should be familiar, as it is possible for a battery to become discharged when no auxilliary power is available. *It is important that no person ever attempt to prop an aircraft until he has been thoroughly checked out in the procedure.* This is something that cannot be learned from a book. There are, however, a few pointers that are helpful.

Be sure that the person in the cockpit is thoroughly familiar with the aircraft and that clear communication is established between him and the person doing the propping. Check to be sure the brakes are actually holding, by attempting to move the aircraft. When ready to pull the propeller to start the engine, call out the word "CONTACT." If the person in the cockpit has all of the controls properly set, he replies "CONTACT" and then turns the ignition switch on.

When actually pulling on the propeller, be sure that your footing is secure. Do not stand on wet grass. Stand close enough to the propeller

Fig. 1-39 It is extremely important that the person starting a turbine engine be thoroughly familiar with the procedure.

that you will be slightly overbalanced *away* from the engine rather than into it. Grasp the propeller by laying your palm over the blade and do not grip your finger around the trailing edge of the blade, as there is danger of the engine kicking back and pulling you into the propeller.

As soon as the engine starts, check the oil pressure gage. If there is no indication of oil pressure within the time limit specified in the aircraft operator's manual, the engine must be shut down and the cause of no pressure determined. If the oil pressure is good, the engine RPM should be adjusted to the point that the engine runs smoothest and is allowed to warm up at this speed.

B. Turbine Engines

Turbine engine aircraft usually have enough automatic sequencing done by the fuel controls so that starting is less involved than starting a reciprocating engine of comparable power. But the extremely high cost of turbine aircraft and the possibility of severe damage in the event of an improper start make it doubly important that only qualified personnel start these engines.

One of the problems that can occur when starting a turbine engine is the lack of sufficient electrical power to get the engine up to proper speed for it to become self-accelerating in its operation. Improper starts are usually categorized as one of two types.

1. Hot start

A hot start is one in which ignition occurs when there is an excess of fuel and insufficient air. The exhaust gas temperature will exceed the allowable limit and the engine will be damaged. The exhaust gas temperature gage must be monitored during a start and if it exceeds the limit, the fuel to the engine should be shut off immediately.

2. Hung start

A hung start is one in which the engine starts but does not accelerate enough for the compressor to supply sufficient air for the engine to become self-accelerating. The start is normal and the exhaust gas temperature is all right, but the engine will not accelerate to normal operating speed. If a hung start occurs, the engine should be shut down, the cause for insufficient starting speed corrected and another attempt made.

PART II

Maintenance Publications

I. Federal Aviation Regulations

The regulations, as we have mentioned, are the actual law, and it is the responsibility of an A&P mechanic to know what he can and what he cannot legally do. The FAR's are publishing by the Superintendent of Documents, U.S. Government Printing Office in Washington, D.C., and some are sold on a single-sale basis, with others only as a subscription. An Advisory Circular AC 00-44 is issued and periodically updated to give the current status of the Federal Aviation Regulations. The latest revision of this Advisory Circular should be used when ordering the FAR's.

Fig. 2-1 Federal Aviation Regulations are the guidelines by which the Federal Aviation Administration controls aviation.

II. Advisory Circulars

The Advisory Circulars are non-regulatory information issued by the FAA to enable us to understand and comply with the Federal Aviation Regulations. Their numbering system is keyed to the FAR's to which they relate:

Subject Number	Subject
00	General
10	Procedural Rules
20	Aircraft
60	Airmen
70	Airspace
90	Air Traffic and General Operating Rules
120	Air Carriers, Air Travel Clubs, and Operators for Compensation or Hire: Certification and Operations
140	Schools and other Certificated Agencies
150	Airports
170	Navigational Facilities
180	Administrative Regulations
190	Withholding Security Information, War Risk Insurance; Aircraft Loan Guarantee Program
210	Fight Information

Three times each year an Advisory Circular Checklist AC 00-2 is issued that lists all of the Advisory Circulars and gives the directions for

Fig. 2-2 Advisory circulars help us understand the Federal Aviation Regulations.

obtaining them. Most AC's are free and a copy may be obtained by writing to FAA in Washington, D.C. Other Advisory Circulars are more lengthy and there is a charge for them. Any AC for which there is a charge must be ordered from the Superintendent of Documents, U.S. Government Printing Office in Washington, D.C., or may be purchased at any of the GPO bookstores that are located throughout the United States in many of the major cities.

III. Type Certificate Data Sheets, Aircraft Specifications, and Aircraft Listings

When an aircraft is inspected, it must be determined that it meets the specifications under which it was certificated. Aircraft Specifications are available to the A&P for all aircraft that were certificated under the Civil Air Regulations (superseded by the Federal Aviation Regulations), and Type Certificate Data Sheets are issued for all aircraft that have been certificated under the Federal Aviation Regulations. Any of the older aircraft of which there are less than 50 still on the FAA registry, have their specifications included in the book of Aircraft Listings.

All of this information is available in six volumes that may be purchased from the Superintendent of Documents. The first five volumes are issued on a subscription basis and the sixth on a single-sale basis.

Volume	Contents
I	Single-Engine Airplanes
II	Small Multi-engine Airplanes
III	Large Multi-engine Airplanes
IV	Rotorcraft, Gliders, and Balloons
V	Aircraft Engines and Propellers
VI	Aircraft Listings and Aircraft Engine and Propeller Listing

IV. General Aviation Airworthiness Alerts

These alerts are issued on a subscription basis from the Superintendent of Documents as Advisory Circular AC 43-16. When the subscription is bought, a summary is sent that includes all of the pertinent alerts, and as new alerts are issued, they are sent to the holder of the subscription.

V. Airworthiness Directives

This vital information is sold in two subscriptions; Volume I is the *Summary of Airworthiness Directives for Small Aircraft*, and Volume II is for large aircraft. These are both published by the Government Printing Office, but because of their importance, they are sold and distributed by the FAA in Oklahoma City. A subscription for either volume includes a summary of all of the effective AD's and bi-weekly updates for a two-year period. The latest address and price is included in the Advisory Circular Checklist (AC 00-2).

VI. Aerofiche as a System of Keeping Publications Current

The mass of information that is required to keep current in the field of aviation maintenance is truly staggering, and it is difficult if not almost impossible to obtain the needed publications from the U.S. Government Printing Office. These problems have made the use of Aerofiche a practical means of keeping all of our information current. One publisher has put on microfiche all of the government publications that are applicable for aviation maintenance, as well as maintenance publications from most of the aircraft, engine, and component manufacturers. The information

Fig. 2-3 The tremendous amount of information needed for aviation maintenance is now available on microfiche.

from the industry includes service manuals, parts manuals, and catalogs.

An Aerofiche card is a piece of transparent film, approximately 4" × 6", which includes up to 88 pages of information. These cards are read in special Aerofiche readers, and many of these readers have a printer built into them that will provide a photocopy of the image you want.

VII. A.T.A Specification 100

The Air Transportation Association of America, an organization of air carriers in America, have established many standards and procedures that make airline operation more effective and efficient. One of the specifications of value to us in the maintenance field is A.T.A. Specification 100, which is a standard for the presentation of technical information. Because of this specification, maintenance information from any of the manufacturers of transport aircraft is arranged with the same numbers used to locate information in their technical publications.

For example, regardless of the aircraft manufacturer, all of the technical information on passenger oxygen would be indexed under the number 3520. In Fig. 2-4 we have a brief listing of the system numbers and the subsystem numbers of A.T.A. Spec. 100.

General Aviation aircraft manufacturers are in the process of standardizing their maintenance information, and A.T.A. Specification 100 will be used as the format for this standardization.

System	Sub	Title	System	Sub	Title	System	Sub	Title
21	00	AIR CONDITIONING		20	Extinguishing		70	Supplementary Gear
	10	Compression	27	00	FLIGHT CONTROLS	33	00	LIGHTS
	20	Distribution		10	Aileron & Tab		10	Flight Compartment
	30	Pressurization Control		20	Rudder & Tab		20	Passenger Compartment
	40	Heating		30	Elevator & Tab		30	Cargo & Service Compartment
	50	Cooling		40	Horiz. Stabilizer Control		40	Exterior
	60	Temperature Control		50	Flaps		50	Emergency Lighting
	70	Humidity Regulation		60	Spoiler & Drag	34	00	NAVIGATION
22	00	AUTO PILOT		70	Gust Lock & Dampener		10	Air Data Instrumentation
	10	Amplification		80	Lift Augmenting		20	Altitude & Direction Inst.
	20	Actuation	28	00	FUEL		30	Radio Navigation
	30	Controlling		10	Storage		40	Radar Navigation
	40	Indicating		20	Distribution		50	Proximity Warning
	50	Sensing		30	Dump		60	Position Computing
	60	Coupling		40	Indications	35	00	OXYGEN
23	00	COMMUNICATIONS	29	00	HYDRAULIC POWER		10	Crew
	10	HF		10	Main		20	Passenger
	20	VHF		20	Auxiliary		30	Portable
	30	PA & Pass. Entertainment		30	Indicating	36	00	PNEUMATIC
	40	Interphone	30	00	ICE & RAIN PROTECTION		10	Distribution
	50	Audio Integrating		10	Airfoil		20	Indicating
	60	Static Discharging		20	Air Intakes	37	00	VACUUM
	70	Voice Recorders		30	Pitot and Static		10	Distribution
24	00	ELECTRICAL POWER		40	Windows & Windshields		20	Indicating
	10	Generator Drive		50	Antennas & Radomes	38	00	WATER/WASTE
	20	AC Generation		60	Propellers/Rotors		10	Potable
	30	DC Generation		70	Water Lines		20	Wash
	40	External Power		80	Detection		30	Waste Disposal
	50	Elect. Load Distribution	31	00	INSTRUMENTS		40	Air Supply
25	00	EQUIP./FURNISHINGS		10	Panels	49	00	AIRBORNE AUX. POWER
	10	Flight Compartment		20	Independent Instruments		10	Power Plant
	20	Passenger Compartment	32	00	LANDING GEAR		20	Engine
	30	Buffet/Galley		10	Main & Doors		30	Engine Fuel & Control
	40	Lavatory		20	Nose & Doors		40	Ignition Starting
	50	Cargo & Accessory Compartment		30	Extension & Retraction		50	Air
	60	Emergency		40	Wheels and Brakes		60	Engine Controls
26	00	FIRE PROTECTION		50	Steering		70	Indicating
	10	Detection		60	Position & Warning		80	Exhaust

Fig. 2-4 The A.T.A. Specification 100 system

System	Sub	Title
	90	Oil
51	00	STRUCTURES
52	00	DOORS
	10	Passenger/Crew
	20	Emergency Exit
	30	Cargo
	40	Service
	50	Fixed Interior
	60	Entrance Stairs
	70	Door Warning
	80	Landing Gear
53	00	FUSELAGE
	10	Main Frame
	20	Auxiliary Structure
	30	Plates, Skins
	40	Attach Fitting
	50	Cones & Fillets/Fairings
54	00	NACELLES/PYLONS
	10	Main Frame
	20	Auxiliary Structure
	30	Plates/Skin
	40	Attach Fittings
	50	Fillets/Fairings
55	00	STABILIZERS
	10	Horizontal
	20	Elevator
	30	Vertical
	40	Rudder
	50	Attach Fittings
56	00	WINDOWS
	10	Flight Compartment
	20	Cabin
	30	Door
	40	Inspection & Observation
57	00	WINGS
	10	Main Frame
	20	Auxiliary
	30	Plates/Skin
	40	Attach Fittings
	50	Flight Surfaces
61	00	PROPELLERS-General
	10	Propeller Assembly
	20	Controlling
	30	Braking
	40	Indicating
65	00	ROTORS
	10	Main Rotor
	20	Anti-Torque Rotor Assy.
	30	Accessory Driving
	40	Controlling
	50	Braking
	60	Indicating
71	00	POWER PLANT
	10	Cowling
	20	Mounts
	30	Fireseals
	40	Attach Fittings
	50	Electrical Harness
72	00	ENGINE TURBINE
	10	Reduction Gear & Shaft Sect.
	20	Air Inlet Section
	30	Compressor Section
	40	Combustion Section
	50	Turbine Section
	60	Accessory Drives
	70	By-Pass Section
73	00	ENGINE FUEL & CONTROL
	10	Distribution
	20	Controlling
	30	Indicating
74	00	IGNITION
	10	Electrical Power Supply
	20	Distribution
	30	Switching
75	00	AIR
	10	Engine Anti-Icing
	20	Accessory Cooling
	30	Compressor Control
	40	Indication
76	00	ENGINE CONTROLS
	10	Power Control
	20	Emergency Shutdown
77	00	ENGINE INDICATING
	10	Power
	20	Temperature
	30	Analyzers
78	00	EXHAUST
	10	Collector
	20	Noise Suppressor
	30	Thrust Reverser
79	00	OIL
	10	Storage
	20	Distribution
	30	Indicating
80	00	STARTING
	10	Cranking
	20	Igniting
81	00	TURBINES
	10	Power Recovery
	20	Turbo-Supercharger
82	00	WATER INJECTION
	10	Storage
	20	Distribution
	30	Dumping & Purging
	40	Indicating
83	00	ACCESSORY GEAR BOXES
	10	Drive Shatt Section
	20	Gearbox Section

Fig. 2-4 The ATA Specification 100 system (continued)

DEPARTMENT OF TRANSPORTATION
FEDERAL AVIATION ADMINISTRATION

A12CE
Revision 11
BEECH
60
A60

May 7, 1973

TYPE CERTIFICATE DATA SHEET NO. A12CE

This data sheet which is part of type certificate No. A12CE prescribes conditions and limitations under which the product for which the type certificate was issued meets the airworthiness requirements of the Federal Aviation Regulations.

Type Certificate Holder — Beech Aircraft Corporation
Wichita, Kansas 67201

I - Model 60, 4 or 6 PCLM (Normal Category), Approved February 1, 1968
Model A60, 4 or 6 PCLM (Normal Category), Approved January 30, 1970

Engines	Lycoming TIO-541-E1A4 or TIO-541-E1C4 (2 of either or 1 of each)
Fuel	100/130 minimum grade aviation gasoline
Engine limits	For all operations, 2900 r.p.m. (380 b. hp.)
Propeller and propeller limits	(a) Two (in any combination) Hartzeil three-blade propellers Diameter: 74 in., (Normal) Minimum allowable for repair 73 1/2 in. (No further reduction permitted) Pitch settings at 30 in. sta.: low 14°, high 81.7° HC-F 3 YR-2/C7479-2R or HC-F-3 YR-2/C7497B-2R or HC-F 3 YR-2F/FC 7479-2R or HC-F 3 YR-2F/FC 7479B-2R (b) Beech 60-389000-3 governor

Airspeed limits

Never exceed	235 knots
Maximum structural cruising	208 knots
Maneuvering	161 knots
Maximum flap extension speed	
Approach position 15°	175 knots
Full down position 30°	135 knots
Landing gear extended	174 knots
Landing gear operating	174 knots

C.G. range (landing gear extended)
(+134.2) to (+139.2) at 6725 lb.
(+128.0) to (+139.2) at 51[illegible] lb. or less
Straight line variation between points given
Moment change due to retracting landing gear (+857 in.-lb.)

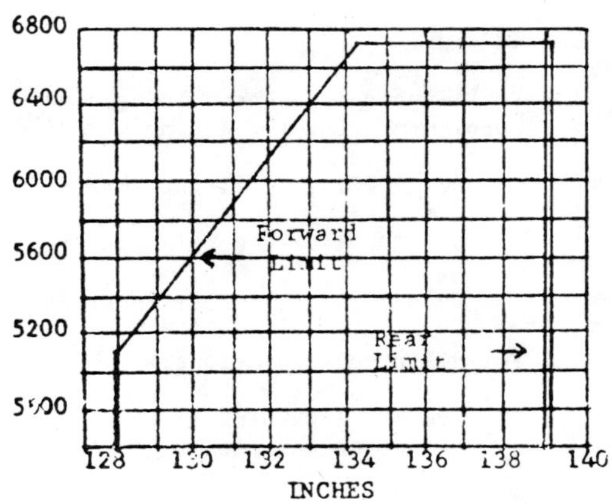

Empty wt. C.G. range — None
Maximum weight — Takeoff and landing — 6725 lb.
Ramp weight — 6819 lb.

Fig. 2-5 An Airplane Type Certificate Data Sheet

No. of seats	4 (2 at +141, 2 at +173) (add 2 at +205)
Maximum baggage (structural limit)	500 lb. at +75 (nose compartment 655 lb. at +212 (aft area of cabin)
Fuel capacity	142 gal. (+138) comprising two interconnected cells in each wing or 204 gal. (+139) comprising three cells in each wing and one cell in each nacelle (four cells interconnected) See NOTE 1 for data on system fuel
Oil capacity (wet sump)	26 qt. (+88)
Max. operating limit	30,000 ft. pressure altitude

Control surface movements

Wing flaps			Maximum	30°
Aileron	Up	25°	Down	15°
Aileron tab (L.H. only)	Up	10°	Down	10°
Aileron tab anti-servo	Up	12°	Down	7°
Elevator	Up	17°	Down	15°
Elevator tab (L.H. only)	Up	10°	Down	30°
Elevator tab servo	Up	6°	Down	7°
Rudder	Right	33°	Left	28°
Rudder tab	Right	26°	Left	26°

Serial Nos. eligible	Model 60: P-3 thru P-126 (except P-123) Model A60: P-123, P-127 and up (see NOTE 3)
Datum	Located 100 in. forward of front pressure bulkhead
Leveling means	Drop plumb line between leveling screws in cabin door frame rear edge
Certification basis	Part 23 of the Federal Aviation Regulations effective February 1, 1965 as amended by 1, 2, 3, and 12; and Special Conditions dated May 16, 1967, forwarded with FAA letter dated June 1, 1967; approved for flight into known icing conditions when equipped as specified in the approved airplane flight manual. Application for Type Certificate dated December 22, 1965. Type Certificate No. A12CE issued February 1, 1968, obtained by the manufacturer under delegation option procedures.
Production basis	Production Certificate No. 8 issued and Delegation Option Manufacturer No. CE-2 authorized to issue airworthiness certificates under delegation option provisions of Part 21 of the Federal Aviation Regulations.

Equipment

The basic required equipment as prescribed in applicable airworthiness regulations (see Certification basis) must be installed in the aircraft for certification. This equipment must include, for all operations, Airplane Flight Manual P/N 60-590000-5D dated January 15, 1971, amended July 1, 1971, or later issue.

In addition:

1. For flights into known icing conditions, these flight manual supplements and the equipment noted therein:

   60-590001-17 Flight Into Known Icing Conditions.

   60-590001-11 Continuous Pressure Operated Surface Deice System.

   60-590001-13 Goodrich Electrothermal Propeller Deice System.

2. For all other operations:

   Pre-stall warning indicator P/N 151-6, 151-7, or 190-2 (Safe Flight Corporation).

Fig. 2-5 An Airplane Type Certificate Data Sheet (continued)

NOTE 1. Current weight and balance data including list of equipment included in certificated empty weight and loading instructions when necessary must be provided for each aircraft at the time of original certification.

The certificated empty weight and corresponding center of gravity locations must include unusable fuel of 24 lb. at (+135).

NOTE 2. The following placard must be displayed in front of and in clear view of the pilot:

"This airplane must be operated in the normal category in compliance with the operation limitations stated in the form of placards, markings and manuals."

NOTE 3. Fuselage pressure vessel structural life limit - refer to the latest revision of the Airplane Flight Manual for mandatory retirement time.

NOTE 4. Model 60 (S/N P-3 thru P-126 except P-123) when modified to Beech dwg. 60-5008 and Model A60 (S/N P-123, P-127 and up) eligible for a maximum weight of 6775 lb.

NOTE 5. A landing weight of 6435 lb. must be observed if 10 PR tires are installed on aircraft not equipped with 60-810012-15 (LH) or 60-810012-16 (RH) shock struts.

. . . END . . .

Fig. 2-5 An Airplane Type Certificate Data Sheet (continued)

DEPARTMENT OF TRANSPORTATION
FEDERAL AVIATION ADMINISTRATION

E-269-15
CONTINENTAL
0-470-4, 0-470-11
0-470-11B
0-470-11-CI
0-470-11B-CI (see NOTE 5)
0-470-13, 0-470-13A
0-470-15

June 1, 1973

TYPE CERTIFICATE DATA SHEET NO. E-269

Engines of models described herein conforming with this data sheet (which is part of type certificate No. 269) and other approved data on file with the Federal Aviation Administration, meet the minimum standards for use in certificated aircraft in accordance with pertinent aircraft data sheets and applicable portions of the Federal Aviation Regulations provided they are installed, operated and maintained as prescribed by the approved manufacturer's manuals and other approved instructions.

Type Certificate Holder — Teledyne Continental Motors, Mobile, Alabama 36601

Model	0-470-11,-11B	0-470-13,-13A	0-470-4	0-470-15
Type	6HOA	- -	- -	- -
Rating, ICAO or ARDC standard atmosphere				
Max. continuous hp., r.p.m., full throttle, at sea level pressure altitude	190-2300	225-2600	- -	190-2300
Takeoff hp., 5 min., r.p.m., full throttle, at sea level pressure altitude	213-2600	225-2600	- -	213-2600
Fuel (min. grade) aviation gasoline	80/87	- -	- -	- -
Lubricating oil				
Ambient air temperature	Oil grade			
Above 40°F.	SAE 50	- -	- -	- -
Below 40°F.	SAE 30	- -	- -	- -
Bore and stroke, in.	5.00 x 4.00	- -	- -	- -
Displacement, cu. in.	471	- -	- -	- -
Compression ratio	7:1	- -	- -	- -
Weight (dry), lb.	391	415	- -	405
C.G. location (with starter, generator and fuel pump)				
Fwd. of rear face, engine crankcase, in.	8.8	9.8	- -	- -
Below crankshaft, centerline, in.	.8	.2	- -	- -
Beside crankshaft, centerline toward 1-3-5 side, in.	—	.1	- -	- -
Propeller shaft, SAE No.	4 flange	Special integral flange 4-7/8 in. O.D. with six 1/2 in. bolt holes in 4 in. dia. circle	- -	- -
Carburetion	Bendix-Stromberg Model PS-5C (CMC No. 531857)	- -	Bendix-Stromberg Model PS-5CD (CMC No. 538685)	Bendix-Stromberg Model PS-5C (CMC No. 625098)
Ignition, dual magnetos	Scintilla S6LN-21 (CMC No. 531676)	Scintilla S6LN-23 (CMC No. 533173)	- -	Scintilla S7LN-21 (CMC No. 531676)

Fig. 2-6 An Engine Type Certificate Data Sheet

Model	0-470-11,-11B	0-470-13,-13A	0-470-4	0-470-15
Timing, °BTC	Right 26, left 26	- -	- -	- -
Spark plugs	(See NOTE 6)	- -	- -	- -
Oil sump capacity, qt.	10; 7 usable	Dry	- -	10; 7 usable
NOTES	1,2,3,4,5,6	1,2,3,4,6	1,2,3,4,6	1,2,3,4,6

"- -" indicates "same as preceding model"
"——" indicates "does not apply"

Certification basis Part 13 of the Civil Air Regulations effective August 1, 1949 as amended by 13-1
Type Certificate No. 269 issued January 19, 1951
Application for Type Certificate dated October 23, 1950

Production basis P.C. 7 (All models except 0-470-11B and -11B-C1)
P.C. 508 (All models with "R" suffix on serial number except 0-470-11B and-11B-C1)

NOTE 1. Maximum permissible temperatures:

Cylinder head
<u>0-470-11,-11B,-15</u>
525°F. (spark plug gasket)
450°F. (bayonet thermocouple downdraft cooling)
<u>0-470-4,-13,-13A</u>
465°F. (bayonet thermocouple downdraft cooling)

Cylinder barrel 315°F.
Oil inlet 225°F.

NOTE 2. <u>0-470-11,-11B,-13,-13A,-4,-15</u>
Fuel inlet pressure limits:
9 to 15 p.s.i.g.

Oil pressure limits:
30 - 60 p.s.i.a.

<u>0-470-11-CI,-11B-CI</u>
Fuel inlet pressure limits:
Min. minus 2.25 p.s.i.g.
Max. plus 10 p.s.i.g.
Oil pressure limits:
30 to 60 p.s.i.a.

NOTE 3. The following accessory drive or mounting provisions are available:

Original Accessory	Type	Dwg. No.	Direction of Rotation*	Speed Ratio to Crankshaft	Maximum (in.-lb.) Continuous	Torque (in.-lb.) Static	Max. Overhang Moment (in.-lb.)
Starter	XIV-A	AND 20004	C	1.0	——	3600	300
Generator (0-470-11, -11B,-15)	XII-A	AND 20002	C	2.37	500	2200	400
(0-470-13, -13A,-4)	XII-A	AND 20002	C	3.094	350	1000	400
Fuel pump	XIII-A	AND 20003	CC	.94	25	450	——
Tachometer	XV-A	AND 20005	CC	.50	7	50	——
Fluid P. Pump	X	AND 20000	C	1.43	100	800	25
Fluid P. Pump	XI-A	AND 20001	C	1.45	250	1650	75
Spare	XV-A	AND 20005	CC	.50	7	50	——
Governor	XX	AND 20010	C	1.03	125	825	——

*C - Clockwise viewing drive pad; CC - Counterclockwise.

NOTE 4. Model 0-470-11B is identical to the 0-470-11 except that it incorporates the model 0-470-15 cylinder and piston assemblies.

Model 0-470-11 incorporates two 6th order dampers.

Model 0-470-13A is identical to the 0-470-13 except that an additional tachometer drive is provided through the camshaft gear and incorporates one 5th and one 6th order crankshaft damper.

Model 0-470-15 is identical to the 0-470-11 except that a four 6th order damper crankshaft is used with the same size propeller flange specified on the model 0-470-13. Model 0-470-15 incorporates propeller control provisions. Also, revised engine mount brackets and long skirt pistons are used with this engine.

Fig. 2-6 An Engine Type Certificate Data Sheet (continued)

NOTE 4. (con.)

0-470-13 engines incorporating crankshaft with one 5th and one 6th order dampers have the numeral 5 stamped on the outer circumference of the crankshaft propeller flange. 0-470-13 engines with two 6th order dampers have no numeral stamped on the flange.

Model 0-470-4 which was previously designated as 0-470-13B is identical to the Model 0-470-13A except for the carburetor.

NOTE 5. Models 0-470-11 and -11B are eligible for incorporation of CMC continuous flow fuel injection equipment No. 5843 replacing carburetion system, with a weight increase of approximately 3 lb. When this modification is accomplished the suffix letters "CI" are to be added to the model designation.

NOTE 6. The following spark plugs are approved for use in these engines:

AC	SR83P, HSR83P, S88D, HS88, HSR88, SR88D
Auto Lite	SH2M, SH15, SH15R, SH20, SH20A, SH200A
BG	RB485S, 706S, 706SR, 919SR, 919SR5, RB955S, RB963S, HRB963S
Champion	RC26S, C27S, REM38P, RHM38P, RED39N, REM39N, RHD39N, RHM39N, RED40E, REM40E, RHM40E, ED41N, EM41N, EM42E
Red Seal	SE190, SJ190, SE230, SJ230

NOTE 7. Model 0-470-11 engines which have been modified in accordance with U.S. Army Aviation and Surface Material Command Work Requirement No. WR-55-2810-112-B(1) issued 5 April 1963 and revised 13 February 1964, and designated Model 0 470-11A, are considered to be equivalent to and interchangeable with Model 0-470-11.

... END ...

Fig. 2-6 An Engine Type Certificate Data Sheet (continued)

DEPARTMENT OF TRANSPORTATION
FEDERAL AVIATION ADMINISTRATION

P-888-6
HARTZELL
HC-13X

November 1, 1968

TYPE CERTIFICATE DATA SHEET NO. P-888

Propellers of models described herein conforming with this data sheet (which is a part of type certificate No. P-888) and other approved data on file with the Federal Aviation Administration, meet the minimum standards for use in certificated aircraft in accordance with pertinent aircraft data sheets and applicable portions of the Federal Aviation Regulation provided they are installed, operated and maintained as prescribed by the approved manufacturer's manuals and other approved instructions.

Manufacturer	Hartzell Propeller, Inc. Piqua, Ohio 45356
Type	Constand speed; hydraulic (See NOTES 3 and 4)
Engine shaft	SAE #20 spline
Hub material	Alloy steel
Blade material	Aluminum Alloy
No. of blades	Three
Hub models eligible	HC-13X20-5, -6, -8. (see NOTES 1 and 4)

Blades Eligible (See NOTE 2)	Continuous HP	Continuous RPM	Takeoff HP	Takeoff RPM	Diameter Limits	Approx. Max. Wt. Complete (For reference only) (See NOTE 3)
7636D-0 to 7636D-8	280	2600	280	2600	76 5/8" - 68 5/8" (-0 to -8)	94 lb.
8433-0 to 8433-12	280	2600	280	2600	84 5/8" - 72 5/8" (-0 to -12)	97 lb.

Certification basis	CAR 14 effective March 5, 1952 as amended by 14-1 and 14-2. Type Certificate No. 888 issued February 1, 1955. Date of Application for Type Certificate January 14, 1955.
Production basis	Production Certificate No. 10

NOTE 1. Hub Model Designation

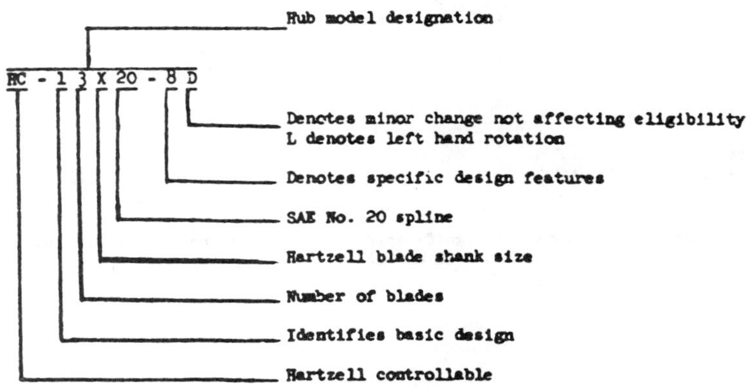

Fig. 2-7 A Propeller Type Certificate Data Sheet

NOTE 2. Blade Model Designation

L 84 33 H - 3

- Blade model designation
- 3: Number of inches cut off from basic diameter
- H: H when used indicates Hard Alloy; D when used indicates dimensional modification from original design
- 33: Basic model or template
- 84: Basic diameter in inches
- L: L when used denotes left hand blade

Diameter limits shown are nominal diameters of the assembled propeller and do not include the + 1/4 inch to - 1/8 inch manufacturing tolerance permissible.

NOTE 3. Pitch Control. Eligible with Hartzell manual or constant speed control. Constant speed control includes the Hartzell governor Model A-1-1. Additional weight of governor is 4.5 lb.
(Note: Some installations require a T-drive adapter for the governor. Additional weight of the Hartzell Model C-253-2 adapter is 2.7 lb.)

NOTE 4. (a) Feathering. Not applicable.
(b) Reversing. The -5 model incorporates a reversing feature.

NOTE 5. Left Hand Models. The left hand version of an approved model propeller is eligible at the same rating and diameter as listed for the right hand model. (See NOTES 1 and 2).

NOTE 6. Interchangeability.
(a) Blades. Only blades listed in the same group of the following listed groups are sufficiently similar aerodynamically and vibrationwise to permit interchangeability in the same diameter without a flight test.
Group (a) 8433, 8433H*
*Hard alloy blades interchangeable only on seaplanes and amphibious aircraft.
(b) Propellers. Only propellers listed in this data sheet may be replaced by corresponding propellers listed in Type Certificate Data Sheet No. P1EA provided the propeller model designations are the same except that the identifying basic digit "1" (See NOTE 1) is replaced by the letter "D".

Note 7 and 8. Not applicable.

NOTE 9. Table of Propeller-Engine Combinations
Approved Vibrationwise for Use on Normal Category Single-Engine Tractor Aircraft

The maximum and minimum propeller diameters that can be used from a vibration standpoint are shown below. No reduction below the minimum diameter listed is permissible since this figure includes the diameter reduction allowable for repair purposes.

Blade Model	Engine Model	Max. Dia. (Inches)	Min. Dia. (Inches)	Placards
8433	Lycoming GO-435	84	84	None
8433	Franklin 6A8-215-B8F	84	80	None
	Franklin 6A8-215-B9F	84	80	None

NOTE 10. The word "eligible" as used herein does not signify approval. For approval compliance with the applicable aircraft airworthiness requirements is necessary.

...END...

Fig. 2-7 A Propeller Type Certificate Data Sheet (continued)

ADVISORY CIRCULAR
AC 43–16

U.S. DEPARTMENT OF TRANSPORTATION
FEDERAL AVIATION ADMINISTRATION
WASHINGTON, D.C. 20590

GENERAL AVIATION AIRWORTHINESS ALERTS

JULY 1980

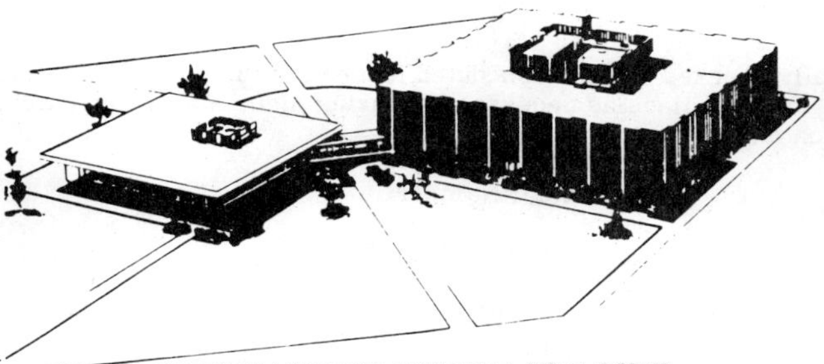

FLIGHT STANDARDS NATIONAL FIELD OFFICE
MIKE MONRONEY AERONAUTICAL CENTER

The General Aviation Airworthiness Alerts provide a common communication channel through which the aviation community can economically interchange service experience and thereby cooperate in the improvement of aeronautical product durability, reliability, and safety. This publication is prepared from information submitted by those of you who operate and maintain civil aeronautical products. The contents include items that have been reported to be significant, but which have not been evaluated fully by the time the material went to press. As additional facts such as cause and corrective action are identified, the data will be published in subsequent issues of the Alerts. This procedure gives Alerts readers prompt notice of conditions reported via Malfunction or Defect Reports. Your comments and suggestions for improvement are always welcome. Send to: Flight Standards National Field Office, Attn: AFO-580, P. O. Box 25082, Oklahoma City, Oklahoma 73125.

AIRCRAFT

AGUSTA

Agusta Model A109A — Engine Mount, P/Ns 109-0605-06-13 and -14

The engine mount flange attached to the airframe was found broken on each engine. The pilot had reported higher than normal vibration. The aft end of the engines would move 1/2 inch vertically when moderate hand force was applied. Total time - 259 hours.

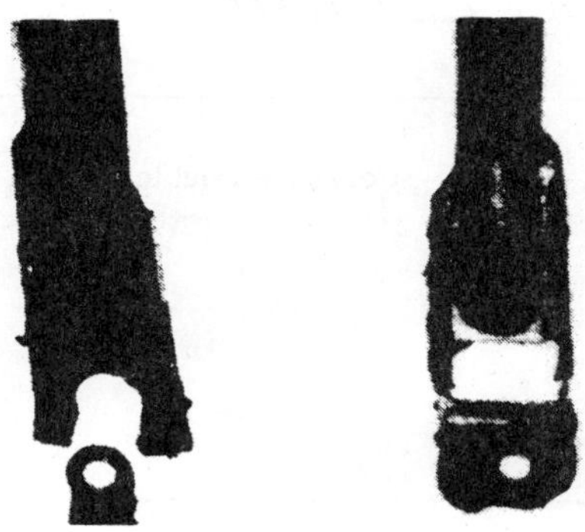

1

Fig. 2-8 A General Aviation Airworthiness Alert

BEECH

Beech — Fuel Cell Inspection

Beech issued Class I Service Instruction No. 0632-280 to provide inspection criteria for bladder type fuel cells because numerous reports have been received of leaks being found at the drain nipples and fuel vents. Wing lower surfaces and under the nacelles should be examined each 100 hours for fuel stains or other evidence of possible fuel seepage. If stains are found, remove the access plates from the wing lower surfaces and inspect the areas between the fuel cells and the wing skin. Aircraft with fuel cells over four years old, or which are continually subjected to high temperature and/or high humidity should be checked more frequently. If fuel seepage is found or if the fuel cells appear to be hard and brittle, they should be replaced.

Beech — Fuel Lines in Wheel Wells

Chafing between the landing gear retract rods and fuel lines routed through the wheel well areas has been found on some in-service airplanes.

It is recommended that fuel lines in the wheel wells be visually checked at each scheduled inspection for possible chafing on the landing gear retract rod. Reposition or replace the fuel lines as necessary. During maintenance activities, avoid using fuel lines as "hand holds" to preclude distortion and resultant chafing.

Aircraft affected:	Models	Serial Numbers
	95-B55 and 95-B55A	TC 1970, TC 2003 - 2260
	E55 and E55A	TE 1081, 1084 - 1165
	56TC and A56TC	TG 2 - TG 94
	58 and 58A	TH 1 - TH 1060
	58P and 58PA	TJ 3 - TJ 215
	58TC and 58TCA	TK 1 - TK 105

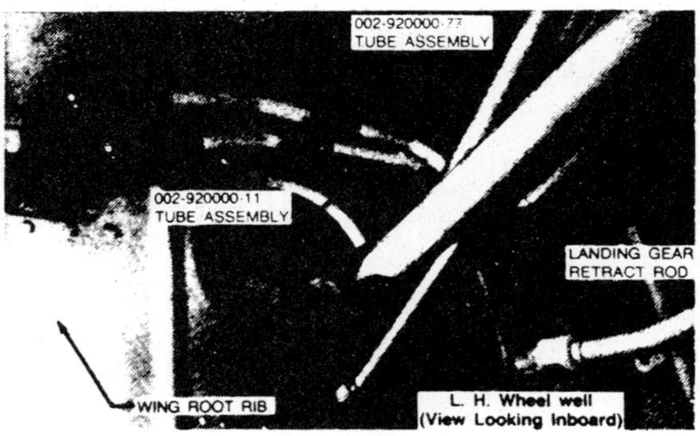

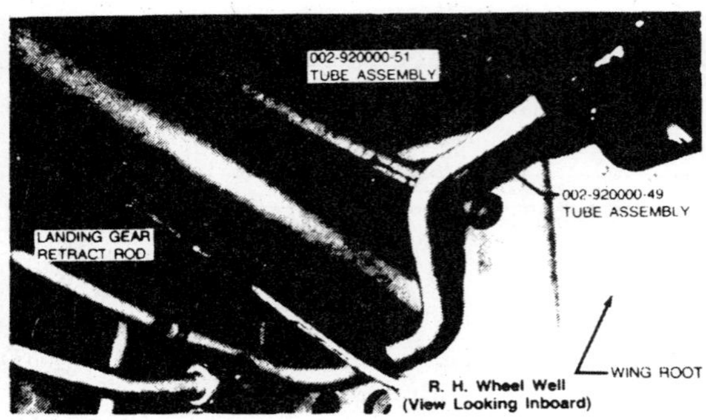

Beech Model A23A — Corroded Doubler

The rear seat air vent flex duct passes through the longitudinal stringer in the fuselage just aft of station 132. The doubler is installed at the cutout in the stringer. The flex duct had worn through and the duct wire was in contact with the aluminum doubler. Moisture, apparently condensate, and dissimilar metals caused severe corrosion of the doubler on the left side of the fuselage. Corrosion was beginning to develop on the doubler on the right side of the fuselage. Total aircraft time - 2,817 hours.

In order to ensure continued receipt of the "Alerts," an address change must be reported to the FAA district office under which you exercise your certificate.

2

Fig. 2-8 A General Aviation Airworthiness Alert (Continued)

DEPARTMENT OF TRANSPORTATION
FEDERAL AVIATION ADMINISTRATION
AERONAUTICAL CENTER
P. O. BOX 25082
OKLAHOMA CITY, OKLAHOMA 73125

Airworthiness Directive

Revision September 21, 1973

The following Airworthiness Directive Revision issued by the Federal Aviation Administration in accordance with the provisions of Federal Aviation Regulation, Part 39, applies to an aircraft model of which our records indicate you may be the registered owner. Airworthiness Directives affect aviation safety. They are regulations which require immediate attention. You are cautioned that no person may operate an aircraft to which an Airworthiness Directive applies, except in accordance with the requirements of the Airworthiness Directive (FAR 39.3).

72-19-4 Nickel-Cadmium Battery. Amdt. 39-1521 as amended by Amendment 39-1598 is further amended by Amendment 39-1723. Applies to all aircraft having a primary electrical system that includes a nickel-cadmium battery that is capable of being used to start the aircraft's engine or APU, except those aircraft that have the charging rate of such a battery automatically controlled so as to prevent battery overheating, or that have a battery temperature sensing and over-temperature warning system with means and operating procedures for disconnecting the battery from its charging source in the event of an over-temperature condition, or that have a battery failure sensing and warning system with means and operating procedures for disconnecting the battery from its charging source in the event of a battery failure, and except Learjet Models 23, 24, and 25 airplanes.

Compliance is required as indicated.

To prevent a possible battery fire that may result from overheating caused by an undetected battery failure, accomplish the following:

(a) For any battery rated at less than 50 amp-hours, containing any polystyrene cell cases, that is installed on a turbine engine powered aircraft, within the next 10 hours' time in service after the effective date of Amendment 39-1521, either —

(1) Replace each cell having a polystyrene cell case with an equivalent cell having a polyamide cell case; or

(2) Replace any battery containing any polystyrene cell cases with a battery containing all polyamide cell cases that is approved by the Chief, Engineering and Manufacturing Branch of an FAA Region (or in the case of the Western Region, the Chief, Aircraft Engineering Division).

(b) For all batteries, within the next 10 hours' time in service after the effective date of Amendment 39-1521 unless already accomplished within the last 50 hours' time in service, and thereafter within the next seven days or 30 hours' time in service, whichever is later, following use of the battery for an engine or APU start or attempted start, visually inspect the battery, including the cell links and tops, for evidence of heat damage, or accomplish a procedure that is found by the Chief, Engineering and Manufacturing Branch of an FAA Region (or in the case of the Western Region, the Chief, Aircraft Engineering Division), to provide evidence of heat damage that is equivalent to that provided by physical inspection of the battery.

(c) If a battery is found to have evidence of heat damage during an inspection or procedure required by paragraph (b), before further flight either —

(1) Replace the battery with an equivalent serviceable battery; or

(2) On aircraft approved for operation without the affected battery being operational, mechanically disconnect the battery at the battery terminal.

(d) Upon request of the operator, an FAA maintenance/avionics inspector may adjust the repetitive inspection intervals specified in this AD to permit compliance at an established inspection period of the operator if the request contains substantiating data that show that:

Fig. 2-9 An Airworthiness Directive

(1) The aircraft is being maintained and inspected in accordance with a continuous airworthiness maintenance program under Parts 121 or 127, or is being inspected in accordance with a continuous inspection program under Subpart D of Part 91, or Part 123, or Part 135; and

(2) The battery system has been maintained for at least the past year in accordance with the operators' maintenance program, and battery service on the aircraft has been free of heat associated problems during that time.

(3) The substantiating data required by paragraphs (d)(1) and (2) must include at least the battery inspection and reconditioning frequency, the name and address of the facility/operator maintaining the battery, and a log book or other records showing the number of battery removals and reasons for removal.

(e) Report evidence of battery heat damage found during inspections or procedures required by this AD, in writing, within 10 days of the inspection, to the Chief, Engineering and Manufacturing Division, Attention: AFS-130, Federal Aviation Administration, 800 Independence Ave., S.W., Washington, D.C. 20591. Each report must include the aircraft model, serial, and registration numbers, the battery make and model numbers, battery hours' time in service, and a description of the heat damage. (Reporting approved by the Bureau of the Budget under BOB No. 04-R0174).

Note: A pre-addressed Malfunction or Defect Report, FAA Form 8330-2 (available at any General Aviation District Office) may be used for a report required by this AD, if reference to the AD is made on the form.

(f) Before February 1, 1974, accomplish at least one of the following:

(1) Install a battery charging rate control system, approved by an FAA maintenance/avionics inspector, or by the Chief, Engineering and Manufacturing Branch of an FAA Region (or in the case of the Western Region, the Chief, Aircraft Engineering Division).

(2) Install a battery temperature sensing and over-temperature warning system and provide a means and an operating procedure for disconnecting the battery from the charging source in the event of battery over-temperature warning, that are approved by an FAA maintenance/avionics inspector, or by the Chief, Engineering and Manufacturing Branch of an FAA Region (or in the case of the Western Region, the Chief, Aircraft Engineering Division).

(3) Install a battery failure sensing and warning system and provide a means and an operating procedure for disconnecting the battery from the charging source in the event of battery failure, that are approved by an FAA maintenance/avionics inspector, or by the Chief, Engineering and Manufacturing Branch of an FAA Region (or in the case of the Western Region, the Chief, Aircraft Engineering Division).

(g) The inspections or procedures required by paragraph (b) may be discontinued after compliance with paragraph (f) of this AD.

This amendment supersedes Amendment 39-1302 (36 F.R. 19075), AD 71-21-5, as amended by Amendment 39-1333 (36 F.R. 21581).

Amendment 39-1521 became effective September 20, 1972.

Amendment 39-1598 became effective March 6, 1973.

This Amendment 39-1723 became effective September 19, 1973.

NOTE: Address all inquiries regarding this AD to:

DOT, Federal Aviation Administration
Systems Branch, AFS-130
800 Independence Ave., S.W.
Washington D.C. 20591

Fig. 2-9 An Airworthiness Directive (continued)

PART III

Federal Aviation Regulations

I. *The Federal Aviation Administration*

When airplanes contributed little to our national economy, there was little reason for their regulation, but as the number of airplanes in the air increased and flight became of more concern to our nation, the federal government stepped in to regulate aviation and keep it from getting out of hand.

In 1926, only 23 years after the first flight by the Wright brothers, the Air Commerce act was passed that charged the Department of Commerce with the responsibility of promoting air commerce by regulating it and by providing aids to navigation. The Department of Commerce, and later the Civil Aeronautics Authority, regulated civilian aviation as the airspace became more crowded and airplanes became more and more vital to our national economy. In 1958 the Federal Aviation Agency was created to give aviation a place of importance. This agency answered directly to the president. The regulating body was again shuffled in 1966 when the Federal Aviation Administration was created which has less stature than the Agency. It was placed under the Department of Transportation and answers to the Secretary of Transportation.

We who are involved in maintaining and operating aircraft are well aware of the place of government regulations in our everyday business. We must be licensed to work on aircraft and to fly them. The aircraft themselves must have both their design and their production approved by the FAA, and the individual aircraft must be certificated. They must be maintained and inspected according to federal regulations, and the federal government keeps a watch to determine if any dangerous problems show up in the aircraft, its engine, or any of its components or accessories when the aircraft is operational in the field.

Because of all of this government control, we must be thoroughly informed concerning the Federal Aviation Administration so we can know what we must do to keep all of our operations both safe and legal.

The Federal Aviation Administration is headquartered in Washington, D.C., and has its Aeronautical Center, the place of most interest to us, in Oklahoma City, Oklahoma. All of the records for both airmen and aircraft are maintained in Oklahoma City, and here the FAA has its training academy where all of the FAA personnel are trained.

A. *Operational Levels of the FAA*

For geographic coverage, the United States is divided into nine domestic FAA regions, and there are three international regions that provide FAA surveillance for all U.S. registered aircraft operating outside of the U.S.

The regions are divided into districts, with Air Carrier District Offices (ACDOs) set up to provide assistance to the airlines and General Aviation District Offices (GADOs) to help the non-airline segment of civilian aviation.

The GADO is our normal point of contact with the FAA. It is here that we take our examinations for ratings and here we can go for any assistance we require from the FAA. If we have any problem that cannot be handled on the district level, the personnel in the GADO will direct the level that has the appropriate jurisdiction.

B. *The Guidelines of the FAA*

Any organization as large as the FAA must operate within clearly defined guidelines, and these are provided by the Federal Aviation Regulations which are documents defining the law in every situation.

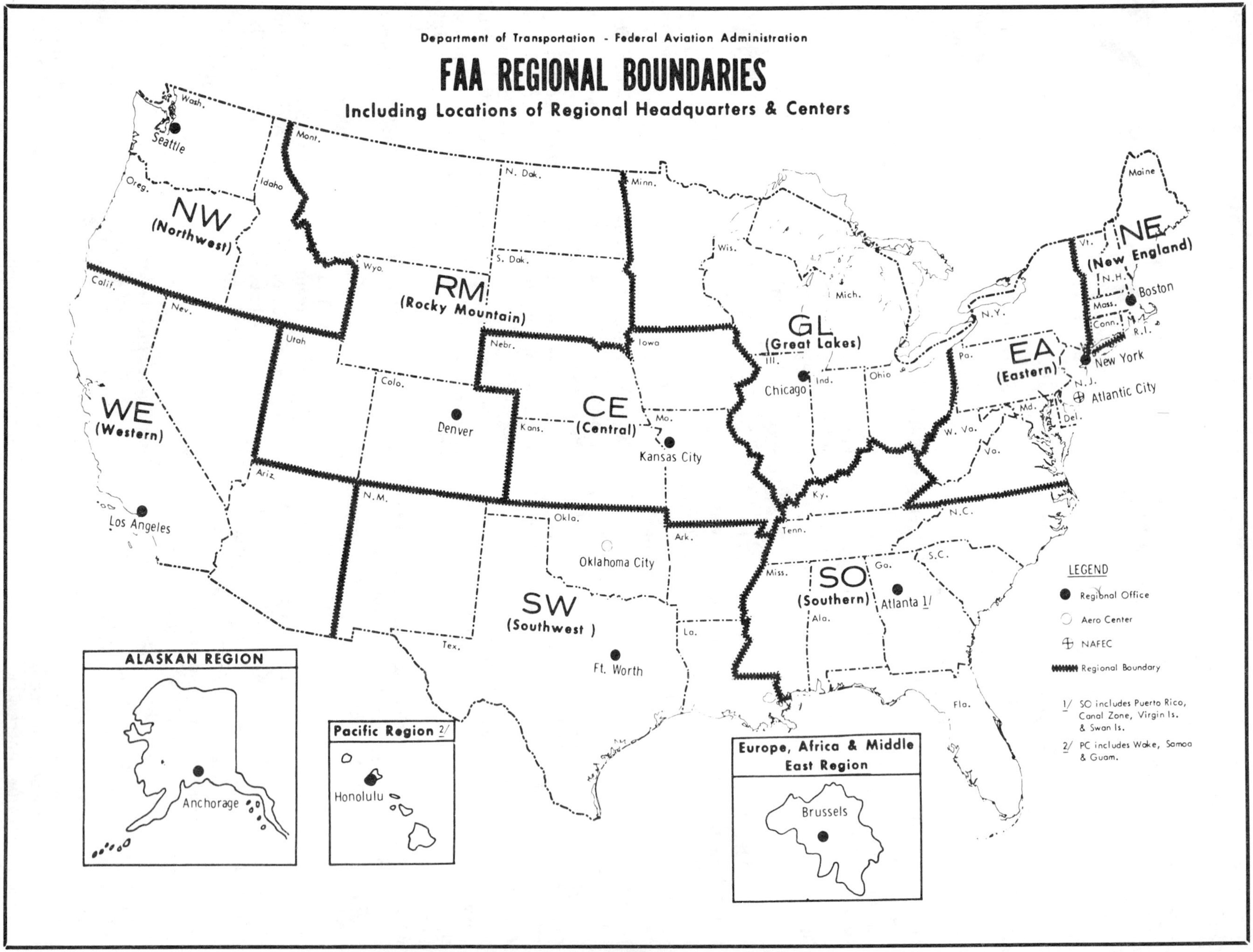

Fig. 3-1 Regions of the Federal Aviation Administration

These regulations are written in legal terminology, which is not easily understood, so the FAA has published a large collection of Advisory Circulars which are keyed to the regulations and interpret them or help us in other ways to abide by the regulations.

For example, FAR Part 43 contains the regulations regarding "Maintenance, preventive maintenance, rebuilding, and alteration." This is the legal description of what we can and cannot do in these areas of aviation operation. The entire regulation is less than 30 pages in length, so it cannot give us any details. But there are hundreds of pages of advisory information published by the FAA that are keyed to FAR Part 43 and help us comply with these regulations. The advisory circular most familiar to us is AC 43.13-1A, which is entitiled "Acceptable Methods, Techniques, and Practices - Aircraft Inspection and Repair." This book is over 300 pages long and gives us detailed methods of complying with FAR Part 43. The methods described are not the only means of complying, but if the inspections and repairs are performed according to this information and it is referenced as approved data, the FAA inspectors will accept the method of repair.

In Fig. 3-2 we have a list of the parts of the Federal Aviation Regulations that we will likely be involved with but the parts of most concern to an A&P are these:

Part 1 gives us the legal definitions and abbreviations that pertain to all aviation operations, both flight and maintenance.

Part 13 tells us how the FAA enforces the regulations.

Part 21 describes the different types of certificates that are issued for the design, manufacture, and operation of products and parts.

Part 23 includes detailed specifications for the design and testing of airplanes that are certificated in the normal, utility, or acrobatic category, and which seat nine people or less, excluding the pilot.

Part 25 gives the same type of information for transport category airplanes.

Part 33 has detailed specifications for the design and testing of reciprocating and turbine engines.

Part 35 covers the design and testing of propellers.

Part 37 deals with the requirements for issuing a Technical Standard Order. A TSO authorization specifies the design and testing requirements for a component that is to be used on certificated aircraft. For example, fabric, seat belts, and instruments that are built according to TSO specifications can be used on certificated aircraft.

Part 39 includes a description of the Airworthiness Directive system. When an aircraft is certificated, it must meet all of the certification requirements. But after it is in service and some deficiency shows up, an airworthiness directive (AD) is issued that describes the way to correct the deficiency. Compliance with AD's is mandatory.

Part 43 gives us the legal requirements for maintenance, preventive maintenance, rebuilding, and alteration of certificated aircraft.

Part 45 shows the details of the markings that are required for aircraft identification.

Part 47 gives us the information we need to register an aircraft, and describes the paperwork needed to transfer the title of the aircraft.

Part 65 is one of the personnel requirement portions of the FAR's, and it deals with the certification of mechanics and repairmen as well as air traffic control tower operators, aircraft dispatchers, and parachute riggers.

Part 91 is the most widely used part of the FAR's and it contains the general operating and flight rules. This regulation includes the requirements for maintenance, inspections, and record keeping. For details, it specifies other parts of the FAR's.

Part 145 gives the requirements for repair station certification and lists the different classes of domestic repair stations.

Part 147 provides information on the requirements for Aviation Maintenance Technician schools.

FEDERAL AVIATION REGULATIONS

Part	Title
1	Definitions and abbreviations
11	General rulemaking procedures
13	Enforcement procedures
21	Certification procedures for products and parts
23	Airworthiness standards: Normal, utility, and acrobatic category airplanes.
25	Airworthiness standards: Transport category airplanes
27	Airworthiness standards: Normal category rotorcraft
29	Airworthiness standards: Transport category rotorcraft
31	Airworthiness standards: Manned free balloons
33	Airworthiness standards: Aircraft engines
35	Airworthiness standards: Propellers
36	Noise standards: Aircraft type certification
37	Technical standard order authorizations
39	Airworthiness directives
43	Maintenance, preventive maintenance rebuilding, and alteration
45	Identification and registration marking
47	Aircraft registration
49	Recording of aircraft titles and security documents
61	Certification: Pilots and flight instructors
63	Certification: Flight crewmembers other than pilots
65	Certification: Airmen other than flight crewmembers
67	Medical standards and certification
71	Designation of Federal airways, area low routes, controlled airspace, and reporting points
73	Special use airspace
75	Establishment of jet routes and high area routes
77	Objects affecting navigable airspace
91	General operating and flight rules
93	Special air traffic rules and airport traffic patterns
95	IFR altitudes
97	Standard instrument approach procedures
99	Security control of air traffic
101	Moored ballons, kites, unmanned rockets, and unmanned free balloons
103	Transportation of dangerous articles and magnetized materials
105	Parachute jumping
107	Airport security
121	Certification and Operations: Domestic, flag, and supplemental air carriers and commercial operators of large aircraft
123	Certification and operations: Air travel clubs using large airplanes
127	Certification and operations of scheduled air carriers with helicopters
129	Operations of foreign air carriers
133	Rotorcraft external-load operations
135	Air taxi operators and commercial operators of small aircraft
137	Agricultural aircraft operations
139	Certification and operations: Land airports serving CAB-certificated scheduled air carriers operating large aircraft [other than helicopters]
141	Pilot schools
143	Ground instructors
145	Repair Stations
147	Aviation maintenance technician schools
149	Parachute lofts
151	Federal aid to airports
152	Airport aid program
153	Acquisition of U.S. land for public airports
154	Acquisition of U.S. land for public airports under the Airports and Airway Act of 1970
155	Release of Airport property from surplus property disposal
157	Notice of construction, alteration, activation, and deactivation of airports
159	National Capital airports
169	Expenditure of Federal funds for nonmilitary airports or air navigational facilities thereon
171	Non-Federal navigation facilities
183	Representatives of the Administrator
185	Testimony by employees and production of records in legal proceedings and service of legal process and pleadings
187	Fees
189	Use of Federal Aviation Administration communication system

Fig. 3-2 The Federal Aviation Regulations

Since an FAR becomes part of the law of the land once it is approved, the FAA is required to give anyone affected by the regulation an opportunity to participate in the formulation of these regulations. Before any new regulations can be made, or any existing regulation can be changed, the FAA must issue a Notice of Proposed Rulemaking (NPRM) and make it available to all interested parties. The NPRM includes a description of the proposed change and the reason for the change. A date is set in which anyone interested can make their comments and these comments are to be considered before the proposed rule becomes law. All mechanics are urged to express their opinions on any proposal that would affect aircraft maintenance.

C. The FAA Involvement in Civilian Aircraft Development and Operation

We will follow a typical aircraft through the various stages of operation in which the FAA is involved from the time the idea is conceived up through the time the aircraft is operational.

A market survey is made by the manufacturer to determine the size and configuration of the aircraft that is needed, and if it is to be, for example, a single-engine aircraft with four to six seats, the engineering department will be guided in their design and testing by FAR Part 23. When the aircraft is built to these specifications, it is licensed in the experimental category until it

Fig. 3-3 Before an aircraft is certificated, it must undergo a rigid flight test program.

completes all of its flight tests and is issued a certificate in the normal or utility category. A Type Certificate is issued for the design, and a Type Certificate Data Sheet is prepared so the mechanics will have a copy of the specifications the aircraft must meet on each of its inspections.

The FAA examines the manufacturing facilities and if they pass the requirements for production, the manufacturer is awarded a production certificate which allows him to issue the airworthiness certificate for each aircraft as they are built.

When the aircraft is delivered to the customer and becomes part of our general aviation fleet, it must be flown by FAA-licensed pilots, and operated according to the general operating and flight rules of FAR Part 91. Every 12 calendar months the aircraft must be given a thorough inspection by an FAA-licensed A&P mechanic holding an Inspection Authorization, and if it is operated for hire, it must be given the same inspection by an A&P mechanic every 100 hours of operation.

If the mechanic finds a malfunction or a defect during an inspection, he should send a Malfunction or Defect Report to the FAA in Oklahoma City, where it is entered into a computer with all other M and D reports. If this is an isolated problem, nothing is done but if the discrepancy shows up often enough, it is then described in the General Aviation Airworthiness Alerts. These are made available to all A&P mechanics so they can keep a sharp eye for this same problem on aircraft they are inspecting. If the problem is serious, an Airworthiness Directive will be issued, and all of the affected aircraft will be grounded until the prescribed fix is made.

As maintenance personnel, our involvement with the FAA is primarily through the recently

Fig. 3-4 Before an aircraft is certificated, it must undergo a complete vibration survey.

established Office of Airworthiness which includes Engineering, Manufacturing, and Maintenance divisions. The maintenance division which is of primary concern to us, is further divided into the Air Carrier and the General Aviation Branches.

II. Personnel Certification

Perhaps the best way to understand the FAR's is to examine the ones that most intimately affect our lives and discuss their provisions in plain English.

A. The Mechanic Certificate

1. Requirements for issuance

FAR Part 65 regulates the certification of airmen other than flight crewmembers. This includes mechanics and repairmen and gives us the requirements for the issuance of an Inspection Authorization.

The FAA issues a Mechanic Certificate with two ratings, Airframe and Powerplant. To be eligible for this certificate, the applicant must provide documented evidence that he has had at least 18 months of practical experience in the area of the raing for which he is applying, that is, 18 months as an airframe mechanic or 18 months as a powerplant mechanic. If he is applying for both ratings, he needs to show evidence of only 30 months combined experience. If the applicant is a graduate of an Aviation Maintenance Technician school that is certificated under FAR Part 147, the Certificate of Completion issued by the school will serve as evidence of the applicant having met the experience requirements.

After the applicant satisfies the personnel at the FAA GADO that he has met the experience requirement, he is allowed to take the written examinations. These examinations are given in five sections: General, Airframe Structures, Airframe Systems and Components, Powerplant Theory and Maintenance, and Powerplant Systems and Components.

When all five sections have been passed with a grade of 70% or more, the applicant can arrange with a Designated Mechanic Examiner to take his oral and practical examination. The DME is not employed by the FAA, but is a private individual whom the FAA has authorized to conduct these examinations. When the oral and practical examinations in the same five sections as the written exams are passed to the satisfaction of the examiner, the applicant is issued a temporary certificate that is valid for 120 days. By this time he should have his permanent certificate.

2. Privileges and limitations

A person holding a Mechanic Certificate with both the Airframe and Powerplant ratings is allowed to conduct a 100-hour inspection on an aircraft and approve it for return to service. He can also *perform* both minor and major repairs and alterations to an aircraft or engine. But this authorization does not include major repairs or alterations to propellers or *any* repairs or alterations to instruments. An A&P cannot authorize the return-to-service of an aircraft after a major repair or alteration unless he also holds an Inspection Authorization.

Unless we think that the A&P certificate authorizes a person to do any maintenance he would like, there is a stipulation that prevents a mechanic from supervising the maintenance or alteration of, or approving for return-to-service, any aircraft or engine for which he is rated unless he has satisfactorily performed the work concerned at an earlier date. He is also restricted from exercising the priviledges of his certificate and rating unless he understands the current instructions of the manufacturer for the specific operation.

In order to keep his A&P certificate current, a mechanic must work for at least six months out of the preceding 24 months, either actually using his certificate or supervising other mechanics.

B. The Inspection Authorization

An A&P mechanic who has held his certificate for at least three years and has been actively engaged in inspecting and maintaining certificated aircraft for the two year period preceding the application, may apply for an Inspection Authorization. He must take written and oral examinations given by the FAA GADO inspector covering the requirements for inspection and the paperwork involved in an inspection and the paperwork required to return an aircraft to service after a major repair or alteration.

Every Inspection Authorization expires on March 31 of each year, and in order to renew it, the holder must attend a meeting set up by the FAA GADO inspector. At this meeting he must show that within the last year he has conducted the required number of inspections and has returned to service the proper number of aircraft after major repairs or alterations. Provisions are made in this requirement for a person who has not held his IA for a full year to require a proportionally smaller number of inspections and returns-to-service.

The mechanic who holds an Inspection Authorization is allowed to conduct an annual or a progressive inspection, and is allowed to return-to-service an aircraft after it has undergone a major repair or alteration as long as the repair or alteration has been done according to approved data.

C. The Repairman Certificate

Much of the specialized work performed in aircraft maintenance is done in the shops of the airlines or in approved repair stations. These facilities operate under FAR Part 121 or 125 which allow the use of repairmen for specialized work.

A Repairman Certificate may be issued by the FAA to a person recommended by the airline or repair station who has had at least 18 months practical experience in the work he is to be doing. This work is normally of such a nature as the overhaul of fuel control units, propellers, instruments, or other components or appliances whose overhaul or repair require specialized knowledge and equipment.

The Repairman Certificate is valid for as long as the holder is employed by the facility that recommended him. But as soon as he leaves that employ, he loses his rating. It can naturally be reissued when he goes to work for another approved repair station.

A recent exception has been made to the experience requirements for issuance of a Repairman Certificate. One manufacturer has been granted a deviation from the regulations that allows a person who has gone through their propeller repairman's course to be granted a Repairman Certificate for the overhaul of the particular models of propellers that were included in the training.

There has been much discussion about adding a rating to the Mechanic Certificate that would allow a person to work on avionics or instruments. Those advocating the creation of new mechanic ratings claim a better qualified technician would be provided by the new certification system. The argument against the creation of this additional rating is that the Repairman Certificate has proven adequate to assure qualification for doing this type of work.

III. Maintenance and Inspection Requirements

A. Types of Maintenance

A description of the maintenance we are authorized to perform on a certificated aircraft is covered in FAR Part 43, and we must be aware of all of its provisions. No textbook can be relied on to keep current with the changes that are continually being made in the regulations, but here we will examine a few of the basic provisions of FAR 43. Be sure that you have a copy of the latest regulations and comply with them for specific details.

Appendix A of FAR 43 lists the basic types of maintenance and gives examples of each. In order to best understand aircraft maintenance, we should examine each of these types.

1. Major alterations

When an aircraft is initially certificated, it is issued an Approved Type Certificate (ATC) that covers its specific configuration, and in order to maintain this certification the aircraft must retain this configuration. But if the operator of an airplane needs better short-field performance than the original design provides and must alter the design of the wing to change its lift characteristics, the airplane no longer conforms to the original Type Certificate and special approval must be granted for the alteration.

All of the data used to make this alteration must be approved by the FAA engineering department. When the data is approved, an A&P mechanic can follow it to make the alteration, but an A&P with an Inspection Authorization must inspect the alteration to assure that it actually conforms to the approved data before the aircraft can be returned to service.

Part 65—Certification: Airmen Other Than Flight Crewmembers

Subpart A—General

§ 65.1 Applicability.

This Part prescribes the requirements for issuing the following certificates and associated ratings and the general operating rules for the holders of those certificates and ratings:

(a) Air traffic control tower operators.

(b) Aircraft dispatchers.

(c) Mechanics.

(d) Repairmen.

(e) Parachute riggers.

§ 65.11 Application and issue.

(a) An application for a certificate and rating, or for an additional rating, under this Part is made on a form and in a manner prescribed by the Administrator.

(b) An applicant who meets the requirements of this Part is entitled to an appropriate certificate and rating.

(c) Unless authorized by the Administrator, a person whose air traffic control tower operator, mechanic, or parachute rigger certificate is suspended may not apply for any rating to be added to that certificate during the period of suspension.

(d) Unless the order of revocation provides otherwise—

(1) A person whose air traffic control tower operator, aircraft dispatcher, or parachute rigger certificate is revoked may not apply for the same kind of certificate for one year after the date of revocation; and

(2) A person whose mechanic or repairman certificate is revoked may not apply for either of those kinds of certificates for one year after the date of revocation.

Ch. 2 (Amdt. 65–23, Eff. 6/26/78)

§ 65.12 Offenses involving narcotic drugs, marihuana, and depressant or stimulant drugs or substances.

(a) No person who is convicted of violating any Federal or State statute relating to the growing, processing, manufacture, sale, disposition, possession, transportation or importation of narcotic drugs, marihuana, and depressant or stimulant drugs or substances, is eligible for any certificate or rating issued under this Part for a period of 1 year after the date of final conviction.

(b) No person who commits an act prohibited by § 91.12(a) of this chapter is eligible for any certificate or rating issued under this Part for a period of 1 year after the date of that act.

(c) Any conviction specified in paragraph (a) of this section, or the commission of the act referenced in paragraph (b) of this section, is grounds for suspending or revoking any certificate or rating issued under this Part.

§ 65.13 Temporary certificate.

A certificate and ratings effective for a period of not more than [120] days may be issued to a qualified applicant, pending review of his application and supplementary documents and the issue of the certificate and ratings for which he applied.

§ 65.15 Duration of certificates.

(a) Except for repairman certificates, a certificate or rating issued under this Part is effective until it is surrendered, suspended, or revoked.

[(b) Deleted.]

(c) The holder of any certificate issued under this Part that is suspended or revoked shall, upon the Administrator's request, return it to the Administrator.

Fig. 3-5 FAR Part 65

(d) Unless it is sooner surrendered, suspended, or revoked, a repairman certificate is effective until the holder is relieved from the duties for which he was employed and certificated. The holder of such a certificate that is no longer effective shall return it to the Administrator.

§ 65.15a Reissuance: Expired certificates.

(a) Any parachute rigger certificate or rating that was issued after September 4, 1950, and any other certificate or rating covered by this Part (except a repairman certificate) that was issued after September 26, 1950, bearing an expiration date, to a person who was not a citizen of the United States may be reissued to that person without an expiration date.

(b) The holder of a certificate or rating with an expiration date may not after that date exercise the privileges of the certificate or rating without obtaining its reissuance.

§ 65.16 Change of name; replacement of lost or destroyed certificate.

(a) An application for a change of name on a certificate issued under this Part must be accompanied by the applicant's current certificate and the marriage license, court order, or other document verifying the change. The documents are returned to the applicant after inspection.

(b) An application for a replacement of a lost or destroyed certificate is made by letter to the Department of Transportation, Federal Aviation Administration, Airman Certification Branch, P.O. Box 25082, Oklahoma City, Okla. 73125. The letter must—

(1) Contain the name in which the certificate was issued, the permanent mailing address (including zip code), social security number (if any), and date and place of birth of the certificate holder, and any available information regarding the grade, number, and date of issue of the certificate, and the ratings on it; and

(2) Be accompanied by a check or money order for $2.00, payable to the Federal Aviation Administration.

(c) An application for a replacement of a lost or destroyed medical certificate is made by letter to the Department of Transportation, Federal Aviation Administration, Civil Aeromedical Institute, Aeromedical Certification Branch, P.O. Box 25082, Oklahoma City, Okla. 73125, accompanied by a check or money order for $2.00.

(d) A person whose certificate issued under this Part or medical certificate, or both, has been lost may obtain a telegram from the FAA confirming that it was issued. The telegram may be carried as a certificate for a period not to exceed 60 days pending his receiving a duplicate certificate under paragraph (b) or (c) of this section, unless he has been notified that the certificate has been suspended or revoked. The request for such a telegram may be made by prepaid telegram, stating the date upon which a duplicate certificate was requested, or including the request for a duplicate and a money order for the necessary amount. The request for a telegraphic certificate should be sent to the office prescribed in paragraph (b) or (c) of this section, as appropriate. However, a request for both at the same time should be sent to the office prescribed in paragraph (b) of this section.

§ 65.17 Tests: general procedure.

(a) Tests prescribed by or under this Part are given at times and places, and by persons, designated by the Administrator.

(b) The minimum passing grade for each test is 70 percent.

§ 65.18 Written tests: cheating or other unauthorized conduct.

(a) Except as authorized by the Administrator, no person may—

(1) Copy, or intentionally remove, a written test under this Part;

(2) Give to another, or receive from another, any part or copy of that test;

(3) Give help on that test to, or receive help on that test from, any person during the period that test is being given;

(4) Take any part of that test in behalf of another person;

(5) Use any material or aid during the period that test is being given; or

(6) Intentionally cause, assist, or participate in any act prohibited by this paragraph.

(b) No person who commits an act prohibited by paragraph (a) of this section is eligible for any airman or ground instructor

Fig. 3-5 FAR Part 65 (continued)

certificate or rating under this chapter for a period of one year after the date of that act. In addition, the commission of that act is a basis for suspending or revoking any airman or ground instructor certificate or rating held by that person.

[§ 65.19 Retesting after failure.

[An applicant for a written, oral, or practical test for a certificate and rating, or for an additional rating under this Part, may apply for retesting—

[(a) After 30 days after the date the applicant failed the test; or

[(b) Before the 30 days have expired if the applicant presents a signed statement from an airman holding the certificate and rating sought by the applicant, certifying that the airman has given the applicant additional instruction in each of the subjects failed and that the airman considers the applicant ready for retesting.]

§ 65.20 Applications, certificates, logbooks, reports, and records: falsification, reproduction, or alteration.

(a) No person may make or cause to be made—

(1) Any fraudulent or intentionally false statement on any application for a certificate or rating under this Part;

(2) Any fraudulent or intentionally false entry in any logbook, record, or report that is required to be kept, made, or used, to show compliance with any requirement for any certificate or rating under this Part;

(3) Any reproduction, for fraudulent purpose, of any certificate or rating under this Part; or

(4) Any alteration of any certificate or rating under this Part.

(b) The commission by any person of an act prohibited under paragraph (a) of this section is a basis for suspending or revoking any airman or ground instructor certificate or rating held by that person.

§ 65.21 Change of address.

Within 30 days after any change in his permanent mailing address, the holder of a certificate issued under this Part shall notify the Department of Transportation, Federal Aviation Administration, Airman Certification Branch, P.O. Box 25082, Oklahoma City, Okla. 73125 in writing, of his new address.

Subpart D—Mechanics

§ 65.71 Eligibility requirements: general.

(a) To be eligible for a mechanic certificate and associated ratings, a person must—

(1) Be at least 18 years of age;

(2) Be able to read, write, speak, and understand the English language, or in the case of an applicant who does not meet this requirement and who is employed outside of the United States by a U.S. air carrier, have his certificate endorsed "Valid only outside the United States";

(3) Have passed all of the prescribed tests within a period of 24 months; and

(4) Comply with the sections of this subpart that apply to the rating he seeks.

(b) A certificated mechanic who applies for an additional rating must meet the requirements of § 65.77 and, within a period of 24 months, pass the tests prescribed by §§ 65.75 and 65.79 for the additional rating sought.

§ 65.73 Ratings.

(a) The following ratings are issued under this subpart:

(1) Airframe.

(2) Powerplant.

(b) A mechanic certificate with an aircraft or aircraft engine rating, or both, that was issued before, and was valid on, June 15, 1952,

Fig. 3-5 FAR Part 65 (continued)

is equal to a mechanic certificate with an airframe or powerplant rating, or both, as the case may be, and may be exchanged for such a corresponding certificate and rating or ratings.

§ 65.75 Knowledge requirements.

(a) Each applicant for a mechanic certificate or rating must, after meeting the applicable experience requirements of § 65.77, pass a written test covering the construction and maintenance of aircraft appropriate to the rating he seeks, the regulations in this subpart, and the applicable provisions of Parts 43 and 91 of this chapter. The basic principles covering the installation and maintenance of propellers are included in the powerplant test.

(b) The applicant must pass each section of the test before applying for the oral and practical tests prescribed by § 65.79. A report of the written test is sent to the applicant.

§ 65.77 Experience requirements.

Each applicant for a mechanic certificate or rating must present either an appropriate graduation certificate or certificate of completion from a certificated aviation maintenance technician school or documentary evidence, satisfactory to the Administrator, of—

(a) At least 18 months of practical experience with the procedures, practices, materials, tools, machine tools, and equipment generally used in constructing, maintaining, or altering airframes, or powerplants appropriate to the rating sought; or

(b) At least 30 months of practical experience concurrently performing the duties appropriate to both the airframe and powerplant ratings.

§ 65.79 Skill requirements.

Each applicant for a mechanic certificate or rating must pass an oral and a practical test on the rating he seeks. The tests cover the applicant's basic skill in performing practical projects on the subjects covered by the written test for that rating. An applicant for a powerplant rating must show his ability to make satisfactory minor repairs to, and minor alterations of, propellers.

§ 65.80 Certificated aviation maintenance technician school students.

Whenever an aviation maintenance technician school certificated under Part 147 of this chapter shows to an FAA inspector that any of its students has made satisfactory progress at the school and is prepared to take the oral and practical tests prescribed by § 65.79, that student may take those tests during the final subjects of his training in the approved curriculum, before he meets the applicable experience requirements of § 65.77 and before he passes each section of the written test prescribed by § 65.75.

§ 65.81 General privileges and limitations.

(a) A certificated mechanic may perform or supervise the maintenance or alteration of an aircraft or appliance, or a part thereof, for which he is rated (but excluding major repairs to, and major alterations of, propellers, and any repair to, or alteration of, instruments), and may perform additional duties in accordance with §§ 65.85, 65.87, and 65.95. However, he may not supervise the maintenance or alteration of, or approve and return to service, any aircraft or appliance, or part thereof, for which he is rated unless he has satisfactorily performed the work concerned at an earlier date. If he has not so performed that work at an earlier date, he may show his ability to do it by performing it to the satisfaction of the Administrator or under the direct supervision of a certificated and appropriately rated mechanic, or a certificated repairman, who has had previous experience in the specific operation concerned.

(b) A certificated mechanic may not exercise the privileges of his certificate and rating unless he understands the current instructions of the manufacturer, and the maintenance manuals, for the specific operation concerned.

§ 65.83 Recent experience requirements.

A certificated mechanic may not exercise the privileges of his certificate and rating unless, within the preceding 24 months—

(a) The Administrator has found that he is able to do that work; or

Fig. 3-5 FAR Part 65 (continued)

(b) He has, for at least 6 months—

(1) Served as a mechanic under his certificate and rating;

(2) Technically supervised other mechanics;

(3) Supervised, in an executive capacity, the maintenance or alteration of aircraft; or

(4) Been engaged in any combination of subparagraphs (1), (2), or (3) of this paragraph.

§ 65.85 Airframe rating: additional privileges.

A certificated mechanic with an airframe rating may approve and return to service an airframe, or any related part or appliance, after he has performed, supervised, or inspected its maintenance or alteration (excluding major repairs and major alterations). In addition, he may perform the 100-hour inspection required by Part 91 of this chapter on an airframe, or any related part or appliance, and approve and return it to service.

§ 65.87 Powerplant rating: additional privileges.

A certificated mechanic with a powerplant rating may approve and return to service a powerplant or propeller or any related part or appliance, after he has performed, supervised, or inspected its maintenance or alteration (excluding major repairs and major alterations). In addition, he may perform the 100-hour inspection required by Part 91 of this chapter on a powerplant or propeller, or any part thereof, and approve and return it to service.

§ 65.89 Display of certificate.

Each person who holds a mechanic certificate shall keep it within the immediate area where he normally exercises the privileges of the certificate and shall present it for inspection upon the request of the Administrator or an authorized representative of the National Transportation Safety Board, or of any Federal, State, or local law enforcement officer.

§ 65.91 Inspection authorization.

(a) An application for an inspection authorization is made on a form and in a manner prescribed by the Administrator.

(b) An applicant who meets the requirements of this section is entitled to an inspection authorization.

[(c) To be eligible for an inspection authorization, an applicant must—

[(1) Hold a currently effective mechanic certificate with both an airframe rating and a powerplant rating, each of which is currently effective and has been continuously in effect for not less than the three year period immediately before the date of application;]

(2) Have been actively engaged, for at least the 2-year period before the date he applies, in maintaining aircraft certificated and maintained in accordance with this chapter;

(3) Have a fixed base of operations at which he may be located in person or by telephone during a normal working week, but it need not be the place where he will exercise his inspection authority;

(4) Have available to him the equipment, facilities, and inspection data necessary to properly inspect airframes, powerplants, propellers, or any related part or appliance; and

(5) Pass a written test on his ability to inspect according to safety standards for returning aircraft to service after major repairs and major alterations and annual and progressive inspection performed under Part 43 of this chapter.

An applicant who fails the test prescribed in subparagraph (5) of this paragraph may not apply for retesting until at least 90 days after the date he failed the test.

[(d) [Reserved]

[§ 65.92 Inspection authorization: duration.

[(a) Each inspection authorization expires on March 31 of each year. However, the holder may exercise the privileges of that authorization only while he holds a currently effective mechanic certificate with both a currently effective airframe rating and a currently effective powerplant rating.

[(b) An inspection authorization ceases to be effective whenever any of the following occurs:

Fig. 3-5 FAR Part 65 (continued)

(1) The authorization is surrendered, suspended, or revoked.

(2) The holder no longer has a fixed base of operation.

(3) The holder no longer has the equipment, facilities, and inspection data required by § 65.91(c)(3) and (4) for issuance of his authorization.

(c) The holder of an inspection authorization that is suspended or revoked shall, upon the Administrator's request, return it to the Administrator.

§ 65.93 Inspection authorization: renewal.

(a) To be eligible for renewal of an inspection authorization for a one-year period an applicant must present evidence at an FAA General Aviation District Office or an International Field Office during the month of March that he still meets the requirements of § 65.91(c)(1) through (4), and by showing that during the current period that he held the inspection authority he—

(1) Has performed at least one annual inspection for each 90 days that he held the current authority;

(2) Has performed inspections of at least two major repairs or major alterations for each 90 days that he held the current authority; or

(3) Has performed or supervised and approved at least one progressive inspection in accordance with standards prescribed by the Administrator.

(b) The holder of an inspection authorization that has been in effect for less than 90 days before the expiration date need not comply with paragraph (a)(1) through (3) of this section.

§ 65.95 Inspection authorization: privileges and limitations.

(a) The holder of an inspection authorization may—

(1) Inspect and approve for return to service any aircraft or related part or appliance (except any aircraft maintained in accordance with a continuous airworthiness program under Part 121 or 127 of this chapter) after a major repair or major alteration to it in accordance with Part 43 of this chapter, if the work was done in accordance with technical data approved by the Administrator.

(2) Perform an annual, or perform or supervise a progressive inspection according to §§ 43.13 and 43.15 of this chapter.

(b) When he exercises the privileges of an inspection authorization the holder shall keep it available for inspection by the aircraft owner, the mechanic submitting the aircraft, repair, or alteration for approval (if any), and shall present it upon the request of the Administrator or an authorized representative of the National Transportation Safety Board, or of any Federal, State, or local law enforcement officer.

(c) If the holder of an inspection authorization changes his fixed base of operation, hv may not exercise the privileges of the authorization until he has notified the FAA General Aviation District Office or International Field Office for the area in which the new base is located, in writing, of the change.

Subpart E—Repairmen

§ 65.101 Eligibility requirements: general.

[(a)] To be eligible for a repairman certificate a person must—

[(1)] Be at least 18 years of age;

[(2)] Be specially qualified to perform maintenance on aircraft or components thereof, appropriate to the job for which he is employed;

[(3)] Be employed for a specific job requiring those special qualifications by a certificated repair station, or by a certificated commercial operator or certificated air carrier, that is required by its operating certificate or approved operations specification to provide a continuous airworthiness maintenance program according to its maintenance manuals;

[(4)] Be recommended for certification by his employer, to the satisfaction of the Administrator, as able to satisfactorily maintain aircraft or components, appropriate to the job for which he is employed;

Fig. 3-5 FAR Part 65 (continued)

[(5)] Have at least 18 months of practical experience in the procedures, practices, inspection methods, materials, tools, machine tools, and equipment generally used in the maintenance duties of the specific job for which he is to be employed and certificated; and

[(6)] Be able to read, write, speak, and understand the English language, or, in the case of an applicant who does not meet this requirement and who is employed outside the United States by a certificated repair station, a certificated U.S. commercial operator, or a certificated U.S. air carrier, described in paragraph (c) of this section, have his certificate endorsed "Valid only outside the United States."

[(b) This section does not apply to the issuance of repairman certificates (experimental aircraft builder) under § 65.104.]

§ 65.103 Repairman certificate: privileges and limitations.

(a) A certificated repairman may perform or supervise the maintenance of aircraft or components thereof, appropriate to the job for which he was employed and certificated, but only in connection with duties for the repair station, commercial operator, or air carrier, by whom he was employed and recommended.

(b) A certificated repairman may not perform or supervise duties under his certificate unless he understands the current instructions of the air carrier, commercial operator, or manufacturer of the article on which he is to perform or supervise those duties and the maintenance manuals relating to the specific operations concerned.

[§ 65.104 Repairman certificate (experimental aircraft builder)—eligibility, privileges and limitations.

[(a) To be eligible for a repairman certificate (experimental aircraft builder), an individual must—

[(1) Be at least 18 years of age;

[(2) Be the primary builder of the aircraft to which the privileges of the certificate are applicable;

[(3) Show to the satisfaction of the Administrator that the individual has the requisite skill to determine whether the aircraft is in a condition for safe operations; and

[(4) Be a citizen of the United States or an individual citizen of a foreign country who has lawfully been admitted for permanent residence in the United States.

[(b) The holder of a repairman certificate (experimental aircraft builder) may perform condition inspections on the aircraft constructed by the holder in accordance with the operating limitations of that aircraft.

[(c) Section 65.103 does not apply to the holder of a repairman certificate (experimental aircraft builder) while performing under that certificate.]

§ 65.105 Display of certificate.

Each person who holds a repairman certificate shall keep it within the immediate area where he normally exercises the privileges of the certificate and shall present it for inspection upon the request of the Adminstrator or an authorized representative of the National Transportation Safety Board, or of any Federal, State, or local law enforcement officer.

Fig. 3-5 FAR Part 65 (continued)

When a major alteration is made, a Form 337 "Major Repair and Alteration" must be filled in to describe the work that was done, and one copy of this form is sent to the FAA GADO and one copy is kept with the aircraft.

If the person making an alteration to a certificated aircraft chooses, he can have the alteration approved and a Supplemental Type Certificate (STC) issued. The description of the STC serves as approved data for anyone wanting to make a similar alteration. A summary of Supplemental Type Certificates with periodic updates is available from the U.S. Government Printing Office on a subscription basis.

2. Minor Alterations

When an aircraft is issued its Approved Type Certificate, a list of all of the accessories that are approved for installation is included. If the aircraft comes from the factory equipped with, for example, a Slick magneto, but a Bendix magneto is also approved, the removal of the Slick magneto and installation of the Bendix constitutes a minor alteration and may be done by an A&P mechanic without any additional approval. The only record that is required in this case is an entry in the engine records telling what was done, when it was done, who did it, and noting the authorization, which was the inclusion of both magnetos in the approved equipment list.

3. Major repairs

A repair can be considered a maintenance function that restores a damaged or worn out part to a serviceable condition. If the repair is of such a nature that it requires special procedures or equipment, or if its being improperly done would appreciably affect the flight or strength characteristics of the aircraft, it is considered to be a major repair.

If an aircraft is damaged to the extent that some of the skin must be replaced, or if the engine mount or landing gear requires welding to be done, the repair is major. It must be done according to approved data, and when the repair is completed, an A&P with an IA must examine it to assure that it does conform with the approved data. He can then approve the aircraft for return to service. A Form 337 must be completed to describe the repair, and one copy sent to the FAA GADO and one copy retained with the aircraft records.

The overhaul of a non-supercharged, direct-drive aircraft piston engine is considered to be a minor repair, but if the crankcase of an engine having either a geared propeller or a geared supercharger is opened, the operation becomes a major repair.

4. Minor repairs

The definition given in FAR Part 1 of a minor repair is simply a repair that is not a major repair. And by working the definition of a major repair backward, we can define a minor repair as one that can be done by simple operations. When any definition is as ambiguous as this, the interpretation is left to the local FAA GADO. In order to determine whether or not a particular operation is considered to be a minor or a major repair, you should get the interpretation of the person who will ultimately have to approve the repair, the GADO inspector.

Generally speaking, a minor repair is one that can be accomplished without requiring any welding, riveting, or gluing, and one which does not change the characteristics or configuration of the aircraft.

An example that includes both minor and major repair involves the replacement of a wing. If a wing has been damaged and rebuilt, the rebuilding constitutes a *major* repair and a Form 337 is required, and the wing must be approved for return to service by an A&P with an IA. But this repaired wing may be installed on an airplane by an A&P without an IA, as its installation is considered to be a minor repair.

B. Preventive Maintenance

The large number of airplanes used by private pilots for non-commercial flying, coupled with the high cost of shop time, have caused many pilots to want to perform their own maintenance. To authorize this and at the same time have a framework in which some semblance of control can be maintained, the FAA has included in FAR 43, Appendix A, a list of about 25 items they considered to be preventive maintenance. A person holding a pilot's license can perform these on the aircraft he owns or operates, as long as it is not used in air carrier service. There is no requirement that the pilot make any maintenance record of the preventive maintenance that he has done.

C. *Types of Inspections*

One of the most important functions of an A&P mechanic is that of looking for trouble. Inspections make up one of the most important jobs of the mechanic, and because of their importance we must consider them in detail.

When an airplane is issued its Approved Type Certificate from the Federal Aviation Administration, its design is approved as meeting all of the requirements for certification under the appropriate parts of the Federal Aviation Regulations. The airplane is constructed under the provisions of a Production Certificate, which means that the production facilities have been closely monitored so that we can be assured that each airplane that comes off of the production line will conform to the provisions of its Type Certificate.

As soon as the airplane leaves the factory, it becomes the responsibility of its owner or operator to determine that it continues to meet all of the requirements for its certification. For this reason, a schedule of inspections has been set up by the manufacturer and is implemented by the aviation maintenance technician.

1. *Preflight inspection*

The FAA requires that before each flight, the pilot must assure himself that the airplane is in condition to fly. This means that there is sufficient fuel for the flight, the oil quantity is adequate, and there is no indication of damage to the aircraft. The manufacturer of each aircraft publishes a comprehensive walk-around inspection, and since this is an approved procedure, it is the responsibility of the pilot to use this method, or to contuct an inspection that is equally comprehensive.

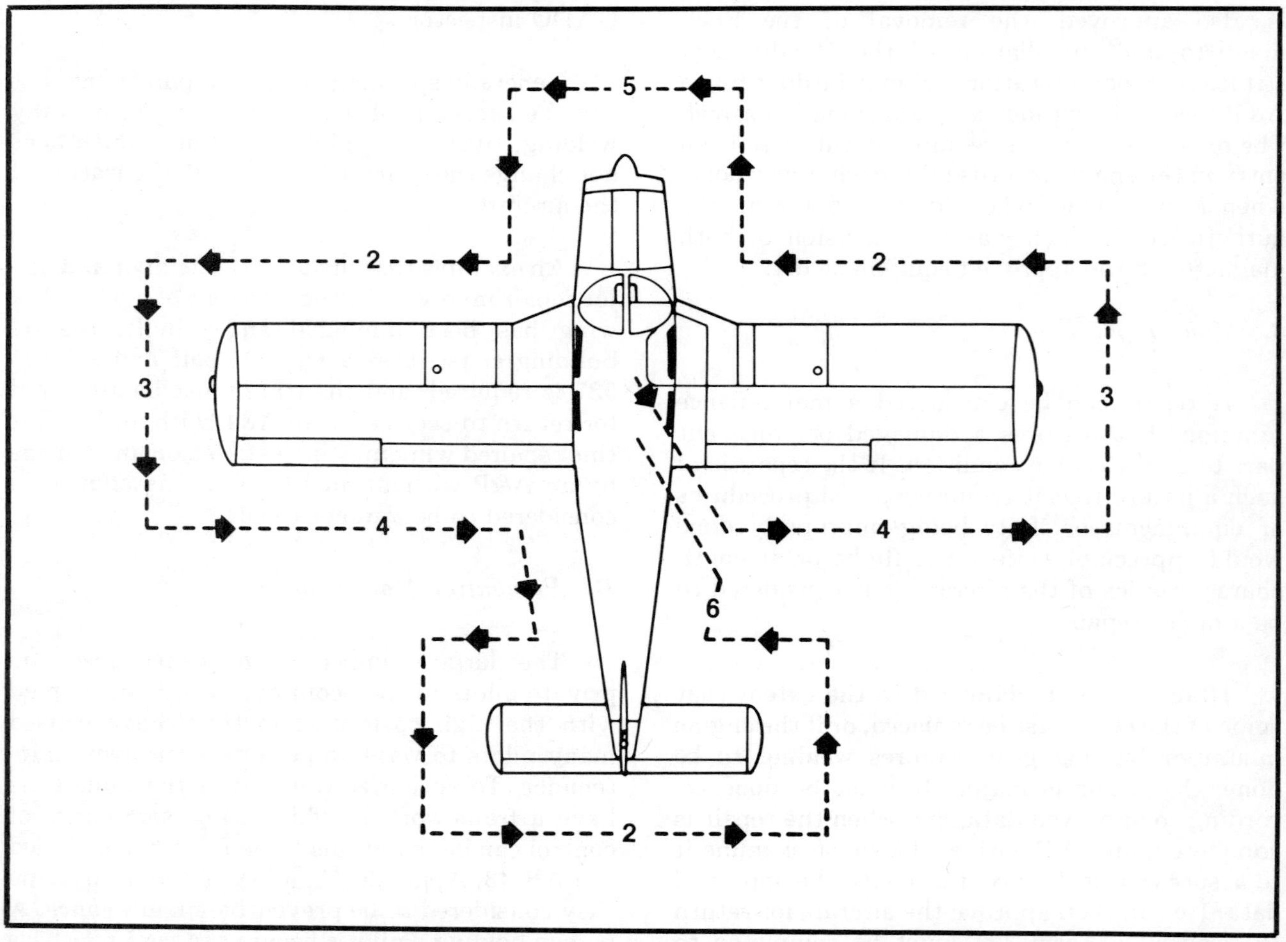

Fig. 3-6 A walk-around inspection is required before each flight.

Part 43—Maintenance, Preventive Maintenance, Rebuilding, and Alteration

§ 43.1 Applicability.

(a) Except as provided in paragraph (b), this Part prescribes rules governing the maintenance, preventive maintenance, rebuilding, and alteration of any—

(1) Aircraft having a U.S. airworthiness certificate; and

(2) Airframe, aircraft engine, propeller, or appliance of such an aircraft.

(b) This Part does not apply to any aircraft for which an experimental airworthiness certificate has been issued, unless a different kind of airworthiness certificate had previously been issued for that aircraft.

§ 43.3 Persons authorized to perform maintenance, preventive maintenance, rebuilding, and alterations.

(a) Except as provided in this section, no person may maintain, rebuild, alter, or perform preventive maintenance on an aircraft, airframe, aircraft engine, propeller, or appliance to which this Part applies. Those items, the performance of which is a major alteration, a major repair, or preventive maintenance, are listed in appendix A.

(b) The holder of a mechanic certificate may perform maintenance, preventive maintenance, and alterations as provided in Part 65.

(c) The holder of a repairman certificate may perform maintenance and preventive maintenance as provided in Part 65.

(d) A person working under the supervision of a holder of a mechanic or repairman certificate may perform the maintenance, preventive maintenance, and alterations that his supervisor is authorized to perform, if the supervisor personally observes the work being done to the extent necessary to ensure that it is being done properly and if the supervisor is readily available, in person, for consultation. However, this paragraph does not authorize the performance of 100–hour or annual inspections, nor inspections performed after a major repair or alteration.

(e) The holder of a repair station certificate may perform maintenance, preventive maintenance, and alterations as provided in Part 145.

(f) An air carrier may perform maintenance, preventive maintenance and alterations as provided in Part 121, 127, or 135 of this chapter, as applicable.

(g) The holder of a commercial operator certificate issued under Part 121 may perform maintenance, preventive maintenance, and alterations as provided in that Part.

(h) The holder of a pilot certificate issued under Part 61 may perform preventive maintenance on any aircraft owned or operated by him that is not used in air carrier service.

(i) A manufacturer may—

(1) Rebuild or alter any aircraft, aircraft engine, propeller, or appliance manufactured by him under a type or production certificate;

(2) Rebuild or alter any appliance or part of aircraft, aircraft engines, propellers, or appliances manufactured by him under a Technical Standard Order Authorization, an FAA-Parts Manufacturer Approval, or Product and Process Specification issued by the Administrator; and

(3) Perform 100-hour, annual, and progressive inspections on aircraft manufactured by him, while operating under a production certificate or under an approved production inspection system for such aircraft.

1

Fig. 3-7 FAR Part 43

§ 43.5 Return to service after maintenance, preventive maintenance, rebuilding, or alteration.

(a) No person may return to service any aircraft, airframe, aircraft engine, propeller, or appliance, that has undergone maintenance, preventive maintenance, rebuilding, or alteration unless—

(1) It has been approved for return to service by a person authorized under § 43.7;

(2) The maintenance record entry required by § 43.9 has been made;

(3) The repair or alteration form authorized by or furnished by the Administrator has been executed in a manner prescribed by the Administrator; and

(4) If a repair or an alteration results in any change in the aircraft operating limitations or flight data contained in the approved aircraft flight manual, those operating limitations or flight data are appropriately revised and set forth as prescribed in § 91.31.

(b) This section does not apply to preventive maintenance performed by a certificated pilot under § 43.3(h).

§ 43.7 Persons authorized to approve aircraft, airframes, aircraft engines, propellers, and appliances for return to service after maintenance, preventive maintenance, rebuilding, or alteration.

(a) Except as provided in this section, no person, other than the Administrator, may approve any aircraft, airframe, aircraft engine, propeller, or appliance for return to service after it has undergone maintenance, preventive maintenance, rebuilding, or alteration.

(b) The holder of a mechanic certificate or an inspection authorization may approve an aircraft, airframe, aircraft engine, propeller, or appliance for return to service as provided in Part 65.

(c) The holder of a repair station certificate may approve an aircraft, airframe, aircraft engine, propeller, or appliance for return to service as provided in Part 145.

(d) A manufacturer may approve for return to service any aircraft, airframe, aircraft engine, propeller, or appliance that he has worked on under § 43.3(i). However, except for minor alterations, the work must have been done in accordance with technical data approved by the Administrator.

(e) An air carrier may approve an aircraft, airframe, aircraft engine, propeller, or appliance for return to service as provided in Part 121, 127, or 135 of this chapter, as applicable.

(f) The holder of a commercial operator certificate issued under Part 121 of this chapter may approve an aircraft, airframe, aircraft engine, propeller, or appliance for return to service as provided in that Part.

§ 43.9 Content, form, and disposition of maintenance, rebuilding, and alteration records (except 100-hour, annual, and progressive inspections).

(a) *Maintenance record entries.* Except as provided in paragraphs (b) and (c) of this section, each person who maintains, rebuilds, or alters an aircraft, airframe, aircraft engine, propeller, or appliance shall make an entry in the maintenance record of that equipment containing the following information:

(1) A description (or reference to data acceptable to the Administrator) of the work performed.

(2) The date of completion of the work performed.

(3) The name of the person performing the work.

(4) If the aircraft, airframe, aircraft engine, propeller, or appliance is approved for return to service, the signature (and if a certificated mechanic, the certificate number) of the person who approved it.

(5) If the work performed is an inspection required under § 91.217 of this chapter for a large airplane, or a turbojet or turbopropeller powered multiengine airplane, the entry must name the kind of inspection conducted (continuous airworthiness inspection program, approved inspection program, etc.) and include a statement that—

(i) The inspection was performed in accordance with the instructions and pro-

Fig. 3-7 FAR Part 43 (continued)

cedures for the kind of inspection program selected by the owner or operator of the airplane; and

(ii) A signed and dated list of the defects, if any, found during the inspection was given to the owner or operator of the airplane.

In addition to the entry required by this paragraph, major repairs and major alterations shall be entered on a form, and the form disposed of, in the manner prescribed in Appendix B, by the person performing the work.

(b) Each holder of an air carrier or commercial operator certificate that is required by its operating certificate or by approved operations specifications to provide for a continuous airworthiness maintenance program, shall make a record of the maintenance, rebuilding, and alteration, on aircraft, airframes, aircraft engines, propellers, appliances, or parts thereof, which it operates, in accordance with the provisions of Part 121 or Part 127, of this chapter as appropriate.

(c) This section does not apply to persons performing the annual, 100-hour, and progressive inspections required by Part 91.

§ 43.11 Content, form, and disposition of annual, 100–hour, and progressive inspection records.

(a) *Maintenance record entries.* The person approving or disapproving for return to service an aircraft, airframe, aircraft engine, propeller, or appliance after any annual, 100-hour, or progressive inspection required by Part 91 shall make an entry in the maintenance record of that equipment, containing the following information:

(1) The type of inspection (and for progressive inspections, a brief description of the extent of the inspection).

(2) The date of the inspection and aircraft time in service.

(3) The signature (and if a certificated mechanic, the certificate number) of the person approving or disapproving for return to service, the aircraft, airframe, aircraft engine, propeller, or appliance.

(4) For annual or 100-hour inspections if the aircraft is approved for return to service, the following or a similarly worded statement—"I certify that this aircraft has been inspected in accordance with (insert type) inspection and was determined to be in airworthy condition".

(5) For annual inspections, if the aircraft is not approved for return to service because of needed maintenance, noncompliance with applicable specifications, airworthiness directives or other approved data, the following or a similarly worded statement—"I certify that this aircraft has been inspected in accordance with an annual inspection and a list of discrepancies and unairworthy items dated (date) has been provided for the aircraft owner or lessee".

(6) For progressive inspections, the following or a similarly worded statement—"I certify that in accordance with a progressive inspection program, a routine inspection of (identify whether aircraft, or components) and a detailed inspection of (identify components) were performed in accordance with a progressive inspection and the aircraft is approved for return to service".

(b) *Listing of discrepancies.* If the person performing an annual inspection finds that the aircraft is unairworthy or does not meet the applicable type certificate data, airworthiness directives, or other approved data upon which airworthiness depends, he shall give the owner or lessee a signed and dated copy of a list of discrepancies. If the aircraft is not approved for return to service, he shall send the list of discrepancies to the local FAA District Office, within 48 hours after completing the inspection.

[§ 43.12 Maintenance records: falsification, reproduction, or alteration.

[(a) No person may make or cause to be made:

[(1) Any fraudulent entry in any record or report that is required to be kept, made, or used to show compliance with any requirement under this Part;

Fig. 3-7 FAR Part 43 (continued)

[(2) Any reproduction, for fraudulent purpose, of any record or report under this Part; or

[(3) Any alternation, for fraudulent purpose, of any record or report under this Part.

[(b) The commission by any person of an act prohibited under paragraph (a) of this section is a basis for suspending or revoking the applicable airman, operator, or production certificate, Technical Standard Order Authorization, FAA-Parts Manufacturer Approval, or Product and Process Specification issued by the Administrator and held by that person.]

§ 43.13 Performance rules (general).

(a) Each person maintaining or altering, or performing preventive maintenance, shall use methods, techniques, and practices acceptable to the Administrator. He shall use the tools, equipment, and test apparatus necessary to assure completion of the work in accordance with accepted industry practices. If special equipment or test apparatus is recommended by the manufacturer involved, he must use that equipment or apparatus or its equivalent acceptable to the Administrator.

(b) Each person maintaining or altering, or performing preventive maintenance, shall do that work in such a manner and use materials of such a quality, that the condition of the aircraft, airframe, aircraft engine, propeller, or appliance worked on will be at least equal to its original or properly altered condition (with regard to aerodynamic function, structural strength, resistance to vibration and deterioration, and other qualities affecting airworthiness).

(c) *Special provisions for air carriers and commercial operators.* Unless otherwise notified by the Administrator, the methods, techniques, and practices contained in the maintenance manual or the maintenance part of the air carrier manual of a certificated air carrier or commercial operator (that is required by its operating certificate or approved operating specifications to provide a continuous airworthiness maintenance and inspection program) constitute acceptable means of compliance with this section.

(d) Each person performing an inspection required by § 91.217 for a large airplane, or a turbojet- or turbopropeller-powered multiengine airplane shall do that work in accordance with the standards prescribed in subparagraphs (1) through (5) of this paragraph for the applicable inspection program.

(1) For a continuous airworthiness ininspection program (§ 91.217(b)(1) of this chapter), the standards prescribed in paragraph (c) of this section apply.

(2) For an approved aircraft inspection program (§ 91.217(b)(2) of this chapter), the standards prescribed in paragraph (a) of this section apply.

(3) For an approved continuous inspection program (§ 91.217(b)(3) of this chapter), the standards prescribed in § 43.15(a) of this Part apply.

(4) For an inspection program recommended by a manufacturer (§ 91.217(b)(4) of this chapter), the standards contained in the recommendations and instructions of the aircraft, engine, propeller, or appliance manufacturer apply.

(5) For an approved inspection program (§ 91.217(b)(5) of this chapter), the standards prescribed in paragraph (a) of this section apply, except when the inspection program for the particular airplane includes other standards.

§ 43.15 Additional performance rules for inspections.

(a) *General.* Each person performing a 100-hour, annual, or progressive inspection required by Part 91 of this chapter or an inspection required under Part 123 of this chapter, shall perform those inspections in such a manner as to determine whether the aircraft concerned meets all applicable airworthiness requirements.

(b) *Rotorcraft.* Each person performing a 100-hour, annual, or progressive inspection of a rotorcraft shall inspect the following systems in accordance with the maintenance manual of the manufacturer concerned:

(1) The drive shafts or similar systems.

(2) The main rotor transmission gear box for obvious defects.

Fig. 3-7 FAR Part 43 (continued)

(3) The main rotor and center section (or the equivalent area).

(4) The auxiliary rotor on helicopters.

(c) *Annual and 100-hour inspections.*

(1) Each person performing an annual or 100-hour inspection shall use a checklist while performing the inspection. The checklist may be of the person's own design, one provided by the manufacturer of the equipment being inspected, or one obtained from another source. This checklist must include the scope and detail of the items contained in appendix D to this Part and paragraph (b) of this section.

(2) Each person approving an aircraft for return to service after an annual or 100-hour inspection shall, before that approval, run the aircraft engine or engines to determine satisfactory performance, in accordance with the manufacturer's recommendations, of—

(i) Power output (static and idle r.p.m.);

(ii) Magnetos;

(iii) Fuel and oil pressure; and

(iv) Cylinder and oil temperature.

(d) *Progressive inspection.*

(1) Each person performing a progressive inspection shall, at the start of a progressive inspection system, inspect the aircraft completely. After this initial inspection, routine and detailed inspections must be conducted as prescribed in the progressive inspection schedule. Routine inspections consist of visual examination or check of the appliances, the aircraft, and its components and systems, insofar as practicable without disassembly. Detailed inspections consist of a thorough examination of the appliances, the aircraft, and its components and systems, with such disassembly as is necessary. For the purposes of this subparagraph, the overhaul of a component or system is considered to be a detailed inspection.

(2) If the aircraft is away from the station where inspections are normally conducted, an appropriately rated mechanic, a certificated repair station, or the manufacturer of the aircraft may perform inspections in accordance with the procedures and using the forms of the person who would otherwise perform the inspection.

§ 43.16 Rotorcraft Maintenance Manual: "Airworthiness Limitations" section.

For rotorcraft for which a Rotorcraft Maintenance Manual containing an "Airworthiness Limitations" section has been issued, each person performing an inspection or other work specified in that section of the manual shall perform the inspection or work in accordance with that section of the manual.

§ 43.17 Mechanical work performed on U.S. registered aircraft by certain Canadian persons.

(a) A person holding a valid mechanic certificate of competence (Aircraft Maintenance Engineer license) and appropriate ratings issued by the Canadian Government, or a person who is an authorized employee (Approved Inspector) performing work for a company whose system of quality control for the inspection and maintenance of aircraft has been approved by the Canadian Department of Transport may, in connection with aircraft of U.S. registry in Canada:

(1) Perform maintenance, preventive maintenance and alterations if those operations are done in accordance with § 43.13 and the maintenance record entries are made in accordance with § 43.9.

(2) Perform the 100-hour inspection required by Part 91 of this chapter if that inspection is done in accordance with § 43.15 and the maintenance record entries are made in accordance with § 43.11.

(3) Approve (certify) maintenance, preventive maintenance, and alterations performed under this section except that a Canadian Aircraft Maintenance Engineer may not approve a major repair or major alteration.

(b) A Canadian Department of Transport Airworthiness Inspector, or an authorized employee (Approved Inspector) performing work

Fig. 3-7 FAR Part 43 (continued)

for a company approved by the Canadian Department of Transport, may approve (certify) a major repair or major alteration performed under this section if the work was done in accordance with technical data approved by the Administrator.

(c) No person may operate in air commerce an aircraft, airframe, aircraft engine, propeller, or appliance on which maintenance, preventive maintenance, or alteration has been performed under this section unless it has been approved by a person authorized in this section.

NOTE.—The recordkeeping and reporting requirements contained herein have been approved by the Bureau of the Budget in accordance with the Federal Reports Act of 1942.

Fig. 3-7 FAR Part 43 (continued)

Appendix A

Major Alterations, Major Repairs, and Preventive Maintenance

(a) *Major alterations.*

(1) *Airframe major alterations.* Alterations of the following parts and alterations of the following types, when not listed in the aircraft specifications issued by the FAA, are airframe major alterations:

(i) Wings.

(ii) Tail surfaces.

(iii) Fuselage.

(iv) Engine mounts.

(v) Control system.

(vi) Landing gear.

(vii) Hull or floats.

(viii) Elements of an airframe including spars, ribs, fittings, shock absorbers, bracing, cowlings, fairings, and balance weights.

(ix) Hydraulic and electrical actuating system of components.

(x) Rotor blades.

(xi) Changes to the empty weight or empty balance which result in an increase in the maximum certificated weight or center of gravity limits of the aircraft.

(xii) Changes to the basic design of the fuel, oil, cooling, heating, cabin presssurization, electrical, hydraulic, de-icing, or exhaust systems.

(xiii) Changes to the wing or to fixed or movable control surfaces which affect flutter and vibration characteristics.

(2) *Powerplant major alterations.* The following alterations of a powerplant when not listed in the engine specifications issued by the FAA, are powerplant major alterations:

(i) Conversion of an aircraft engine from one approved model to another, involving any changes in compression ratio, propeller reduction gear, impeller gear ratios or the substitution of major engine parts which requires extensive rework and testing of the engine.

(ii) Changes to the engine by replacing aircraft engine structural parts with parts not supplied by the original manufacturer or parts not specifically approved by the Administrator.

(iii) Installation of an accessory which is not approved for the engine.

(iv) Removal of accessories that are listed as required equipment on the aircraft or engine specification.

(v) Installation of structural parts other than the type of parts approved for the installation.

(vi) Conversions of any sort for the purpose of using fuel of a rating or grade other than that listed in the engine specifications.

(3) *Propeller major alterations.* The following alterations of a propeller when not authorized in the propeller specifications issued by the FAA are propeller major alterations:

(i) Changes in blade design.

(ii) Changes in hub design.

(iii) Changes in the governor or control design.

(iv) Installation of a propeller governor or feathering system.

(v) Installation of propeller de-icing system.

(vi) Installation of parts not approved for the propeller.

(4) *Appliance major alterations.* Alterations of the basic design not made in accordance with recommendations of the appliance manufacturer or in accordance with an FAA Airworthiness Directive are appliance major alterations. In addition, changes in the basic design of radio communication and

Fig. 3-7 FAR Part 43 (continued)

navigation equipment approved under type certification or a Technical Standard Order that have an effect on frequency stability, noise level, sensitivity, selectivity, distortion, spurious radiation, AVC characteristics, or ability to meet environmental test conditions and other changes that have an effect on the performance of the equipment are also major alterations.

(b) *Major repairs.*

(1) *Airframe major repairs.* Repairs to the following parts of an airframe and repairs of the following types, involving the strengthening, reinforcing, splicing, and manufacturing of primary structural members or their replacement, when replacement is by fabrication such as riveting or welding, are airframe major repairs:

(i) Box beams.

(ii) Monocoque or semimonocoque wings or control surfaces.

(iii) Wing stringers or chord members.

(iv) Spars.

(v) Spar flanges.

(vi) Members of truss-type beams.

(vii) Thin sheet webs of beams.

(viii) Keel and chine members of boat hulls or floats.

(ix) Corrugated sheet compression members which act as flange material of wings or tail surfaces.

(x) Wing main ribs and compression members.

(xi) Wing or tail surface brace struts.

(xii) Engine mounts.

(xiii) Fuselage longerons.

(xiv) Members of the side truss, horizontal truss, or bulkheads.

(xv) Main seat support braces and brackets.

(xvi) Landing gear brace struts.

(xvii) Axles.

(xviii) Wheels.

(xix) Skis, and ski pedestals.

(xx) Parts of the control system such as control columns, pedals, shafts, brackets, or horns.

(xxi) Repairs involving the substitution of material.

(xxii) The repair of damaged areas in metal or plywood stressed covering exceeding six inches in any direction.

(xxiii) The repair of portions of skin sheets by making additional seams.

(xxiv) The splicing of skin sheets.

(xxv) The repair of three or more adjacent wing or control surface ribs or the leading edge of wings and control surfaces, between such adjacent ribs.

(xxvi) Repair of fabric covering involving an area greater than that required to repair two adjacent ribs.

(xxvii) Replacement of fabric on fabric covered parts such as wings, fuselages, stabilizers, and control surfaces.

(xxviii) Repairing, including rebottoming, of removable or integral fuel tanks and oil tanks.

(2) *Powerplant major repairs.* Repairs of the following parts of an engine and repairs of the following types, are powerplant major repairs:

(i) Separation or disassembly of a crankcase or crankshaft of a reciprocating engine equipped with an integral supercharger.

(ii) Separation or disassembly of a crankcase or crankshaft of a reciprocating engine equipped with other than spurtype propeller reduction gearing.

(iii) Special repairs to structural engine parts by welding, plating, metalizing, or other methods.

(3) *Propeller major repairs.* Repairs of the following types to a propeller are propeller major repairs:

(i) Any repairs to, or straightening of steel blades.

(ii) Repairing or machining of steel hubs.

(iii) Shortening of blades.

(iv) Retipping of wood propellers.

(v) Replacement of outer laminations on fixed-pitch wood propellers.

(vi) Repairing elongated bolt holes in the hub of fixed-pitch wood propellers.

(vii) Inlay work on wood blades.

(viii) Repairs to composition blades.

(ix) Replacement of tip fabric.

(x) Replacement of plastic covering.

Fig. 3-7 FAR Part 43 (continued)

(xi) Repair of propeller governors.

(xii) Overhaul of controllable pitch propellers.

(xiii) Repairs to deep dents, cuts, scars, nicks, etc., and straightening of aluminum blades.

(xiv) The repair or replacement of internal elements of blades.

(4) *Appliance major repairs.* Repairs of the following types to appliances are appliance major repairs:

(i) Calibration and repair of instruments.

(ii) Calibration of radio equipment.

(iii) Rewinding the field coil of an electrical accessory.

(iv) Complete disassembly of complex hydraulic power valves.

(v) Overhaul of pressure type carburetors, and pressure type fuel, oil, and hydraulic pumps.

(c) *Preventive maintenance.* Work of the following type is preventive maintenance:

(1) Removal, installation, and repair of landing gear tires.

(2) Replacing elastic shock absorber cords on landing gear.

(3) Servicing landing gear shock struts by adding oil, air, or both.

(4) Servicing landing gear wheel bearings, such as cleaning and greasing.

(5) Replacing defective safety wiring or cotter keys.

(6) Lubrication not requiring disassembly other than removal of nonstructural items such as cover plates, cowlings, and fairings.

(7) Making simple fabric patches not requiring rib stitching or the removal of structural parts or control surfaces.

(8) Replenishing hydraulic fluid in the hydraulic reservoir.

(9) Refinishing decorative coating of fuselage, wings, tail group surfaces (excluding balanced control surfaces), fairings, cowling, landing gear, cabin, or cockpit interior when removal or disassembly of any primary structure or operating system is not required.

(10) Applying preservative or protective material to components where no disassembly of any primary structure or operating system is involved and where such coating is not prohibited or is not contrary to good practices.

(11) Repairing upholstery and decorative furnishings of the cabin or cockpit interior when the repairing does not require disassembly of any primary structure or operating system or interfere with an operating system or affect primary structure of the aircraft.

(12) Making small simple repairs to fairings, nonstructural cover plates, cowlings, and small patches and reinforcements not changing the contour so as to interfere with proper airflow.

(13) Replacing side windows where that work does not interfere with the structure or any operating system such as controls, electrical equipment, etc.

(14) Replacing safety belts.

(15) Replacing seats or seat parts with replacement parts approved for the aircraft, not involving disassembly of any primary structure or operating system.

(16) Trouble shooting and repairing broken circuits in landing light wiring circuits.

(17) Replacing bulbs, reflectors, and lenses of position and landing lights.

(18) Replacing wheels and skis where no weight and balance computation is involved.

(19) Replacing any cowling not requiring removal of the propeller or disconnection of flight controls.

(20) Replacing or cleaning spark plugs and setting of spark plug gap clearance.

(21) Replacing any hose connection except hydraulic connections.

(22) Replacing prefabricated fuel lines.

(23) Cleaning fuel and oil strainers.

(24) Replacing batteries and checking fluid level and specific gravity.

(25) Removing and installing glider wings and tail surfaces that are specifically designed for quick removal and installation and when such removal and installation can be accomplished by the pilot.

Fig. 3-7 FAR Part 43 (continued)

Appendix B

Recording of Major Repairs and Major Alterations

(a) Except as provided in paragraphs (b) and (c), each person performing a major repair or major alteration shall—

(1) Execute FAA Form 337 at least in duplicate;

(2) Give a signed copy of that form to the aircraft owner; and

(3) Forward a copy of that form to the local FAA District Office within 48 hours after the aircraft, airframe, aircraft engine, propeller, or appliance is approved for return to service.

(b) For major repairs made in accordance with a manual or specification acceptable to the Administrator, a certificated repair station may, in place of the requirements of paragraph (a)—

(1) Use the customer's work order upon which the repair is recorded;

(2) Give the aircraft owner a signed copy of the work order and retain a duplicate copy for a least 2 years from the date of approval for return to service of the aircraft, airframe, aircraft engine, propeller, or appliance;

(3) Give the aircraft owner a maintenance release signed by an authorized representative of the repair station and incorporating the following information:

(i) Identity of the aircraft, airframe, aircraft engine, propeller, or appliance.

(ii) If an aircraft, the make, model, serial number, nationality and registration marks, and location of the repaired area.

(iii) If an airframe, aircraft engine, propeller, or appliance, give the manufacturer's name, name of the part, model, and serial numbers (if any); and

(4) Include the following or a similarly worded statement—

"The aircraft, airframe, aircraft engine, propeller, or appliance identified above was repaired and inspected in accordance with current Regulations of the Federal Aviation Administration and is approved for return to service.

"Pertinent details of the repair are on file at this repair station under Order No. ______. Date ______

Signed ______________________ for
(signature of authorized representative)

______________________ ______________
(repair station name) (certificate no.)

______________________."
(address)

(c) For a major repair or major alteration made by a person authorized in § 43.17, the person who performs the major repair or major alteration and the person authorized by § 43.17 to approve that work shall execute a FAA Form 337 at least in duplicate. A completed copy of that form shall be—

(1) Given to the aircraft owner; and

(2) Forwarded to the Federal Aviation Administration, Aircraft Registration Branch, P.O. Box 25082, Oklahoma City, Oklahoma 73125, within 48 hours after the work is inspected.

PART 43 11

Fig. 3-7 FAR Part 43 (continued)

Appendix D

Scope and Detail of Items (as Applicable to the Particular Aircraft) to be Included in Annual and 100–Hour Inspections

(a) Each person performing an annual or 100-hour inspection shall, before that inspection, remove or open all necessary inspection plates, access doors, fairing, and cowling. He shall thoroughly clean the aircraft and aircraft engine.

(b) Each person performing an annual or 100-hour inspection shall inspect (where applicable) the following components of the fuselage and hull group:

(1) Fabric and skin—for deterioration, distortion, other evidence of failure, and defective or insecure attachment of fittings.

(2) Systems and components—for improper installation, apparent defects, and unsatisfactory operation.

(3) Envelope, gas bags, ballast tanks, and related parts—for poor condition.

(c) Each person performing an annual or 100-hour inspection shall inspect (where applicable) the following components of the cabin and cockpit group:

(1) Generally—for uncleanliness and loose equipment that might foul the controls.

(2) Seats and safety belts—for poor condition and apparent defects.

(3) Windows and windshields—for deterioration and breakage.

(4) Instruments—for poor condition, mounting, marking, and (where practicable) for improper operation.

(5) Flight and engine controls—for improper installation and improper operation.

(6) Batteries—for improper installation and improper charge.

(7) All systems—for improper installation, poor general condition, apparent and obvious defects, and insecurity of attachment.

(d) Each person performing an annual or 100-hour inspection shall inspect (where applicable) components of the engine and nacelle group as follows:

(1) Engine section—for visual evidence of excessive oil, fuel, or hydraulic leaks, and sources of such leaks.

(2) Studs and nuts—for improper torquing and obvious defects.

(3) Internal engine—for cylinder compression and for metal particles or foreign matter on screens and sump drain plugs. If there is weak cylinder compression, for improper internal condition and improper internal tolerances.

(4) Engine mount—for cracks, looseness of mounting, and looseness of engine to mount.

(5) Flexible vibration dampeners—for poor condition and deterioration.

(6) Engine controls—for defects, improper travel, and improper safetying.

(7) Lines, hoses, and clamps—for leaks, improper condition, and looseness.

(8) Exhaust stacks—for cracks, defects, and improper attachment.

(9) Accessories—for apparent defects in security of mounting.

(10) All systems—for improper installation, poor general condition, defects, and insecure attachment.

(11) Cowling—for cracks, and defects.

(e) Each person performing an annual or 100-hour inspection shall inspect (where applicable) the following components of the landing gear group:

(1) All units—for poor condition and insecurity of attachment.

PART 43 15

Fig. 3-7 FAR Part 43 (continued)

(2) Shock absorbing devices—for improper oleo fluid level.

(3) Linkage, trusses, and members—for undue or excessive wear, fatigue, and distortion.

(4) Retracting and locking mechanism—for improper operation.

(5) Hydraulic lines—for leakage.

(6) Electrical system—for chafing and improper operation of switches.

(7) Wheels—for cracks, defects, and condition of bearings.

(8) Tires—for wear and cuts.

(9) Brakes—for improper adjustment.

(10) Floats and skis—for insecure attachment and obvious or apparent defects.

(f) Each person performing an annual or 100-hour inspection shall inspect (where applicable) all components of the wing and center section assembly for poor general condition, fabric or skin deterioration, distortion, evidence of failure, and insecurity of attachment.

(g) Each person performing an annual or 100-hour inspection shall inspect (where applicable) all components and systems that make up the complete empennage assembly for poor general condition, fabric or skin deterioration, distortion, evidence of failure, insecure attachment, improper component installation, and improper component operation.

(h) Each person performing an annual or 100-hour inspection shall inspect (where applicable) the following components of the propeller group:

(1) Propeller assembly—for cracks, nicks, binds, and oil leakage.

(2) Bolts—for improper torquing and lack of safetying.

(3) Anti-icing devices—for improper operations and obvious defects.

(4) Control mechanisms—for improper operation, insecure mounting, and restricted travel.

(i) Each person performing an annual or 100-hour inspection shall inspect (where applicable) the following components of the radio group:

(1) Radio and electronic equipment—for improper installation and insecure mounting.

(2) Wiring and conduits—for improper routing, insecure mounting, and obvious defects.

(3) Bonding and shielding—for improper installation and poor condition.

(4) Antenna including trailing antenna—for poor condition, insecure mounting, and improper operation.

(j) Each person performing an annual or 100-hour inspection shall inspect (where applicable) each installed miscellaneous item that is not otherwise covered by this listing for improper installation and improper operation.

Fig. 3-7 FAR Part 43 (continued)

Appendix E

Altimeter System Test and Inspection

Each person performing the altimeter system tests and inspections required by § 91.170 shall comply with the following:

(a) Static pressure system:

(1) Ensure freedom from entrapped moisture and restrictions.

(2) Determine that leakage is within the tolerances established in § 23.1325 or § 25.1325, whichever is applicable.

(3) Determine that the static port heater, if installed, is operative.

(4) Ensure that no alterations or deformations of the airframe surface have been made that would affect the relationship between air pressure in the static pressure system and true ambient static air pressure for any flight condition.

(b) Altimeter:

(1) Test by an appropriately rated repair facility in accordance with the following subparagraphs. Unless otherwise specified, each test for performance may be conducted with the instrument subjected to vibration. When tests are conducted with the temperature substantially different from ambient temperature of approximately 25 degrees C., allowance shall be made for the variation from the specified condition.

(i) *Scale error*—With the barometric pressure scale at 29.92 inches of mercury, the altimeter shall be subjected successively to pressures corresponding to the altitude specified in Table I up to the maximum normally expected operating altitude of the airplane in which the altimeter is to be installed. The reduction in pressure shall be made at a rate not in excess of 20,000 feet per minute to within approximately 2,000 feet of the test point. The test point shall be approached at a rate compatible with the test equipment. The altimeter shall be kept at the pressure corresponding to each test point for at least one minute, but not more than ten minutes, before a reading is taken. The error at all test points must not exceed the tolerances specified in Table I.

(ii) *Hysteresis* — The hysteresis test shall begin not more than 15 minutes after the altimeter's initial exposure to the pressure corresponding to the upper limit of the scale error test prescribed in subparagraph (i); and while the altimeter is at this pressure, the hysteresis test shall commence. Pressure shall be increased at a rate simulating a descent in altitude at the rate of 5,000 to 20,000 feet per minute until within 3,000 feet of the first test point (50 percent of maximum altitude). The test point shall then be approached at a rate of approximately 3,000 feet per minute. The altimeter shall be kept at this pressure for at least 5 minutes, but not more than 15 minutes, before the test reading is taken. After the reading has been taken, the pressure shall be increased further, in the same manner as before, until the pressure corresponding to the second test point (40 percent of maximum altitude) is reached. The altimeter shall be kept at this pressure for at least one minute, but not more than 10 minutes, before the test reading is taken. After the reading has been taken, the pressure shall be increased further, in the same manner as before, until atmospheric pressure is reached. The reading of the altimeter at either of the two test points shall not differ by more than the tolerance specified in Table II from the reading of

Fig. 3-7 FAR Part 43 (continued)

the altimeter for the corresponding altitude recorded during the scale error test prescribed in subparagraph (i).

(iii) *After effect*—Not more than 5 minutes after the completion of the hysteresis test prescribed in subparagraph (ii), the reading of the altimeter (corrected for any change in atmospheric pressure) shall not differ from the original atmospheric pressure reading by more than the tolerance specified in Table II.

(iv) *Friction*—The altimeter shall be subjected to a steady rate of decrease of pressure approximating 750 feet per minute. At each altitude listed in Table III, the change in reading of the pointers after vibration shall not exceed the corresponding tolerance listed in Table III.

(v) *Case leak*—The leakage of the altimeter case, when the pressure within it corresponds to an altitude of 18,000 feet, shall not change the altimeter reading by more than the tolerance shown in Table II during an interval of one minute.

(vi) *Barometric scale error*—At constant atmospheric pressure, the barometric pressure scale shall be set at each of the pressures (falling within its range of adjustment) that are listed in Table IV, and shall cause the pointer to indicate the equivalent altitude difference shown in Table IV with a tolerance of 25 feet.

[(2) Altimeters which are the air data computer type with associated computing systems, or which incorporate air data correction internally, may be tested in a manner and to specifications developed by the manufacturer which are acceptable to the Administrator.]

(c) Records: Comply with the provisions of §43.9 of this chapter as to content, form and disposition of the records. The person performing the altimeter tests shall record on the altimeter the date and maximum altitude to which the altimeter has been tested and the persons approving the airplane for return to service shall enter that data in the airplane log or other permanent record.

TABLE I

Altitude (feet)	Equivalent pressure (inches of mercury)	Tolerance ± (feet)
−1,000	31.018	20
0	29.921	20
500	29.385	20
1,000	28.856	20
1,500	28.335	25
2,000	27.821	30
3,000	26.817	30
4,000	25.842	35
6,000	23.978	40
8,000	22.225	60
10,000	20.577	80
12,000	19.029	90
14,000	17.577	100
16,000	16.216	110
18,000	14.942	120
20,000	13.750	130
22,000	12.636	140
25,000	11.104	155
30,000	8.885	180
35,000	7.041	205
40,000	5.538	230
45,000	4.355	255
50,000	3.425	280

TABLE II—TEST TOLERANCES

Test	Tolerance (Feet)
Case Leak Test	±100
Hysteresis Test:	
First Test Point (50 percent of maximum altitude)	75
Second Test Point (40 percent of maximum altitude)	75
After Effect Test	30

TABLE III—FRICTION

Altitude (Feet)	Tolerance (Feet)
1,000	±70
2,000	70
3,000	70
5,000	70
10,000	80
15,000	90
20,000	100
25,000	120
30,000	140
35,000	160
40,000	180
50,000	250

Fig. 3-7 FAR Part 43 (continued)

TABLE IV—PRESSURE-ALTITUDE DIFFERENCE

Pressure (Inches of Hg)	Altitude Difference (Feet)
28.10	−1727
28.50	−1340
29.00	−863
29.50	−392
29.92	0
30.50	+531
30.90	+893
30.99	+974

Appendix F

ATC Transponder Tests and Inspections

[The ATC transponder tests required by § 91.177 of this chapter may be conducted using a bench check or portable test equipment and must meet the requirements prescribed in paragraphs (a) through (d) of this Appendix. If portable test equipment with appropriate coupling to the aircraft antenna system is used, operate the test equipment at a nominal rate of 235 interrogations per second to avoid possible ATCRBS interference. An additional 3db loss is allowed to compensate for antenna coupling errors during receiver sensitivity measurements conducted in accordance with paragraph (c)(1) when using portable test equipment.

[(a) For reply radio frequency, interrogate the transponder and verify that the reply frequency of the system is 1090 ±3 MHZ.

[(b) Suppression: When the transponder is interrogated on Mode 3/A at an interrogation rate between 230 and 1000 interrogations per second for Class 1B and 2B transponders or between 230 and 1200 interrogations per second for Class 1A and 2A transponders:

[(1) Verify that the transponder does not respond to more than 1 percent of the interrogations when the amplitude of P_2 pulse is equal to the P_1 pulse.

[(2) Verify that the transponder replies to at least 90 percent of the interrogations when the amplitude of the P_2 pulse is 9db less than P_1 pulse.

[If the test is conducted with a radiated test signal, the interrogation rate shall be 235 ±5 interrogations per second unless a higher rate has been approved for the test equipment used at that location.]

(c) Receiver sensitivity:

(1) Verify that receiver sensitivity of the system is −73±4dbm by use of a test set—

(i) Connected to the antenna end of the transmission line;

(ii) Connected to the antenna terminal of the transponder with a correction for transmission line loss; or

(iii) Utilizing a radiated signal.

(2) Verify that the difference in mode 3/A and mode C receiver sensitivity does not exceed 1db.

(d) Records:

Comply with the provisions of § 43.9 of this chapter as to content, form, and disposition of the records.

Fig. 3-7 FAR Part 43 (continued)

2. Annual inspection

Most aircraft operating under FAR Part 91 must have a complete inspection every 12 calendar months, and this inspection must be done by an A&P holding an Inspection Authorization. The exceptions to this rule are aircraft operating under FAR Part 91, Subpart D, which are inspected according to a continuous airworthiness inspection program, or aircraft operated according to a progressive inspection. We will discuss both of the inspections later.

FAR 43, Appendix D, includes a list of items that are typically included in an annual inspection. A checklist must be used when conducting these inspections, and the list may be compiled by the inspector or may be a list that is supplied by the manufacturer of the aircraft.

If the aircraft that is undergoing an annual inspection does not conform to the airworthiness requirements, the inspector cannot return it to service, but he can make an entry in the aircraft records to the effect that this aircraft has been given an annual inspection and found to be in unairworthy condition "because of the following items . . ." A list of the unairworthy items must be supplied to the aircraft owner and a copy sent to the FAA GADO. When these items have been corrected by an A&P mechanic, he can enter in the records the fact that he has corrected the discrepancies, and he can return the aircraft to service.

3. 100-hour inspection

When an aircraft is operated for compensation or hire, it must be given, in addition to its annual inspection, an inspection that is identical to the annual every 100 hours of operation. The only difference between an annual and a 100-hour inspection is the fact that an Inspection Authorization is not required for an A&P to conduct a 100-hour inspection, and there is no requirement that a list of discrepancies be furnished to the owner of the aircraft or to the FAA if the aircraft fails the inspection.

4. Progressive inspection

When an aircraft is operated every day and having it out of service for several days for an annual or hundred-hour inspection would be uneconomical, the aircraft can be put on a progressive inspection program. An A&P with an IA must furnish the FAA GADO with a schedule that will include a complete inspection within the required time frame, either 100-hours of operation or 12 calendar months. The aircraft must have been given a complete annual inspection before the progressive is started, and the portions of the inspection must be performed according to the schedule that the FAA has approved.

The advantage of a progressive inspection is that it allows the aircraft to be decommissioned for only a few hours at a time for the inspection to be performed. For example, the engine may be inspected at one time, the landing gear at another time, and the structure at another time. The total inspection is the same as that of an annual inspection.

5. Continuous airworthiness inspection program

Large aircraft—that is, aircraft which have a maximum certificated takeoff weight of more than 12,500 pounds—or multi-engine turbojet or turboprop airplanes must be inspected under a continuous airworthiness inspection program specified in FAR 91.217. This program is tailored to the specific operation of the aircraft and will vary with the same type of aircraft operated under different conditions.

The inspection schedule adopted by an operator may be one of five types: It may be the same program as is used by airlines operating under FAR Part 121; it may be the same program as is approved for an Air Taxi and Commercial Operator working under FAR Part 135; or a program that is approved for and used by a Part 123 Air Travel Club. It may even be an inspection schedule recommended by the aircraft manufacturer; or it may be one that is established by the operator and is approved by the FAA GADO.

6. Special inspections

a. Altimeter and static system check

Federal Aviation Regulation 91.170 requires that every aircraft operating under instrument flight rules have its altimeter and static system checked every twenty-four calendar months. The inspection of the altimeter must be conducted by the manufacturer of the aircraft or by a repair station approved for this operation, and it consists

ANNUAL INSPECTION FORM

Make & Model ______________________ N ______________ Date ______________

Serial No. ______________________ Year ________ Tach Time ______________

Owner ______________________ Airframe Total ______________

Address ______________________ Engine Total ______________

______________________ Engine TSMOH ______________

ITEM	MAKE	MODEL/PART NO.	SERIAL NO.	APPROVED
Engine				
Prop				
Carb				
Mags: Left				
Right				
Generator / Alternator				
Starter				
Misc. Acces.				
ELT				
Altimeter				
Seat Belts				

II

AD NO.	DESCRIPTION	COMPLIANCE	ADDITIONAL COMPLIANCE

Fig. 3-8 Annual inspection form

III Flight Manual required: Yes ____ No ____

Special Markings/Placards required: ________________________________

____

____

____

____

____

____

IV Aircraft papers in order:

	Yes	No
Registration		
Airworthiness Certificate		
Radio License		
Flight Manual		
Special Markings		

V Altimeter Static Test [FAR 91.170] Date of Compliance ________________________________

VI Special Inspections [Service Manuals, Bulletins, Inspec. Aids, etc.]

Reference	Inspection	Completed

Fig. 3-8 Annual Inspection Form (continued)

INSPECTION REPORT

This form meets requirements of FAR Part 43

Work Order No. ________

Make	Model	Serial No.	Registration No.
Owner		Date	
Type of Inspection		Tach Time	

A. PROPELLER GROUP

	L	R	100	500	insp.
1. Inspect spinner and back plate					
2. Inspect blades for nicks and cracks					
3. Inspect hub for cracks and corrosion					
4. Check for grease and oil leaks					
5. Check mounting bolts and safety					
6. Constant speed — check blades for tightness in hub pitot tube					
* 7. Constant speed — remove prop, remove sludge					
* 8. Lubricate as per manual					
9. Inspect complete assembly					
10. Replace spinner					

B. ENGINE GROUP

	L	R	100	500	insp.
1. Remove engine cowls					
2. Clean cowling, check for cracks, missing fasteners, etc.					
3. Compression check: /80 L. #1 #2 #3 #4 #5 #6 R. #1 #2 #3 #4 #5 #6					
* 4. Drain oil					
5. Check oil screens and clean					
6. Replace oil filter element					
7. Check oil temp sender unit for leaks and security					
8. Clean and check oil radiator fins					
* 9. Remove and flush oil radiator					
10. Check and clean fuel screens					
11. Drain carburetor					
*12. Service fuel injector nozzles					
13. Check fuel system for leaks					
14. Check oil lines for leaks and security					
15. Check fuel lines for leaks and security					
*16. Service air cleaner					
17. Check induction air and heat ducts					
18. Check condition of carb heat box					
19. Check mag points for proper clearance					
20. Check mags for oil seal leakage					
21. Check breaker felts for lubrication					
22. Check distributor block for cracks, burned areas, corrosion, height of contact springs					
23. Check ignition harness and insulators					

Fig. 3-8 Annual Inspection Form (continued)

Item	L	R	100	500	insp.
24. Check mag to engine timing:					
Left ________					
Right ________					
*25. Service or replace spark plugs					
26. Check condition of generator or alternator					
27. Check condition of starter					
28. Check condition and tension of drive belts					
29. Check hydraulic pump and strainer					
30. Check vacuum pump and lines					
31. Inspect exhaust stacks, gaskets, etc.					
32. Inspect muffler and shrouds					
34. Check engine baffles					
35. Check breather tube for obstructions, security					
36. Check crankcase for leaks, cracks, etc.					
37. Check engine mounts for cracks, loose mounts					
38. Check engine mount bushings					
39. Check firewall seals					
40. Check throttle, carb heat, mixture, and prop governor controls for travel and operating condition					
41. Check cowl flap condition and operation					
42. Inspect engine for general condition, loose parts, chafing, proper safeties, proper installations, etc.					
*43. Fill engine with oil					
44. Clean engine					
45. Lubricate all controls					
46. Reinstall engine cowl					

C. CABIN GROUP

1. Inspect cabin doors for damage and operation			
2. Check windows for general condition			
3. Check upholstery for general condition			
4. Check seats, seat belts, and mountings			
5. Check trim operation			
6. Check flap operation			
7. Check fuel selector valve operation			
8. Check operation of fuel drain			
9. Check landing, navigation, cabin, and instrument lights			
10. Check rudder pedals			
11. Check brake cylinders for operation and leaks			
12. Check brake fluid level			
13. Check control wheels, column, pulleys, cables			
14. Check instruments, lines, and attachments			
*15. Service filters for gyro instruments			
16. Check condition in general under panel for loose wires, loose equipment, chafing, etc.			
17. Check condition of heater controls and ducts, air vents, and air conditioning ducts			
18. Lubricate controls			
19. Check condition of instrument panel shock mounts			
20. Check altimeter calibration			

D. FUSELAGE AND EMPENNAGE GROUP

1. Remove inspection plates and panels			
2. Check general condition of aircraft skin			
3. Check baggage door[s] latches, and hinges			

Fig. 3-8 Annual Inspection Form (continued)

	100	500	insp.
* 4. Service battery			
5. Check bulkheads and stringers for damage			
6. Check wiring for damage and security			
7. Check security of all lines [fuel, hydraulic, etc.]			
8. Check cables, turnbuckles, guides, and pulleys for safeties, damage, and operation			
9. Check radio antennas for mounting and electrical connections			
10. Check rotating beacon or strobes for security and operation			
11. Check empennage surfaces for damage			
12. Check rudder hinges, horn, and attachments			
13. Check vertical fin attachments			
14. Check elevator or stabilator hinges, horn, and attachments			
15. Check horizontal stabilizer attachments			
16. Check trim mechanism			
*17. Service hydraulic and brake systems			
*18. Service cabin heater			
*19. Lubricate as per manual			
20. Reinstall inspection plates and panels			

E. WING GROUP

	100	500	insp.
1. Remove inspection plates and fairings			
2. Check surfaces and tips for damage, loose rivets, etc.			
3. Check ailerons, hinges, cables, pulleys and bellcranks for damage and operation			
4. Check flaps for damage and operation			
5. Check fuel tanks for leaks and water			
6. Check fuel tank vents			
7. Fuel tanks marked for octane and capacity			
8. Check wing attachment bolts			
9. Check pitot tube for security			
10. Check stall warning indicator for operation			
11. Lubricate as per manual			
12. Reinstall inspection plates and fairings			

F. LANDING GEAR GROUP

	100	500	insp.
1. Put airplane on jacks			
2. Check tires for wear and damage			
* 3. Remove wheels for cracks, corrosion, broken bolts			
5. Check tire pressure N M			
6. Check brake linings and drum or disc			
7. Check brake lines for security and condition			
8. Check oleo struts for leaks and scoring			
9. Check gear forks, torque links, etc., for general condition			
10. Check gear attachment bolts			
11. Check nose gear steering control and travel			
12. Check shimmy dampener			
13. Check gear doors and attachments			
14. Retract gear, check for proper clearances and operation			
15. Check warning horn and lights			
16. Check safety "squawk" switch			
17. Check oleo fluid level and air pressure			
18. Inspect all hydraulic lines and electrical wires for security, routing, and general condition			
*19. Lubricate as per manual			
20. Remove airplane from jacks			

Fig. 3-8 Annual Inspection Form (continued)

G. OPERATIONAL INSPECTION	L	R	100	500	insp.
1. Check fuel pump					
2. Check fuel quantity and pressure- or flow-gage					
3. Check alternator or generator output					
4. Check vacuum gage					
5. Check gyros for noise and roughness					
6. Check cabin heater operation					
7. Check electronic equipment operation					
8. Check air conditioner operation					
9. Check parking brake					
10. Check oil pressure and temperature					
11. Check manifold pressure					
12. Check alternate air					
13. Check propeller smoothness					
14. Check prop governor action					
15. Check mag RPM variation L. Engine R. Engine Mag L. Mag R.					
16. Check static RPM ______________					
17. Check idle RPM ______________					
18. Check mag switch operation					
19. Check throttle and mixture operation					
20. Check idle mixture					

H. GENERAL	100	500	insp.
1. Aircraft conforms to FAA specifications			
2. All AD notes complied with			
3. Manufacturers' service letters, bulletins, instructions complied with			
4. Aircraft papers in proper order a. Registration b. Airworthiness Certificate c. Radio license d. Flight Manual			

Signature of Mechanic or Inspector ______________________________

Certificate No. ______________________________

* Items not required by FAA as part of 100-hour inspection are marked with a star.

Fig. 3-8 Annual Inspection Form (continued)

of an accuracy test and other tests that determine the condition of the instrument. These tests are described in FAR Part 43, Appendix E.

A mechanic holding an Airframe rating can conduct the static system check, which consists of checking the system for leaks and distortion and for the presence of water in the lines.

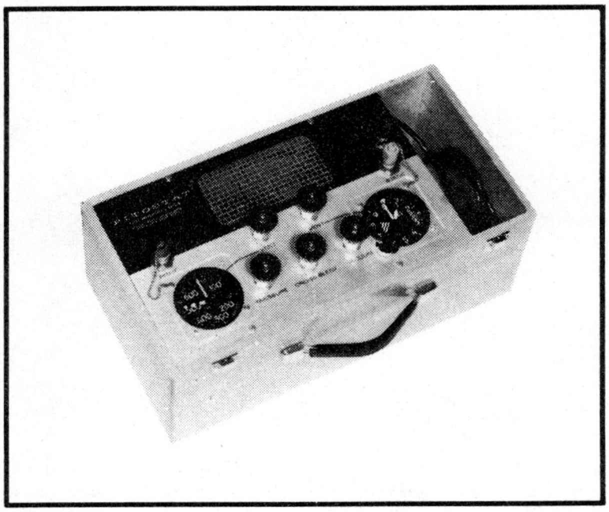

Fig. 3-9 The altimeter test prescribed by FAR 91.170 can be performed with test equipment of this type.

b. Transponder check

The radar beacon transponder that replies to the traffic control radar must be checked every twenty-four calendar months by a repair station certificated for this inspection. A notation must be made in the aircraft records of this inspection. This inspection is detailed in FAR Part 43, Appendix F.

c. Emergency Locator Transmitter (ELT) check

The Emergency Locator Transmitter (ELT) is a small, totally self-contained transmitter usually located in the tail of the airplane. An impact will trigger it to continuously transmit a downsweeping tone on two of the emergency frequencies, 121.5 and 243 MHz. An ELT must be inspected at each regular aircraft inspection for the condition of its batteries, which must be replaced at a specific time interval. The replacement date must be recorded on the outside of the ELT case where the technician can see it.

d. Inspection following an overweight or hard landing

An aircraft is designed and built to withstand the stresses imposed by normal operation, but sometimes a pilot may misjudge his sink rate on a landing and make much harder contact with the ground than he should. The load imposed on the structure by the landing is due not only to the weight of the aircraft, but to the acceleration which is determined by the vertical velocity at touchdown. Also, some aircraft are certificated with a higher allowable takeoff weight than they are allowed to have for landing. The reason is that a takeoff imposes less load on the landing gear than a landing, and the weight of the aircraft is reduced by either burning off some of the fuel, or by dumping it through a special dump system built into the aircraft for this purpose. It is possible, however, for an emergency condition to arise in which the pilot must land the aircraft while it is loaded to a weight above its allowable landing gross weight. If either of these two types of landing have occured, the mechanic must give the aircraft a careful inspection to ascertain whether or not any damage has been done.

The wing skin should be carefully examined for indications of wrinkles or popped rivets. The skin on the underside of the wing will be subjected to a compressive stress outboard of the landing gear, and to a tensile stress between the main landing gears. Either type of stress can cause rivets to shear.

The tires must be removed from the wheels and inspected to be sure that none of the ply cords were broken by the impact. The wheel should be inspected by the eddy current inspection method to determine whether there are any cracks that may have occured when the tire contacted the ground. Sometimes these cracks, if quite small, will close up so much that they cannot be found by a liquid penetrant inspection, but are large enough to seriously weaken the wheel.

e. Inspection following flight into severe turbulence

When an aircraft is flown into severe turbulence, it is possible for it to be stressed so highly that the skin can be wrinkled and rivets popped from seams. Where there is an indication of this type of damage, several rivets on each end of the damage should be removed and examined for any sign of shear failure, as would be indicated by the rivet shank being distorted by being pushed to the side.

Another type of damage evidence that will appear after an overstressed condition is fuel leakage from integral tanks. Most large aircraft have part of the structure sealed off with a flexible sealant around all of the riveted seams so they will hold fuel. When the structure is subjected to extremes of flexing from either turbulence or from a hard landing, the sealant may be pulled away from the structure and fuel will seep out along the seam. While all fuel leaks are not an indication of extreme flexing, this possibility may be considered when a leak is found.

Subpart C—Maintenance, Preventive Maintenance, and Alterations

§ 91.161 Applicability.

(a) This subpart prescribes rules governing the maintenance, preventive maintenance, and alteration of U.S. registered civil aircraft operating within or without the United States.

(b) Sections 91.165, 91.169, 91.170, 91.171, 91.173, and 91.174 of this subpart do not apply to an aircraft maintained in accordance with a continuous airworthiness maintenance program as provided in Part 121, 127, or 135 of this chapter.

§ 91.163 General.

(a) The owner or operator of an aircraft is primarily responsible for maintaining that aircraft in an airworthy condition, including compliance with Part 39 of this chapter.

(b) No person may perform maintenance, preventive maintenance, or alterations on an aircraft other than as prescribed in this subpart and other applicable regulations, including Part 43.

(c) No person may operate a rotorcraft for which a Rotorcraft Maintenance Manual containing an "Airworthiness Limitations" section has been issued, unless the replacement times, inspection intervals, and related procedures specified in that section of the manual are complied with.

§ 91.165 Maintenance required.

Each owner or operator of an aircraft shall have that aircraft inspected as prescribed in Subpart D or § 91.169 of this Part, as appropriate, and § 91.170 of this Part and shall, between required inspections, have defects repaired as prescribed in Part 43 of this chapter. In addition, he shall ensure that maintenance personnel make appropriate entries in the aircraft and maintenance records indicating the aircraft has been released to service.

§ 91.167 Carrying persons other than crewmembers after repairs or alterations.

(a) No person may carry any person (other than crewmembers) in an aircraft that has been repaired or altered in a manner that may have appreciably changed its flight characteristics, or substantially affected its operation in flight, until it has been approved for return to service in accordance with Part 43 and an appropriately rated pilot, with at least a private pilot's certificate, flies the aircraft, makes an operational check of the repaired or altered part, and logs the flight in the aircraft's records.

(b) Paragraph (a) of this section does not require that the aircraft be flown if ground tests or inspections, or both, show conclusively that the repair or alteration has not appreciably changed the flight characteristics, or substantially affected the flight operation of the aircraft.

§ 91.169 Inspections.

(a) Except as provided in paragraph (c) of this section, no person may operate an aircraft unless, within the preceding 12 calendar months, it has had—

(1) An annual inspection in accordance with Part 43 of this chapter and has been approved for return to service by a person authorized by § 43.7 of this chapter; or

(2) An inspection for the issue of an airworthiness certificate.

No inspection performed under paragraph (b) of the section may be substituted for any inspection required by this paragraph unless it is performed by a person authorized to perform annual inspections, and is entered as an 'annual' inspection in the required maintenance records.

(b) Except as provided in paragraph (c) of this section, no person may operate an aircraft carrying any person (other than a crewmember) for hire, and no person may give flight instruction for hire in an aircraft which that person provides, unless within the preceding 100 hours of time in service it has received an annual or 100-hour inspection and been approved for return to service in accord-

Fig. 3-10 FAR Part 91

ance with Part 43 of this chapter, or received an inspection for the issuance of an airworthiness certificate in accordance with Part 21 of this chapter. The 100-hour limitation may be exceeded by not more than 10 hours if necessary to reach a place at which the inspection can be done. The excess time, however, is included in computing the next 100 hours of time in service.

(c) Paragraphs (a) and (b) of this section do not apply to—

(1) Any aircraft for which its registered owner or operator complies with the progressive inspection requirements of § 91.171 and Part 43 of this chapter;

(2) An aircraft that carries a special flight permit or a current experimental or provisional certificate;

(3) Any airplane operated by an air travel club that is inspected in accordance with Part 123 of this chapter and the operator's manual and operations specifications; or

(4) An aircraft inspected in accordance with an approved aircraft inspection program under Part 135 of this chapter and so identified by the registration number in the operations specifications of the certificate holder having the approved inspection program.

(5) Any large airplane, or a turbojet- or turbopropeller-powered multiengine airplane, that is inspected in accordance with an inspection program authorized under Subpart D of this Part.

§ 91.170 Altimeter system tests and inspections.

(a) No person may operate an airplane in controlled airspace under IFR unless, within the preceding 24 calendar months, each static pressure system and each altimeter instrument has been tested and inspected and found to comply with Appendix E of Part 43. The static pressure system and altimeter instrument tests and inspections may be conducted by—

(1) The manufacturer of the airplane on which the tests and inspections are to be performed;

(2) A certificated repair station properly equipped to perform these functions and holding—

(i) An instrument rating, Class I;

(ii) A limited instrument rating appropriate to the make and model altimeter to be tested;

(iii) A limited rating appropriate to the test to be performed;

(iv) An airframe rating appropriate to the airplane to be tested; or

(v) A limited rating for a manufacturer issued for the altimeter in accordance with § 145.101(b)(4) of this chapter; or

(3) A certificated mechanic with an airframe rating (static pressure system tests and inspections only).

(b) [Revoked.]

(c) No person may operate an airplane in controlled airspace under IFR at an altitude above the maximum altitude to which an altimeter of that airplane has been tested.

§ 91.171 Progressive inspection.

(a) Each registered owner or operator of an aircraft desiring to use the progressive inspection must submit a written request to the Flight Standards District Office having jurisdiction over the area in which the applicant is located, and shall provide—

(1) A certificated mechanic holding an inspection authorization, a certificated airframe repair station, or the manufacturer of the aircraft, to supervise or conduct the progressive inspection;

(2) A current inspection procedures manual available and readily understandable to pilot and maintenance personnel containing, in detail—

(i) An explanation of the progressive inspection, including the continuity of

Fig. 3-10 FAR Part 91 (continued)

inspection responsibility, the making of reports, and the keeping of records and technical reference material;

(ii) An inspection schedule, specifying the intervals in hours or days when routine and detailed inspections will be performed and including instructions for exceeding an inspection interval by not more than 10 hours while en route and for changing an inspection interval because of service experience;

(iii) Sample routine and detailed inspection forms and instructions for their use; and

(iv) Sample reports and records, and instructions for their use;

(3) Enough housing and equipment for necessary disassembly and proper inspection of the aircraft; and

(4) Appropriate current technical information for the aircraft.

(b) The frequency and detail of the progressive inspection shall provide for the complete inspection of the aircraft within each 12 calendar months and be consistent with the manufacturer's recommendations, field service experience, and the kind of operation in which the aircraft is engaged. The progressive inspection schedule must insure that the aircraft at all times will be airworthy and will conform to all applicable FAA aircraft specifications, type certificate data sheets, airworthiness directives, and other approved data.

(c) If the progressive inspection is discontinued, the owner or operator shall immediately notify the local General Aviation District Office, in writing, of the discontinuance. After the discontinuance, the first annual inspection under § 91.169(a) is due within 12 calendar months after the last complete inspection of the aircraft under the progressive inspection. The 100-hour inspection under § 91.169(b) is due within 100 hours after that complete inspection. A complete inspection of the aircraft, for the purpose of determining when the annual and 100-hour inspections are due, will require a detailed inspection of the aircraft and all its components in accordance with the progressive inspection. A routine inspection of the aircraft and a detailed inspection of several components is not considered to be a complete inspection.

§ 91.173 Maintenance records.

[(a) Except for work performed in accordance with §§ 91.170 and 91.177, each registered owner or operator shall keep the following records for the periods specified in paragraph (b) of this section:]

(1) Records of the maintenance and alteration, and records of the 100-hour, annual, progressive, and other required or approved inspections, as appropriate, for each aircraft (including the airframe) and each engine, propeller, rotor, and appliance of an aircraft. The records must include—

(i) A description (or reference to data acceptable to the Administrator) of the work performed;

(ii) The date of completion of the work performed; and

(iii) The signature and certificate number of the person approving the aircraft for return to service.

(2) Records containing the following information:

(i) The total time in service of the airframe.

(ii) The current status of life-limited parts of each airframe, engine, propeller, rotor, and appliance.

(iii) The time since last overhaul of all items installed on the aircraft which are required to be overhauled on a specified time basis.

(iv) The identification of the current inspection status of the aircraft, including the times since the last inspections required by the inspection program under which the aircraft and its appliances are maintained.

[(v) The current status of applicable airworthiness directives (AD) including, for each, the mthod of compliance, the AD number, and revision date. If the AD involves recurring action, the time and date when the next action is required.]

Ch. 27 (Amdt. 91–152, Eff. 6/26/78)

Fig. 3-10 FAR Part 91 (continued)

(vi) A list of current major alterations to each airframe, engine, propeller, rotor, and appliance.

[(b) The owner and operator shall retain the following records for the periods prescribed:]

(1) The records specified in paragraph (a)(1) of this section shall be retained until the work is repeated or superseded by other work or for one year after the work is performed.

(2) The records specified in paragraph (a)(2) of this section shall be retained and transferred with the aircraft at the time the aircraft is sold.

[(3) A list of defects furnished to a registered owner or operator under § 43.9 of this chapter, shall be retained until the defects are repaired and the aircraft is approved for return to service.]

(c) The owner or operator shall make all maintenance records required to be kept by this section available for inspection by the Administrator or any authorized representative of the National Transportation Safety Board (NTSB).

§ 91.174 Transfer of maintenance records.

Any owner or operator who sells a U.S. registered aircraft shall transfer to the purchaser, at the time of sale, the following records of that aircraft, in plain language form or in coded form at the election of the purchaser, if the coded form provides for the preservation and retrieval of information in a manner acceptable to the Administrator:

(a) The records specified in § 91.173(a)(2).

(b) The records specified in § 91.173(a)(1) which are not included in the records covered by paragraph (a) of this section, except that the purchaser may permit the seller to keep physical custody of such records. However, custody of records in the seller does not relieve the purchaser of his responsibility under § 91.173(c), to make the records available for inspection by the Administrator or any authorized representative of the National Transportation Safety Board (NTSB).

§ 91.175 Rebuilt engine maintenance records.

(a) The owner or operator may use a new maintenance record, without previous operating history, for an aircraft engine rebuilt by the manufacturer or by an agency approved by the manufacturer.

(b) Each manufacturer or agency that grants zero time to an engine rebuilt by it shall enter, in the new record—

(1) A signed statement of the date the engine was rebuilt;

(2) Each change made as required by Airworthiness Directives; and

(3) Each change made in compliance with manufacturer's service bulletins, if the entry is specifically requested in that bulletin.

(c) For the purposes of this section, a rebuilt engine is a used engine that has been completely disassembled, inspected, repaired as necessary, reassembled, tested, and approved in the same manner and to the same tolerances and limits as a new engine with either new or used parts. However, all parts used in it must conform to the production drawing tolerances and limits for new parts or be of approved oversized or undersized dimensions for a new engine.

§ 91.177 ATC transponder tests and inspections.

(a) After January 1, 1976, no person may use an ATC transponder that is specified in §§ 91.24(a), 121.345(c), 127.123(b), or 135.143(c) of this chapter, unless, within the preceding 24 calendar months, that ATC transponder has been tested and inspected and found to comply with Appendix F of Part 43 of this chapter.

(b) The tests and inspections specified in paragraph (a) of this section may be conducted by—

(1) A certificated repair station properly equipped to perform those functions and holding—

(i) A radio rating, Class III;

Fig. 3-10 FAR Part 91 (continued)

(ii) A limited radio rating appropriate to the make and model transponder to be tested;

(iii) A limited rating appropriate to the test to be performed; or

(iv) A limited rating for a manufacturer issued for the transponder in accordance with § 145.101(b)(4) of this chapter; or

(2) A certificate holder authorized to perform maintenance in accordance with § 121.379 or § 127.140 of this chapter: or

(3) The manufacturer of the aircraft on which the transponder to be tested is installed, if the transponder was installed by that manufacturer.

Subpart D—Large and Turbine-Powered Multiengine Airplanes

§ 91.217 Inspection program.

(a) No person may operate a large airplane, or a turbojet- or turbopropeller-powered multiengine airplane, unless the replacement times for life-limited parts specified in the aircraft data sheets or other documents approved by the Administrator are complied with and the airplane, including the airframe, engines, propellers, appliances, survival equipment, and emergency equipment is inspected in accordance with an inspection program selected under the provisions of this section.

(b) The registered owner or operator of each airplane governed by this subpart must select and must use one of the following programs for the inspection of that airplane:

(1) A continuous airworthiness inspection program that is a part of a continuous airworthiness maintenance program currently in use by a person holding an air carrier or commercial operator certificate under Part 121 of this chapter.

(2) An approved aircraft inspection program currently in use by a person holding an ATCO certificate under Part 135 of this chapter.

(3) An approved continuous inspection program currently in use by a person certificated as an Air Travel Club under Part 123 of this chapter.

(4) A current inspection program recommended by the manufacturer.

(5) Any other inspection program established by the registered owner or operator of that airplane and approved by the Administrator under paragraph (e) of this section.

(c) Notice of the inspection program selected shall be sent to the local FAA District Office having jurisdiction over the area in which the airplane is based. The notice must be in writing and include—

(1) Make, model, and serial number of the airplane;

(2) Registration number of the airplane;

(3) The inspection program selected under paragraph (b) of this section; and

(4) The name and address of the person responsible for scheduling the inspections required under the selected inspection program.

(d) The registered owner or operator may not change the inspection program for an airplane unless he has given notice thereof as provided in paragraph (c) of this section and the new program has been approved by the FAA, where appropriate.

Fig. 3-10 FAR Part 91 (continued)

(e) Each registered owner or operator of an airplane desiring to establish an approved inspection program under paragraph (b)(5) of this section must submit the program for approval to the local FAA District Office having jurisdiction over the area in which the airplane is based. The program must include the following information:

(1) Instructions and procedures for the conduct of inspections for the particular make and model airplane, including necessary tests and checks. The instructions and procedures must set forth in detail the parts and areas of the airframe, engines, propellers, and appliances, including emergency equipment required to be inspected.

(2) A schedule for the performance of the inspections that must be performed under the program expressed in terms of the time in service, calendar time, number of system operations, or any combination of these.

(f) [Deleted]

§ 91.219 Availability of inspection program.

Each owner or operator of an airplane shall make a copy of the inspection program selected under § 91.217 available to—

(a) The person responsible for the scheduling of the inspections;

(b) Any person performing inspections on the airplane; and

(c) Upon request, to the Administrator.

Fig. 3-10 FAR Part 91 (continued)

Part 45—Identification and Registration Marking

Subpart A—General

§ 45.1 Applicability.

This Part prescribes the requirements for—

(a) Identification of aircraft, and identification of aircraft engines and propellers that are manufactured under the terms of a type or production certificate;

(b) Identification of certain replacement and modified parts produced for installation on type certificated products; and

(c) Nationality and registration marking of U.S. registered aircraft.

Subpart B—Identification of Aircraft and Related Products

§ 45.11 General.

(a) *Aircraft and aircraft engines.* Aircraft covered by § 21.182 must be identified, and each person who manufactures an aircraft engine under a type or production certificate shall identify his engine, by means of a fireproof plate that has the information specified in § 45.13 marked thereon by etching, stamping, engraving, or other approved method of fireproof marking. The identification plate for aircraft must be secured in such a manner that it will not likely be defaced or removed during normal service, or lost or destroyed in an accident; and it must be secured to the aircraft at an accessible location near an entrance, except that it it is legible to a person on the ground it may be located externally on the fuselage near the tail surfaces. For aircraft engines, the identification plate shall be affixed to the engine at an accessible location, in such a manner that it will not be likely to be defaced or removed during normal service, or lost or destroyed in an accident.

(b) *Propellers and propeller blades and hubs.* Each person who manufactures a propeller, propeller blade, or propeller hub under the terms of a type or production certificate shall identify his product by means of a plate, stamping, engraving, etching, or other approved method of fireproof identification that is placed on it on a non-critical surface, contains the information specified in § 45.13, and will not be likely to be defaced or removed during normal service or lost or destroyed in an accident.

§ 45.13 Identification data.

(a) The identification required by § 45.11 shall include the following information:

(1) Builder's name.

(2) Model designation.

(3) Builder's serial number.

(4) Type certificate number, if any.

(5) Production certificate number, if any.

(6) For aircraft engines, the established rating.

(7) Any other information the Administrator finds appropriate.

[(b) Except as provided in paragraph (d)(1) of this section, no person may remove, change, or place identification information required by paragraph (a) of this section, on any aircraft, aircraft engine, propeller, propeller blade, or propeller hub, without the approval of the Administrator.

[(c) Except as provided in paragraph (d)(2) of this section, no person may remove or install any identification plate required by § 45.11 of this Part, without the approval of the Administrator.

[(d) Persons performing work under the provisions of Part 43 of this chapter may, in accordance with methods, techniques, and practices acceptable to the Administrator—

[(1) Remove, change, or place the identification information required by paragraph

Fig. 3-11 FAR Part 45

(a) of this section on any aircraft, aircraft engine, propeller, propeller blade, or propeller hub; or

[(2) Remove an identification plate required by § 45.11 when necessary during maintenance operations.

[(e) No person may install an identification plate removed in accordance with paragraph (d)(2) of this section on any aircraft, aircraft engine, propeller, propeller blade, or propeller hub other than the one from which it was removed.]

§ 45.14 Identification of critical components.

Each person who produces a part for which a replacement time, inspection interval or related procedure is specified in the "Airworthiness Limitations" section of a Rotorcraft Maintenance Manual shall mark that component with a part number (or equivalent) and with a serial number (or equivalent).

§ 45.15 Replacement and modification parts.

(a) Except as provided in paragraph (b) of this section, each person who produces a replacement or modification part under a Parts Manufacturer Approval issued under § 21.303 of this chapter shall permanently and legibly mark the part with—

(1) The letters "FAA-PMA";

(2) The name, trademark, or symbol of the holder of the Parts Manufacturer Approval;

(3) The part number; and

(4) The name and model designation of each type certificated product on which the part is eligible for installation.

(b) If the Administrator finds that a part is too small or that it is otherwise impractical to mark a part with any of the information required by paragraph (a) of this section, a tag attached to the part or its container must include the information that could not be marked on the part.

Subpart C—Nationality and Registration Marks

§ 45.21 General.

(a) Except as provided in § 45.22, no person may operate a United States registered aircraft unless that aircraft displays nationality and registration marks in accordance with the requirements of this section and §§ 45.23 through 45.33.

(b) Unless otherwise authorized by the Administrator, no person may place on any aircraft a design, mark, or symbol that modifies or confuses the nationality and registration marks.

(c) Aircraft nationality and registration marks must—

(1) Except as provided in paragraph (d) of this section, be painted on the aircraft or affixed by any other means insuring a similar degree of permenence;

(2) Have no ornamentation;

(3) Contrast in color with the background; and

(4) Be legible.

(d) The aircraft nationality and registration marks may be affixed to an aircraft with readily removable material if—

(1) It is intended for immediate delivery to a foreign purchaser;

(2) It is bearing a temporary registration number: or

(3) It is marked temporarily to meet the requirements of § 45.22(c) (1).

§ 45.22 Exhibition, antique, and other aircraft: special rules.

(a) When display of aircraft nationality and registration marks in accordance with §§ 45.21 and 45.23 through 45.33 would be inconsistent with exhibition of that aircraft, a United States registered aircraft may be operated without displaying those marks anywhere on the aircraft if—

(1) It is operated for the purpose of exhibition, including a motion picture or television production, or an airshow;

(2) Except for practice and test flights necessary for exhibition purposes, it is oper-

Fig. 3-11 FAR Part 45 (continued)

ated only at the location of the exhibition, between the exhibition locations, and between those locations and the base of operations of the aircraft; and

(3) For each flight in the United States—

(i) It is operated with the prior approval of the General Aviation District Office, in the case of a flight within the designated airport control zone of the takeoff airport, or within 5 miles of that airport if it has no designated control zone; or

(ii) It is operated under a flight plan filed under § 91.83 of this chaper describing the marks it displays, in the case of any other flight.

(b) When it was built at least 30 years ago or has the same external configuration as an aircraft built at least 30 years ago, a United States registered aircraft may be operated without displaying marks in accordance with §§ 45.21 and 45.23 through 45.33 if—

(1) It displays in accordance with § 45.21(c) marks at least 2 inches high on each side of the fuselage or vertical tail surface consisting of the Roman capital letter "N" followed by—

(i) The U.S. registration number of the aircraft; or

(ii) The symbol appropriate to the airworthiness certificate of the aircraft

Fig. 3-11 FAR Part 45 (continued)

("C", standard; "R", restricted; "L", limited; or "X", experimental) followed by the U.S. registration number of the aircraft; and

(2) It displays no other mark that begins with the letter "N" anywhere on the aircraft, unless it is the same mark that is displayed under subparagraph (1) of this paragraph.

(c) No person may operate an aircraft under paragraph (a) or (b) of this section—

(1) In an ADIZ or DEWIZ described in Part 99 of this chapter unless it temporarily bears marks in accordance with §§ 45.21 and 45.23 through 45.33;

(2) In a foreign country unless that country consents to that operation; or

(3) In any operation conducted under Part 121, 127, 133, 135, or 137 of this chapter.

(d) If, due to the configuration of an aircraft, it is impossible for a person to mark it in accordance with §§ 45.21 and 45.23 through 45.33, he may apply to the Administrator for a different marking procedure.

[§ 45.23 Display of marks; general.

[(a) Each operator of an aircraft shall display on that aircraft marks consisting of the Roman capital letter "N" (denoting United States registration) followed by the registration number of the aircraft. Each suffix letter used in the marks displayed must also be a Roman capital letter.]

(b) When marks that include only the Roman capital letter "N" and the registration number are displayed on limited or restricted category aircraft or experimental or provisionally certificated aircraft, the operator shall also display on that aircraft near each entrance to the cabin or cockpit, in letters not less than 2 inches nor more than 6 inches in height, the words "limited", "restricted", "experimental", or "provisional airworthiness", as the case may be.

[§ 45.25 Location of marks on fixed-wing aircraft.

[(a) The operator of a fixed-wing aircraft shall display the required marks on either the vertical tail surfaces or the sides of the fuselage, except as provided in § 45.29(f).

[(b) The marks required by paragraph (a) of this section shall be displayed as follows:

[(1) If displayed on the vertical tail surfaces, horizontally on both surfaces of a single vertical tail or on the outer surfaces of a multivertical tail. However, on aircraft on which marks at least 3 inches high may be displayed in accordance with § 45.29(b)(1), the marks may be displayed vertically on the vertical tail surfaces.

[(2) If displayed on the fuselage surfaces, horizontally on both sides of the fuselage between the trailing edge of the wing and the leading edge of the horizontal stabilizer. However, if engine pods or other appurtenances are located in this area and are an integral part of the fuselage side surfaces, the operator may place the marks on those pods or appurtenances.]

§ 45.27 Location of marks; nonfixed-wing aircraft.

(a) *Rotorcraft.* Each operator of a rotorcraft shall display on that rotorcraft the marks required by § 45.23 on the—

(1) Bottom surface of the fuselage or cabin with the top of the marks toward the left side of the fuselage; and

(2) Side surfaces of the fuselage below the window lines, as near the cockpit as possible.

(b) *Airships.* Each operator of an airship must display on that airship the marks required by § 45.23, horizontally on—

(1) The upper surface of the right horizontal stabilizer and on the under surface of the left horizontal stabilizer with the top of the marks toward the leading edge of each stabilizer; and

(2) Each side of the bottom half of the vertical stabilizer.

Fig. 3-11 FAR Part 45 (continued)

(c) *Spherical balloons.* Each operator of a spherical balloon shall display the marks required by § 45.23 in two places diametrically opposite and near the maximum horizontal circumference of that balloon.

(d) *Nonspherical balloons.* Each operator of a nonspherical balloon shall display the marks required by § 45.23 on each side of the balloon near its maximum cross section and immediately above either the rigging band or the points of attachment of the basket or cabin suspension cables.

§ 45.29 Size of marks.

(a) Except as provided in paragraph (f) of this section, each operator of an aircraft shall display marks on the aircraft meeting the size requirements of this section.

(b) *Height.* The character marks must be of equal height and on—

[(1) Fixed-wing aircraft—must be at least 12 inches high, except that marks at least 3 inches high may be displayed on small aircraft when none of the following exceeds 180 knots TAS:

[(i) V_C;

[(ii) V_{MO};

[(iii) V_{NO}; and

[(iv) If V_C, V_{MO}, or V_{NO} has not been determined for the aircraft, the speed shown to be the maximum cruising speed of the aircraft;]

(2) Airships, spherical balloons, and nonspherical balloons—must be at least 20 inches high; and

(3) Rotorcraft fuselage or cabin—

(i) Bottom surfaces—must be at least four-fifths as high as the fuselage is wide or 20 inches high, whichever is less; and

(ii) Side surfaces—must be as high as practicable, but not less than 2 inches high.

However, subparagraph (3)(ii) does not require the use of marks more than 6 inches high.

(c) *Width.* Characters must be two-thirds as wide as they are high, except the number "1", which must be one-sixth as wide as it is high, and the letters "M" and "W" which may be as wide as they are high.

(d) *Thickness.* Characters must be formed by solid lines one-sixth as thick as the character is high.

(e) *Spacing.* The space between each character may not be less than one-fourth of the character width.

(f) If either one of the surfaces authorized for displaying required marks under § 45.25 is large enough for display of marks meeting the size requirements of this section and the other is not, full-size marks shall be placed on the larger surface. If neither surface is large enough for full-size marks, marks as large as practicable shall be displayed on the larger of the two surfaces. If any surface required to be marked by § 45.27 is not large enough for full-size marks, marks as large as practicable shall be displayed on that surface.

[(g) *Uniformly.* The marks required by this Part for fixed-wing aircraft must have the same height, width, thickness, and spacing on both sides of the aircraft.]

§ 45.31 Marking of export aircraft.

A person who manufactures an aircraft in the United States for delivery outside thereof may display on that aircraft any marks required by the State of registry of the aircraft. However, no person may operate an aircraft so marked within the United States, except for test and demonstration flights for a limited period of time, or while in necessary transit to the purchaser.

§ 45.33 Sale of aircraft to a non-U.S. citizen; removal of marks.

When an aircraft that is registered in the United States is sold to a person who is not a citizen of the United States, the United States registered owner shall, before its delivery to the purchaser, remove all United States marks from that aircraft.

Fig. 3-11 FAR Part 45 (continued)

PART IV

Maintenance Forms And Records

I. *Permanent Records*

The accepted method of keeping aircraft maintenance records has been in a bound airframe and powerplant logbook, in which the flight time is recorded and a record made of all maintenance and inspections.

The requirements for a bound logbook have been changed so the records can now be kept in a loose-leaf format, and the records that are needed have been clearly defined and divided into two categories: those that must be kept with the aircraft as long as it is operating and those that may be destroyed after a given period of time.

FAR 91.173 describes the records that must be kept. Permanent records that must stay with the aircraft and be transferred with it when it is sold are:

a. The total time in service of the airframe.

b. The current status of any life-limited parts.

c. The time since last overhaul of all items which are required by the FAA to be overhauled on a given time basis.

d. The identification of the current inspection status, including the time since the last required inspection.

e. The current status of applicable Airworthiness Directives. This must include the method of their compliance.

f. A list of the current major alterations that have been made on the airframe, engine, propeller, rotor, or appliance.

The allowable operational life of life-limited parts is specified by the FAA, and on each inspection we must determine the status of these parts. We do this by finding the total time on the aircraft and subtracting from this the total time on the component. Let's consider the way we can determine the status of the transmission installed in a helicopter.

For example, the helicopter airframe has a total (TT) of 3,456.2 hours, and a transmission was installed at an airframe time of 2,956.7 hours; this transmission, however, already had 500.7 hours of service in another machine. We can determine from this that the transmission has a total operational time of 991.2 hours.

Total airframe time	**3,456.2 hours**
Transmission installed	**2,965.7 hours**
Transmission time this machine	**490.5 hours**
Transmission time—other	**500.7 hours**
Total transmission time	**991.2 hours**

If this transmission must be overhauled every 1,200 hours, it must be overhauled when the airframe total time reaches 3,665.0 hours.

Time for overhaul	**1,200.0 hours**
Time on transmission	**991.2 hours**
Time remaining	**208.8 hours**

Airframe total time	**3,456.2 hours**
Transmission time remaining	**208.8 hours**
Airframe time at transmission overhaul	**3,665.0 hours**

We must be careful when determining the time remaining before inspection or overhaul, because some of the larger components may have parts installed inside of them that must be inspected at a shorter time interval.

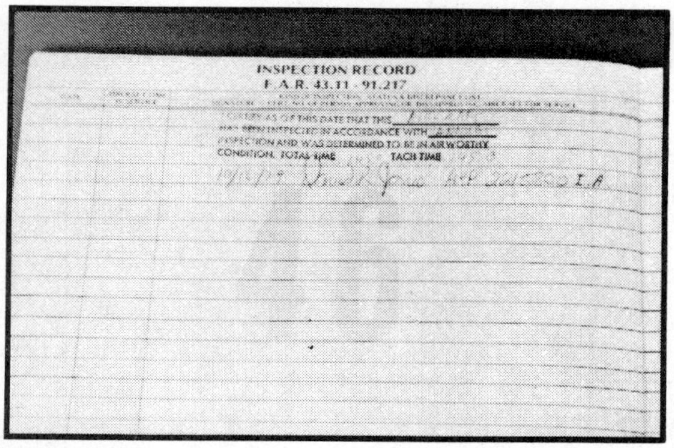

Fig. 4-1 An airframe logbook can be used as part of the required maintenance records.

II. Temporary Records

In addition to the records that must be permanently maintained, the owner or operator of the aircraft must keep the following records for a period of one year, or until the operation is repeated or superseded:

a. Records of the maintenance that has been done to the airframe, engine, propeller, rotor, or appliance

b. Records of 100-hour, annual, or progressive inspections

These records must include:

a. Description of the work performed

b. Date of completion of the work performed

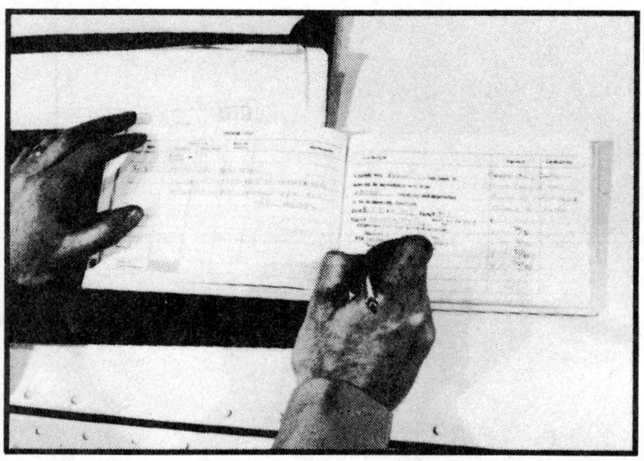

Fig. 4-2 It is extremely important that records be kept of all maintenance done to an aircraft.

c. Signature and certificate number of the person approving the aircraft for return to service

III. Forms

A. Malfunction or Defect Report Form 8330-2

If a malfunction or defect shows up on an aircraft, powerplant, propeller, or appliance that is not caused by poor maintenance practices, a malfunction or defect report should be filled out and mailed to the FAA GADO. In Fig. 4-3 we have an example of this form with spaces for the make, model, and serial number of the unit that had the difficulty and a place for the identification of the specific part that failed or caused the problem. There is space on the report for recommendations for a fix that would prevent any recurrence of the problem.

These forms are forwarded by the GADO to the FAA Aeronautical Center in Oklahoma City, where all of the reports that have been received are analyzed and any trends that would indicate a dangerous situation are noted. The information is made available to A&P's in the form of General Aviation Airworthiness Alerts, and if the problem has occurred frequently enough and is of a serious enough nature, an Airworthiness Directive is issued to require the fix to be made before a similar failure can cause an accident.

B. Major Repair and Alteration Form 337

When a major repair or major alteration is made on an airframe, powerplant, propeller, or appliance, a Form 337 must be filled out to record the work that has been done.

Two copies of the form are required, but it is always a good policy to make three. The original form stays with the aircraft records where it is to be kept for at least one year, and the duplicate copy must be sent to the FAA GADO within 48 hours after the aircraft is approved for return to service. Although it is not required by the FAA, it is a good idea to make an extra copy of the form for the mechanic who approved the aircraft for return-to-service.

The information required on the front side of the form is quite straightforward, but the back side is just a big blank space in which to describe

1. REGISTRATION NO. N–	DEPARTMENT OF TRANSPORTATION FEDERAL AVIATION ADMINISTRATION **MALFUNCTION OR DEFECT REPORT**			Form Approved Budget Bureau No. 04–R0003	8. DATE SUB.	FOR FAA USE ONLY CONTROL NO.
	A. MAKE	B. MODEL	C. SERIAL NO.	7A. COMMENTS (*Describe the malfunction or defect and the circumstances under which it occurred. State probable cause and recommendations to prevent recurrence.*)		
2. AIRCRAFT						
3. POWERPLANT						
4. PROPELLER						

5. APPLIANCE/COMPONENT (*assy. that includes part*)			
A. NAME	B. MAKE	C. MODEL	D. SERIAL NO.

6. SPECIFIC PART (*of component*) CAUSING TROUBLE		
A. NAME	B. NUMBER	C. PART/DEFECT LOCATION

Continue on reverse

SUBMITTED BY

FAA USE D. ATA CODE	E. PART TT	F. PART TSO	G. PART CONDITION	B. REP. STA.	C. OPER.	D. MECH.	E. AIR TAXI	F. MFG.	G. FAA	H. OTHER	I.

FAA FORM [illegible]-2 (9–70) SUPERSEDES PREVIOUS EDITIONS

8110-4

USE THIS SPACE FOR ADDITIONAL COMMENTS IF NEEDED

DEPARTMENT OF TRANSPORTATION
FEDERAL AVIATION ADMINISTRATION
WASHINGTON, D.C. 20590

OFFICIAL BUSINESS
PENALTY FOR PRIVATE USE, $300

POSTAGE AND FEES PAID
FEDERAL AVIATION ADMINISTRATION

Fig. 4-3 Malfunction or Defect report form

DEPARTMENT OF TRANSPORTATION FEDERAL AVIATION ADMINISTRATION **MAJOR REPAIR AND ALTERATION** **(Airframe, Powerplant, Propeller, or Appliance)**	*Form Approved* *Budget Bureau No. 04–R060.1* *FOR FAA USE ONLY* OFFICE IDENTIFICATION

INSTRUCTIONS: Print or type all entries. See FAR 43.9, FAR 43 Appendix B, and AC 43.9–1 (or subsequent revision thereof) for instructions and disposition of this form.

1. AIRCRAFT	MAKE	MODEL
	SERIAL NO.	NATIONALITY AND REGISTRATION MARK
2. OWNER	NAME *(As shown on registration certificate)*	ADDRESS *(As shown on registration certificate)*

3. FOR FAA USE ONLY

4. UNIT IDENTIFICATION / 5. TYPE

UNIT	MAKE	MODEL	SERIAL NO.	REPAIR	ALTERATION
AIRFRAME	(As described in item 1 above)				
POWERPLANT					
PROPELLER					
APPLIANCE	TYPE MANUFACTURER				

6. CONFORMITY STATEMENT

A. AGENCY'S NAME AND ADDRESS	B. KIND OF AGENCY	C. CERTIFICATE NO.
	☐ U.S. CERTIFICATED MECHANIC	
	☐ FOREIGN CERTIFICATED MECHANIC	
	☐ CERTIFICATED REPAIR STATION	
	☐ MANUFACTURER	

D. I certify that the repair and/or alteration made to the unit(s) identified in item 4 above and described on the reverse or attachments hereto have been made in accordance with the requirements of Part 43 of the U.S. Federal Aviation Regulations and that the information furnished herein is true and correct to the best of my knowledge.

DATE	SIGNATURE OF AUTHORIZED INDIVIDUAL

7. APPROVAL FOR RETURN TO SERVICE

Pursuant to the authority given persons specified below, the unit identified in item 4 was inspected in the manner prescribed by the Administrator of the Federal Aviation Administration and is ☐ APPROVED ☐ REJECTED

BY			
	☐ FAA FLT. STANDARDS INSPECTOR	☐ MANUFACTURER	☐ INSPECTION AUTHORIZATION	OTHER *(Specify)*
	☐ FAA DESIGNEE	☐ REPAIR STATION	☐ CANADIAN DEPARTMENT OF TRANSPORT INSPECTOR OF AIRCRAFT	

DATE OF APPROVAL OR REJECTION	CERTIFICATE OR DESIGNATION NO.	SIGNATURE OF AUTHORIZED INDIVIDUAL

FAA Form 337 (7-67) (8320)

Fig. 4-4 Major Repair And Alteration Form

NOTICE

Weight and balance or operating limitation changes shall be entered in the appropriate aircraft record. An alteration must be compatible with all previous alterations to assure continued conformity with the applicable airworthiness requirements.

8. DESCRIPTION OF WORK ACCOMPLISHED *(If more space is required, attach additional sheets. Identify with aircraft nationality and registration mark and date work completed.)*

☐ ADDITIONAL SHEETS ARE ATTACHED

Fig. 4-4 Major Repair And Alteration Form (continued)

the work that has been done. Some basic rules to follow when filling out this space are in order:

Be sure that your illustrations and figures are neat and clear.

Don't say anything that does not need to be said. This form is a record of only the major repair or alteration, and any work that was done at the same time but is not part of the repair or alteration should not be mentioned on the form.

Locate the repair on the airplane. Someone may later make a similar repair, and if the location of your work is not pinpointed with the fuselage or wing station, or referenced from some other recognizable location, you may be held responsible for the work someone else did.

Identify the materials that were used in the repair. If you have replaced some factory parts, use the part numbers. If you have made the parts, give the specifications of the materials that were used.

List the approved data that was used for the work. If it was a manufacturer's service bulletin or STC, give the number. If it is to comply with an AD note, be sure to reference the AD number. If the work is done according to some description or drawing in AC 43.13-1A, reference the paragraph or the drawing number.

Make a statement to the effect that changes in the weight and balance have been noted in the weight and balance statement of the aircraft records, or that no weight and balance change was effected.

When the description is finished, draw a line across the bottom of the form and write the word END in the center of the line.

C. Certificate of Registration and Bill of Sale, Form 8050

When an aircraft is sold, the seller must return the Aircraft Registration form to the FAA in Oklahoma City, Oklahoma, and the purchaser must file his application for a new Registration.

The form 8050-1 is filled out in triplicate, and the original, which is the white copy, and the green copy are sent to the FAA in Oklahoma City, along with the registration fee and proof of ownership. The pink copy of the form is kept in the aircraft until the new registration form is returned by the FAA.

Proof of ownership may be substantiated by the Bill of Sale Form 8050-2 or by any other form of proof that the purchaser may want to use. Before buying an aircraft, the purchaser should have the title of the aircraft verified by a title search that can be conducted by any of a number of companies in Oklahoma City that provide this service. If a search is not made, it is possible that the person that sells the aircraft may not be the same as the person listed on the FAA records as the legal owner of the aircraft, and the ownership can be disputed at some later date.

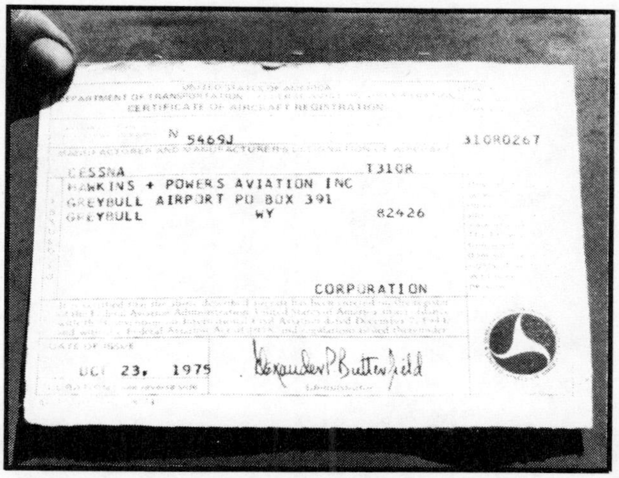

Fig. 4-5 Certificate of Registration

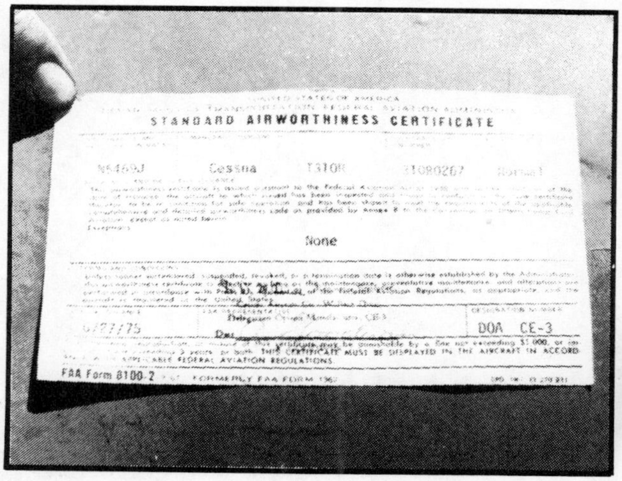

Fig. 4-6 Airworthiness Certificate

D. Standard Aircraft Airworthiness Certificate From 8100-2

When an aircraft is built and is found to meet all of the requirements for its certification, it is issued an Airworthiness Certificate, Form 8100-2. This certificate must be displayed in the aircraft and is transferred with the aircraft when it is sold, unless it is sold to a foreign purchaser. A standard Airworthiness Certificate is issued to aircraft licensed under the normal, utility, acrobatic, or transport category, or manned free balloons. These standard certificates remain in effect as long as the aircraft receives the required maintenance and is registered in the United States.

E. Special Airworthiness Certificate Form 8130-7

Aircraft that are certificated under any category other than those of the standard classifications are issued a Special Airworthiness Certificate. This includes aircraft that are operating under the Experimental, Restricted, Limited, and Provisional categories.

NOTES

NOTES